Seventeenth Symposium on Biotechnology for Fuels and Chemicals

Presented as Volumes 57 and 58 of
Applied Biochemistry and Biotechnology

Proceedings of the Seventeenth Symposium
on Biotechnology for Fuels and Chemicals
Held May 7–11, 1995, at Vail, Colorado

Sponsored by
US Department of Energy
(Office of Alternative Fuels, Office of Industrial Processes)
National Renewable Energy Laboratory
Oak Ridge National Laboratory
Idaho National Engineering Laboratory
Argonne National Laboratory
A. E. Staley Manufacturing Company
Archer Daniels Midland Company
Bio-Technical Resources, L.P.
Chronopol, Inc.
Colorado Institute for Research in Biotechnology
Dow Chemical
DuPont Company
Enzyme Bio-Systems Ltd.
FermPro Manufacturing L.P.
Gist-brocades
Golden Technologies Company, Inc.
Grain Processing Corporation
Martin Marietta Energy Systems
National Corn Growers Association
New Energy Company of Indiana, L. P.
Raphael Katzen Associates International, Inc.
South Point Ethanol
St. Lawrence Technologies Inc.
Weyerhaeuser
American Chemical Society (Division of Biochemical Technology)

Editors
Charles E. Wyman
National Renewable Energy Laboratory
Brian H. Davison
Oak Ridge National Laboratory

Springer Science+Business M

ISBN 978-1-4612-6669-3 ISBN 978-1-4612-0223-3 (eBook)
DOI 10.1007/978-1-4612-0223-3

Applied Biochemistry and Biotechnology

Volumes 57 and 58, Complete, Spring 1996

Originally published by Humana Press Inc in 1996
Softcover reprint of the hardcover 1st edition 1996

Applied Biochemistry and Biotechnology is abstracted or indexed regularly in *Chemical Abstracts, Biological Abstracts, Current Contents, Science Citation Index, Excerpta Medica, Index Medicus,* and appropriate related compendia.

For product safety concerns and information please contact our
[illegible]
[illegible] Germany

Introduction to the Proceedings of the Seventeenth Symposium on Biotechnology for Fuels and Chemicals

CHARLES E. WYMAN
National Renewable Energy Laboratory

BRIAN H. DAVISON
Oak Ridge National Laboratory

The application of microorganisms to produce valuable products dates from antiquity with the fermentation of various plant materials to alcoholic beverages. These techniques were extended to manufacture industrial-based products, but chemical routes to produce such commodities from fossil sources displaced many of these biologically based technologies following World War II. With concerns about oil shortages, high petroleum prices, and air quality, fermentation processes again became the basis for the emergence of major industries for production of ethanol from starch crops in the United States and from cane sugar in Brazil in the early 1980s. Mounting environmental issues of waste disposal, the potential for global climate change, and pollution prevention have driven the need for development of new biological processes to produce fuels and chemicals from renewable cellulosic biomass feedstocks, such as agricultural and forestry residues, municipal wastes, and woody and herbaceous energy crops as well as existing crops, coal, carbon dioxide, and other resources. Similar technologies can also be applied to remedy environmental problems. However, widespread use of biological processes depends on achieving technical advances that make them profitable.

The vast potential of metabolic engineering coupled with modern advances in bioprocessing methods opens up a wide range of new possibilities and unprecedented uses for biotechnology in the production of fuels and chemicals that more traditional technologies could not likely achieve. Recent advances in biotechnology have reduced the estimated cost of ethanol production from cellulosic biomass to the point that it is competitive for blending with gasoline, and continued reductions in the cost of producing ethanol from cellulosic biomass will lead to its widespread use as a pure automotive fuel at a scale well beyond anything we have seen before. Such low-cost ethanol could become viable as a feedstock for ethylene synthesis, opening up a large segment of the chemical market to biotechnology-based products. Biotechnologies are emerging for directly producing other chemicals and intermediates, such as lactic and succinic acid and polyhydroxybutyrate, that are environmentally friendly. Continued advances in biotechnology will lead to development of algae and higher plants that can economically produce oils for transformation into diesel fuel substitutes, chemicals, and materials. Biotechnology is also being applied to the removal of sulfur from petroleum, the liquefaction of coal, and the processing of waste streams.

These and numerous other applications of biotechnology to the production of fuels and chemicals and environmental cleanup, as well as yet unknown developments that will emerge in the future, will result in tremendous benefits to society. They will create new jobs, lower the dependence on oil imports, improve the trade balance, improve our air quality, cut greenhouse gas emissions, reduce the hazards of producing fuels and chemicals, and prevent pollution. They will also provide profitable enterprises that fuel our economy. Through the coordinated efforts of universities, government research laboratories, and industry, biological processes can be taken from conception to commercial uses that realize this enormous potential.

In the tradition of the annual symposium series, the Seventeenth Symposium on Biotechnology for Fuels and Chemicals continues to provide a forum for the exchange of information and ideas on advances in biotechnology that facilitates the emergence of a large-scale industry for the competitive production of fuels and chemicals. This year's program included 35 oral and 115 poster presentations covering topics in thermal, chemical, and biological processing; applied biological research; bioprocessing research; process economics and commercialization; and environmental biotechnology. Formal discussion sessions were organized on moving ethanol and ethyl tertiary butyl ether into the marketplace, biotechnology for the forest products industry, moving biodiesel into the marketplace, and trends in commercial cellulases, such as production, specifications, and cost. Coupled with various tours, the banquet, informal discussions, and a tranquil setting, the Seventeenth Symposium provided a congenial atmosphere for productive interactions among the 225 attendees from academia, industry, and government.

The Seventeenth Symposium was sponsored by the Office of Alternative Fuels and Office of Industrial Processes of the US Department of Energy (DOE); the National Renewable Energy Laboratory; Oak Ridge National Laboratory; Idaho National Engineering Laboratory; Argonne National Laboratory; A.E. Staley Manufacturing Company; Archer Daniels Midland Company; Bio-Technical Resources, L.P.; Chronopol, Inc.; Colorado Institute for Research in Biotechnology; Dow Chemical; DuPont Company; Enzyme Bio-Systems, Ltd; FermPro Manufacturing LP; Gist-brocades; Golden Technologies Company, Inc.; Grain Processing Corporation; Martin Marietta Energy Systems; National Corn Growers Association; New Energy Corporation of Indiana; Raphael Katzen Associates International, Inc.; South Point Ethanol; St. Lawrence Technologies Inc.; Weyerhaeuser; and the American Chemical Society, Division of Biochemical Technology.

Organization of the Symposium was as follows:

Organizing Committee

Charles E. Wyman, *Chairman, National Renewable Energy Laboratory*
Brian H. Davison, *Cochairman, Oak Ridge National Laboratory*
Kathleen Clarkson, *Genencor International*
James Doncheck, *Bio-Technical Resources, L.P.*
Barbara Goodman, *National Renewable Energy Laboratory*
Nancy Ho, *Purdue University*
Thomas Jeffries, *USDA Forest Products Laboratory*
Donald Johnson, *Grain Processing Corporation*
LaMar Johnson, *Idaho National Engineering Laboratory*
Raphael Katzen, *Rafael Katzen Associates International, Inc.*
Richard Moorer, *U.S. Department of Energy*
David Ollis, *North Carolina State University*
Jack Saddler, *University of British Columbia*
Linda Schilling, *National Institute for Standards and Technology*
Jonathan Woodward, *Oak Ridge National Laboratory*

Session Chairpersons and Cochairpersons

Session 1: Thermal, Chemical, and Biological Processing
Marion Bradford, *A.E. Staley*
Cynthia Riley, *National Renewable Energy Laboratory*
Session 2: Applied Biological Research
Ting Carlson, *Cargill, Inc.*
William Apel, *Idaho National Energy Laboratory*
Session 3: Bioprocessing Research
David Ollis, *North Carolina State University*
George Philippidis, *National Renewable Energy Laboratory*
Session 4: Process Economics and Commercialization
Harry Folster, *Weyerhaeuser*
Karl Sanford, *Genencor International*
Session 5: Environmental Biotechnology
Kerry Sublette, *University of Tulsa*
K. Thomas Klasson, *Oak Ridge National Laboratory*
Poster Session
Antonios Antonopoulos, *Argonne National Laboratory*
Karel Grohmann, *USDA Citrus and Subtropical Products Research Laboratory*

Acknowledgments

The submitted manuscript has been authored by a contractor of the US Government under contract DE-AC36-83CH10093. Accordingly, the US Government retains a nonexclusive, royalty-free license to publish or reproduce the published form of this contribution, or allow others to do so, for US Government purposes.

The National Renewable Energy Laboratory is operated by Midwest Research Institute, for the US Department of Energy under contract DE-AC36-83CH190093.

Oak Ridge National Laboratory is managed by Lockheed Martin Energy Systems, Inc., for the US Department of Energy under contract DE-AC05840R21400.

Other Proceedings in This Series

1. "Proceedings of the First Symposium on Biotechnology in Energy Production and Conservation" (1978), *Biotechnol. Bioeng. Symp.* **8.**
2. "Proceedings of the Second Symposium on Biotechnology in Energy Production and Conservation" (1980), *Biotechnol. Bioeng. Symp.* **10.**
3. "Proceedings of the Third Symposium on Biotechnology in Energy Production and Conservation" (1981), *Biotechnol. Bioeng. Symp.* **11.**
4. "Proceedings of the Fourth Symposium on Biotechnology in Energy Production and Conservation" (1982), *Biotechnol. Bioeng. Symp.* **12.**
5. "Proceedings of the Fifth Symposium on Biotechnology for Fuels and Chemicals" (1983), *Biotechnol. Bioeng. Symp.* **13.**
6. "Proceedings of the Sixth Symposium on Biotechnology for Fuels and Chemicals" (1984), *Biotechnol. Bioeng. Symp.* **14.**
7. "Proceedings of the Seventh Symposium on Biotechnology for Fuels and Chemicals" (1985), *Biotechnol. Bioeng. Symp.* **15.**
8. "Proceedings of the Eighth Symposium on Biotechnology for Fuels and Chemicals" (1986), *Biotechnol. Bioeng. Symp.* **17.**
9. "Proceedings of the Ninth Symposium on Biotechnology for Fuels and Chemicals" (1988), *Appl. Biochem. Biotechnol.* **17,18.**
10. "Proceedings of the Tenth Symposium on Biotechnology for Fuels and Chemicals" (1989), *Appl. Biochem. Biotechnol.* **20,21.**
11. "Proceedings of the Eleventh Symposium on Biotechnology for Fuels and Chemicals" (1990), *Appl. Biochem. Biotechnol.* **24,25.**
12. "Proceedings of the Twelfth Symposium on Biotechnology for Fuels and Chemicals" (1991), *Appl. Biochem. Biotechnol.* **28,29.**
13. "Proceedings of the Thirteenth Symposium on Biotechnology for Fuels and Chemicals" (1992), *Appl. Biochem. Biotechnol.* **34,35.**
14. "Proceedings of the Fourteenth Symposium on Biotechnology for Fuels and Chemicals" (1993), *Appl. Biochem. Biotechnol.* **39,40.**
15. "Proceedings of the Fifteenth Symposium on Biotechnology for Fuels and Chemicals" (1994), *Appl. Biochem. Biotechnol.* **45,46.**
16. "Proceedings of the Sixteenth Symposium on Biotechnology for Fuels and Chemicals" (1995), *Appl. Biochem. Biotechnol.* **51/52.**

This symposium has been held annually since 1978. We are pleased to have the proceedings of the Seventeenth Symposium currently published in this special issue to continue the tradition of providing a record of the contributions made.

The Eighteenth Symposium is planned for May 5–9, 1996 in Gatlinburg, Tennessee. We encourage comments or discussions relevant to the format or content of that meeting.

CONTENTS

*For papers with multiple authorship, the asterisk identifies the author to whom correspondence and reprint requests should be addressed.

SESSION 1

Thermal, Chemical, and Biological Processing

MARION BRADFORD[1] AND CYNTHIA RILEY[2]

[1]*A. E. Staley, Decatur, IL;*
and [2]*National Renewable Energy Laboratory, Golden, CO*

Historically, this session has dealt primarily with thermal, chemical, and biological processing of biomass to produce fuels and chemicals. There have been presentations on the downstream processing as well. This year's program has been chosen to look at current practices, new processes for production of chemicals, and the future technology that could be reached with genetic engineering.

Pretreatment of biomass is an essential step in the preparation of a fermentable substrate. This year's session includes papers on the traditional dilute-acid process, methods of monitoring inhibitory factors produced in the dilute acid process, and a new water-only process that shows promise for the future. Production of useful chemicals requires efficient, low-energy recovery processes due to the dilute nature of substrates that come from hydrolysis of biomass. This year's session looks at three recovery processes that are state-of-the-art for efficiency and offer economical recovery of valuable products from dilute process streams. Novel genetic engineering has shown it possible to produce standard petrochemicals through fermentation technology with or without the use of some conventional chemical steps. This year's program includes one such approach.

The overall intent of this year's session is to examine the state of reduction to practice in the conversion of biomass into usable intermediates.

0273-2289/96/5758—0003/$9.50

Design and Initial Operation of a High-Solids, Pilot-Scale Reactor for Dilute-Acid Pretreatment of Lignocellulosic Biomass

TEH-AN HSU,*,[1] MIKE HIMMEL,[1] DAN SCHELL,[1] JODY FARMER,[1] AND MARK BERGGREN[2]

[1]*Alternative Fuels Division, National Renewable Energy Laboratory, Golden, CO 80401-3393; and* [2]*Hazen Research, Inc., Golden, CO*

ABSTRACT

A recently installed 100-L horizontal shaft custom-fabricated mixer/reactor, made of Carpenter 20 Cb-3 stainless steel and designed for high-solids, dilute-acid pretreatment of lignocellulosic biomass, is described. Several preliminary runs with a hardwood and a herbaceous feedstock are reported; the results show the operation to be reliable with good reproducibility. No significant differences between comparable pretreatments in the high-solids reactor and a 1-L laboratory-scale reactor are apparent. Pretreatment efficiency is not affected by 10–15% (w/w) solids loading or biomass particle size of up to –10 mm. Mass balances are closed reasonably well, with improvement in reactor contents handling and biomass analytical methods desirable.

Index Entries: Dilute sulfuric acid pretreatment; high-solids reactor; pilot-scale reactor; ethanol from biomass; prehydrolysis.

INTRODUCTION

A pretreatment step is necessary to overcome the recalcitrant nature of the native cellulose to facilitate subsequently enzymatic conversion of cellulose in lignocellulosic biomass into glucose that can be fermented into alternative fuels and chemicals. The numerous pretreatment techniques reported in the literature can generally be grouped into physical, chemical, and biological methods *(1–7)*. Some consider thermal pretreatments to form a fourth family *(8)*, which is considered by others as either physical *(1,3–5,7)* or chemical *(2,6)* pretreatment. Chemical pretreatment has received the most attention by far *(2)*, probably because physical pretreatment is generally too costly and not very effective, whereas biological pretreatment is in its developmental infancy *(9)*.

*Author to whom all correspondence and reprint requests should be addressed.

Among various chemical techniques, dilute-acid pretreatment has been studied for the past two decades, including sulfuric *(10–20)*, hydrochloric *(21–23)*, nitric (D. L. Brink, Forest Products Laboratory, University of California, Richmond, personal communication, 1993; *24,25*), phosphoric *(21)*, and peracetic *(26)* acid. Sulfuric acid, apparently because of its low cost and high effectiveness in removing the recalcitrant nature of native cellulose, has been most extensively investigated. Laboratory studies *(11,12,17–19,27,28)* of a dilute sulfuric acid pretreatment process have shown that it significantly enhances the enzymatic digestibility of many forms of lignocellulosic feedstocks, including hardwoods, herbaceous crops, and agricultural residues. At the same time, hemicellulosic sugar yields are reasonably high from pretreatment. Yields of xylose (or, more precisely, soluble xylose equivalents), for example, have reached as high as 80% of theoretical. Therefore, although it suffers from the need for corrosion-resistant construction materials for reactors, generation of gypsum during postpretreatment neutralization, and formation of saccharification and fermentation inhibitors during pretreatment, dilute sulfuric acid pretreatment is still one of the most, if not the most, promising techniques.

All except two *(12,16)* of the dilute sulfuric acid pretreatment studies cited used reactor dry solids loadings of 10% or lower. However, the dry solids loading needs to be substantially higher to reduce reactor size, limit acid use, and reduce steam consumption. For example, the ethanol production cost estimates by the National Renewable Energy Laboratory (NREL) *(29)* employ a dry solids loading of 35%, because ethanol production costs increase prohibitively with lower solids loadings (R. Elander, NREL, personal communication, 1995). Nevertheless, literature data on high-solids, dilute-acid pretreatment are extremely limited. Grohmann et al. *(12)* extended the solids loading to 40% using small (13-mm [1/2-in.] od × 100-mm [4-in.] long) pipe reactors with indirect heating, and Schell et al. *(16)* extended it to 30% using open trays in a cylindrical reactor with direct steam injection aided by indirect heating. In both studies, the biomass was completely static during pretreatment, and heat and mass transfers would likely be severely uneven at a large scale.

To gather higher solids data in a system that can be applied at a commercial scale, we designed and fabricated a 100-L horizontal shaft mixer/reactor. This reactor was designed to demonstrate the feasibility of the dilute sulfuric acid pretreatment technique at high-solids levels, provide scale-up data, and produce large quantities of pretreated solids for fermentation research. The reactor was recently installed and tested with a hardwood and a herbaceous feedstock. This article describes the high-solids, pilot-scale pretreatment reactor and its initial operation, and compares the results to those obtained from a bench-scale reactor.

MATERIALS AND METHODS

Materials

The woody and herbaceous lignocellulosic feedstocks were obtained by the NREL Alternative Fuels Program feedstock interface project. The woody material was primarily hardwood hybrid poplar DN34 *(Populus eugenei)* (mixed with minor other clones), whole tree, purchased from Domtar Paper Company (Cornwall, Ontario, Canada) as commercial chips. The chips were then air-dried, knife-milled, and sieved to –2-mm to +0.25-mm (–1/16-in. to +60-mesh) size range. The herbaceous material was switchgrass (*Panicum virgatum L.*, Alamo cultivar), purchased from Auburn University (Auburn, AL). It was field-dried and baled before being shipped to NREL. On receipt, it was knife-milled to either –3-mm (–1/8-in.) or –10-mm (–3/8-in.) particles for use in this study.

Chemicals used for pretreatment and sample analysis were all of reagent grade. The cellulase enzyme used was a Genencor Cytolase CL preparation.

Pretreatment Reactor

The pilot-scale reactor (dubbed the "paddle reactor," Fig. 1) was designed based on classical paddle-blender designs, with mixing elements (arms) placed along a horizontal shaft in such a manner as to ensure that all surfaces inside the reactor shell are contacted by the element tips. Teflon™ wiping blades are attached to each mixing element (including special arms designed to wipe the end plates) to ensure the biomass is effectively removed from the hot inner walls. This system of gentle, positive mixing combined with extensive wall scraping ensures minimal biomass overheating and maximum overall heat transfer of high-solids slurries.

The reactor vessel was constructed (by Sterling Machine Company, Sterling, CO) entirely of Carpenter 20 Cb-3 stainless steel to provide resistance to hot dilute sulfuric acid. The reaction vessel, about 100 L, is approx 610-mm (24 in.) long × 460-mm (18 in.) in diameter and is surrounded axially by a 25-mm (1-in.) wide steam/cooling fluid jacket with one 19-mm (3/4-in.) bottom port and two 19-mm (3/4-in.) top ports. All penetrations into the reaction vessel are made through the primary head plate, a 250-mm (10-in.) diameter manway. The vessel is constructed with only one removable end plate, located on the hand-crank side of the vessel, and this plate is dismantled during major repair only. All routine cleaning is done through the manway. A unique vessel handling system allows the vessel to be rotated in the support frame from the operating/loading position with the manway facing up to a position 180° around axially with the manway facing down for cleaning and product removal. To allow the vessel to rotate, a dual support system was employed. Whereas the reactor shaft and electric motor are one independent unit driven from one side of the framework, a hand wheel and gear box drive a lug attached to the reactor housing on the other (the hand-crank) side. This reactor housing lug is bolted to the framework during normal operation. Thus, the electric motor-side train drives the reactor shaft, and the hand wheel rotates the reactor vessel.

The reactor shaft is sealed by two annular stuffing box seals packed with 13-mm (1/2-in.) GFO® packing rings. This type of seal was found to be very practical for biomass pretreatment, in which the mixed material is often abrasive as well as corrosive. The drive shaft is Carpenter 20 Cb-3 stainless steel, 56 mm (2–3/16 in.) in diameter, and is supported by four "Seal Master"-type, double-row ball bearing blocks. The drive motor is a U.S. Motors 750-W (1-HP), 208 maximum rpm, adjustable-speed gear motor. The overall unit footprint and weight are 1.6 (65 in.) × 1.0 m (40 in.) and 680 kg (1500 lb), respectively. The reactor is insulated where practical.

Originally, the reactor was built for a maximum allowable working pressure (MAWP) of 720 kPa (90 psig), but later altered (by Eaton Metal Products Company, Denver, CO) to 1.1 MPa (150 psig). Assuming that the vapor pressure for dilute-acid pretreatment could be estimated from the vapor pressure of saturated steam, the original 720-kPa (90-psig) pressure limit constrained the operating tempera-

A

B

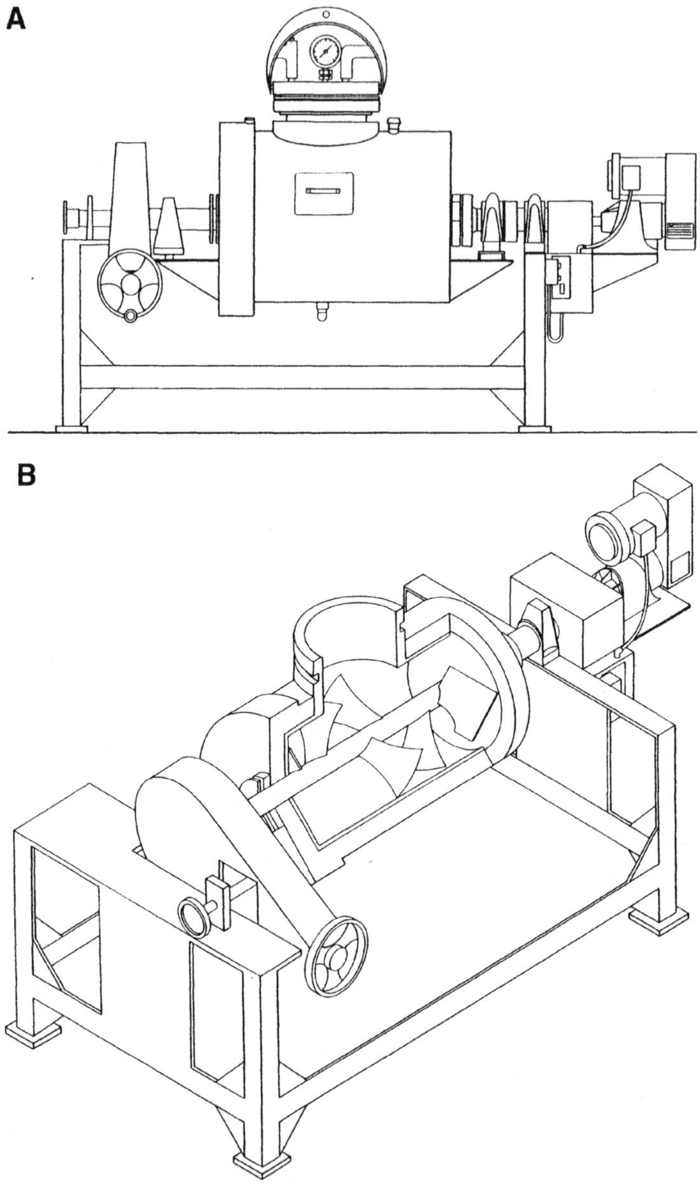

Fig. 1. Schematics of paddle reactor: **(A)** front view and **(B)** cut-away view showing inside of reactor.

ture for dilute-acid prehydrolysis to about 166°C (331°F) or below. According to our previous observations (data not shown), however, the actual reactor pressure was consistently slightly higher than the corresponding saturated steam pressure—possibly caused by a small amount of residual air in the reactor, volatile gases generated during pretreatment, or both. With this consideration and the consideration of the installation of a rupture disk, prior to the pressure rating alteration, we were planning on limiting pretreatment temperature in the paddle reactor to approx 160°C (320°F).

The modification of the MAWP was necessary, because, as study of the dilute sulfuric acid pretreatment technique progressed (using a laboratory-scale reactor), it became desirable to operate the paddle reactor at temperatures higher than the originally projected maximum of 160°C. Early NREL studies showed that temperatures up to 160°C were effective *(11,12)*. More recent studies, however, suggest that higher temperatures (as high as 180°C) might be desirable for some feedstocks *(17,19)* or certain temperature strategies *(13)*. Without the modification and with the installation of a rupture disk having a typical tolerance margin of 5% of the rated pressure to ensure the vessel not to be overpressured accidently, the maximum pretreatment temperature allowed of the reactor would be approx 160°C (320°F). As a result, the vessel was modified to an MAWP of 1.1 MPa (150 psig). The reactor can now be safely operated at 175°C. Without the alteration, the original design would have restricted operation of the vessel to temperatures possibly below those necessary for effective pretreatment of many of the feedstocks of interest. The pressure rating alteration involved changing of the removable end plate to a heavier flange and redoing the welds of the reactor. The reactor jacket is now rated at 960 kPa (125 psig).

To our knowledge, this is the only reactor of its kind ever reported in the literature for high-solids, dilute-acid pretreatment of lignocellulosic biomass.

Pretreatment Reactor Utility Skid

A utility skid was built by a subcontractor (Interpro [now Colorado Minerals Research Institute], Golden, CO) to allow the paddle reactor to be operated. The skid provides a centralized and portable location for all support equipment and external utility connections (e.g., steam, water). Flexible stainless-steel hoses are used to make connections between the skid and the reactor using quick-disconnect and Swagelok fittings. A flow diagram for the system is shown in Fig. 2. Not shown in the figure are the rupture disk and a pressure gage that are tapped to the manway lid, and a tee fitting between the lid and the quick coupling in the reactor condensate valve (RCV) line that houses a thermocouple (the tip of which protrudes into the reactor head space). The skid is equipped with a high-pressure metering pump (Cheminjector Diaphragm Metering Pump, Hydroflo Corporation, Plumsteadville, PA, 55.3 L/h [14.6 gal/h] maximum flow rate and 1.1 MPa [150 psig] maximum delivery pressure), steam pressure regulator, pressure relief valve, steam traps, heat exchanger, and (regular) valves. All connectors, tubing, tankage, and valves used to construct the system are either Carpenter 20 Cb-3 or 304 stainless steel.

Pretreatment Runs

To conduct a dilute-acid pretreatment run, a known amount of biomass is loaded into the paddle reactor through the manway, and then mixed with a predetermined amount of deionized water to form a slurry of the desired solids content. The manway lid is then closed and connections to the utility skid made. Next, the biomass in the reactor is deaerated by heating by steam to 75°C for 10 min. Following deaeration, the reactor is sealed by closing the vent valve (the reactor condensate valve, RCV) and then heated to the target pretreatment reaction temperature, either by injecting steam into the reactor jacket or directly into the reactor. For this study, we applied indirect heating of the jacket. The steam used is filtered to remove par-

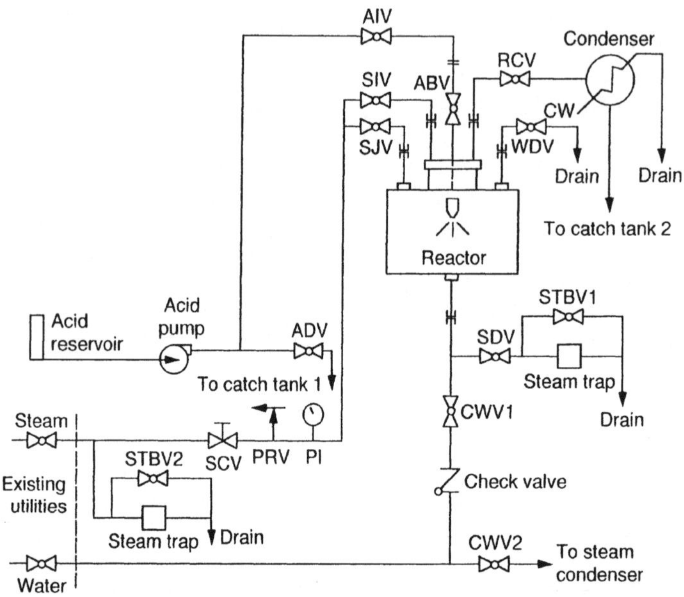

Fig. 2. Paddle reactor process diagram. ⋈, ball valve; needle valve; H, quick coupling; = , swagelok connector; ABV, acid block valve; ADV, acid drain valve; AIV, acid injection valve; CWV, cold water valve; PI, pressure indicator; PRV, pressure relief valve; RCV, reactor condensate valve; SCV, steam control valve; SDV, steam drain valve; SIV, steam injection valve; SJV, steam jacket valve; STBV, steam trap bypass valve; WDV, water drain valve.

ticulates and reduced to an appropriate pressure by a control valve, and then further controlled by a steam control valve at the utility skid. Steam traps drain condensate from the steam lines on the skid and from the reactor jacket.

When the target reaction temperature is reached, a calculated amount of 72 wt% sulfuric acid is injected quickly (in 20–40 s, depending on the amount of acid injected) into the reactor by the metering pump through a spray nozzle (Fig. 2). At the completion of the pretreatment time, cooling water is injected into the bottom port of the jacket to cool the reactor rapidly. When the reactor temperature drops to 95°C, the vapors are vented through a condensing heat exchanger and collected in a separate tank (Catch Tank 2). At 80°C, the manway lid is opened. Although the reactor can also be cooled by venting it immediately after the predetermined pretreatment time is reached, this cooling method was not used in this study. To discharge the reactor, it is rotated about 180° and the contents removed by gravity. A filter pan under the reactor collects the discharged reactor contents, which can be filtered to separate the prehydrolysate from the pretreated solids and the prehydrolysate sampled for compositional analysis. For runs to determine dry biomass solids yield, the wet filter cake of pretreated solids and residual prehydrolysate is washed until the filtrate pH rises to 5.0 or higher and then partially air-dried before determining the dry solids weight. A solids sample is collected for compositional analysis before air-drying. If the dry biomass solids yield is not needed, the filter cake is partially washed until the filtrate pH turns to 4.0 or higher and a sample is then collected for analysis.

For initial operation of the paddle reactor, a total of two hybrid poplar and six switchgrass runs was conducted. The pretreatment conditions are summarized in Table 1, and include both 10 and 15% solids loadings. The higher acid concentrations for switchgrass were needed to compensate for the higher buffering capacity of switchgrass or the higher solids loading, or both, so that the prehydrolysate pH was controlled to about 1.3. Throughout the runs, the rotational speed of the reactor shaft was maintained at 90 rpm. During loading and discharging, the shaft speed was set at a much lower level of approx 25 rpm (the lowest speed without incurring undesirable noise, possibly owing to belt slippage, in the gear box) to facilitate loading and discharging. The reactor temperature during heating was manually controlled by the steam control valve (SCV, Fig. 2); the steam supply upstream of the steam control valve was regulated to 930 kPa (120 psig).

For investigating scale-up differences, the results obtained from a laboratory-scale (1-L) reactor were compared to the paddle reactor results under the same pretreatment conditions. The laboratory-scale reactor operation has been previously described *(19)*, except that the acid injection was achieved by using a high-performance liquid chromatography pump.

Analytical Methods

All samples generated were refrigerated until analyzed. To analyze the composition of prehydrolysates and solids, NREL standard analytical procedures as described by Ehrman and Himmel *(30)* were followed. All pretreated solids were washed extensively to ensure the soluble sugars and acids were removed before analysis for enhanced accuracy *(30,31)*. Solids were analyzed for glucan, xylan, arabinan, galactan, mannan, Klason lignin, acid-soluble lignin, and ash. The extractives content was also determined for raw switchgrass. Prehydrolysates were analyzed for glucose, xylose, arabinose, galactose, mannose, acetic acid, furfural, 5-hydroxymethylfurfural (HMF), and acid-soluble lignin.

For enzymatic digestibility tests, a modified NREL standard analytical procedure was followed. The NREL standard protocol (Laboratory Analytical Procedure no. 009, Ethanol Project, Alternative Fuels Division, NREL) calls for using 1% cellulose loading with 60 filter paper units (FPU) cellulase/g cellulose in pH 4.8 sodium citrate buffer at 50°C and determining digestibility by following glucose concentrations. In the modified procedure, instead of following only glucose concentrations that take, at times, 6–7 d to reach the final values, 3-d samples were analyzed for glucose and cellobiose, and total glucose equivalents were calculated by summing total glucose units in glucose and cellobiose to represent the ultimate digestibility. Some digestibility tests were conducted after the pretreated solids were size-reduced in a laboratory blender (Waring Commercial Lab Blender, Dynamics Corporation of America, New Hartford, CT). The blending procedure has been described elsewhere *(32)*.

Reactor Mass Balances

Overall mass balances were estimated for three paddle reactor runs. The entire reactor discharge was filtered immediately following discharge. The total weight of the reactor contents after pretreatment was obtained by adding the weights of filtrate (prehydrolysate), wet filter cake, vent condensate, and reactor clean-out (residual material in reactor on discharging), and compared to the total weight of materials (biomass, water, and acid) loaded into the reactor. The ratio, expressed as a percentage of total reactor charge, is referred to as the reactor mass recovery. The filter

Table 1
Paddle Reactor and 1-L Laboratory-Scale Reactor Pretreatment Runs[a]

Run no.	Sulfuric acid concentration,[b] wt%	Dry solids loading %	Dry solids loading g	Biomass particle size	Reactor mass recovery, %	Dry biomass solids yield, %
Hybrid poplar DN34						
Paddle reactor run 1	0.73	10	6000	–2 mm +0.25 mm (–1/16 in. +60 mesh)	nd	nd
Paddle reactor run 2	0.73	10	5830	–2 mm +0.25 mm (–1/16 in. +60 mesh)	nd	nd
1-L lab-scale reactor	0.73	10	50	–2 mm +0.25 mm (–1/16 in. +60 mesh)	nd	nd
Switchgrass						
Paddle reactor run 1	0.88	10	6000	–10 mm (–3/8 in.)	nd	nd
Paddle reactor run 2	0.88	10	6000	–10 mm (–3/8 in.)	nd	nd
Paddle reactor run 3	0.88	10	6000	–3 mm (–1/8 in.)	nd	nd
Paddle reactor run 4	0.88	10	6000	–3 mm (–1/8 in.)	101.0	54.0
Paddle reactor run 5	0.88	10	6000	–3 mm (–1/8 in.)	93.3	53.8
Paddle reactor run 6	1.11	15	9000	–3 mm (–1/8 in.)	99.7	56.0
1-L lab-scale reactor	0.88	10	60	–3 mm (–1/8 in.)	nd	52.0

[a]Pretreatment temperature was 170°C for all runs; pretreatment time was 10 min for all hybrid poplar runs and 5 min for all switchgrass runs. nd, not determined.
[b]Acid concentration calculated based on weight of liquid phase of reactor contents.

cake dry weight, plus reactor clean-out dry weight, relative to the dry weight of the feedstock biomass, expressed as a percentage, is referred to as the dry biomass solids yield.

Furthermore, the mass balance for major biomass components, including glucan, xylan, and Klason lignin, was established for these runs. In doing so, the generation of the small amounts of volatiles during pretreatment was ignored (most or all of the volatiles would result in the vent condensate, which was observed to be negligible in all cases—data not shown). Moreover, HMF and furfural were assumed to be generated from glucose and xylose, respectively.

RESULTS AND DISCUSSION

All paddle reactor runs conducted thus far have operated satisfactorily. We have encountered only very minor and predictable operating issues, such as the need for tightening the packing seals and fittings during a run. Figure 3 shows the temperature profiles of the two hybrid poplar DN34 runs. Apparently, except for a time shift, the two profiles are very similar and can be superimposed almost exactly on each other. The time shift resulted from the difference in the starting points when the first temperature point was recorded. This observation of reproducibility of heating course was also true for the switchgrass runs (data not shown). Thus, the control of the steam heating was quite reproducible.

The compositions of the pretreated solids and prehydrolysates resulting from the hybrid poplar runs, along with that of the raw feedstock, are presented in Table 2. Comparison of the duplicate paddle reactor runs (runs 1 and 2) shows good reproducibility, including solids and prehydrolysate compositions and pretreated solids enzymatic digestibility. Hemicellulose was removed effectively, and the cellulose digestibility of the pretreated solids was approx 80% (based on available cellulose in the pretreated solids) for both runs.

In comparison with the laboratory-scale reactor, the solids digestibility seems to be significantly higher, and more glucan seems to have been solubilized for the smaller reactor. These results, however, may be somewhat misleading—because, as noted in Table 2, the pretreated solids sample from the laboratory-scale reactor was only partially washed before the composition was analyzed. This inadequate washing probably resulted in the carbohydrate (specifically, glucan) content being underestimated *(31)*. Thus, the significantly lower pretreated solids glucan content (58.8%) for the laboratory-scale reactor was likely higher than that reported, and accordingly, the digestibility of the solids is likely to be somewhat lower than that (89%) shown in the table.

Table 3 shows the compositions of raw and pretreated switchgrass and the enzymatic digestibility of the pretreated solids, as well as the corresponding prehydrolysate compositions. Based on these data, two points are worth noting. First, the enzyme digestibility shown for Run 2 is based on the glucan content reported in the table, but these values are somewhat biased, because they were obtained from a partially washed solids sample *(31)*. Thus, the true glucan content was somewhat higher, and the true digestibility lower, perhaps in the neighborhood of 60%. As a result, the digestibilities of runs 1 and 2 are considered comparable to each other. Second, the laboratory-scale reactor run represents the average of six runs of the laboratory-scale reactor, and pertinent statistical analysis results of these six runs are included in the table to allow meaningful data comparison.

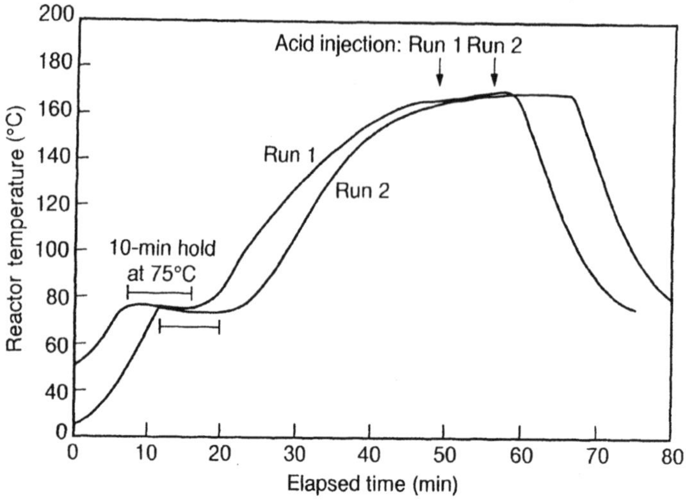

Fig. 3. Temperature profiles of hybrid poplar DN34 pretreatment runs.

It is apparent from Table 3 that the 10% solids loading runs agree well for the paddle and the laboratory-scale reactors. The difference in feedstock particle sizes (Table 1) did not significantly affect the pretreatment efficiency. The enzyme digestibility, before size reduction of the pretreated solids, of the larger (–10-mm or –3/8-in.) particles was significantly lower than that of the smaller (–3-mm or –1/8-in.) particles (cf runs 1 and 2 vs runs 3–6 and laboratory-scale reactor run). However, the paddle reactor 10 and 15% solids loadings and the laboratory-scale reactor, all being of the same smaller particle size, showed comparable digestibilities between 78 and 83% (cf runs 3–6 and laboratory-scale reactor run). For the smaller particle size, size reduction after pretreatment showed no effect (cf digestibilities of run 6 solids with and without blending). Because the solids from runs 1 and 6 showed similar digestibilities after laboratory blending, we conclude that the same pretreatment efficiency was achieved for both feedstock particle sizes.

In comparison, the two reactors resulted in comparable pretreatments (cf runs 1–5 vs laboratory-scale reactor run). Only the Klason lignin content appears to differ in product obtained from the two reactor designs. This small difference, however, is probably not statistically significant. A larger data base will have to be established for a definitive answer.

Because the preliminary paddle reactor runs showed acceptable reproducibility, agreed reasonably well with the laboratory-scale reactor runs, and resulted in no significant differences for low (10%) and high (15%) solids loadings, we have placed the reactor into service for producing large quantities of dilute-acid pretreated biomass for fermentation research. The reactor is also ready for pretreatment research. Solids loadings higher than 15% are now being tested.

Interestingly, Tables 2 and 3 indicate a tendency for the total accounting of mass to be slightly (up to 6%) more than 100% for pretreated solids. This is suspected to be largely owing to the built-in variabilities and weaknesses in the analytical methods employed (C. I. Ehrman, NREL, personal communication, 1995). For raw feedstocks, the approx 5% composition unaccounted for is likely to be a combination of acetyl groups and uronic acids that were not specifically analyzed.

Table 2
Results of Dilute Sulfuric Acid Pretreatment of Hybrid Poplar DN34

	Pretreated solids composition[a] (% dry weight)									
Run no.	Glucan	Xylan	Galactan	Arabinan	Mannan	Klason lignin	Acid-soluble lignin	Ash	Total	Enzymatic digestibility, %
Feedstock	43.7	16.0	1.0	1.8	3.1	25.2	2.7	1.4	94.9	nd[b]
Paddle reactor run 1	65.1	2.6	0.0	0.0	0.0	32.6	1.3	0.2	101.9	78
Paddle reactor run 2	65.0	2.5	0.0	0.0	0.0	33.8	1.7	0.2	103.2	79
1-L lab-scale reactor[c]	58.8	2.5	0.9	0.6	1.2	33.7	1.9	0.2	100.4	89[d]

	Prehydrolysate composition, g/L							
Run no.	Glucose[e]	Xylose[e]	Galactose[e]	Arabinose[e]	Mannose[e]	Acetic acid	HMF	Furfural
Paddle reactor run 1	4.4	16.2	0.0	0.7	2.1	4.3[e]	0.1	1.4
Paddle reactor run 2	4.2	13.8	0.0	0.4	1.8	4.5[e]	0.1	1.2
1-L lab-scale reactor	6.0	15.5	0.0	3.2	2.9	3.0	0.2	1.7

[a]Except for run noted as feedstock.
[b]nd, not determined.
[c]Compositional analysis performed on a partially washed solids sample.
[d]Determined with NREL standard procedure.
[e]Analyzed after hydrolysis (4% H_2SO_4 at 121°C for 1 h) of prehydrolysates; sugars reported should therefore be interpreted as monomeric sugar equivalents, such as glucose equivalents.

Table 3
Results of Dilute Sulfuric Acid Pretreatment of Switchgrass

	Pretreated solids composition,[b] % dry weight										Enzymatic digestibility, %	
Run no.	Extractives	Glucan	Xylan	Galactan	Arabinan	Mannan	Klason lignin	Acid-soluble lignin	Ash	Total	Without blending	With blending
Feedstock	12.2	33.6	20.0	2.0	3.3	0.0	17.7	2.5	3.1	94.4	nd	nd
Paddle reactor run 1	nd	59.8	3.2	0.0	0.0	0.0	38.2	1.5	2.9	105.6	63	84
Paddle reactor run 2	nd	53.3[c]	3.3[c]	0.0[c]	0.0[c]	0.0[c]	37.7[c]	1.0[c]	4.1[c]	99.8	64[d]	nd
Paddle reactor run 3	nd	57.5	3.5	0.0	0.0	0.0	37.9	1.4	3.7	104.0	81	nd
Paddle reactor run 4	nd	nd	nd	nd	nd	nd	nd	nd	nd	—	nd	nd
Paddle reactor run 5	nd	57.5	4.6	0.0	0.4	0.0	35.6	1.6	3.7	103.3	83	nd
Paddle reactor run 6	nd	54.6	5.1	0.0	0.0	0.0	32.5	0.9	4.8	97.9	78	80
1-L lab-scale reactor[e]	nd	57.2	3.3	0.0	0.6	0.2	34.2	1.9	4.3	101.1	78	nd

Run no.	pH	pH Prehydrolysate composition, g/L								
		Glucose[f]	Xylose[f]	Galactose[f]	Arabinose[f]	Mannose[f]	Acetic acid	HMF	Furfural	Acid-soluble lignin
Paddle reactor run 1	nd	5.2[g]	20.8[g]	nd	nd	nd	3.0	0.6	0.9	nd
Paddle reactor run 2	nd	5.2[g]	20.7[g]	nd	nd	nd	2.8	0.5	0.8	nd
Paddle reactor run 3	1.3	5.5	20.7	2.1	3.7	0.7	0.6	0.4	0.9	3.7
Paddle reactor run 4	1.3	nd	nd	nd	nd	nd	nd	nd	nd	nd
Paddle reactor run 5	1.4	5.8	21.6	1.8	3.7	0.7	2.8	0.5	0.7	5.0
Paddle reactor run 6	1.4	6.6	26.6	0.6	4.6	1.4	1.2	0.3	0.5	5.3
1-L lab-scale reactor[h]	1.3–1.4	5.7	19.7	2.6	3.9	1.2	2.7	0.4	0.6	5.1

[a]nd, not determined.

[b]Except for run noted as feedstock.

[c]Analysis performed on a partially washed solids sample.

[d]Obtained based on glucan content analyzed from a partially washed solids sample.

[e]Data presented are averages of six duplicate runs; statistical analyses for glucan, Klason lignin, acid-soluble lignin, and ash resulted in coefficients of variation (CVs) <10%; CV for xylan was 29.6%.

[f]Analyzed after hydrolysis (4% H_2SO_4 at 121°C for 1 h) of prehydrolysates; sugars reported should therefore be interpreted as monomeric sugar equivalents, such as glucose equivalents.

[g]Analyzed without further (secondary) hydrolysis of prehydrolysate.

[h]Data presented are averages of six duplicate runs; statistical analyses resulted in coefficients of variation <10% for all components, except for HMF and furfural (CVs for HMF and furfural were 23.6 and 26.3%, respectively).

Table 4
Paddle Reactor Pretreatment Mass Balances
for Switchgrass Runs[a]

Run no.	Mass balance closure, %		
	Glucan	Xylan	Klason lignin
Paddle reactor run 3	107.5	101.5	115.5
Paddle reactor run 5	108.4	106.7	108.2
Paddle reactor run 6	102.5	87.8	103.1
1-L lab-scale reactor[b]	104.4	94.8	100.5

[a]The small amounts of volatiles generated during pretreatment were neglected, and HMF and furfural were assumed to be generated from glucose and xylose, respectively.

[b]Data presented are averages of six duplicate runs; statistical analyses resulted in CVs equal to or lower than 5% for all components.

Table 4 summarizes the mass balance closures for the major components: glucan, xylan, and Klason lignin. For run 3 (dry biomass solids yield not available, *see* Table 1), calculations were made based on the average dry biomass solids yield of runs 4 and 5. It shows that comparisons between the two duplicate runs (runs 3 vs 5), between the 10 and 15% solids loading runs (runs 3 and 5 vs run 6), and between the paddle and laboratory-scale reactor runs (runs 3, 5, and 6 vs laboratory-scale reactor run) all appear to agree reasonably well. However, in general, there was an overaccounting of glucan (by up to 8%) and Klason lignin (by up to 16%). Xylan recovery was not as consistent, in that the two 10% solids loading runs resulted in an average 104% recovery, whereas the 15% solids loading run resulted in only 88%, and the laboratory-scale reactor run achieved 95%.

It should be noted that three assumptions were used to derive the numbers in Table 4:

1. The volatiles generated in pretreatment were negligible;
2. HMF and furfural were generated from glucose and xylose, respectively; and
3. Regardless of the reactor mass recoveries (Table 1) determined, the total amount of prehydrolysate in each run was calculated by subtracting the weight of the dry biomass solids recovered after pretreatment from the total reactor loading.

The first assumption was justified from our consistent observations of the negligible amount of vent condensates. The second should not introduce any significant errors because of the low levels of the minor sugars. The third was also considered justifiable, because handling the large quantities of the paddle reactor contents was difficult, and the reactor mass recovery numbers that showed some scattering served only to provide an indication of how consistent this reactor discharge handling was.

The consistent mass balance overaccounting is likely to be, at least in part, owing to the overaccounting of mass discussed. It thus appears desirable to improve the mass balance closures by improving the handling of the reactor contents (before and after pretreatment) and, possibly, the various biomass analytical methods. We are currently working on improving reactor contents handling.

Improving biomass analytical methods is a continual goal of the NREL ethanol project chemical analysis and testing (CAT) task (C. I. Ehrman, NREL, personal communication, 1995).

CONCLUSIONS

A 100-L, custom-fabricated paddle pretreatment reactor has been constructed, which provides a valuable tool for high-solids operation. The reactor is useful for both producing pretreated biomass and allowing study of high-solids pretreatment. No significant scale-up differences in pretreatment efficiency are apparent between a 1-L laboratory-scale reactor and the paddle reactor. The paddle reactor, designed to be able to handle high-solids dilute-acid pretreatment, has thus far been successfully tested for solids loadings up to 15%, and different solids loadings do not appear to affect pretreatment efficiency. Moreover, different biomass particle sizes, up to –10 mm (–3/8 in.), do not appear to affect pretreatment efficiency.

Material balances estimated for the reactor resulted in reasonably good closures, although it is desirable to have the overall accuracy further improved.

ACKNOWLEDGMENTS

We thank C. Wyman for providing valuable guidance during a modification phase of the paddle reactor and A. Wiselogel for providing relevant information on the sources of the lignocellulosic feedstocks used in this study. We also thank the NREL ethanol project CAT task and Y. Y. Lee's laboratory (under subcontract No. XAW-4-14170-01) at Auburn University for analytical support. Further, we thank C. Ehrman and R. Elander for their valuable input on chemical analysis and process economics aspects, respectively, during the course of the study. Five of the six laboratory-scale, switchgrass runs were conducted by K. Connors, R. Elander, K. Evans, S. Hall, and J. Hora. This work was funded by the Biochemical Conversion Element of the Biofuels Program of the US Department of Energy.

REFERENCES

1. Chang, M. M., Chou, T. Y. C., and Tsao, G. T. (1981), *Adv. Biochem. Eng.* **20,** 15–42.
2. Dale, B. E. (1985), *Ann. Rep. Fermentation Processes* **8,** 299–323.
3. Datta, R. (1981), *Process Biochem.* **June/July,** 16–19.
4. Fan, L. T., Lee, Y., and Gharpuray, M. M. (1982), *Adv. Biochem. Eng.* **23,** 157–187.
5. Han, Y. W. (1978), *Adv. Appl. Microbiol.* **23,** 119–153.
6. Millett, M. A., Baker, A. J., and Satter, L. D. (1976), *Biotechnol. Bioeng. Symp.* **6,** 125–153.
7. Shimizu, K. (1981), *JARQ* **14,** 4244–4248.
8. Ghosh, P. and Singh, A. (1993), *Adv. Appl. Microbiol.* **39,** 295–333.
9. Grohmann, K., Wyman, C. E., and Himmel, M. E. (1992), *ACS Symp. Ser.* **476,** 354–392.
10. Grethlein, H. E., Allen, D. C., and Converse, A. O. (1984), *Biotechnol. Bioeng.* **26,** 1498–1505.
11. Grohmann, K., Torget, R., and Himmel, M. (1985), *Biotechnol. Bioeng. Symp.* **15,** 59–80.
12. Grohmann, K., Torget, R., and Himmel, M. (1986), *Biotechnol. Bioeng. Symp.* **17,** 135–151.
13. Grohmann, K. and Torget, R. W. (1992), US Patent No. 5,125,977.
14. Knappert, D., Grethlein, H., and Converse, A. (1980), *Biotechnol. Bioeng.* **22,** 1449–1463.
15. Knappert, D., Grethlein, H., and Converse, A. (1981), *Biotechnol. Bioeng. Symp.* **11,** 67–77.
16. Schell, D. J., Water, P. J., and Johnson, D. K. (1992), *Appl. Biochem. Biotechnol.* **34/35,** 659–665.
17. Torget, R., Werdene, P., Himmel, M., and Grohmann, K. (1990), *Appl. Biochem. Biotechnol.* **24/25,** 115–126.

18. Torget, R., Walter, P., Himmel, M., and Grohmann, K. (1991), *Appl. Biochem. Biotechnol.* **28/29,** 75–86.
19. Torget, R., Himmel, M., and Grohmann, K. (1992), *Appl. Biochem. Biotechnol.* **34/35,** 115–123.
20. Torget, R. and Hsu, T. (1994), *Appl. Biochem. Biotechnol.* **45/46,** 5–21.
21. Israilides, C. J., Grant, G. A., and Han, Y. W. (1978), *Appl. Environ. Microbiol.* **36(1),** 43–46.
22. Mehlberg, R. L. (1977), MS Thesis, Purdue University, West Lafayette, IN.
23. Mehlberg, R. L. and Tsao, G. T. (1979), Paper presented at the 178th ACS National Meeting, Washington, DC.
24. Brink, D. L. (1993), US Patent No. 5,221,357.
25. Brink, D. L. (1994), US Patent No. 5,366,558.
26. Fan, L. T., Gharpuray, M. M., and Lee, Y. (1981), *Biotechnol. Bioeng. Symp.* **11,** 29–45.
27. Grohmann, K., Himmel, M., Rivard, C., Tucker, M., Baker, J., Torget, R., and Graboski, M. (1984), *Biotechnol. Bioeng. Symp.* **14,** 137–157.
28. Torget, R., Himmel, M., Wright, J. D., and Grohmann, K. (1988), *Appl. Biochem. Biotechnol.* **17/18,** 89–104.
29. Hinman, N. D., Schell, D. J., Riley, C. J., Bergeron, P. W., and Walter, P. J. (1992), *Appl. Biochem. Biotechnol.* **34/35,** 639–649.
30. Ehrman, C. I. and Himmel, M. E. (1994), *Biotechnol. Tech.* **8(2),** 99–104.
31. Hsu, T. and Nguyen, Q. (1995), *Biotechnol. Tech.* **9(1),** 25–28.
32. Torget, R., Hatzis, C., Hayward, T. K., Hsu, T., and Philippidis, G. P. (1995), Paper presented at the 17th Symposium on Biotechnology for Fuels and Chemicals, Vail, CO, May 7–11.

0273-2289/96/5758—0019$7.50

Downstream Processing of Acetate Fermentation Broths by Nanofiltration

IN SOO HAN AND MUNIR CHERYAN*

University of Illinois, Agricultural Bioprocess Laboratory, 1302 West Pennsylvania Avenue, Urbana, IL 61801

ABSTRACT

Acetate can be separated from fermentation broths and partially purified by nanofiltration (NF). Membrane performance was a function of pressure, pH, concentration of acetate, temperature, and the presence of other media components. With Nitto-Denko's NTR729 membrane, average acetate rejection was 60%, glucose rejection was 99%, and flux was 15 L/m^2/h at 200 psig, 30°C, pH 5.6, and 20 g/L acetic acid. The best downstream strategy is to clarify the fermentation broth by microfiltration (MF), recycle the cells for improving fermenter productivity, nanofilter the cell-free broth, and then evaporate the permeate. The cost of ceramic MF and NF purification is about $66/t acetate, compared to the cost of potassium acetate ($950/t) or calcium-magnesium acetate ($700/t).

Index Entries: Acetate; membrane technology; nanofiltration; separation.

INTRODUCTION

Acetic acid is one of the most important commodity chemicals, with a US production of 3.6 billion lb/yr, representing a market value of $1.1 billion. Much of it is petroleum-derived and converted into acetates, e.g., polyvinyl acetate, sodium acetate, potassium acetate, calcium magnesium acetate (CMA), and acetate esters. Acetate can also be produced by the anaerobic fermentation of dextrose by *Clostridium thermoaceticum*. This organism, however, produces acetate at low concentrations (2–5%) and over long periods (3–5 d), unless the medium is supplemented with expensive nutrients *(1–3)*.

The feedstock (e.g., dextrose) and downstream processing constitute the largest costs in the manufacture of acetates. Membrane technology can be used in most phases of downstream processing. Microfiltration (MF) and ultrafiltration (UF) are effective for separation and recycling of the cells *(4–6)*. Nanofiltration (NF) is a relatively new membrane technique that can separate organic acids (depending on the degree of dissociation) and monovalent salts from sugars and other organic compounds *(7,8)*. NF should allow us to purify the acetate and recycle the nutrients simultaneously, thus lowering the cost of production. Furthermore, a combination of high-rejection (HR) and low-rejection (LR) membranes could be used to obtain

*Author to whom all correspondence and reprint requests should be addressed.

very high concentrations of acetic acid. This article reports on the potential applicability of membrane technology for the separation, purification, and concentration of acetates from fermentation broths. Membrane performance data obtained with a CMA fermentation broth were combined with earlier studies on model sodium acetate solutions to obtain a preliminary cost estimate of applying membrane technology in downstream processing.

MATERIALS AND METHODS

Flux and rejection characteristics of several membranes were studied using a stirred filtration cell as described earlier *(7)*. The fermentation broth was from our CMA fermentation studies in which dextrose was fermented in the presence of yeast extract and other trace elements by *C. thermoaceticum (5,6)*. Dolime (CaO·MgO) was used as the neutralizing agent during the fermentation. Prior to use in the NF test cell, the broth was clarified using a 0.2-μ ceramic MF module (CeraMem Separations, Waltham, MA) and its pH adjusted using pure glacial acetic acid, if necessary.

Several reverse osmosis (RO) and NF membranes were evaluated with model acetate solutions. From this study *(7)*, two membranes were selected for the first NF-purification stage: NTR729 (Nitto-Denko, Japan) and MX07 (Osmonics, Minnetonka, MN). Rejection (R) of components is calculated as:

$$R = 1 - C_p/C_R \tag{1}$$

where C_p is concentration of a component in the permeate and C_R is concentration of the component in the retentate. Flux data can be correlated by the following equation:

$$J = A\,(P_T - \Delta\pi) \tag{2}$$

where J is flux (LMH), P_T is the applied pressure, $\Delta\pi$ is the osmotic pressure difference across the membrane, and A is a membrane permeability coefficient.

Acetate and glucose were analyzed by high-performance liquid chromatography (HPLC) using the HPX-87 column (Bio-Rad, Hercules, CA) as described earlier *(5–7)*. Unless otherwise mentioned, the concentration of acetates is reported in terms of acetic acid. An acetic acid concentration of 1 g/L is equivalent to 1.225 g of CMA and 1.375 g of sodium acetate. Minerals (Na, K, Ca, Mg) were measured by atomic absorption as practiced in the Microanalytical Laboratory, Department of Chemistry, University of Illinois, Urbana.

RESULTS AND DISCUSSION

Process Design

A diagram of a possible downstream process with membranes is shown in Fig. 1. The process is designed in three stages: clarification, purification, and concentration. After the fermentation, the whole fermentation broth is pumped to a microfiltration unit that serves two purposes: to clarify the broth prior to NF and to recycle cells to the fermenter to improve productivity *(5,6)*. The clarified broth is then sent to the first of two NF stages. The first is for separation/purification of the acetate; it consists of membranes with low acetate rejection and high glucose rejection (glucose was used as an index of nutrients). Since the acetate passage was significantly better at low pH *(7)*, the fermentation broth was adjusted to pH 5.6. This pH

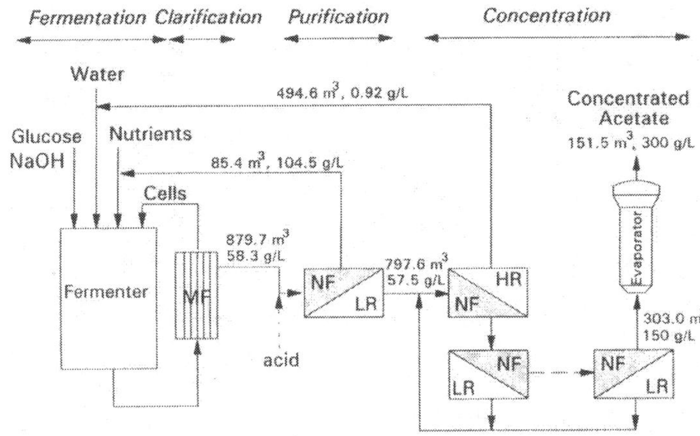

Fig. 1. Process for downstream processing of acetate fermentation broths. Data refer to volumetric flow rate (m^3/h) and concentration of sodium acetate (g/L).

value was a compromise based on the rejections of acetate vs glucose, and the cost of acidifying the broth. This required 3.8 L of full-strength glacial acetic acid/m^3 of sodium acetate fermentation broth, based on titration curves.

The retentate from the NF-purification stage contains the nutrients and some acetate, which are recycled to the fermenter. The permeate from this stage is a dilute and substantially purified acetate solution. This permeate stream is sent to the second NF operation, a series of HR and LR membranes. The HR membrane (e.g., FT-30, Dow, Midland, MI or TLC, Fluid Systems, San Diego, CA) should produce a permeate that is mostly water, which can be recycled to the fermenter. The retentate, containing partially concentrated acetate, is sent to an LR membrane (e.g., NF40, Dow). Owing to the partial passage of acetate through the LR membrane, the $\Delta\pi$ will be low, thus keeping the driving force (and flux) at reasonably high levels, according to Eq. (2). The retentate from this LR membrane, containing higher acetate concentration than the feed and the permeate (since it is partially rejected), is sent to more LR units until the desired concentration of acetate is obtained. The permeate from these particular LR modules contains dilute acetate, which can be recycled back to the first HR membrane in this stage. Thus, in the ideal case, the product from this series of HR–LR membranes is water (from the HR membrane) and concentrated acetate (from the last LR membrane stage).

The six-effect evaporator (for final concentration up to 300 g/L) shown in Fig. 1 has been included in the economics. Depending on the nature of the product, a subsequent drying step may be needed (e.g., for CMA) or further evaporation (e.g., for sodium or potassium acetates). This last moisture removal step has not been included in this analysis.

Membrane Performance

Initial studies were conducted with low-strength CMA fermentation broth, which contained 17 g/L acetic acid (20.8 g/L CMA). When adjusted from pH 6.8 to 5.6 with concentrated acetic acid, the acetic acid concentration of the tested sample

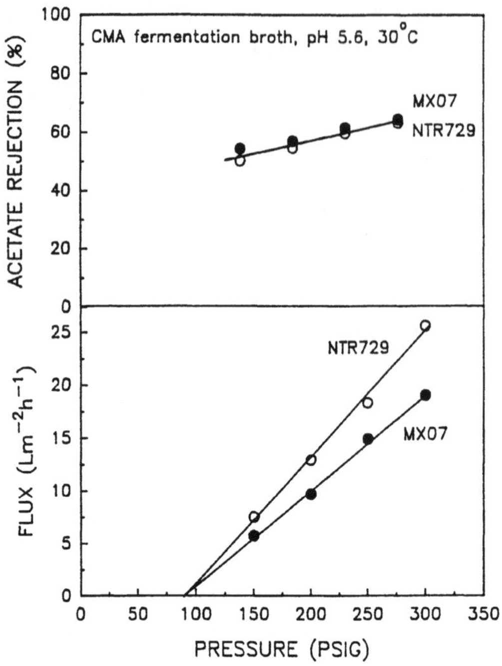

Fig. 2. Performance of NF membranes with CMA fermentation broth. The pH of the CMA broth was adjusted from 6.8 to 5.6, resulting in a final concentration of 19 g/L acetic acid.

increased to 19 g/L. Other components included about 26 g/L other total solids, mostly salts and vitamins (the Na^+ and K^+ ions are in the culture medium). As shown in Fig. 2, flux was 10–15 L/m^2/h (LMH) at 200 psi, whereas the rejection was 60%. In contrast, model sodium acetate solution had a flux of 45 LMH and 40% rejection under the same conditions with the same membranes *(7)*. The different ionic conditions were probably responsible for the deviation from model solution behavior. For example, the observed osmotic pressure difference ($\Delta\pi$) was about 90 psi, and permeability constant (A) was 1.81 LMH/bar for the NTR729 membrane at 30°C. In contrast, with model solutions at 20 g/L, $\Delta\pi$ was 60 psi and A = 4.03 LMH/bar. The higher osmotic pressure of the fermentation broth was owing to other salts and vitamins in the culture medium.

The behavior of the minerals (K^+, Na^+, Ca^{2+}, Mg^{2+}) during NF of CMA fermentation broth is shown in Fig. 3 with the NTR729 membrane (this was a different batch of CMA broth: This tested sample contained 23 g/L acetate). The concentration of the elements in the retentate did not change much as the pressure was increased. However, higher pressure caused a higher volumetric flux, which resulted in a dilution of the components in the permeate, as shown in the upper graph of Fig. 3. This is the main reason for the observed phenomenon of higher apparent rejection at higher pressures.

In addition, as expected of NF membranes *(8)*, rejections of Ca^{2+} and Mg^{2+} were higher than Na^+ and K^+ rejections. Although the Na^+ and K^+ concentrations were about the same in the retentate and permeate, indicating low rejection of these ions,

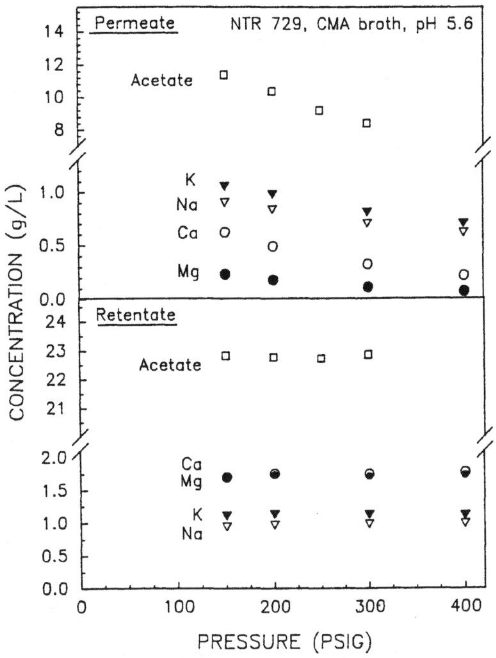

Fig. 3. Mineral composition of permeate and retentate from the NF of CMA fermentation broth. The pH of the CMA broth was adjusted from 6.8 to 5.6, resulting in a final concentration of 23 g/L acetic acid.

the permeate contained much lower levels of Ca^{2+} and Mg^{2+} than the retentate. At 200 psi, rejections of Na^+ and K^+ were 14%, whereas Ca^{2+} rejection was 72% and Mg^{2+} was 90%. At pH 5.6, acetic acid was 87% ionized *(7)*, and its separation was dependent on the cation composition of the feed. Thus, the nature of the cations and the pH influence the degree of separation of organic acids.

Figure 4 shows a batch concentration with the NTR729 membrane. Glucose was added to the CMA broth to 10 g/L and acetic acid was 20 g/L. The concentration of acetic acid and glucose is plotted against the volume concentration ratio (VCR) for both permeate and retentate. As the concentration of acetate increases in the feed/retentate, the concentration of acetate in the permeate increases. Until a VCR of ~3 (corresponding to 40 g/L of acetate and 30 g/L of glucose), rejections were constant at ~95 and 55%, respectively. Above VCR 3, the rejections decreased for both compounds, indicating perhaps that the ion-exchange capacity of the membrane had been exceeded. Similar effects (higher acetate concentration causing a decrease in acetate rejection) were reported for model systems *(7)*.

The product (permeate) from the NTR729 membrane was clear and colorless, in contrast to the dark-colored feed and retentate. The HPLC chromatograms also showed a substantial degree of purification *(9)*.

Economics

The basis for the economic analysis is a 1000 t/d sodium acetate plant. The cost estimates presented in this article are for downstream processing from the MF to

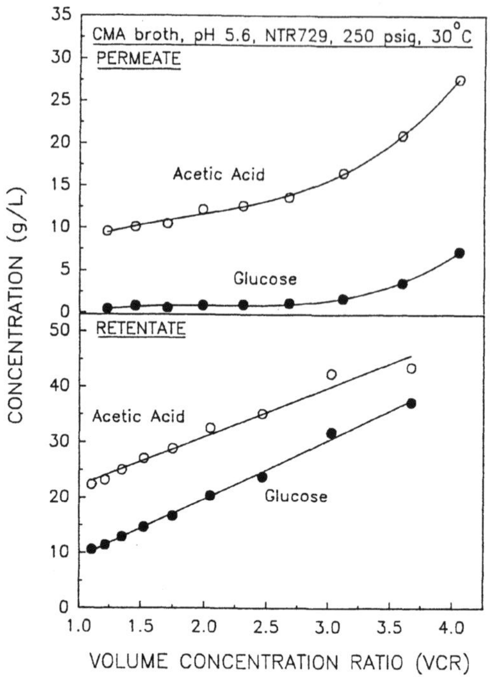

Fig. 4. Effect of concentrating CMA fermentation broth with NTR729 membrane on retentate and permeate concentrations. The pH of the CMA broth was adjusted from 6.8 to 5.6, resulting in a final concentration of 20 g/L acetic acid. Glucose (10 g/L) was added to the broth.

the evaporation step as shown in Fig. 1. The target concentration of the product from the evaporator is 300 g/L. For comparison purposes, three options have been considered: a process as shown in Fig. 1 (labeled "Option 1—MF, NF, NF"), a process where there is no NF concentration ("Option 2—MF, NF") and a process that bypasses both NF-concentration and NF-purification stages, so that the MF-clarified broth goes directly to the evaporator (labeled "Option 3—MF"). The membrane units are operated for 22 h/d, with 2 h for cleaning. Broth from the fermenter is at 60°C, acetic acid concentration of 58.3 g/L, and pH of 6.8.

MF

Two types of MF membrane modules were considered: hollow fibers of 2–3 mm internal diameter (available from Koch-Romicon, A/G Technology, or Dow), and ceramic (alumina) membrane modules (e.g., with 3–4 mm internal diameter from US Filter, Warrendale, PA or with 2-mm channel height from CeraMem). The broth flow rate from the MF system is 879,737 L/h, based on a material balance as shown in Fig. 1. From performance studies done in our laboratory and the relative capital and replacement costs for these membrane modules, the CeraMem ceramic membrane system was chosen (*see* Table 1 for operating parameters, concentrations, flows, and unit costs).

Table 1
Specifications and Operating Levels of Membrane Processes
Used for Downstream Processing of Acetate Fermentation Broths

Operating parameter	MF	NF purification	NF concentration
Module type	Ceramic	Spiral[a]	Spiral[a]
Operating pressure (psi)	75	500	700
Temperature (°C)	60	50	50
pH	6.8	5.6	5.6
Pressure drop/module (psi)	20	15	15
Recirculation (GPM/module)	300	200	200
Energy (W/m^2)	250	55	55
Membrane area (m^2)	2513	20,712	201,338
System cost (\$/m^2)	2150	225	225
Membrane replacement cost (\$/m^2/yr)[b]	80	50	50
Feed flow rate (m^3/h)	—	883.1	797.6
Permeate flow rate (m^3/h)	879.7	797.6	494.6
Acetate in feed (g/L)	58.3	62.1	57.5
Acetate in retentate (g/L)	58.3	104.5	150
Acetate in permeate (g/L)	58.3	57.5	0.92

[a] 8 × 40 in. modules, with 60-mil spacers, 25 m^2 each.
[b] Five-year life for ceramic membranes, 1-yr life for NF polymeric membranes.

NF Purification

The permeate from the MF unit is the feed to the NF-purification step. The operating parameters and flows for this step are given in Table 2. The NF area is determined as follows:

$$\text{NF membrane area (m}^2\text{)} = \text{NF permeate flow (L/h)/NF flux (LMH)} \quad (3)$$

Because the concentration of acetate and glucose is continuously increasing during a batch process or with each stage in a feed-and-bleed process *(4)*, there is a simultaneous increase in osmotic pressure, which causes flux to decrease and eventually become zero when the osmotic pressure of the retentate is the same as the applied transmembrane pressure. To design this stage, a computer program was written to take flux and rejection data from our previous work *(7,9)* and perform calculations until the final product stream met the requirements for the process: In this case, it is a product stream (permeate) concentration of 57.5 g/L acetate at a total permeate flow rate of 797,617 L/h (these parameters come from the design of the NF-concentration stage: *see next section*). The outer loop varied the amount of area to meet the requirements.

A batch process cannot achieve 100% recovery of the acetate owing to the buildup of the osmotic pressure. Some acetate remains in the retentate and will have to be recycled back to the fermenter with the other nutrients. A high recovery of acetate (i.e., a high VCR) in the purification stage is desirable, because less acetate recovery in this step means a higher feed rate into the NF unit, which then increases the MF membrane area. At maximum VCR, recovery of acetate is maximized, MF

Table 2
Capital Investment ($) for the Downstream Processing of Acetate Fermentation Broths
(Basis: 320,000 t/yr of Acetate Production)

Item	Option 1, MF, NF, NF	Option 2, (MF, NF)	Option 3, (MF)
Membranes			
MF (CeraMem)	4,091,200	4,691,200	4,158,400
NF purification	4,538,900	4,538,900	0
NF concentration	45,319,000	0	0
Installation (10% of 1)	5,454,910	923,010	415,840
Evaporator[a] (installed)	2,199,300	4,850,100	4,850,101
Total fixed capital	62,203,310	15,003,210	9,424,341

[a]From Tejayadi and Cheryan *(10)*.

area is minimized, but NF area is maximized. Thus, there is an optimum VCR at which the NF-purification step should be operated to obtain a net minimum cost. For this case, the optimum VCR was 10.31 based on economic analysis *(9)*, resulting in a net 82.93% recovery of the acetate, where recovery is defined as follows:

$$\text{Acetate recovery (\%)} = [\text{acetate (kg) from last stage of NF concentration/ acetate (kg) in the feed to the NF-purification stage}] \times 100 \quad (4)$$

The average NF flux was 39.5 LMH, resulting in an NF-purification area of 20,172 m^2, and an MF area of 2513 m^2. Table 1 shows flow rates and acetate concentrations in this step.

CONCENTRATION BY NF

The criteria in this step are:

1. The acetate concentration in the retentate of the final stage should be 150 g/L;
2. The permeate from the first stage in this step should be as close to zero acetate as possible; and
3. The weight of acetate produced should be 1000 t/d, which translates into an output of 303,030 L/h of product stream containing 150 g/L of acetate.

The type of modules used here are essentially the same as for the NF purification step, except for the membrane within the modules. The operating parameters and costs are shown in Table 1.

The first module in this operation is used only for water removal, using an HR membrane. This permeate could be recycled to the fermenter as make-up water. The flux-concentration data for this stage was obtained from our earlier

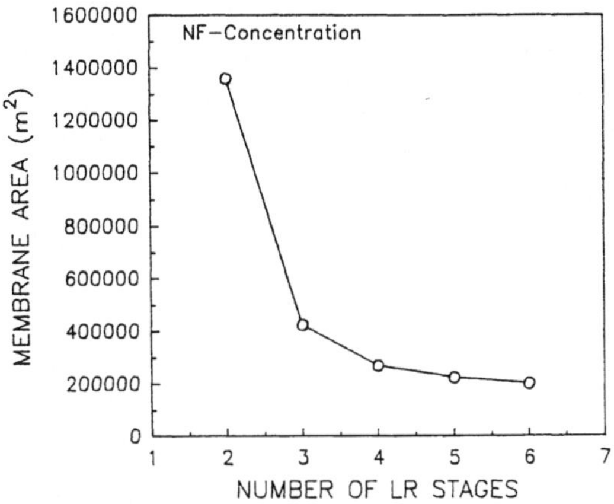

Fig. 5. Effect of number of LR stages in the NF-concentration operation on the membrane area required in the NF-concentration operation.

study *(7,9)*. By increasing the number of LR membranes, higher flux could be achieved in each membrane owing to the smaller concentration gradient. The permeates from LR membranes are recycled back to the feed stream of the first (HR) membrane. By trial and error, the intermediate retentate concentration of each stage was obtained. The effect of number of stages on area is shown in Fig. 5. At a minimum, one HR membrane and two stages of LR membranes are required. A six-stage NF-concentration system was selected as a compromise.

Unit Costs

The most important parameter for determining capital cost is the membrane area in each stage. Table 1 shows some of these unit costs. Table 2 shows fixed capital costs for the process. There is a considerable difference in cost between Option 1 and Option 2, owing to high cost of using NF for concentration to 150 g/L acetate. However, the difference in capital cost between Options 2 and 3 may be justified if (1) there is a need for a purified acetate product or (2) if the recycle of the nutrients can partially offset the cost of NF purification. There are no data available as yet on the feasibility of recycling the spent nutrient broth. Thus, we have assumed only a 50% credit for the nutrients, i.e., a reuse of the nutrients only once or a continuous recycle of the nutrients with a 50% make-up of the nutrients.

Table 3 shows the annual operating cost with the three options. There is a substantial saving in evaporation cost with Option 1, but this is more than offset by the costs of membrane replacement, cleaning, and electric power needed for NF concentration. With Options 2 and 3, the net downstream cost (excluding drying) adds only \$37–66/t of acetate. This is small compared to the cost of CMA (\$700/t) or potassium acetate (\$950/t). A large portion of the operating costs is for purchasing acetic acid to lower the pH of the broth for NF purification. Efforts are under way in our laboratory to adapt the *C. thermoaceticum* culture to operate at lower pH. If successful, this would significantly lower the downstream processing cost.

Table 3
Annual Operating Cost ($/yr) for Downstream Processing

Item	Option 1 (MF,NF,NF)	Option 2 (MF,NF)	Option 3 (MF)
Electric power for membranes			
MF ceramic	258,016	258,016	228,448
NF purification	203,084	203,084	0
NF concentration	2,029,749	0	0
Energy for evaporation	2,413,551	10,165,774	10,165,774
Membrane replacement			
MF ceramic	201,040	201,040	201,040
NF purification	1,008,645	1,008,645	0
NF concentration	10,068,897	0	0
Membrane cleaning ($18/m²/yr)	4,039,974	415,890	46,782
Labor ($24/h)	30,000	15,000	15,000
Maintenance (3% of fixed capital)	1,866,101	450,096	282,730
Depreciation (10% of fixed capital)	6,220,336	1,500,321	942,434
Acetic acid for pH adjustment ($0.68/L)[a]	16,003,610	16,003,610	0
Total cost	44,343,003	30,221,476	11,882,208
Nutrient credit[b]	(8,960,000)	(8,960,000)	0
Net downstream cost:			
$/yr	35,383,003	21,261,476	11,882,208
$/t of acetate	110.6	66.4	37.1

[a]Anon. *(11)*.
[b]Assuming 50% recycle of nutrients, based on data of Witjitra et al. *(3)*.

ACKNOWLEDGMENTS

Funds were provided by the Illinois Corn Marketing Board, US Department of Agriculture through the NRICGP program, Minnesota Corn Promotion and Research Council, and the Illinois Agricultural Experiment Station. We are grateful for contributions of membranes from Desalination Systems, Inc., Dow, Fluid Systems, Nitto-Denko, and Osmonics.

REFERENCES

1. Parekh, S. R. and Cheryan, M. (1991), *Appl. Microbiol. Biotechnol.* **36,** 384–387.
2. Parekh, S. R. and Cheryan, M. (1994), *Biotechnol. Lett.* **16,** 139–142.
3. Witjitra, K., Shah, M. M., and Cheryan, M. (1996), *Enzyme Microb. Technol.*, in press.
4. Cheryan, M. (1986), *Ultrafiltration Handbook*. Technomic, Lancaster, PA.
5. Parekh, S. R. and Cheryan, M. (1994), *Enzyme Microb. Technol.* **16,** 104–109.
6. Shah, M. M. and Cheryan, M. (1995), *Appl. Biochem. Biotechnol.* **51–52,** 413–422.
7. Han, I. S. and Cheryan, M. (1995), *J. Membrane Sci.* **107,** 107–113.
8. Raman, L. P., Rajagopalan, N., and Cheryan, M. (1994), *Chem. Eng. Prog.* **90(3),** 68–74.
9. Han, I. S. (1994), MS thesis, University of Illinois, Urbana.
10. Tejayadi, S. and Cheryan, M. (1995), *Appl. Microbiol. Biotechnol.* **43,** 242–248.
11. Anon. (1994), *Chem. Marketing Reporter* **245,** 28–36.

0273-2289/96/5758—0029$7.50

Chemical Interactions Between Aqueous and Organic Phases in a Reactive Extraction Process

CHRISTIAN SCHELER,[1] MILAN POPOVIC,[1]
EUGENE IANNOTTI,[2] RICK MUELLER,[2] AND RAKESH BAJPAI*,[3]

[1]*Department of Biotechnology, Technische Fachhochschule, Berlin, Germany;* [2]*Department of Agricultural Engineering, University of Missouri—Columbia;* and [3]*Department of Chemical Engineering, University of Missouri—Columbia, MO 65211*

ABSTRACT

Partitioning of lactic acid between aqueous phase and kerosene containing tri-*n*-octyl phosphine oxide (TOPO) as a reactive agent was measured and analyzed in terms of an association model between lactic acid and TOPO. Equilibrium association was calculated to involve 1.1 mol of lactic acid/mol of TOPO. ^{31}P-NMR and FT-IR analyses found no evidence of covalent bond formation between lactic acid and TOPO in solutions in kerosene, and pointed to hydrogen bond formation between the two chemicals. In a hydrophobic, microporous, hollow-fiber-membrane module, the simultaneous presence of lactic acid, TOPO, and the back-extractant NaOH resulted in drastic membrane fouling as evidenced by the loss of kerosene flux through the membrane. The membrane could be restored back to the original kerosene flux after extensive washing.

Index Entries: Partition coefficient; NMR; FT-IR; lactic acid; Tri-*n*-octyl phosphine oxide.

INTRODUCTION

A number of commercially important products of biological origin express toxicity towards the biological agents that produce them *(1,2)*. This results in dilute solutions of the products, reduced observed yields, low production rates, and increased costs of product recovery. Hence, on-line methods of product recovery are being aggressively explored. Of these, solvent extraction is particularly attractive owing to the possibility of engineering high selectivity for desired compounds and the potential for not only minimizing the product toxicity but also reducing the need for extensive downstream processing *(4,5)*. Unfortunately a number of solvents with high product specificity and good partitioning characteristics, are toxic towards the microorganisms that produce the biological products *(5)*. Solvent toxicity has been

*Author to whom all correspondence and reprint requests should be addressed.

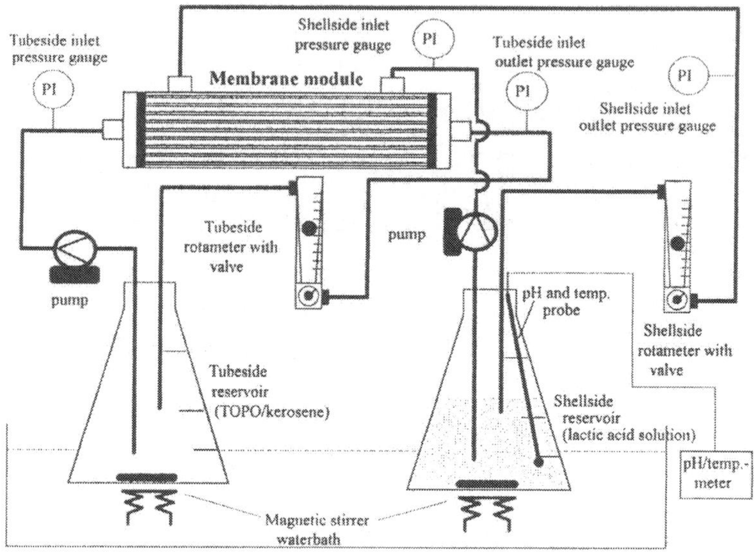

Fig. 1. A schematic of a membrane-associated reactive extraction system for recovery of lactic acid from fermentation broths.

demonstrated to stem from the solubilized solvents in the aqueous phase and from the exposure of the microorganisms to the solvent–aqueous phase interface *(5,6)*. In this respect, solvent extractions in which microporous membranes are used to separate the solvent and aqueous phases are promising, since they eliminate the problems of interfacial toxicity while reducing the costs associated with creation and maintenance of dispersions *(7–9)*. A schematic of such a process for lactic acid fermentation has been presented in Fig. 1. This process involves reactive extraction of lactic acid from the fermentation broth into a mixture of kerosene and tri-*n*-octyl phosphine oxide (TOPO). Additionally, the presence of a membrane barrier between the aqueous and solvent phases has been further exploited by using a dispersion of alkali droplets in solvent. These alkali droplets back-extract lactic acid from the solvent phase and enhance extraction of the organic acid from the fermentation broth. Separation of the neutralized alkali droplets from the solvent phase has been shown to produce a concentrated stream of purified sodium lactate *(9)*.

Incorporation of such reactive systems in a scheme for on-line product recovery poses a number of questions regarding their ramifications on the fermentation and the extraction process. The objective of the present research was to address these questions. The specific objectives were to study: (1) the interactions between the reactive agent TOPO and lactic acid, and (2) the effect of TOPO and other components present in the aqueous and solvent phases on membrane fouling.

MATERIALS AND METHODS

TOPO was obtained as 98% pure chemical from Aldrich Chemical Co. (Milwaukee, WI). It was dissolved in kerosene obtained from Phillips Petroleum Co. (Bartlesville, OK). Kerosene was washed twice with equal amounts of deionized water to remove water-soluble chemicals and passed through a 0.2-μm filter before use. The solubility of TOPO in kerosene was found to be 19.7% (w/w) at 20 ± 1°C. Lactic acid solutions were prepared by diluting 88% lactic acid stock solutions (Pfalz and Bauer) to either 8 or 16% concentration. Measurements of equilibrium partitioning coefficients of lactic acid in kerosene with or without TOPO were conducted at 37°C in Erlenmeyer flasks. After contacting the two phases for 24 h at an agitation rate of 200 rpm in a rotary incubator, the two phases were separated. Lactic acid concentration in the aqueous phase was determined directly by HPLC using a procedure described previously *(5)*. The concentration of lactic acid in the organic phase was similarly determined after back-extraction of lactic acid in sodium hydroxide solution.

The experimental setup shown in Fig. 1 was procured from Hoechst-Celanese Corporation (Charlotte, NC). It consisted of Liqui-Cel Laboratory Liquid–Liquid Extraction system 5PCS-100 and the Liqui-Cel Laboratory Membrane Contactor 5PCM-106. It contained 2100 Celgard X-10 microporous polypropylene hollow fibers (internal diameter 210 μm, external diameter 300 μm, length 16 cm, effective pore diameter 0.05 μm, porosity 30%, and effective surface area 0.23 m^2). Changes in kerosene flux through the membrane at a transmembrane pressure of 29 psi were used as measures of membrane fouling. Membrane permeability was independent of the transmembrane pressure. Membrane extractions were conducted at 37 ± 0.5°C by countercurrent flow of organic phase through the tube side (250 mL/min) and aqueous phase on the shell side at the same flow rate. Details of these measurements have been presented by Scheler *(10)*.

RESULTS AND DISCUSSION

Interactions Between Lactic Acid and TOPO in Kerosene

Equilibrium partitioning of lactic acid between water and TOPO in kerosene was measured at different concentrations of lactic acid and TOPO. The distribution coefficients (m) and partition coefficients (m^*) were calculated from the concentrations of lactic acid in the two phases according to the following definitions:

$$\text{Distribution coefficient } (m) = (C_{LAorg}/C_{LAaq}) \tag{1}$$

and

$$\text{Partition coefficient } (m^*) = (C_{HAorg}/C_{HAaq}) = m^* \, (1 + 10^{\,pH - pK}) \tag{2}$$

where LA represents the concentration of total lactic acid and HA the concentration of undissociated lactic acid. pH is the negative logarithm of hydrogen ion concentration in aqueous medium at equilibrium, and pK is the negative logarithm of equilibrium dissociation constant of lactic acid.

The partition coefficient data have been presented in Fig. 2. These decrease linearly with increasing total lactic acid concentration and increase with increasing TOPO concentration. Similar results were also obtained by Golob et al. *(11)* and by Kuo and Gregor *(12)*. Experiments were also conducted with technical-grade TOPO and at room temperature (25°C). However, no significant differences were found between the partition coefficients with technical and analytical grades of TOPO. Within the temperature range, the effect of temperature was also minimal.

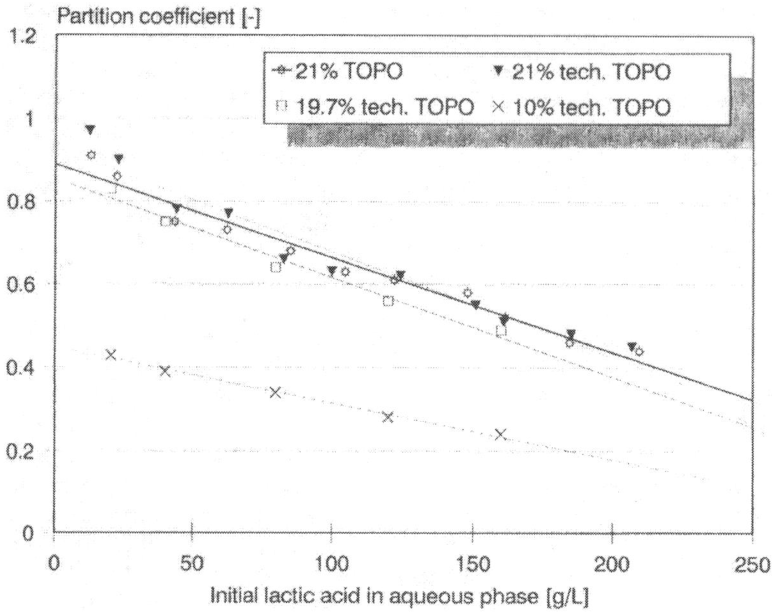

Fig. 2. Measured partition coefficients for lactic acid between aqueous phase and kerosene/TOPO.

A phase equilibrium model of the type proposed by Fahim et al. *(13)* was used to analyze the partition coefficient data. According to this model, if the interactions between lactic acid in aqueous phase and TOPO in the solvent phase are written as:

$$n\mathrm{HA} + \mathrm{S_o} \rightleftharpoons [\mathrm{HA}]^n S_o \quad (3)$$

the partition coefficient can be derived to be:

$$\text{Partition coefficient} = m^* = n\mathrm{K_1}\,(\mathrm{HA_{aq}})^{n-1}[(\mathrm{S_{oT}}) - (\mathrm{HA_o}/n)]$$

Here, n is the number of molecules of undissociated acid HA that are associated with 1 molecule of TOPO in the organic phase, and K_1 is the association constant of the organic acid with the reactive agent. (S_{oT}) is the total amount of lactic acid per unit volume of aqueous phase. Thus, a Log-Log plot of m^* and $[HA_{aq}]^{n-1}\,[(So_T) - (HA_{aq})/n]$ should lead to a straight line of slope 1.0. Since the number (n) of acid molecules associating with 1 molecule of TOPO is not known, different values of n were chosen with the objective of achieving a slope of 1.0 for the line of best fit between the experimental data presented in Fig. 2 and the predictions of Eq. *(3)*. Based on such calculations, a value of $n = 1.1$ was obtained. The plot of partition coefficient against the function mentioned above has been presented in Fig. 3. Such values for n for association of organic acids with reactive agents are in agreement with the observations of Fahim et al. *(13)*, who reported a value of 1.4 for acetic acid—TOPO interactions. Similar numbers were found for propionic acid—TOPO also.

Noninteger numbers associated with reactive agents are difficult to comprehend. Further investigations of the interactions between lactic acid and TOPO in kerosene were conducted. These methods involved investigation of TOPO–lactic acid complexes with the help of FT-IR and ^{31}P-NMR techniques.

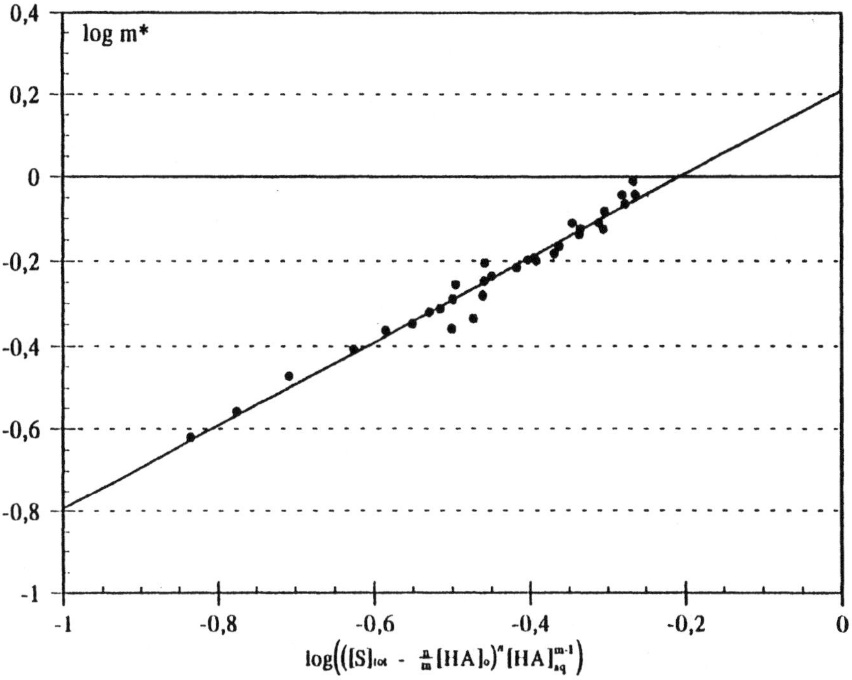

Fig. 3. Representation of partition coefficient data according to equation *(3)*.

^{31}P-NMR and FT-IR Studies of TOPO–Lactic Acid Complexes

^{31}P-NMR studies were conducted at different concentrations of lactic acid to study the nature of the lactic acid–TOPO complex in kerosene. In all the cases, only one peak was observed suggesting a rapid, reversible, and equilibrium association between the two constituents of the complex. The absolute chemical shift at which the peak occurred was found to change with the concentration of lactic acid in the kerosene–TOPO solution (Fig. 4). Yet the changes were small compared to the expected shift if any covalent bonds were formed between TOPO and lactic acid. Owing to the small changes in the absolute chemical shift, it is concluded that ≡P=O group in TOPO remains essentially unaffected by the presence of lactic acid molecules, and the nature of this interaction is mostly through hydrogen bonding. Kuo and Gregor *(12)* have also reported formation of hydrogen bond between the TOPO and acids. Similar conclusions were also made from the results of FT-IR analysis. FT-IR spectra of TOPO in kerosene, aqueous solution of lactic acid, and of lactic acid partitioned in kerosene/TOPO are presented in Fig. 5. Neither did the presence of lactic acid influence the position of phosphate group in TOPO, nor was there any effect of TOPO on the wavelength for the carbonyl group.

Effect of the Reactive Solvent System on Kerosene Flux Through the Membrane

A number of experiments were conducted where different organic (tube side) and aqueous (shell side) phases were pumped through the membrane system at a flow rate of 250 mL/min for 5 h. After each experiment, the kerosene flux through

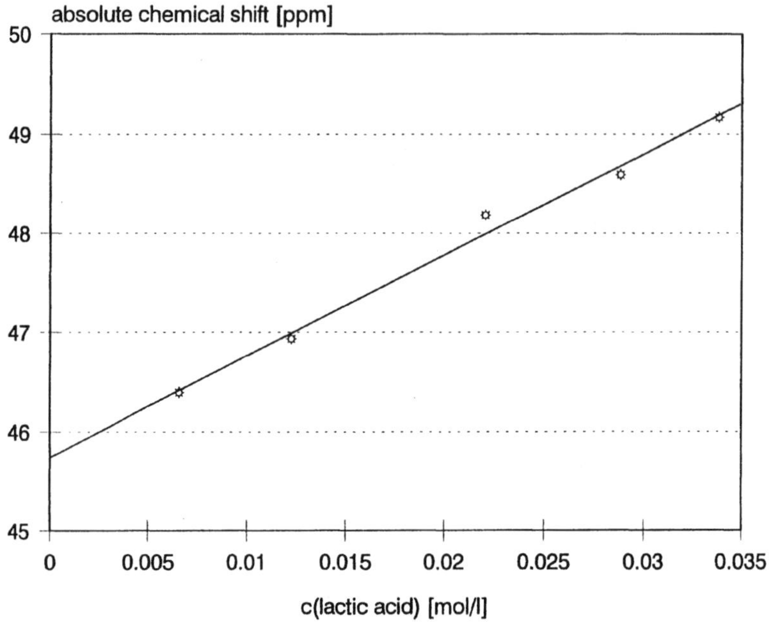

Fig. 4. Absolute chemical shift as a function of lactic acid concentration in ^{31}P-NMR.

the membrane pores was measured at 29 psi and compared with the flux before exposure to the extractants. The results are presented in Table 1 where positive fouling was indicated for cases where the kerosene flux decreased by at least 80% as a result of the extraction process. Clearly, a severe reduction in kerosene flux was observed when TOPO, NaOH, and lactic acid were simultaneously present in the system. A ^{31}P-NMR spectrum of TOPO–lactic acid–NaOH in kerosene did not reveal any additional complex formation at the phosphate bond. It is possible that the presence of these three chemicals together causes a physical change, such as precipitation or formation of stable emulsions, which has been reported earlier *(11)*. Wang et al. *(14)* have reported that crystals of TOPO were observed on the membranes that were exposed to air after the experiments. Therefore, the membrane was not exposed to air after coming in contact with TOPO.

Figure 6 shows electron micrographs of a fresh membrane and of one in which the kerosene flux had dropped. These observations support the formation of a precipitate when the three chemicals (TOPO, NaOH, and lactic acid) are present together in the membrane system. This fact is also supported by the measurements conducted during an attempt to clean the fouled membrane with the objective of restorinf the kerosene flux. These procedures and the results have been presented in Table 2. The extensive cleaning restored the membrane to its original condition. Whether the formation of the precipitate results in a corresponding loss of extractive flux of lactic acid is an open question. Our earlier measurements *(14)* and some of the recently conducted measurements suggest that the extractive flux of lactic acid does not drop as rapidly.

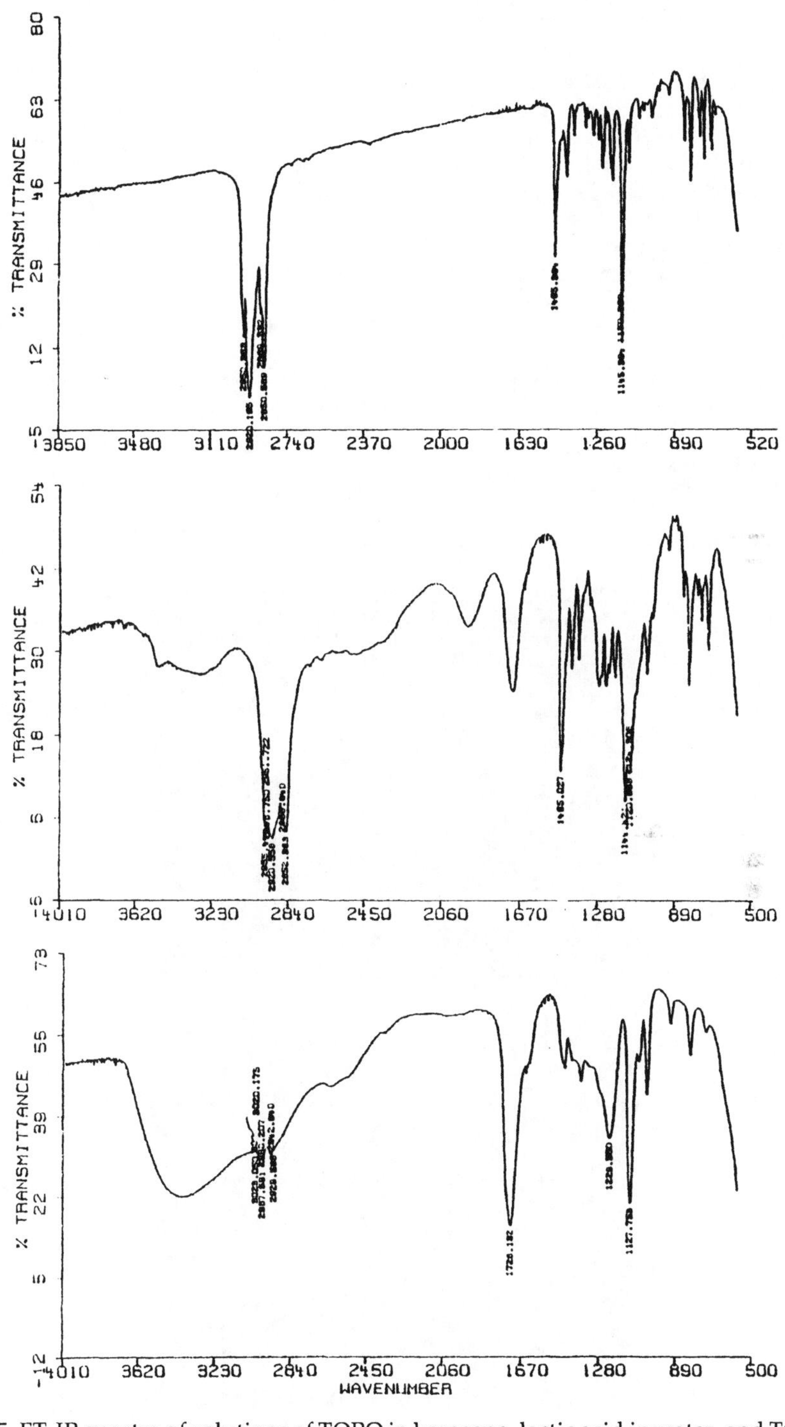

Fig. 5. FT-IR spectra of solutions of TOPO in kerosene, lactic acid in water, and TOPO-lactic acid in kerosene.

Table 1
Operational Conditions That Result in Fouling of Membranes During Extractive Recovery of Lactic Acid from Aqueous Phase

Tube side	Shell side	Fouling % flux change
10% TOPO in kerosene with 20 vol% 2*N* NaOH dispersed	8% Lactic acid, pH 4	+ (–80 ± 1)
Supported liquid membrane formed with 10% TOPO in kerosene in membrane pores and 2*N* NaOH on shell side	8% Lactic acid, pH 4	+ (–85 ± 1)
10% TOPO in kerosene	8% Lactic acid, pH 4	– (+8)
Kerosene	8% Lactic acid, pH 4	– (–1)
20 vol% 2*N* NaOH in kerosene	8% Lactic acid, pH 4	(+1)
10% TOPO in kerosene	2*N* NaOH	– (+1)
10% TOPO in kerosene with 20 vol% 2*N* NaOH dispersed	Distilled water	– (+1)

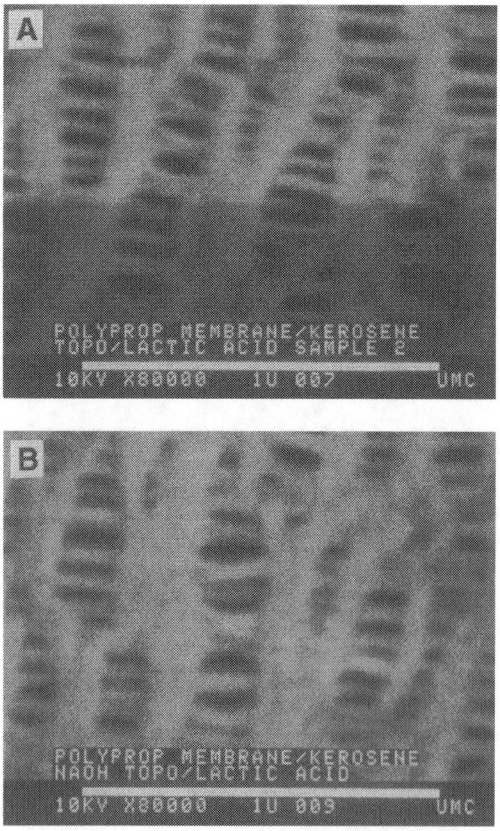

Fig. 6. Electron micrographs of clean and fouled membranes.

Table 2
Kerosene Flux Through Membrane Before and After an Extraction Experiment and During Successive Cleaning Steps

Process	Kerosene flux in L/(m^2h) at 29 psi TMP Before	After
Extraction with kerosene/TOPO/NaOH	42	7.8
20 h Storage in kerosene	7.8	7.8
1 h Washing with kerosene at 55°C	7.8	24
3 h Storage in kerosene at room temperature	24	13.2
1 h Washing with 50% isopropanol at 54°C	13.2	13.2
Storage in kerosene for 20 h at room temperature	13.2	41.4

CONCLUSIONS

Partition coefficient of lactic acid between aqueous and organic phase (kerosene + TOPO) increases linearly with increase in the concentration of TOPO in kerosene and decreases with increase in the concentration of lactic acid. On the average, 1.1 mol of lactic acid can associate with each molecule of TOPO in the organic phase. There is no covalent bond formation between TOPO and lactic acid in the organic phase. NMR and FT-IR data suggest that a hydrogen-bond formation is more likely. The presence of TOPO, lactic acid, and NaOH together in the membrane system causes a rapid drop in the transmembrane flux of kerosene. The kerosene flux could be restored to its original value only after extensive cleaning.

ACKNOWLEDGMENTS

Christian Scheler was supported during these studies by a Deutscher Akademischer Austausch Dienst (DAAD) fellowship, which is gratefully acknowledged. We are also very grateful to R. Glaser of the Department of Chemistry, University of Missouri—Columbia for his help in the NMR and FT-IR measurements and their interpretations. Financial support for the project came from a grant of Missouri Corn Growers' Association.

REFERENCES

1. Harrero, A. A. (1983), *Trends Biotechnol.* **1,** 49–53.
2. Bajpai, R. K. and Iannotti, E. L. (1988), in *Handbook on Anaerobic Fermentations*, Erickson, L. E. and Funk, D. Y., eds., Marcel Dekker, New York, pp. 207–241.
3. Holten, C. H. (1973), *Lactic Acid.* Verlag Chemie GmbH, Weinheim.
4. Kertes, A. S. and King, C. J. (1986), *Biotechnol. Bioeng.* **28,** 269–282.
5. Su, Y. T. (1991), MS thesis, Department of Chemical Engineering, University of Missouri—Columbia, MO.
6. Bar, R. and Gainer, J. L. (1987), *Biotechnol. Progr.* **3,** 109–114.
7. Cho, T. and Shuler, M. L. (1986), *Biotechnol. Progr.* **2,** 53–60.
8. Prasad, R. and Sirkar, K. K. (1988), *AIChEJ* **34,** 177–188.
9. Bajpai, R. K., Iannotti, E. L., Wang, C. J., and Su, B. (1990), Simultaneous Production and Recovery of Lactic Acid by Fermentation, AIChE Spring National Meeting, Orlando, FL.

10. Scheler, C. (1994), Recovery of Lactic Acid from Fermentation of Starch Using Membrane Associated Solvent-Extraction. Diplomarbeit, Technische Fachhochschule Berlin, Fachbereich 3 Chemie und Biotechnologie, Berlin, Germany.
11. Golob, J., Grilc, V, and Zadnik, B. (1981), *Ind. Eng. Chem. Process Design Dev.* **20,** 433–435.
12. Kuo, Y. and Gregor, H. P. (1983), *Sep. Sci. Tech.* **18,** 421–440.
13. Fahim, M. A., Qader, A., and Hughes, M. A. (1992), *Sep. Sci. Tech.* **27,** 1809–1821.
14. Wang, C. J., Bajpai, R. K., and Iannotti, E. L. (1991), *Acid. Appl. Biochem. Biotechnol.* **28/29,** 589–603.

0273-2289/96/5758—0039$7.25

Enzymatic Hydrolysis of Corn Starch After Extraction of Corn Oil with Ethanol

NINGJUN CAO, QIN XU, JIANLONG NI, AND LI-FU CHEN*

Food Science Department, Purdue University, West Lafayette, IN 47907-1160

ABSTRACT

Simultaneous corn oil extraction and alcohol dehydration, a solvent corn-milling process developed in our laboratory, was tested on a pilot scale to recover corn oil. Over 92% of the corn oil was extracted by 95% ethanol with a liquid-to-solid ratio of 0.75 at 65°C. Following the oil extraction, 95% of zein fraction was also extracted from the defatted ground corn by 65% ethanol at 65°C with a liquid-to-solid ratio of 1.0 to produce extracted ground corn. After corn oil and zein extraction by ethanol, the ability of the amylases to hydrolyze the starch fraction in extracted ground corn was studied. Results showed the starch fraction can be hydrolyzed to glucose by α-amylase and glucoamylase without separatio of fiber and residual protein. The glucose yield in hydrolysis was a function of temperature, enzyme concentration, and solid-to-liquid ratio. With a two-step heating hydrolysis, corn starch was converted to glucose with a 97.2% yield from extracted ground corn. After filtration and washing of the hydrolyzed mass, the final glucose concentration was 24.3% glucose (w/v).

Index Entries: Corn starch; enzymatic hydrolysis; oil extraction; anhydrous ethanol; corn oil.

INTRODUCTION

Glucose is the largest end use of corn starch, which can serve as a feedstock for microbial fermentation into industrial solvents and fuel ethanol *(1)*. Generally, corn is processed by either wet milling or dry milling to produce corn oil, corn starch, and corn meal to facilitate the processing of starch by hydrolysis *(2)*.

Wet milling has been the preferred method for recovering starch from corn for many years. It has experienced enormous growth since the 1970s because of the increased demand for starch as raw material for the production of high-fructose corn syrup as a sweetener and ethanol as an alternative fuel source. Because current wet-milling methods require considerable amounts of capital investment and the process involves high energy consumption, including the steeping and separation of starch from protein processes, many researchers have been searching for new low-cost approaches for fractionating corn to produce novel byproducts with higher values than those obtained by current milling processes *(3)*.

*Author to whom all correspondence and reprint requests should be addressed.

In an effort to develop an energy-efficient alternative method to wet milling, a solvent corn milling process that simultaneously carries out oil extraction and alcohol dehydration has been developed *(4–8)*. This process has the following advantages over the current wet-milling methods:

1. Elimination of the time-consuming steps that adversely affect the functional and edible properties of the protein. As a result, a food-grade protein concentrate is produced that is expected to have a higher nutritional value than gluten meal because the germ proteins are also extracted;
2. Replacement of the ternary distillation for dehydration of asotropic ethanol with a simple water-absorption step with dried corn as the adsorbent; and
3. Realization of a higher oil yield than the wet-milling process.

In order to illustrate the large-scale feasibility of the new process, the objective of this project is to complete a series of experiments to study starch hydrolysis after extraction of corn oil and protein by ethanol. Since the use of enzymes has several advantages over acid hydrolysis of starch, such as fewer byproducts and higher yield *(9)*, we have chosen enzymatic hydrolysis as the starch hydrolysis method.

MATERIALS AND METHODS

Materials

Ground No. 2 yellow dent corn was obtained from a local feed station. Anhydrous and 95% ethanol were purchased from McCormick Distilling Co., Inc. (Weston, MO). Thermostable α-amylase (EC 3.2.1.1, Termamyl 120L) and glucoamylase (EC 3.2.1.3, AMG 200L) were generously provided by Novo Nordisk Bioindustrials, Inc. (Danbury, CT); cellulase TAP was a gift of International Enzyme Co., Inc. (Fallbrook, CA).

Methods

Characterization of Ground Corn

For the determination of the particle size distribution, dried corn was fractionated on a sieve shaker (Ro-tap, Model B, Tyler Industrial Products) and shaken for 10 min. The retained corn on each sieve was weighed. The moisture content of the ground corn was determined by drying in an air oven at 135°C for 2 h (AACC 1983). The total oil content of the ground corn was determined in accordance with a standard procedure (AACC 1983).

Pilot Plant Oil Extraction

The pilot-scale experiments of oil extraction were carried out in a countcurrent extractor (Crown Iron Work, Minneapolis, MN) by 95% ethanol at 65°C and 0.75 of liquid-to-solid ratio for 30–45 min retention time. Extraction residue was desolventized by Down Draft Desolventizer at 80°C to produce a defatted dried ground corn containing 4% moisture. Oil–ethanol solution was distilled to yield crude corn oil and anhydrous ethanol. Zein protein was then extracted from defatted ground corn by 65% ethanol at 65°C at an equal ratio of solid to liquid with a residence time of 30 min in the column (6 × 80 cm).

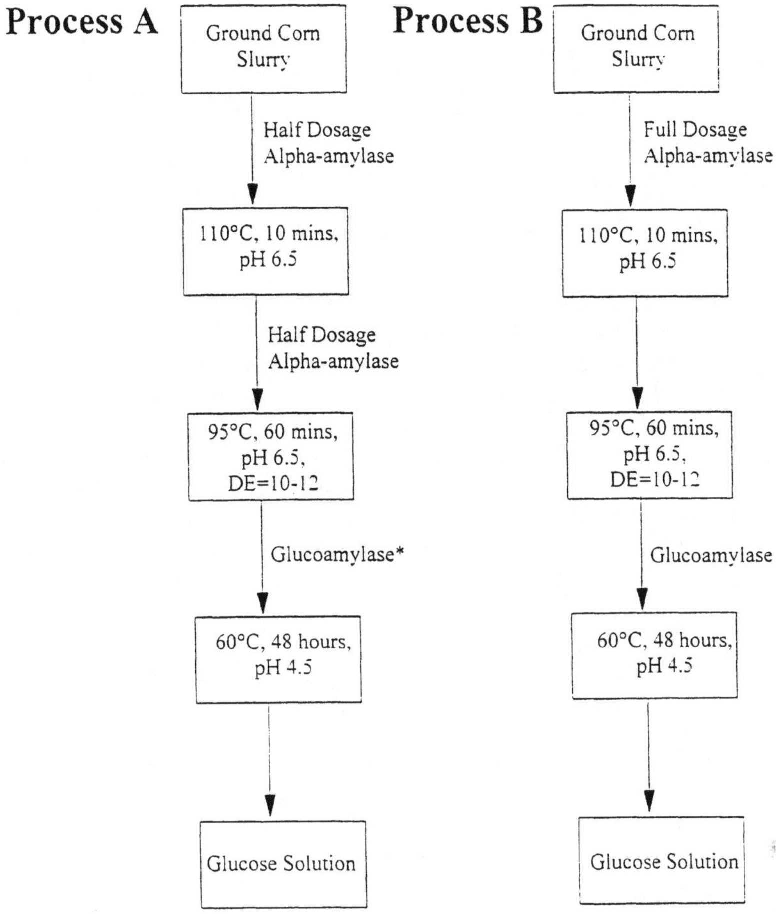

Fig. 1. Processes of defatted ground corn hydrolysis. (*) In process C an amount of cellulose was added with glucoamylase.

Gelatinization Temperature of Starch

Corn starch was prepared from blending extracted ground corn with water in a Waring Blender for 5 min with the blade turned upside down. The slurry was centrifuged in a gradually increased gravitational field. All centrifuged starch samples were combined. Corn starch (0.2%) was dispersed in a water and ethanol solution ranging from 0–60% of ethanol (v/v). Samples were determined according to the microscopic hot-stage technique described by MacMasters and Chien *(10,11)*.

Starch Hydrolysis

Slurry was prepared by mixing the extracted ground corn with 100 ppm calcium chloride aqueous solution (pH 6.5) at different liquid-to-solid ratios *(12)*. Different dosages of α-amylase were added in different processes (Fig. 1). The slurry was heated to 110°C and kept for 10 min in an autoclave. After flash cooling, the slurry was kept at 95°C until no blue color reaction occurred with iodine

Table 1
Composition of Ground Corn, dry base, g/100 g

Main component	Whole corn	Extracted ground corn
Lipids	4.9	0.35
Proteins	10.3	4.15
Starch and sugars	74.6	84.2
Fiber	10.2	11.27

solution. After liquefaction, the pH of the slurry was adjusted to pH 4.5 using 2.0*N* HCl solution. The saccharification step was completed by adding 4 AGU of glucoamylase (AMG 200L)/g corn starch in processes A and B, or 4 AGU of glucoamylase (AMG 200L)/g corn starch and 5 U of cellulase (TAP)/g corn starch in process C. The reactions were kept at 60°C for 48 h. The slurry was filtered and washed to collect glucose hydrolysate. Glucose concentration was determined by high-performance liquid chromatography (HPLC) methods (Aminex HPX-87C column, Bio-Rad Laboratories, Hercules, CA), and dextrose equivalent (DE) value was calculated on a dry wt basis by determining the relative reducing power of the liquefact with anhydrous dextrose given the value 100.

Enzyme Activity

α-Amylase activity was determined by the Novo analytical method AF-9. One KNU was the amount of enzyme that breaks down 5.26 g starch/h at pH 5.6 at 37°C *(13)*. Glucoamylase activity was determined by the Novo analytical method AF-22. One AGU was defined as the amount of enzyme that hydrolyzes 1 μmol maltose/min at pH 4.3 at 25°C *(13)*. One cellulase unit was defined as that amount of activity liberating 1.0 μmol of glucose from α-cellulose in 1 h at pH 5.0 at 37°C.

RESULTS AND DISCUSSION

Pilot Plant Oil Extraction

In this pilot-scale testing, over 92% of the corn oil was recovered, which was about 15–17% over the oil yield of a typical wet-milling process *(12)*. When the cracked corn was dried to about 1% moisture, each kilogram of the cracked corn was able to absorb water in 95% ethanol to produce 500 mL 99.6% (w/w) ethanol. Table 1 shows the composition change of ground corn before and after extraction. Compared with industrial corn starch containing 0.65% of fat, the extraction residue of the current method contained only 0.35% of fat. Furthermore, with 95% of zein protein removal, the residue after desolventization exhibited a very loose structure that facilitated corn starch hydrolysis without the need for further separation of starch and fiber. A glucose yield of 95% was obtained by enzymatic hydrolysis of defatted ground corn.

Gelatinization and Liquefaction

Extracted ground corn was subjected to enzymatic hydrolysis to study the effect of ethanol treatment on starch hydrolysis. Starch gelatinization is a precondition of efficient hydrolysis. Once the starch has been gelatinized, the molecule must be

Table 2
Effect of Ethanol Treatment
on Initial and Final Gelatinization Temperature of Corn Starch

	Gelatinization temperature, °C[a]			
	Sample 1		Sample 2	
% EtOH, v/v	Initial	Final	Initial	Final
0	59.3	66.0	61.0	68.0
20	59.3	66.1	61.0	68.0
40	68.0	75.5	70.0	77.5
45	70.5	78.8	68.0	77.0
50	76.4	81.0	78.5	>81.0
60	>81.0	>85.0	>81.0	>85.0

[a]Sample 1, the undefatted ground corn; sample 2, extracted ground corn; deviation of temperature = ±0.5°C.

rapidly hydrolyzed or retrogradation will occur, particularly if the substrate is cooled during liquefaction. On cooling, the starch chains reassociate with each other, with the linear outer branches of amylopectin, or with lipids to reform the lipid complex, forming semicrystalline precipitate, which is highly resistant to hydrolysis, leading to filtration problems as well as loss of yield *(2)*. Table 2 shows the initial and final gelatinization temperatures of extracted ground corn. This indicates that gelatinization of starch granules can be affected by high concentration of ethanol and the gelatinization does not occur during the extraction and desolventization processes.

The liquefaction of extracted ground corn is a function of α-amylase concentration. The progress of liquefaction can be monitored by measuring the DE value of the slurry and its ability to develop a blue color with iodine. Increasing enzyme concentration can accelerate the liquefaction rate in the same condition (Fig. 2, process A). Comparing the effect of three different enzyme concentrations, the DE values for two lower enzyme dosages were approximately linear with time in the first hour at 95°C. When the enzyme dosage was over 0.8 KNU/g ground corn, which was slightly higher than the amount used in the commercial practice of 0.7 KNU/g *(13)*, the DE went up to a value of about 12, which was ideal for the subsequent saccharification process. Therefore, a 0.8 KNU/g corn of α-amylase dosage was used in all the following experiments.

Substrate composition also affected the DE progress curve of liquefaction (Fig. 3). Using 30% starch slurry as control, which was equal to the starch content in extracted ground corn in a liquid-to-solid ratio of 3.5, results showed that the removal of oil and zein protein enhanced the enzyme's capability of accessing starch molecules in the extracted residue. The hydrolysis characteristic of extracted ground corn is similar to industrial starch. Because fat exerts a considerable influence on the starch hydrolysis process, whole ground corn shows a totally different hydrolysis characteristic. Ethanol extraction reduced the effect of fat on starch hydrolysis.

Dextrose Yield

High yield and high concentration of glucose are the key issues for an economically viable practice to utilize corn starch. Although extracted ground corn had a characteristic of loose structure after oil and zein protein extraction, the hydrolysis

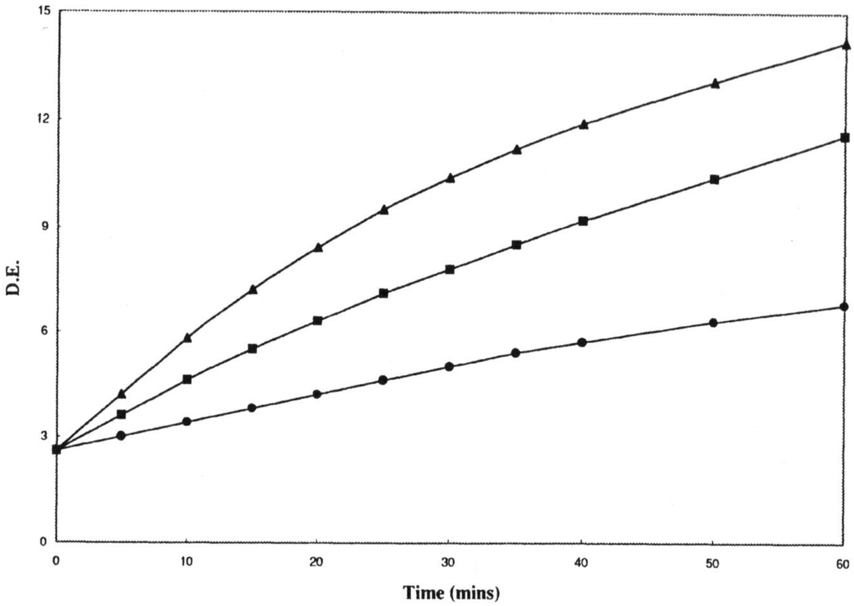

Fig. 2. Liquefaction time-course of defatted ground corn at different amylase concentration. Temperature: 95°C; pH = 6.5; liquid-to-solid ratio: 3.5; size of ground corn: 32–60 mesh, process A. ●, 0.4 KNU/g corn; ■, 0.8 KNU/g corn; ▲, 1.6 KNU/g corn.

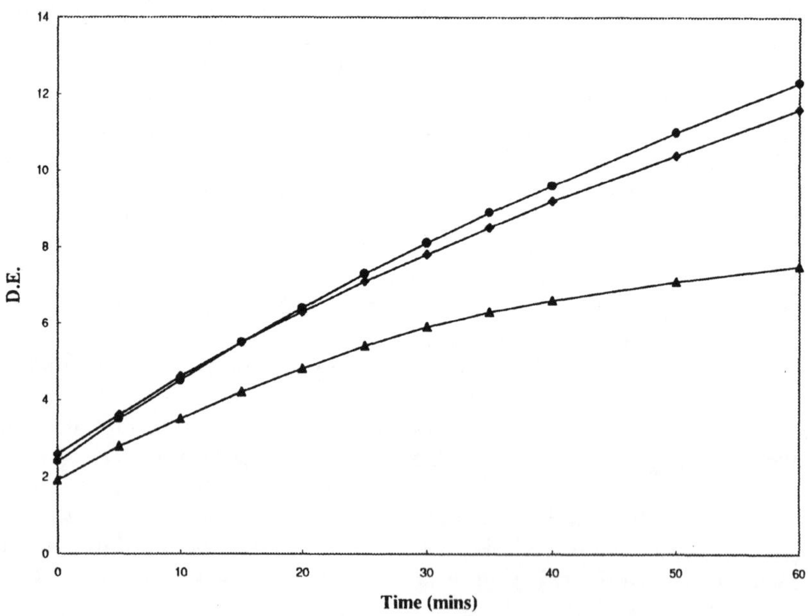

Fig. 3. Liquefaction time course of different substrates. Temperature: 95°C; pH = 6.5; liquid-to-solid ratio: 3.5; size of ground corn: 32–60 mesh; α-amylase concentration: 0.8 KNU/g corn; control: 30% starch slurry. ◆, defatted ground corn; ●, 30% starch control; ▲, ground corn.

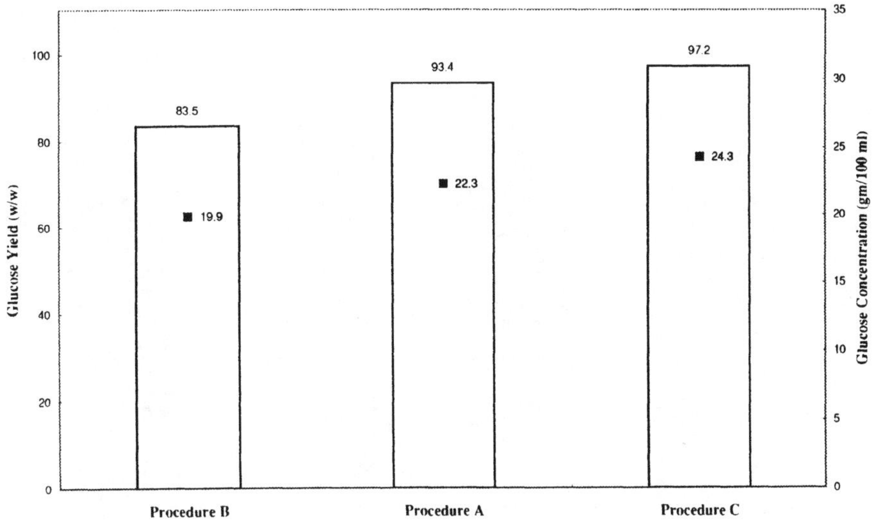

Fig. 4. Effect of processes on glucose yield (bar) and glucose concentration (■). Liquefaction condition: temperature: 95°C; pH = 6.5; Liquid-to-solid ratio: 3.5; size of ground corn: 32–60 mesh; α-amylase concentration: 0.8 KNU/g corn; time: 60 min. Saccharification condition: temperature: 60°C; pH = 4.5; enzyme concentration: glucoamylase, 4 AGU/g corn for processes A and B, and 4 AGU/g corn plus 5 U cellulase/g corn for process C; time: 48 h.

condition of starch fraction is not very different from the commercial starch process. The existence of cellulose fiber and residual tissue structure may hinder the diffusion of enzyme to starch. We have observed that starch can be hydrolyzed more efficiently at a temperature of 110°C than at 105°C. Since 110°C is the maximum limit of thermostable temperature for α-amylase Termamyl 120L, a relatively large part of activity has been lost by the process. In process B, therefore, only an 83.5% glucose yield was achieved (Fig. 4). To avoid enzyme deactivation at this temperature, enzyme was added in two stages: one-half of the enzyme dosage was added before heating at 110°C and the rest after the first step and before the second step of heating of 95°C (process A). As a result, 93.4% of the glucose yield was obtained. A modified process suggested by the Novo company *(13)*, process C, was used to enhance the hydrolysis of extracted ground corn. In this process, 5 IU of cellulase/g of corn were added with glucoamylase during the saccharification stage at 60°C, and the glucose yield was increased to 97.2%. After filtering and washing of the hydrolysis residue, a glucose concentration of 24.3% (w/v) was obtained from process C.

Particle size affects the yield of oil extraction as well as glucose yield in the subsequent starch hydrolysis. Figure 5 shows that glucose yield increased from 80 to 98% following the particle size changes from 16 to 60 mesh. Decreasing particle size of ground corn can enhance the diffusion rate of extraction solvent, water, and enzyme.

Adsorption of water is the most important factor in controlling the rate of starch hydrolysis. The nonstarch materials, such as cellulose fiber, are hydrophilic and exist in the exterior of the extracted ground corn. It holds water and competes with starch for water; therefore, the amount of water added to the residue has to be compensated. It is expected that the liquid-to-solid ratio will affect the glucose yield as well as glucose concentration of hydrolysate (Fig. 6). In our experiments, a liquid-to-solid

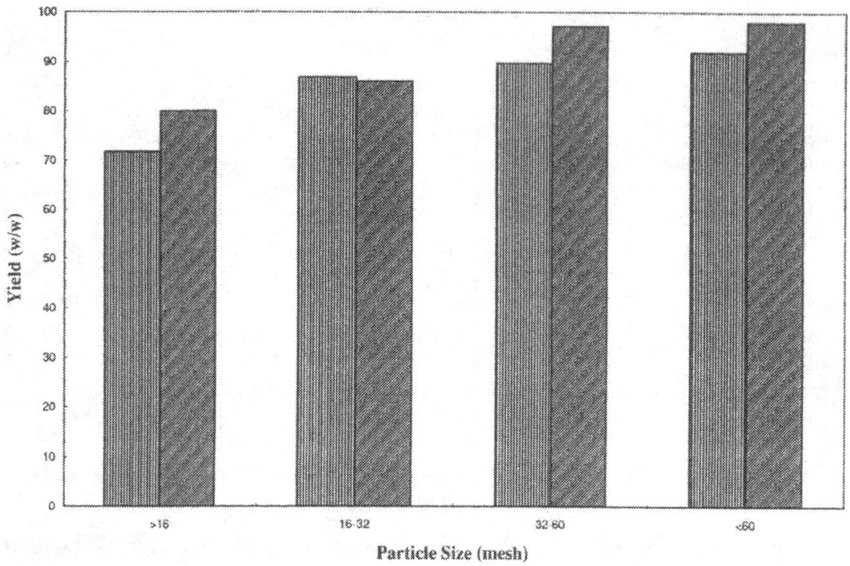

Fig. 5. Effect of particle size of ground corn on oil extraction yield (▥) and glucose yield (▨). Condition of oil extraction with countcurrent extractor: extraction solvent: 95% ethanol; temperature: 65°C; retention time: 30–45 min; ratio of liquid to solid: 0.75; hydrolysis condition: same as Fig. 4.

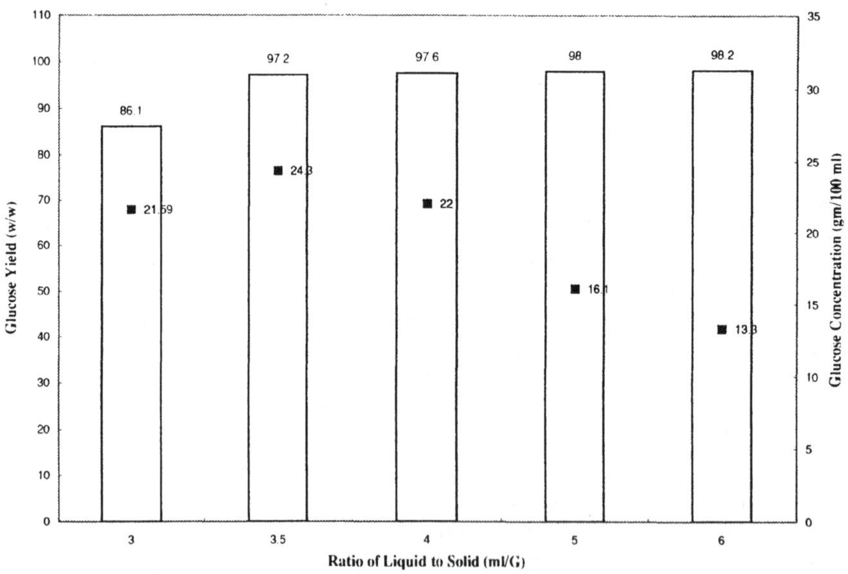

Fig. 6. Effect of liquid-to-solid ratio on glucose yield (bar). Liquefaction condition: temperature: 95°C; pH = 6.5; Size of ground corn: 32–60 mesh; α-amylase concentration: 0.8 KNU/g corn; time: 60 min. Saccharification condition: temperature: 60°C; pH = 4.5; enzyme concentration: 4 AGU/g corn plus 5 U cellulase/g corn; time: 48 h. ■, glucose concentration.

ratio of 3.5 gave a 97.2% glucose yield and 24.3% of glucose hydrolysate. This glucose concentration is suitable for commercial fermentation processes.

CONCLUSION

A new solvent corn-milling process could be an economical alternative process to utilize corn in the production of high fructose corn syrup and ethanol. It not only produced up to 92% oil recovery and 500 mL anhydrous ethanol/kg dried corn, but also provided a kind of extracted ground corn having loose matrix structure. The removal of corn oil can reduce the amount of oil present on starch hydrolysis, and it would simplify the liquefaction and saccharification processes. The defatted ground corn produced can be efficiently hydrolyzed by thermostable α-amylase and glucoamylase combined with a small amount of cellulase. At optimum conditions, without separation and purification of starch, a hydrolysate has been produced with 97.2% yield of glucose.

REFERENCES

1. Galliard, T. (1985), *Ann. Proc. Phytochem. Soc. Eur.* **26,** 103–115.
2. Reeve, A. (1992), in *Starch Hydrolysis Products,* Schenck, F. W. and Hebeda, R. E., eds., VCH Publishers, New York, pp. 79–83.
3. Hojilla-Evangelista, M. P., Johnson, L. A., and Myers, D. J. (1992), *Cereal Chem.* **69,** 643–647.
4. Chen, L. F. and Hoff, J. E. (1987), US Patent 4,716,218.
5. Chien, J. T., Hoff, J. E., and Chen, L. F. (1988), *Cereal Chem.* **65,** 484–486.
6. Chien, J. T., Hoff, J. E., Lee, M. J., Lin, H. M., Chen, Y. J., and Chen, L. F. (1990), *Chem. Eng. J.* **43,** B103–B113.
7. Robertson, G. H. and Pavlath, A. E. (1986), *Energy Agricult.* **5,** 295–308.
8. Hojilla-Evangelista, M. P., Myers, D. J., and Johnson, L. A. (1992), *JAOCS* **69,** 199–204.
9. Kennedy, J. F., Cabral, J. M. S., Sa-Correia, I., and White, C. A. (1987), *Crit. Repts. Appl. Chem.* **13,** 115–121.
10. MacMaster, M. M. (1964), in *Methods in Carbohydrate Chemistry,* vol. 4, Whistler, R. L., ed., Academic, New York, pp. 233–235.
11. Chien, J. T. (1986), Ph.D. Thesis, pp. 71,72.
12. Johnson, L. A. (1991), in *Handbook of Cereal Science and Technology,* Lorenz, K. J. and Kulp, K., eds., Marcel Dekker, New York, pp. 55–131.
13. Novo Nordisk Bioindustrials, Inc. (1994), Technical Bulletin, Danbury, CT.

0273-2289/96/5758—0049$7.00

Pretreatment of Sugar Cane Bagasse Hemicellulose Hydrolysate for Xylitol Production by Yeast

JOSE M. DOMINGUEZ,[1]
CHENG S. GONG,[2] AND GEORGE T. TSAO*,[2]

[1]Department of Chemical Engineering. University of Vigo (Campus Orense), Las Lagunas, Orense, Spain; and [2]Laboratory of Renewable Resources Engineering, Purdue University, West Lafayette, IN 47907

ABSTRACT

A total of six known xylitol-producing yeast strains were screened for production of xylitol from xylose. *Candida* sp. 11-2 proved to be the best producer. It was chosen to study its ability to produce xylitol from hemicellulose hydrolysate derived from sugar cane bagasse. The hydrolysate was prepared by dilute sulfuric acid (2–3% [w/v]) hydrolysis, with a high-solid, low-liquid ratio followed by leaching. Owing to the inhibitors present in the hydrolysate, different treatments were studied to overcome its effect. In order to reach higher xylitol productivity, treated hydrolysates were concentrated by vacuum evaporation in rotavapor to provide a higher initial xylose concentration. After treatment, *Candida* sp. 11-2 was able to ferment xylose in hemicellulose hydrolysate to produce xylitol.

Index Entries: Xylose; xylitol; sugar cane bagasse hemicellulose hydrolysate; *Candida* sp.

INTRODUCTION

In the past years, the number of works related to obtaining high sweetening power products has increased. One of such product is xylitol, a naturally occurring five-carbon polyalcohol. This alcohol has other important properties in addition to its sweetening effect that make it desirable for sugar-free confections: It is completely safe for the teeth because of it anticaries properties, it is tolerated by diabetics, and it has a high negative heat of solution *(1,2)*. Xylitol is present in certain fruits and vegetables, but the small amounts render its extraction uneconomical. Producing xylitol by microbiological means can overcome this problem. Many yeast species, especially those belonging to *Candida* species, are good xylitol producers *(3)*. Xylitol is formed as a metabolic intermediate product of D-xylose fermentation: D-xylose can be converted to xylitol by NADPH-dependent aldehyde reductase, or can be isomerized to D-xylose by D-xylose isomerase, and then reduced to xylitol by NADH-dependent xylitol dexhydrogenase *(4)*.

*Author to whom all correspondence and reprint requests should be addressed.

A good source of D-xylose is sugar cane bagasse, the fibrous residue obtained after the extraction of sugar from sugar cane. This waste has been used as raw material for hydroxymethyl furfural production, paper pulp, acoustical boards, press woods, and agriculture mulch. Sugar cane bagasse can be hydrolysed using dilute acid with high-solid, low-liquid ratio to obtain a mixture of fermentable sugars with xylose as the major component *(5)*.

In the hydrolysate, some byproducts generated in the hydrolysis, such as acetic acid, furfural, phenolic compounds, or lignin-degradation products can be present. These are potential inhibitors of the microbial metabolism. In order to provide hydrolysates suitable as fermentation substrate, different treatments, such as acidified activated charcoal or cation-exchange resins, can be carried out to remove these toxic substances *(3)*.

The initial xylose concentration is an important factor to obtain a high xylitol production. This initial xylose concentration can be increased by concentrating the hydrolysate using vacuum evaporation. During the concentrating of hydrolysate, there is also an increase in toxic substances that can act as inhibitors. Therefore, it is important to pretreat concentrated hydrolysate to remove these toxic substances. In this article, we investigate the effect of different pretreatments of sugar cane bagasse hydrolysate on the production of xylitol from hydrolysate by *Candida* sp. 11-2.

MATERIALS AND METHODS

Microorganisms

The yeast cultures were purchased from American Type Culture Collection (ATCC; Rockville, MD). The yeast cultures chosen were *Candida parapsilosis* ATCC 28474, *Candida guilliermondii* ATCC 20118, *C. guilliermondii* ATCC 42050, *Pachysolen tannophilus* ATCC 32691, *Candida* sp. 11-2 (laboratory strain), and the yeast *Debaryomyces hansenii* NRRL Y-7426 (obtained from the Northern Regional Research Laboratory, USDA, Peoria, IL).

Growth of Yeast Cells

The yeast cells were incubated for 3 d at 32°C in an incubator shaker at 200 rpm (New Brunswick). The liquid medium contains 1% glucose, 1% xylose, 3 g/L Bacto-yeast extract, 3 g/L Bacto-malt extract, and S g/L Bacto-peptone.

Fermentation

Fermentation with Pure Xylose

The fermentation batches to compare different yeasts was performed at 28°C in 50-mL Erlenmeyer flasks (containing 10 mL of culture media) placed in a gyratory shaker at 180 rpm for 4 d. The culture media was prepared with pure xylose (15% [w/v]) and supplemented with nutrients with the following composition per liter: 3 g Bacto-yeast extract, 3 g Bacto-malt extract, and 5 g Bacto-Peptone. Yeast cells from growth medium were harvested by centrifugation and transferred into fermentation medium.

Fermentation with Hydrolysates

The fermentation of hydrolysates was carried out under the same conditions as described for with pure xylose. The same amounts of nutrients were added in the hydrolysate to eliminate the effect of nutrients that might affect the fermentation of different concentrations of substrate.

Acid Hydrolysate

Sugar cane bagasse hemicellulose hydrolysate was obtained by hydrolysis of ground bagasse in dilute sulfuric acid g 2–3% [w/v]) at 100°C followed by downflow leaching using water according to the method described by Ladisch *(5)*. Hydrolysate has a low pH value and was kept at room temperature.

Treatments

1. Neutralization: The pH of the hydrolysate was increased to 4.5–6 by the addition of calcium carbonate. Calcium sulfate of approx 15 g/L formed during neutralization was removed from liquid by filtration.
2. Activated charcoal and neutralization: Activated charcoal (Savannah Foods & Industrials, Inc., Savannah, GA) was washed and equilibrated with HCl 0.4*N*. After being packed into a column (1.5 × 25 cm), charcoal was washed with water, and the acid hydrolysate was applied to the column. The hydrolysate collected was neutralized by addition of calcium oxide slurry and calcium carbonate to pH 4.5–6.0. The calcium sulfate was removed by filtration.
3. Cation–exchange resins and neutralization: Acid cation-exchange resins (Dowex 50 WX4, 200–400 mesh) were washed with NaOH (0.1*N*) followed by water. The resin was packed into a column (1.5 × 25 cm) and after washing with water, the acid hydrolysate was applied to the column. The hydrolysate collected was treated with calcium oxide slurry and calcium carbonate to increase the pH to 4.5–6.0. Calcium sulfate formed was removed by filtration.

Analytical Methods

Sugar and alcohol were analyzed using a Hitachi high-performance liquid chromatographic system consisting of an AS-4000 Intelligent Auto Sampler, a Hitachi L3350 refractive index monitor, a Hitachi L-6000 pump, and a Hitachi D-2500 chromatointegrator. Separation was achieved using an organic acid column (Aminex HPX-87 H Ion Exclusion Column 300 × 7.8 mm, Bio-Rad, Hercules, CA) at 60°C with 0.008*N* sulfuric acid as eluant at 0.8 mL/min over a 18-min period. Figure 1 shows the liquid chromatography of acid hydrolysate without treatment.

RESULTS AND DISCUSSION

Comparison of Yeast

Table 1 shows the fermentation batches carried out with pure xylose under microaerobic conditions for the different yeast chosen. Only *C. guilliermondii* 20118, yeast mutant 11-2, *D. hansenii* NRRL Y-7426 showed substantial xylose consumption and xylitol production (0.72, 0.99, and 0.90g/L/h, respectively). The other strains were poor D-xylose utilizers under these fermentative conditions. The yeast mutant, *Candida* sp. 11-2, was chosen to perform the further experimentation.

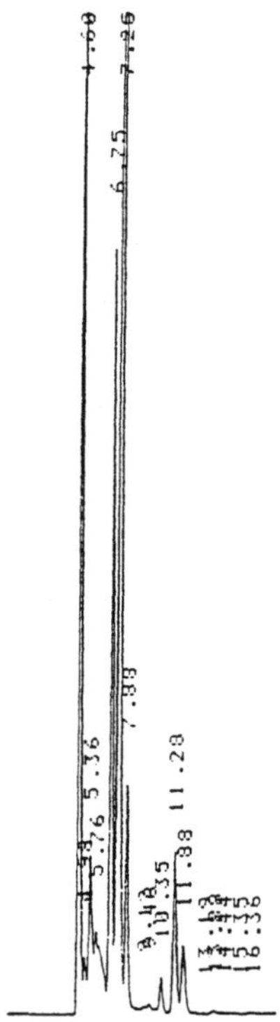

Fig. 1. Liquid chromatogram of acid hydrolysate. 4.60, phytic acid + sulfuric acid; 6.75, glucose; 7.25, xylose; 7.88, arabinose; 9.48, lactic acid; 10.35, glycrol; 11.28, acetic acid.

Comparison of Different Treatments and Different Dilutions

Table 2 shows the typical chemical composition of bagasse acid hydrolysate. Glucose and xylose are the fermentable sugars. Xylose (6.69%) can be fermented to xylitol by yeast, and glucose (2.39%) can be fermented to ethanol. Phytic acid present (4.87%) can be removed from hydrolysate by an alkaline treatment. Acetic acid (1.39%) can become an inhibitor when it is present in significant amounts. In order to increase the xylitol production, the simple neutralization of this hydrolysate was compared to other treatments, like activated charcoal and cation-exchange resins. The effect of the initial xylose concentration on yeast fermentation after each treatment was also studied. The results after 48 h of fermentation are shown in Table 3.

The increase in initial xylose concentration resulted in the decrease in the amount of xylose utilized in the "only neutralization" sample, indicating the presence of an inhibitor(s) in the hydrolysate that inhibits the utilization of xylose. The inhibitory effect can be removed by activated charcoal treatment. From our experience, approx 1 g of charcoal can treat about 20 mL of original sugar cane bagasse hemicellulose hydrolysate. In all the cases, xylitol yield from xylose is low when hydrolysates with different treatment were used as substrate. The reason for low xylitol yield is not known. We believe the extraneous chemicals present in the hydrolysate could inhibit the release of xylitol into the medium. This possibility is currently under investigation.

Table 1
Xylitol Production from Pure Xylose by Different Yeasts

	48 h				96 h		
Yeast	Initial xylose, g/L	Xylose used, %	Xylitol, g/L	Qp, g/Lh	Xylose used, %	Xylitol, g/L	Qp, g/Lh
C. parapsilosis 28474	152.72	26.94	6.72	0.14	88.32	17.84	0.19
C. guillierrnondii 20118	155.97	40.11	33.28	0.69	80.33	68.96	0.72
C. guilliermondii 42050	148.98	8.11	5.91	0.12	25.24	13.79	0.14
P. tannophilus 32691	152.78	14.42	0.91	0.02	45.91	0.88	0.01
Mutant of *Candida* sp. 11-2	152.47	54.77	56.35	1.17	97.33	95.35	0.99
D. hansenii NRRL Y-7426	149.50	93.89	105.82	2.21	100	86.00	0.90

Table 2
Typical Chemical Composition of Acid Hydrolysate

Constituent	Wt%
Moisture	83.64
Xylose	6.69
Glucose	2.39
Arabinose	0.79
Acetic acid	1.39
Phytic acid	4.87

Table 3
Xylitol Production after 48 h of Fermentation
for the Different Treatments and with Different Initial Xylose Concentrations

Treatment	Initial xylose, g/L	Xylose used, %	Xylitol, g/L	Qp, g/Lh	Ethanol, g/L)
Only neutralization	58.38	5.07	2.55	0.053	0
	43.25	8.38	4.19	0.073	1.38
	27.99	95.94	4.32	0.066	9.06
Activated charcoal and neutralization	42.96	94.75	10.54	0.205	12.18
	34.66	96.77	7.32	0.137	6.13
	22.70	96.31	4.62	0.071	1.83
Cation-exchange resins and neutralization	45.80	83.94	10.10	0.195	7.185
	32.58	90.31	8.32	0.158	2.011
	24.33	89.49	5.96	0.066	0.878

The lowest xylose consumption corresponded to the hydrolysate only neutralized. This is owing to the fact that there are substances in the fermentation media that make the fermentation processes difficult. Consequently, the xylitol production was very poor, especially when the media was not diluted, making the

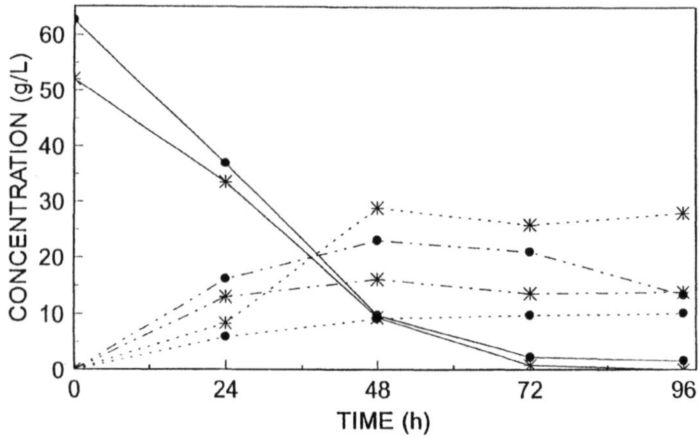

Fig. 2. Comparation of hydrolysates treated with activate charcoal. Without passing through a cation-exchange resin: (—●—) xylose concentration, (- -●- -) xylitol concentration, (—●- -) ethanol concentration. Passing through a cation-exchange resin: (—✳—) xylose concentration, (-✳ -) xylitol concentration, (—✳- -)ethanol concentration.

effect of these substances stronger. However, the results show that the hydrolysates treated with activated charcoal or cation-exchange resins produced xylitol readily. This confirms that the samples without a pretreatment have significant inhibition. During the activated charcoal treatment, the acetic acid is removed. Some toxic byproducts formed during acid hydrolysis of cellulosic materials, such as 5-hydroxymethyl furfural and furfural (with significant fermentation inhibition *[6]*) and other chemicals toxic to microorganisms (like phenolic compounds, especially in its oxidized form *[7]*) are removed during these treatments. The highest xylitol productivity (0.205 g/L/h), corresponding to 10.54 g/L, was obtained from hydrolysate treated with activated charcoal. For this reason, it was selected to be the best treatment. Another reason is its low cost as compared to the cation-exchange resins.

Glucose and other sugars that can be present in the hydrolysate in insignificant amounts, like mannose and galactose, were utilized by the yeast for growth and ethanol production, but not for their corresponding polyol production. The amounts of ethanol found in some experiments were even higher than the xylitol concentration. This is because *Candida* sp. 11-2 is able to produce both ethanol and xylitol from xylose. The highest ethanol concentration (12.183 g/L) appeared under the same conditions that led to the highest xylitol concentration. Another sugar present, arabinose (0.79%), was hardly consumed. A similar behavior was observed by Meyrial et al. *(8)*.

Concentration of the Hydrolysate Treated with Activated Charcoal

The hydrolysate treated with activated charcoal was concentrated by vacuum evaporation in a rotavapor. An additional treatment with cation-exchange resins and neutralization was conducted. The xylitol production from that treated with cation-exchange resins was compared with that without the resin treatment. Figure 2 shows the results after 96 h of incubation. After 48 h, the highest xylitol production

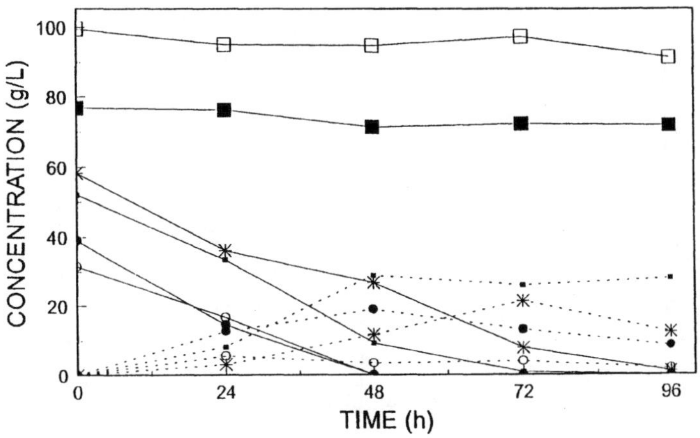

Fig. 3. Effect of the initial xylose concentration. Concentration of xylose with the following initial xylose concentrations: (—○—) 30 g/L, (—●—) 40 g/L, (—▪—) 50 g/L, (—✳—) 60 g/L, (—■—)75 g/L, (—⊟—) 100 g/L. Concentration of xylitol with the following initial xylose concentration: (- -○- -) 30 g/L, (- -●- -) 40 g/L, (- -▪- -) 50 g/L, (-✳ -) 60 g/L.

was reached. When the hydrolysate was not treated with cation-exchange resins, the xylitol productivity and the final xylitol concentration after 48 h were 0.192 g/L/h and 9.22 g/L, respectively, whereas when it was treated, the xylitol productivity and the final xylitol concentration increased to 0.602 g/L/h and 28.90 g/L, respectively. When the hydrolysate was treated with resins, the ethanol concentration was lower (16.10 g/L) than the xylitol production (28.90 g/L), but when the hydrolysate was not treated with resins, the ethanol production was higher (23.03 g/L) than the xylitol production (9.22 g/L).

Effect of the Initial Xylose Concentration

The hydrolysate treated with activated charcoal was concentrated to obtain different concentrations of initial xylose. After being applied to a column with cation-exchange resins, the hydrolysates were neutralized and supplemented with nutrients. Figure 3 shows the fermentation results after 96 h with different initial xylose concentrations. The rate of xylitol production increased with increasing initial xylose concentration from 30 to 50 g/L, reaching a maximum of 28.90 g/L after 48 h of fermentation. The decrease in xylitol production is dramatic with a further increase in the initial xylose concentration. With an initial xylose concentration of 60 g/L the xylose was mostly utilized and reached the maximum xylitol production of 21.35 g/L after 72 h. With initial xylose concentration higher than 75 g/L, xylose was not consumed at all.

Figure 4 shows the ethanol production from different initial xylose concentrations. Ethanol increased with increasing initial xylose concentration to 60 g/L, reaching a maximum of 26.87 g/L after 72 h of fermentation. Ethanol production was very low when the initial xylose concentration in the hydrolysate was 75 g/L or higher.

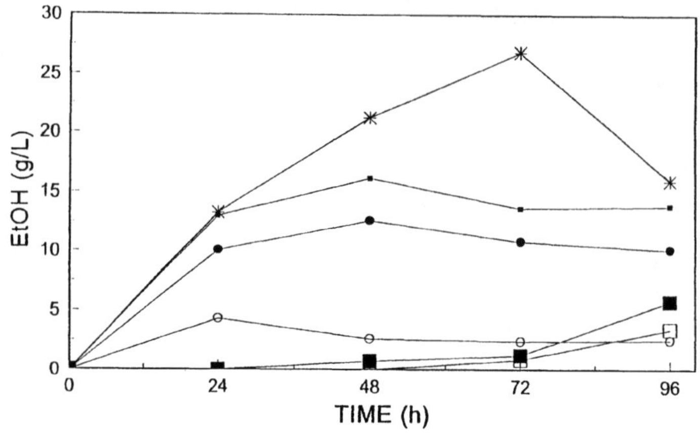

Fig. 4. Ethanol production with the following initial xylose concentrations: (—○—) 30 g/L, (—●—) 40 g/L, (—▪—) 50 g/L, (—✳—) 60 g/L, (—■—)75 g/L, (—□—) 100 g/L.

REFERENCES

1. Pepper, T. and Olinger, P. M. (1988), *Food Technol.* **42,** 98–106.
2. Makinen, K. K. and Joderling, E. (1980), *J. Food Sci.* **45,** 367–371.
3. Gong, C. S., Chen, C. S., and Chen, L. F. (1993), *Appl. Biochem. Biotechnol.* **39/40,** 83–89.
4. Hofer, M., Betz, A., and Kotyk, A. (1971), *Biochem. Biophys. Acta.* **252,** 1–12.
5. Ladisch, M. R. (1979), *Process Biochem.* **14,** 21–25.
6. Vitriskaya, A. M. and Soboleva, G. A. (1975), *Appl. Biochem. Microbiol.* **11,** 579–585.
7. Cole, M. (1958), *Nature* **181,** 1596,1597.
8. Meyrial, V., Delgenes, J. P., Moletta, R., and Navarro, J. M. (1991), *Biotechnol. Lett.* **13,** 281–286.

0273-2289/96/5758—0057$7.25

The Effect of External Reflux on the Foam Fractionation of Proteins

SÉRGIO LUIZ DE LUCENA,[1] EVERSON ALVES MIRANDA,[2] AND CESAR COSTAPINTO SANTANA*,[1]

[1]Departamento de Termofluidodinâmica and [2]Departamento de Processos Quìmicos, Faculdade de Engenharia Quìmica, Universidade Estadual de Campinas Caixa Postal 6066, CEP 13081-970, Campinas, SP, Brazil

ABSTRACT

Foam fractionation of proteins is especially advantageous in treating dilute solutions where other separation methods run into economical limitations. Improvement in separation efficiency can be obtained by adding external reflux to the column. This work studied operational variables, such as superficial gas velocity and solution concentration, on the enrichment of the foam phase when external reflux was added to the separation. Bovine serum albumin, cytochrome C, and barley malt α-amylase were the proteins fractionated in a column 3.2 cm in diameter and 105 cm in height operated by bubbling nitrogen in pH-controlled solutions. The determination of solute enrichment and purification factor for several external reflux ratios demonstrated the importance of this operational variable on the efficiency of foam fractionation of proteins.

Index Entries: Foam fractionation; bubble separation; reflux; protein separation; proteins.

INTRODUCTION

Foam fractionation is a technique that has long been used in the laboratory to recover and purify dilute solutions of proteins, including enzymes. Also, this technique can be an inexpensive way to separate two proteins present in an aqueous solution. This technique is based on the ability of surface-active compounds, such as proteins, to adsorb and concentrate in a gas–liquid interface. Therefore, a gas bubble rising through a pool of aqueous protein solution can have its interface enriched in protein. The formation of foam as the bubble reaches the top of the liquid phase provides a large gas–liquid interface where the adsorption of proteins can occur. Also, the liquid draining down through the foam bed is another important mechanism in which the proteins are exposed once more to the gas–liquid interface. This internal reflux is a second chance for proteins to adsorb at the interface of bubbles that are rising through the foam bed.

*Author to whom all correspondence and reprint requests should be addressed.

Foam fractionation was first used by Ostwald and Sierh in 1937 to separate albumin from potato and beet juices *(1)*. The possibility of protein denaturation owing to foaming was always a drawback for broader use of foam fractionation of proteins, especially enzymes. However, although some proteins are very sensitive to foaming, others are not and can be processed by this technique with high activity recovery *(2)*. Recently, researchers have revisited this technique owing to the need to develop cost-effective downstream separation techniques of valuable products and to remove pollutants from waste streams *(3–5)*.

Environmental factors that affect protein structure and, consequently, its surface activity, like pH, ionic strength, and the presence of organic solvents, are important operational variables. Other important variables are superficial gas flow rate, temperature, length of the liquid pool, length of the foam bed, and type of gas.

As mentioned, the draining of liquid from bubbles that have collapsed inside the column allows proteins a renewed contact with the gas phase. It gives the adsorption process a chance to move toward completion, resulting in higher loadings and selectivity. An enhancement of this phenomenon can be achieved by the introduction of external reflux into the column: The liquid resulting from the foam that collapses outside the column is partially returned to the column. It drains down the column, adding to the internal reflux. Studies of the introduction of external reflux have been done for the case of the fractionation of surfactants *(6,7)*. In an extensive literature search, no work on the effect of external reflux on the foam fractionation of proteins was found. This work had the objective of studying the effect of external reflux on the foam fractionation of different proteins, including an enzyme.

MATERIALS AND METHODS

Materials

Bovine serum albumin (BSA) and a crude preparation of barley malt α-amylase (BMA) were purchased from Sigma (St. Louis, MO). All other chemicals were reagent-grade. Solutions of BSA (50–100 mg/L) and cytochrome C (100 mg/L) were prepared in 0.1*M* phosphate buffer, pH 7.0. BMA solutions (2 g/L) were prepared with 0.027*M* acetate buffer, pH 5.1. Since BMA preparation had an insoluble fraction, its solutions were filtered through a 1.2-μm pore membrane prior to use.

Column Setup and Runs

A scheme of the experimental setup is shown in Fig. 1. The system was composed of a column, a foam breaker, and a reservoir for protein solution. The gas for the separation (nitrogen) was provided from a gas cylinder and humidified before entering the column. Peristaltic pumps were used to introduce external reflux at the top of the column and to recirculate protein solution from the external reservoir to the liquid pool in the column.

The glass column was 105 cm in height and 3.2 cm in diameter. Temperature was controlled by a water jacket and measured at two points: at the top of the column (foam) and at the external reservoir (protein solution). Bubbles were formed when humidified nitrogen gas flowed through a fritted glass disk at the bottom of the column. The foam breaker consisted of a stirrer whose paddles were located just at the column exit.

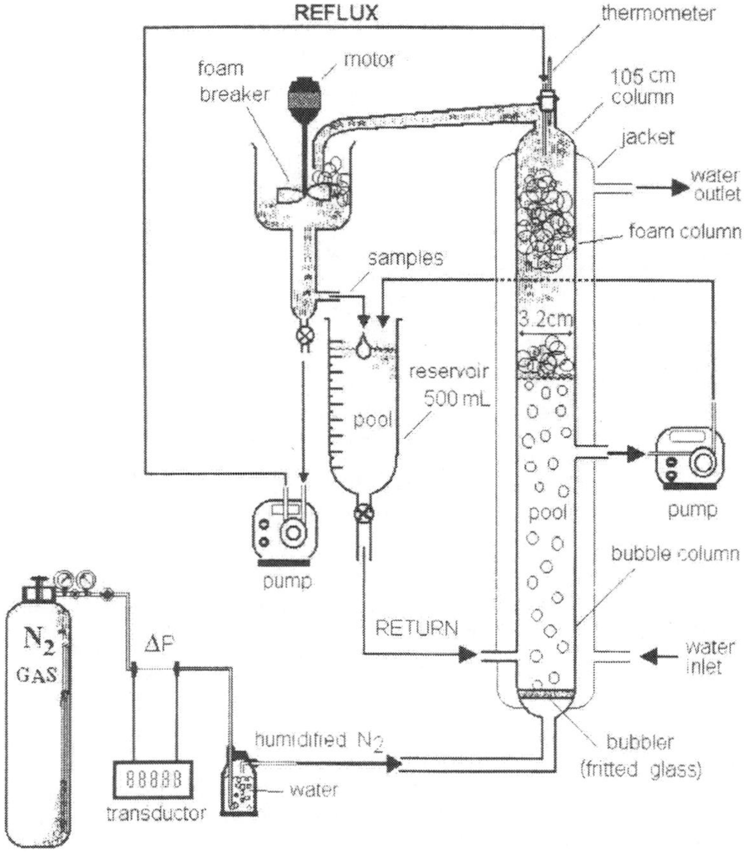

Fig. 1. Scheme of the experimental setup for fractionation of proteins in foam columns with external reflux.

Prerun operations consisted of turning the temperature control system on (recirculation of water in the column jacket), loading the column with 1 L of protein solution through the external reservoir, and turning the protein solution recirculating pump on. When the solution temperature at the external reservoir achieved 22°C, gas was introduced into the column at a fixed flow rate. Timing of the actual run was started when the first bubbles achieved the top of the liquid pool and started forming foam. When the foam reached the top of the column (initial height of foam bed of 48 cm), the foam breaker was turned on. External reflux was introduced at the top of the column by a peristaltic pump as soon as liquid overflowed from the foam breaker to the external reservoir. Samples of the foamate (collapsed foam) drawn at different times and samples of protein solution before and after fractionation were also taken for protein concentration determination and activity analysis. Protein enrichment was defined as the ratio of protein concentration in the sample to the protein concentration in the initial solution. Similarly, activity ratio was defined as the ratio of the activity of the sample to the activity of the initial enzyme solution (activity in U/mL).

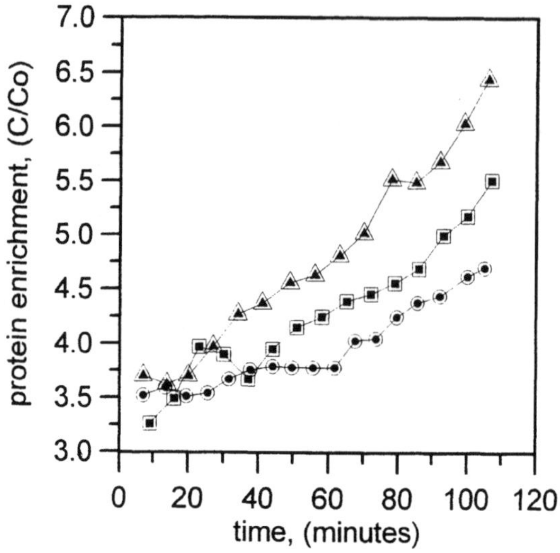

Fig. 2. Effect of external reflux ratio on protein enrichment from BSA solution with superficial gas velocity of 0.30 cm/s. BSA solution, 100 mg/L. ●, without external reflux; ■, external reflux ratio, 0.31; ▲, external reflux ratio, 1.10.

Analytical Methods

Protein concentration was determined by the Coomassie blue assay using a commercial kit based on the method developed by Bradford in 1976 (Bio-Rad [Hercules, CA] LIT33 Rev B) with insulin as the standard protein in the case of fractionation of α-amylase. BSA concentration was determined by the difference in absortion at 280 and 320 nm. When measuring protein in the binary mixture of cytochrome C and BSA, solution absorption was taken at 280, 320, and 409 nm. This procedure enables the setup of two algebraic equations relating the extinction coefficients of proteins with the concentration and the absorbances at 280 and 409 nm. Measurements at 320 nm were used as the baseline to subtract turbidity interference possibly present in the protein solutions. Enzyme activity was measured with 1% soluble starch solution, as substrate, at 20°C for 30 min, and the total reducing sugar released during reaction was measured by the 2,3-dinitrosalicilic acid method *(8)*. One unit of activity was defined as 1 mg of reducing sugar released as maltose equivalent/min at the assay conditions.

RESULTS AND DISCUSSION

Figures 2 and 3 show the results of BSA concentration by foaming at two different superficial gas velocities (0.30 and 0.45 cm/s, respectively). In both cases, the introduction of the external reflux increased the protein concentration in the foamate up to two times when compared with the case with no reflux. Also, the larger the external reflux ratio, the larger the enrichment in protein. In the case of no reflux, the increase in the superficial gas flow rate resulted in a lower enrichment of the foamate. Certainly, a shorter residence time of the gas in the column resulted in a

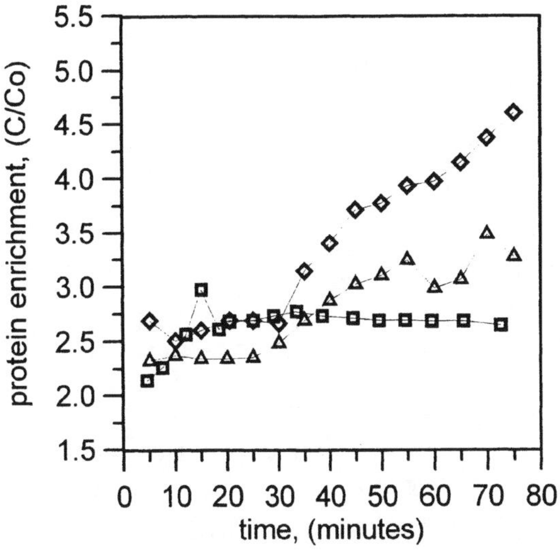

Fig. 3. Effect of external reflux ratio on protein enrichment from BSA solution with superficial gas velocity of 0.45 cm/s. BSA solution, 100 mg/L. External reflux started after 30 min of run. ◇, external reflux ratio, 0.68; △, external reflux ratio, 0.21; □, without external reflux.

lower loading of protein at the gas–liquid interface. Although the reflux ratios used at the two different superficial gas velocities were not the same, the overall enrichment seemed higher for lower superficial gas velocities when external reflux was present than it was for the case of no external reflux.

The effect of the initial protein concentration in the protein enrichment was evaluated at two different superficial gas velocities (Fig. 4). Protein enrichment showed an inverse relationship with initial protein concentration: Higher enrichments were obtained at lower protein concentrations. This is an advantageous feature of foam fractionation, since it enables the processing of diluted streams with high efficiency. The effect of superficial gas velocity on protein enrichment at different initial protein concentration was the same as verified in Figs. 2 and 3: The higher the superficial gas velocity, the lower the enrichment.

The fractionation of BSA in the presence of a second protein, cytochrome C, resulted in the same magnitude of BSA enrichment as before (Fig. 5). Also, the external reflux increased the enrichment of BSA in the foamate. However, cytochrome C enrichment was not strongly affected by the introduction of the external reflux. The fact that this protein is strongly hydrophilic may be the reason for this behavior: It does not strongly interact with the hydrophobic gas phase.

The fractionation of the α-amylase contained in the BMA preparation resulted in an increase in protein enrichment much higher than the increase in activity ratio (Fig. 6). However, the total activity recovery was only from 60 to 85%. Experiments showed that around 15% of the total activity loss was owing to the pumping necessary for the recirculation of protein solution between the external reservoir and the liquid pool in the column. Therefore, up to 25% of the initial activity was lost owing to foaming and foam breaking. The protein preparation used in these experiments was a crude α-amylase preparation where the enzyme was a small fraction of

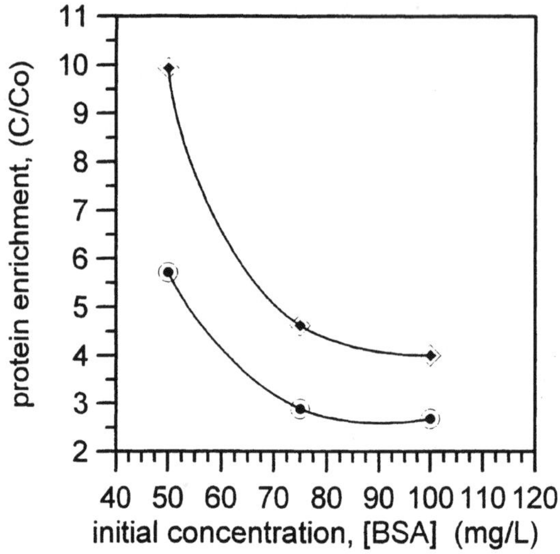

Fig. 4. Effect of initial protein concentration on protein enrichment from BSA solutions at two superficial gas velocities (V_{gas}) without external reflux. Processing time 60 min. Without external reflux: ◆, V_{gas}, 0.30 cm/s; ●, V_{gas}, 0.45 cm/s.

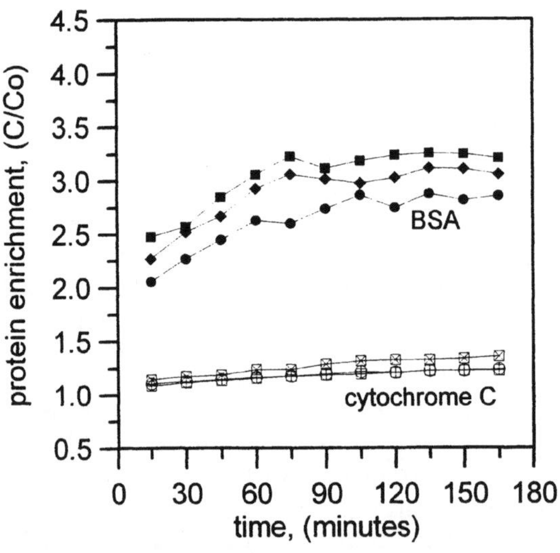

Fig. 5. Effect of external reflux on the BSA and cytochrome C enrichments from a solution 100 mg/L in both proteins. Superficial V_{gas}, 0.45 cm/s. ●, ⊕, without external reflux; ◆, ⊞, external reflux ratio, 0.15; ■, ⊠, external reflux ratio, 0.29.

the total protein (SA 1–3 U/mg of solids, whereas pure α-amylases have SAs in the range of thousands of U/mg of protein). Consequently, the denatured α-amylase contribution for the total protein enrichment in the foam was assumed small and negligible. Therefore, the results suggested that contaminant proteins

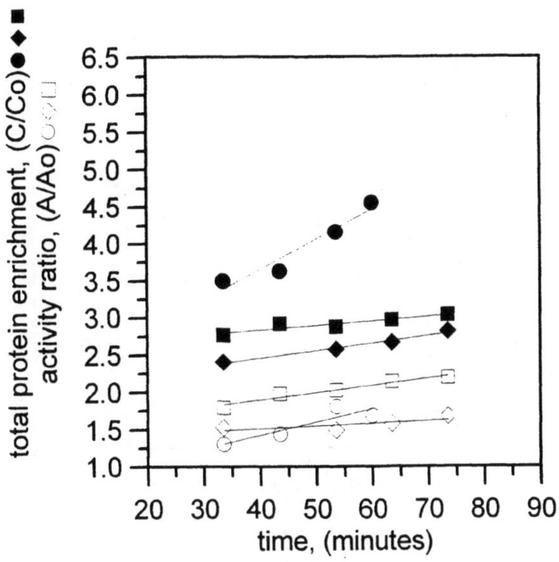

Fig. 6. Effect of superficial V_{gas} on the fractionation of BMA without external reflux. Without external reflux: ●, ○, V_{gas}, 0.20 cm/s; ■, □, V_{gas}, 0.30 cm/s; ◆, ◇, V_{gas}, 0.45 cm/s.

were preferentially adsorbed in the gas–liquid interface in comparison with α-amylase. The contaminants seems to be more hydrophobic than the α-amylase contained in the BMA solution.

The foaming of BMA with no reflux confirmed that total protein enrichment is a function of the superficial gas flow rate (higher enrichment for lower velocities). However, the effect of changing the superficial gas velocity on the activity ratio (ratio between activity of the foamate and activity of the initial solution) was not well defined by the experimental results; no significant changes in the activity ratio were verified with the change in the superficial gas velocity. The same behavior was observed when external reflux was introduced in the column (Figs. 7–9). The lack of sensitivity of the activity enrichment for changes in superficial gas velocity is probably owing to the lower hydrophobic character of the α-amylase at the experimental conditions as already mentioned. Since we expect no strong interaction between the gas phase and the enzyme based on its hydrophobicity, changes in the former are not to affect the latter.

CONCLUSION

The introduction of external reflux in the foam fractionation of proteins resulted in an increase in the protein enrichment in the foamate. This increase was more evident for proteins with strong hydrophobic characteristics, like BSA. The superficial gas velocity had the same effect on the protein enrichment with external reflux as for the case of no reflux. The larger the superficial gas velocity, the smaller the protein enrichment. The external reflux proved to be an important variable in the search for optimum operational conditions in the foam separation of proteins.

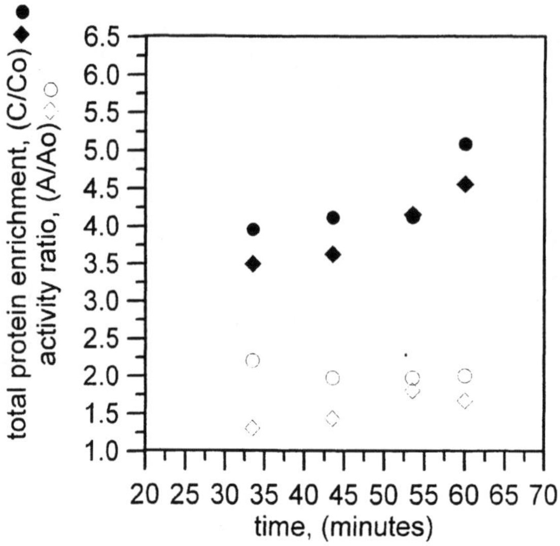

Fig. 7. Effect of external reflux on the foam fractionation of BMA. Superficial V_{gas}, 0.20 cm/s. ◆, ◇, without external reflux; ●, ○, external reflux, 0.69 mL/min.

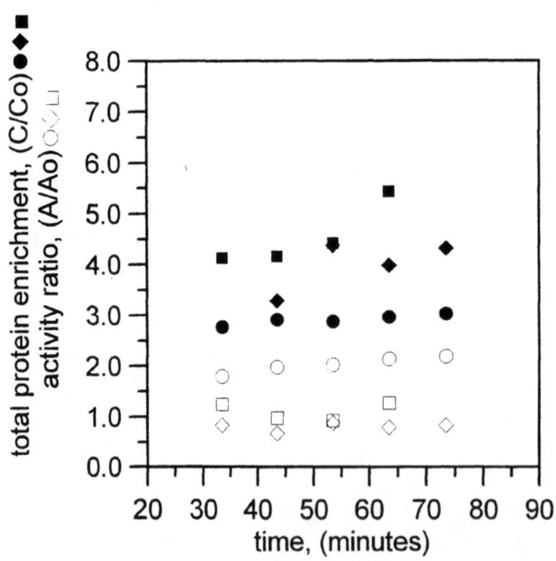

Fig. 8. Effect of external reflux on the foam fractionation of BMA. Superficial V_{gas}, 0.30 cm/s. ■, □, external reflux, 1.25 mL/min; ◆, ◇, external reflux, 0.78 mL/min; ●, ○, without external reflux.

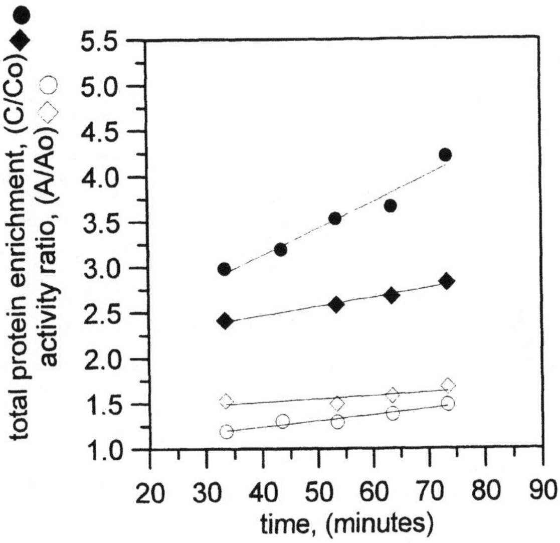

Fig. 9. Effect of external reflux on the foam fractionation of BMA. Superficial V_{gas}, 0.45 cm/s. ●, ○, external reflux, 0.95 mL/min; ◆, ◇, without external reflux.

REFERENCES

1. Santana, C. C. (1994), *Revista Brasileira de Engenharia* **5(Oct.),** 1–74.
2. Uraizee, F. and Narsimhan, G. (1990), *Enzyme Microb. Technol.* **12(March),** 232,233.
3. Uraizee, F. and Narsimhan, G. (1995), *Separation Sci. Technol.* **30(6),** 847–881.
4. Bhattacharya, P., Ghosal, S. K., and Sen, K. (1994), *Separation Sci. Technol.* **29(7),** 855–865.
5. DeSouza, A. H. G., Tanner, R. D., and Effler, W. T. (1991), *Appl. Biochem. Biotechnol.* **28/29,** 655–666.
6. Lemlich, R. and Lavi, E. (1961), *Science* **134,** 191.
7. Harper, D. O. and Lemlich, R. (1965), *Ind. Eng. Chem.: Process Design Dev.* **4(1),** 13–16.
8. Bailey, J. M. (1988), *Appl. Microbiol. Biotechnol.* **29,** 494–496.

0273-2289/96/5758—0067$7.00

Biosorption of Precious Metal Ions by Chicken Feather

Kyozo Suyama,*[,1] Yoshitaka Fukazawa,[1] and Hiroshi Suzumura[2]

[1]*Department of Applied Biological Chemistry, Faculty of Agriculture, Tohoku University, Sendai 981, Japan; and* [2]*Rheology Laboratory, Hiroshima Research and Development Center Technical Headquarters, Mitsubishi Heavy Industries, Ltd., Hiroshima 450, Japan*

ABSTRACT

Chicken feather (C-feather) is an intricate network of stable and water-insoluble protein fibers with high surface area and is an abundant bioresource. C-feather protein was found to accumulate various precious metal ions (gold and platinum metals) selectively from their dilute aqueous solutions in high yield and in short contact time, depending on pH and characteristics of the individual precious metal ions. Under certain condition, the sequestering level of precious ions, Au(III), Pt(II), and Pd(II) approaches about 17, 13, and 7% of dry wt of C-feather, respectively. Gold(III) potassium cyanade was also accumulated up to 5.5% at pH 2.0. The presence of 100-fold (mol) of coexisting cations, such as Na^+, Fe(III), Cu(II), and Ni(II), did not show a discernible effect on the precious metal uptake rate and capacity of C-feather. Experiment suggested C-feather is promising for use in the removal/recovery of precious metals as well as water pollution control. A qualitative discussion is given about the excellent adsorption behavior of C-feather.

Index Entries: Chicken feather; precious metal; biosorption.

INTRODUCTION

Gold and platinum metal ions are precious metal that have been in high demand for centuries. Despite continued threats of material substitution, consumption of these precious metals by the industry grew and is expected to increase further in the years to come. Gold is used extensively in many high-technology areas, particularly in computer applications, that demand the highest reliability. Platinum metals are used as a catalyst in organic synthesis and combustion gas treatment. Precious metals recovery from secondary sources, such as electronic scrap and catalyst, and waste electroplating solution, is therefore an important technology as well as good recovery from primary resources, such as leach solutions.

The recovery and collection of these precious metals from dilute solutions generally involves either zinc dust precipitation, carbon adsorption, solvent extraction, and usage of ion-exchange resin *(1)*. However, in most cases, these techniques have low selectivity and are likely to become increasingly expensive.

*Author to whom all correspondence and reprint requests should be addressed.

Recently, some success has been achieved in the development of biosorbents for precious metals recovery. Certain types of biomaterials, such as microbial cells *(2–4)*, chitosan *(5)*, marine algae *(6–8)*, egg shell membrane *(9)*, possess the potential to sequester and accumulate metal cations, including precious metals in aqueous solution. Metal accumulation by biomaterials is termed "biosorption." In the present investigation, chicken feather (C-feather) was found to accumulate precious metal ions from their dilute aqueous solutions in high yield and in short contact time, but is dependent on the pH and characteristics of the individual metal ion. Biosorption is generally considered to be a rapid physical/chemical process and not expensive. C-feather is an abundant resource (byproduct of chicken industry), highly stable, water-insoluble, homogeneous, and has high surface area. It allows metal ions, as well as extremely large organic nutrient crosslinked protein molecule, to pass freely through it.

This article outlines some basic process parameters in the potential use of a new biomaterial, C-feather as a biosorbent, for sequestering and accumulating precious metals from aqueous solution.

MATERIALS AND METHODS

C-feather sample was obtained from the waste of a local chicken industry of Japan. The feather was washed twice with water and detergent by an electric washing machine, then washed with deionized distilled water, dried in air at room temperature, and then desiccated over phosphorus pentoxide under reduced pressure at room temperature. The powdered product was obtained by grinding the feather in the blender for approx 2 min.

Two procedures were utilized in experiments reported here. In the "batch" procedure, approximately the same weight of feather (around 30–40 mg) was precisely weighed, then placed directly in contact with aqueous solutions containing various metal ions. The feather was placed directly and soaked in the aqueous solutions containing metal ions in 40-mL stoppered glass test tubes. To prevent metal contamination during experiments, the test tubes were soaked with 10% HCl overnight. Then, they were washed and rinsed with distilled deionized water. This cleaning procedure was used throughout the experiments. The suspension volume was always 20 mL, unless otherwise specified. Precious metal solutions of desired concentrations (initial concentration of metal solution was always $3.0 \times 10^{-3}M$, unless otherwise specified) were prepared by dissolving Au(III) as $AuCl_4^-$, Pt(II) as $PtCl_4^{2-}$ and Pd(II) as $PdCl_2$ in deionized distilled water and shaken at 60 rpm with a shaker in a thermostated water bath at 25°C. The pH was adjusted with 0.1*N* HCl or 0.1*N* NaOH. Measurement of pH of solution was done by glass electrode pH meter type F-12 (Horiba Inc., Kyoto, Japan). The following chemicals—$NiCl_2{\cdot}6H_2O$, NaCl, $FeCl_3{\cdot}4H_2O$, and $CuCl_2{\cdot}H_2O$—were chosen for coion effect experiments considering their high solubility in solutions at room temperature. All the reactions were terminated by the separation of the feather from the solution with tweezers, and then the feather was washed with the metal-ion-free aqueous solution of the same pH.

The kinetics of precious ion biosorption by the feather was investigated in a series of contact test tubes at the stated initial concentration at 25°C and at the desired pH. After desired durations of contact, the reactions were terminated, then the residual precious metal concentrations were determined, and the result was a time-dependent precious metal uptake profile. The biosorptive uptake capacity was determined as a function of residual metals concentration or by measurement of dry weight of feather complexed with metals. In order to confirm the retention of precious metal on the feather, the exposed feather was analyzed for precious metal content. For this purpose, the metal-laden feather was digested with an aqua regia solution.

Experimental scale of selective collection of gold and palladium from aqua regia digest of anode slime was carried out by feather powder-packed minicolumn (0.5×3.0 cm). The slurry of powdered feather was poured into a column, then aqua regia solutions of anode slime was passed through the column, and the effluent was analyzed for metal ions. For preparation of the aqua regia solution of metal ions from anode slime, 10 g of dried anode slime were digested with 30 mL of aqua regia. Gold and platinum adsorption was accomplished by passing the 10-times-diluted aqua regia solution through a column of the feather powder biosolvent.

Metal ion analysis was performed by using inductively coupled plasma atomic spectrometry (ICP AES: SPS 1200A, Seiko Instrument Inc., Tokyo, Japan).

RESULTS AND DISCUSSION

The feather exhibited a certain degree of biosorption of all metal ions tested. These included Ni(II), Zn(II), Cu(II), Ag^+, Hg(II), Au(III) as $AuCl_4^-$, Pt(II) as $PtCl_4^{2-}$, and Pd(II) as $PdCl_2$ (data for all not shown). The metal ions not specified were added as sulfate, nitrate, or chloride salts.

Figures 1A–C demonstrates the pH dependence of gold, platinum, and palladium at different soaking times of 1, 3, and 5 h at 25°C, respectively. These metal ions are bound to the feather much more strongly than any other metal ions tested, e.g., under certain conditions the feather has a binding capacity for gold(III) that is at most 17% of the dry wt of feather (about 0.7 mmol of gold / g of dry wt feather, for platinum that is 9%, and for palladium that is at most 6%. There are significant differences in behavior for each metal. For example, the maximum uptake values for gold , platinum, and palladium were obtained at pH 3.0, 1.8, and 2.2, respectively. This variation would enable one to perform selective elution of bound metal ions from feather for these ions.

In contrast to the behavior of most metal ions under these conditions, the binding of precious elements is not completely reversed by lowering the pH of the solution. We found a considerable amount of biosorption of all these precious elements to the feather takes place under the condition of 2% HCl solution (data not shown) and 10-times-diluted aqua regia solution with distilled water as described later. However, exposure of precious metal-bound feather, which was prepared by the treatment of the feather in the solution of precious metals at pHs ranging from 2–3, to 2% HCl solution was effective in stripping these metal ions from the feather. For example, roughly 55% of Au loaded at pH 3.0 was removed by the 2% HCl solution. Under the condition of 2% HCl at room temperature, the feather is stable and can be used repeatedly. However, exposure of precious metal-bound feather to high concentrations of sulfur-containing ligands, such as thiourea or mercaptoethanol, are effective at certain pHs (not shown) in stripping these metal ions from the feather surface.

Figure 2 illustrates the pH dependence of uptake of dicyanoaurate, $Au(CN)^{2-}$ at 25°C for 1 h soaked in $3.0 \times 10^{-3}M$ solution. And Figure 3 shows the relationships between contact time and uptake of $Au(CN)^{2-}$ on the feather at pH 2.5 and 25°C. The

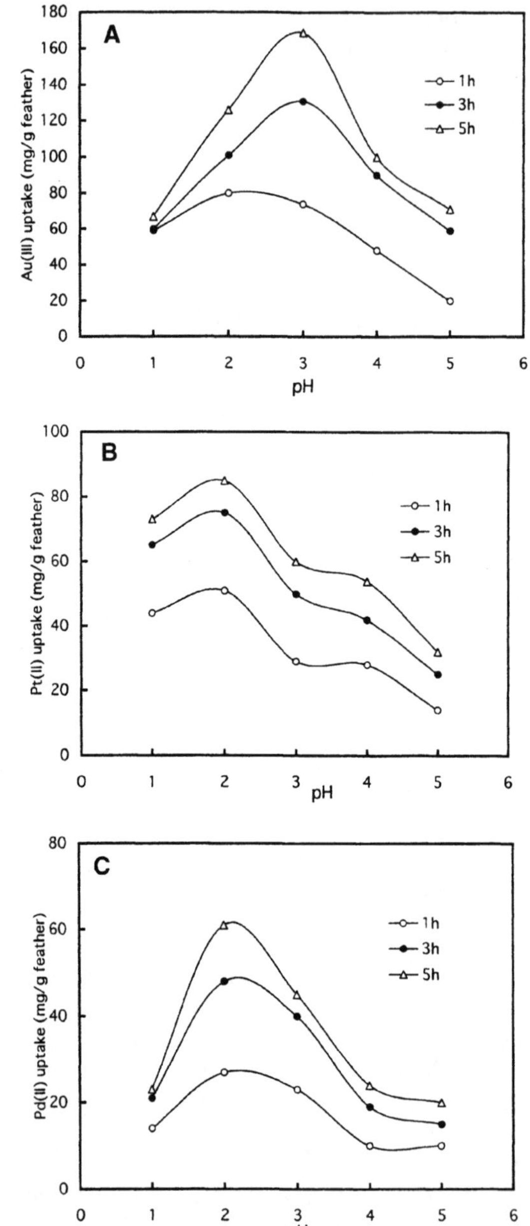

Fig. 1. **(A)** Effect of pH and soaking time on uptake of Au(III). Au(III): $3.0 \times 10^{-3}M$, temp.: 25°C. **(B)** Effect of pH and soaking time on uptake of Pt(II). Pt(II): $3.0 \times 10^{-3}M$, temp.: 25°C. **(C)** Effect of pH and soaking time on uptake of Pd(II). Pt(III): $3.0 \times 10^{-3}M$, temp.: 25°C.

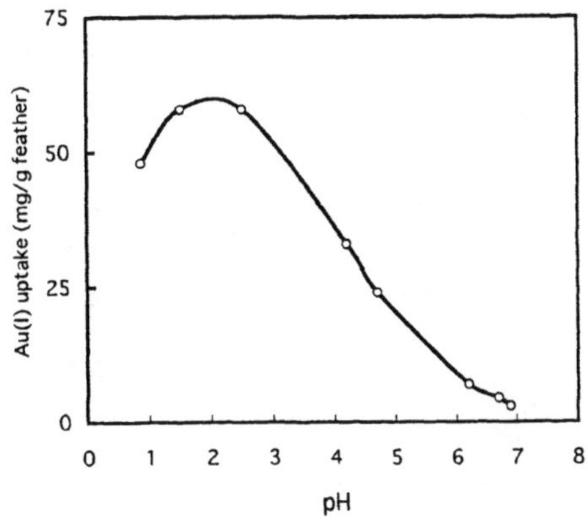

Fig. 2. Effect of pH on uptake of $Au(CN)^{2-}$. Au(III): $3.0 \times 10^{-3}M$, temp.: 25°C. Soaking time: 1 h.

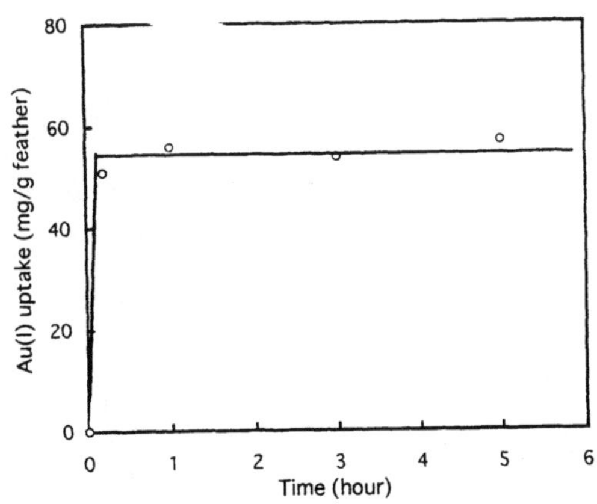

Fig. 3. Relationships between contact time and uptake of $Au(CN)^{2-}$ on chicken feather. Au(III): $3.0 \times 10^{-3}M$, ph: 2.5 at 25°C.

solution pH considerably affected the $Au(CN)^{2-}$ uptake. Interestingly, the feather shows unique behavior for $Au(CN)^{2-}$ to have high binding capacity and uptake equilibrated in extremely short contact time, i.e., under pH 2.5 the feather has binding capacity that is 5% of the dry wt of the feather within 15 min. Also the binding of $Au(CN)^{2-}$ is completely reversible when pH is increased to 6.0. From this behavior, we show that the feather could be used to the remove and/or recover gold from the effluent electroplating solutions.

The effect of the presence of coions in the solution on the precious metals biosorptive capacity of the feather was examined. Figure 4 shows the effect on palladium uptake, for example. Experiments were performed on $3.0 \times 10^{-3}M$ palladium solution with five times the concentration of coions, Fe(III), Cu(II), Ni(II), and Na^+ as cations, which were tested because of their high probability of occurrence in solutions from industrial sources, at pH 2.0 for 5 h of soaking time. It was observed that no uptake of coions took place and there was a slight change in the uptake capacity of palladium. Gold and platinum also showed slight change

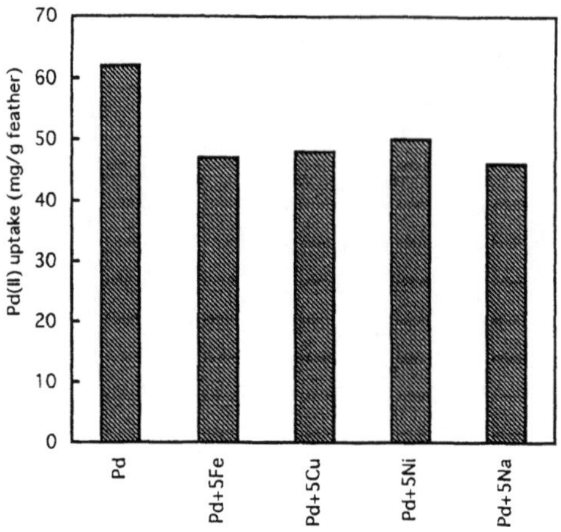

Fig. 4. Effect of cometal ions on uptake of Pd(II). Pd(III): $3.0 \times 10^{-3}M$, cometal ions Fe(III), Cu(II), Ni(II), Na^+: $1.5 \times 10^{-2}M$.

in the uptake rate and capacity in the presence of five times more (by mol) of these coexisting metal ions (data not shown).

Figure 5 shows the adsorption of palladium(II) on feather from the dilute (16 ppm) palladium solutions at pH 2.0 without and with 100 times more (by mol) coions, Fe(III), Cu(II), Ni(II), and Na^+. The results show no detectable amount of palladium found in all the aqueous solutions tested after 1 h of contact with the feather.

The different affinities of metal ions for binding to the feather can be used to develop a scheme whereby metal ions can be removed from solutions and can be selectively recovered. To demonstrate this, a solution that contained mixture of palladium ($1 \times 10^{-4}M$) and 10 times more (by mol) Cu(II) at pH 2.0 was applied to a column packed with feather powder. Copper ions eluted freely through the column, and palladium was retained completely. The result suggested that selective separation and recovery of precious metal ions may be possible. To confirm this, an aqua regia digestive solution of anode slime (Au: 0.4749, Ag: 9.4389, Cu: 0.7700, Se: 5.4, Pb: 46.17, and Pd: 0.0183 wt%) was charged on a column. Eluting solution was simply prepared by diluting with 10 times more (by volume) distilled water to the aqua regia solution. Figure 6 shows the elution profile of gold and palladium. The results suggested that precious metals can be separated and collected by feather biosorption.

The identities of the functional groups responsible for metal ion binding to the chicken feather are not known at this time. From the amino acid composition, sulfhydryl groups are involved in the binding for metal, but they are not the only groups responsible for precious metal binding. Precious metal ions have larger ionic radii than the coions tested in this research. This could conceivably be a part of the reason why they are so selectively sorbed from a solution of mixed cation. Further works are necessary to investigate the regeneration and multiple reuse potential of the biosorbent in many subsequent application cycles. Also further experiments are in progress to characterize for the binding sites and mechanisms of the binding of the precious metal ions. Whatever the nature of the binding sites, C-feather has a high affinity for the binding of precious metal ions.

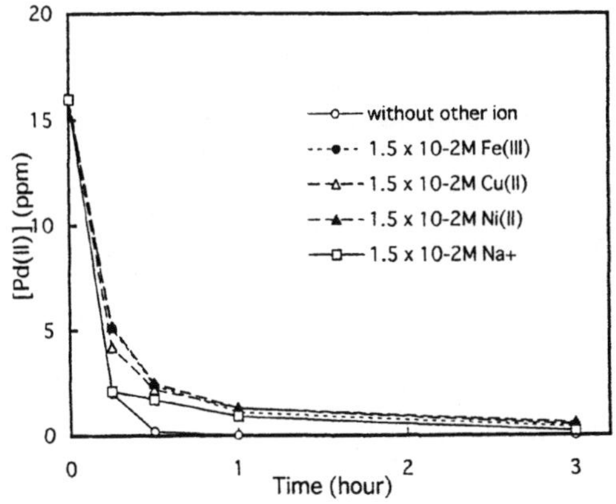

Fig. 5. Relationships between Pd(II) concentration and soaking times treated by batch method from the dilute Pd(II) solution in the presence of 100 times more (by mol) of cometal ions. Initial Pd(II) concentration and volume: $1.5 \times 10^{-4}M$ and 20 mL. Feather: 50 mg, pH: 2.0 temp.: 25°C.

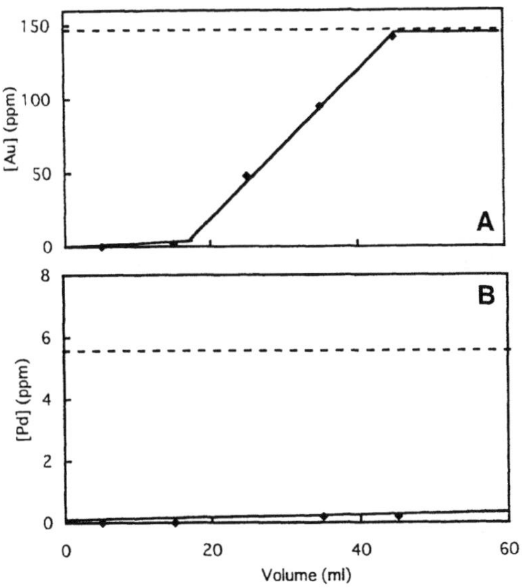

Fig. 6. Elution profiles of precious metal ions by minicolumn of the aqua regia solution of anod slime. **(A)** au, **(B)** pd. Anode slime was digested with three-fold increase (w/v) of aqua regia, then aqua regia solution was diluted 10-fold increase with water. Dilute solution was charged to feather minicolumn (5 × 30 mm: P 115 mg of feather powder). pH was not adjusted. -------: Initial concentration of Au and Pd.

REFERENCES

1. Tobin, J. M., Cooper, D. G., and Neufeld, R. J. (1984), *Appl. Environ. Microbiol.* **47,** 821–824.
2. Wood, J. M. and Wang, H. K. (1984), *Environ. Sci. Technol.* **18,** 106–110.
3. Akser, M., Wan, R. Y., and Miller, J. D. (1986), *Solvent Extr. Ion Exch.* **4,** 531–546.
4. Finnegan, I., Toerien, S., Abbott, L., and Raubenheimer, H. G. (1991), *Appl. Microbiol. Biotechnol.* **35,** 274–276.

5. Inoue, K., Yoshizaki, K., and Baba, Y. (1991), in *New Dev. Ion Exchange* (Proc. ICIE, 1991, Tokyo), pp. 2–4.
6. Kuyucak, N. and Volesky, B. (1989), *Biorecovery* **1,** 189–204.
7. Darnell, D. W., Green, B., Henzle, M. T., Hosea, J. M., McPhason, R. A., Sneddon, J. and Alexander, M. D. (1986), *Environ. Sci. Technol.* **20,** 206–208.
8. Hosea, M., Green, B., McPhason, R., Henzl, M., Alexander, M. D. (1986), *Inog. Chimica Acta* **123,** 161–165.
9. Suyama, K., Fukazawa, Y. and Umetsu, Y. (1994), *Appl. Biochem. Biotechnol.* **45/46,** 871–879.

0273-2289/96/5758—0075$7.50

The Effect of Fermentation (Retting) Time and Harvest Time on Kudzu *(Pueraria lobata)* Fiber Strength

Sibel Uludag,[1] Veara Loha,[2] Ales Prokop,[2] and Robert D. Tanner*,[2]

[1]*Civil and Environmental Engineering Department; and* [2]*Chemical Engineering Department, Vanderbilt University, Nashville, TN 37235*

ABSTRACT

The noncommercial kudzu plant has been growing wild in the southern United States since the 1930s. In this article, the kudzu fibrous vine is investigated for possible economic applications. The feasibility of removing the fibers by microbial retting is investigated as a low-cost method for recovery of these fibers. The harvest season, vine diameter, and the natural retting time are the three main variables investigated to determine the optimal fiber tensile strength. To extract these light yellow, unidirectional, and multicellular kudzu fibers, retting using a culture screened from a naturally occurring mixed culture was investigated. Late fall, winter, and spring retted fibers were subjected to tensile strength tests. The kudzu fibers were harvested from the wild in both Nashville, TN and Muscle Shoals, AL.

Index Entries: Kudzu; retting; natural fibers; fiber strength; tensile strength.

INTRODUCTION

Although kudzu *(Pueraria lobata)* was deliberately grown in the southern United States by the US Soil Conservation Service in 1930s and 1940s *(1)*, today it is considered to be an "escaped plant" *(2)*. Kudzu can reach a growth rate of up to 0.3 m (1 ft) in a day in late spring/early summer, and its vines can grow up to 18–30 m (60–100 ft) in a single season. The vines can climb upward and cover nearly everything in their path (such as fields and trees). Hence, it is considered to be an "unwanted plant" particularly by farmers, foresters, and park naturalists.

Kudzu provides herbal medicines that are used in China and Japan. For example, such medicines have proven to be effective in relaxing tight and painful muscles in the neck, and pain in the shoulders and back caused by "wind-heat" injuries *(3)*. The kudzu root is also very rich in starch, and the kudzu vine has a

*Author to whom all correspondence and reprint requests should be addressed.

fibrous structure. Although natural fiber production from kudzu vines has largely been replaced by synthetic fibers, other traditional natural fibers, such as cotton, flax, hemp, silk, and wool, continue to retain their popularity owing to their unique physical properties. In China, kudzu fiber is still being woven into cloth and is occasionally used to make baskets, packing materials, and suspension bridges *(4)*. The most famous area of handicraft cloth production is located near the headwaters of the Yangtze River. On the other hand, one of the last industrial-scale weaving plants located in Japan using twisted kudzu filaments closed in 1975 because of competitive pressures from synthetic fibers *(4)*. Today, coarse kudzu vine fibers are imported to the United States from Korea, where the kudzu plants are cultivated for use as "grass-cloth" wall coverings in interior design.

Kudzu fibers extracted from the woody stem (vines) of the plant have a multiple cellular and unidirectional structure. Kudzu fibers are classified as bast fibers (fibers extracted from the interior part of the plant sheath) *(5)*. When chemically treated, such fibers can be converted to long staple fibers (yarns).

The remarkable combination of properties—durability, strength, natural light yellow color, and purity, and the absence of noncellulosic foreign material (except in the winter fibers)—makes the kudzu fibers attractive and desirable. Natural fibers are obtained from the retting process as ready yarn, but of variable widths, rather than individual fibers. These natural fibers can be easily spun into bulky yarns in large-scale production.

METHODS

Retting Procedure

Kudzu vines of various diameters were harvested locally in both Tennessee (Nashville area) and Alabama (Muscle Shoals area). These cut vines were then classified according to diameter size into groups: seven groups of varying diameters for the late fall vines, four groups for the winter vines, and three groups for the early spring vines. The largest diameter vine tested was 0.008 m (0.32 in.). Afterward, vines of approx 0.15 m (6 in.) in length were submerged in tap water (ca. pH = 5.9) along with the microbial culture. They were kept submerged with the aid of weights on the top of container. This retting process then degraded the vine at room temperature (ca. 21°C) in the water solution (i.e., the fibers underlying the vine sheath were exposed for later recovery by the action of sheath degradation enzymes, which were contributed by the growing culture). After the fibers were removed, the underlying straw stalk was then recovered. Vine samples were taken from the water chamber until this solid-state (with respect to the "solid" vines) fermentation process was terminated. The stopping point was reached when the outer sheath was easily removed from the vine under flowing tap water without mechanical assistance. Mechanical assistance in the laboratory was provided by a spatula, used as needed in the retting process both to remove the outer sheath and extract the fibers. Retting action was followed in all of these experiments by the fiber recovery step. In that step, the separated fibers were washed with tap water and then dried at room temperature. In addition to all of the physical measurements related with the fiber itself, the pH profile of the developed microbial culture was observed using a Chemcadet pH meter to determine the optimal retting time.

Preparation of Fibers for the Tensile Strength Test

The thickness of each weighed fiber was measured by a micrometer with ±2.54 × 10^{-6} m (±0.0001 in.) tolerance, whereas the width and length were measured using an ordinary ruler. To measure the thickness of a fiber, the fiber was placed between two microscopic slides, each of which were 0.001 m (0.04 in.) thick. Since the kudzu fibers come from the retting process as flat strips, these measurements were easily carried out. To make a fiber thickness measurement, a flat kudzu fiber strip was placed between two glass microscope slides to create a "fiber sandwich." The flat side of one of the glass slides of the "fiber sandwich" was then placed against the anvil of the micrometer, and the micrometer handle was rotated clockwise until contact was made with the flat side of the other glass slide. The handle was turned until it could turn no more (the instrument prevented glass breakage with an internal "slip gear"). The final contact measurement of the "fiber sandwich" was easily replicated for each measurement, because the micrometer contained a ratchet that provided a replicable limit to the force applied on the glass plate (the "slip gear"). The flat strips, incidentally, came sectioned from the retting process (like narrow peel-off labels on the same page). This multisectioned kudzu strip of fibers was generally clear, soft, and nearly uniform in appearance, although comprised of fibers of apparently random widths. These textural properties of the fibers make them similar to today's modern fibers. The tensile strengths reported here are expressed in several different forms to fit the common units used in the textile industry, such as denier and texture. Denier is a direct numbering system for expressing linear density of a fiber, equal to mass in grams (g_f) per 9000 m of yarn or filament *(6)*. Although denier is used widely, the texture (tex) system has been adopted by the ASTM as the standard unit of linear density of textile fibers, and it is accepted as a part of the SI unit system. The texture is the "mass" of a fiber in grams force (g_f) per length of 1000 m (1 denier = 0.1111 tex) *(7)*.

Tensile Strength Test

Tensile force effects were tested using a D-Dillon Weigh-Tronix Inc. (Fairmont, MN) Tensile Testing Machine Model DTM and a Model P-3500 Instruments Division Strain Indicator. Since the expected strength was near the lowest detection limit of the machine, the experiments were run with the specified strain gage factor for low-load applications of +0.252. Fibers of approx 0.06 m (2.36 in.) in length were used in this breaking test. The tensile strength measurements were replicated in approximately one-third of the total fiber samples. The total number of fiber tests was around 130.

RESULTS

Our focus here was to determine for a given year three effects on the tensile strength of the fiber:

1. The season in the southeastern US in which the vine was harvested;
2. The effect of vine size; and
3. The effect of retting time.

The literature indicated that in temperate Japan, the optimal strength of the fibers came from vines harvested in early June to late July *(4)*.

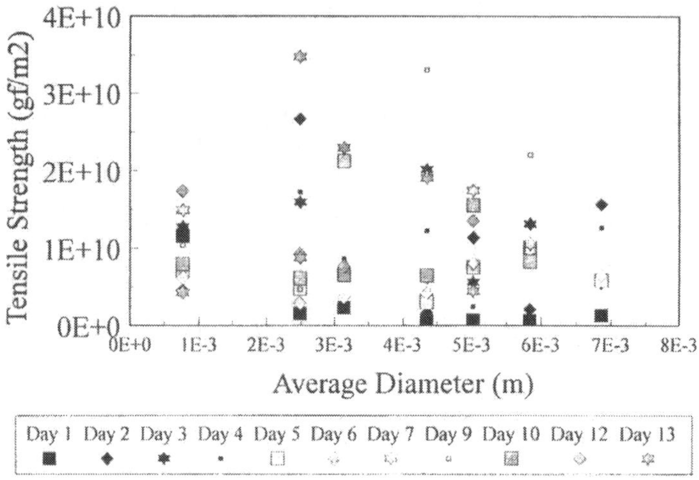

Fig. 1. The effect of kudzu vine diameter on the tensile strength (late fall, 1994).

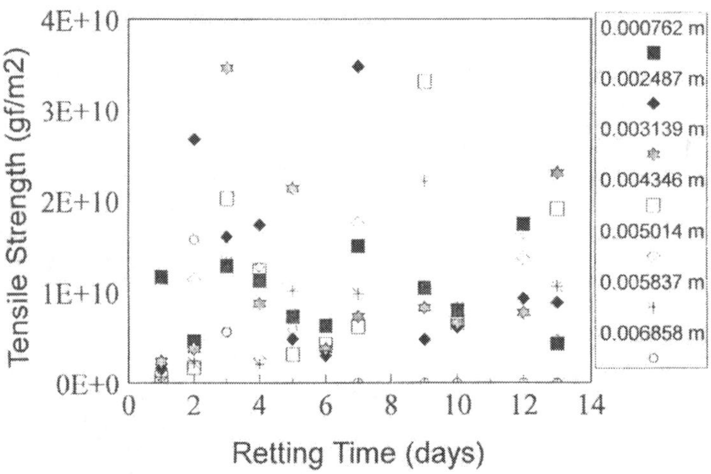

Fig. 2. Late fall kudzu fiber's tensile strength change with retting time.

Late Fall Harvest (November 19, 1994)

The effects of vine diameter and retting time on fiber tensile strength are shown in Figs. 1 and 2. The fiber strength reaches a maximum for a vine diameter of 0.002487, 0.003139, and 0.004346 m (0.0979, 0.1236, and 0.1711 in.) for d 7, 3, and 9, respectively (retting times). The kudzu fiber tensile strength reached a maximum limit around 3.5×10^{10} g_f/m^2 (50,000 lb_f/in^2), for the late fall, as shown in Fig. 1. Fibers extracted from vines between 0.002487 and 0.004346 m (0.0979 and 0.1711 in.) in diameter seem to be stronger than the others. The optimum retting time for the late fall harvest seems to fall within the range of the three highest values shown in Fig. 2 (3–9 d, with 6 d being the average).

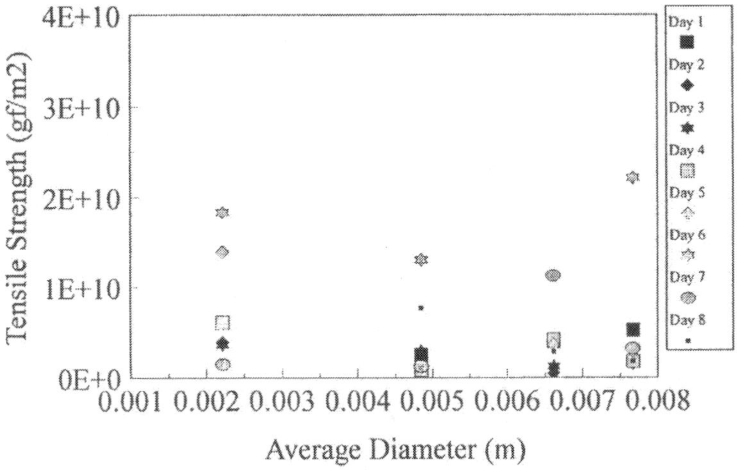

Fig. 3. The effect of kudzu vine diameter on the tensile strength (winter, 1995).

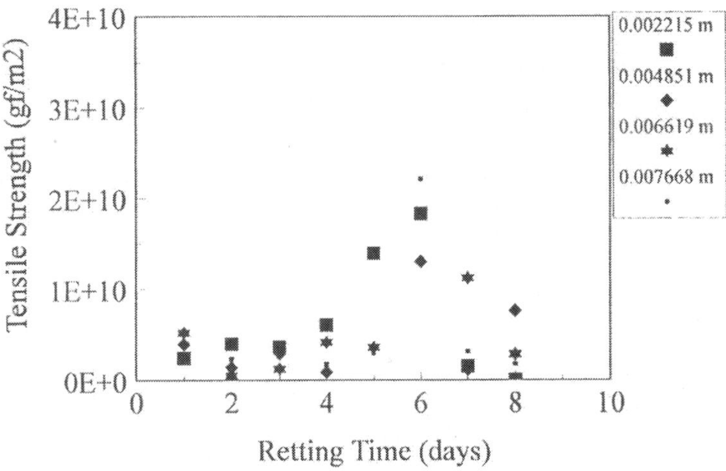

Fig. 4. Winter kudzu fiber's tensile strength change with retting time.

Winter Harvest (January 26, 1995)

The major result obtained from the fibers that were exposed to cold weather is that they are not as strong as late fall fibers. The fibers tested in this set all have a value below 2.0×10^{10} g_f/m^2 (28,421 lb_f/in^2), as shown in Fig. 3. The strongest fibers are the fibers of 0.002215 m (0.0872 in.) diameter vines. The optimum retting time, based on the maximum value obtained, is determined as 6 d in winter from the tensile strength vs retting time data (Fig. 4). Most of the kudzu fibers from the winter vines were gray in color (vs clear or light brown from the late fall vines), possibly because of fungal or bacterial contamination in the winter.

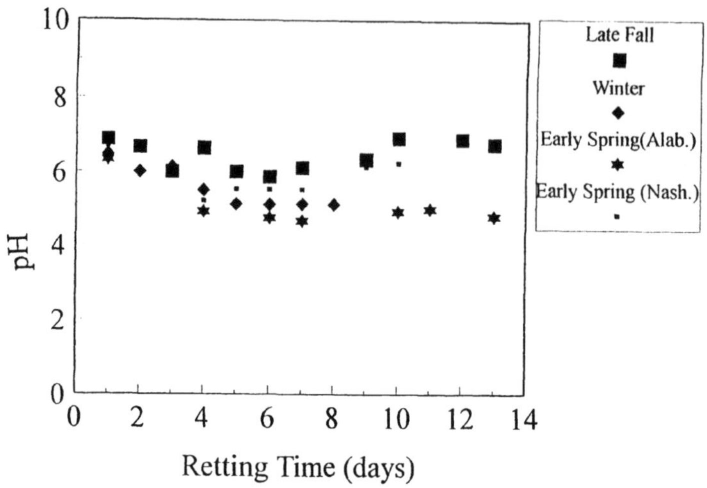

Fig. 5. pH change of retting solution with time.

pH During the Retting Process

In all seasons studied, the pH of the uncontrolled retting solution decreased with time up to 5–6 d and then usually increased (Fig. 5). The global pH value of the retting solutions changed between 4.8 and 7.0, when all of the seasons were considered. The reason for this change in pH may have been the dominance of acid-producing bacteria in the solution during the 5–6 d retting time.

DISCUSSION

In this section, we will compare the quality of the kudzu vine fiber between seasons and with other natural fibers. The averages of our experimental results are graphed in Figs. 6 and 7 and summarized in Table 1. In this project, the number of samples of early spring kudzu vine fibers were limited, since there were very few new-growth kudzu vines available at the harvesting location in Nashville. The kudzu vines available for harvest in Alabama in the early spring were thicker than those previously studied for late fall and winter, again because of the limited number of new-growth vines. Therefore, the discussion part of this article emphasizes mainly the late fall and winter fiber test results.

Late fall kudzu fibers are stronger than the winter kudzu fibers as shown in Figs. 6 and 7 and in Table 1. The early spring fibers from Alabama (harvested on March 20, 1995) were about the same strength as the late fall fibers, but the early spring fibers from Tennessee (harvested on May 8, 1995) were about twice as strong, as shown in Table 1. It is interesting that although late fall fibers are 1.6 times stronger in g_f/m^2 U than the winter fibers, this ratio rises to 2.4 in terms of g_f/denier and g_f/tex U. This follows primarily because of linear density differences between late fall and winter fibers. In winter, the fiber weight per unit length was different owing to the noncellulosic (contaminating) material, such as fungus, as previously mentioned. It appears that the absolute strength remains about constant, whereas the weight increases owing to the contaminants. This contaminating material might also lead to

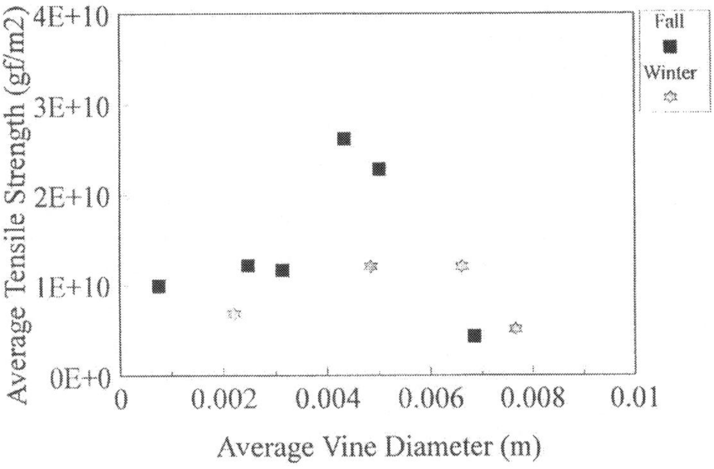

Fig. 6. Average tensile strength change based on average vine diameter in late fall and winter.

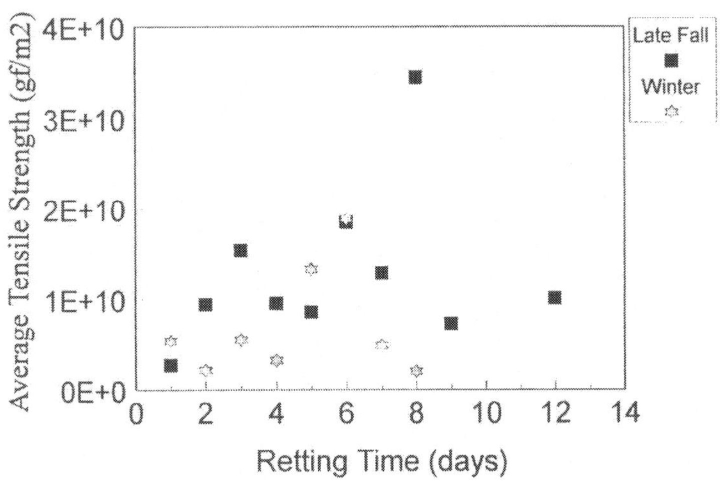

Fig. 7. Average tensile strength change based on retting time in late fall and winter.

greater or lesser measurements of the fiber thickness, leading to false thickness readings. In addition, since the kudzu vines harvested in Nashville, TN came from a public park the plant might have been exposed to an eradicating herbicide. If so, this exposure might have adversely affected the nature of the kudzu vine and its fiber.

The optimum harvest season for collecting strong kudzu fibers is from early June to late July in Japan *(4)*. Only 1-yr-old vines were reported to have been used to produce fibers in Japan. Here, the 0.002-m vines appear to be 1 yr old, the 0.004-m vines 2 yr old, and so on. Since the spring comes earlier to the southern United States than to Japan, the optimum harvesting time seems to be earlier in this study than the June or July optimum previously observed in Japan. There are no known reported

Table 1
Average Seasonal Tensile Strength Values of Kudzu Fiber for the Given Dimensions of Tested Fiber

	Season			
	Late fall	Winter	Early spring	
Property	Nashville, TN	Nashville, TN	Muscle Shoals, AL	Nashville, TN
Average thickness, m	0.0002	0.0001	0.00006	0.000046
Average width, m	0.002	0.002	0.002	0.002
Average length, m	0.06	0.06	0.06	0.06
Diameter of vines, m	0.00076–0.00584	0.00220–0.00767	0.00726–0.00779	0.00254–0.00559
Average linear density, g_f/m	0.128	0.182	0.145	0.075
Average tensile strength				
g_f/m^2	2.07×10^{10}	1.30×10^{10}	1.88×10^{10}	4.27×10^{10}
g_f/denier	2.400	0.995	2.654	4.395
g_f/tex	21.384	8.965	23.890	39.556

tensile strength measurements for kudzu fibers in the literature to compare with our results. However, we can compare our findings with other natural fibers. It is interesting that even kudzu fibers harvested out of the expected optimum season (which is spring or early summer), as summarized in Table 1, are comparable with wool fibers, as summarized in Tables 2 and 3. Sunhemp seems to be the most comparable natural fiber with kudzu fiber. This may be the case because they are both bast fibers, and retting methods used for their fiber extraction are similar. Sunhemp fibers retted in pond water for 7 d can reach a value as high as 76 g_f/tex tensile strength *(8)*, which is several times higher than our kudzu fibers.

The tensile strength specification of an ideal textile fiber has been set at 5–8 g_f/denier *(9)* in order to meet standard fiber spinning and weaving conditions. Therefore, wild kudzu fiber quality of late fall and winter seasons determined in this study fall well below the strength desired for use by the textile industry, but the early spring Nashville untreated fibers do approach that standard. Building on the results determined here, with the aid of enormous versatility of fiber-manufacturing processes (e.g., a mixture of natural and synthetic fibers, twisting techniques, coating the fibers with oil, and so forth), kudzu fibers appear to be viable for commercial use by the textile industry.

The type of retting employed to process the vines to fibers may be just as an important variable as the retting time to obtain good-quality fibers. Microbial retting with a given cultural strain can be used as a baseline case to reference other retting methods. Improvements in retting processes can be achieved using techniques, such as the controlled warm-water procedure, and microbial retting by a specific group of bacteria *(10)*. The techniques of retting can affect both the fiber quality and the retting time.

The optimum retting time is defined in terms of the desired objectives. It is usually defined as the shortest time required to recover the strongest fibers. Indirect indicators for monitoring the retting appear to be the pH of the solution used in

Table 2
Tensile Strength (g_f/denier) of Various Natural Fibers and Kudzu Vine Fiber

Natural fiber	Description	Tensile strength, g_f/denier	Reference
Cotton	Wet	3.3-6.4	*7*
Wool	Wet	0.76-1.6	*7*
Abaca	Not twisted	6.710	*5*
Abaca	Oiled, combed	6.900	*5*
Sisal	Oiled, combed	4.400	*5*
Henequen	Oiled, combed	3.300	*5*
Kudzu vine fiber	Unprocessed		
	Late fall	2.400	This study
	Winter	0.995	This study
	Early spring		
	Muscle Shoals	2.654	This study
	Nashville	4.395	This study

Table 3
Tensile Strength (g_f/tex) of Sunhemp and Kudzu Vine Fiber

Natural fiber	Description	Tensile strength, g_f/tex	Reference
Sunhemp	Retted in village (6–8 wk)	18	*8*
	Sample retted in pond water (7 d)	76	*8*
Kudzu vine fiber	Unprocessed		
	Late fall	21.384	This study
	Winter	8.965	This study
	Early spring		
	Muscle Shoals	23.890	This study
	Nashville	39.556	This study

retting and the color change of the vine, usually from a yellow-brown to a gray-blue. The pH of the retting solution typically decreases until a lower pH plateau of 4.6–4.9 is reached, and then it rises. This increase appears to indicate the start of overretting, which is the deterioration of the fiber strength *(11)*. The optimum retting time of kudzu fiber in late fall, winter, and early spring seems to be indicated by the minimum plateau value in the relationship between pH and retting time, shown in Fig. 5.

It is noted that the kudzu vine fibers extracted from the early spring vines were the highest in quality in terms of tensile strength. As summarized in Tables 1, 2, and 3, the tensile strength of the kudzu vine fiber increased significantly in the early spring such that it may fit the specifications for direct use in the textile industry.

CONCLUSIONS

It is concluded from this study that kudzu, a wild plant endemic to the southeastern part of the United States, can provide quality fibrous materials to the textile industry. Average tensile strength of the kudzu vine fiber was measured as 2.07×10^{10} g_f/m^2, and 1.3×10^{10} g_f/m^2 in the late fall and winter, respectively, in Nashville, TN. In the early spring it rose to 1.88×10^{10} g_f/m^2 and 4.27×10^{10} g_f/m^2 from samples obtained at Muscle Shoals, AL and Nashville, TN, respectively. Based on this study, but without tests between June and November, the optimum kudzu harvest time of the year in the southeastern United States for the extraction of strong fibers appears to be in the spring. Kudzu fiber tensile strength was found to be of sufficiently high quality for the plant fibers to compete with other natural fibers that are presently used in the textile industry. The optimum retting time of approx 6 d also appeared to be marked by the 6-d minimum in the pH vs time curve describing the kudzu retting solution. The tensile strength was a weak function of the vine diameter of kudzu.

ACKNOWLEDGMENTS

Support from the Vanderbilt University Research Council for this project is gratefully acknowledged. The authors thank James C. Hightower for setting up the tensile strength testing apparatus for our kudzu fiber tests.

REFERENCES

1. Hipps, C. B. (1994), *Horticulture* **June–July,** 36–39.
2. Tanner, R. D., Hussain, S. S., Hamilton, L. A., and Wolf, F. T. (1979), *Econ. Botany* **33,** 400–412.
3. Reid, D. P. (1993), *Chinese Herbal Medicine* Shambhala Publications Inc., Boston, MA.
4. Shurtleff, W. and Aoyagi, A. (1985), *The Book of Kudzu, A Culinary and Healing Guide.* Avery Publishing Group, Wayne, NJ, pp. 72–78.
5. Batra, S. K. (1985), *Other Long Vegetable Fibers, Fiber Chemistry. Handbook of Fiber Science and Technology,* vol. 4, *International Fiber Science and Technology,* series 7 (Lewin, M. and Sello, S. B., eds.), Marcel Dekker, New York, pp. 727–807.
6. American Society for Testing and Materials (1990), *Compilation of ASTM Standard Definitions.* PCN:03-001090-42, 7th ed., American Society for Testing and Materials, Baltimore, MD.
7. Kirk, R. E. and Othmer, D. F. (1983), *Explosives and Propellants to Flame Retardants for Textiles, Encyclopedia of Chemical Technology,* vol. 10, 14th ed. (Kroschwitz, I. and Howe-Grant, M., eds.), Wiley, New York, pp. 539–559.
8. Maheshwari, G. Syamala Devi, G., Bisaria, V. S., and Vasudevan, P. (1994), *Bioresource Technol.* **47,** 177–179.
9. Mark H. F. (1987), *Chem. Eng. Prog.* **83,** 44–54.
10. Grayson, M. (1983), *Encyclopedia of Composite Materials and Components,* Encyclopedia Reprint Series, Wiley, New York, pp. 497–512.
11. Clark J. G. (1990), *Chem. Eng. Prog.* **86,** 45–50.

0273-2289/96/5758—0083$9.25

Optimization of Reverse-Flow, Two-Temperature, Dilute-Acid Pretreatment to Enhance Biomass Conversion to Ethanol

ROBERT TORGET,* CHRISTOS HATZIS, TAMMY KAY HAYWARD, TEH-AN HSU, AND GEORGE P. PHILIPPIDIS

Alternative Fuels Division, National Renewable Energy Laboratory, Golden, CO 80401-3393

ABSTRACT

A reverse-flow, two-temperature dilute-acid prehydrolysis process of commercial yellow poplar sawdust using two percolation reactors was designed to simulate countercurrent flow of the biomass solids and prehydrolysis liquor, and to exploit the xylan biphasic kinetics. Lower temperatures (150–174°C) are initially applied to hydrolyze the easily hydrolyzable xylan, and higher temperatures (180–204°C) are applied to hydrolyze the remaining xylan. Two reactors were used to optimize each temperature range, using varying concentrations of sulfuric acid from 0.073–0.73 wt% and reaction times. Yields of soluble xylose, as high as 97% of theoretical, expressed as monomeric and oligomeric xylose, have been achieved with only 2.9% of the xylan being degraded to furfural, at concentrations of total potential sugar between 2.4 and 3.7 wt% before flashing. Depending on the combined severity of the acid concentration, residence time of the solids and liquor, and temperature of prehydrolysis, 81–100% of the hemicellulose, 3–32% of the glucans, and up to 46% of the Klason lignin could be solubilized. The lignocellulosic substrate produced from the pretreatment is readily converted to ethanol at a yield of approx 91% of theoretical, with ethanol concentrations of up to 4.0 wt% in 55 h via a simultaneous saccharification and fermentation (SSF) process. In terms of xylose recovery and ethanol production level and rate, the present results are far superior to those previously reported using a single-temperature, dilute-acid pretreatment.

Index Entries: Counterconcurrent pretreatment; dilute-acid pretreatment; ethanol; xylan kinetics; percolation reactor.

INTRODUCTION

Energy sources alternative to petroleum, especially those that are renewable, have been a subject of intense research interest in recent years. Ethanol production from lignocellulosic biomass is one area that has received considerable attention. Ethanol can be used as a transportation fuel in at least three forms: as hydrous

*Author to whom all correspondence and reprint requests should be addressed.

ethanol (95% ethanol and 5% water), in ethanol–gasoline blends (e.g., 10% ethanol and 90% gasoline), and as a feedstock material for producing the high-octane, oxygenated fuel additive ethyl tertiary butyl ether (ETBE) *(1)*. These uses may displace the use of petroleum-derived gasoline, boost the octane number of gasoline, improve the quality of automobile emissions, and result in a net reduction in the current CO_2 accumulation in the atmosphere. These effects appear to be particularly important for the US because:

1. With the adoption of the 1990 Clean Air Act Amendments (CAAA) that mandate many urban areas to have oxygenates added to gasoline, ethanol can be used to meet the CAAA requirements;
2. Enough ethanol might be produced from lignocellulosic biomass to replace all gasoline used *(2)*;
3. Energy use by the transportation sector accounts for more than 60% of the total petroleum consumption; and
4. Imported oil accounts for a significant fraction of the US foreign trade deficit *(3)*.

Lignocellulosic biomass consists of cellulose fibers embedded in a sheath of hemicellulose and lignin. Dense binding among cellulose chains presents additional constraints to the hydrolysis of cellulose. To convert the carbohydrates in lignocellulosic biomass to ethanol, a dilute-acid pretreatment effectively removes the lignin-hemicellulose protective shield and renders the remaining cellulose amenable to enzymatic conversion to glucose *(4–11)*. The acid-catalyzed removal of hemicellulose in hardwoods and agricultural residues follows biphasic kinetics *(4,10,12–14)*. As a result, xylan in lignocellulosic biomass probably consists of an easy-to-hydrolyze and a hard-to-hydrolyze fraction *(4,12–15)*, and can thus be modeled as two parallel pseudo-first-order reactions of xylan-solubilization yielding oligomeric xylose. These are followed by xylose formation with subsequent degradation, and have been modeled accordingly *(16,17)*. The biphasic nature of hardwood xylan solubilization suggests using a reactor configuration that exploits these kinetics, which should produce higher yields of the solubilized carbohydrates. Additionally, because the hydrolysis reactions are all pseudo-first-order, countercurrent operation will produce higher yields of the desired carbohydrates than does a cocurrent mode of operation *(18)*.

Researchers have reported that yields of solubilized xylose equivalents of 92% of theoretical can be achieved by applying a nonuniform temperature policy to address the biphasic xylan *(11,19,20)*. In this study, further improvements and optimization of the performance parameters using percolation reactors in a "reverse-flow" mode of operation to simulate countercurrent flow was used to increase further the solubilized xylose equivalent yield and concentration. The effects of particle size, dry vs wet packings before prehydrolysis, temperature, acid concentrations, and residence time of the acid catalyst and the solids were all studied to optimize the process in terms of sugar yield and concentration in the prehydrolyzate.

MATERIALS AND METHODS

Materials

Liriodendron tulipifera L. (yellow poplar) sawdust substrate was provided by Eric Miller of Saw Miller Inc., Haydenville, OH, and was shipped in 1-t parcels from

Lannes Williamson Pallets, Inc., Southside, WV, as fresh green sawdust with a moisture content of 43 wt% when received at the National Renewable Energy Laboratory (NREL). The sawdust was then double-bagged in plastic and stored frozen at –20°C. The chemical composition of a representative sample of the sawdust after being milled to pass through a 2-mm screen (*see below* for a description of the mill) was 41.9 wt% glucan, 15.5 wt% xylan, 2.7 wt% mannan, 0.8 wt% galactan, 0.6 wt% arabinan, 23.4 wt% Klason lignin (determined on unextracted sawdust), 3.2 wt% acid-soluble lignin, 0.6 wt% ash, 3.4 wt% acetyl groups, and 7.9 wt% others (most likely extractives and uronic acids). For pretreatment runs, the sawdust was thawed at room temperature and separated into two size fractions using a portable sieve shaker (Tyler Industrial Products, Mentor, OH) equipped with a 4-mm mesh screen (USA Standard Testing Sieves). The material that passed through a 4-mm screen (87% of the feed material) was set aside to be mixed with the milled overs. The overs (13% of the feed material) were milled using a laboratory knife mill (Thomas-Wiley Laboratory Mill, Arthur H. Thomas Co., Philadelphia, PA) equipped with a 6-mm rejection screen. These milled overs were manually mixed with the 4-mm screened material and used for experiments without further processing; thus, 100% of the commercial sawdust was thus used.

A liquid cellulase preparation (CPN International, Jupiter, FL) was stored at 4°C until used. The specific activities of the cellulase and β-glucosidase enzymes were 88 filter paper units (FPU)/mL and 230 IU/mL, respectively *(21)*. The yeast used in the SSF experiments was *Saccharomyces cerevisiae* D_5A *(22)*. Cellulose powder (α-cellulose, Sigma Chemical Co., St. Louis, MO) was used as a control substrate. The remaining chemicals were purchased from national laboratory supply houses.

Analytical Methods

Compositional analyses of the pretreated biomass and the prehydrolyzate were carried out by standard methods as previously described *(4-7,11,23–25)*. Before enzymatic saccharification and SSF testing of the pretreated sawdust, the particle size had to be reduced for maximum conversion of the glucan to glucose *(26,27)*. Thawed pretreated chips and 3 vol of deionized (DI) water were blended for 5 min in a single-speed laboratory Waring blender, which reduced the chips to a very fine slurry. The solids were separated from the water by filtration.

Enzymatic hydrolysis was performed as previously described *(4,5,9)*, except that cellulase enzyme loading was approx 60 FPU/g cellulose with β-glucosidase at approx 157 IU/g cellulose. This level of cellulase is higher than the FPU/g cellulose used in the SSF experiments described below.

In addition to testing the pretreated substrate for the efficacy of the pretreatment in terms of rates and the extent of enzymatic saccharification, an SSF protocol was used to provide additional information on the fermentability of the pretreated substrate. SSF was carried out in 250-mL flasks outfitted with stoppers constructed to vent CO_2 through a water trap as previously described *(28)*, with minor modifications. The cellulase preparation was used at a concentration of 25 FPU/g cellulose (16.8 FPU/g bone dry pretreated wood) with a substrate loading of approx 8 wt% cellulose (11.9 wt% solids). All SSF experiments were run at 38°C, in a laboratory rotating incubation at 150 rpm, an initial pH of 5.0, and with an α-cellulose control. Glucose and ethanol concentrations were monitored by using a YSI model 2700 glucose analyzer (Yellow Springs Instruments, Yellow Springs, OH). Ethanol levels were confirmed by gas chromatography as previously described *(28)*.

EXPERIMENTAL

The present study was undertaken (1) to further optimize process parameters for scale-up of a previously described two-temperature, dilute-acid percolation pretreatment of hardwood sawdust *(11)*, and (2) to investigate a reported alternative reactor configuration (reverse-flow, two-stage, dilute-acid prehydrolysis), which has been shown through modeling to generate higher concentrations of xylose equivalents in the prehydrolysis liquor and in higher sugar yields *(29)*. We sought to address concerns of the limited size of the previous pretreatment reactor (2 in length × 1 in diameter), the small particle size (–60 mesh to +80 mesh), and the limited amount of pretreated biomass that could be generated from our prior reactor system. We therefore chose a reactor of a length/diameter (L/D) ratio of 6 with reactor volume 24-fold larger than our previous reactor, and used particle sizes (–4 mm mesh) that would require less energy for milling to obtain more realistic data to be used in the NREL biomass-to-ethanol economic analysis model *(30)* to assess the viability of this process. Before pretreatment experiments could begin, we initiated a study to quantify the flow behavior of the acid catalyst through the sawdust bed as a function of temperature and liquid flow rate. We then assessed the quality of the pretreated residue by both enzymatic saccharification and converting the glucan to ethanol via SSF.

Reverse-Flow, Two-Temperature Percolation Reactor and Liquor Collection System

The design of the percolation reactor (Fig. 1) is a modified 2 × 12-in. Hastelloy C276 Modcolumn (Anspec Inc., Ann Arbor, MI) with titanium dispersion (60-μ pore size) and retention (20-μ pore size) frits. Three 1/8-in. thermocouples (Hastelloy C sheaths, Omega Engineering Co., Stamford, CT) were installed equidistant along the length of the reactor to monitor temperature. Hastelloy C276 tubing (1/8-in. [1.6-mm] od × .03-in. [0.8-mm] id, 1/4-in. [1.6-mm] od × .03-in. [0.8-mm] id) was used to connect the reactor with other components of the system as well as for the preheating coil (Colorado Springs Windustrial Co., Colorado Springs, CO). The preheating coil was long enough to allow the incoming prehydrolysis liquor at a flow rate of 200 mL/min to be at the desired prehydrolysis temperature. The three-way Hastelloy C valves were obtained from Valco Valve Corporation, Houston, Texas (the third port of the three-way valves was not used in the current study). Teflon-lined 304 stainless steel collection cylinders (1- and 1/2-gal capacities) were obtained from Denver Valve and Fitting, Denver, CO. A schematic of the flow system and equipment design used for the reverse-flow experiments is shown in Fig. 2. For the flow characteristic studies, only one reactor was used as previously described *(11)*.

To determine the void volume of the packed bed, the reactor was charged with 191.7 g (114.7 g bone dry basis) of thawed, 4°C refrigerated screened sawdust, and the void volume was determined as previously described *(11)* using a Dynamax Model SD-1 HPLC pump (Rainin Instrument, Inc., Ridgefield, NJ). The total void volume (interparticle and intraparticle) of the water-saturated, deaerated, packed reactor was 470 mL.

Flow Characterization Studies

To characterize the flow in the reactor (i.e., how the observed flow varies from ideal plug flow), we used the NaCl step-change method and recorded the resi-

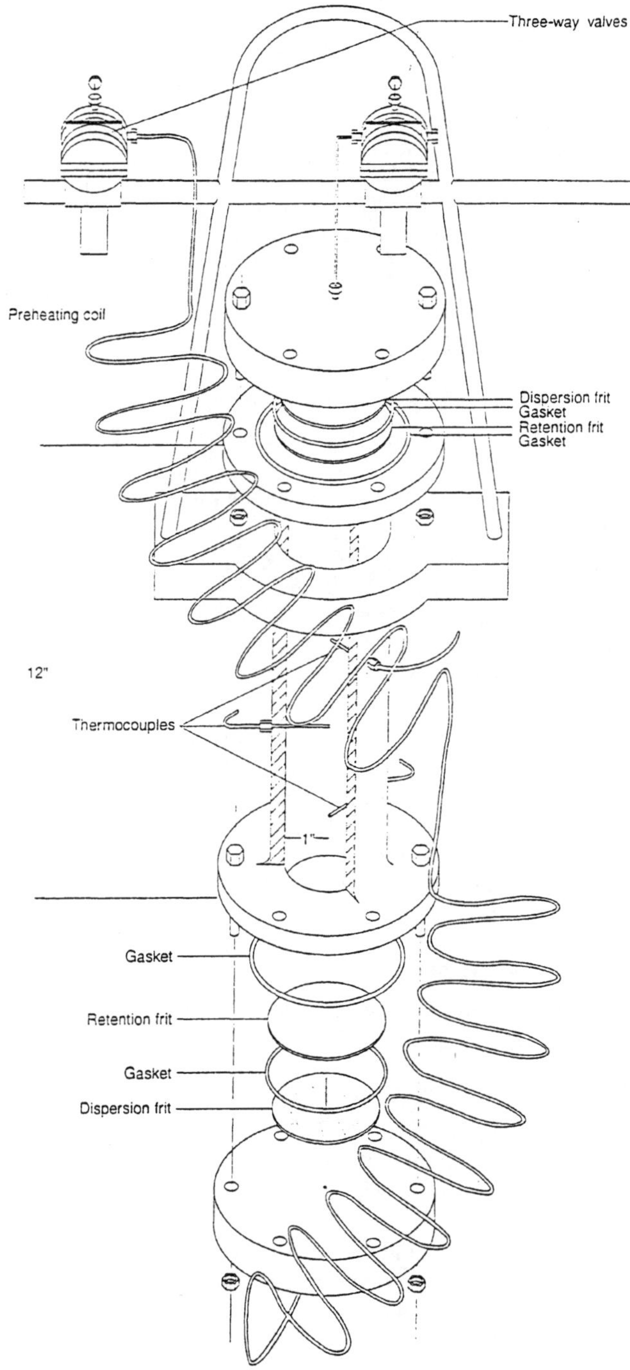

Fig. 1. Schematic diagram of the percolation reactor.

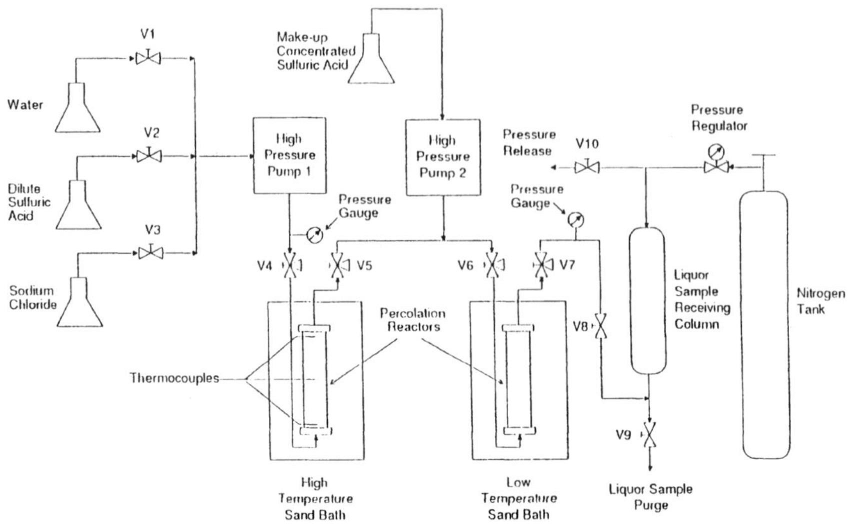

Fig. 2. Schematic diagram of the laboratory reverse-flow, two-temperature pretreatment reactor apparatus.

dence-time distribution (RTD) of the eluant NaCl in a similar manner as previously described *(11)*. We used 160°C and a flow rate of 100 mL/min measuring the NaCl (0.25*M*) RTD response via a density differential (Biotage model mPDS, Charlottesville, VA).

A computer reproduction of a tracing of the 0.25*M* NaCl concentration distribution as a function of void volumes pumped is shown in Fig. 3. Significant deviation from a plug flow plot is observed, indicating nonideal dispersion of the salt front. This plot was used to calculate the make-up acid pumping rate in the reverse-flow dilute-acid pretreatment runs described below under step 2 (*see* more detailed explaination under step 2: "Steady-State" Reverse-Flow, Two-Temperature Pretreatment).

Dilute-Acid Pretreatment Studies: Three Steps in Generating Simulated Steady-State, Reverse-Flow, Two-Temperature Samples

Step 1: Generating Partially Pretreated Sawdust at Low Temperature to be Used for the Reverse-Flow High-Temperature Reactor (Fig. 4)

Two percolation reactors were charged with biomass and placed in two sand baths with valves to the reactors closed. (Note: The packed bed was not prewet or deaerated to pretreat the biomass.) One sand bath was set at the "low" temperature mode (150 or 174°C), and the other at the "high" temperature mode (180, or 204°C). Once the reactors were at the "low" temperature, reactor 1 was connected on line as previously described *(11)*, whereas reactor 2 was set aside. The entire system was overpressurized by 10–20 psig above the steam pressure. Three voids of acid (0.735, 0.404, or 0.0735 wt%, one void to wet the reactor contents, and two voids for collection) were pumped through reactor 1 and collected in the desired solids residence

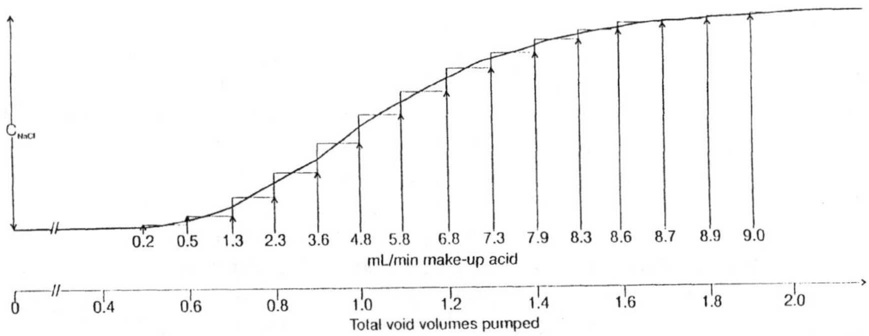

Fig. 3. Percolation flow characteristics using density detection and NaCl as tracer at 160 C and a flow rate of 100 mL/min as a function of total void volumes pumped. Also depicted is a representative step change make-up acid pump rate as a function of total void volumes pumped.

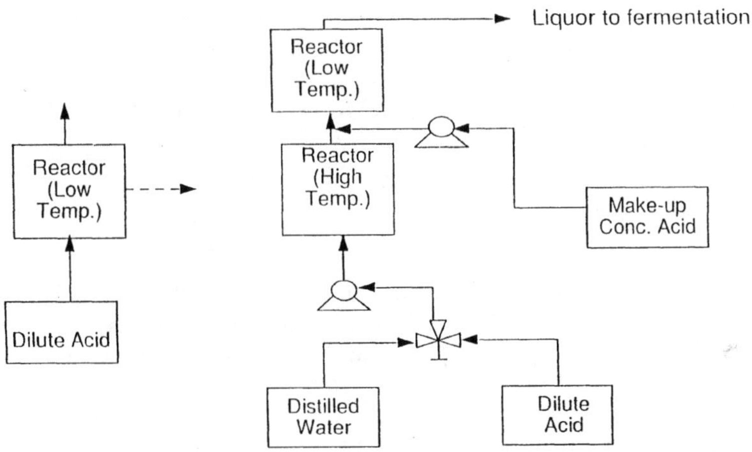

Fig. 4. Schematic of the concept of reverse-flow, two-temperature pretreatment.

time. This prehydrolysis liquor (step 1 liquor) was expelled from the collection column and saved for compositional analysis. V5 was then closed, and V10 opened to release all pressure downstream of reactor 1.

Step 2: "Steady-State" Reverse-Flow, Two-Temperature Pretreatment (Fig. 4)

Reactor 1 was moved to the high-temperature sand bath, while reactor 2 was placed in the low-temperature sand bath and the system assembled (Fig. 2). The entire system was over-pressurized by 10–20 psig. One void of acid (0.735, 0.404, or 0.0735 wt%, depending on the experiment) was then pumped into reactor 1 in 1/3 residence time of the solids (i.e., one void of liquor was pumped into reactor 1 in 3.33 min for the 10-min solids cook; in 5 min for the 15-min solids cook, and 6.66 min for the 20-min solids cook), which simultaneously wetted reactor 2 with the displaced acid from reactor 1, since the reactors are in series (Fig. 2). Two voids of DI water were then pumped through reactor 1 in 2/3 residence time of the solids (i.e. two voids of liquor were pumped into reactor 1 in: 6.66 min for the 10-min solids cook, 10 min for the 15-min solids cook, and 13.3 min for the 20-min solids cook) to totally wash the contents of reactor 1. To the diluted acid eluant from reactor 1 was added concentrated make-up acid (0.649, 1.224, 4.595, 6.49, or 12.24 wt% sulfuric acid, depending on the experiment) using increased step change pump rates dictated by the RTD curve (an example of pump rates is seen in Fig. 3), in which after about 0.48 voids of water had been pumped into the high-temperature reactor, the effluent from this reactor, which fed directly to the low-temperature reactor, had to have 0.2 mL/min of concentrated make-up acid added at the "T" to keep the concentration of acid being fed to the low-temperature reactor constant. Then after approx 0.6 voids of water had been added to the high-temperature reactor, 0.5 mL/min of concentrated make-up acid was added to keep the concentration of acid being fed to the low temperature reactor constant. This procedure of increasing the pumping rate of make-up acid as a function of total voids of wash water pumped through the high-temperature reactor was continued until 2.0 voids of wash water had been pumped through the system). Thus, reactor 2 during the entire solids residence time always saw a constant concentration of acid catalyst. The two total voids of liquor collected from this "reverse-flow" stage were expelled from the collection column and set aside for compositional analysis (step 2 liquor). Both reactors were isolated by valving, and reactor 1 was disconnected and cooled. The sugar concentrations and yields in this liquor fraction represented the actual "steady-state" liquor from a semicontinuous countercurrent reactor system.

Step 3: Completing the Prehydrolysis of Reactor 2 for Mass Balance Closure

Reactor 2 was submerged in the high-temperature sand bath and connected as in step 1 above (without the addition "T"). One void of dilute acid was pumped through reactor 2 in 1/3 residence time for the solids, followed by two voids of DI water in 2/3 residence time for the solids to wash out all the acid and solubilized sugars. Reactor 2 was disconnected and cooled.

The liquor from step 3 was saved for analysis. Solid contents of both reactors were expelled, combined, and collected in a glass-sintered funnel of medium porosity. The solid residue was then weighed and chemically analyzed; the three liquor fractions were individually analyzed for sugars, hydroxyl-methyl-furfural (HMF), furfural, acetic acid, and acid-soluble lignin. The additive quantities of the components from all three liquors were then used for total mass balance calculations.

RESULTS AND DISCUSSION

In the present study, a reverse-flow, two-temperature pretreatment using two percolation reactors (which simulates a semicontinuous, countercurrent mode of operation) packed with commercial yellow poplar sawdust was investigated. To define better the experimental conditions that would give high xylose equivalent yields, we used the mathematical modeling of the reverse-flow percolation process using the two-temperature pretreatment of hybrid poplar xylan studied by Auburn University, Auburn, AL (through an NREL subcontracted study). Mathematical simulations *(29)* and our prior study *(11)* have demonstrated that, after the biomass is cooked at temperatures between 135 and 150°C using 0.73 wt% sulfuric acid for residence times necessary to hydrolyze 60% of the xylan, a step-change increase in prehydrolysis temperature of between 25 and 35°C to hydrolyze the remaining xylan results in maximum yields of xylose (92% of theoretical xylose equivalents) in the prehydrolyzate. We sought to optimize these results not only in terms of total yield, but also in terms of concentration of fermentable sugar in the prehydrolyzate using the concept of reverse flow. The biomass is first treated at a low temperature in the percolation mode. Then it is treated again at a higher temperature. The throughput stream from the high-temperature treatment is again put through a percolation reactor packed with fresh biomass at low temperature. The reacted residue from this low-temperature reactor is then treated with fresh acid at high temperature (Fig. 4). This process is repeated and simulates countercurrent movement of the solids relative to the liquor *(29)*. A further refinement of the process uses process water to wash the acidic medium containing solubilized sugars from the high-temperature reactor after the desired extent of acidic prehydrolysis. To the effluent from the high-temperature reactor (before entering the low-temperature reactor) is added make-up acid to the desired concentration for prehydrolyzing the biomass in the low-temperature reactor. Thus, the contents of the high-temperature reactor (which have been washed *in situ*) can now be added directly to a fermentation vessel to produce ethanol.

The flow characteristics of the catalyst in percolation reactors, which influence observed rates of reactions and thus yields of desired products, are a function of temperature, packing density, flow rate, particle size, and porosity, as well as other parameters, such as moisture content. Additionally, since in a reverse-flow mode of operation with a water-wash step in the high-temperature stage, a make-up acid loop before the low-temperature stage would be required, it was essential to have a method available to quantitate the amount of make-up acid to add to achieve constant acid concentration throughout the lower-temperature stage. We therefore initiated our laboratory investigation with a study of the flow characteristics in the packed-bed reactor under experimental prehydrolysis conditions. However, because solubilized sugar yields and flow patterns are influenced by sawdust particle size, we conducted a literature search to define the largest particle size that could be used without incurring significant losses in sugar yield.

Particle Size Selection

Among the most important parameters to be considered when scaling up from our previous reactor was the selection of the nominal screened chip size to be used in the experiments. Milling sawdust to a size amenable for the enzymatic conversion of the glucan fraction of pretreated hardwood sawdust *(26,27)* requires substantial energy *(31)*. Therefore, our approach was to use the largest critical chip size possible for which diffusion of the solubilized sugar products out of the porous wood will not contribute to degradation losses.

Additionally, bulk packing density was a concern, as was the rate of "wetting" or impregnation of the hydrolyzing liquor as dictated by diffusion and capillary action. Capillary absorption proceeds significantly faster than diffusionary liquid penetration in wood *(32)*, and the rate of diffusion for a solute in wood in the longitudinal direction is more than an order of magnitude greater than either in the radial or in the tangential direction *(32)*. Also, for hybrid poplar, the critical chip size for diffusing hydrolyzed sugars is from 0.29–0.4 cm for two-temperature pretreatments that use 150 or 140°C as the lower temperature, respectively *(33)*. Additionally, the pulping industry uses wood chips that are 3–6 mm thick *(34)*, which agrees well with the above-mentioned critical chip size. Thus, based on the above considerations and on the fact that 87% of the commercial sawdust passed through a 4-mm laboratory screen and thus only 13% overs would require milling (data not shown), sawdust passing through a 4-mm screen was therefore used for all experiments.

Flow Characteristics in the Percolation Reactor

It has been previously shown that NaCl at low concentrations (10 m*M*) can be used instead of sulfuric acid as a tracer in RTD studies up to 140°C to determine the flow characteristics of the H_3O^+ ion *(11)*. Because our lower prehydrolysis temperatures were 150 and 174°C at flow rates from 91–141 mL/min, we chose to measure the RTD function at 160°C at a flow rate of 100 mL/min (Fig. 3) using 0.25*M* NaCl, which gave a good response using a densitometer to measure the concentration profile. However, the Na^+ ion (at elevated temperatures) causes partial hydrolysis of polymeric carbohydrates *(35)*, and therefore might not be considered an inert tracer. We therefore performed a prehydrolysis at 160°C for 10 min using 0.25*M* NaCl and a dry-packed column and passing three void volumes of liquor through in 10 min to quantitate the catalytic effect of Na^+ ions. Under these conditions, 54% of the xylan is hydrolyzed along with 30% of the lignin, with only 2% of the glucan being hydrolyzed. Therefore, indeed, the observed RTD curve reflects a partially hydrolyzed bed. This may be a fortuitous degree of hydrolysis to simulate what rates of make-up acid to add in the reverse-flow mode, since the xylan in the lower temperature reactor is partially hydrolyzed with one void of acid, which totally wetted the column, and an additional 0.5 void of acid before make-up acid needed to be added (step 2, Materials and Methods). Therefore, although this method of calculating the amount of make-up acid is not totally quantitative, it offers a reasonable approximation of the required mode of addition of make-up acid. An affirmation that the RTD curve is reasonably close to what is experimentally required is the fact that the measured pH of the effluent of the reverse-flow mode of operation (liquor from step 2, Materials and Methods) was nearly identical to that measured from the liquor from step 1 (Materials and Methods) in which no make-up acid was added. The dispersion effects seen in the present study using –4-mm screened material were comparable to those in the previous study using –60 to +80 mesh material.

Scale-Up Effects from Our Prior Reactor

Several concerns from our prior study were addressed in scaling up the reactor 24-fold in total capacity. First, we chose an L/D ratio of 6 in the present reactor, instead of 2, to examine whether a different L/D ratio affected the flow characteristics and channeling. Despite the different particle size (–4 mm instead of –60 to + 80 mesh), we obtained similar RTD curves with the larger reactor packed with larger mesh particles (data not shown). Therefore, wall effects and channeling do not significantly affect the flow patterns in the packed sawdust bed; rather, the flow characteristics are predominantly dictated by the intraparticle flow.

Second, we not only sought to reproduce the previously published results using the larger reactor, but also to compare pretreatment results using a dry-packed bed (just before adding acid) to results using a totally water-saturated and deaerated bed. Using a dry-packed bed allows higher sugar concentrations in the prehydrolysis liquor, which is desirable for fermentation and downstream processing. Table 1 lists the prehydrolysis conditions used to reproduce our prior results and further optimize the process parameters, and Table 2 tabulates the mass balance results of

Table 1
Prehydrolysis Parameters

Run, (code)	Temp 1, C°	Temp 2, C°	Reaction time 1, min	Reaction time 2, min	Acid concentration, wt%
Scale-up reproducibility					
1 (inhpop 01) (water-saturated)	140	170	30	20	0.735
2 (inhpop 05) (dry-packed)	140	170	30	20	0.735
Factorial optimization of reverse-flow pretreatment					
3 (inhpop 08)	150	180	10	10	.0735
4 (inhpop 15)	150	200	10	10	.0735
5 (inhpop 13)	150	180	20	20	.0735
6 (inhpop 09)	150	200	20	20	.0735
7 (inhpop 14)	150	180	10	10	0.735
8 (inhpop 12)	150	200	10	10	0.735
9 (inhpop 11)	150	180	20	20	0.735
10 (inhpop 10)	150	190	15	15	.4015
11 (inhpop 16)	150	190	15	15	.4015
Verification of optimization					
12 (inhpop 17)	174	204	10	10	0.0735

the scale-up study. As can be calculated from Table 2, the mass balance closure on the xylan fraction (%soluble + %insoluble + %furfural) from both run 1 (the water-saturated, deaerated reactor that represents the method used in our prior study *[11]*) and run 2 (dry chip loading) averages only 91.5%, which is well below our average closures of 99.7% from runs 3–12 in Table 2. Nevertheless, similar amounts of xylan are solubilized irrespective of the wetting mode (86.5% for the water-saturated reactor and 83.0% for the dry packing). Moreover, the 86.5% xylose equivalents solubilized is similar to the previously reported 92% of the xylose equivalents solubilized using the small reactor. This demonstrates that under reaction conditions, the acid catalyst reaches sites of reaction via capillary or diffusion mechanisms at rates that cannot be differentiated in the two experimental setups.

Reverse-Flow, Two-Temperature Pretreatment Optimization Experiments

To increase the solubilized sugar concentrations and reduce the required reaction times of the above pretreatment regime, a reverse-flow, two-temperature mode of operation (Fig. 4), which simulates countercurrent operation *(29)*, was employed using 150°C as the low prehydrolysis temperature. A factorial experimental design involving nine runs was developed. Acid concentrations varied from 0.0735–0.735 wt%, the high prehydrolysis temperature from 180–200°C, and the solids residence time in each stage from 10–20 min, as shown in Table 1 (runs 3–11 under Factorial optimization of reverse-flow pretreatment). As shown in Table 2, all three major polymers are solubilized to varying extents, with xylan being preferentially hydro-

Table 2
Mass Balance Data

Run,code	Xylan				Glucan				Lignin	
	%Insoluble	%Soluble	%Monomer	%Furfural	%Insoluble	%Soluble	%Monomer	%HMF	%Insoluble	%Soluble
Scale-up reproducibility										
1 (inhpop 01) (water saturated)	0.8	86.5	92.0	4.4	90.1	10.3	96.6	0.2	59.7	35.8
2 (inhpop 05) (dry packing)	0.6	83.0	95.2	7.7	87.5	12.2	95.2	0.5	64.6	41.0
Factorial optimization of reverse-flow pretreatment										
3 (inhop 08)	12.6	86.0	32.1	1.8	88.7	2.8	88.9	0.1	71.2	40.8
4 (inhpop 15)	4.6	86.1	58.5	2.8	94.8	6.0	56.9	0.1	66.2	29.7
5 (inhpop 13)	13.3	80.8	44.3	1.0	98.4	3.2	36.8	0	73.6	26.3
6 (inhpop 09)	2.1	94.0	47.4	3.6	89.6	8.1	87.9	0.3	60.9	46.5
7 inhpop 14)	6.2	90.4	N.R.[a]	4.9	89.2	12.3	100	0.2	69.8	32.9
8 (inhpop 12)	0	98.4	73.4	4.4	61.8	31.7	91.9	1.5	65.1	52.5
9 (inhpop 11)	0.7	101.3	79.6	5.8	72.9	17.7	92.3	0.6	68.6	40.6
10 (inhpop 10)	1.0	91.8	63.8	3.6	83.2	14.0	78.4	0.5	59.8	46.3
11 (inhpop 16)	2.4	94.4	68.2	3.7	81.7	14.0	91.0	0.3	68.1	33.3
Verification of optimization										
12 (inhpop 17)	0.5	97.0	36.0	2.9	91.6	9.2	77.0	0.3	57.7	35.7

[a]N.R. not reported.

lyzed. Mass balance closures for each component (%insoluble + %soluble + % as a degradation product) varies from run to run and does not necessarily depend on the combined severity *(36)* of pretreatment. Mass balance closures for xylan vary from a low of 93.5% to a high of 107.8%, for glucan from 91.2–101.7%, and for lignin from 95.9–117.6%. The respective means and standard deviations of these closures are 99.7 ± 4.3%, 97.1 ± 4.0%, and 105.8 ± 6.6%, indicating that experimental reproducibility was at a satisfactory level in this study. Additionally, when pretreatment conditions are severe, as in run 8, we have observed lignin closures routinely above 100%, probably because of insoluble complexes of humic substances and sugar degradation products, which are quantitated as Klason lignin.

As shown in Table 2, when the combined severity was relatively low (runs 3–7), only 3–12% of the glucan (recovered both as glucose and oligomeric glucose in the prehydrolyzate) was hydrolyzed. This correlates well with the reported 10% of the cellulose that is assumed to hydrolyze at 50 times the rate of the rest of the cellulose *(37)*. However, as the combined severity increased, up to 32% of the glucan fraction was hydrolyzed (run 8), but with very low conversion to HMF. Except for two runs (run 4 and 5), most hydrolyzed glucan was recovered in the monomeric form. Lignin solubilization in all runs was relatively high, varying from 26% (run 5) to a high of approx 40%–46% (run 6 and 10), depending on which substrate (insoluble vs soluble fractions) is quantitated. This is significantly higher than that previously reported for batch dilute-acid pretreatment *(4–7)*, but similar to the hydrothermolysis experiments using a percolation reactor *(38)*. Not only would this solubilized lignin be very toxic to fermentative organisms *(39)*, but the lignin would have to be recovered for its energy content either by producing methane in the waste-water treatment facility or by separating, drying, and burning to generate power in the proposed NREL wood-to-ethanol plant design *(30)*.

One promising aspect of this pretreatment configuration is the very high yields of xylose equivalents recovered in the prehydrolysis liquor. Yields of xylose equivalents (released as oligomers and monomers) range from 81% (run 5) to approx 100% (run 9), with yields of more than 90% being common (run 6, 7, 8, 10, and 11). The concentration of total sugar equivalents in these liquors vary from 2.4 wt% for run 3, to 3.7 wt% for run 8 before any process flashing of the liquor would take place (data not shown). The high xylose yields obtained in the present study compared to those previously reported for a batch reactor *(4,5,7,8)* are highly desirable for maximizing ethanol yields. Also small amounts of furfural (5.8% of the xylan in the worst case, run 9), HMF (1.5% of the glucan in the worst case, run 8), and acetic acid (<0.4 wt%, data not shown) accumulated in the prehydrolyzate.

The fraction of monomeric xylose in the hydrolyzate relative to the oligomeric xylose, as shown in Table 2, varies from a low of 32.1% (run 3) to a high of nearly 80% (run 9). Further studies are needed to elucidate the degree of polymerization (DP) profile of the released oligomers. The impact of the presence of oligomeric xylose in the prehydrolyzate on its fermentability to ethanol needs to be addressed. The dimers and trimers, but not higher DP, of xylose can be fermented by certain microorganisms, such as ethanologenic strains of *Klebsiella oxytoca (40)*. Therefore, if higher DP values of xylose are obtained, a posthydrolysis of the prehydrolyzate must be performed. From a process standpoint, this could possibly be accomplished by holding the temperature below the hydrolysis temperature for a residence time necessary to form the fermentable low DP oligomers (monomers, dimers, and trimers) of xylose without further degrading the xylose. Alternatively, the use of xylanases can accomplish the same task.

Another promising aspect of this pretreatment configuration is that by selecting a set of different operational parameters (which all give high yields of xylose equivalents in the liquor), different ratios of the solubilized sugars in the prehydrolyzate can be achieved. This option of varying ratios of sugars can be extremely important in effectively converting the C_5 fraction to ethanol *(41)*.

Verification of Optimal Reverse-Flow Process

From the results of the factorial design, we sought to optimize the process around three criteria: low acid, minimal residence time of both the liquor and the solids, and high yield of xylose equivalents in the prehydrolysis liquor. We therefore chose a 10-min residence time of the solids at each stage, 0.0735 wt% sulfuric acid, 174°C for the lower temperature, and 204°C *(29,33)* for the higher temperature. Using these conditions, near-quantitative recovery (97.0%) of the xylose equivalents was seen in the prehydrolyzate with 36% hydrolyzed to monmers and only 2.9% of the xylan being recovered as furfural, and a total mass balance closure around xylan of 100.4% (run 12, Table 2). Slightly less than 10% of the glucan was solubilized and 36% of the lignin was solubilized (Table 2).

Enzymatic Saccharification and SSF Performance of Pretreated Samples

The enzymatic digestibility of cellulose in the pretreated sawdust substrates reached 84–91% completion in 6 h, except for runs 3 and 5, and all substrates reached quantitative conversion (97.0% to 100% of theoretical) in 24 h (Fig. 5). In comparison, the α-cellulose control at 24 h was 76%. The two substrates that exhibited slower initial hydrolysis rates, but did ultimately digest to completion, had incomplete xylan hydrolysis (Table 2) compared to the rest of the substrates. This slow initial rate might be explained by the degree of porosity found in the two groups of substrates, i.e., the faster hydrolyzing substates could have had larger pores from the dilute-acid pretreatment, thus allowing the cellulase enzymes easier access to the cellulose backbone *(42)*.

To investigate whether this increased rate of digestibility would translate into an increased rate and extent of fermentation of the glucan into ethanol, we conducted an SSF investigation. In Fig. 6, the rates of ethanol production from selected pretreated sawdust substrates are compared to the performance of the control substrate, α-cellulose. The rates and conversion of the pretreated sawdust substrates were better than α-cellulose. With one substrate, run 12, 91% of the cellulose in the pretreated sawdust was converted to ethanol in 55 h at an ethanol concentration in the fermentation broth of 4.1 wt%. The other two substrates achieved slightly more than 90% conversion in 72 h at nearly identical ethanol concentrations of 4.0 wt%. These performance results, both in terms of yield and productivity, exceed those obtained with three other poplars pretreated using a batch reactor and similar cellulase and β-glucosidase loadings *(28)*. The straight enzymatic saccharification (Fig. 5) results agreed well with the SSF results in that more than 90% of the available glucan in the pretreated substrate was converted in the former test to glucose and in the latter to ethanol.

In addition to the significantly increased xylose equivalent yield and SSF rate and yield, the reverse-flow, two-temperature pretreatment technique appears to present opportunities for other process improvements. It inherently produces a

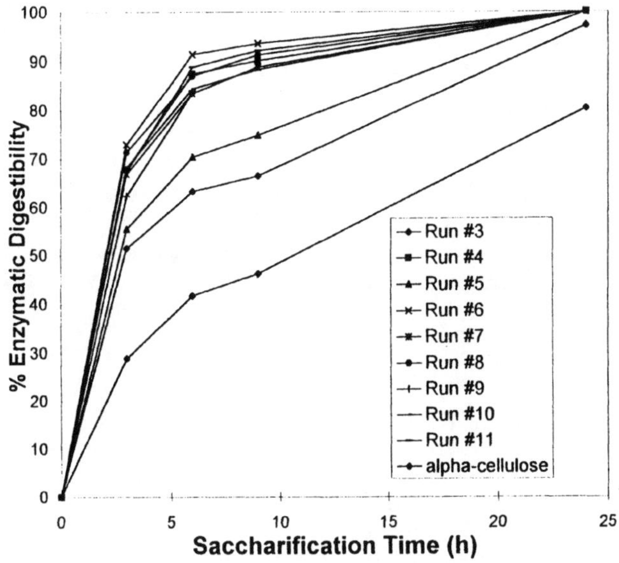

Fig. 5. Enzymatic digestibility of pretreated sawdust substrates as a function of time.

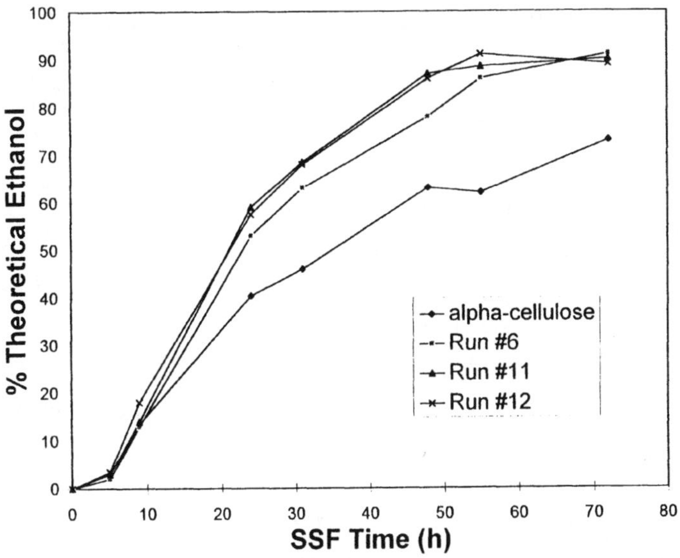

Fig. 6. SSF of selected pretreated sawdust substrates as a function of time.

xylose stream separate from the pretreated solid, with the pretreated solid being washed free of any fermentation inhibitors. Our SSF test, using pretreated wood discharged from the reactors without further washing, demonstrated the fermentation compatibility of the pretreated wood. Also, the byproduct gypsum produced from neutralizing the xylose stream can be removed readily to produce a xylose stream free of any solids. This stream may be more amenable to any detoxification process that addresses the fermentation inhibitors. Currently, we are looking closely into these potential benefits, and an engineering study is under way to project the possible impact of the proposed pretreatment method on the economics of the NREL reference case biomass-to-ethanol process *(30)*.

ACKNOWLEDGMENTS

The authors wish to thank Christine Ehrman and members of the chemical analysis and testing task of the NREL Alternative Fuels Division for analyzing the numerous samples generated during the course of this study. We also thank K. Connors of NREL for his support with the enzymatic saccharification experiments. This work was funded by the Biochemical Conversion Element of the Biofuels Program of the US Department of Energy.

REFERENCES

1. Wyman, C. E. and Hinman, N. D. (1990), *Appl. Biochem. Biotechnol.* 735–753.
2. Wyman, C. E. (1993), *Proc. 1st Biomass Conference of the Americas: Energy, Environment, Agriculture, and Industry*, Burlington, VT, August 30–September 2, 1010–1031.
3. Lynd, L. R., Cushman, J. H., Nichols, R. J., and Wyman, C. E. (1991), *Science* **251,** 131.
4. Grohmann, K., Torget, R., and Himmel, M. (1985), *Biotechnol. Bioeng. Symp.* **15,** 59–80.
5. Grohmann, K.,Torget, R. , and Himmel, M. (1986), *Biotechnol. Bioeng. Symp.* **17,** 135–151.
6. Torget, R., Werdene, P., Himmel, M., and Grohmann, K. (1990), *Appl. Biochem. Biotechnol.* **24,** 115–126.
7. Torget, R., Walter, P., Himmel, M., and Grohman, K. (1991), *Appl. Biochem. Biotechnol.* **28/29,** 75–86.
8. Torget, R., Himmel, M., and Grohman, K. (1992), *Appl. Biochem. Biotechnol.* **34/35,** 115–123.
9. Grohmann, K., Himmel, M., Rivard, C., Tucker, M., Baker, J., Torget, R., and Graboski, M. (1984), *Biotechnol. Bioeng. Symp.* **14,** 137–157.
10. Torget, R. (1985), MS thesis, Colorado School of Mines, Golden, CO.
11. Torget, R. and Hsu, T. (1994), *Appl. Biochem. Biotechnol.* **45/46,** 5–22.
12. Springer, E. L., Harris, J. F., and Neill, W. K. (1963), *TAPPI* **46,** 551–555.
13. Springer, E. L. and Zoch, L. L. (1968), *TAPPI* **51,** 214–218.
14. Kobayashi, T. and Sakai, Y. (1956), *Bull. Agr. Chem. Soc. Jpn* **20,** 1–7.
15. Maloney, M. T., Chapman, T. W., and Baker, A. J. (1985), *Biotechnol. Bioeng.* **27,** 355–361.
16. Conner, A. H. (1984), *Wood Fiber Sci.* **16,** 268–277.
17. Carrasco, F. and Roy, C. (1992), *Wood Sci. Technol.* **26,** 189–208.
18. Wright, J. D., Bergeron, P. W. and Werdene, P. J. (1987), *Ind. Eng. Chem. Res.* **26,** 699–705.
19. Grohmann, K. and Torget, R. W. (1992), US Patent #5,125,977.
20. Torget, R., Hsu, T., Kim, B. J., and Lee, Y. Y. (1992), AICHE National Meeting, Miami, FL.
21. Ghose, T. K. (1987), *Pure Appl. Chem.* **59,** 257–268.
22. Spindler, D. D., Wyman, C. E., Grohmann, K. and Philippidis, G. P. (1992), *Biotech. Lett.* **14,** 403–407.
23. Official Test Methods (1983), Atlanta, GA.
24. Moore, W. E. and Johnson, D. B. (1967), *Procedures for the Chemical Analysis of Wood and Wood Products*, USDA Forest Products Laboratory, Madison, WI.
25. Technical Association of the Pulp and Paper Industry Standard Method T 250, *TAPPI*, Acid-Soluble Lignin in Wood and Pulp.
26. Torget, R., Himmel, M., Wright, J., and Grohmann, K. (1988), *Appl. Biochem. Biotechnol.* **17,** 89–104.
27. Hayward, T. K. (1995), personal communication, NREL, Golden, CO.
28. Philippidis, G., and Smith, T. K. (1995), *Appl. Biochem. Biotechnol.* (in press).
29. Kim, B. J., Lee, Y. Y., and Torget, R. (1994), *Appl. Biochem. Biotechnol.* **45/46,** 113–129.
30. Hinman, N. D., Schell, D. J., Riley, C. J., Bergeron, P. W., and Walter, P. J. (1992), *Appl. Biochem. Biotechnol.* **34/35,** 639–649.
31. Schell, D. J. and Harwood, C. (1994), *Appl. Biochem. Biotechnol.* **45/46,** 159–168.
32. Kraev, L. N. and Lorolkov, I. I. (1968), Sb Tr Vses Nauchno-Issled Inst Gidroliza Rastit Mater. **17,** 13–25.

33. Kim, B. J., Lee, Y. Y., and Torget, R. (1993), *Appl. Biochem. Biotechnol.* **39/40,** 119–129.
34. Smook, G. A. (1992), *Handbook for Pulp and Paper Technologist*, 2nd ed., p. 30.
35. Zarauyika, M. F., Moses, P., and Mavunganidze, T. (1990), *J. Polymer Sci. Part A Polym. Chem.* **28,** 3565–3574.
36. Chum, C. L., Johnson, D. K., Black, S. K., and Overend, R. P. (1990), *Appl. Biochem. Biotechnol.* **24/25,** 1–14.
37. Conner, A. H., Wood, B. F., Hill, C. G., Jr., and Harris, J. F. (1985), *J. Wood Chem. Technol.* **5(4),** 461–489.
38. Mok, W. S. and Antal, M. J., Jr. (1992), *Ind. Eng. Chem. Res.* **31(4),** 1157–1161.
39. Cole, M. (1958), *Nature* **181,** 1596.
40. Burchhardt, G. and Ingram, L. O. (1992), *Appl. Environ. Microb.* **58,** 1128–1133.
41. Takahashi, D. F., Carvalhal, M. L., and Alterthum, F. (1994), *Biotechnol. Lett.* **16,** 747–750.
42. Grethlein, H. E. (1985), *Bio. Technology* **3,** 155–160.

0273-2289/96/5758—0103$9.25

Sorptive Recovery of Dilute Ethanol from Distillation Column Bottoms Stream

Manish Gulati,[1,2] Paul J. Westgate,[1,2,3] Mark Brewer,[1] Rick Hendrickson,[1] and Michael R. Ladisch*,[1,2]

[1]*Laboratory of Renewable Resources Engineering and* [2]*Department of Agricultural Engineering, Purdue University, West Lafayette, IN 47907; and* [3]*W. R. Grace & Company, Columbia, MD*

ABSTRACT

Modern ethanol distillation processes are designed to ensure removal of all ethanol from the column bottoms, i.e., to levels <100 ppm ethanol, and utilize substantial stripping steam to achieve this result. An alternate approach using sorption was attempted as a means to reduce energy requirements in the stripping section, and thereby reduce cost. Adsorbents tested for use in such an application showed that carbonaceous supports, in particular Ambersorb XEN 572, gave alcohol-free water as effluent when a 1% (w/w) starting ethanol concentration was passed downflow at 1 bed vol/h over a fixed-bed adsorber at 70°C. Regeneration was readily achieved at 70–90°C using hot air, vacuum, superheated steam, or hot water to strip the ethanol from the column, and yielded ethanol streams containing a maximum of 5.9% alcohol, with average concentrations of 2.5–3.5% depending on the regeneration method used. These experimentally determined operating conditions combined with distillation energy calculations have enabled development of a process concept for sorptive concentration of dilute ethanol which is more energy efficient than distillation alone. The combination of existing distillation and corn grit drying technologies, with sorptive recovery of dilute ethanol (from the column bottoms) shows promise of recovering a fuel grade, 99.4% ethanol product from a 4.5% ethanol broth with an energy requirement of 23,100 BTU/gal. The potential energy saving of 3600 BTU/gal over distillation alone corresponds to 1.8¢/gal, and provides motivation for further examination of this approach in reducing costs of ethanol production from biomass.

Index Entries: Adsorption; dilute ethanol; Ambersorb; stripping steam; distillation energy.

Abbreviations: C_p, specific heat capacity of water; D, mass flow rate of overhead product; $\dot{m}_c$, mass flow rate of cooling water in overhead condenser; $\dot{m}_S$, mass flow rate of steam in reboiler; R_D, R, reflux ratio for distillation column; R_{Dmin}, R_{min}, minimum reflux ratio; l, latent heat of vaporization of liquid in bottoms; l_S, latent heat of vaporization of steam; ΔT, rise in temperature of cooling water in overhead condenser.

*Author to whom all correspondence and reprint requests should be addressed.

Table 1
Approximate Cost Ranges for Conversion of Corn to Ethanol *(2)*

Item	Approximate cost range, ¢/gal
Net corn	40–60
Operating	
Energy fuel (steams)	10–20
Electricity	5–10
Enzymes	3–4.5
Chemicals	1.5–2.5
Other	18.5–38
Capital	30–60
Total	108–195

INTRODUCTION

A modern integrated alcohol plant has energy requirements in the range of 34,000–41,000 BTU/gal, and includes alcohol stripping, rectification and dehydration steps, as well as evaporating water from soluble components and drying fibrous material *(1)*. The energy cost is about 10–20% of the overall cost of ethanol production from corn (Table 1), with a reduction in energy costs often requiring higher capital investment *(2)*. Operating costs are distributed principally among energy, enzyme, chemical, and other costs.

Energy has been a significant operating cost, and therefore, innovations have focused on conserving energy. Although the distillation of ethanol consumed up to 50% of the overall energy used in some types of grain alcohol plants *(3,4)*, process improvements and plant design have significantly reduced this. This area has accounted for most of the savings in the last 5 yr, and further large savings in energy costs are going to be more difficult to achieve *(5)*.

The concentration of fermentation-strength ethanol involves an atmospheric distillation step that removes ethanol from water to obtain an overhead concentration of 90–95.6% (by weight) ethanol. This is followed by adsorption or azeotropic or extractive distillation to obtain fuel-grade ethanol. The energy required to distill fermentation-strength ethanol to its azeotropic composition (95.6% by weight at atmospheric pressure) is a function of ethanol concentration in the feed as well as the reflux ratio *(6,7)*.

Various energy-efficient approaches have been implemented to reduce energy requirements for removing the last 5–10% of water. These include the use of ground corn or molecular sieves in fixed-bed sorption processes, and staged distillation systems that use various organic agents for removing the water in an extractive or azeotropic distillation. Ethanol-water mixtures can be separated by passing the vapors over various biomass materials, such as corn *(6,8,9)*, corn meal *(10–14)*, and corn grits *(9,12,15,16)*; cellulosics *(6,11,17)*; and starch *(6,9,11,18,19)*. The sorption system, which utilizes corn grits as the sorbent currently processes about 50% of the alcohol produced in the United States (750 million gal/yr) *(20)*. Laboratory data show that this process readily accepts 85% ethanol with a minimal energy penalty compared to a 90% ethanol feed *(12)*.

Removing the last traces of alcohol in the bottom or stripping section of the distillation column can also consume a significant fraction of energy, particularly for 4–6% ethanol feed concentrations, which are likely to be typical for cellulose conversion processes. Modern ethanol distillation processes are designed to achieve <100 ppm ethanol in the bottoms composition. Ethanol removal, by use of an adsorbent, has the potential to reduce energy requirements in the stripping section. For example, if the stripping section gives a 1% ethanol stream, the remaining ethanol could be recovered by adsorption. A concentrated aqueous ethanol would be recovered on regeneration of the sorbent and then be reintroduced into the column at or above the feed plate.

This article investigates the potential of achieving significant energy savings using an adsorption step at the bottom of the column, in preference to achieving a comparable concentration by distillation alone, since less water is boiled up to recover the same amount of ethanol. Selection of the sorbent, regeneration conditions, and regeneration media are described.

The energy tradeoffs of a combined distillation/adsorption system, and experimental testing of the process concept, are presented in the context of a first estimate of the material and energy balance for a distillation process with sorptive ethanol recovery.

MATERIALS AND METHODS

Selection of the Sorbent

The sorbents examined in this work were Amberlite XAD-2, XAD-4, Amberchrom CG-71, Ambersorb 563, and Ambersorb XEN 572, corn grits, solka floc, and Avicel (Table 2). The carbonaceous, Ambersorb materials are pyrolyzed, highly sulfonated, styrene/divinyl benzene macroreticular ion exchange resins with surface areas 550 m^2/g (for Ambersorb 563) and 1100 m^2/g (for 572).

Sorption capacities of these materials were measured by adding 4 mL of 1% ethanol to 1 g of each material in a culture tube, capping and sealing the tube, shaking the mixture for 5–30 min at ambient temperature, and then measuring the ethanol removed from the liquid based on a gas chromatography (GC) analysis of the liquid. The ethanol present was measured by GC analysis of 3.5 μL of sample using a Hayes Sep D (Supelco, Bellefonte, PA) column in a Shimadzu GC-14A instrument. This combination of the column and GC instrument gives a baseline separation of water from ethanol with helium as the carrier gas, and detection by a thermal conductivity detector (TDC). The column temperature was 180°C, the injector 120°C, and the detector 135°C for this analysis.

A follow up study at 70°C was conducted with the adsorbent Ambersorb XEN 572, since it was found to have the best properties with respect to removing ethanol from water efficiently. This sorbent has a particle size of 0.3–0.84 mm with a density of 0.49 g/mL, and micro-, meso and macroporosities of 0.41, 0.19, and 0.24, respectively.

Adsorption and Regeneration Methods and Conditions

Apparatus

Testing of adsorption and regeneration conditions was carried out in an experimental apparatus (Fig. 1) consisting of a 25 mm id × 60 cm long, jacketed column

Table 2
Comparison of Ethanol Removed by Different Sorbents at Ambient Temperature

Sorbent	Available from	Chemical structure/ composition	Wt, air-dry, g	Vol 1% EtOH added, mL	Contact time, min	% EtOH removed	Wt EtOH removed, g/g of sorbent
Ambersorb 572	Supelco (Bellefonte, PA) developed by Rohm and Haas	Carbonaceous, produced by pyrolysis of a highly sulfonated styrene-divinyl-benzene macro-reticular ion-exchange resin; surface area = 1100 m^2/g	1	4	30	87	0.0344
			1.0017	4	5	88	0.0347
Ambersorb 563	Supelco	Similar to Ambersorb 572, surface area = 550 m^2/g	1	4	30	87.5	0.0346
			1.0022	4	5	81	0.0319
Amberlite XAD-2	Supelco	Polyaromatic, surface area = 300 m^2/g	1.0008	4	30	0.3	0.0001
Amberlite XAD-4	Supelco	Polyaromatic, surface area = 725 m^2/g	1.0244	4	30	22.1	0.0085
Amberchrom CG-71	Supelco	Highly crosslinked methacrylic resin, surface area = 450–550 m^2/g	1.005	4	30	24.7	0.0097
Corn grits	J. R. Short Milling Co., Kankakee, IL	75% starch, the balance is protein, oil, and fiber	1.0055	8	30	11.9	0.0094
Solka floc	James River Corporation, Berlin, NH	Pulping cellulose	1.0072	8	30	10.5	0.0082
Avicel	FMC Corporation	Microcrystalline cellulose	1.0015	8	30	11.3	0.0089

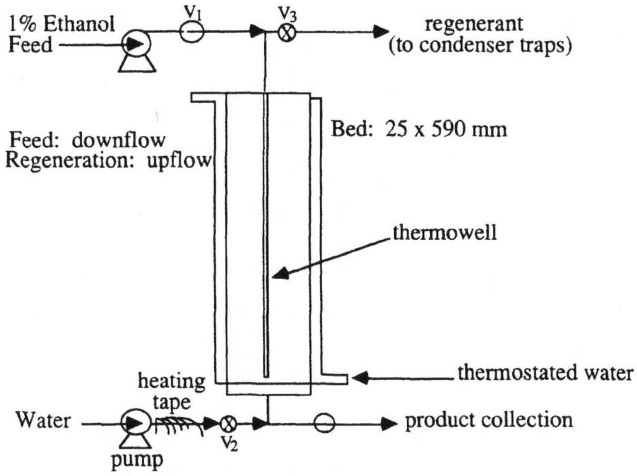

Fig. 1. Schematic representation of adsorption apparatus (not to scale).

containing the Ambersorb XEN 572 adsorbent with a bed height of 59 cm and a bed volume of 288 mL (corrected for the volume displaced by the thermowell). The thermowell (2-mm id) was located in the center of the column, and three thermocouple wires were fitted inside and connected to an automatic signal scanner (Omega Engineering Inc., Stamford, CT) to read the temperature. The wire tips were placed 5, 30, and 55 cm from the bottom of the column. Two circular wire screens were placed on the ends to retain the sorbent. The column (Ace Glass Inc., Vineland, NJ) was inverted and packed with the adsorbent by the tap and fill method. The polypropylene parts were then screwed in. The jacket was heated to the desired temperature using circulating water. Heated 1% aqueous ethanol was passed downflow at 1 bed vol/h.

The connecting tubes of the apparatus were made of steel, except for the tube between valves V_1 and V_3, which was plastic. The tube leading from the preheater to the bottom of the column was covered with insulated electric heating tape. This was connected to a powerstat to control the power and, hence, the temperature of the tape. A setting of 130 was chosen for the powerstat to give a temperature >110°C, to heat the water/ethanol mixture. The pump (Milton Roy Company, Riviera Beach, FL) used was rated at 29–290 mL/h flow rate. Owing to the small size of the apparatus, significant heat losses occurred, thus dropping the temperature to below 100°C.

Operating Procedures

During the adsorption cycle, the valve V_1 was connected to the ethanol feed reservoir, and V_3 was adjusted to make all the ethanol feed enter the column from the top. Samples collected at the column outlet were used to generate breakthrough profiles. The sorbent was regenerated by one of four methods: air, vacuum, steam stripping, and hot water. Each method of regeneration was evaluated with respect to the time of regeneration, percentage of ethanol in the condensate, and the degree of regeneration of the bed. The effectiveness of the regeneration step was assessed by running an adsorption cycle after regeneration and comparing the breakthrough profiles obtained at different regeneration conditions.

During regeneration, valve V_2 was adjusted to allow upward flow, and valve V_3 was switched to collect the stripped vapors. The preheater was kept at 95°C, and the electric heater tape was connected to a powerstat set at 130. The condenser vessel was cooled in a mixture of dry ice and acetone. This facilitated the capture of all condensable vapors for purposes of obtaining accurate mass balances. Regeneration by hot air was carried out with the column jacket maintained at 70°C. Air, heated by the preheater and heating tape to 70°C, was passed upflow at a pressure of 1.5 psig.

Upflow vacuum regeneration was carried out by pulling a vacuum with the jacket at 93°C. During vacuum stripping, the condenser outlet was connected to a vacuum pump, which gave a vacuum of 660 mmHg below atmospheric pressure.

Steam stripping was approximated by pumping water upflow at 5.5 mL/min through a tube covered with heating tape, which vaporized the water. A 660-mmHg vacuum was also applied to prevent condensation of water vapor at the column temperature of 89°C.

The apparatus for runs where regeneration was carried out with pressurized hot liquid water included a 12 mm x 14.5 cm steel column packed with 6.6 g of the Ambersorb XEN 572. Liquid chromatography end fittings with frits were used to retain the sorbent. The column was loaded with 1% ethanol at room temperature at a flow rate of 3.03 mL/min. For regeneration, the entire system, except the condenser, check valve, and pump, was immersed in a sand bath maintained at 120°C. Water was pumped through the column at a back pressure of 350 psig and a flow rate of 3.03 mL/min. In case of water regeneration, a vacuum at 660 mmHg was applied for 10 s between collection of regeneration samples to clear all lines. Vacuum was also applied for 10 min at the end of each run to clear the system of vapors.

Once feasibility was established based on the quick runs with the small column, regeneration with hot water was carried out by pumping at 5.5 mL/min with the larger column at 80°C and atmospheric pressure using the apparatus of Fig. 1. Collection of the effluent was the same as before.

Effect of Glucose on Ethanol Sorption

The dilute fermentation stream that enters a distillation column will have sugars in it. A batch study was conducted with a dilute solution of 4 mg/mL glucose to quantify the adsorption behavior of Ambersorb in the presence of glucose. The procedure consisted of soaking 5.0 g of Ambersorb in 30 mL of water for 60 min. The water was then drained off (9 mL of interstitial water remained) and 50 mL of glucose solution at 3.96 mg/ mL were added. The mixture was allowed to stand with no agitation. The mobile phase glucose concentration at different times was measured by a Beckman Glucose Analyzer (Beckman Instruments Inc., Fullerton, CA).

The effect of glucose on the breakthrough volume for a 1% ethanol feed was also carried out using the column adsorption system shown in Fig. 1. Two runs were carried out with 1% alcohol containing 4 mg/mL glucose. Collection of the sample and analysis of the ethanol content were carried out as described previously. Regeneration was by the hot water method. Vacuum was used between samples to clear the system as previously described.

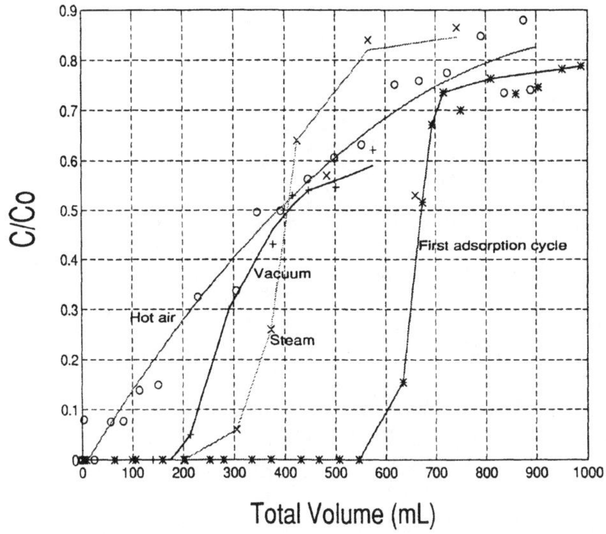

Fig. 2. Ethanol breakthrough profiles for Ambersorb XEN 572 at 70°C for dry, never-used sorbent (first adsorption cycle) and after sorption/regeneration/sorption sequence with regeneration using hot air, steam regeneration, and vacuum.

RESULTS

Sorbent Selection

Ambersorb XEN 572 and 563 removed more than 80% of the ethanol in 30 min of batch contacting. A shorter contact time of 5 min gave similar ethanol uptake, although Ambersorb XEN 572 performed marginally better than 563. The capacities of polymeric, starch (corn grits), and cellulosic (solka floc, Avicel) materials were lower (Table 2). Hence, Ambersorb 572 was selected as the adsorbent to be used in further runs.

Ambersorb 572 had lower capacity at 70°C (81% ethanol uptake) than at room temperature (88% uptake). The difference was small enough, however, that the adsorption step was carried out at 70°C. Sorption at the higher temperature is beneficial since the bottom stream from an ethanol column is hot (typically close to 100°C). Processing of the bottom stream to 70°C (or higher) for the sorption step would leave more of the thermal content of this stream to be used elsewhere in the plant.

Fixed-Bed Adsorption Runs

Dry Ambersorb XEN 572, evaluated in the apparatus shown in Fig. 1, initially gave the sharp breakthrough profile as shown in Fig. 2. Breakthrough occurred at 600 mL. Ethanol uptake, corrected for an estimated column void volume of 115 mL (40%), was 4.85 g of ethanol, or a loading of 16.8 g/L column volume. After several adsorption and regeneration cycles, the breakthrough occurred at 200–220 mL instead of 600 mL. Ethanol uptake, corrected for column void volume, now corresponded to 1 g of ethanol, or 3.4 g/L column volume. Significant, nonreversible

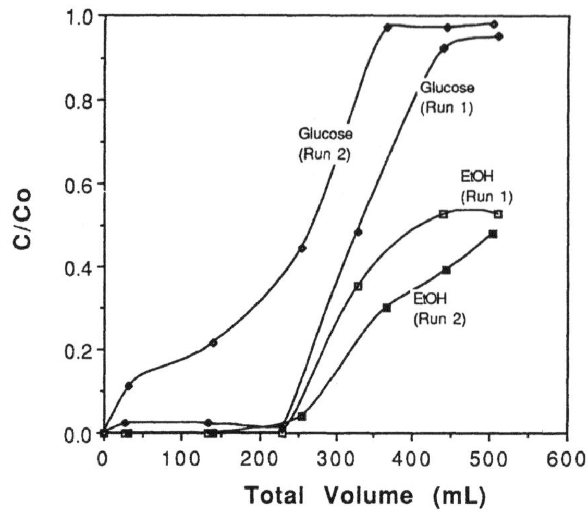

Fig. 3. Ethanol sorption in the presence of glucose—two replicate runs. Glucose breakthrough curves showing lower capacity for glucose uptake after the first run.

ethanol sorption occurs during the initial phases of the adsorption cycles at the regeneration conditions used in this work.

Effect of Glucose on Alcohol Adsorption

The Ambersorb takes up glucose, with 16% of the glucose (at 4 mg/mL) being adsorbed in 40 min. Nearly 70% was removed in 24 h. Since the sorption of glucose is slower than ethanol uptake, a short contact time with a mixture of ethanol and water should favor ethanol sorption.

Ethanol sorption runs with glucose present gave breakthrough volumes for ethanol that were the same as for runs carried out in the absence of glucose (compare Fig. 3 to Fig. 2). The regeneration stream contained 0.1–1.0 mg/mL glucose when regeneration was carried out using hot water at 93°C, with a decrease in glucose uptake occurring after the first adsorption/ regeneration cycle (Fig. 3). This indicates that a background adsorption of glucose occurs. Nonetheless, the adsorbent retains its effectiveness to sorb ethanol in the presence of dilute glucose.

Regeneration

The results of the regeneration run are summarized in Table 3. The degree of regeneration of the bed was measured by loading the column after a regeneration run and recording the breakthrough profile thus obtained. The profiles obtained after regeneration with hot air, steam, and vacuum are shown in Fig. 2. Regeneration with hot air is slow and incomplete, as indicated by the elution profile obtained after regeneration with hot air (Fig. 2). Steam stripping is faster and gives a more complete regeneration of the bed (the breakthrough volume is close to 225 mL). The results for the vacuum regeneration are similar (Fig. 2).

Both methods are relatively expensive and hence a regeneration by hot liquid water was attempted by passing hot water under a 50-psig back pressure through the column. Hot water was found to give a near-complete regeneration of the bed at

Table 3
Results of Regeneration Runs

Method	Temperature, °C	Pump flow rate, mL/min	Volume of condensate, mL	Ethanol in condensate, wt%	Time of regeneration, min	Maximum ethanol concentration obtained wt%
Hot air	70	—	81	4.9	415	6.2
Steam stripping	70	5.50	270	2.6	53	4.1
Vacuum	70	—	120	4.0	75	NA[c]
Hot water column	80	5.50	151	2.3	35	2.6
	93[a]	5.50	83	2.6	30	2.7
	120[b]	3.03	16	2.2	9	2.3
	93[a]	3.06	88	3.2	41	4.7
	93[a]	2.80	70	3.5	41	5.9

[a]Column jacket insulated.
[b]Small column.
[c]Not available.

80°C, but only gave a stripped ethanol concentration of 2.3%. Stripping with hot water at 93°C and 120°C gave almost the same concentration of ethanol in the condensate (2.2%), as expected, since earlier batch studies had indicated only a small temperature dependence for adsorption. When the flow rate of regenerating water was decreased from 5.5 to 3.06 mL/min, the average concentration of ethanol increased to 3.2%. With a still lower flow rate of 2.8 mL/min, an average concentration of 3.5% ethanol was obtained (with a maximum concentration of 5.9%) (Table 3). Sorptive concentration of a 1% ethanol stream using hot water as the regeneration media is thus shown to be possible. This result also indicates that 4% ethanol could be recovered from a hot water regeneration step, if multiple adsorber columns are properly staged for countercurrent flow of hot water. These results justify use of the basis of recovering 3.2% ethanol from a 1% feed for calculating material and energy balances.

Material Balance

The capacity of 3.4 g/L column volume would require a process designed to carry out adsorption/desorption in 5-min cycles. On this basis, the amount of sorbent required for a 50 million gal/yr ethanol plant would be on the order of 250,000 L. Hence, a higher capacity is needed. Significant further research, development, and testing are needed before large-scale use would be achieved. The current work presents the energy consequence of such a process if it were developed and implemented.

A process concept for combined distillation/adsorption was thus developed based on the experimental results, and the material balance generated (Fig. 4). The fermentation broth entering the recovery section is 4.5 wt%. A 90% ethanol overhead product is assumed for the distillation tower. The minimum reflux for such a system is 4.23. A practical operating reflux ratio of 6.35 (=1.5 × R_{Dmin}) is used. The bottoms at 1 wt% alcohol are directed to the adsorber.

DISCUSSION

Description of Distillation/Sorption System

A standard ethanol stripping system splits the feed into an ethanol rich overhead and a dilute bottoms stream. An approach that combines distillation and adsorption would use adsorption processes both overhead and at the bottom of the column as energy-efficient alternatives to remove the remaining small amounts of water and ethanol, respectively. The use of adsorption with stripping entails addition of adsorption columns at the bottom of the distillation tower as illustrated in the simplified diagram of Fig. 5. Energy-efficient integrated distillation systems have been studied and developed, including a staged distillation system using an alkali-extracting agent and distillation with intermediate heat pumps and optimal sidestream return *(21–23)*. The analysis here is for an atmospheric distillation column with recouping of heat from the column bottoms through preheating of the distillation column feed, and condensation of the overhead distillation column vapors to preheat partially the water or gas used to regenerate the sorbent on which the dilute ethanol is captured and concentrated.

Ethanol/Water Vapor-Liquid Equilibrium Data for Energy Calculations

The equilibrium data used for evaluating the energy tradeoffs were obtained from Carey and Lewis *(24)*, and fitted to polynomials by least-squares regression

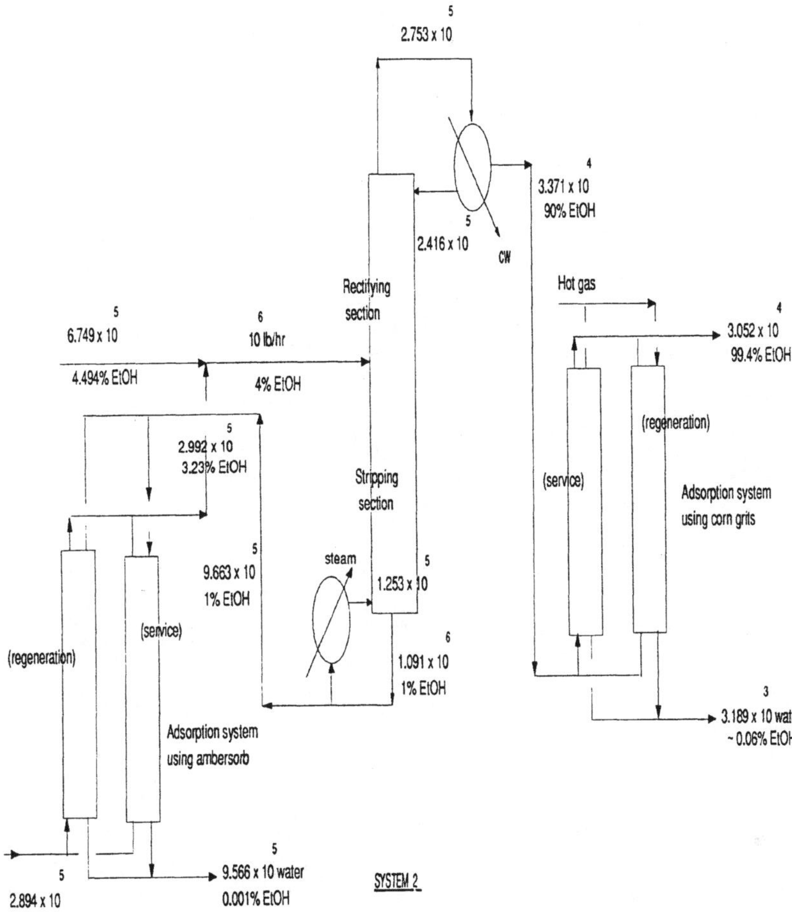

Figure 4. Process schematic with material balance for a combined distillation/adsorption system for ethanol/water, with an ethanol fermentation broth concentration of 4.5%, and a column feed concentration of 4% ethanol. Flows are in lb/h for a 50,000,000 gal/yr plant.

using the statistical package, SAS. The best fit was given by the three polynomials, shown in Table 4. The fit over the ranges of 0.0–0.12, 0.12–0.52, and 0.52–0.9 mol fraction ethanol is shown in the combined plot in Fig. 6. Data from other sources *(25–27)* generally coincided with that of Carey and Lewis as shown in Fig. 6.

Calculation of Energy Requirements

Several cases were calculated with respect to energy requirements. For a thermally saturated liquid feed, the steam required per hour ($\dot{m}_S$) is given by *(28)*:

$$\dot{m}_S = D \times (R_D + 1) \times (\lambda/\lambda_S) \tag{1}$$

The cooling water requirement is given by:

$$\dot{m}_c = D \times (R_D + 1) \times (\lambda/\Delta T) \times C_p \tag{2}$$

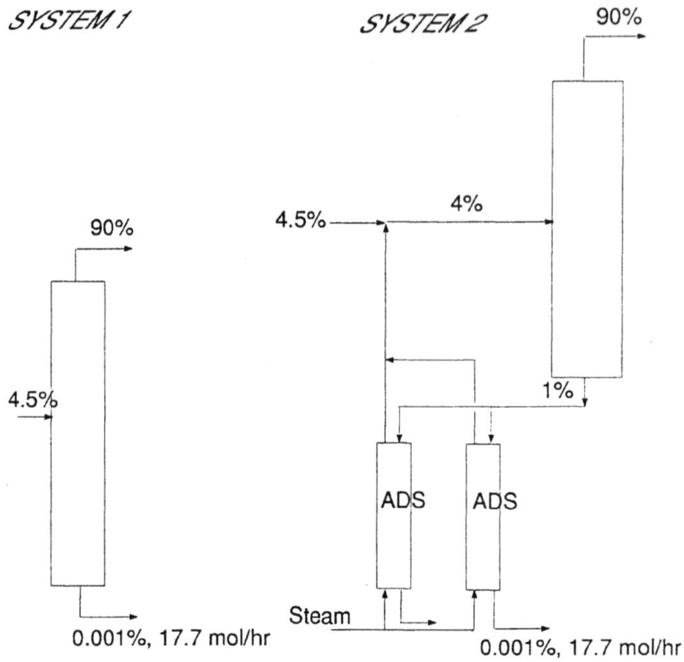

Fig. 5. Simplified schematic: comparing distillation (system 1) vs distillation combined with adsorption (system 2). Overhead condensers, reflux, and reboiler not shown for purposes of this illustration.

The minimum reflux is a function of the wt% ethanol in the feed (Fig. 7). For a 90% (by weight) overhead product, feed compositions of 20% or lower avoid the pinch region between the equilibrium curve and operating line. At 92% ethanol, the pinch region (where the rectifying operating line is tangent to the equilibrium) corresponds to about 17.7 wt% ethanol. Prior calculations have shown that a pinch can occur for 12 wt% ethanol as the azeotrope is approached. The calculation is approximate given the sensitivity of the pinch region (and corresponding feed concentration) to the exact shape of the equilibrium curve. The optimum reflux ratio for the two systems will be 1.2–1.5 times the minimum reflux ratio (R_{Dmin}). We chose $1.5 \times R_{Dmin}$ as the basis of our calculations, together with the assumptions that the feed entering the column is thermally saturated.

The reboiler duty sharply increases for feed concentrations below 3% and under these conditions decreases slightly when the ethanol in the overhead product decreases from 90 to 85%. The energy requirement decreases when the feed concentration increases from 4 to 5% but changes little with respect to changes in the ethanol bottom concentration, assuming that the number of theoretical plates can be adjusted to meet the compositional constraints.

The additional theoretical plates required to increase overhead composition from 85 to 90% is small as long as the reflux ratio is allowed to increase accordingly. However, the number of plates increases dramatically as the reflux ratio decreases (Fig. 8). Figure 8 illustrates the limitations posed: by specifying the bottom composition at 0.001 (rather than 1%) in terms of theoretical plates at reflux ratios of

Table 4
Polynomials Used to Fit Ethanol/Water Vapor/Liquid Equilibria Data from Carey and Lewls *(24)*

Range, mole fraction	Polynomial	R^2
$x \leq 0.1238$	$11.07x - 126.9x^2 + 833.21x^3 - 2279.7x^4$	0.99
$0.1238 \leq x \geq 0.5198$	$0.15 + 4.90x - 27.2x^2 + 86.98x^3 - 159.1x^4 + 162.65x^5 - 74.06x^6$	0.99
$x \geq 0.5198$	$0.98 - 2.72x + 6.82x^2 - 6.67x^3 + 2.61x^4$	0.99

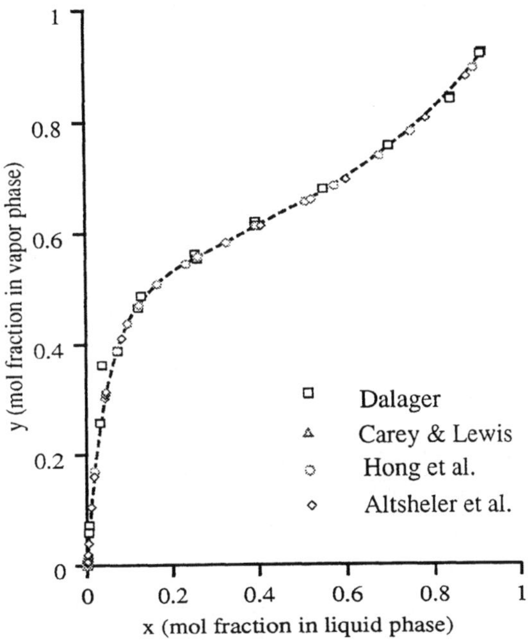

Fig. 6. Comparison of polynomials given in Table 4 to ethanol water equilibrium data atmospheric pressure for ethanol concentrations ranging from 0.0–0.92 mol fraction ethanol. Vapor liquid equilibrium curve based on combined equilibrium data text from several sources *(24–27)*. △, Carey and Lewis *(24)*; ○, Hong et al. *(25)*; □, Dalager *(26)*; ◇, Altsheler *(27)*.

10 or less. A column originally designed for 1% ethanol in the bottoms would be impractical to operate, if 0.01% ethanol were the goal since a very high reflux ratio (much >50) would be needed.

Potential for Energy Savings

The tradeoff between the number of theoretical plates and steam requirements is shown in Fig. 9. An energy usage (as reboiler duty) as low as 20,000 BTU/gal is practical for a 4% feed and 85% overhead product, and about 17,000 BTU/gal for a 5% feed, if the bottoms were to contain 1% ethanol and the overhead product 85% ethanol. The base case of a 90% overhead product with 4% feed shows that a

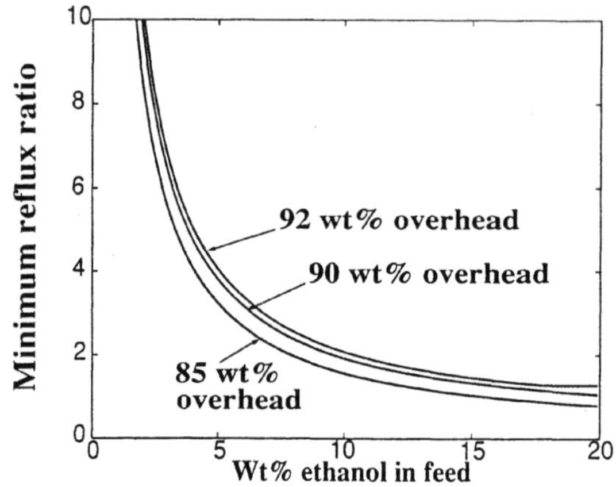

Fig. 7. Minimum reflux as function of ethanol concentration in feed, for 85, 90, and 92% (by weight) ethanol in the overhead.

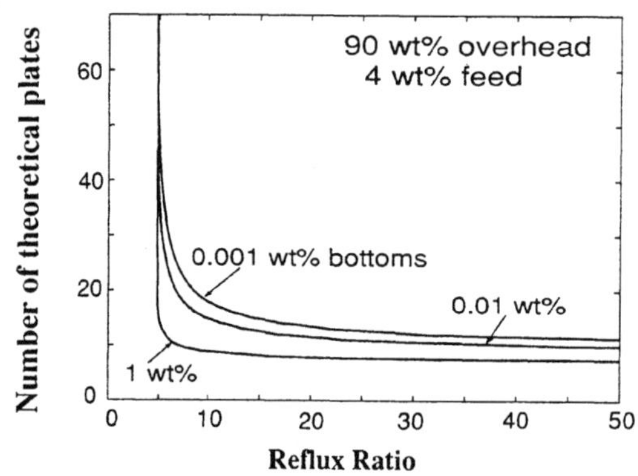

Fig. 8. Theoretical plates as function of reflux ratio for 90% ethanol in overhead product and 4% ethanol in feed.

column with 20 theoretical trays incurs a significant energy penalty as the ethanol in the bottoms decreases from 1 to 0.001 wt% (Figure 9). For a 40-tray column, the difference in energy requirements is <8%. If an overhead product of 90% can be processed by the downstream dehydration system, then the difference in energy required for recovering 0.001% alcohol in the bottoms vs 1% becomes smaller. An additional energy benefit is gained if 5% ethanol feed is available and 85% overhead product can be processed in the dehydration section to give the final fuel grade product. The energy and theoretical plate requirements in Fig. 9 show the potential benefits of stopping distillation at 1% bottoms and then recovering that ethanol by an alternative method.

Energy Requirements

A hypothetical process in which a 4.5% ethanol broth results in a 4% thermally saturated ethanol feed to the distillation column is shown in Fig. 4. A distillation tower

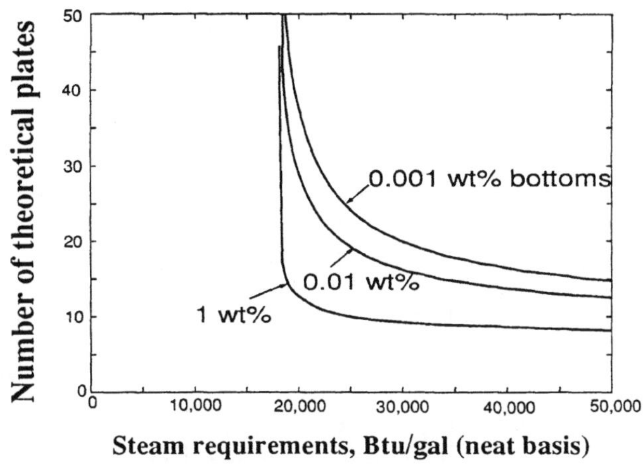

Fig. 9. Comparison of theoretical trays required as a function of steam requirements for 0.001, 0.01, and 1% ethanol (by weight) in the bottom product for 90% ethanol in overhead and 4% feed.

with 23 theoretical trays and a reboiler load of 23,100 BTU/gal ethanol (neat basis) would result. If a feed of 4.5% ethanol, an overhead product of 90%, and a bottom product of 0.001% is specified, the combination of distillation to 1% and adsorption to process the 1% bottom product would result in about 19,500 BTU/gal to obtain a 0.001% bottom stream. Of this energy requirement, about 18,300 BTU/gal are for distillation and about 1200 BTU/gal for regeneration (Fig. 10). The distillation column for such a process could be specifically designed to produce 1% ethanol, thereby offering capital cost savings since only 10 plates would be needed (instead of 21) with a minimal energy penalty. However, once such a column is built, it could not be used to recover 0.01 or 0.001% ethanol directly, since the energy costs would be prohibitive, requiring nearly 127,000 BTU/gal for a 0.01% bottoms and more than 160,000 BTU/gal for a 0.001% bottom.

The energy savings associated with an 90% overhead product are attractive, if the 90% ethanol vapors are dried using a corn grits sorption system. The energy required to obtain a fuel grade, dried ethanol product would be about 2000 BTU/gal for an 90% ethanol vapor stream in a properly designed and integrated unit *(12)*. The data and calculations show that a 4.5% fermentation broth (4% feed to the distillation column) has the potential of yielding a fuel-grade ethanol product using about 21,500 BTU/gal ethanol (as 99.4%). Hence, energy-efficient recovery of an ethanol stream derived from a biomass or cellulose process is shown to be technically feasible if sorptive, liquid-phase, recovery of the ethanol from distillation column bottoms is carried out. A closer study to optimize sorbent characteristics, capacity, and cost is now needed.

CONCLUSIONS

Recovery of ethanol from dilute distillation column bottoms by adsorption is feasible, and benefits can be derived from a reduced distillation tower size or energy savings. Cost savings may be possible if a column is specifically designed for cou-

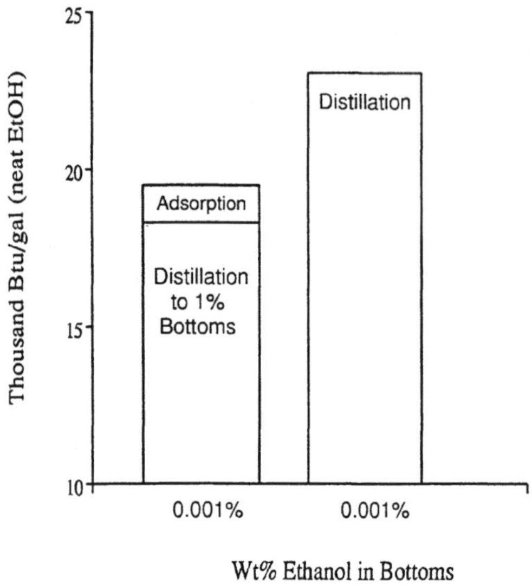

Fig. 10. Comparison of energy requirements of distillation vs distillation/sorption for ethanol recovery for a fermentation broth containing 4.5% ethanol with column designed for 0.001 wt% ethanol in bottom product. 90 wt% overhead; $R = 1.5 \times R$ min; 23 plates.

pling with an adsorptive ethanol recovery system, since the number of theoretical plates is significantly reduced, compared to a process that utilizes only distillation. This could decrease capital cost for a given reflux ratio. The potential energy savings for the case presented in Fig. 10 exceed 3500 BTU/gal. This assumes that a fermentation broth containing 4.5% ethanol (resulting in a distillation column feed containing 4% ethanol) is recovered by a combination of distillation and adsorption, rather than distillation alone, to obtain a 90% overhead product and 0.001% ethanol in the distillation column bottoms. At $5 per million BTU, this corresponds to 1.8¢/gal fuel-grade ethanol product. The benefit of combining distillation with adsorption lies not only in energy savings, but also in the options it offers for designing separation systems that will ultimately be needed for alcohol recovery from biomass conversion processes.

ACKNOWLEDGMENTS

This work was supported by NREL subcontract No. XCC-3-13388-01. We thank Nandan Padukone, Franki Parson, and Brian Rieper of NREL for assistance during the course of this project.

We thank also Phillip Wankat, Joe Weil, and Kyle Beery of Purdue University and Rod Bothast and Bruce Dien of USDA (Peoria) for their comments and suggestions during preparation of this manuscript.

REFERENCES

1. Ladisch, M. R. and Westgate, P. J. (1994), *Improvements in Separation and Recovery Technology For Lower Biofuel Costs,* Proceedings of Corn Utilization Conference V Corn Growers Association, St. Louis, MO.

2. Ladisch, M. R. and Schwandt, R. (1992), in *Technology for Expanding the Biofuels Industry.* Proceedings of workshop sponsored by DOE, USDA and Renewable Fuels Association in Chicago, p. II-4.
3. David, M. L., Hamnzaker, G. S., Buzenberg, R. J., and Wagner, J. P. (1978), in *Gasohol Economic Feasibility Study*, Development Planning and Research Assoc., Inc., Manhattan, KS.
4. Ghose, T. K. and Tyzgi, R. D. (1979), *Biotechnol. Bioeng.* **21,** 1387.
5. Hohmann, N. and Rendleman, C. M. (1993), in *Emerging Technologies in Ethanol Production*, USDA Agriculture Information Bulletin No. 663, p. 1.
6. Ladisch, M. R. and Dyck, K. (1979). *Science* **205,** 898.
7. Phillips, J. A. and Humphrey, A. E. (1983), in *Wood and Agricultural Residues: Research in Use for Feed, Fuels, and Chemicals*, Soltes, E. J., ed., Academic, New York, p. 503.
8. Robertson, G. H., Doyle, L. R., and Pavlath, A. E. (1983), *Biotechnol. Bioeng.* **25,** 3133.
9. Crawshaw, J. P. and Hills, J. H. (1990), *Ind. Eng. Chem. Res.* **29,** 307.
10. Voloch, M., Ladisch, M. R., Bienkowski, P., and Tsao, G. T. (1984), *Cereal Polysaccharides in Technology and Nutrition*, Rasper, V. F. ed., American Association of Cereal Chemists, St. Paul, MN, p. 103.
11. Hong, J., Voloch, M., Ladisch, M. R., and Tsao, G. T. (1982), *Biotechnol. Bioeng.* **24,** 725.
12. Ladisch, M. R., Voloch, M., Hong, J., Bienkowski, P., and Tsao, G. T. (1984), *Ind. Eng. Chem. Process Design Dev.* **23,** 437.
13. Hills, J. H. and Pirzada, I. M. (1989), *Chem. Eng. Res. Des.* **67,** 442.
14. Hassaballah, A. A. and Hills, J. H. (1990), *Biotechnol. Bioeng.* **35,** 598.
15. Bienkowski, P. R., Barthé, A., Voloch, M., Neuman, R. N., and Ladisch, M. R. (1986), *Biotechnol. Bioeng.* **28,** 960.
16. Neuman, R., Voloch, M., Bienkowski, P., and Ladisch, M. R. (1986), *Ind. Eng. Chem. Fundam.* **25,** 422.
17. Walsh, P. K., Liu, C. P., Findley, M. E., Liapis, A. I., and Siehr, D. J. (1983), *Biotechnol. Bioeng. Symp.* **13,** 629.
18. Lee, J. Y., Westgate, P. J., and Ladisch, M. R. (1991), *AIChE J.* **37,** 1187.
19. Westgate, P. J. and Ladisch, M. R. (1993), *Ind. Eng. Chem. Res.* **32,** 1676.
20. Goetz, R. J. (1995), In Forefront, Goetz, R., ed. AGAD, Purdue University, West Lafayette, IN.
21. Katzen, R., Diebold, V. B., and Moon, G. B. (1969), US Patent 3, 445, 345.
22. Katzen, R. and Diebold, V. B. (1976), US Patent 3, 990, 952.
23. Lynd, L. R. and Grethlein, H. E. (1986) *AIChE J.* **32(8),** 1347.
24. Carey, J. S. and Lewis, W. K. (1932), *J. Ind. Eng. Chem.* **(24),** 882.
25. Hong, J., Ladisch, M. R., and Tsao, G. T. (1981), *J. Chem. Eng. Data* **26(3),** 305–307.
26. Dalager, P. (1969), *J. Chem. Eng. Data* **14(3),** 298–301.
27. Altsheler, W. B., Unger, E. D., and Kolachov, P. (1951), *J. Ind. Eng. Chem.* **43(11),** 2559–2564.
28. McCabe, W. L., Smith, J. C., and Harriott, P. (1985), in *Unit Operations of Chemical Engineering.* McGraw-Hill, New York, p. 489.

0273-2289/96/5758—0119$8.00

Ammonia Recycled Percolation Process for Pretreatment of Herbaceous Biomass

PRASHANT V. IYER, ZHANG-WEN WU, SUNG BAE KIM,[†] AND YOON Y. LEE*

Department of Chemical Engineering, Auburn University, Auburn, AL 36849

ABSTRACT

An ammonia-based pretreatment method termed ammonia recycled percolation (ARP) was developed for pretreating herbaceous biomass like corn cobs/stover mixture (CCSM) and switchgrass. The process involves treatment of biomass with aqueous ammonia through a percolation reactor (packed-bed, flow-through type). The effects temperature, reaction time, and ammonia concentration were studied. The extent of delignification in the ARP process was in the range of 60–80% for CCSM and 65–85% for switchgrass. The ARP process solubilized significant amounts of the hemicellulose fraction into the pretreatment effluent, yet left most of the glucan fraction intact. The experimental data on CCSM and switchgrass indicate that the ARP is a highly effective pretreatment method. Near-complete conversion of cellulose to glucose was obtained by enzymatic hydrolysis of ARP-treated solid samples of CCSM, whereas conversion was slightly lower for switchgrass. The rate of enzymatic hydrolysis of ARP-treated samples was substantially higher than that of α-cellulose. The ARP effluents were evaluated for fermentability/toxicity by the xylose-fermenting yeast *Pichia stipitis* (NRRL Y-11545). The adaptability of ARP-treated solid samples to simultaneous saccharification and fermentation (SSF) was tested for ethanol production using cellulase enzyme and the yeast, *Saccharomyces cerevisiae* (NREL, D5A).

Index Entries: Pretreatment; ammonia; delignification; switchgrass; corn cobs/stover.

INTRODUCTION

Pretreatment is a necessary element in bioconversion of lignocellulosics to fuels and chemicals. The primary purpose of pretreatment is to make the cellulosic biomass amenable to the action of cellulase enzymes. A number of pretreatment methods have been proposed and investigated *(1–7)*. Although most of these pretreatments do increase the yields of sugars by enzymatic hydrolysis, the reported yields are usually below the theoretical maximum *(1–5)*. There are some exceptional cases where quantitative conversion of cellulose into glucose has been reported *(6,7)*. These pretreatment methods, however, require a high-energy input or utilize

*Author to whom all correspondence and reprint requests should be addressed.
†On leave from Gyongsong National University, Chinju, Korea.

Table 1
Initial Composition of CCSM and Switchgrass

	Percentage	
Components identified[a]	CCSM	Switchgrass
Glucan	38.10	35.20
Xylan	20.00	17.76
Arabinan	3.10	3.72
Galactan	1.20	1.92
Mannan	0.60	3.40
Klason lignin[b]	15.95	19.88
Acid-soluble lignin	3.10	3.30
Ash	4.29	5.25
Extractives	6.70	6.09
Others	6.96	3.48

[a]Based on oven-dry untreated biomass.
[b]For Klason lignin determination, biomass was treated with 95% ethanol to remove extractives.

expensive and/or toxic reagents. There are other problems associated with pretreatments, such as loss of sugars through decomposition and generation of toxic products that inhibit microbial action in the subsequent fermentation *(8,9)*. The economic and environmental constraints further limit the applicability of these known methods. Recently, a novel pretreatment method based on aqueous ammonia has been developed in our laboratory *(10)*. This process was termed the ammonia recycled percolation (ARP) process. The ARP when applied to woody biomass brought about not only pretreatment effects, but also a partial fractionation of biomass into cellulose, hemicellulose, and lignin. Lignin generated from the ARP process is sulfur- and sodium-free, unlike the lignin generated from conventional pulping processes. It is quite conceivable that the uncontaminated lignin can be a marketable byproduct, enhancing the overall economics of the bioconversion process.

In this study, we extended our investigation on the ARP as a pretreatment method to cover herbaceous biomass feedstocks, including corn cobs/stover mixture and switchgrass. Switchgrass has a number of desirable characteristics as an energy crop *(11)*. The scope of this work included the technical factors, primarily the reaction conditions, effectiveness as a pretreatment, and fermentability/toxicity of the effluents to simultaneous saccharification and fermentation (SSF).

MATERIALS AND METHODS

Materials

Corn cobs/stover mixture (CCSM) and switchgrass feedstocks were supplied from National Renewable Energy Laboratory (NREL, Golden, CO). The CCSM was milled and screened to the nominal size of 0.25–2.00 mm. Switchgrass was used as supplied (the size being very fine—5 mm). The initial compositions of CCSM and switchgrass are shown in Table 1. The cellulase enzyme, Cytolase CL, Lot No. 17-92262-09, was obtained from Environmental Biotechnology (Santa Rosa, CA). The average specific activity of the enzyme as determined by the supplier is as follows: Filter paper activity = 95.9 FPU/mL, β-glucosidase activity = 80.6 p-NPGU/mL, endoglucanase activity = 613 CMCU/mL. A culture of xylose-fermenting yeast *Pichia stipitis* (NRRL Y-11545) was obtained from USDA and was used for fermentability and toxicity studies on ARP effluents. An NREL culture of *Saccharomyces cerevisiae*, D_5A strain, was used for SSF studies. The Sigma (St. Louis, MO) α-cellulose was supplied from NREL.

Experimental Setup and Operation

The details of the experimental apparatus are described elsewhere *(10)*. A slight modification was made in that a temperature-programmable oven replaced a sand bath. In operation of ARP, 5–7 g of air-dried biomass sample were packed into the reactor, soaked with ammonia solution, and left overnight. The ratio of liquid to solid inside the reactor was kept in the range of 7–10. The preheating time to reach the desired reaction temperature was about 15 min. Nitrogen back pressure was applied to the reaction system at 325 psi to prevent the ammonia vaporization. After the reaction, water was pumped to remove residual sugars and ammonia trapped in biomass. The reactor was then flushed with nitrogen to remove the excess water. The biomass samples discharged from the reactor were divided into two portions. One portion of the wet solid residues was oven-dried at 105°C overnight to measure moisture content and then subsequently the weight loss on pretreatment. It was further subjected to chemical composition analysis. The other remaining portion of solid residues was stored in a walk-in cool room (4°C) under wet condition for carrying out the digestibility test. The ARP effluent collected in a holding tank was transferred into an air-tight sample bottle and stored in the cool room for further composition analysis.

Digestibility Test

Enzymatic hydrolysis of pretreated substrate was performed at 50°C and pH 4.8 (0.05*M* sodium citrate buffer) on a shaker bath (New Brunswick Scientific, Edison, NJ) agitated at 150 rpm. The enzyme loading was 60 IFPU/g glucan, and the initial glucan concentration was 1% (w/v). Samples were taken periodically and analyzed for glucose and cellobiose content using high-performance liquid chromatography (HPLC). The total glucose content after 72 h of hydrolysis was taken to calculate the enzymatic digestibility. Untreated switchgrass, CCSM, and α-cellulose were also subjected to the same digestibility test as a control and as a reference.

Analytical Methods

The solid samples were analyzed for sugars, Klason lignin, acid-soluble lignin, and ash following the procedures of NREL-CAT Standard Procedures *(12)*. Prior to Klason lignin determination, the feedstocks were put through ethanol reflux to remove the extractives. The effluents from the ARP process were boiled until all free ammonia was evaporated. Since most of the sugars contained in the effluents were oligomers, a secondary hydrolysis was carried out at 121°C with 4% (w/w) sulfuric acid for 1 h of reaction time to convert oligomers into monomers. Sugars and decomposition products were measured by HPLC using Bio-Rad (Hercules, CA) Aminex HPX-87H column. The total amount of xylan, mannan, arabinan, and galactan was used to represent the hemicellulose content in this study. The fermentation samples were analyzed using the H-column and/or YSI 2300 Stat Plus sugar analyzer.

Fermentation of ARP Effluents and ARP Solid Samples

Fermentability/toxicity tests of ARP effluents containing xylose were done following the procedure obtained from the USDA *(13)*. The fermentation was performed at 30°C, pH 6.0 with 3% (w/v) xylose. The adaptability of ARP-treated solid samples for ethanol production by SSF was determined by following NREL-CAT Standard Procedure No. 00 8 *(14)*. The SSF conditions were: 38°C, pH 5.0, 3% (w/v) glucan, and 25 IFPU/g glucan enzyme loading.

Determination of Fermentation Parameters

For the ARP effluent fermentability/toxicity tests, the yield was calculated as the ratio of maximum ethanol concentration detected minus the concentration of ethanol at zero time to the initial concentration of sugars (glucose + xylose) present. Volumetric productivity (g/L/h) was calculated as the maximum ethanol concentration detected minus the concentration of initial ethanol divided by the fermentation time (time at which the ethanol concentration reaches maximum). For the SSF of ARP-treated solid substrates, the ethanol yield was defined as the ratio of the amount of net ethanol produced to the amount of the initial glucose content in the pretreated substrate divided by 0.51.

RESULTS AND DISCUSSION

Effect of Ammonia Concentration

Ammonia concentrations were varied over 2.5–20% (w/w) keeping other conditions at 170°C, reaction time of 1 h, flow rate = 1 mL/min. Figure 1 shows the weight remaining, amounts of glucan, hemicellulose, lignin remaining, and cellulose enzymatic digestibility on the ARP pretreatment. During the ARP, CCSM lost 43–47% of its weight. The weight loss for switchgrass was slightly higher at 49–56%. Lignin and hemicellulose accounted for the major part of the weight loss. In the case of CCSM, as high as 56% of the original amount of hemicellulose of CCSM was extracted into ARP effluent, whereas <8% of the total glucose content was extracted. These numbers were slightly higher for switchgrass, 64 and 14%, respectively. The amount of glucan and hemicellulose solubilized remained essentially constant showing little effect of ammonia concentration over this range. The lignin remaining, however, decreased by 48 to 19% for CCSM (32 to 16% for switchgrass) as the concentration of ammonia was increased from 2.5 to 20%. The increase in ammonia concentration beyond 10% had negligible effect on the degree of delignification. Figure 1 also shows that the enzymatic digestibility of ARP samples exceeded 90% for CCSM and 87% for switchgrass in all cases. The digestibility generally increased as the ammonia concentration was increased. However, above 5% for CCSM, the increase of digestibility was insignificant. Results were similar for switchgrass, but the leveling off occurred at 10%. An ammonia concentration of 10% is therefore regarded as a near-optimum level for pretreatment of switchgrass. For CCSM, ammonia concentration of 5% was sufficient to achieve high digestibility. Considering that an ammonia concentration of 10% gives a higher degree of delignification (79%) and that almost all of it is recycled, it was taken as the optimum level for CCSM as well.

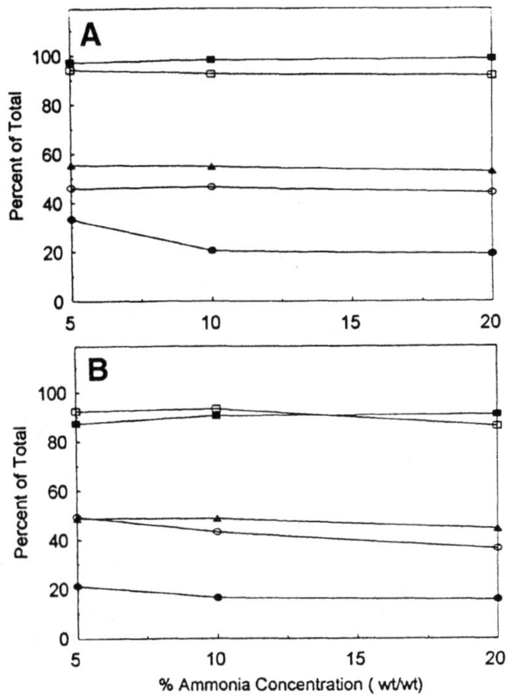

Fig. 1. Compositional changes and enzymatic digestibility in ARP at various ammonia concentrations. **(A)** CCSM; **(B)** switchgrass. Pretreatment conditions: 170°C, flow rate = 1 mL/min. Enzymatic hydrolysis conditions: 60 IFPU/g glucan, pH 4.8, 50°C. ▲, solid remaining; □, glucan remaining; ○, HC remaining; ●, lignin remaining; ■, enzymatic digestibility.

Effect of Temperature

ARP on CCSM retained 44% of hemicellulose (43% for switchgrass) at 170°C, 10% ammonia, treatment time of 60 min, flow rate = 1 mL/min (Fig. 1). Complete retention or complete removal of the hemicellulose (depending on the subsequent fermentation strategy) is desirable for efficient utilization of hemicellulose sugars. Increased solubilization of the hemicellulose is expected with increase in the reaction temperature. Ensuing runs were made over 160–200°C for CCSM, maintaining other conditions at 10% ammonia, 1-h reaction time, flow rate = 1 mL/min. Table 2 summarizes the composition data for the hydrolysate and the remaining solid residue after pretreatment. From the data on CCSM, it is seen that the amounts of glucan and hemicellulose solubilized increased from 8 to 15% and 46 to 80%, respectively, as the pretreatment temperature was increased from 160 to 200°C. However, the maximum amount of hemicellulose recovered as oligomers in the ARP effluent was only about 72% of the original amount and occurred at 200°C. The ARP treatment of CCSM gave a near-complete sugar balance at temperatures below 200°C, indicating sugar decomposition is insignificant. With regard to delignification, temperature effect gave a complicated picture. As the temperature was increased from 160 to 180°C, lignin remaining in the solid decreased to 19% of the original. As the temperature was increased from 180 to 200°C, however, lignin remaining increased to 35%.

Table 2
Effect of Reaction Temperature on Composition of Hydrolysates and Solid Residues in ARP[a]

Temperature, °C	% Solid remaining	% Lignin, Klason	% Glucan			% Hemicellulose		
			Solid	Liquid	Total	Solid	Liquid	Total
CCSM								
Untreated	100.00	15.95	38.10	—	38.10	24.90	—	24.90
160	55.56	5.03	35.10	1.89	36.99	12.53	11.04	23.57
170	54.90	3.26	35.30	2.04	37.34	11.01	12.65	23.66
180	51.26	3.10	34.78	2.11	36.89	9.51	14.53	24.04
190	49.70	5.32	33.41	2.36	35.77	7.96	16.51	24.47
200	43.24	5.63	32.18	2.39	34.57	5.09	18.07	23.16
Switchgrass								
Untreated	100.00	19.88	35.20	—	35.20	26.80	—	26.80
160	52.15	4.84	33.34	1.58	34.92	13.24	12.26	25.50
165	50.22	4.26	32.02	1.28	33.30	11.60	13.30	24.90
170	48.57	3.26	32.88	1.27	34.15	10.94	14.34	25.28
175	45.12	3.53	31.94	1.11	33.05	8.51	15.40	23.91

[a]All sugar and lignin contents based on the oven-dry untreated biomass. Hemicellulose: the total amount of xylan, mannan, arabinan, and galactan. Pretreatment conditions: 10% (w/w) ammonia, 60 min, flow rate = 1 mL/min. Secondary hydrolysis conditions: 121°C, 45 min, 4% sulfuric acid.

We do not have a clear explanation for this unusual behavior at this time. The data in Tables 2 and 3 show that the material balance of the treated solids do not close out because ash content and extractable and gaseous degradation components (if any) were not measured, and were thus unaccounted for. The enzymatic digestibility was investigated for all samples of CCSM treated at various temperatures (Fig. 2). The maximum cellulose enzymatic digestibility after 72 h was 98% of theoretical maximum for CCSM, and it occurred with the ARP conducted at 170°C. With one exception of ARP-treated sample (CCSM at 160°C), all treated samples are seen to exhibit high digestibility, most of them showing above 90% in 36 h. At the 36-h point, the digestibility of α-cellulose was only 56%. Although the reaction temperatures of both 170 and 180°C offer good pretreatment performance for CCSM, reaction temperature of 170°C would be preferred to 180°C from process viewpoint.

A 10°C increment was used in the study of temperature effect on CCSM. After a finding that the optimum temperature for CCSM is near 170°C, a 5°C increment was then used for switchgrass, applying four levels over 160–175°C. The composition data of treated switchgrass are also included in Table 2. The enzymatic digestibilities for ARP-treated switchgrass samples are shown in Fig. 2. As the temperature was increased from 160 to 175°C, there was a substantial increase in hemicellulose removal. However, the lignin content in the solid remained essentially constant. The glucan removal was below 10% for all the temperatures. Again the initial rate of hydrolysis of ARP-treated samples was much higher than that of α-cellulose and the digestibility at 72 h of the best-case ARP-treated switchgrass was above 90% as opposed to 77% for α-cellulose. From the data in Table 2 and Fig. 2, 170–175°C is judged to be the optimum region for treatment of switchgrass in ARP.

Table 3
Effect of Reaction Time on Composition of Hydrolyzates, Solid Residues, and Cellulose Enzymatic Digestibility in ARP

Reaction	% Solid	% Lignin,	% Glucan			% Hemicellulose			% Digestibility
time, min	remaining	Klason	Solid	Liquid	Total	Solid	Liquid	Total	72 h
CCSM									
Untreated	100.00	15.95	38.10	—	38.10	24.90	—	24.90	22.90
15	56.07	4.16	35.48	1.82	37.30	11.95	12.61	24.56	94.10
30	55.68	3.82	34.43	1.88	36.31	11.15	13.01	24.16	95.28
45	55.28	3.44	34.98	2.16	37.14	11.39	13.35	24.74	96.45
60	54.90	3.26	35.30	2.04	37.34	11.01	12.65	23.66	98.30
Switchgrass									
Untreated	100.00	19.88	35.20	—	35.20	26.80	—	26.80	17.30
30	50.10	5.76	31.07	1.88	32.95	12.88	12.72	25.60	87.30
60	48.57	3.26	32.88	1.27	34.15	10.94	14.34	25.28	90.46

Notes are the same as for Table 2.

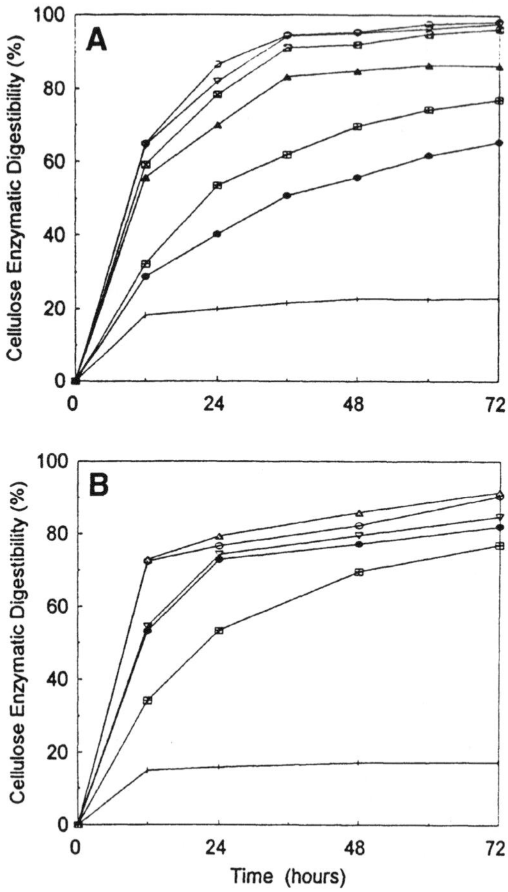

Fig. 2. Enzymatic digestibility of substrates treated in ARP at various temperatures. **(A)** CCSM; **(B)** switchgrass. Pretreatment conditions: 10% ammonia, 60 min, flow rate = 1 mL/min. Enzymatic hydrolysis conditions: 60 IFPU/g glucan, pH 4.8, 50°C. (A) ●, 160°C; ○, 170°C; ▽, 180°C; ⊠, 190°C; ▲, 200°C; +, untreated; ⊞, alpha-cellulose. (B) ●, 160°C; ▽, 165°C; ○, 170°C; △, 175°C; +, untreated; ⊞, alpha-cellulose.

Effect of Reaction Time

The time effect on the ARP process was first studied by taking on-line samples of the ARP effluent of switchgrass. The conditions were: 10% ammonia, 170°C, reaction time of 1 h, flow rate = 1 mL/min. These samples were analyzed for sugars and lignin. Lignin contents were determined by the UV at 280 nm using 10% ammonia solution as a reference. The data indicate that most of the hemicellulose removal and delignification occur within first 30 min of the ARP operation (Fig. 3). In the ensuing runs, reaction time was varied over 15–60 min for CCSM, keeping other conditions at 10% ammonia, 170°C, flow rate = 1 mL/min. The composition data for the hydrolysate and the remaining solid residues after the ARP pretreatment and the enzymatic digestibility for both switchgrass and CCSM are presented in Table 3. From the data, it is quite evident that the reaction time of 15 min is sufficient to increase the 72-h digestibility from the control level of 23–94% for CCSM. For switch-

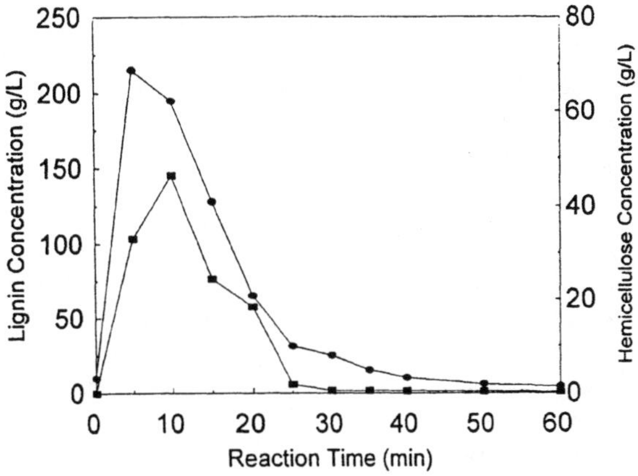

Fig. 3. Time-course of lignin and hemicellulose in ARP effluent of switchgrass. Pretreatment conditions: 170°C, 60 min, flow rate = 1 mL/min. Secondary hydrolysis conditions: 121°C, 4% sulfuric acid, 1 h; HC, hemicellulose. ●, lignin; ■, hemicellulose.

grass, a reaction time of 30 min was required. The digestibilities were somewhat lower at 17.3% for control and 87.3% for treated samples. Further increases in reaction time showed only a marginal increase in digestibility for both CCSM and switchgrass. As the reaction time was increased from 15 to 60 min for CCSM (30–60 min for switchgrass), the remaining lignin decreased from 26 to 20% of original for CCSM (29 to 16% for switchgrass). However, the hemicellulose and glucan remained relatively constant (Table 3). Data in Table 3 collectively indicate that the acceptable range of reaction time under the previously described reaction conditions is 15–60 min.

ARP Effluent and Lignin Separation

ARP treatment of CCSM and switchgrass solubilized about 80% of original lignin into the ARP effluent at the pretreatment condition of 170°C, 1 h, 10% ammonia, 1 mL/min. At the same time, about 55% of the original hemicellulose was extracted into the ARP effluent, mostly in the form of oligomers. The ARP effluents were boiled until all the free ammonia evaporated (or until pH reached 7.0 from 11.5) and then subjected to a secondary acid hydrolysis. The acidic condition of secondary hydrolysis induced precipitation of lignin from the ARP effluent. The effluent was cooled to room temperature and filtered. The precipitated lignin was further washed with deionized (DI) water, and then dried at 105°C overnight and weighed. During the pretreatment, 80% of the original Klason lignin was extracted into the liquid effluent for CCSM. It was 84% for switchgrass. However, only 50–60% of the original Klason lignin was precipitated and recovered. The balance of the lignin is believed to be broken down into low-mol-wt lignin fragments and remained soluble even under acidic conditions.

Ammonia Consumption

From an economic standpoint, the consumption of ammonia is an important issue in the ARP process. Ammonia material balance was calculated using data on Kjeldahl nitrogen analysis on both solids and liquids. The results have shown that the recovery of ammonia is in excess of 99%, essentially the same as that for hybrid poplar *(10)*. The ammonia consumption was 0.02 g/g dry biomass, not a significant cost factor. Most of the unrecovered ammonia is believed to be in the form of ammonium acetate in the ARP effluent. Only a small fraction is affixed to the treated solid residue as a result of pretreatment. This fraction may even be utilized as a source of nitrogen in the subsequent fermentation *(15)*.

Fermentability/Toxicity Tests on ARP Effluents

The ARP treatment of CCSM and switchgrass under near-optimum conditions solubilized 50–56% of total hemicellulose into the ARP effluent. Extraneous components, such as extractives, acetic acid, furfural, and dissolved lignin components, present in the hydrolysates are potential inhibitors of microorganism fermenting the xylose. A toxicity test was therefore conducted to determine the maximum effluent loading for efficient fermentation of xylose to ethanol. The effluent loading is defined as the volume percentage of the ARP hydrolysate in the total fermentation liquid.

Multiple runs of ARP were made at 170°C, 10% ammonia, flow rate = 1 mL/min, at the reaction time of 15 min for CCSM (30 min for switchgrass) to generate a sufficient amount of liquor to be tested. Effluent was boiled to remove ammonia, and its final volume was readjusted to initial volume adding DI water. Since ARP effluent contained a large amount of oligomeric sugars (mainly xylan), a secondary hydrolysis was conducted to convert them into monomers. In this process, lignin was precipitated and filtered out before subjecting the effluent to fermentability/toxicity tests.

Figure 4 shows the yield (g/g) and volumetric productivity (g/L/h) at various effluent loadings for CCSM and switchgrass. The yield at 40 and 60% effluent loading for CCSM was essentially the same as that of the control (0% loading). The volumetric productivity, however, has decreased significantly with effluent loading of 60%. Fermentation with 80% effluent loading was not completed within 72 h, leading to lower concentration of ethanol than other runs. For switchgrass, the results were quite similar to those of CCSM in that 60% effluent loading showed little effect on the product yield, whereas the volumetric productivity was adversely affected.

SSF

ARP-treated solid samples used in the SSF test were prepared under the conditions of 170°C, 10% ammonia, flow rate = 1 mL/min, 15-min reaction time for CCSM (30 min for switchgrass). The glucan content of pretreated biomass was 65% (dry basis) for CCSM and 66% for switchgrass. The glucan content of α-cellulose, which was used as a control, was 94%.

The ethanol yield data for both ARP-treated CCSM and switchgrass are presented in Fig. 5. The ethanol yield for pretreated CCSM reached 83% of theoretical maximum at 168 h. It was essentially same as that of α-cellulose. The pretreated switchgrass gave about the same level of ethanol yield, 84% at 168 h. At the 24-h point, the ethanol yield of the pretreated CCSM reached 60% and switchgrass, 52%. These values were substantially higher than the 35% yield observed for α-cellulose (Fig. 5). The SSF results on the ARP-treated CCSM and switchgrass are deemed quite satisfactory in view of the fact that a certain fraction of the substrate is channeled into cell growth.

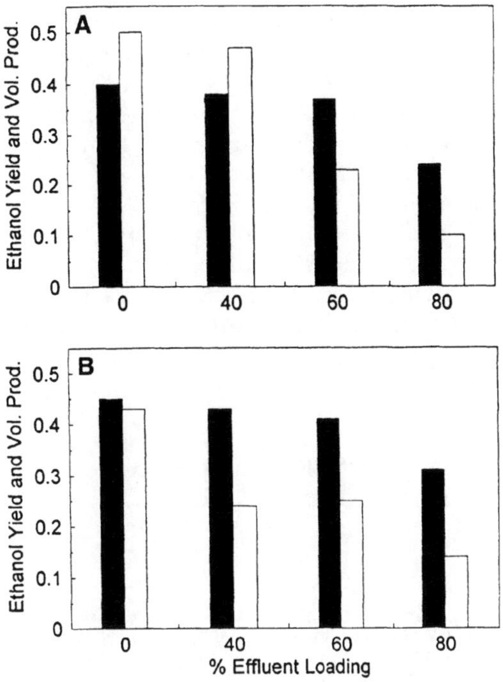

Fig. 4. Ethanol yield (■, g/g) and volumetric productivity (□, g/L/h) at various effluent loadings. **(A)** CCSM, 15 min in ARP; **(B)** switchgrass, 30 min in ARP. Pretreatment conditions: 170°C, 10% ammonia, flow rate = 1 mL/min; 15 min for CCSM, 30 min for switchgrass. Fermentation conditions: 30°C, pH 6.0, 72 h, 3% (w/v) xylose.

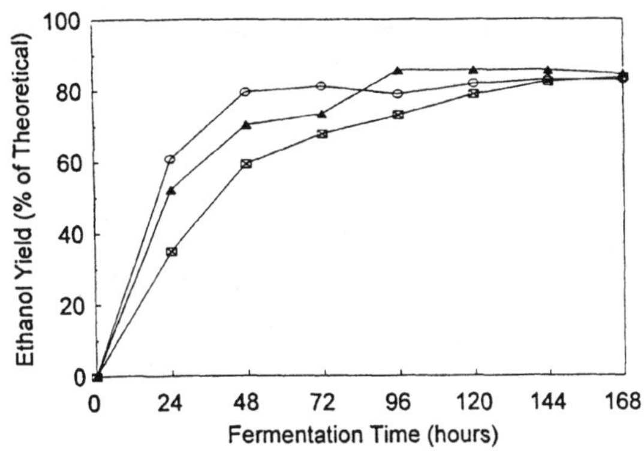

Fig. 5. Ethanol yield substrates pretreated in ARP at different fermentation times. Pretreatment conditions: 170°C, 10% ammonia, flow rate = 1 mL/min, 15 min for CCSM, 30 min for switchgrass. SSF condition: 38°C, pH 5.0, 3% (w/v) glucan, enzyme loading 25 IFPU/g glucan. ○, pretreated CCSM; ▲, pretreated switchgrass; ⊠, alpha-cellulose.

CONCLUSIONS

ARP as tested against CCSM and switchgrass is a highly effective pretreatment method. The enzymatic digestibility of the pretreated substrates was found to be in the vicinity of 90%. Hemicellulose and lignin removal and modification of physical structure are attributed to the increase in the enzymatic hydrolysis. The optimum condition identified for ARP on CCSM is: 170°C, 10 wt% ammonia concentration, flow rate = 1 mL/min, and reaction time 15–60 min. For switchgrass, it is: 170°C, 10 wt% ammonia concentration, flow rate = 1 mL/min, and reaction time 30–60 min. The extent of delignification was 74–80% for CCSM and 71–84% for switchgrass. Under the optimum conditions, 50–56% of the total hemicellulose in CCSM and switchgrass, but <8% of the total glucan in CCSM (10% for switchgrass), was solubilized. The toxicity of the ARP effluent is such that its loading up to 60% could be employed without adversely affecting the ethanol yield. ARP-treated solid samples of CCSM and switchgrass are easily adaptable to SSF for ethanol production.

ACKNOWLEDGMENTS

This research was as a part of NREL subcontract-XAW-4-14170-01. Additional equipment support was provided by the College of Engineering, Auburn University.

REFERENCES

1. Cahela, D. R., Lee, Y. Y., and Chambers, R. P. (1983), *Biotechnol. Bioeng.* **25,** 3–18.
2. McGinnis, G. D., Wilson, W. W., Mullen, C. E. (1983), *Ind. Eng. Chem. Prod. Res. Dev.* **22,** 352–357.
3. Gould, J. M. (1984), *Biotechnol. Bioeng.* **27,** 225–231.
4. Vallander, L. and Eriksson, K.-E. (1985), *Biotechnol. Bioeng.* **27,** 650–659.
5. Ropars, M., Marchal, R., Pourquie, J., and Vandecasteele, J. P. (1992), *Bioresource Technol.* **42,** 197–204.
6. Holtzapple, M. T., Lundeen, J. E., Sturgis, R., Lewis, J. E., and Dale, B. E. (1992), *Appl. Biochem. Biotechol.* **34/35,** 5–21.
7. Torget, R. and Hsu, T.-A. (1994), *Appl. Biochem. Biotechnol.* **45/46,** 5–22.
8. Detroy, R. W., Lindenfelser, L. A., Sommer, S., and Orton, W. L. (1981), *Biotechnol. Bioeng.* **23,** 1527.
9. Averinos, G. C. and Wang, D. I. C. (1983), *Biotechnol. Bioeng.* **25,** 67.
10. Yoon, H. H., Wu, Z. W., and Lee, Y. Y. (1995), *Appl. Biochem. Biotechnol.* **51/52,** 5–19.
11. Turhollow, A. F. (1991), *Bioresource Technol.* **36,** 247–252.
12. Ehrman, T. and Magill, K. (1992), NREL-CAT Standard Procedure No. 001-003, Golden, CO.
13. Philippidis, G. P., Smith, T. K., and Schmidt, S. L. (1993), NREL-CAT Standard Procedure No. 008, Golden, CO.
14. Kurtzman, C. P. (1994), *P. stipitis*-Based Pretreatment Liquor Fermentability/Toxicity Test Protocol (USDA), Peoria, IL.
15. Dale, B. E., Henk, L. L., and Striang, M. (1985), *Dev. Ind. Microb.* **26,** 223–233.

0273-2289/96/5758—0133$8.50

Kinetic and Modeling Investigation on Two-Stage Reverse-Flow Reactor as Applied to Dilute-Acid Pretreatment of Agricultural Residues

RONGFU CHEN,[1] YOON Y. LEE,*[,1] AND ROBERT TORGET[2]

[1]Department of Chemical Engineering, Auburn University, AL 36849; and [2]National Renewable Energy Laboratory, Golden, CO

ABSTRACT

The kinetics of dilute-acid pretreatment/hydrolysis of the hemicellulose in a mixture of corn cobs and corn stover was investigated. The kinetic data confirmed that the hemicellulose in this feedstock is of a biphasic nature. The kinetic model recognizes the presence of soluble xylose oligomers, xylose monomer, and the decomposition of xylose. The kinetic parameters were determined over the conditions of 120–150°C, and sulfuric acid concentration of 0.44–1.90%.

The biphasic nature of the kinetics brings about an additional flexibility in the reactor design and the operation strategy, since different reaction conditions can be applied to each of the two different fractions of hemicellulose in the feedstock. With incorporation of the kinetic data, a percolation reactor operated under various modes, uniform temperature, temperature step change (along with or without flow rate step change), and two-stage reverse-flow operation, was modeled and investigated for its performance. The modeling results affirmed that a step-change/reverse-flow operation is advantageous for biphasic substrates, including agricultural residues. The optimum temperature difference in the step-change operation was determined to be 30°C over a wide range of reaction temperature. Temperature step change alone (without use of reverse-flow mode) increased the product yield by 3–11% (depending on the reaction conditions) over that of uniform temperature operation. The most significant improvement, however, was seen with application of a two-stage reverse-flow reactor arrangement with temperature step change employing different conditions at each stage. This operation essentially doubled the sugar concentration over that of the temperature step change operation.

Index Entries: Dilute-acid pretreatment; kinetics, modeling; percolation; corn cobs/stover mixture.

Abbreviations: A, acid concentration; C_0, initial xylan concentration in percolation; E_i, activation energy for k_i; F1,F2, fast and slow fraction of hemicellulose; H, hemicellulose concentration; k_{oi}, frequency factor for k_i; k_i, rate constant = $A^{ni},k_{oi}\exp(E_i/RT)$, min^{-1}; L, reactor length, cm; n_i, acid concentration exponent; O, oligomer concentration; R, universal gas constant; S, dimensionless concentration;

*Author to whom all correspondence and reprint requests should be addressed.

t, time, min; T, absolute temperature, °K; u, velocity inside percolation reactor; x, distance coordinate along reactor length, cm; X, xylose concentration; Y, yield; z, x/L;

Greek: α_i, k_3/k_i, I = 1, 2; β_i, $k_i L/u$; γ_i, k_4/k_i, I = 1, 2; ρ, τ_1/τ; τ, tu/L; τ_1, τ_2, dimensionless time before and after temperature shifting in step-change mode; ω, ratio of velocity at high temperature to low temperature;

Subscript: 0, value at $t = 0$; 1, fast-hydrolyzing hemicellulose; 2, slow-hydrolyzing hemicellulose; O, oligomer; T, total; X, xylose.

INTRODUCTION

Dilute-acid hydrolysis of hemicellulose is one of the major process options applicable for biomass pretreatment. In this process, it is essential to consider the recovery of hemicellulose sugars as well as the effectiveness of the pretreatment. One common problem found in the acid pretreatment is that the sugars are decomposed under high temperature and low pH conditions to form undesirable components that are toxic to subsequent fermentation. Under the circumstances, it is critically important to choose proper reaction conditions, type of reactor, and mode of reactor operation. In our previous studies along these lines *(1–4)*, we have established that the percolation reactor (packed-bed flow through type) is one of the reactor types most appropriate for biomass pretreatment. In the application of this reactor, the sugar product is discharged from the reactor as it is formed. This permits the process to attain high sugar yield by minimizing the sugar decomposition.

The scope of our recent study on this subject was extended to cover agricultural residues including corn cobs/stover mixture (CCSM). Although some of the characteristics are similar among various biomass feedstocks, we found that the kinetic behavior of the hemicelluloses in various species differ significantly. We have reported that hemicellulose in hardwoods can be modeled as a biphasic substrate *(4,5)*, i.e., it is composed of two different fractions, fast-hydrolyzing and slow-hydrolyzing ones. We also found it to be true with CCSM. However, the reactions with CCSM are much more rapid than those of hardwoods, such that much less stringent conditions are required, i.e., lower temperature and/or acid levels. Under the lower temperature/acid conditions, the reaction pattern also changes so that soluble xylose oligomers appear as a recognizable component in the overall reaction of dilute-acid hydrolysis of the hemicellulose. These observations led us to modify our reaction model to account for the different kinetic pattern, including xylose oligomers as an additional component. This kinetic model is introduced here as an improved version over that of Kim et al. *(6)* in which the oligomer formation is neglected.

This investigation was undertaken first to establish a kinetic model that accounts for xylose oligomers as an additional entity and to determine the associated reaction kinetic parameters specifically applicable for acid hydrolysis of CCSM hemicellulose based on our experimental data of the kinetic study. The experimental conditions of interest were 120–150°C, and 0.44–1.90 wt% of sulfuric acid, substantially lower than those applied to hardwoods in both accounts. The mathematical modeling of a percolation reactor operated under various modes was then developed considering the kinetic pattern as well as the newly obtained kinetic parameters of CCMS. The modeling work covered operations under uniform temperature, two-stage step change, and two-stage step change reverse flow. The latter is a process scheme originally developed at the National Renewable Energy Laboratory (NREL) *(7)*.

EXPERIMENTAL METHODS

Materials

The CCSM feedstock was supplied in the form of milled fine particles (20–60 mesh) by NREL and used without any modification. The Cellulase Enzyme, Cytolase CL (Lot No 17-92262-09) was obtained from Environmental Biotechnologies, Inc. (Santa Rosa, CA).

Batch Reactions

For the kinetic study, reactions were carried out in Pyrex™ glass tube reactors (11-mm id). The reactors were packed with 0.4 g biomass and 6 mL acid solution, and sealed at both ends under natural gas–oxygen flame. To initiate the reaction, the glass reactor ampules were placed into an oil bath (Haake FS model) in which the temperature was preadjusted to a level 50°C higher than the desired point. After 50 s, the ampules were transferred into another oil bath preset at the desired reaction temperature. The two-oil-bath procedure was done to minimize the preheating time. Temperature measurement within a reactor has shown that the center section of the reactor reached the set point in 50 s. After being subjected to specified reaction times, the reactors were quenched in a cold water bath.

Analytical Methods

Sugar analyses for solid samples were performed according to the NREL CAT Standard Procedure No. 02 using high-performance liquid chromatography (HPLC) equipped with a refraction index (RI) detector, and Bio-Rad's (Hercules, CA) Aminex HPX-87P column. The liquid samples were analyzed for sugars by HPLC, RI detector, and Bio-Rad's Aminex HPX-87H column, with mobile phase of 0.005*M* sulfuric acid. The amount of soluble xylose oligomer in the hydrolysate (from primary hydrolysis) was measured by secondary hydrolysis with Cytolase CL enzyme. The increase in xylose after secondary hydrolysis was taken as xylose oligomers. All of the kinetic data analysis was done on the basis of xylose equivalent.

MODEL DEVELOPMENT

Kinetic Model

The sharp breakage in the semilog plot for the remaining xylan (Fig. 1) confirms that the xylan in CCSM is biphasic. This finding is in agreement with the kinetics of hemicellulose hydrolysis in other herbaceous biomass feedstocks, such as switchgrass *(8)*. With the recognition of xylose oligomer as a separate entity, the overall kinetic pattern of dilute-acid treatment of hemicellulose is now described as that shown in Fig. 2. For the proposed kinetic model, the variation of each individual component can be theoretically determined by the following set of differential equations:

$$(dH_1/dt) = -k_1H_1 \tag{1}$$

$$(dH_2/dt) = -k_2H_2 \tag{2}$$

$$(dO/dt) = k_1H_1 + k_2H_2 - k_3O \tag{3}$$

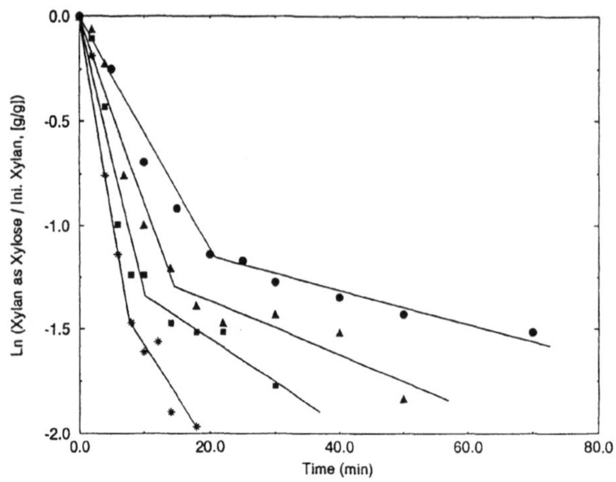

Fig. 1. Decay of xylan content in CCSM hemicellulose during hydrolysis (temperature = 140°C, solid:liquid = 1:16.4). ●, 0.47 wt% acid, ▲, 0.73 wt%, ■, 1.22 wt%, *, 1.95 wt%.

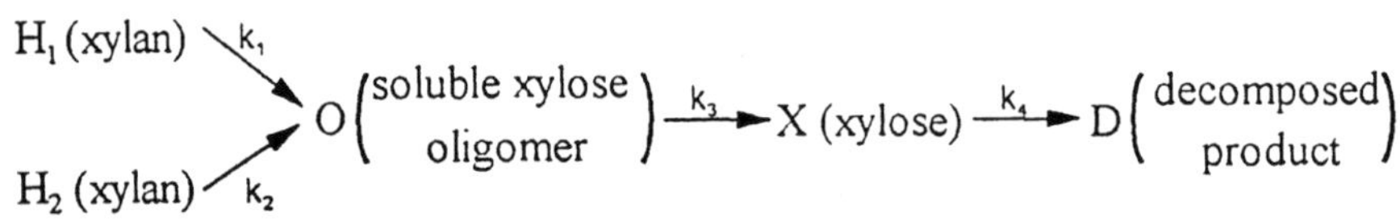

Fig. 2. Acid hydrolysis pathway of CCSM hemicellulose.

$$(dX/dt) = k_3O - k_4X \tag{4}$$

with initial conditions at $t = 0$, $H_1 = F_1 H_0$, $H_2 = F_2 H_0$, $O = 0$, $X = 0$.

The analytical solutions for eqs. *(1–4)* are obtained as follows:

$$O = \left(\frac{F_1H_0k_1}{k_3-k_1}\right)\left(e^{-k_1t} - e^{-k_3t}\right) + \left(\frac{F_2H_0k_2}{k_3-k_2}\right)\left(e^{-k_2t} - e^{-k_3t}\right) \tag{5}$$

$$\mathbf{X} = \frac{F_1H_0k_1k_3}{(k_3-k_1)(k_4-k_1)}\left(e^{-k_1t} - e^{-k_4t}\right) + \frac{F_2H_0k_2k_3}{(k_3-k_2)(k_4-k_2)}\left(e^{-k_2t} - e^{-k_4t}\right)$$

$$-\left(\frac{k_3}{k_4-k_3}\right)\left(\frac{F_1H_0k_1}{k_3-k_1} + \frac{F_2H_0k_2}{k_3-k_2}\right)\left(e^{-k_3t} - e^{-k_4t}\right) \tag{6}$$

Process Model of Percolation Reactor

The percolation reactor is a packed-bed, flow-through-type reactor, as shown in Fig. 3. In the following modeling procedure, it is assumed that the axial heat transfer after temperature step change and the internal and external diffusion effects are negligible.

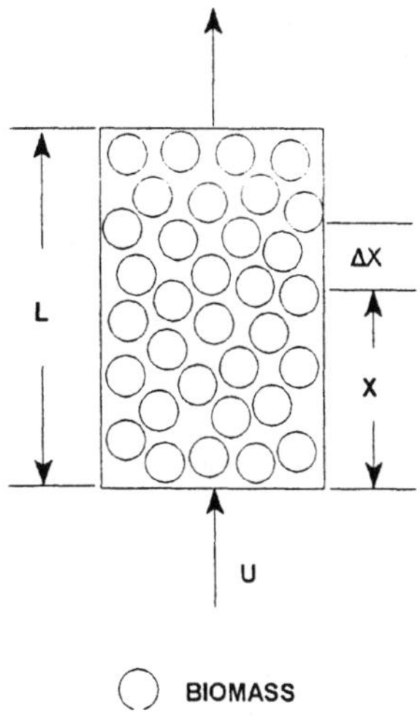

Fig. 3. Conceptual sketch of percolation reactor.

Uniform Temperature Operation

Material balance on component O (oligomers) leads to the following:

$$(\partial O/\partial \tau) = (k_1 H_1 + k_2 H_2 - k_3 O) - u[(\partial O/\partial x)] \quad (7)$$

where

$$H_1 = F_1 H_0 \exp\{-k_1[t-(x/u)]\},\ t > (x/u)$$
$$H_2 = F_1 H_0 \exp\{-k_2[t-(x/u)]\},\ t > (x/u) \quad (8)$$

The pertinent initial and boundary conditions are: $x = 0, O = 0$, and $t = 0, O = 0$. The equation regarding fast hemicellulose is:

$$(\partial O_1/\partial \tau) = (k_1 H_1 + k_3 O_1) - u(\partial O_1/\partial x) \quad (9)$$

where H_1 is shown in Eq. (8). In dimensionless form, the PDE is expressed as:

$$(\partial S_{O1}/\partial \tau) = \beta_1 \exp[-\beta_1(\tau - z)] - \alpha_1\beta_1 S_{O1} - (S_{O1}/\partial z) \quad (10)$$

with pertinent initial and boundary conditions of $t = 0, S_{O1} = 0$, and $z = 0, S_{O1} = 0$.

By the Laplace Transform method, one obtains the following solution for Eq. (10):

$$S_{O1} = (1/\alpha_1)(1 - \exp{-\alpha_1\beta_1 z})\exp{-\beta_1(\tau - z)} \quad (11)$$

The yield of oligomer from fast hemicellulose is then calculated by integrating S_{X1} value at the reactor exit point ($z = 1$) over the range of operation time $1 \sim 1 + \tau$.

$$Y_{O1} = \int_1^{\tau+1} (S_{O1})_{z=1} d\tau = [1 - \exp(-\alpha_1\beta_1)]/(\alpha_1\beta_1)\,[1 - \exp(-\beta_1\tau)] \quad (12)$$

Oligomer yield from slow hemicellulose Y_{O2} is identical to that shown in Eq. (12) with subscript value of 2 replaced by 1. The combined yield of soluble xylose oligomer from both fast and slow hemicellulose can be expressed as:

$$Y_O = F_1 Y_{O1} + F_2 Y_{O2} \quad (13)$$

Similarly, material balance for xylose from fast hemicellulose leads to the following:

$$(\partial S_{X1}/\partial \tau) = \alpha_1\beta_1 S_{O1} - \gamma_1\beta_1 S_{X1} - (\partial S_{X1}/\partial z) \tag{14}$$

where S_{O1} is given by Eq. (11).

The solution for Eq. (14), xylose from fast hemicellulose, is expressed as:

$$\begin{aligned} S_{X1} = {} & \frac{1}{\gamma_1} e^{\beta_1 z}(e^{-\beta_1\tau} - e^{-\beta_1(1+\gamma_1)\tau}) + \\ & \frac{1}{\gamma_1} e^{-\gamma_1\beta_1 z}(e^{-\beta_1(1+\gamma_1)(\tau-z)} - e^{-\beta_1(\tau-z)}) + \\ & \frac{e^{-\beta_1\tau}}{\gamma_1-\alpha_1}(e^{\beta_1(1-\gamma_1)z} - e^{\beta_1(1-\alpha_1)z}) \end{aligned} \tag{15}$$

The yield of xylose from fast hemicellulose is obtained as:

$$\begin{aligned} Y_{X1} = \int_1^{\tau+1}(S_{X1})_{z=1} = {} & \frac{1}{\gamma_1} e^{\beta_1}\left(\frac{e^{-\beta_1} - e^{-\beta_1(\tau+1)}}{\beta_1} + \frac{e^{-\beta_1(1+\gamma_1)(\tau+1)} - e^{-\beta_1(1+\gamma_1)}}{\beta_1(1+\gamma_1)}\right) + \\ & \frac{1}{\gamma_1} e^{-\gamma_1\beta_1}\left(\frac{1-e^{-\beta_1(1+\gamma_1)\tau}}{\beta_1(1+\gamma_1)} + \frac{e^{-\beta_1\tau}-1}{\beta_1}\right) + \\ & \frac{e^{\beta_1(1-\gamma_1)} - e^{\beta_1(1-\alpha_1)}}{\gamma_1-\alpha_1}\left(\frac{e^{-\beta_1} - e^{-\beta_1(\tau+1)}}{\beta_1}\right) \end{aligned} \tag{16}$$

The yield of xylose from slow hemicellulose Y_{X2} is identical to Eq. (16) with subscript value of 2 replaced by 1. The total yield of sugar including soluble xylose oligomer and xylose monomer, is expressed as:

$$Y_T\,(\text{uniform}) = F_1(Y_{O1} + Y_{X1}) + F_2(Y_{O2} + Y_{X2}) \tag{17}$$

Two-Stage Step-Change Operation

The term τ_1 represents the dimensionless reaction time up to the temperature shifting point, and τ_2 represents the rest of the reaction time. Total yield consists of four parts, namely, fast fraction reacting for the duration of τ_1, slow for τ_1, fast for τ_2, and slow for τ_2. The total yield is then expressed as:

$$\begin{aligned} Y_T(\text{step}) = {} & F_1[Y_{O1,T1} + Y_{X1,T1} + R_1(Y_{O1,T2} + Y_{X1,T2})] + \\ & F_2[Y_{O2,T1} + Y_{X2,T1} + R_1(Y_{O2,T2} + Y_{X2,T2})] \end{aligned} \tag{18}$$

where:

$$\begin{aligned} R_1 &= \exp(-\beta_1\tau_1) \\ R_2 &= \exp(-\beta_2\tau_1) \end{aligned} \tag{19}$$

The yield is now expressed as a function of reaction time and a number of dimensionless parameters including $\beta_1(=k_1L/u)$. In actual computation of yields, τ

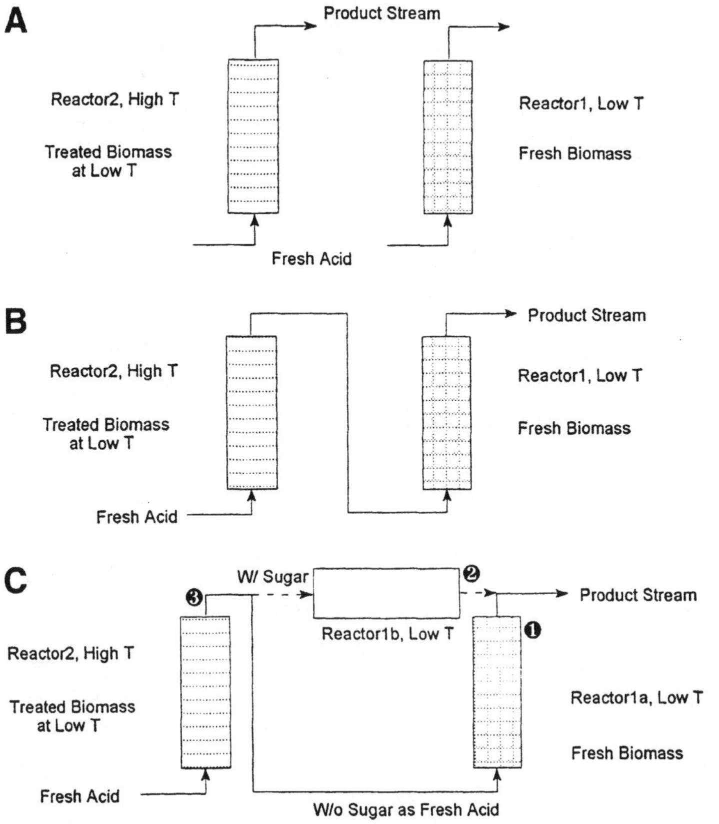

Fig. 4. Schematics of temperature step change and two stage reverse flow in modeling study. **(A)** Two-stage step change, **(B)** two-stage reverse flow, **(C)** two-stage reverse flow (artificial).

value is first set, and then the optimum β_1 is computed (to maximize the yield value) using Eq. (20) before the integration process.

$$\left(\frac{\partial Y_T}{\partial \beta_1}\right)_{\alpha,\tau} = \left(\frac{\partial Y_T}{\partial \beta_1}\right)_{\alpha,\tau,\beta_2} + \left(\frac{\partial Y_T}{\partial \beta_2}\right)_{\alpha,\tau,\beta_1}\left(\frac{\partial \beta_2}{\partial \beta_1}\right)_{\alpha,\tau} = 0 \qquad (20)$$

where $\beta_2 = (k_2/k_1)\,\beta_1$.

Two-Stage Reverse-Flow Operation

As depicted in Fig. 4C, a two-stage reverse-flow operation can be simulated as a two-stage step-change operation with a bypass reactor (Reactor 1b). The results from Reactor 1a are the same as those from the first stage reactor in two-stage step-change operation, Reactor 1 in Fig. 4A. The results from Reactor 2 are the same as those from the second-stage reactor in two-stage step-change operation, Reactor 2 in Fig. 4A. The yields from Reactor 1b can be calculated by the following equations:

Xylose yield = xylose from Reactor 2 + xylose generated
from oligomer in Reactor 1b - xylose decomposed in Reactor 1b (21)

Table 1
Activation Energy for Each Reaction

k_i	k_{0i} min^{-1} (%w/w)$^{-ni}$	n_i	E_i, kcal/g mol
1	1.998×10^{10}	1.0	20.6
2	1.237×10^{13}	1.0	27.7
3	1.046×10^{14}	1.2	27.5
4[a]	8.990×10^{11}	1.0	28.2

[a]After Kim and Lee *(5)*.

Oligomer yield = oligomer
from Reactor 2 – oligomer converted to xylose in Reactor 1b (22)

RESULTS AND DISCUSSION

The kinetic model was set to follow the pattern of parallel-serial reactions shown in Fig. 2. The kinetic information was then used in the simulation of a percolation reactor performing dilute-acid pretreatment/hemicellulose hydrolysis. For simulation purposes, a computer program was developed on the basis of theoretical results described in the previous section. The program incorporates experimentally determined kinetic data as well as the solution of modeling equations. The optimization process involved repeated calculation of yields changing the operation parameters. In the optimization part of the program, the Fortran subroutines from IMSL, Inc. (Houston, TX) were employed *(9)*.

Determination of Kinetic Parameters

Equations (8) and (9) were used to determine the kinetic parameters. For this purpose, the experimental data taken at 120, 130, 140, and 150°C were put through a nonlinear regression using the SAS NLIN Program *(10)*. The xylose decomposition data (k_4) were taken from Kim and Lee *(5)*. The regression for parameter estimation was done in two tiers. For a given set of four parameters of F_f (the fast hydrolyzable fragment) n_1, n_2, and n_3, the remaining 12 parameters (four each for k_1, k_2, and k_3 at four different temperature levels) were determined by nonlinear regression analysis. This process was repeated varying the four aforementioned parameters until the minimum of the sum of square of errors was attained. The best fitting values of F_f, n_1, n_2, and n_3 were found to be 0.65, 1.0, 1.0, and 1.2 respectively. The rest of the kinetic parameters are shown in Table 1.

The comparison between predicted reaction progress based on the kinetic model and the actual experimental data is shown in Fig. 5. The model prediction is seen to be in good agreement with the experimental data. Hence, it is confirmed that the proposed kinetic model and the associated kinetic parameters are valid.

Optimum Temperature Step Change

Kim et al. *(4)* reported that the performance in a percolation reactor can be improved by adopting a step change in operating conditions over that of uniform operating condition. The step-change temperature strategy was further refined to

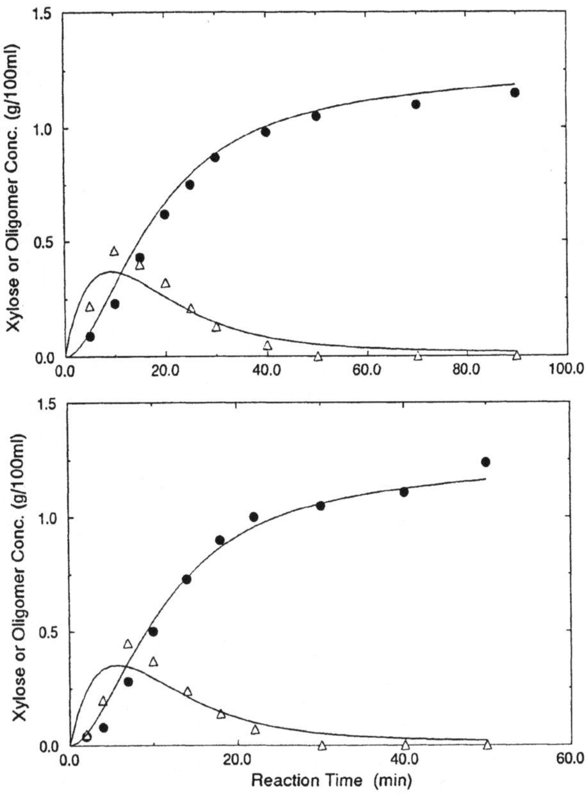

Fig. 5. Comparison between the predicted and experimental value on reaction progression in CCSM hydrolysis at 140°C, acid conc.=0.44 and 0.68% (point = experimental values, line = calculated kinetic model values). **(A)** Acid concentration, 0.44%, solid:liquid, 1:16.4. **(B)** Acid concentration, 0.68%, solid:liquid, 1:16.4. –, model predicted value; ●, xylose (experimental data); △, oligomer (experimental data).

obtain an optimum level of temperature difference. Temperature differences of 10, 20, 30, and 40°C were applied in the study with low temperatures set at 120, 130, and 140°C. The highest yield occurred with temperature differences of 30°C for low temperatures of 130 and 140°C, and 40°C for a low temperature of 120°C (Fig. 6). The temperature difference of 30°C was then taken as the optimum, ignoring the results at 120°C, a temperature unacceptable for this substrate because of low reaction rate and yield. The step-change operation brought about 8.8–10.9% improvement in yield over that of uniform low-temperature operation, and 2.9–3.5% over uniform high-temperature operation.

Optimum Flow Rate Step Change (Along with Temperature Step Change)

The preceding simulation of temperature step change was conducted with a constraint that the liquid flow rate was kept constant for the entire operation. Since β is an operational parameter related not only to temperature, but also the flow rate,

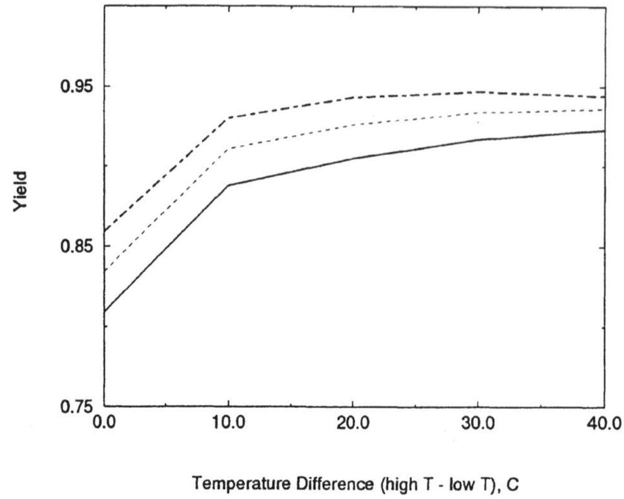

Fig. 6. Modeling results on effect of temperature difference on yield in step-change operation (acid conc. = 0.8%, $\tau = 1.5$, $\rho = 0.5$). (---) low $T = 140°C$; (—) low $T = 130°C$; (-) low $T = 120°C$.

it is probable that the β value shift (change of liquid velocity and/or temperature) may also influence the sugar yield. To verify this, the velocity ratio ω was introduced as an additional adjustable parameter in the percolation reactor simulation program. The velocity ratio ω was defined as the velocity in high-temperature operation to the velocity in the low-temperature operation. Figure 7 shows the effect of ω on product yield. It is seen that the maximum yield occurs at $\omega = 2.0$ for all three cases. The additional improvement of yield owing to ω adjustment was 0.4–0.6%.

Two-Stage Reverse-Flow Operation

Two-stage reverse-flow percolation reactor design was proven to be advantageous in obtaining high product concentration theoretically *(6)* and experimentally *(11)*. The concept of the process is shown in Fig. 4B. The biomass is first treated at a low temperature in percolation mode. It is then treated again at a high temperature. Up to this point, the procedure is identical to the two-stage process with a step change in temperature. The difference is that the stream from the high-temperature treatment is again put through a reactor packed with fresh biomass at low temperature. The reacted solid residue in this reactor is then treated with fresh acid at high temperature. This process is repeated. The difference between two-stage step-change and two-stage reverse-flow configurations is depicted in Fig. 4. For computation purpose, it is postulated that Reactor 1 in Fig. 4B is composed of two artificial reactors (superposition), namely Reactor 1a containing fresh biomass with input stream as fresh acid without sugar content and Reactor 1b containing no biomass with sugar stream input. The process composed of Reactor 1a and Reactor 2 (Fig. 4C) is in fact identical to the two-stage step-change operation shown in Fig. 4A.

There are two important points in the role of the two-stage reverse-flow reactor. One is the advantage of superposition, which can combine two sugar streams together (from Reactor 1a and Reactor 1b), thus increasing the product concentra-

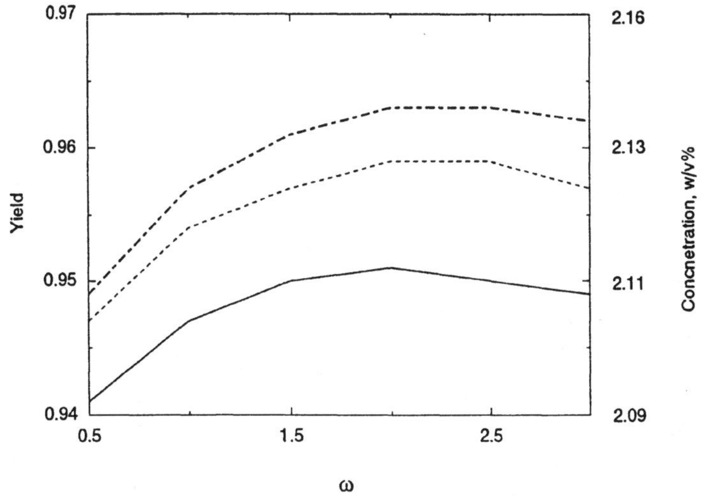

Fig. 7. Modeling results on effect of flow rate ratio on product yield (acid conc. = 0.8%, τ = 1.5, ρ = 0.5). (--) 140–170°C; (—) 130–160°C; (-), 120–150°C.

tion. Another is the disadvantage of decomposing the sugar products inside Reactor 1b. The key issue in the two-stage reverse-flow mode is that because of the low temperature applied in Reactor 1b (Reactor 1), the sugar decomposition inside Reactor 1b is in fact minimized. There are three different flow arrangements that are available for the two-stage step-change process as indicated in Fig. 8. Depending on the acid throughput applied at each stage of the process, the process stream for low-temperature processing is either supplemented with fresh acid or partially bypassed to the sugar product. In Fig. 8B, make-up fresh acid is introduced to the system. The product concentration is therefore expected to decrease. Furthermore, the bypass stream in Fig. 8C not being subjected to superposition, the product concentration is also expected to be lower than that of Fig. 8A. Thus, Fig. 8A would generally be regarded as a preferred mode of operation.

There is an important aspect of the percolation reactor that needs to be brought up in assessing its performance. In percolation reactor operation, the yield generally increases with τ, whereas the product concentration decreases with it. An inverse relationship, or a trade-off from a practical viewpoint, thus exists between the yield and the product concentration. The yield value alone without the corresponding concentration will not serve as a meaningful performance indicator.

The results of the simulation with optimized process parameters are summarized in Fig. 9. The yield values are indeed shown to follow the inverse relationship with product concentration. When the yield is compared at a fixed product concentration level, the two-stage reverse-flow mode gives highest yield, 96.5% at $\tau = 2$ and 98% at $\tau = 3$, followed by step change, uni-high, and uni-low. The advantage of reverse flow in attaining high product concentration is phenomenal, almost doubling the product concentration of step-change mode. The advantage of the superposition (the product stream coming out of one reactor being put through a second reactor) is clearly demonstrated here. The concentrations from the other three modes are in the order of step change > uni-high > uni-low. It is also noteworthy that

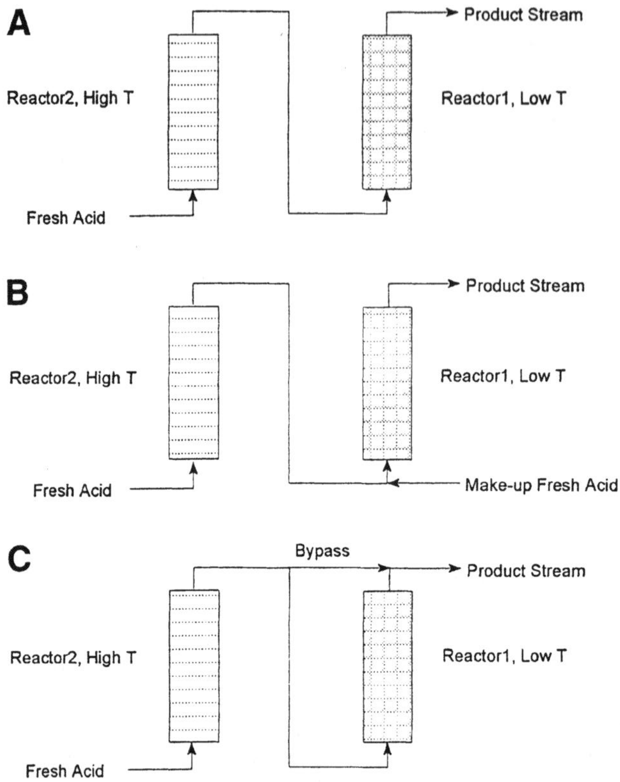

Fig. 8. Flow configuration in two-stage reverse-flow percolation reactor. **(A)** In(reactor 1) = out(reactor 2), ρ = 0.5); **(B)** in(reactor 1) = out(reactor 2) + fresh acid, $0.5 < \rho < 1.0$, **(C)** in(reactor 1) = out(reactor 2) - bypass, $0 < \rho < 0.5$.

for a given τ value (fixed amount of liquid throughput), step-change gives the highest yield, reverse flow the second highest, and uni-low the lowest yield. For the step-change mode, an optimum ρ was taken for a specific τ value. This is done so as to compare the optimum step-change mode with reverse flow. Although the reverse-flow mode gives highest product concentration, it also pays a price for it: additional sugar decomposition in Reactor 1. Sugar decomposition in Reactor 1 depends on the input sugar concentration. Yield loss in reverse flow can be as high as 2.8% at τ = 1.0 with product concentration of 6.02 %(w/v), but as low as 0.2% at τ = 3.0 with product concentration of 2.18%(w/v). Here again is a clear case of the trade-off between the product concentration and product yield. The true optimum operating condition can only be determined from consideration of overall process economics, which must include the cost of down stream processing.

Sugar Distribution in Reverse-Flow Mode

It is important to delineate the functions of the two separate reactors employed in the two-stage reverse-flow operation mode. Table 2 lists the sugar concentration at various points of the process (*see* Fig. 4C) from each fraction. Under the listed process condition, it is seen that 86.2% of fast-hydrolyzing hemicellulose, and only 19.4% of slow-hydrolyzing hemicellulose were recovered from a low-temperature

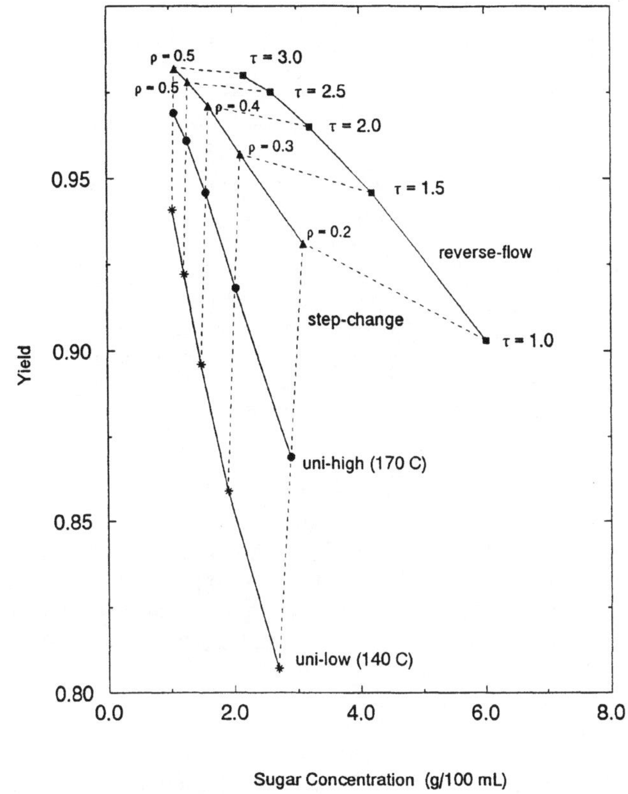

Fig. 9. Modeling results on yield vs sugar product concentration under various operation modes of percolation reactor (acid conc. = 0.8%, C_0 = 3.333 w/v%) (ρ: optimum values for step change, 0.5 for reverse flow).

reactor (Reactor 1a), whereas 67.4% of slow-hydrolyzing hemicellulose and only 12.9% of fast-hydrolyzing hemicellulose were recovered from a high-temperature reactor (Reactor 2). These results prove that the main reason for the existence of variational optimal temperature is the biphasic nature of the hemicellulose in the CCSM. The sugar stream passing through the Reactor 1b has decreased from 0.084 + 0.236 = 0.320 to 0.0006 + 0.315 = 0.3156 owing to decomposition, a mere 1.4% sugar loss. As we expected, oligomer formation was high in the low-temperature reactor (Reactor 1) and low in the high-temperature reactor (Reactor 2). This observation also agreed well with the results of the kinetic study.

CONCLUSIONS

The hemicellulose in CCSM is biphasic when it is subjected to dilute-acid hydrolysis. The biphasic kinetic model, including xylose oligomer and monomer, has shown good agreement with experimental data. On the basis of this kinetic data, a mathematical model was established, and process simulations were conducted for a percolation reactor operated under various modes. The optimum temperature difference in temperature step-change operation was found to be 30°C. The product

Table 2
Modeling Results of Product Sugar Distribution from Two-Stage Reverse-Flow Operation (low T = 140°C, high T = 170°C, τ = 1.5, and ρ = 0.5)

Location	Sugar source	Oligomer yield	Xylose yield	Total yield	Sugar recovered[a]
Reactor 1a	Fast fraction	0.179	0.381	0.560	0.862
	Slow fraction	0.022	0.046	0.068	0.194
Reactor 1b	Oligomer	0.0006	—	0.0006	—
	Xylose	—	0.315	0.315	—
Reacto 2	Fast fraction	0.003	0.081	0.084	0.129
	Slow fraction	0.009	0.227	0.236	0.674

[a]Sugar recovered = sugar received/sugar available from each fraction = total yield/F_i, where i = 1, 2.

yield in temperature step-change operation was substantially higher than that of uniform temperature operation at either limit. The application of flow rate step-change along with temperature step change brought about additional improvement of yield over the case with temperature step change alone. Use of a two-stage reverse-flow percolation reactor arrangement has brought about a significant improvement in the performance in that it has doubled the product concentration over that of two-stage step-change operation.

ACKNOWLEDGMENT

This research was supported by NREL through a Subcontract-XAW-3-13441-01.

REFERENCES

1. Lee, Y. Y., Lin, C. M., Johnson, T., and Chambers, R. P. (1978), *Biotechnol. Bioeng. Symp.* **8,** 75–88.
2. Limbaugh, M. L. (1980), MS Thesis, Auburn University, AL.
3. Cahela, D. R., Lee, Y. Y., and Chambers, R. P. (1983), *Biotechnol. Bioeng.* **25,** 3–17.
4. Kim, B. J., Lee, Y. Y., and Torget, R. (1993), *Appl. Biochem. Biotechnol.* **39,** 119–129.
5. Kim, S. B. and Lee, Y. Y. (1987), *Biotechnol. Bioeng. Symp.* **17,** 71–84.
6. Kim, B. J., Lee, Y. Y., and Torget, R. (1994), *Appl. Biochem. Biotechnol.* **45,** 113–129.
7. Torget, R. W., Hsu, T.-A., Kadam, K., and Phillippidis, G. P., and Wyman, C. E. (1993), Prehydrolysis of Lignocellulosic Materials, US Patent applied for.
8. Lee, Y. Y., Kim, B. J., and Chen, R. (1993), Final Report for Subcontract-NREL-XD-1-11121-1.
9. FORTRAN Subroutines for Mathematical Applications, User's Manual, IMSL, Inc., Houston, TX, (1991), p. 1096–1102.
10. Freund, R. and Littell, R. (1992), *SAS System for Regression,* 2nd ed., SAS Institute Inc., Cary, NC, p. 169–188.
11. Torget, R. W., Hayward, T. K., Hatzis, C., and Philippidis, G. P. (1995), presented at the 17th Symposium on Biotechnology for Fuels and Chemicals, Vail, CO.

0273-2289/96/5758—0147$7.50

Fractionation of Herbaceous Biomass by Ammonia–Hydrogen Peroxide Percolation Treatment

SUNG BAE KIM[†] AND YOON Y. LEE*

Department of Chemical Engineering, Auburn University, Auburn, AL 36849

ABSTRACT

Treatment with ammonia and hydrogen peroxide was investigated as a means of fractionating herbaceous biomass. The main feature of this process is that aqueous forms of these reagents are pumped simultaneously into a packed-bed flow-through-type reactor (percolation reactor) under a semibatch mode with ammonia being recycled. Experimental tests on corn cobs/stover mixture (CCSM) and switchgrass feedstocks have proven that a high degree of fractionation of biomass into three major components is attainable under this process scheme. The extent of delignification was 94–99% It was achieved at a representative condition of 170°C, 0.28 g loading of H_2O_2/g biomass, and 10 wt% ammonia concentration. At the same time, about 80% of total hemicellulose in the biomass was separated out into the effluent primarily in the form of xylose oligomers. Decomposition of sugar components was insignificant. The remaining solids had a composition of 80–93% glucan, 5–10% xylan, and 1–6% lignin. Selected solid samples. obtained under near-optimum conditions, exhibited a chemical composition close to that of commercial α-cellulose The enzymatic digestibilities of these solid samples were substantially higher than that of α-cellulose.

Index Entries: Biomass, herbaceous; ammonia; hydrogen peroxide; fractionation; pretreatment.

INTRODUCTION

Ammonia has a number of characteristics suitable for processing of lignocellulosic substrates. It is a proven delignification reagent *(1,2)*. It also carries other reactive properties, including hydrolysis of glucuronic acid ester crosslinks in biomass *(3)*, cleaving of the lignin–hemicellulose bonds *(4)*, and change of cellulose fiber structure *(5)*. High volatility of ammonia in comparison to water makes it easy to separate from aqueous mixture. To our knowledge, there is no harmful byproduct formation from ammonia–lignin–carbohydrate interaction at elevated temperatures. For these reasons, the use of aqueous ammonia in biomass pretreatment has been a subject of our laboratory investigation. Recently, we have reported on a novel biomass pretreatment process termed ammonia recycled percolation (ARP) process

*Author to whom all correspondence and reprint requests should be addressed.
†On leave from Gyeongsang National University, Chinju, Korea.

(6). Its main technical features are: (1) aqueous ammonia ($NH_3 \cdot H_2O$) is used as the pretreatment reagent, and (2) a packed-bed flow-through-type (percolation) reactor is employed and operated under a recirculation mode The concept of percolation process applies well to chemical processing of lignocellulosic substrates. A distinctive feature of the percolation process in comparison to a straight batch process is that the process stream is continuously fed and withdrawn from the reactor. In connection with the biomass pretreatment, it offers a unique advantage that the lignin and other extraneous components are separated from the biomass structure, preventing recondensation of lignin within the biomass. In this study, the ARP process has indeed shown a great potential as a pretreatment method in that it provided the digestibility exceeding that of α-cellulose, high delignification, and negligible sugar decomposition. One result from the study was that a substantial amount of hemicellulose was also solubilized into the effluent along with lignin. In this article, we investigate the fractionation of herbaceous biomass by adding hydrogen peroxide to the ARP process to improve the performance of hemicellulose recovery and delignification. The process is henceforth referred to as ARP-H, H standing for hydrogen peroxide. There have been a number of pretreatment studies based on hydrogen peroxide reported in literature in which a high degree of delignification and enzymatic digestibility was claimed *(7–13)*. The observation by Gould and Freer *(10)* that 80% of the hemicellulose in agricultural residues was solubilized in the alkaline peroxide solution was particularly supportive of this approach. The scope of the work covered the technical factors pertaining to the overall assessment of ARP-H as a fractionation method.

MATERIAL AND METHODS

Materials

Milled corn cobs/stover mixture (CCSM), switchgrass, and hybrid poplar were supplied from National Renewable Energy Laboratory (NREL). They were screened to the nominal size of 2 mm ~ 60 mesh. The composition of untreated CCSM was 38.1 wt% glucan, 20 wt% xylan, 3.1 wt% arabinan, 1.2 wt% galactan, 0.6 wt% mannan, 16.0 wt% klason lignin, 4.7 wt% acid soluble lignin, 6 9 wt% ash, 6.7 wt% extractives, and 2.7 wt% unaccounted for. The cellulase enzyme, Cytolase CL (Lot No. 17–92262 09), was supplied from Environmental Biotechnology, Inc. (Santa Rosa, CA). The average activities of the enzymes as determined by the supplier are: 95.9 IFPU/mL, β-glucosidase activity = 80.6 p-NPGU/mL, endoglucanase activity = 613 CMCU/mL.

Experimental Setup and Operation

A schematic diagram of the reactor setup is shown in Fig. 1. The system consists of a stock solution reservoir, pump, programmable drying oven, reactor, and liquid holding tank, which also served as a back-pressure vessel. Aqueous ammonia and hydrogen peroxide solutions were pumped simultaneously by a duplex metering pump (LDC minipump) to a packed-bed reactor through a preheating coil. The flow rate of ammonia solution was monitored by flow meter and hydrogen peroxide solution by a buret. The reactor was constructed out of SS 316 tubing to the dimension of 5/8 in. od × 6 in. L (33 cm^3 of internal volume). The reactor temperature was controlled in a temperature-programmable oven. An autoclave (600 mL, Parr Instrument, Moline, IL) was used as a liquid holding tank to which a nitrogen cylinder

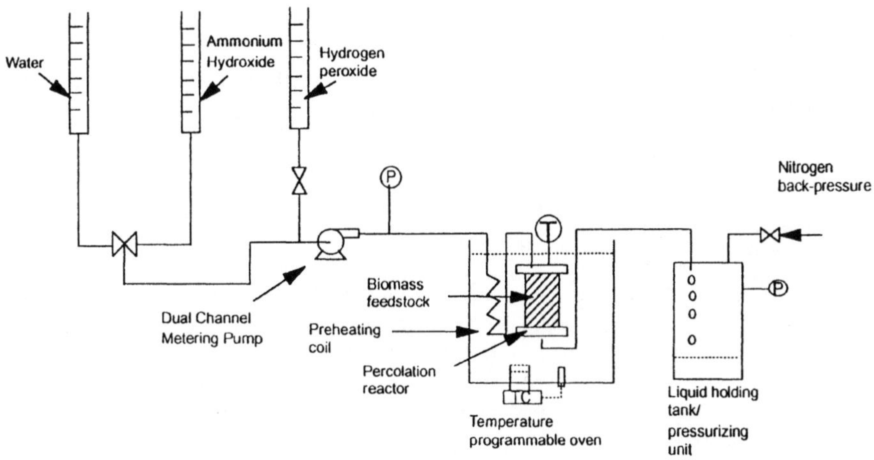

Fig. 1 . Schematics of ARP-H reactor system.

was connected to apply back pressure, preventing evaporation of reactant fluid. The biomass feed was presoaked with ammonia solution overnight. In an ARP-H experiment, 6 g of biomass sample were packed into the reactor. To carry out the reaction, oven temperature was initially set 10°C higher than the desired set point. When the reactor temperature reached within 6°C of the set point, it was reset to the original set point. This operation reduced the preheating time to 15 min. At the completion of a run, the reactor was pumped with water to remove the residual sugar and ammonia trapped in the treated biomass.

The effluent collected in the holding tank was filtered and analyzed for composition. The wet solids discharged from the reactor were separated into two portions. One was oven-dried at 105°C overnight for measurement of weight loss and further subjected to composition analysis. The other was used in the enzymatic digestibility test.

Digestibility Test

Enzymatic hydrolysis of pretreated substrate was performed at 50°C and pH 4.8 (0.05*M* sodium citrate buffer) on a shaker bath agitated at 150 rpm. The cellulase enzyme loading was 60 IFPU/g glucan. The initial glucan concentration was 1% (w/v). Samples were taken periodically and analyzed for glucose and cellobiose content using high-performance liquid chromatography (HPLC). Total glucose plus cellobiose content after 72 h of hydrolysis was taken for calculation of the enzymatic digestibility.

Analytical Methods

The solid biomass samples were analyzed for sugars, klason lignin, acid-soluble lignin, and ash, and extractives of solid samples were analyzed following the procedures described in NREL Chemical Analysis & Testing Standard Procedure (No 002 and 010). All experiments were done in duplicate. The hydrolysates from the ARP-H process were boiled until all free ammonia was evaporated. Since the hydrolysates contained oligomers, a secondary hydrolysis was carried out with 4 wt% sulfuric acid at 121°C for 1 h. Sugars and decomposition products were measured by HPLC using Bio-Rad (Hercules, CA), Aminex HPX-87H column. Since Aminex HPX-87H column does not resolve xylose, mannose, and galactose, the combined value of xylan + mannan + galactan is used in part of this article, expressed as XMG.

RESULTS AND DISCUSSION

Effect of Hydrogen Peroxide on Delignification and Hemicellulose Recovery

The reaction of hydrogen peroxide with lignin goes through a rather complex mechanism *(14,15)*. Nonetheless, the net reaction can be expressed as the reaction between lignin and two lignin-oxidizing species, hydroxy radical (·OH) and perhydroxy anion (HO_2^-) formed during the decomposition of H_2O_2. The perhydroxyl anion is a mild oxidant that is not strong enough for prompting delignification, whereas the hydroxyl radical is a powerful lignin oxidant *(16)*. The latter also nonspecifically attacks carbohydrates resulting in depolymerization of carbohydrates *(16,17)*. The formation of both species can thus be expressed by the following two reactions:

$$H_2O_2 = H^+ + HOO^-, \qquad H_2O_2 + HOO^- = \cdot OH + O_2^{-}\cdot + H_2O$$

These reactions are strongly pH-dependent. Alkaline condition promotes formation of perhydroxyl anion by neutralizing the H^+ ions, whereas the perhydroxyl anion accelerates formation of hydroxyl radical through the second reaction. Table 1 shows various reaction schemes for ARP-H treatment of CCSM under a fixed level of H_2O_2 loading. The control reaction, No. 1, was conducted at the reaction condition of 170°C, 10 wt% ammonia without H_2O_2. In this run, about 60% hemicellulose removal and 85% delignification was achieved, an indication that the ARP treatment alone is quite effective in the delignification of CCSM. To improve delignification further, only ammonia solution was pumped initially, and then H_2O_2 was added into the ammonia stream to remove the residual lignin (run nos. 2–4). This scheme has improved the fractionation to achieve 77% hemicellulose removal, a phenomenal 98% delignification. From run 2–4, it seems clear that ammonia concentration is an important variable in hemicellulose recovery. In runs 5–7, H_2O_2 was added from the beginning of the experiment. This has increased hemicellulose recovery to 81% and delignification close to 100%. It is to be noted that the solubilization of cellulose portion increased slightly with increase in hemicellulose recovery. In view of the fact that commercial α-cellulose contains a considerable amount of xylan (about 6%), it seem to be extremely difficult to fractionate all of the xylan content selectively. We consider 80% of hemicellulose recovery as an upper limit of fractionation in ARP-H. The extent of fractionation achieved in this method is substantially higher than those reported for acid hydrolysis and two-stage organosol treatment of corn stover *(18)*, as well as those from similar studies on agricultural residues using H_2O_2. The degree of fractionation in these studies was reported to be 40–60% in delignification and 40–80% in hemicellulose removal *(7,11–13)*.

Effect of Reaction Parameters on Hemicellulose Recovery and Delignification

The temperature effect on ARP-H was investigated over the range of 165–180°C. The results are summarized in Table 2. The isolated effect of temperature was

Table 1
Effect of Ammonia Stream Concentration and Hydrogen Peroxide Feeding Mode on ARP-H Treatment of CCSM[a]

No.	reaction time, min	Ammonia stream conc., wt%	Duration of hydrogen peroxide input		Glucan, %	xmg,[b] %	Delignification, %	Solid remaining, %
1[d]	90	10	No addition	L[c]	2.6	13.1		
				S	34.9	8.6	85.1	49.2
2	90	10	60–90	L	3.0	13.8		
				S	33.5	7.6	96.2	40.6
3	90	15	60–90	L	3.4	15.3		
				S	32.3	6.3	95.3	39.9
4	90	20	60–90	L	3.6	16.7		
				S	31.8	5.1	98.2	38.6
5	60	15	0–90	L	3.5	15.7		
				S	32.2	5.6	97.9	37.5
6	60	20	0–90	L	3.9	16.9		
				S	32.1	5.0	98.6	38.5
7	90	20	0–90	L	4.0	17.6		
				S	30.7	4.3	98.3	37.1

[a] All sugar contents are based on oven-dried untreated biomass, and expressed as glucan, xylan, mannan, and galactan equivalents. Reaction condition: 170°C, hydrogen peroxide loading 0.84 g/g biomass, flow rate of ammonia and hydrogen peroxide stream 1 mL/min.
[b] xylan + mannan + galactan.
[c] L refers to hydrolysate after acid secondary hydrolysis, and S refers to the solid residue.
[d] flow rate of ammonia solution = 2 mL/min.

Table 2
Effect of Temperature on the Compositions of CCSM Hydrolysate and Solid Residue[a]

Temperature, °C		Glucan, %	XMG,[c] %	Delignification, %	Solid remaining, %
165	L[b]	4.0	17.1		
	S	32.4	4.8	98.6	38.9
170	L	4.0	17.6		
	S	30.7	4.3	98.3	37.1
175	L	4.0	17.4		
	S	30.8	4.2	98.7	35.4
180	L	4.1	17.3		
	S	30.5	4.4	98.6	34.8

[a] All sugar contents are based on oven-dried untreated biomass, and expressed as glucan, xylan, mannan, and arabinan equivalents. Reaction condition: 20 wt% ammonia, 90 min. (the hydrogen peroxide loading 0.84 9/9 dry biomass, the flow rate of ammonia and hydrogen peroxide 1 mL/min.

[b] L refers to hydrolysate after acid secondary hydrolysis, and S refers to the solid residue.

[c] xylan + mannan + galactan.

Table 3
Effect of Reaction Time on the Compositions of CCSM Hydrolysate and Solid Residue[a]

Reaction time, min		Glucan, %	XMG,[c] %	Delignification, %	Solid remaining, %
30	L[b]	3.5	15.2		
	S	33.7	6.6	96.8	42.2
60	L	3.8	16.8		
	S	31.8	4.8	98.0	38.5
90	L	4.0	17.6		
	S	30.7	4.3	98.3	37.1

[a] All sugar contents are based on oven-dried untreated biomass, and expressed as glucan, xylan, mannan, and arabinan equivalents. Reaction condition: 170°C, 20 wt% ammonia, 90 min, the hydrogen peroxide loading 0.84 9/9 dry biomass, the flow rate of ammonia and hydrogen peroxide 1 mL/min.

[b] L refers to hydrolysate afler acid secondary hydrolysis, and S refers to the solid residue.

[c] xylan + mannan + galactan.

such that the amount of solubilization of cellulose and delignification was insensitive to temperature over this range, remaining relatively constant at 4 and 98%, respectively. The hemicellulose recovery increased slightly as temperature was raised from 165 to 170°C. Further increase in temperature did not result in any improvement in hemicellulose recovery. For this reason, a reaction temperature of 170°C was chosen in most of the subsequent experiments.

In order to reduce total liquid input in the ARP-H, the reaction time was reduced from 90 min to 30 and 60 min. The results are shown in Table 3. The degree of delignification with lower reaction times was almost at the same level as that of 90 min. The hemicellulose recovery, however, has decreased from 81 to 70% (30 min) and 77%

Table 4
Effect of Hydrogen Peroxide Input on the Compositions of CCSM Hydrolysate and Solid Residue[a]

Hydrogen peroxide loading, g/g biomass		Glucan, %	XMG,[c] %	Delignification, %
1.12	L[b]	4.4	17.9	
	S	29.3	3.9	98.2
0.84	L	4.0	17.6	
	S	30.7	4.3	98.3
0.56	L	4.2	17.3	
	S	29.6	4.3	98.8
0.28	L	4.2	17.4	
	S	29.7	4.5	99.1
0.14	L	3.5	16.2	
	S	32.9	5.7	97.8
0.06	L	2.7	13.5	
	S	34.9	8.1	95.9

[a] All sugar contents are based on oven-dried untreated biomass, and expressed as glucan, xylan, mannan, and arabinan equivalents. Reaction condition: 170°C, 20 wt% ammonia, 90 min, the flow rate of ammonia, and hydrogen peroxide 1 mL/min.
[b] L refers to hydrolysate after acid secondary hydrolysis, and S refers to the solid residue.
[c] xylan + mannan + galactan.

(60 min). For a process in which delignification is the primary purpose, it seems logical to choose a reaction time of 30 min. However, if the fractionation aims at maximum recovery of each biomass component, the reaction time should be above 90 min.

Increase of H_2O_2 from 0.84 to 1.12 g/g caused a slight increase in solubilization of hemicellulose and cellulose, but no significant change in the delignification (Table 4). A decrease of H_2O_2 from 0.84 to 0.28 g/g again showed little effect on delignification and only a slight decrease in hemicellulose recovery. A further decrease of H_2O_2 down to 0.14 caused a slight decrease in delignification, but rather significant reduction in hemicellulose recovery. These results collectively indicate that 0.28 H_2O_2 g/g biomass is the minimum loading for maximum delignification and hemicellulose recovery within the conditions applied in this study.

Effect of Flow Pattern on Hemicellulose Recovery

The data in Table 4 confirms that a near-complete delignification is achievable by way of ARP-H treatment. Attaining hemicellulose recovery of above 80%, however, seems to be a difficult task, unless an extremely high loading H_2O_2 is applied. In our attempt to improve the ARP-H further, we have looked into the possibility of reducing the liquid throughput. It is a way of saving energy input and increasing the concentration of hemicellulose in the ARP-H effluent. The effect of flow rate and flow pattern on the ARP-H performance was therefore studied. As the flow rate, and thus the cumulative liquid throughput, was reduced to half, most of the performance indices remained unchanged, although the hemicellulose recovery decreased from 81 to 75% (Table 5). The gain here is the increase in the concentration level of hemicellulose sugars in the ARP-H effluent and, more importantly, the reduction in the cost in the downstream processing.

Table 5
Effect of Ammonia/Hydrogen Peroxide Delivery Flow Pattern on the Compositions of CCSM Hydrolysate and Solid Residue Under Fixed Total Hydrogen Peroxide Input[a]

Flow rate, mL/min						
Ammonia	Hydrogen peroxide		Glucan, %	XMG[c] %	Delignification, %	Lignin recovery, %
1	1	L[b]	4.0	17.6		
		S	30.7	4.3	98.3	59.8
1	0.5	L	3.8	16.9		
		S	30.5	4.8	97.5	—
0.5	0.5	L	3.8	16.4		
		S	31.4	4.8	98.2	61.8
1	1[d]	L	3.5	16.2		
		S	32.9	5.7	97.8	68

[a] All sugar contents are based on oven-dried untreated biomass, and expressed as glucan, xylan, mannan, and arabinan equivalents. Reaction condition: 170°C, 20 wt% ammonia, 90 min, the hydrogen peroxide loading 0.28 g/g biomass.
[b] L refers to hydrolysate after acid secondary hydrolysis, and S refers to the solid residue.
[c] Xylan + mannan + galactan.
[d] The hydrogen peroxide loading 0.14 g/g biomass.

Lignin Separation from Spent Liquid

Followup experiments were conducted to see how much of the lignin generated from the ARP-H could actually be separated from spent liquid. The hydrolysate collected from ARP-H reactor was evaporated until the pH dropped to 7.0 from the initial value of 11.5. Part of the lignin precipitates at this point. As the effluent was subjected to a secondary hydrolysis, lignin was further precipitated. The lignin precipitate was filtered, washed, and dried overnight at 105°C. The lignin content thus determined is summarized in Table 5. As H_2O_2 loading was doubled from 0.14 to 0.28 g/g, the lignin precipitated (based on the total lignin in the untreated biomass) decreased from 68 to 60%. The decrease in lignin recovery can be attributed to the fact that higher H_2O_2 loading increases the low-mol-wt fraction in the ARP-H effluent, which is not precipitated in acidic solution *(9,19,20)*. According to Gould and Freer *(10)*, these lignin degradation products are not toxic in either enzymatic hydrolysis or in subsequent fermentation.

Application of ARP-H Treatment to Switchgrass and Hybrid Poplar

The performance of the ARP-H was further tested using additional substrates of switchgrass and hybrid poplar. For these substratest the ARP-H runs were conducted at 170°C, for 90 min and 0.28 g H_2O_2/g biomass. The performance indices are depicted in Fig. 2. The hemicellulose recovery obtained from switchgrass was almost the same as that of CCSM. The degree of delignification, however, was lower. The ARP-H treatment was less effective on the woody substrate on both accounts, hybrid poplar yielding 50% hemicellulose removal and 80% delignification. Interestingly, the cellulose solubilized in the ARP-H effluent of hybrid poplar was only about one-fifth of the herbaceous substrates. This may be

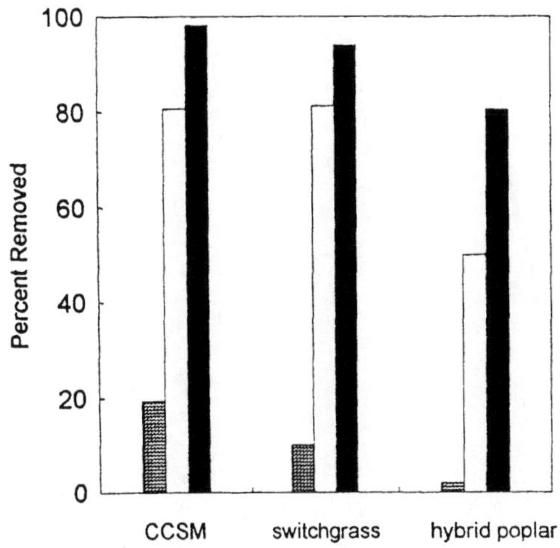

Fig. 2. Sugars and lignin removed from biomass after ARP-H treatment of CCSM, switchgrass, and hybrid poplar. Reaction conditions: 170°C, 20 wt% ammonia, 90 min, the H_2O_2 loading 0.28 g/g biomass, the flow rate of ammonia stream and H_2O_2 stream = 1 mL/min. ▦, glucan; □, xmg; ■, lignin.

explained by the fact that wood cellulose in general has a more rigid structure, higher crystallinity, and higher molecular weight than cellulose in herbaceous feedstock.

Enzymatic Digestibility

The enzymatic digestibilities of ARP-H-treated CCSM, switchgrass, hybrid poplar, untreated filter article (Whatman No. 1), α-cellulose and CCSM are shown in Fig. 3. The 72-h digestibilities of ARP-H-treated CCSM and switchgrass were 95 and 93%, respectively. These values were much higher than that of α-cellulose, which stands at 83% and is essentially in the same range of filter article (94%). The digestibility of hybrid poplar was 90%. The digestibility of the treated CCSM represents a 3- to 4-fold increase over the untreated one. The 24-h digestibility, 84%, of CCSM shown in Fig. 3 was much higher than that previously reported for prehydrolyzed and solvent extracted corn stover (20–37%) *(18)*.

Other Pertinent Comments

Although the process economics is beyond the scope of this work, we bring up a few relevant points. The use of hydrogen peroxide is undoubtedly a major cost factor in this process. From ammonia material balance based on Kjeldahl nitrogen analysis, it was found that the recovery factor of ammonia is 99.2%. The actual consumption of ammonia was 0.02 g NH_3/g dry biomass, a rather insignificant cost factor. The ammonia consumption was owing to reaction with the acetic acid/ acetates formed during the ARP-H process. The lignin generated in this process is sulfur- and sodium-free, unlike the ones generated from pulping processes. It should receive due attention as such. A point of caution or limitation of this process is to be mentioned. It has been reported that the decomposition products of hydrogen

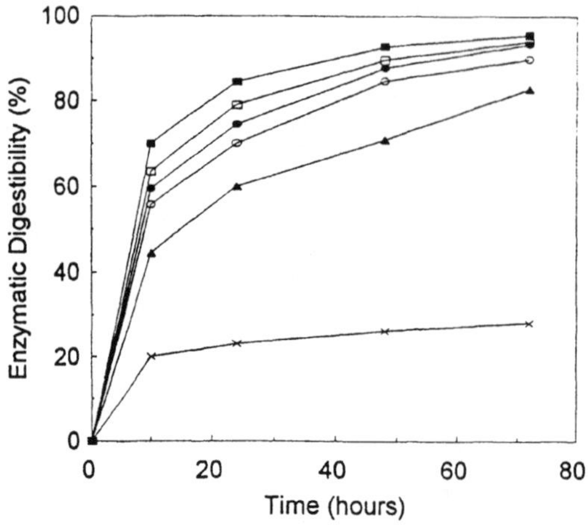

Fig. 3. Enzymatic digestibility of ARP-H treated and untreated CCSM, switchgrass, and hybrid poplar. Reaction conditions: 170°C, 20 wt% ammonia 90 min, the H_2O_2 loading 0.28 g/g biomass, the flow rate of ammonia stream and H_2O_2 stream = 1 mL/min. Enzymatic hydrolysis condition: 60 IFPU/g glucan, pH4.8, 50°C. ■, CCSM; □, switchgrass; ●, untreated filter paper; ○, hybrid poplar; ▲, alpha-cellulose; x, untreated CCSM.

peroxide in an alkaline solution not only act as lignin oxidants, but also attack carbohydrates, resulting in depolymerization and decomposition of carbohydrates *(16,17)*. We have paid particular attention to this matter and found that the significant decomposition of glucan occurs at temperatures above 170°C, a region that can be avoided for processing of herbaceous biomass.

REFERENCES

1. Jangalgi, N. R. (1983), *Chem. Abstr.* **98(14),** 109,117t.
2. Dale, B. E. and Moreira, M. J. (1982), *Biotechnol. Bioeng. Symp.* **12,** 31–43.
3. Han, Y. W. (1978), *Adv. Appl. Microbiol.* **23,** 119.
4. Wang, P. Y., Bolker, H. I., and Purves, C. B. (1967), *Tappi* **50,** 123,124.
5. Lewin, M. and Roldan, L. G. (1971), *J. Poly. Sci.* **36,** 213.
6. Yoon, H. H., Wu, Z. W., and Lee, Y. Y. (1994), *Appl. Biochem. Biotechnol.* **51/52,** 5–19.
7. Gould, J. M. (1984), *Biotechnol. Bioeng.* **26,** 46–52.
8. Holtzapple, M. T., Joseph, E. L., Sturgis, R., Lewis, J. E., and Dale, B. E. (1992), *Appl. Biochem. Biotechnol.* **34,** 5–21.
9. Abbott, T. and Peterson, R. (1985), *Biotechnol. Bioeng.* **27,** 1073–1076.
10. Gould, J. M. and Freer, S. N. (1984), *Biotechnol. Bioeng.* **26,** 628–631.
11. Gould, J. M. (1985), *Biotechnol. Bioeng.* **27,** 225–231.
12. Wei, C. and Cheng C. (1985), *Biotechnol. Bioeng.* **27,** 1418–1426.
13. Cunningham, R. L. and Carr, M. E. (1984), *Biotech. Bioeng. Symp.* **14,** 95.
14. Gierer, J. (1981), *Proc. Int. Symp. Wood and Pulping Chem.*, The Ekman-Days, Stockholm, vol. 2, p. 12.
15. Dencc, C. W. (1980), *Chemistry of Delignification with Oxygen, Ozone, and Peroxides,* Uni Publishers, Tokyo, Japan, p. 199.
16. Sjostrom, E. (1983), *Wood Chemistry—Fundamentals and Applications,* 2nd ed., Academic, San Diego, CA.
17. Samuelson, O. (1981), *Proc. Int. Symp. Wood and Pulping Chem.*, The Ekman-Days, Stockholm, vol. 2, p. 78.
18. Lee, Y., Robinson, C. W., and Moo-Young, M. (1987), *Biotechnol. Bioeng.* **29,** 572–581.
19. Lachenal, D., de Choudens, C., and Monzie, P. (1980), *Tappi* **63,** 119.
20. Bailey, C. W. and Dence, C. W. (1975), *Tappi* **58,** 104.

0273-2289/96/5758—0157$8.50

Conversion of Lignocellulosics Pretreated with Liquid Hot Water to Ethanol

G. PETER VAN WALSUM,[1] STEPHEN G. ALLEN,[2] MARK J. SPENCER,[2] MARK S. LASER,[1] MICHAEL J. ANTAL, JR.,[2] AND LEE R. LYND*[,1]

[1]*Thayer School of Engineering, Dartmouth College, Hanover, NH 03755; and* [2]*Hawaii Natural Energy Institute, University of Hawaii at Manoa, Honolulu, HI*

ABSTRACT

Lignocellulosic materials pretreated using liquid hot water (LHW) (220°C, 5 MPa, 120 s) were fermented to ethanol by batch simultaneous saccharification and fermentation (SSF) using *Saccharomyces cerevisiae* in the presence of *Trichoderma reesei* cellulase. SSF of sugarcane bagasse (as received), aspen chips (smallest dimension 3 mm), and mixed hardwood flour (–60 +70 mesh) resulted in 90% conversion to ethanol in 2–5 d at enzyme loadings of 15–30 FPU/g. In most cases, 90% of the final conversion was achieved within 75 h of inoculation. Comminution of the pretreated substrates did not affect the conversion to ethanol. The hydrolysate produced from the LHW pretreatment showed slight inhibition of batch growth of *S. cerevisiae*. Solids pretreated at a concentration of 100 g/L were as reactive as those pretreated at a lower concentration, provided that the temperature was maintained at 220°C.

Index Entries: Liquid hot water; pretreatment; SSF; inhibition; particle size reduction.

INTRODUCTION

In order to harness biological processes to convert plant biomass into fuels on a scale large enough to displace conventional fuels significantly, lignocellulosic materials must be utilized *(1)*. Conversion of naturally occuring lignocellulosic materials to ethanol requires a pretreatment of some kind *(2)*: that is, processing to make biomass sufficiently accessible and reactive to allow high rates and yields on enzymatic hydrolysis. Unfortunately, pretreatment is one of the most expensive *(3,4)*, and poorly understood *(2,5–7)* unit operations in the conversion of biomass to ethanol.

Determinants of Pretreatment Efficacy

From an applied perspective, the principal determinants of pretreatment efficacy are the degree of fiber reactivity, recovery of pentosans, hydrolysate

*Author to whom all correspondence and reprint requests should be addressed.

inhibition of microbial growth, extent of size reduction required, cost of associated equipment and utilities, and waste production. These metrics are discussed in the following.

Fiber Reactivity

Increased fiber reactivity is accomplished through a variety of mechanisms: solubilization of hemicellulose *(5)*; removal of lignin *(8)*; reduction of particle size *(2)*; and alteration of the cellulose characteristics, such as degree of polymerization, abundance of cellulose chain ends *(2)*, and crystallinity *(9,10)*. Effective pretreatments in general approach or exceed 80% of theoretical cellulose conversion on subsequent hydrolysis of a representative hardwood feedstock (e.g., poplar) using moderate (e.g., 10–15 FPU/g cellulose) cellulase loadings. Such conversions are achieved in a period on the order of 5 d, although this is highly feedstock dependent *(1)*.

Pentosan Recovery

For the production of a commodity fuel such as ethanol, product yield has a strong influence on process economics *(11,12)*. With the increasing availability of organisms capable of converting pentoses to ethanol *(13–16)*, high recovery of pentoses is a key feature of leading pretreatment technologies. A recovery of 80% of theoretical is representative of dilute-acid hydrolysis. *(17)*.

Extent of Hydrolysate Inhibition

The organisms used for the fermentation of ethanol are often inhibited by the degradation products produced during pretreatment. Inhibitory compounds originate from *(6)* the hydrolysis of extractive components, organic and sugar acids esterified to hemicellulose (e.g., acetic, formic, glucuronic, galacturonic), and solubilized phenolic lignin derivatives; the degradation of solubilized compounds (e.g., furfural, hydroxymethyl furfural); and the release of corrosion products (e.g., metal ions). The production of inhibitors has been documented for dilute-acid *(6)*, steam-explosion *(18,19)*, and acid-hydrolysis *(20,21)* pretreatments.

Extent of Size Reduction Required

Different pretreatment processes differ widely in the extent of size reduction required. Costs and energy requirements for particle size reduction increase geometrically with decreasing particle size and can be very significant *(5)*. For dilute-acid pretreatment, grinding down to a particle size on the order of 1 mm is required *(17)*. This can account for one-third of the power requirements of the entire process *(17)*.

Cost Associated with Equipment and Waste Production

Reactor size is determined by residence time and solids concentration. High reactivity and solids concentration reduce the equipment size and, thus, the investment necessary. Materials requirements are primarily a function of the corrosivity of the process conditions and, secondarily, of the pressure at which the process operates. Process residues arise principally from neutralization of acids, e.g., formation of gypsum when limestone is used to neutralize sulfuric acid. Such residues are largely, if not entirely inert, but they do require disposal, most likely via landfill.

Pretreatment Processes

The most thoroughly characterized pretreatment process options include dilute-acid pretreatment *(22–25)*, steam explosion *(26–31)* including acid-catalyzed steam explosion *(32,33)*, ammonia fiber explosion *(34–36)*, and treatment with organic solvents *(37,38)*. Additional processes that have been proposed, but have received less attention to date include use of milling *(39)*, supercritical fluids *(5)*, irradiation *(40)*, biological delignification *(10)*, oxidizing agents *(23)*, alkali *(41)*, and liquid hot water (LHW) (*see* "LHW Pretreatment" below).

Both steam-explosion and dilute-acid pretreatments have been the object of over a decade of research and development, much of which has been specifically targeted at fuel production from biomass. Steam explosion has been studied on a pilot scale, and a pilot plant for dilute-acid pretreatment has recently been constructed *(42)*. Steam explosion offers the advantage of less corrosive operating conditions. The chief advantage of dilute-acid hydrolysis relative to steam explosion is higher recoveries of hemicellulose sugars. Specifically, a pentosan recovery following hardwood pretreatment of 80% is representative of dilute-acid pretreatment *(20)*, whereas we know of no study on steam explosion reporting recoveries higher than that of 65% observed by Heitz et al. *(28)*.

In the context of fuel production, cost estimates for ammonia fiber explosion (AFEX) *(36)* and catalytic pretreatment with organic solvents (organosolv), (R. Katzen, personal communication) support the view that in the absence of future breakthroughs, these two processes are somewhat more costly than dilute-acid pretreatment. Milling and other purely mechanical pretreatment processes are considered not to be cost-competitive and are not under active consideration to our knowledge.

LHW Pretreatment

A much smaller body of literature exists with respect to the use of LHW to fractionate biomass, with but a portion of this literature directed toward pretreatment. Table 1 summarizes results for biomass fractionation using LHW. In general, this approach results in very high and often complete solubilization of hemicellulose, significant solubilization of both lignin and overall biomass, and rather low solubilization of cellulose. Recent work by Antal and coworkers *(43,49)* is differentiated from previous work by pentosan recovery that is higher than reported values and, in most cases, by the use of much shorter reaction times. Apart from the authors, only two of the groups represented in Table 1 have examined the use of LHW as a pretreatment: those of Ladisch and Bobleter *(10,50,51)*. Using a 30-min pretreatment reaction time, Hormeyer et al. obtained enzymatic hydrolysis yields of about 40% theoretical from poplar and straw *(50)*, with significant hydrolysis of the same materials also achieved when mediated by *Clostridium thermocellum* in a direct microbial conversion (DMC) system *(51)*. Enzymatic hydrolysis of hemicellulose from beech bark and bagasse using various xylanases was also reported *(47)*. Kohlmann et al. *(10)* employed a reaction time of 50–60 min in the presence of added caustic to pretreat rapeseed stems and soybean hulls, with near-theoretical yields reported on subsequent enzymatic hydrolysis.

Based on the results just summarized, the thermochemical aspects of LHW accord it outstanding potential as a pretreatment process. Relative to steam explosion, LHW pretreatment offers the potential advantage of high pentosan recovery.

Table 1
Solublization Profiles for Aqueous Fractionation of Biomass[a]

System type	Feedstock	Pretreatment conditions: T, oC	t, min	Log Ro[b]	Solubilized material, wt.%: Biomass	Lignin	Cellulose	Hemicellulose	% Pentosan recovery	Ref.
Batch	10 Different	200–230	1-15		40–60	35–60	4–22	100	90	*(43)*
Batch	Aspen	230	2	4.1	47	37	5	100	88	*(43)*
Batch	Cane bagasse	230	2	4.1	41	38	5	100	99	*(43)*
Short percolation	Cane bagasse	220	2	3.8	43	65	7	98	84	([c])
Short flow	Aspen	235	2	4.3	32	<15	6	>90	67	*(44)*
Long percolation	Cane bagasse	195	65	4.6	50	68	nd	100	nd	*(45)*
Long percolation	Wheat straw	195	25	4.2	50	<60	nd	100	nd	*(46)*
Long batch	Aspen	170	120	4.1	29	25	21	83	nd	*(47)*
Long batch	Rapeseed stems	180	50–60	4.1	nd	nd	nd	nd	nd	*(10)*
Long batch	Soybean hull	180	50–60	4.1	nd	nd	nd	nd	nd	*(10)*
Steam explosion	Aspen	230	2	4.1	27	<10	<10	90	55	*(21)*
Steam explosion	Cane bagasse	210	2	3.5	28	25	0	85	62	*(28)*
Steam explosion	Wheat straw	230	2	4.1	54	104	12	88	37	*(29)*

[a] nd, Not determined.
[b] Log Ro refers to the severity parameter defined by Overend and Chornet *(48)*.
[c] R. Katzer (personal communication).

Relative to dilute-acid pretreatment, LHW offers advantages with respect to materials of construction and waste production. For the production of ethanol, some of the largest cost impacts of the choice of pretreatment technology are manifested in the biologically mediated process operations downstream of the pretreatment step itself *(4)*. The objective of this study was to determine the suitability of lignocellulosic materials pretreated by the process developed by Antal et al. *(43,49)* as substrates for the biologically mediated processes of enzymatic hydrolysis and fermentation in a simultaneous saccharification and fermentation (SSF) system. The effects of enzyme loading, post-pretreatment size reduction, pretreatment solids concentration, and inhibition on the conversion to ethanol were evaluated.

MATERIALS AND METHODS

Feed Materials

Fresh sugar-cane bagasse was obtained from Hawaiian Commercial & Sugar Company (Puunene, HI). The bagasse was stored at <0°C and screened (+14 mesh retained) prior to use. Aspen chips (smallest dimension 3 mm) were supplied by the US Forest Products Laboratory (Madison, WI) and also stored at <0°C. Mixed hardwood flour was stored at ambient conditions and screened (+70 mesh retained) prior to use.

LHW Pretreatment

The lignocellulosic feed materials were pretreated with LHW at the University of Hawaii using a custom-built 250-mL immersed percolation reactor *(43)*. Pretreatment was typically a single-stage processing of 10–15 g (oven dry basis) of feed material at 220°C for 120 s. In the case of aspen, three such stages were employed consecutively. The reaction pressure (5 MPa) was always high enough to prevent formation of a vapor phase, resulting in an immersed percolation. The ability to displace either hot or cold liquid water into the preheated reactor also resulted in rapid heating/cooling. A detailed description of the LHW pretreatment process is given elsewhere *(43)*.

Pretreatments were conducted at both low and high solids concentrations, with solids concentration defined as the dry mass of feed material pretreated per unit volume of liquid product. Subsequent flushing of the reactor with cold water cooled the lignocellulosic residue and removed any residual solubilized material.

Two types of operating cycles were used, batch and continuous. During a batch cycle, the displacement of hot water into the reactor was stopped on filling of this vessel. In a continuous cycle, flow was maintained at 0.5 L/min throughout the reaction time. For this study, the continuous cycle was used for the aspen and low concentration (10 g/L) bagasse. The batch cycle was used for the high solids concentration pretreatment and for the hardwood flour.

Fermentation

Conversion to ethanol was done by simultaneous SSF carried out in 250-mL serum vials filled to a working volume of 75–100 mL. The vials were incubated at 37°C and agitated at 100 rpm on an orbital shaker. Growth was allowed to continue to completion before the experiments were terminated. *Trichoderma reesei* cellulase (Genencor Cytolase) was used for hydrolysis of the substrate and was augmented with β-glucosidase (Novozyme 188) in a ratio of 5 U of β-glucosidase/FPU of cellulase. The fermenting organism was *Saccharomyces cerevisiae* strain D_5A supplied by the National Renewable Energy Laboratory (NREL), Golden, CO.

Pretreatment hydrolysate inhibition was tested growing *S. cerevisiae* on glucose. These experiments were carried out in sealed anaerobic test tubes (20 mL filled to 10 mL) or serum vials (125 mL filled to 50 mL) and incubated at 37°C. The test tubes were not agitated. The vials were agitated at 100 rpm on an orbital shaker.

Medium Preparation

For most of the SSF experiments, the concentration of solids in the batch bottles was set to give 20 g/L cellulose. The pretreated solids were stored frozen or refrigerated until ready for use. The pretreated solids were used either as delivered or with the particle size reduced through comminuting for several minutes in a domestic blender set at high speed. For hydrolysate inhibition experiments, the carbon source used was 20 g/L glucose in varying dilutions of pretreatment hydrolysate (0–100% hydrolysate). For all experiments, the medium also included: 10 g/L yeast extract and 20 g/L peptone (both Difco). When necessary, the pH was adjusted to between 4.0 and 4.5 with sulfuric acid or potassium hydroxide. Prior to adding cellulase, the medium was sterilized by autoclaving for 45 min. The cellulase mixture was sterilized through a 0.2-μm filter (Gelman). The inoculum was 2 mL (2% v/v) of yeast broth grown on yeast extract, peptone, and 20 g/L glucose.

Sampling

Samples were taken throughout the fermentations using a sterile syringe through the top of the pressure bottles. CO_2 accumulation in the bottle head space was released at the time of sample extraction. The reaction was allowed to continue until gas production ceased.

Analysis

Dry weights of samples were determined by drying small portions of the samples at 72°C. The cellulose content of solids was determined by quantitative saccharification via acid hydrolysis (72% H_2SO_4 at 30°C for 60 min followed by 2.5% H_2SO_4 at 121°C for 60 min) and then an enzymatic glucose assay on the resulting hydrolysate (Sigma [St. Louis, MO] # 16-UV). Samples of known cellulose content (avicel or hardwood flour) were hydrolyzed alongside each batch of samples to correct for minimal glucose degradation that occurred during the two-step hydrolysis. High-performance liquid chromatography (HPLC) (Bio-Rad [Hercules, CA] HP-87H column) was used to measure ethanol and glucose concentrations in the fermentation broth. Equations (1) and (2) were used to calculate the conversion based on final ethanol concentration and on final residual cellulose, respectively.

$$\text{Conversion to ethanol} = [(\text{ethanol g/L})_{\text{final}} - (\text{ethanol g/L})_{\text{initial}} / (\text{initial cellulose g/L}) \times 0.51] \times 100 \quad (1)$$

$$\text{Cellulose conversion} = 100 - [(\text{final cellulose g/L}) / (\text{initial cellulose g/L})] \times 100 \quad (2)$$

In Eq. (1), the number 0.51 in the denominator results from the stoichiometry of the hydrolysis and fermentation reactions. One mole of a cellulose mono-

Table 2
Solubilization of Different LHW-Pretreated Lignocellulosic Substrates

Substrate	Solids concentration, g/L	Temp, °C	Time, s	Solubilization, %
Bagasse	10	220	120	40
Aspen[a]	10	220	120	32
Hardwood flour	6	220	120	40
Bagasse	80	210	120	33
Bagasse	100	220	120	32

[a] The aspen samples were pretreated three times at the given conditions.

mer hydrolyzes to glucose and is then fermented to ethanol, CO_2, and cell mass according to:

$$1\ C_6H_{10}O_5 + 1\ H_2O \rightarrow 1\ C_6H_{12}O_6 \rightarrow 2\ C_2H_5OH + 2\ CO_2 + \text{cells} \quad (3)$$

The value of carbohydrate allocated to cell production was taken as 0.2 g dry cell material produced/g ethanol produced.

RESULTS

LHW Pretreatment

Table 2 lists results from the pretreatment of three different lignocellulosic substrates. These pretreatments were done at low (<10 g/L) solids concentration. Solubilization of the solids was extensive and gave nearly complete hemicellulose solubilization. Pentosan recovery was consistently above 80%, whereas solubilization of cellulose was low, consistently <10% *(43)*.

SSF Conversion

When pretreated at low solids concentration, all the substrates tested reached a final conversion of about 90% based on cellulose concentrations of the initial and final solids. Figure 1 shows conversion to ethanol vs time for different substrates: LHW-pretreated hardwood flour, bagasse, and aspen chips, as well as raw bagasse that was not pretreated. It can be seen that all of the pretreated materials represented in Fig. 1 achieve 90% of the final conversion within 75 h, except for the unpretreated bagasse, which is minimally reactive, and the bagasse pretreated at high solids concentration, but low temperature (*see* Effect of Pretreatment Solids Concentration on Conversion). The similarity between the conversion curves testifies to the broad applicability of LHW pretreatment. For most of the experiments undertaken for this study, 90% of the final conversion was achieved in 75 h. Table 3 lists the results of those experiments plus additional experiments on comminuted bagasse at varying enzyme loadings and comminuted aspen. It can be seen from Table 3 that conversion calculated with the final ethanol concentration is usually higher than conversion based on the final residual cellulose. This may have been owing to one or a combination of low measurements for initial cellulose content in the solids, low values for the glucose content in the enzyme mixture, or too high an estimate for carbohydrate

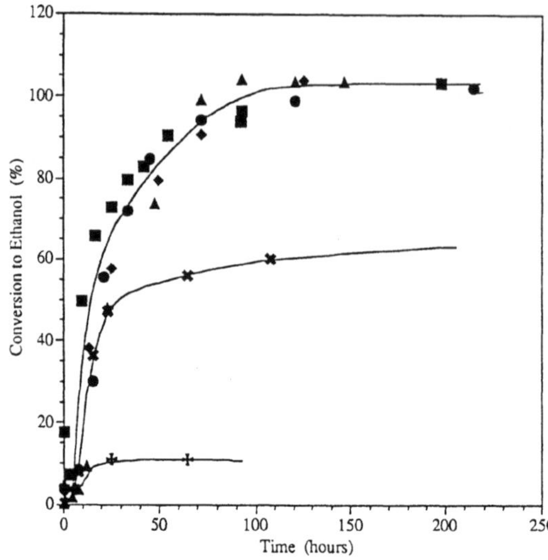

Fig. 1. Percent of theoretical conversion to ethanol in batch SSF of different LHW-pretreated materials (pretreatment: 5–10 g/L, 220°C unless noted otherwise): (■) bagasse, 20 g cellulose/L at 15 FPU/g cellulose; (▲) mixed hardwood flour, 35 g cellulose/L at 20 FPU/g; (●) Aspen chips, 22 g cellulose/L at 14 FPU/g; (✚) raw bagasse (unpretreated), 20 g cellulose/L at 17.5 FPU/g; (◆) bagasse (pretreatment 100 g solids/L, 220°C), 28 g cellulose/L at 13 FPU/g; (✖) bagasse (pretreatment: 80 g solids/L, 210°C), 20 g cellulose/L at 15 FPU/g. Bagasse and Aspen symbols each represent averages of duplicate experiments.

allocated to cell production. Note that the conversion data represented in Fig. 1 are based on ethanol concentrations.

Effect of Enzyme Loading

Figure 2 shows the ethanol conversion obtained from batch SSF experiments on comminuted bagasse at different cellulase loadings of 7, 15, and 30 FPU cellulase/g of cellulose. For all three experiments, the cellulase was supplemented with β-glucosidase. The rate is seen to increase with increasing cellulase content, and at 7 FPU/g, the time to reach 90% of final conversion is up toward 100 h. Results from experiments carried out on bagasse with and without supplemental β-glucosidase show that the absence of supplement β-glucosidase slows down the initial conversion rate, but does not substantially diminish the final conversion (data not shown).

Effect of Post-Pretreatment Particle Size Reduction

Side-by-side experiments were carried out on both aspen and bagasse to compare the effect of post-pretreatment comminution on conversion. Both the bagasse and the aspen showed no difference in reaction rate or extent as a result of comminution. Table 3 lists final conversion numbers for these experiments. Particle size after comminution was not measured, but for both aspen and bagasse, the change was from a substrate with large, obvious particles (chips up to 1.5 cm for aspen, or coarse fibers up to several cm long for bagasse) down to a homogeneous watery mash.

Table 3
Conversion in Batch SSF of LHW Pretreated Lignocellulosic Materials

Sample pretreatment conditions	Enzyme load, IU/g	Additional size reduction	Cellulose before rxn, g/L	Cellulose after rxn, g/L	Final EtOH, g/L	Conversion%	
						By cellulose	By EtOH
Hard wood flour							
10 g/L at 220°C	20	no	35	—	18	—	100.8
Bagasse							
10 g/L at 220°C	15	no	20	2.55	9.94	87.25	99.4
10g/L at220°C	7	yes	20	2.23	8.5	88	85
10g/L at 220°C	15	yes	20	1.73	10.1	91.35	101
10 g/L at 220°C	30	yes	20	1.53	10.7	92.4	107
Not pretreated	17.6	no	20	—	1.07	—	10.7
10 g/L at 220°C	15[a]	no	20	—	9.73	—	97
80g/L at 210°C	15	no	20	1.18	7.6	41	76.3
80 g/L at 210°C	15	yes	20	1.19	7.6	40.5	76
100 g/L at 220°C	13	no	28	1.94	14.6	93	103.8
Aspen chips							
	14	no	22	1.69	11.5	92.3	104.5
	14	yes	22	1.97	11.4	91.0	103.6

[a] No supplemental β-glucosidase added.

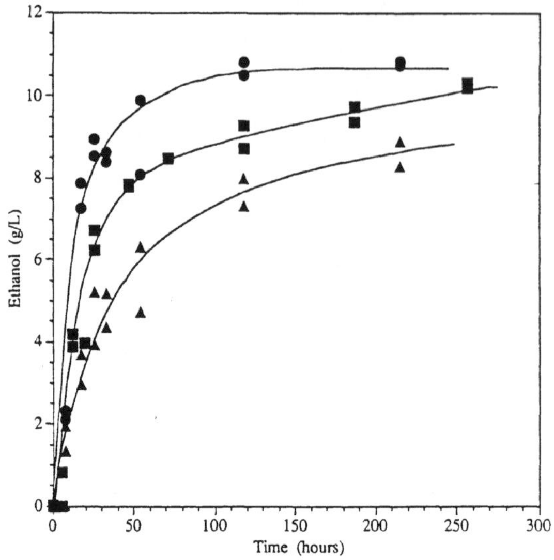

Fig. 2. Effect of enzyme loading on ethanol production in batch SSF of LHW-pretreated bagasse, 20 g cellulose/L. (▲) Enzyme loading of 7 FPU/g cellulose; (■) 15 FPU/g; (●) 30 FPU/g. Symbols each represent averages of duplicate experiments.

Effect of Pretreatment Solids Concentration on Solubilization

Table 2 lists the pretreatment conditions and solubilization results for the higher concentration substrates tested using SSF.

Effect of Pretreatment Solids Concentration on Conversion

Figure 1 shows the contrast between the SSF conversion of solids pretreated under different conditions. Bagasse treated at 220°C showed no reduction in reactivity as pretreatment concentration was increased from 10–100 g/L. However, bagasse pretreated at high concentration (80 g/L), but only 210°C reacted both more slowly and less completely than did the bagasse pretreated at high-temperature. The entries in Table 3 show that, as with the low-concentration pretreated material, there was no difference in conversion through comminution of the high-concentration pretreated material.

Hydrolysate Inhibition

Experiments were done to determine the degree of microbial growth inhibition caused by the hydrolysate produced during LHW pretreatment. Two hydrolysates were tested, both produced by the high-concentration pretreatment of bagasse: one at 210°C (80 g/L solids) and the other at 220°C (100 g/L solids). The fermentation was carried out on glucose to remove the effect of potential hydrolysis limitation to growth. The hydrolysate produced at 210°C showed no sign of growth inhibition up to a concentration of 100% hydrolysate in the fermentation medium. The hydrolysate produced at 220°C did show some inhibition. From Fig. 3 it can be seen that the inhibited culture took about twice as long as the uninhibited

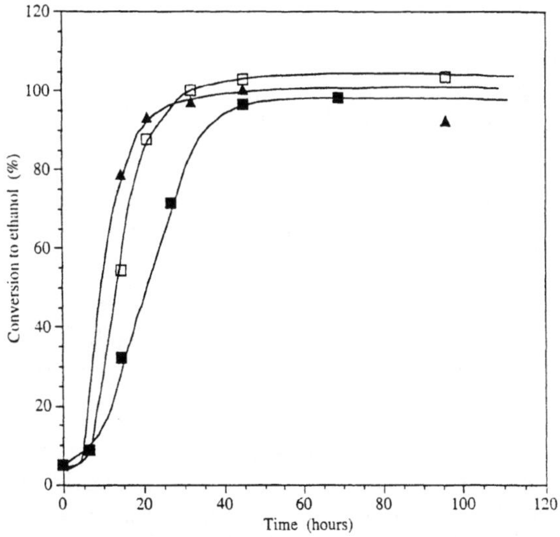

Fig. 3. Effect of presence of L,HW pretreatment hydrolysate on ethanol production in batch fermentation of 20 g/L glucose. (▲) 0% Hydrolysate; (□) 100% hydrolysate from pretreatment at 210°C, 80 g solids/L; (■) 100% hydrolysate from pretreatment at 220 C, 100 g solids/L. Symbols each represent averages of duplicate experiments.

cultures to produce full conversion of the glucose feed. This inhibited growth proceeds at roughly the same rate as the SSF conversion of the pretreated lignocellulosics.

DISCUSSION

Previous work looking at the physical and chemical performance of LHW has suggested that it would be an effective pretreatment method for the conversion of lignocellulosic material to ethanol via SSF. LHW pretreatment offers the promise of high pentosan recovery *(43,49)* without the use of added acid. This translates into less need for corrosion-resistant materials of construction and reduces solid waste disposal requirements. Other metrics for the suitability of LHW to pretreatment for conversion of biomass to ethanol have been addressed in this study.

Figure 1 shows that the LHW pretreatment is effective on a variety of lignocellulosic materials. In most experiments, the substrates tested went to 90% of their final conversion within 75 h at a cellulase loading of 15 FPU/g cellulose. This is comparable to the performance of other available pretreatments. Figure 2 illustrates that there is a definite reduction in the rate of conversion with decreasing enzyme load. However, when examining the conversion numbers (Table 3) based on cellulose concentration, it appears that the decline in final substrate utilization with declining enzyme load is relatively light. If this retention of high extent of reaction at low cellulase loading proves to be a pretreatment-specific effect, rather than a substrate-specific effect, it would be a significant advantage for LHW pretreatment. This effect may result from reduced inactivation of enzyme owing to the low lignin content of the LHW material. This point will require further investigation.

Table 4
Comparison of Pretreatment Processes

Metric of pretreatment effectiveness	Steam explosion	Dilute acid	LHW
Reactive fiber	Yes	Yes	Yes
Pentosan recovery	Low	Medium/high	High
Hydrolysate inhibitory	Yes *(18,19)*	Yes *(6)*	Slightly
Size reduction required	No	Yes	No
Materials of construction	Less costly	More costly	Less costly
Solid residue from neutralization salts	Slight	Significant	Slight

The absence of any difference between comminuted and as-received substrates in both the rate and extent of conversion suggests that LHW pretreatment may require much less by way of size reduction than acid-based pretreatment alternatives. Although steam explosion appears to be similarly robust with respect to size reduction, this process is accompanied by low recovery of hemicellulose sugars.

For a pretreatment to be useful for ethanol production, it is important that the carbohydrate concentration of the substrate medium not be too diluted. The results in Fig. 1 showing effective conversion of solids pretreated at a solids concentration of 100 g/L are encouraging to this end. The striking difference between the samples pretreated at 210°C and 220°C underlines the importance of maintaining effective control over reaction severity at high solids concentration. It has not been determined if the deficit in temperature at 210°C can be compensated for with a longer reaction time.

The data obtained in this study are the first known to us that approach the question of inhibition by pretreatment byproducts associated with LHW. The low level of inhibition by the hydrolysate tested here suggests that LHW may have a decided advantage in this category relative to both acid and steam-explosion pretreatment. The results in Fig. 3 show a marked difference between the degree of inhibition at 210°C and 220°C pretreatment. This difference seems to reflect the observed differences in reactivity of the two substrates. Further study of these two phenomena may help in understanding the mechanisms at work and optimization of this process.

Table 4 summarizes the performance of LHW, dilute-acid hydrolysis, and steam explosion with respect to the metrics that have been developed. Acknowledging that the two other pretreatment methods have been far more extensively studied at this point, including on the pilot-scale level, it can be seen that based on the results of this study, LHW compares favorably with these two other technologies on all counts.

CONCLUSION

LHW is an effective pretreatment for the conversion of lignocellulosic materials to ethanol via SSF. Lack of particle size reduction was seen to pose no disadvantage to hydrolysis. Bagasse pretreated at higher solids concentration performs comparably to that pretreated at lower concentrations provided the pretreatment temperature can be maintained at 220°C. The hydrolysate produced in high-concentration pretreatment showed only slight inhibition of *S. cerevisiae*. Future work should look at performing a definitive experiment in which all of these advantages are observed simultaneously along with the scale-up of the LHW process.

ACKNOWLEDGMENTS

Fermentation work was performed at Thayer School of Engineering and supported by the Linbergh Foundation and the Link Foundation. The authors thank Avinash C. Balbahadur, Hae U. Choi, Scott M. Brady, Hilarie Schubert, and Harley J. McAllister for carrying out experiments, and Colin South for advice on SSF procedures.

The pretreatment work was conducted at the University of Hawaii and supported by the NREL as part of the Hawaii Integrated Biofuels Research Program. The authors thank Ralph Overend (NREL) for insightful discussions, Francis Keany (Hawaiian Commercial & Sugar), and Andy Baker (US Forest Products Laboratory) for supplying feed materials, William Mok for assistance in the design of the pretreatment process, Eric Croiset and Ben Respicio for construction of the process equipment, and Michael Onuma and Lance Kam for help in conducting experiments.

REFERENCES

1. Lynd, L. R., Cushman, J. H., Nichols, R. J., and Wyman, C. E. (1991), *Science* **251,** 1318–1323.
2. Converse, A. O. (1993), in *Bioconversion of Forest and Agricultural Residues,* (Saddler, J. N., ed.), CAB International, Wallingford, UK, pp. 93–106.
3. Chem Systems (1990), Technical and economic evaluation: wood to ethanol process. National Renewable Energy Laboratory, Golden CO.
4. Stone, K. and Lynd, L. R. (1995), Appl. Biochem. Biotechnol. **45/46,** 569–584.
5. McMillan, J. D. (1994), in *Enzymatic Conversion of Biomass for Fuels Production,* (Himmel, M. E., Baker, J. O., and Overend, R. P., eds.), ACS Symposium Series 566, Washington, DC, pp. 292–324.
6. McMillan, J. D. (1994), in *Enzymatic Conversion of Biomass for Fuels Production,* (Himmel, M. E., Baker, J. O., and Overend, R. P., eds.), ACS Symposium Series 566, Washington, DC, pp. 411–437.
7. Weil, J., Westgate, P., Kohlmann, K., and Ladisch, M. R. (1994), *Enzyme Microb. Technol.* **16,** 1002.
8. Wyman, C. E. and Goodman, B. J. (1993), in *Opportunities for Innovation in Biotechnology,* (Busche, R., ed.), US Department of Commerce, National Institute of Standards and Technology, Gaithersburg, MD.
9. Sinitsyn, A. P., Gusakov, A. V., and Vlasenko, E. Y. (1991), *Appl. Biochem. Biotechnol.* **30,** 43–59.
10. Kohlmann, K., Westgate, P. J., Weil, J., and Ladisch, M. R. (1993), SAE Technical Article Series, 932251. SAE, Warrendale, PA.
11. Lynd, L. R. (1990) *Appl. Biochem. Biotechnol.* **24/25,** 695–719.
12. Wright, J. (1988), *Chem. Eng. Prog.* **August,** 62–74.
13. Altherum, F. and Ingram, L. O. (1989), *Appl. Environ. Microbiol.* **55,** 1943–1948.
14. Bealle, D. S., Ohta, K., and Ingram, I. O. (1991), *Biotechnol. Bioeng.* **38,** 296–303.
15. Chen, Z. D. and Ho, N. W. Y. (1993), *Appl. Biochem. Biotechnol.* **39,** 135–147.
16. Zhung, M., Eddy, C., Deand, K., Finkelstein, M., and Picatagio, S. (1995), *Science* **267,** 240–243.
17. Assessment of costs and benefits of flexible and alternative fuel use in the US transportation sector. Technical report 11: Evaluation of a wood-to-ethanol process. (1993), DOE/EP-0004, US Department of Energy, Washington, DC.

18. Mes-Hartree, M. and Saddler, J. N. (1983), *Biotechnol. Lett.* **5(8),** 531–536.
19. Forsberg, C.W., Schellhorn, H. E., Gibbons, L. N., Maine, F., and Mason, E. (1988), *Biotechnol. Bioeng.* **28,** 176–184.
20. Tran, A. V. and Chambers, R. P. (1986), *Enzyme Microb. Technol.* **8,** 439–444.
21. Fraser, F. R. and McKaskey, T. A. (1991), *Enzyme Microb. Technol.* **13,** 110–115.
22. Ooshima, H., Burns, D. S., and Converse, A. O. (1990), *Biotechnol. Bioeng.* **36,** 446–452.
23. Thompson, D.N., Chen, H.-C., and Grethlein, H. E. (1991), *Bioresource Technol.* **39,** 155–163.
24. Torget, R., Himmel, M. E., and Grohmann, K. (1991), *Bioresource Technol.* **35,** 239–246.
25. Torget, R., Walter, P., Himmel, M., and Grohmann, K. (1991), *Appl. Biochem. Biotechnol.* **28/29,** 75–86.
26. Beltrame, P. L., Carniti, P., Visciglio, A., Focher, B., and Marzetti, A. (1991), *Bioresource Technol.* **39,** 165–171.
27. Brownell, H. H. and Saddler, J. N. (1987), *Biotechnol. Bioeng.* **29,** 228–235.
28. Heitz, M., Capek-Menard, E., Koeberle, P. G., Gagne, J., Chornet, E., Overend, R. P., Taylor, J. D., and Yu, E. (1991), *Bioresource Technol.* **35,** 23–32.
29. Jollez, P., Chornet, E., and Overend, R. P. (1994), in *Advances in Thermochemical Biomass Conversion*, vol. 2. (Bridgewater, A. V., ed.), Chapman and Hall, London, pp. 1659–1669.
30. Montane, D., Salvado, J., Farriol, X., and Chornet, E. (1993), *Biomass Bioenergy* **4,** 427–437.
31. Ramos, L. P, Breuil, C., and Saddler, J. N. (1992), *Appl. Biotechnol. Bioeng.* **34/35,** 37–48.
32. Clark, T. A. and Mackie, K. L. (1987) *J. Wood Chem. Technol.* **7,** 373–403.
33. Schell, D., Torget, R., Power, A., Walter, P. J., Grohmann, K., and Hinman, N. D. (1991), *Appl. Biochem. Biotechnol.* **28/29,** 87–97.
34. Dale, B. E. (1985), *Ann. Rep. on Fermentation Processes* **8,** 299–323.
35. Grethlein, H. E. and Dill, T. (1993), The cost of ethanol produced from cellulosic biomass a comparison of selected alternative processes. Report to the USDA for Specific Cooperative Agreement No. 58-1935-2-050.
36. Holtzapple, M. T., Lundeen, J. E., Sturgis, R., Lewis, J. E., and Dale, B. E. (1992), *Appl. Biochem. Biotechnol.* **34/35,** 5–21.
37. Holtzapple, M. T. and Humphrey, A. E. (1984), *Biotechnol. Bioeng.* **26,** 670–676.
38. Kim, S. B. and Lee, Y. Y. (1987), *Biotechnol. Bioeng. Symp.* **17,** 71–84.
39. Dale, B. E., Henk, L. L., and Shiang, M. (1985), *Dev. Ind. Micobiol.* **26(13),** 223–233.
40. Azuma, J.-I., Tanaka, F., and Koshijima, T. (1984), *J. Ferment. Technol.* **62(4),** 377–384.
41. Koullas, D. P., Christakopolous, P., Kekos, D., Macris, B. J., and Koukios, E. G. (1992), *Biotechnol. Bioeng.* **39,** 113–116.
42. Schell, D. and Duff, B. (1994), Review of pilot plant programs for conversion of lignocellulosic biomass to ethanol. Alternative Fuels Division, National Renewable Energy Laboratory, Golden CO.
43. Allen, S. G., Kam, L. C., Zemann, A. J., and Antal, M. J. (1996), Macroscale fractionation of sugar cane with hot compressed liquid water by the aquasolv process. *Ind. Eng. Chem. Res.* (accepted for publication).
44. Bobleter, O., Vidotti, R., Zemann, A., and Prutsch, W. (1990), in 5th E.C. Conference, vol. 2, *Conversion and Utilization of Biomass*, Grassi, G., Gosse, G., and dos Santos, G., eds., Elsevier Applied Science, London pp. 2.31–2.37.
45. Aronovsky, S. I. and Gortner, R. A. (1930), *Ind. Eng. Chem.* **22,** 264.
46. Bouchard, J., Nguyen, T. S., Chornet, E., and Overend, R. P. (1991), *Bioresource Technol.* **36,** 121.
47. Walch, E., Zemann, A., Schinner, F., Bonn, G., and Bobleter, O. (1992), *Bioresource Technol.* **39,** 173–177.
48. Overend, R. P. and Chornet, E. (1987), *Phil. Trans. R. Soc. Lond. A* **321,** 523–526.
49. Mok, W. S.-L. and Antal, M. J. (1992), *Ind. Eng. Chem. Res.* **31,** 1157–1161.
50. Hormeyer, H. F., Bonn, G., Kim, D. W., and Bobleter, O. (1987), *J. Wood Chem. Technol.* **7(2),** 269.
51. Hormeyer, H. F., Tailliez, P., Millet, J., Girard, H., Bonn, G., Bobleter, O., and Aubert, J.-P. (1988) *Appl. Microbiol. Biotechnol.* **29,** 528–535.

0273-2289/96/5758—0171$8.00

Pretreatment of Sugar Cane Bagasse for Enhanced Ruminal Digestion

FRANCISCO C. DESCHAMPS,[1]
LUIZ P. RAMOS,[2] AND JOSÉ D. FONTANA*,[3]

[1]Empresa de Pesquisa e Extensão Agropecuária, EPAGRI, Itajai, SC, Brazil; [2]Department of Chemistry; and [3]Department of Biochemistry, LQBB, Federal University of Paraná, P. O. Box 19.046, Curitiba; Paraná, Brazil 81531-970

ABSTRACT

Crop residues, such as sugar cane bagasse (SCB), have been largely used for cattle feeding However, the close association that exists among the three major plant cell-wall components, cellulose, hemicellulose, and lignin, limits the efficiency by which ruminants can degrade these materials Previously, we have shown that pretreatment with 3% (w/w) phosphoric acid, under relatively mild conditions, increased considerably the nutritional value of SCB. However, in this preliminary study, pretreated residues were not washed prior to *in situ* degradability assays because we wanted to explore the high initial solvability of low-mol-wt substances that were produced during pretreatment We have now studied the suitability of water- and/or alkali-washed residues to *in situ* ruminal digestion. Alkali washing increased substrate cellulose content by removing most of the lignin and other residual soluble substances As a result, the ruminal degradability of these cleaner materials had first-order rate constants five times higher than those substrates with higher lignin content (e.g., stem-exploded bagasse) However, alkali washing also increased the time of ruminal lag phase of the cellulosic residue, probably because of hemicellulose and/or lignin removal and to the development of substrates with higher degree of crystallinity. Therefore, longer lag phases appear to be related to low microbial adherence after extensive water and alkali extraction, as Novell as to the slower process of cellulase induction during ruminal growth. The kinetic data on ruminal digestion were shown to be very well adjusted by a nonlinear model. Although pretreatment enhances substrate accessibility, the occurrence of an exceedingly high amount of lignin byproducts within the pretreated material reduces considerably its potential degradability.

Index Entries: Sugar cane bagasse; autohydrolysis; phosphoric acid; *in situ* degradability.

INTRODUCTION

It has been suggested that the process of ruminal digestion depends on adherence of the rumen biota to the cellulosic substrate *(1)* Three main factors are known

*Author to whom all correspondence and reprint requests should be addressed.

to influence this process: the degree of lignification of the cell wall, the degree of crystallinity of the cellulose component, and the substrate surface area available for microbial and/or enzymatic attack *(2,3)* Therefore, the structure and composition of the substrate must facilitate microbial adherence through interaction with large multicomponent complexes (cellulosomes) *(4,5)*, microbial exopolysaccharides *(6)*, and cellulose binding domains (CBD) of fungal hydrolases *(7,8)*. In general, microbial adhesion occurs rapidly and is strongly influenced by changes in pH, temperature, and ionic strength inside the ruminal tract *(9)*.

Several pretreatment methods have been shown to enhance the susceptibility of lignocellulosic residues to microbial degradation, even at relatively mild severities. This enhancement in degradability is usually triggered by partial hemicellulose solubilization, yielding water-soluble low-mol-wt xylooligomers, which can be readily utilized by the rumen biota *(10)*. However, insoluble pretreated residues often contain most of the metabolic energy available for animal growth and production. Therefore, the utilization of this residue for cattle feeding is of utmost importance, and will depend on pretreatment optimization with regard to substrate accessibility and production of growth inhibitors.

Pretreatment with phosphoric acid under relatively mild conditions increases the initial solubility of sugar cane bagasse (SCB) (fraction that was immediately available to the rumen microorganisms) from 6.8 to 27.5%, without having a similar effect on potential degradability *(10,11)*. This improvement in initial solubility was a result from the hydrolytic effect of pretreatment on the hemicellulose component, yielding low-mol-wt substances, which could be readily utilized by the rumen biota Steam explosion (autohydrolysis) has also been shown to improve both the potential degradability and the initials solubility of cellulosic residues by similar mechanisms *(12–15)*. Finally, the potential degradability of SCB can be considerably increased by alkali washing *(16)*. This has been attributed to the ability of this treatment to reduce both the lignin and hemicellulose content oaf the residue, yielding a cellulosic material that can be readily used for cattle feeding.

In this article, we have tried to identify the substrate-related factors that can be used to evaluate the suitability of several cellulosic residues for cattle feeding. Four pretreatment methods have been assessed: phosphoric acid treatment, alkali washing, steam explosion, and steam explosion followed by alkali washing. All of these residues, except the raw material, were washed thoroughly prior to any ruminal degradability assays in order to allow for the determination of their actual potential degradabilities.

MATERIALS AND METHODS

Pretreatment of SCB

A large batch of SCB was obtained from the processing stages of a sugar mill in southern Brazil (Porto Belo Distilleries, Itajaí, SC). The residue, which contained about 2% (w/w) of contaminating sucrose and had a moisture content of 48% (w/w), was conditioned in plastic bags md stored frozen at –20°C.

The sample named ALK was obtained by treating SCB with hot alkali. Approximately 30 g of SCB were milled to pass a 2-mm screen and treated with 900 mL of NaOH 2*N* for 2 h at room temperature. The suspension was then kept under reflux for l h, filtered through a nylon bag, dialyzed overnight against tap water, and dried for 48 h at 55°C. The alkali-extracted residue was then Willey-milled to pass a 0.5-mm screen and kept for further analysis, yielding approx 40% (w/w) ion relation to the dry weight of the raw material.

Both solvolysis (SOL) and phosphoric acid (OPA) residues were obtained as described previously *(10,11)*. SCB was pretreated at temperatures up to 197°C (13.5 atm) in the presence and absence of OPA as an acid catalyst. OPA was used at 3% (w/w) in relation to the dry weight of the residue. SOL was defined as those experiments in which water was used as a replacement to the dilute-acid solution Both samples were washed thoroughly with ethanol at 70% ([v/v] aqueous) for 1 h at room temperature to eliminate low-mol-wt contaminants. Yields were around 78% for the SOL residue and 60% for the OPA residue, both in relation to their original oven-dried weights

The steam-exploded (SE) SCB was kindly supplied by André Luiz Ferraz (CEBIQ, FAENQUIL—Lorena, SP, Brazil). The SE fraction was prepared at the CEBIQ pilot plant facility by steam treating SCB at 190°C for 15 min. Recovery yield of the water-insoluble fraction was typically around 66.6%. Finally, SE alkali washing (SEALK) was carried out by treating the SE residue with alkali (NaOH 1%, w/w) at 100°C for 1 h and a water-to-solids ratio of 1:20 (dry basis). After alkali washing, the residual material was washed thoroughly with water until neutrality and stored at 4°C for further utilization SEALK had a yield around 39.6% in relation to the dry weight of the original, untreated SCB.

Substrate Analysis

The chemical composition of the residues was determined using the classical methods of neutral detergent fiber (NDF) and acid detergent fiber (A-DF) as described previously *(17)*. By definition, the NDF fraction includes lignin, cellulose, and hemicellulose, whereas the ADF fraction consists of only cellulose and lignin. The hemicellulose content was estimated by subtracting ADF from NDF, whereas cellulose was determined gravimetrically after treatment of the residue with the oxidative acetic/nitric acid reagent *(18)*.

The relative amount of low-mol-wt phenolic compounds within the cellulosic residue was estimated by UV/vis spectroscopy. Approximately 0.25 g of the washed residue was treated with 5 mL of methanol:HCl (1%, [v/v]) for 96 h at room temperature. The supernatant was collected by centrifugation, diluted in methanol (1:50), and scanned in a Zeiss Specord M-500 UV/vis spectrophotometer from 200 to 500 nm in wavelength

Substrate chemical composition was also assessed by Fourier-transformed infrared spectroscopy (FT-IR). After milling, each cellulosic material was compacted into KBr disks at 8 t of pressure for 3 min at room temperature and analyzed using a Bomem FT-IR spectrophotometer within the region of 800–1900/cm. Peak intensities at 897/cm were associated with cellulose *(19)*. Hemicelluloses were monitored by absorption bands at 1734/cm (axial stretch deformations of carboxyl groups) and within the region of 1160–1300 cm (axial stretch deformations of ether bonds) *(19,20)*. Finally, peak intensities at 1590/cm and within the region of 1590–1600/cm were collectively attributed to carbon-to-carbon axial stretch deformations of aromatic compounds found in lignin *(19,20)*.

The degree of crystallinity of the substrates was obtained by X-ray diffraction. Each sample was compressed into compact sheets and analyzed using a Philips X'Pert diffractometer. The wavelength of the Cu/Kaα radiation source was 0.154 nm and the spectra were obtained at 30 mA and 40 kV. Samples were scanned on the automated diffractometer from 8 to 40° of 2Θ (Bragg angle), with an elapsed time of 1 s and data acquisition taken at intervals of 0.02° *(21)*.

The crystallinity index of cellulose was calculated by the empirical method described by Segal et al. *(22)*, using the following equation:

$$\text{CrI}\ (\%) = (\text{I}_{002} - \text{I}_{AM}/\text{I}_{002}) \times 100 \tag{1}$$

where I_{002} is the maximum intensity of the 002 lattice diffraction (reflection attributed to the crystalline region of the sample), and I_{AM} is the intensity of diffraction at 2Θ = 18° (reflection attributed to the amorphous region of the sample).

Determination of *In Situ* Degradability (Ruminal Digestion)

Samples were subjected to *in situ* ruminal digestion as described earlier *(11,23)*. Approximately 0.5 g of biomass (oven-dried weight) were placed in a 9 × 15 cm nylon bag with a pore size of 40 μm and incubated for several residence times in cattle fitted with rumen cannulae. After a given incubation time, the bag was withdrawn from the rumen cannulae and washed thoroughly with tap water. The residual, undigested material was then collected from the bag, air-dried at 55°C, weighed, and kept for further analysis.

Kinetics of ruminal digestion were determined after 0, 6, 10, 16, 24, 32, 48, and 72 h of incubation. Two animals were used in these experiments and each experiment was done in duplicate for each animal. Therefore, each data point was a result of a total of four replicates. Results derived from *in situ* ruminal digestion, expressed as weight loss in relation to the dry weight of the residue, were adjusted according to the following equation *(23)*:

$$p = a + b\,(1 - \exp - c \cdot t) \tag{2}$$

were p is the potential degradability of the substrate, t is the ruminal digestion time, a is the fraction of the substrate that is immediately available to the rumen biota (initial solubility), b is the degradability potential of the insoluble fraction, and c is the expected degradation rate of fraction b.

Based on the model, the *in situ* lag phase of each of the substrates used was calculated with regard to the amount of original material that remained after 6–32 h of incubation (undigested dry matter). For this, the weight loss data for each cellulosic substrate were fitted by linear regression analysis to a discontinuous first-order kinetic equation *(24,25)*:

$$\ln y = \text{a} + k \cdot t \tag{3}$$

were y is the undigested dry matter, α is the intercept at the ordinate axis, k is the first-order rate constant and t is the incubation time. In adjusting the data to the model, k was calculated by regressing $\ln y$ against t. The *in situ* lag phase was then calculated by solving the regression equation for time when $\ln y = \ln 1$.

The effective *in situ* degradability of each residue was determined according to the following equation *(23)*:

$$e = a + (b \times c)/(c + k) \tag{4}$$

where e is the material effectively degraded and k is the rate of passage of the diet through the rumen tract.

Table 1
Chemical Composition of the Various Cellulosic Residues

Sample[b]	NDF[b]	ADF[b]	Hemicellulose[c]	Cellulose[d]	CrI, %[e]
Native	87.0	53.0	34.0	39.5	36.2
ALK	94.7	85.6	9.1	88.3	57.5
SOL	89.7	67.0	22.7	55.5	42.2
OPA	92.0	85.5	6.5	68.1	42.3
SE	78.0	73.3	4.7	65.8	46.3
SEALK	98.5	95.8	2.7	92.3	53.7

[a]ALK, alkali washed; SOL, solvolyzed; OPA, phosphoric acid treated; SE, steam exploded; SEALK, steam exploded and alkali washed. All values are averages of triplicates with a coefficient of variation always lower than 3%.

[b]Neutral detergent fiber (NDF) and acid detergent fiber (ADF) were determined according to Van Soest *(17)*.

[c]Hemicellulose content was determined by subtracting ADF from NDF.

[d]Cellulose was determined gravimetrically after digestion of the residue with the acetic/nitric reagent according to Crampton and Maynard *(18)*.

[e]Crystallinity index was determined according to Segal et al. *(22)*.

RESULTS AND DISCUSSION

The suitability of several cellulosic residues for cattle feeding was evaluated with regard to their chemical composition, potential degradability, accessibility to rumen microorganisms, and time of lag phase during ruminal digestion. In addition, the rate of passage of the diet through the digestive tract was also considered in order to predict the effective degradability of the substrates.

Each of the substrates used in this study was shown to contain very little amount of detergent-extractable materials in their chemical composition (Table 1). Therefore, their NDF values were very high, except for the SE sample, which had a lower NDF content. The high recovery yield of both NDF and cellulose components after alkali washing also indicated that this extraction was very effective in removing both hemicellulose and lignin, leaving a cellulosic residue with nearly 90% of cellulose (ALK and SEALK).

The higher amount of hemicellulose in SOL has demonstrated that hemicellulose is more effectively removed when pretreatment is carried out in the presence of an acid catalyst, such as OPA *(10,11)*. This observation is also valid for the SE fraction, since several organic acids, such as acetic and formic acids, are released during the process of steam explosion (autohydrolysis).

FT-IR spectroscope showed that all pretreatment methods, except solvolysis, were effective in removing most of the hemicellulose component of SCB (Fig. 1). Compared to the native residue, a considerable decrease in intensity of the absorption band at 1734/cm (C = O stretching deformation in acetyl groups and/or 4-*O*-methylglucuronic acid units) was observed in relation to the peak intensities at 1510 and 1600/cm (aromatic compounds of lignin) *(20)*. The effective removal of lignin by alkali washing was further confirmed by the new disappearance of latter absorbance band in both ALK and SEALK spectra. It was also apparent that lignin removal by alkali washing was more effective after steam explosion,

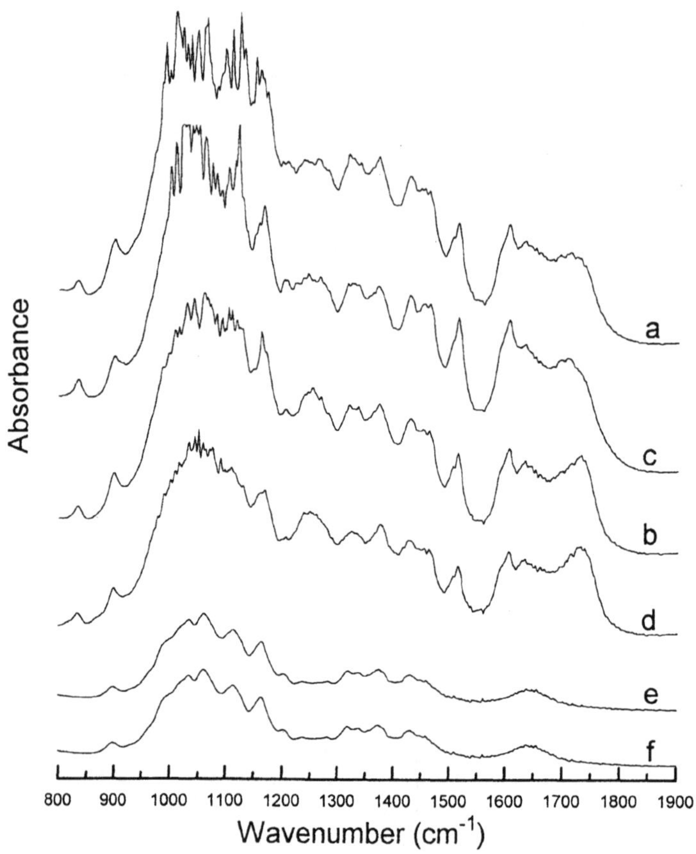

Fig. 1. FT-IR spectra of substrates derived from SCB **(a)** OPA, **(b)** SOL, **(c)** SE, **(d)** Native, **(e)** ALK, and **(f)** SEALK.

probably owing to the higher accessibility of the steam-exploded fibers to the alkaline reagent.

With regard to the kinetic parameters of ruminal digestion, all of the pretreated substrates were shown to have very low *a* values (Table 2). This result was expected, since *a* is usually low for extracted materials, which are mainly composed of polymeric compounds, such as cellulose, hemicellulose, and lignin *(23)*. Nevertheless, even though SE had been water-washed after pretreatment, its initial solubility (3.4%) was slightly higher than that of other lignin-containing residues, such as OPA and SOL. This higher initial solubility SE was consistent with its lower NDF content of 78% (Table 1), indicating that a small amount of detergent-extractable materials (readily accessible, low-mol-wt lignin–carbohydrate complexes) still remained within the sample after water washing. It appeared that these low-mol-wt substances, particularly phenolic compounds, were more effectively removed from the substrates by ethanol/water extraction. In fact, both SOL and OPA were shown to have higher ruminal lag phases than SE (7:00, 6:32, and 1:08 h, respectively), supporting the concept that a simple water washing is not enough to remove most of the low-mol-wt substances from steam-treated pulps (Table 3).

Table 2
Kinetic Parameters of Ruminal Digestion

Sample	a[a]	b[b]	c[c]	r^2	t = 72[d]	k = 0.02[e]
Native	0.068	0.339	0.040	0.97	0.38	0.30
ALK	0.000	0.987	0.037	0.93	0.92	0.64
SOL	0.014	0.815	0.017	0.95	0.59	0.39
OPA	0.004	0.735	0.016	0.96	0.51	0.33
SE	0.034	0.623	0.026	0.96	0.56	0.38
SEALK	0.000	0.905	0.042	0.90	0.86	0.61

[a]Fraction that is immediately available to ruminal digestion.
[b]Degradability potential of the insoluble fraction.
[c]Expected degradation rate of fraction.
[d]Expected degradation after 72 h of ruminal digestion.
[e]Expected degradation when the rate of passage is included.

Table 3
Determination of *In Situ* Lag Phase by Linear Regression Analysis of the Amount of Original Material That Remained After 6–32 h of Ruminal Digestion (Undigested Dry Matter)

Sample	$\ln y = \alpha + k \times t$	r^2	Lag, h[a]	k_N[b]
Native	$0.02025 - 0.01016 \times t$	0.96	1:54	1.00
ALK	$0.31225 - 0.05640 \times t$	0.91	5:32	5.55
SOL	$0.12441 - 0.01773 \times t$	0.85	7:00	1.74
OPA	$0.08779 - 0.01341 \times t$	0.90	6:32	1.32
SE	$0.01516 - 0.01338 \times t$	0.70	1:08	1.32
SEALK	$0.32984 - 0.05189 \times t$	0.92	6:21	5.11

[a]Lag phase as determined according to the linear regression analysis.
[b]Normalized rate in relation to the native residue (k_{sample}/k_{native}).

Compared to the SE residue, both SOL and OPA showed higher potential degradabilities (62.3, 81.5, and 73.5%, respectively). It has been hypothesized that the hydrolytic action of the enzymes can be strongly influenced by lignin hydrophobicity *(3,26)*. Pretreatment methods, such as steam explosion, result in a partial depolymerization of both hemicellulose and lignin, thus exposing the cellulose component to microbial degradation *(13–15)*. However, the higher temperatures and pressures usually developed during steam explosion can modify the lignin component to coalescent spheres of high hydrophobicity. This pseudoplastic behavior of the modified lignin allows its redeposition onto the cellulosic matrix, causing a shielding effect, which translates into a steric hindrance to microbial interaction *(13,27–29)*.

It is also worth mentioning that raw bagasse showed higher initial solubility (6.8%) than any other substrate used. This can be easily explained by the presence of low amounts of sucrose and other soluble materials within the raw fibers. Nevertheless, raw bagasse does not behave as a good substrate for cattle feeding owing to the close association that exists within its three main polymeric components, cellulose, hemicellulose and lignin *(11)*. In fact, raw fibers have a very low potential degradability (parameter *b*) of only 33.9%.

The highest kinetic values of potential degradability (parameter *b*) were observed for the alkali-washed residues, ALK and SEALK. These residues also had high degradation rates (parameter *c*), even at considerably high rates of passage through the ruminal digestive tract (e.g., $k = 0.02$ or 2% passage of the bulk material/h) (Table 3). However, both ALK and SEALK had longer ruminal lag phases, as determined during the *in situ* degradability assay (Table 3). Similar results have been demonstrated for several purified cellulose preparations, which also showed relatively long lag phases *(25,30,31)*. This effect can be attributed to low microbial adherence after extensive water and alkali extraction of the substrate, as well as to the slow process of cellulase induction during ruminal growth.

It has been suggested by several authors that glucose and cellobiose accumulation can impose inhibitory effects on cellulose digestion because of catabolite repression of cellulases *(30,32,33)*. Other authors have attributed this negative effect to a strong interference on the binding capacity of cellulases through their cellulose binding domain *(34)*. Our observations during ruminal digestion of ALK and SEALK did not suggest any strong effect of sugar accumulation on the efficiency by which the rumen biota degraded residues with high cellulose content.

Low molecular weight phenolic compounds, such as vanillin, *p*-coumaric acid, and ferulic acid are also known for their inhibitory effects on the in vitro fiber digestion of pure and mix cultures of rumen microorganisms K35). These and other phenolic compounds appear to compromise microbial adhesion to the substrate *(36,37)*. To investigate this hypothesis, we have determined the relative amount of phenolic compounds within each of the cellulosic residues used in this study by UV/vis spectroscopy. The SE residue was shown to have the highest UV absorbance profile (Fig. 2). This was consistent with the previously observed low NDF values and relatively short lag phase of this residue. Depending on pretreatment conditions, the process of steam explosion produces microbial inhibitors that have to be water-washed prior to their utilization for bioethanol production through in vitro enzymatic hydrolysis and/or simultaneous saccharification and fermentation *(38)*. However, these low-mol-wt substances do not interfere with *in situ* degradability assays because they are either utilized by the diverse rumen microflora or quickly diffused through the digestive tract.

Various researchers have tried to obtain a better understanding of the relative importance of cellulose crystallinity on the ruminal degradability of pretreated cellulosic residues *(39)*. It has been suggested that, at least for pure cultures of *Fibrobacter succinogenes* ssp. succinogenes, crystallinity acts as a barrier for microbial attack, and this barrier must be disrupted before the substrate can be effectively utilized *(34)*. Other authors have suggested that the crystallinity barrier is not a problem for the rumen biota because it can be easily overcome by the strong synergism that occurs among the existing microorganisms *(25)*.

Our results have shown that, regardless of the higher crystallinity index (CrI) of both ALK and SEALK (57.5 and 53.7, respectively), these substrates had higher potential degradabilities than those of lower CrI, that had not been alkali washed (Table 1, Fig. 3). These were the same substrates that were characterized by having a long time of lag phase In light of these results, it seems plausible to assume that longer lag phases might have been required to overcome the stronger crystallinity barrier of both ALK and SEALK residues. However, it must be emphasized that this crystallinity effect may be just apparent, and that lignin and hemicellulose removal could be a factor of a greater importance.

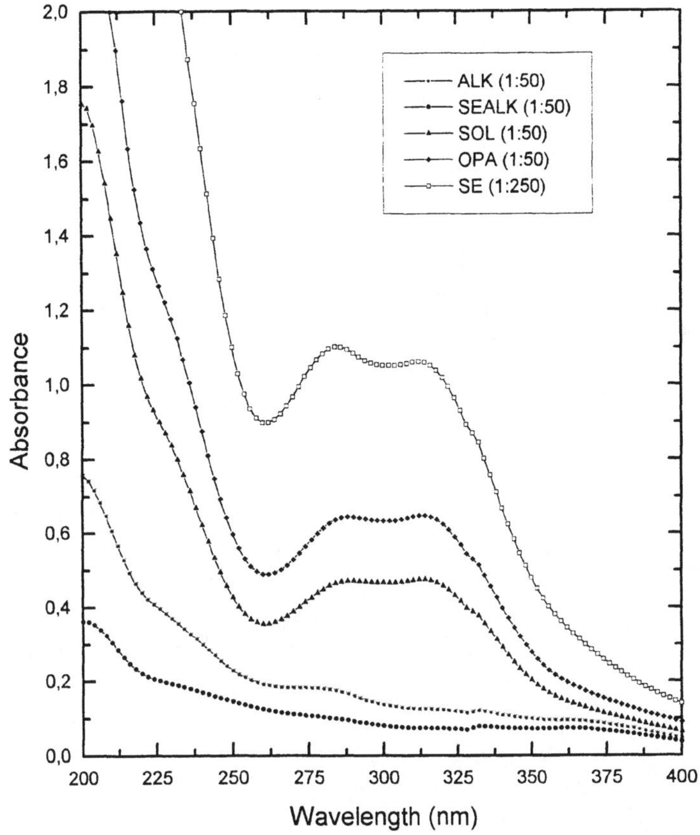

Fig. 2. UV/vis spectra of methanol:HCl extracts obtained from pretreated fractions of SCB. (◆) OPA, (▲) SOL, (□) SE, (*) ALK, (●) SEALK.

The two sharp diffraction intensities, observed at 2Θ (Bragg angles) of 20.89 and 26.65°, were attributed to the occurrence of silica within the residues. This was more apparent for the SE material, which was shown to contain approx 2% (w/w) of ash in its chemical composition.

A linear relationship between crystallinity and the rate of hydrolysis of different cellulose substrates have been often described in the literature *(40,41)*. Iran and coworkers *(41)* suggested that a pretreatment method should decrease the degree of crystallinity to be effective. Other authors *(13–16,40)* have related the improvement in enzyme digestibility of pretreated cellulosic residues to the removal of hemicellulose and lignin, which, according to Grethlein *(42)*, results in an increase in both the accessible pore volume and the specific surface area of the substrate. Stone and coworkers *(43)* showed that the median pore size is also strongly dependent on the degree of swelling. Therefore, it can be seen that the concepts of cellulose crystallinity, hemicellulose removal, substrate swelling (pore volume distribution), and accessible surface area are all strongly related, and any attempt to discuss one of these aspects in isolation can fall into a rough and inac-

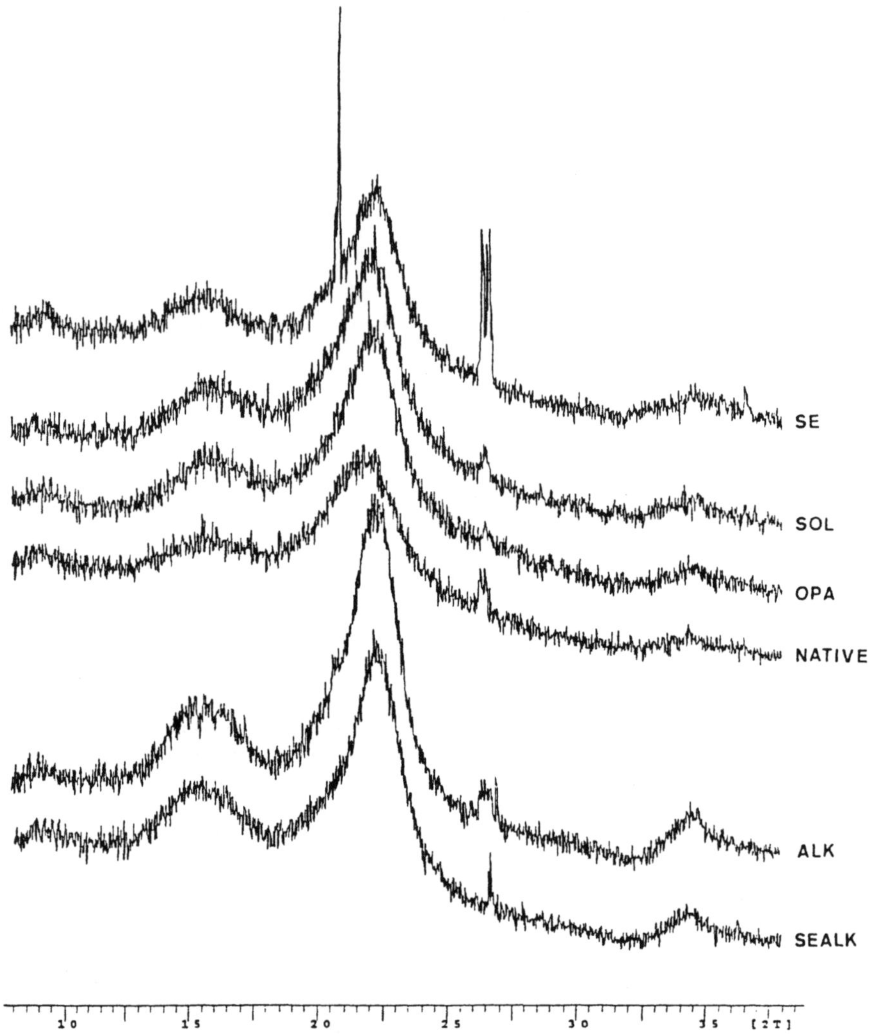

Fig. 3. X-ray diffraction analysis of substrates derived from SCB.

curate simplification of the complexity that the process of cellulose degradation actually represents.

Our work has shown that partial lignin and hemicellulose removal increases the potential degradability of SCB. However, this gain in potential degradability also increases the time of lag phase during ruminal digestion, unless the amount of soluble sugars that is released during pretreatment is not washed out of the substrate prior to their use for cattle feeding. If this is the case, the conditions used for pretreatment have to be optimized in order to avoid production of toxic byproducts and/or growth inhibitors for cattle feeding.

ACKNOWLEDGMENTS

This work was made possible through financial support of the following Brazilian funding agencies: CNPq, CAPES, and FUNCITEC/SC. The authors are grateful to André Luiz Ferraz (Faenquil, Lorena, SP) for the steam-treated bagasse, as well as to Ted Taylor and Cristina Doneda (UFSC, Flonanópolis, SC) for their valuable contribution on X-ray diffraction.

REFERENCES

1. Akin, D. E. (1979), *J. Anim. Sci.* **48,** 701–710.
2. Chesson, A. and Forsberg, C. W. (1988), in *The rumen microbial ecosystem,* Hobson, P. N., ed., Elsevier, New York, pp. 251–284.
3. Kerley, M. S., Fahey, G. C., Gould, J. M., and Ianotti, E. L. (1988), *Food Microstruct.* **7,** 59–65.
7. Lamed, R. and Bayer, E. A. (1988), *Adv. Appl. Microbiol.* **33,** 1–46.
5. Leatherwood, J. M. (1973), *Fed. Proc.* **32,** 1815–1819.
6. Cheng, K. J. and Costerton, J. W. (1980), in *Digestive Physiology and Metabolism in Ruminants,* Ruckebush, Y. and Thivend, P., ed., AVI, Wesport, pp. 227–250.
7. McGavin, M. and Forsberg, C. W. (1989), *J. Bacteriol.* 17, 3310–3315.
8. Gilbert, H. J., Hall, J, Hazlewood, G. P., and Ferreira, L. M. A. (1990), *Mo. Microbiol.* **4,** 759–767.
9. Weimer, P. J. (1993), in *Forage Cell Wall Stucture and Digestibility,* Jung, H. G., Buxton, D. R., Hatfield, R. D., and Ralph, J., eds., Madison, WI, pp. 485–498.
10. Fontana, J. D., Correa, J. B. C., Duarte, J. H., Barbosa, A. M., and Blumel, M. (1984), *Biotechnol. Bioeng. Symp.* **14,** 175–184.
11. Fontana, J. D., Ramos, L. P., and Deschamps, F. C. (1995), *Appl. Biochem. Biotechnol.* **51/52,** 105–116.
12. Playne, M. J. (1984), *Biotechnol. Bioeng.* **26,** 426–433.
13. Brownell, H. H. and Saddler, J. N. (1987), *Biotechnol. Bioeng.* **29,** 228–235.
14. Ramos, L. P., Breuil, C., and Saddler, J. N. (1992), *Appl. Biochem. Biotechnol.* **34/35,** 37–48.
15. Saddler, J. N., Ramos, L. P., and Breuil, C. (1993), in *Bioconversion of Forest and Agricultural Plant Wastes,* Saddler, J. N. ed., C. A. B. International, London, pp. 73–92.
16. Rolz, C., de Arnola, M. C., Valladares, J., and de Cabrera, S. (1987), *Process Biochem.* **23,** 17–23.
17. Van Soest, P. J. (1967), *J. Anim. Sci.* **26,** 119–128.
18. Crampton, E. W. and Maynard, L. A. (1938), *J. Nutr.* **15,** 383–395.
19. Keller, R. J. (1986), The Sigma Library of FT-tR Spectra, Sigma, St. Louis, MO.
20. Reeves, J. B., III (1993), *Vibrational Spectroscopy* **5,** 303–310.
21. Ramos, L. P., Nazhad, M. M., and Saddler, J. N. (1993), *Enzyme Microb. Technol.* **15,** 821–831.
22. Segal, L., Creely, J. J. Martin, A. E., Jr., and Conrad, C. M. (1959), *Textile Res. J.* **27,** 30–41.
23. Orskov, E. R. and McDonald, I. (1979), *J. Agricultural Sci.* **92,** 449–453.
24. Mertens, D. R. (1977), *Fed. Proc.* **36,** 187–192.
25. Weimer, P. J., Guisa-Lopez, J. M., and French, A. D. (1990), *Appl. Environ. Microbiol.* **56,** 2421–2429.
26. Bacic, A., Harris, P. J., and Stone, B. A. (1988), in *The Biochemistry of Plants,* vol. 14, Preiss, J., ed., Academic, New York, pp. 297–371.
27. Donaldson, L. A., Wong, K. K. Y., and Mackie, K. L. (1988), *Wood Sci. Technol.* **22,** 103–114.
28. Wong, K. K. Y., Deverell, K. F., Mackie, K. L., Clark, T. A., and Donaldson, L. A. (1988), *Biotechnol. Bioeng.* **31,** 447–456.
29. Toussain, B., Excoffier, G., and Vignon, M. R. (1991), *Animal Feed Sci. Technol.* **32,** 235–242.
30. Smith, W. R., Yu, I., and Hungate, R. E. (1973), *J. Bacteriol.* **114,** 729–737.
31. Beveridge, R. J. and Richards, G. N. (1975), *Carbohydr. Res.* **43,** 163–172.
32. Russel, J. B. and Baldwin, R. L. (1978), *Appl. Environ. Microhiol.* **36,** 319–329.
33. Roger, V., Fonty, G., Komisarczuk-Bony, S., and Gouet, P. (1990), *Appl. Environ. Microbiol.* **56,** 3081–3087.
34. Huang, L. and Forsberg, C. W. (1990), *Appl. Environ. Microbiol.* **56,** 1221–1228.

35. Chesson, A., Stewart, C. S., and Wallace, R. J. (1982), *Appl. Environ. Microbiol.* **44,** 497–603.
36. Varel, V. H. and Jung, H. G. (1986), *Appl. Environ. Microbiol.* **52,** 275–280.
37. Akin, D. E., Rigsby, L. L., Theodorou, M. K., and Hartley, R. D. (1988), *Anim. Feed Sci. Technol.* **19,** 261–275.
38. Ramos, L. P., Breuil, C., and Saddler, J. N. (1992), *Enzyme Microb. Technol.* **15,** 19–25.
39. Hatfield, R. D. (1993) in *Forage Cell Wall Structure and Digestibility,* Jung, H. G., Buxton, D. R., Hatfield, R. D., and Ralph, J., eds., American Society of Agronomy, Madison, WI, pp. 285–313.
40. Sinitsyn, A. P., Gusakov, A. V., Vlasenko, E. Y. (1991), *Appl. Biochem. Biotechnol.* **30,** 43–59.
41. Fan, L. T., Gharpuray, M. M., and Lee, Y.-H. (1987), *Cellulose Hydrolysis.* Springer-Verlag, Berlin.
42. Grethlein, H. E. (1985), *Bio/Technology* **3,** 155–160.
43. Stone, J., Scallan, A., Donefer, E., and Ahlgren, E. (1969), *Adv. Chem. Ser.* **95,** 219–241.

0273-2289/96/5758—0183$7.25

Scientific Note

Measurement of the Inhibitory Potential and Detoxification of Biomass Pretreatment Hydrolysate for Ethanol Production

CHRISTOPHER J. RIVARD,*[,1] REBECCA E. ENGEL,[1] TAMMY K. HAYWARD,[2] NICHOLAS J. NAGLE,[1] CHRISTOS HATZIS,[2] AND GEORGE P. PHILIPPIDIS[2]

[1]*Applied Biological Sciences Branch and* [2]*Bioprocessing Branch, National Renewable Energy Laboratory, Golden, CO*

ABSTRACT

The Microtox assay represents a rapid, accurate, and reproducible method for determining general microbial toxicity. This assay was used to evaluate the relative toxicity of a variety of hydrolysate samples derived from dilute-acid and alkaline biomass pretreatment. Toxicity is elicited from biomass degradation products, such as furfural, hydroxymethyl furfural, and acetic acid, generated during pretreatment. Microtox results indicate that the pretreatment samples examined ranged from 9 to 71 toxicity units (TU). Correlations of TU and sample absorbance at several wavelengths were evaluated for all sample series. Sample TU values best agreed with absorbance at 230 nm, but the unsatisfactory fit suggests that absorbance should not be used as an absolute measure of sample toxicity.

Microtox data for pretreatment hydrolysate samples were correlated with the inhibition experienced by the ethanologenic yeast, *Saccharomyces cerevisiae* strain D_5A, during the simultaneous saccharification and fermentation (SSF) process of pretreated biomass. None of the alkaline pretreatment conditions produced inhibition during SSF (data not shown). However, the acid pretreatment conditions did produce a wide range of inhibitory and noninhibitory hydrolysates. In general, fermentation was inhibited for acid-pretreated hydrolysate samples with values exceeding 45 TU. Preliminary studies that focused on reducing hydrolysate sample toxicity (detoxification) indicate that adding perlite and zeolite had little effect. However, the use of charcoal, a universal flocculent, or ion-exchange resins significantly reduced sample toxicity, holding promise for the efficient bioconversion of pretreated biomass to ethanol. Moreover, the developed toxicity measurement assay can quickly monitor the quality of the pretreatment process. In this way, biomass conversion operation processes can be reliably controlled at the pilot and commercial scales.

Index Entries: Microtox; toxicity assay; ethanol fermentation; dilute-acid hydrolysis; detoxification.

*Author to whom all correspondence and reprint requests should be addressed.

INTRODUCTION

Numerous pretreatment technologies have been developed for solubilizing portions of biomass feedstocks before simultaneous saccharification and fermentation (SSF) *(1)*. In general, these treatments effectively enhance enzymatic cellulose conversion rates and yields by opening new surface areas for hydrolytic enzyme attack. However, solubilized sugars undergo further degradation reactions, especially under acidic/high-temperature conditions, thus creating unfermentable (and possibly toxic) compounds. Inhibitory compounds produced during dilute sulfuric acid pretreatment of biomass include acetic acid, furfurals, and numerous aromatic compounds. The identification and potential toxicity levels of suspected inhibitors from acid pretreatment processes were recently reviewed by McMillan *(2)*. Indeed, laboratory experiments conducted with selected pretreatment hydrolysates have demonstrated complete inhibition of yeast cell growth and ethanol production.

A rapid biological assay has been developed to determine the relative level of toxicity for waste waters and aqueous extracts (Microtox assay, Microbics Corporation, Carlsbad, CA). This assay measures the light output of luminescent indicator bacteria beforehand and after they are challenged by a sample of unknown toxicity. The degree of light loss (the indication of metabolic inhibition in the test organisms) indicates the sample's degree of toxicity. This type of assay is well suited to determining gross levels of toxicity, which may be a result of multiple variables.

In the present study, the Microtox assay was employed to determine the relative toxicity of pretreatment hydrolysate samples prior to ethanol fermentation. Microtox data were compared to those of sample absorbance and actual fermentation performance. It was also used to examine the efficacy of a variety of methods to detoxify a representative pretreatment hydrolysate sample before it was used in fermentation.

MATERIALS AND METHODS

Pretreatment Hydrolysates

Pretreatment hydrolysate samples were prepared using a mixed waste paper (white office paper, newsprint, coated paper magazines, and cardboard) feedstock. Before pretreatment, this feedstock was knife milled to pass a 3.1-mm (⅛-in.) rejection screen. Parameters including temperature, time, and acid addition are described in Table 1 for batch dilute sulfuric acid pretreatment. Additional pretreatments were performed under alkaline conditions using a temperature range of 25–120°C, 0–4 h, and 0–2 wt% sodium hydroxide. All pretreatments were conducted at 5% (wt/vol) paper solids using a 2-L (.53 gal) Parr stirred reactor constructed from Carpenter 30 cb3. Following pretreatment, hydrolysate samples were filtered through white linen to separate solids from liquids for individual analysis. Separated solids were stored wet and were generally in the range of 20–30% total solids. Liquid and solids fractions were stored at 4°C (39.2°F) until use.

Microtox Analysis

Before Microtox analysis, hydrolysate samples were removed from cold storage and allowed to reach room temperature. For dilute-acid pretreatment samples, the pH was adjusted to 7.0 by adding a commercial grade of solid hydrated lime (Snowflake Lime, Ash Grove Cement Company, Overland Park, KS). Although a

Table 1
Dilute-Acid Pretreatment of Mixed Paper Feedstock

Sample	Temperature, °C	Time, min	Sulfuric acid, wt%
1375-001-01	200 (392°F)	5	0.5
1375-001-02	200 (392°F)	20	0.73
1375-001-03	180 (356°F)	5	0.73
1375-001-04	200 (392°F)	5	0.73
1375-001-05	180 (356°F)	20	0.73
1375-001-06	190 (374°F)	12.5	0.616
1375-001-07	180 (356°F)	20	0.5
1375-001-08	190 (374°F)	12.5	0.616
1375-001-09	180 (356°F)	5	0.5
1375-001-10	200 (392°F)	20	0.5

rapid pH adjustment was observed when the lime was added, the samples slowly readjusted to a lower pH. For alkaline pretreatment samples, the pH was adjusted to 7.0 by adding 2*M* hydrochloric acid. To provide the necessary time for pH equilibration, hydrolysate samples were allowed to stir at room temperature for 2 h before final pH adjustment. The samples were then transferred to 50-mL (1.69-oz.) disposable centrifuge tubes and centrifuged at room temperature for 15 min at 3500 rpm (1400*g*) (Centra 4, International Centrifuge, Needham Heights, MA). The samples were then filtered using sterile 115-mL (3.89 oz.) capacity filter units equipped with a 0.45-μm membrane (Gelman Sciences, Ann Arbor, MI). For samples of known high toxicity or those with a substantially dark appearance, a primary dilution was made (generally 1:10) using distilled water. The Microtox analysis was performed as previously described *(3)* using a model 500 unit equipped with a microcomputer to gather and analyze data. Microtox software (version 6.1) was used to calculate for effective concentration 50 (EC_{50}) and toxicity units (TU) of the samples. Phenol at 90 mg/L in distilled water was used as a standard and analyzed daily to ensure the accuracy of the assay.

Hydrolysate Absorbance Measurements

The absorbance of each sample was analyzed following a 1:100 dilution with distilled water using a Hewlett Packard 8451A Diode Array Spectrophotometer. Samples were assayed at 208, 230, and 274 nm, using quartz cuvets.

Hydrolysate Detoxification

Activated charcoal (Darco-g-60, MCB Manufacturing Chemist, Cincinnati, OH), ion exchange resin (RG501 -XS, Bio-Rad Laboratories, Hercules, CA), flocculent (Universal Flocculent 1002, Nalco Chemical, Denver, CO), zeolite (U.S. Zeolites Inc., Golden, CO), and perlite (Horticulture Perlite, Premier Brands, New Rochelle, NY) were examined for their ability to reduce hydrolysate sample toxicity and color. In general, test materials were added as a dry powder to the pH-adjusted hydrolysate sample at varying concentrations (wt/vol). The sample was then mixed using a magnetic stirrer for 2 h at room temperature. Before analysis, the treated hydrolysate samples were filtered using 115-mL (3.89-oz.) filter units (0.45-μm membrane).

Ethanol Fermentation of Pretreatment Hydrolysates

SSF experiments were performed with acid and alkaline slurry samples as previously described *(4)*. Acid-pretreated hydrolysate liquid was pH adjusted with solid NaOH to 6.5–7.0. Alkaline-pretreated hydrolysate liquid was pH adjusted with 96% phosphoric acid to 4.0. Non-pH-adjusted wet solids were then combined with hydrolysate to achieve a 10% solids loading and a slurry pH of 5.0. Dry nutrients were added so as not to dilute further the hydrolysate sample. Bactopeptone (2 g) and yeast extract (1 g, both from Difco, Detroit, MI) were added to each growth flask. The sample flasks were then autoclaved for 30 min. The flasks were then inoculated with a 10% vol/vol culture of nonhydrolysate adapted *Saccharomyces cerevisiae* strain D_5A and 10 filter paper units/g of cellulosic material of a liquid cellulase preparation (Econase CE, Enzyme Development Corporation, New York, NY). The flasks were incubated at 38°C and 150 rpm using a New Brunswick rotary shaker. Water traps were fitted to each growth flask to vent CO_2 and maintain anaerobic conditions. Whole slurry samples were analyzed every 24 h for free sugars and end products, including ethanol.

Before analysis, slurry samples were centrifuged to remove solids. The supernatant was further filtered using 0.45-µm syringe filters (Gelman Sciences) before HPLC analysis. A Hewlett Packard HPLC (model 1090) equipped with an organic acid column (HPX-87H, Bio-Rad) was used to quantify glucose, cellobiose, glycerol, acetic acid, and ethanol.

Ethanol produced after 7 d of fermentation was compared to the theoretical ethanol yield calculated from the initial six carbon sugars in the substrate.

RESULTS

Two series of pretreatment hydrolysate samples including dilute sulfuric acid pretreatment (#1375-001) and alkaline pretreatment (#1375-004) were evaluated for toxicity. These hydrolysates were stored at 4°C (39.2°F), and the dark components of the liquor impregnated the plastic storage container. The dark color of the hydrolysates correlated well with a strong "phenol-type" odor. In the case of the dilute sulfuric acid hydrolysate samples, pH readjustment was required over time because storage of the sample caused a continual pH drop.

The toxicity of the pH-adjusted and filtered hydrolysate samples was determined using the Microtox assay. The toxicity of each sample was evaluated at least twice (generally on separate days). The data for sample toxicity, expressed in average TU, are given for each series in Tables 2 and 3. In general, the toxicity within a given series varies significantly. Overall, the sample toxicity for the #1375 set ranged from 9 to 71 TU. The toxicity level of the hydrolysate samples may be compared to that of phenol at 90 mg/L, which exhibits a toxicity of 3–4 TU.

Because the toxic nature of the hydrolysate samples may be attributed to a variety of pretreatment by-products, including furfurals and lignin degradation products (i.e., *p*-hydroxybenzaldehyde, *p*-coumaric acid, and vanillic acid), the absorbance optima for these compounds were used to correlate with sample toxicity. Sample absorbance was evaluated for each experimental series at 208, 230, and 274 nm. The correlation of sample TU and absorbance is described in Figs. 1 and 2. In general, the best fit occurs with the correlation of TU and absorbance at 230 nm. However, all correlations of TU and absorbance demonstrate multiple outliers.

Table 2
Microtox Analysis of Pretreatment Hydrolysate Samples #1375-001

	Assay date	Dilution	Total TU	Sample
001-01	3-22-94	1:20	42.6	42.6 ± 0.7
	3-24-94	1:10	43.3	
	3-29-94	1:10	41.9	
001-02	3-22-94	1:20	62.7	64.4 ± 5.4
	3-23-94	1:20	68.0	
	3-24-94	1:10	69.2	
	3-29-94	1:10	57.5	
001-03	6-22-94	1:2	33.4	33.6 ± 0.3
	6-23-94	1:4	33.8	
001-04	6-24-94	1:20	73.6	70.6 ± 4.3
	6-27-94	1:20	67.5	
001-05	6-24-94	1:10	52.6	53.6 ± 1.4
	6-27-94	1:10	54.6	
001-06	6-24-94	1:20	53.9	47.1 ± 6.1
	6-27-94	1:10	45.5	
	9-20-94	1:10	42.0	
001-07	6-24-94	1:10	47.2	46.9 ± 0.4
	6-27-94	1:10	46.6	
001-08	3-22-94	1:20	44.5	45.5 ± 2.5
	3-24-94	1:10	43.7	
	3-29-94	1:10	48.4	
001-09	3-22-94	1:20	20.6	21.1 ± 1.6
	3-23-94	1:10	22.2	
	3-24-94	1:2	19.1	
	3-29-94	1:2	22.6	
001-10	3-22-94	1:20	59.2	62.9 ± 3.2
	3-24-94	1:10	65.2	
	3-29-94	1:10	64.2	

Figure 3 describes the relationship of overall (ethanol) yield and sample TU. In general, the data indicate a sigmoidal curve in which hydrolysate sample TU values in excess of 45 result in severe toxicity to the yeast. In some samples with low TU values, ineffective pretreatment reduced cellulose availability to the hydrolytic enzyme and thus resulted in suboptimal overall yields.

Although some level of sample toxicity may be tolerated in ethanol fermentations, many of the pretreatment hydrolysate samples were demonstrated to be toxic to yeast. The ability to reduce the level of sample toxicity to one that yeast can tolerate was examined using a variety of adsorbents. Adsorbents were added to hydrolysate sample #1375-001-08 at two concentrations (except for the flocculent). The effects on sample toxicity of adding adsorbents are described in Fig. 4. These data indicate that only the flocculent, charcoal, and ion-exchange resins substantially reduced the sample toxicity.

Table 3
Microtox Analysis of Pretreatment Hydrolysate Samples #1375-004

Sample	Assay date	Dilution	Total TU	Average TU
004-01	7-1-94	1:10	32.9	35.2 ± 2.1
	7-5-94	1:5	35.8	
	7-5-94	1:5	37.0	
004-02	7-5-94	1.2	9.8	9.3 ± 0.7
	7-5-94	—	8.8	
004-03	7-1-94	1:10	50.2	52.2 ± 2.0
	7-5-94	1:10	54.2	
	7-5-94	1:10	52.3	
004-04	7-5-94	1:2	9.4	9.0 ± 0.6
	7-6-94	—	8.6	
004-05	7-6-94	1:10	39.6	35.6 ± 3.8
	7-6-94	1:2	32.1	
	7-7-94	1:5	35.0	
004-06	7-6-94	1:10	41.4	39.8 ± 1.4
	7-7-94	1:5	39.0	
	7-7-94	1:5	39.1	
004-07	7-6-94	1:5	15.3	15.2 ± 2.0
	7-7-94	1:2	17.2	
	7-7-94	1:2	12.4	
	7-11-94	1:2	16.0	
004-08	7-7-94	1:10	35.9	37.3 ± 2.0
	7-11-94	1:5	38.7	
004-09	3-24-94	1:2	9.5	10.2 ± 0.7
	3-29-94	1:2	10.3	
	3-29-94	—	10.8	
004-10	3-22-94	1:20	52.6	52.3 ± 2.7
	3-29-94	1:20	49.8	
	3-30-94	1:10	55.9	
	3-30-94	1:10	50.7	

DISCUSSION

Owing to the complex nature of inhibitors produced during pretreatment of biomass and cellulosic wastes, standard methods, such as solution absorbance, are ineffective at predicting the relative toxicity that may be experienced by a fermentation microorganism. The Microtox assay represents a rapid, reproducible, EPA-approved method for evaluating the relative toxicity of a variety of environmental samples, including pretreatment hydrolysates. In this study, pretreatment hydrolysate sample toxicity varied from 9 to 71 TUs. Correlating hydrolysate sample toxicity with absorbance demonstrates the best fit with readings at 230 nm, although substantial line fit variation occurred within series. All correlations of TU and absorbance demonstrate multiple outliers, which may indicate that the toxic byproducts from the pretreatment process are heterogeneous and vary greatly with changes in the pretreatment process. If true, this condition makes assumptions of sample toxicity using absorbance at 230 nm relatively inaccurate.

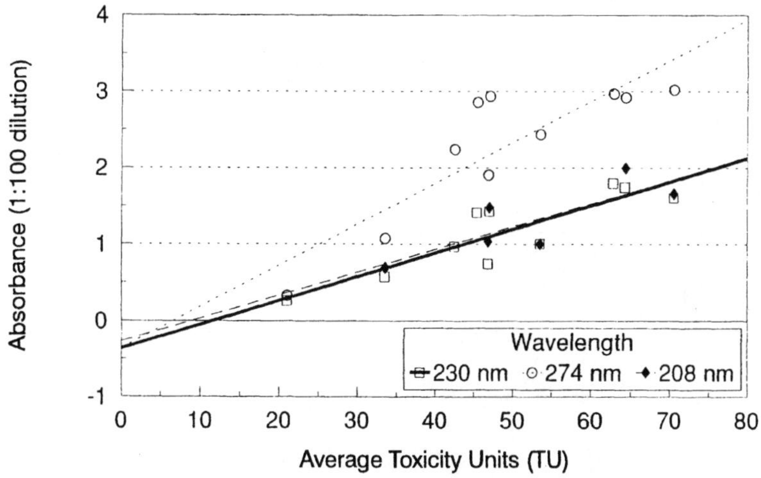

Fig. 1. Correlation of dilute sulfuric acid hydrolysate TU and sample absorbance. Microtox TU was as described in Table 2. Absorbance measurements were performed on samples diluted 1:100 with distilled water.

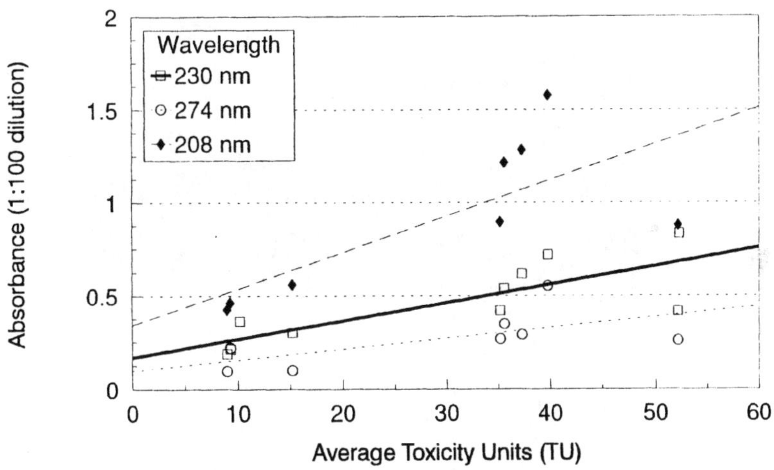

Fig. 2. Correlation of alkaline hydrolysate TU and sample absorbance. Analysis was performed as described in Fig. 1.

The proposed utility of the Microtox assay is to predict the level of toxicity that may be tolerated by a specific fermentation catalyst, under the defined parameters with which this study was performed, i.e., feedstock type, pretreatment protocols, and fermentation catalyst. Although some level of adaptation to toxic components of pretreatment hydrolysates may be anticipated, this study identifies an approximate limitation of 45 TU for effective fermentation under the defined conditions.

Many adsorbents were chosen for study, and the Microtox assay used to determine their effectiveness on reducing overall toxicity. Perlite and zeolite

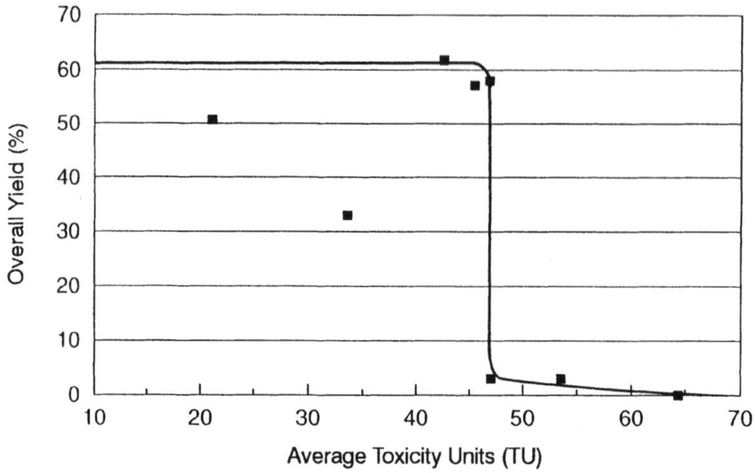

Fig. 3. Comparison of dilute-acid pretreatment sample SSF performance and sample toxicity. Overall yield was determined as described in Materials and Methods.

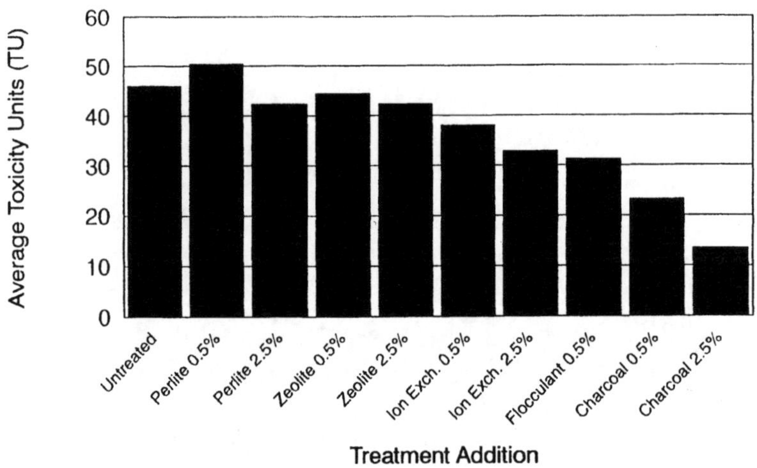

Fig. 4. Changes in measured TU of dilute-acid hydrolysate sample #1375-001-08 following treatment with potential detoxifying agents.

appeared to be relatively ineffective in reducing hydrolysate sample toxicity. However, applying activated charcoal, universal flocculent, or ion-exchange resins substantially reduced hydrolysate toxicity. Additional work will identify alternative detoxifying methods that can provide low cost and minimal obstacles to downstream processing.

ACKNOWLEDGMENT

This work was funded by the Biochemical Conversion Element of the Biofuels Program of the US Department of Energy.

REFERENCES

1. Wright, J. D. (1988), *Economics of Enzymatic Hydrolysis Processes*, SERI/TP-231-3310, Solar Energy Research Institute.
2. McMillan, J. D. (1994), in *Enzymatic Conversion of Biomass for Fuels Production*, Himmel, M. E., Baker, J. O., and Overend, R. P., eds., American Chemical Society Symposium Series 566, pp. 411–437.
3. American Society for Testing and Materials (1995), "The Basic Microtox Test," ASTM Standard D-5660.
4. Vinzant, T.B., Ponfick, L., Nagle, N. J., Ehrman, C. I., Reynolds, J. B., and Himmel, M. E. (1994), *Appl. Biochem. Biotechnol.* **45/46,** 611–626.

Session 2
Applied Biological Research

Ting Carlson[1] and William A. Apel[2]

[1]Cargill, Inc., Dayton, OH;
and [2]Idaho National Laboratory, Idaho Falls, ID

Conversion of biomass to ethanol and other chemicals has been a subject of intensive studies over the last 20 years. The critical elements in this process are hydrolysis of biomass to sugars and microbial fermentation of sugars to ethanol. Recent advances in developing biological catalysts to improve the overall economics of this process are presented in this session.

Being able to convert pentoses and glucose in hydrolysates simultaneously to ethanol is a trait actively pursued by many laboratories. The approach has been to add ethanol-fermentative genes into substrate-versatile microbe or to add sugar-metabolizing genes into good ethanol fermentors. In this session, the results of using a genetically engineered *Zymomonas* to ferment xylose, arabinose, and glucose are presented. The basic physiology of ethanol production in various ethanol microbes is also presented. Another important aspect of this process is the production of efficient cellulase systems to hydrolyze biomass to sugars. The crystalline structure of cellulose and the feedback inhibition of cellulase activity by its reaction products hindered the use of biomass as a sugar source. A presentation on the cellulase system in *Clostridium acetobutylicum* is included in this session.

With advances in genetic engineering such as these, we could expect a commercially viable process that converts inexpensive biomass to ethanol in the near future.

0273-2289/96/5758—0193$6.75

Ethanolic Fermentation of Lignocellulose Hydrolysates

A Minireview

BÄRBEL HAHN-HÄGERDAL

Applied Microbiology, Lund Institute of Technology/ Lund University, PO Box 124, S-221 00 Lund, Sweden

Lignocellulosic raw materials, such as forest products and agricultural residues, constitute a major renewable energy source for the fermentative production of liquid fuels, such as ethanol. Using transportation fuels from renewable resources should not contribute to global warming, generally referred to as the "greenhouse effect," and would thus assure sustainable development of the transportation sector. This argument is based on the assumption that the production and harvest rates of lignocellulosic raw materials used for the production of transportation fuels are balanced by the assimilation rate of combustion carbon dioxide in the photosynthesis of green plants.

Lignocellulosic raw materials are composed of hemicellulose, cellulose, and lignin. The hemicellulosic and cellulosic fractions are made up of the monosaccharides glucose, mannose, galactose, xylose, and arabinose, the composition being dependent on the nature of the raw material. In soft woods and in agricultural residues, the pentose sugars xylose and arabinose account for 5–25 and 0.5–3%, respectively, of the raw material *(1)*. Prior to fermentation, lignocellulosic materials are pretreated and/or hydrolyzed with acids and/or enzymes *(2)* to make the sugars accessible to fermenting microorganisms. The sugars can be fermented to ethanol with bacteria, yeast, and filamentous fungi (for a recent review, *see* ref. 3). During pretreatment and hydrolysis, a number of microbial inhibitors are released and produced. The nature and composition of these compounds are dependent on the pretreatment and hydrolysis conditions, and will have a profound influence on the efficiency of fermentation process. Furthermore, in environmentally sustainable processes, the utilization of fresh water has to be minimized by circulation of process streams, which leads to an accumulation of the inhibitory compounds *(4)*.

Compounds inhibitory to fermentation are both inherent in the lignocellulosic materials and produced during pretreatment, hydrolysis, and fermentation. The degree of inhibition varies with raw material and pretreatment, and the presence of two or more compounds is usually more inhibitory than one single compound alone *(5)*. On pretreatment and/or hydrolysis, acetic acid—commonly used as an antimicrobial food preservative—is released from acetyl groups in the hemicellulose fraction *(6,7)*. Sugar degradation products, such as furfurals *(8)*, and lignin degradation products, such as phenolics *(9–11)*, are formed owing to the chemical reactions in pretreatment and/or hydrolysis. In addition, the product ethanol and the fermentation byproducts acetic acid, acetaldehyde, formic acid, and lactic acid are well-known microbial inhibitors *(12,13)*. Finally, heavy metals released from the equipment have an inhibitory effect *(14)*.

Lignocellulose hydrolysates have been detoxified to enhance their fermentability. Several methods have been suggested: calcium hydroxide in combination with sulfite *(15)*, ion exchange *(16)*, molecular sieves *(10)*, and steam stripping *(17)*. Using calcium hydroxide in combination with sulfite, also called overliming, contributes as much as 22% to the production costs of ethanol produced from a pentose-rich, steam-pretreated willow hydrolysate using the recombinant bacterium *Escherichia coli* KO11 *(18,19)*. The fact that detoxification potentially allows all sugars in a lignocellulose hydrolysate to be fermented to ethanol may justify its relatively high cost. When the liquid stream from steam-pretreated willow was fractionated by evaporation, the fermentation inhibitors were localized to the nonvolatile fraction *(20)*. The nonvolatile fraction also inhibited the enzymatic hydrolysis of the steam-pretreated raw material, however, to a lesser extent. Therefore, in addition to the detoxification methods mentioned, the inhibitors could also be removed by evaporation of the recirculating liquid streams. The volatile fraction would be recirculated, and the nonvolatile fraction containing organic material would be burned in a steam boiler to provide energy for the process. Economic analyses will determine whether or not detoxification should be used and which method to choose.

For optimal process economy, all sugars should be fermented to ethanol, preferentially simultaneously. Microorganisms fermenting all sugars in lignocellulosic hydrolysates to ethanol are either slow or highly susceptible to the inhibitors *(21)*. The efficient fermenting yeast, *Saccharomyces cerevisiae*, ferments only hexoses and the isomeric form of xylose, xylulose. It lacks the two initial enzymatic steps of the xylose metabolism, xylose reductase and xylitol dehydrogenase. The research efforts on the fermentation of all sugars in a lignocellulose hydrolysate have mainly focused on the efficient fermentation of the pentose sugar xylose (for recent reviews, *see* refs. *3,22–25*). A number of different organisms have been considered: filamentous fungi, such as *Mucor sp* 105 *(26)* and *Fusarium oxysporium (27)*; yeasts such as *Pichia stipitis, Candida shehatae, Pachysolen tannophilus* (for a review, *see* ref. 25); Bakers's yeast *S. cerevisiae* in combination with xylose isomerase *(28,29)*; thermophilic anaerobic bacteria, such as *Clostridium thermohydrosulphuricum (30), Clostridium thermosaccharolyticum (31)*, and *Thermoanaerobacter ethanolicus (32)* as well as lactate dehydrogenase mutants of *Bacillus stereothermophilus (33)*. Strains of *P. stipitis (34), C. shehatae (17)*, and *S. cerevisiae (35)*, the latter in combination with xylose isomerase, have been adapted and screened for, respectively, in order to enhance their inhibitor tolerance. Present knowledge suggests that the yeasts together with the recombinant bacteria *(see below)* offer the greatest potential for the development of cost-effective fuel ethanol processes. With the filamentous fungi, the productivity is too low, and with the thermophilic anaerobic bacteria, both productivity and yield are too low.

Recombinant DNA technology makes it possible to introduce new pathways and to block unwanted metabolic pathways, so that novel microorganisms can be engineered for efficient ethanol production from all sugars present in lignocellulose hydrolysates. *E. coli* KO11, which can utilize all lignocellulose sugars, has been engineered to contain the ethanologenic genes from *Zymomonas mobilis (36,37)*. In addition, the acid-producing genes have been deleted. *Klebsiella oxytoca* M5A1 has been similarly genetically engineered *(38)*. In laboratory media, the recombinant strains ferment all sugars present in lignocellulosic raw materials to ethanol with theoretical yields. However, the maximum fermentative performance of *E. coli* KO11 in a lignocellulose hydrolysate is greatly dependent on a cheap and efficient detoxification method *(18,19)*.

Recently, *Z. mobilis* CP4 has been genetically engineered with an entire xylose-utilizing pathway by introducing the *E. coli* genes for xylose isomerase, xylulokinase, transketolase, and transaldolase *(39)*. The fermentative performance of the recombinant strain has been established under selection pressure in laboratory media containing mixtures of glucose and xylose where it ferments both sugars with close to theoretical yields. The main drawback with *Z. mobilis* is that it does not ferment galactose, mannose, and arabinose present in most lignocellulosic raw materials *(1,40)*. The fact that *Z. mobilis* has been under consideration for large-scale fuel ethanol production from sugar feed stocks for over 15 yr without any process being established *(41–43)* throws some doubt on the usefulness of this organism in large-scale industrial applications.

With the aim to combine the xylose-fermenting properties of *P. stipitis* with the inhibitor tolerance of *S. cerevisiae*, the genes for the xylose-utilizing enzymes xylose reductase *(44–46)* and xylitol dehydrogenase *(47–51)* from *P. stipitis* have been cloned and expressed in *S. cerevisiae*. In addition, the genes for the enzyme xylolokinase from *P. tannophilus* and from *S. cerevisiae* have also been cloned and (over)expressed in *S. cerevisiae (52)*. All attempts to express a functional bacterial xylose isomerase in *S. cerevisiae* have so far failed *(53–55)*.

Transformants of *S. cerevisiae* harboring the gene for the enzyme xylose reductase convert xylose to xylitol with theoretical yields in the presence of cosubstrate for the generation of reduced cofactors and maintenance energy *(44,56–58)*. Xylitol is used in the food industry as an anticariogenic sweetener as well as a sugar substitute for diabetics *(59)*. Theoretically, *S. cerevisiae* transformed with the genes for the enzymes xylose reductase and xylitol dehydrogenase should ferment xylose to ethanol. Attempts in this direction met with limited success, so that only low yields of ethanol have been obtained *(49–51)*. However, when the *S. cerevisiae* gene for the enzyme transaldolase was simultaneously overexpressed, mixtures of xylose and glucose were fermented to ethanol *(60)*. Similarly, ethanol formation from xylose was enhanced when the gene for xylulokinase was overexpressed in transformants of *S. cerevisiae* harboring the genes for the enzymes xylose reductase and xylitol dehydrogenase (Ho, personal communication).

ACKNOWLEDGMENTS

The research on xylose fermentation at the Department of Applied Microbiology was financially supported by NUTEK (Swedish National Board for Industrial and Technical Development), NI (Nordic Fund for Technology and Industrial Development), SSEU (Swedish Ethanol Development Foundation), and NFR (Swedish Natural Science Research Council), as well as the Knut and Alice Wallenberg foundation.

REFERENCES

1. Ladisch, M. R., Lin, K. W., Voloch, M., and Tsao, G. T. (1983), *Enzyme Microb. Technol.* **5,** 82–102.
2. von Sivers, M. and Zacchi, G. (1995), *Biores. Technol.* **51,** 43–52.
3. Olsson, L. and Hahn-Hägerdal, B. (1996), *Enzyme Microb. Technol.* (accepted for publication).

4. Galbe, M. and Zacchi, G. (1991), *Appl. Biochem. Biotechnol.* **33/34,** 93–104.
5. Frazier, W. C. and Westhoff, D. C. (1988), *Food Microbiology*, 4th ed., McGraw-Hill, New York, pp. 83–90.
6. Pampulha, M. E. and Loureiro, V. (1989), *Biotechnol. Lett.* **11,** 269–274.
7. van Zyl, C., Prior, B. A., and du Preez, J. C. (1991), *Enzyme Microb. Technol.* **13,** 82–86.
8. Martin, M. Delgenes, J. P., Moletta, R., and Navarro, J. M. (1992), *Proc. Biomass Energy, Ind. Environ., E. C. Conf., 6th Meeting*, Grassi, Colina, Zibetta, eds., Elsevier, London, pp. 1332–1336.
9. Ando, S., Arai, I., Kiyoto, K., and Hanai, S. (1986), *J. Ferm. Technol.* **64,** 567–570.
10. Tran, A. V. and Chambers, R. P. (1986), *Enzyme Microb. Technol.* **8,** 439–444.
11. Nishikawa, N. K., Sutcliffe, R., and Saddler, J. N. (1988), *Appl. Microbiol. Biotechnol.* **27,** 549–552.
12. Brown, S. W., Oliver, S. G., Harrison, D. E. F., and Righelato, R. C. (1981), *Appl. Microbiol. Biotechnol.* **11,** 151–155.
13. Maiorella, B., Blanch, H. W., and Wilke, C. R. (1983), *Biotechnol. Bioeng.* **25,** 103–121.
14. Watson, N. E., Prior, B. A., Lategan, P. M., and Lussi, M. (1984), *Enzyme Microb. Technol.* **6,** 451–456.
15. Leonard, R. H. and Hajny, G. J. (1945), *Ind. Eng. Chem.* **37,** 390–395.
16. Clark, T. A. and Mackie, K. L. (1984), *J. Chem. Technol. Biotechnol.* **34B,** 101–110.
17. Yu, S., Wayman, M., and Parekh, S. K. (1987), *Biotechnol. Bioeng.* **29,** 1144–1150.
18. von Sivers, M., Zacchi, G., Olsson, L., and Hahn-Hägerdal, B. (1994), *Biotechnol. Prog.* **10,** 555–560.
19. Olsson, L., Hahn-Hägerdal, B., and Zacchi, G. (1995), *Biotechnol. Bioeng.* **45,** 356–365.
20. Palmqvist, E., Hahn-Hägerdal, B., Galbe, M., and Zacchi, G. (1996), *Enzyme Microb. Technol.* (accepted for publication).
21. Hahn-Hägerdal, B., Jeppsson, H., Olsson, L., and Mohagheghi, A. (1994), *Appl. Microbiol. Biotechnol.* **41,** 62–72.
22. Schneider, H. (1989), *Crit. Rev. Biotechnol.* **9,** 1–40.
23. Jeffries, T. W. (1990) in *Yeast-Biotechnology and Biocatalysis*. Marcel Dekker, New York, pp. 349–394.
24. Hahn-Hägerdal, B., Hallborn, J., Jeppsson, H., Olsson, L., Skoog, K., and Walfridsson, M. (1993), in, *Bioconversion of Forest and Agricultural Plant Residues*, Saddler, J. N., ed., CAB International, Wallingford, UK, pp. 231–290.
25. Hahn-Hägerdal, B., Jeppsson, H., Skoog, K., and Prior, B. A. (1994), *Enzyme Microb. Technol.* **16,** 933–943.
26. Ueng, P. P. and Gong, C.-S. (1982), *Enzyme Microb. Technol.* **4,** 169–171.
27. Suihko, M.-L. (1983), *Biotechnol. Lett.* **5,** 721–724.
28. Wang, P. Y., Johnson, B. F., and Schneider, H. (1980), *Biotechnol. Lett.* **2,** 273–278.
29. Lindén, T. and Hahn-Hägerdal, B. (1989), *Enzyme Microb. Technol.* **11,** 583–589.
30. Ng, T. K., Ben-Basset, A., and Zeikus, J. G. (1981), *Appl. Environ. Microbiol.* **41,** 1337–1343.
31. Liu, H.-S., Hsu, H.-W., and Saylor, G. S. (1988), *Biotechnol. Prog.* **4,** 40–46.
32. Lacis, L. S. and Lawford, H. G. (1988), *Arch. Microbiol.* **150,** 48–55.
33. San Martin, R., Bushell, D., Leak, D. J., and Hartley, B. S. (1993), *J. Gen. Microbiol.* **139,** 1033–1040.
34. Parekh, S. R., Parekh, R. S., and Wayman, M. (1987), *Process Biochem.* **22,** 85–91.
35. Lindén, T., Peetre, J., and Hahn-Hägerdal, B. (1992), *Appl. Environ. Microbiol.* **58,** 1661–1669.
36. Ohta, K., Beall, D. S., Mejia, J. P., Shanmugam, K. T., and Ingram, L. O. (1991), *Appl. Environ. Microbiol.* **57,** 893–900.
37. Lindsay, S. E., Bothast, R. J., and Ingram, L. O. (1995), *Appl. Microbiol. Biotechnol.* **43,** 70–75.
38. Ohta, K., Beall, D. S., Mejia, J. P., Shanmugam, K. T., and Ingram, L. O. (1991), *Appl. Environ. Microbiol.* **57,** 2810–2815.
39. Zhang, M., Eddy, C., Deande, K., Finkelstein, M., and Picataggio, S. (1995), *Science* **267,** 240–243.
40. Olsson, L. and Hahn-Hägerdal, B. (1993), *Process Biochem.* **28,** 249–257.
41. Rogers, P. L., Lee, K. J., and Tribe, D. E. (1980), *Process Biochem.* **15, Aug./Sept.,** 7–11.

42. Parekh, S. R., Parekh, R. S., and Wayman, M. (1989), *Process Biochem.* **24,** 88–91.
43. Doelle, H. W., Kirk, L., Crittenden, R., and Toh, H. (1993), *Crit. Rev. Biotechnol.* **13,** 57–98.
44. Hallborn, J., Walfridsson, M., Airaksinen, U., Ojamo, H., Hahn-Hägerdal, B., Penttilä, M., and Keränen, S. (1991), *Bio/Technology* **9,** 1090–1095.
45. Amore, R., Kötter, P., Kuester, C., Ciriacy, M., and Hollenberg, C. P. (1991), *Gene Amst.* **109,** 89–98.
46. Takuma, S., Nakashima, N., Tantirungkij, M., Kinoshita, S., Okada, H., Seki, T., and Yoshida, T. (1991), *Appl. Biochem. Biotechnol.* **28/29,** 327–340.
47. Kötter, P., Amore, R., Hollenberg, C. P., and Ciriacy, M. (1990), *Curr. Genet.* **18,** 493–500.
48. Persson, B., Hallborn, J., Walfridsson, M., Hahn-Hägerdal, B., Keränen, S., Penttilä, M., and Jörnvall, H. (1993), *FEBS Lett.* **324,** 9–14.
49. Kötter, P. and Ciriacy, M. (1993), *Appl. Microbiol. Biotechnol.* **38,** 776–783.
50. Tantirungkij, M., Nakashima, N., Seki, T., and Yoshida, T. (1993), *J. Ferment. Bioeng.* **75,** 83–88.
51. Meinander, N., Hallborn, J., Keränen, S., Ojamo, H., Penttilä, M., Walfridsson, M., and Hahn-Hägerdal, B. (1994), in *ECB6: Proc. 6th Congr. Biotechnol.*, Alberghina, L., Frontali, L., and Sensi, P., eds., Elsevier Science, B. V., pp. 1143–1146.
52. Deng, X. X. and Ho, N. W. Y. (1990), *Appl. Biochem. Biotechnol.* **24/25,** 193–199.
53. Ho, N. W. Y., Stevis, P., Rosenfeld, S., Huang, J. J., and Tsao, G. T. (1983), *Biotech. Bioeng. Symp.* **13,** 245–250.
54. Sarthy, A. V., McConaughy, B. L., Lobo, Z., Sundstrom, J. A., Furlong, C. E., and Hall, B. D. (1987), *Appl. Environ. Microbiol.* **53,** 1996–2000.
55. Amore, R., Wilhelm, M., Hollenberg, C. P. (1989), *Applied Microbiol. Biotechnol.* **30,** 35–357.
56. Hallborn, J., Gorwa, M.-L., Meinander, N., Penttilä, M., Keränen, S., and Hahn-Hägerdal, B. (1994), *Appl. Microbiol. Biotechnol.* **42,** 326–333.
57. Meinander, N., Hahn-Hägerdal, B., Linko, M., Linko, P., and Ojamo, H. (1994), *Appl. Microbiol. Biotechnol.* **42,** 334–339.
58. Thestrup, H. N. and Hahn-Hägerdal, B. (1995), *Appl. Environ. Microbiol.* **61,** 2043–2045.
59. Mäkinen, K. K. (1979), *Adv. Food Res.* **25,** 137–158.
60. Walfridsson, M., Hallborn, J., Penttilä, M., Keränen, S., and Hahn-Hägerdal, B. (1995), *Applied Environ. Microbiol.* **61,** 4184–4190.

0273-2289/96/5758—0201$8.00

Peculiarities of the Regulation of Fermentation and Respiration in the Crabtree-Negative, Xylose-Fermenting Yeast *Pichia stipitis*

VOLKMAR PASSOTH,
MARTIN ZIMMERMANN, AND ULRICH KLINNER*

Institut für Biologie IV (Mikrobiologie), RWTH Aachen, Worringer Weg, 52074 Aachen, Germany

ABSTRACT

The respiration of *Pichia stipitis* was not repressed by either high concentrations of fermentable sugars or oxygen limitation. Fermentation was not induced by high sugar concentrations, but was inactivated by aerobic conditions. The activity of pyruvate dehydrogenase was constitutive. In contrast, pyruvate decarboxylase, alcohol dehydrogenase, and aldehyde dehydrogenase were induced by a reduction in the oxygen tension. It was demonstrated that in *P. stipitis* , the pyruvate decarboxylase is not induced by a signal from glycolysis. Contrary to *Saccharomyces cerevisiae*, the pyruvate decarboxylase was not inhibited by phosphate.

Index Entries: *Pichia stipitis*; PDC; PDH; ADH; aldehyde dehydrogenase.

INTRODUCTION

The yeast *Pichia stipitis* converts xylose to ethanol with nearly maximal yield and relatively high rates in comparison to other xylose-fermenting yeasts. However, the rates are rather low in comparison to the glucose fermentation capacity of *Saccharomyces cerevisiae*. This is one of the reasons why *P. stipitis* has not yet been used for the production of ethanol from xylose on an industrial scale *(1,2)*.

Attempts were made to overcome this disadvantage of *P. stipitis* by searching for optimal fermentation conditions, but the results did not lead to substantial improvement *(3)*. A solution could be attained by manipulating the genes of *P. stipitis* coding for enzymes involved in carbohydrate metabolism. The xylose catabolism enables the cells to divide frequently under aerobic conditions. This quick growth is severely retarded under semiaerobic conditions. This retardation does not appear to be caused by a redox imbalance in the cell owing to the reduced respiration, since no xylitol is produced and the decrease in growth rate is also found in glucose growing cells *(4,5)*. Rather, it seems to be the result of a decrease in the metabolic flux through the several steps of the xylose assimilation pathway and glycolysis.

*Author to whom all correspondence and reprint requests should be addressed.

Only little is known about the regulation of the appropriate steps. It was shown that no fermentation occurs under aerobic conditions in medium containing 3% xylose *(6)*. This result suggested that *P. stipitis* is a Crabtree- negative species *(7)*. In *Candida shehatae*, a species closely related to *P. stipitis (8)*, the activity of alcohol dehydrogenase (ADH) during oxygen limitation is much higher than under aerobic conditions *(9)*. The activity of pyruvate decarboxylase (PDC) of *P. stipitis* is strongly influenced by the oxygen concentration in the medium *(10)*. In the Crabtree-positive species *S. cerevisiae*, detailed knowledge exists about the induction mechanism of PDC. At least it is triggered by distinct metabolic signals that originate from the glycolysis *(11)*. However, there is no experimental evidence for the mode of activation of PDC in *P. stipitis*.

The investigations described here deal with the metabolic regulation at the branching points between respirative and fermentative metabolism, including the regulation mechanism for pyruvate dehydrogenase (PDH), PDC, aldehyde dehydrogenase (AldDH), and ADH. An essential aspect of our research was the comparison of the regulation in the Crabtree-negative yeast *P. stipitis* with that described for *S. cerevisiae*.

MATERIALS AND METHODS

Materials

Yeast Strains

P. stipitis NRRL Y-11544 (CBS 5776), kindly provided by C. P. Kurtzman, was used in all experiments. *S. cerevisiae* ATCC 24858 was used as a reference strain for the examinations of the inhibition of PDC by phosphate.

Media

The cells were grown in a mineral salt medium according to ref. *12*, supplemented with 6.7 g/L yeast nitrogen base w/o amino acids (Difco). If not otherwise stated, 30 g/L glucose or xylose or 20 g/L ethanol were added as described below.

Aerobic and Semiaerobic Growth

Aerobic and semiaerobic growth conditions were achieved by varying the volume of the media *(13)* and the speed of shaking. For aerobic growth, the cells were incubated at 25°C and 200 rpm on a gyratory shaker in a Fernbach flask containing 100 mL media. The cells were incubated up to an OD at 623 nm of 1.0. No ethanol was formed below this optical density, indicating a respirative mode of growth.

For semiaerobic growth, the cells were incubated at 25°C and 100 rpm in a 1-L Erlenmeyer flask containing 250 mL media up to an OD of approx 7.0. Under these conditions, the yield of ethanol was estimated to be approx 0.45 g ethanol/g sugar, indicating a preferential fermentative mode of metabolism.

Shift Experiments

The oxygen shift experiments were performed in a Braun Biostat E 1 fermenter, containing 5 L of media at 25°C. The pH was held constant at 5.0 using 1*M* NaOH. The fermenters were inoculated with 50 mL of aerobically grown cells that were grown on the same carbon source as in the fermenter. The oxygen tension was held constantly at 80% by varying both the rotating speed between 300 and 500 rpm and the aeration rate. The cells were grown under these conditions until they reached an OD of about 1.0, which was achieved within 8–10 h. At higher cell densities, the oxygen tension started to decrease and caused irregular conditions. To shift the oxygen tension from 80 to 20% or to 0%, the culture was flushed with nitrogen. The shift was achieved within 30 s. After a shift to 20%, the oxygen tension was regulated by varying the aeration rate. After the shift to 0%, the cells were further gassed with nitrogen to produce anaerobic conditions. The rotation speed was held constant at 200 rpm in both cases.

For glucose induction experiments, the fermenter was inoculated with an aerobic culture grown in ethanol and incubated until the cells reached an OD of 0.6. This was reached within 10 h. After this, glucose was added to a final concentration of 3%. The oxygen tension was held constant at 80%. However, this was only possible for a maximum of 4 h. To estimate the enzyme activities and the concentration of glycolytic metabolites of the cells grown in ethanol, additional yeasts were tested that were grown with ethanol to an OD of about 1.0, to avoid systematic errors owing to the inherent low dry weight and protein concentration.

Enzyme Assays

Samples were taken immediately prior to the oxygen shift (aerobic activity) or to the addition of glucose to cells grown in ethanol. After the oxygen tension shift or the addition of glucose, samples were taken at the times described below (*see* Results). The sample volume was about 100 mL. The cell suspension was quickly cooled by transferring to a tube that had been kept in ice. The crude extracts were prepared from this sample as reported *(14)*. The crude extract buffer was supplemented with the protease inhibitors phenylmethanesulfonyl fluoride, leupeptin, and pepstatin A *(15)*. The samples were stored at 4°C until the measurements were performed, which was within 9 h after the preparation of the crude extract. Within this time, no significant decrease of the activities of the tested enzymes was found. The activities of PDC and PDH were determined as described elsewhere *(15,16)*. For the measurements of PDH, the crude extract was purified prior to the assay using a Sephadex 50 column equilibrated with crude extract buffer. The reaction was initiated using pyruvate instead of Coenzyme A. ADH was assayed with 0.6*M* ethanol and 2.5 m*M* NAD^+ in pyrophosphate-semicarbazide buffer containing 75 m*M* $Na_4P_2O_7$, 21 m*M* glycine, and 2.5 g/L semicarbazide. The pH of the buffer was adjusted to 8.7 using 5*M* NaOH. The AldDH activity was assayed with benzaldehyde as substrate *(17)*. Acetaldehyde was also tested, but the activity was found to be nearly the same as benzaldehyde. All measurements were performed in a Beckmann DU 64 photometer or in a Pharmacia LKB Biochrom 4060 photometer.

The enzyme activities are expressed in U/mg protein. One unit (U) of enzyme activity is defined as the ability to convert 1 μmol of substrate/min at 30°C. The protein concentration was determined by the method of Lowry *(18)* using bovine serum albumin as a standard.

Determination of Kinetic Parameters of PDC and PDH

The kinetic parameters of PDC were estimated in cells grown semiaerobically; those of PDH were determined in aerobically grown cells. The behavior of the enzymes was determined by sampling the activity in the crude cell extract at several concentrations of pyruvate (*see* Results). Inhibition of PDC by phosphate was inves-

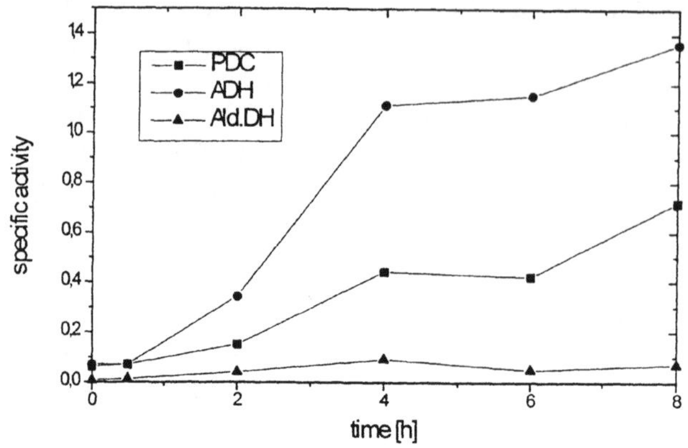

Fig. 1. Specific activities of PDC, ADH, and AldDH of glucose growing cells after a shift of the oxygen tension from 80 to 0%.

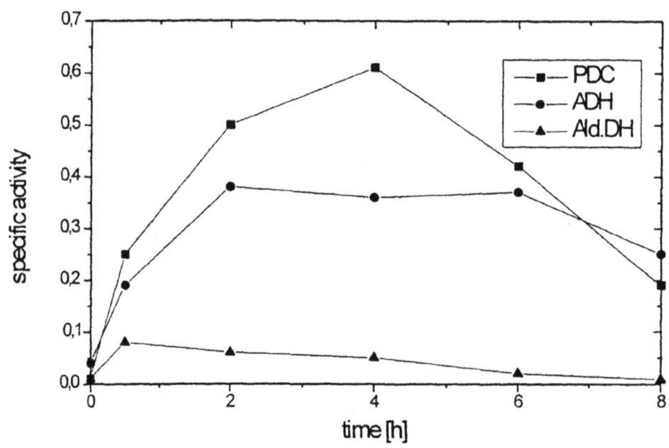

Fig. 2. Specific activities of PDC, ADH, and AldDH of glucose growing cells after a shift of the oxygen tension from 80 to 20%.

tigated in Na phosphate buffer at pH 6.8. As a reference for the inhibition by phosphate, the PDC of *S. cerevisiae* ATCC 24858 pregrown on 2% glucose was measured.

The inhibition of PDH by NADH was estimated against NAD^+ as substrate with 4 m*M* pyruvate. The parameters were calculated as reported *(19)*.

Determination of Glycolytic Metabolites

The samples were prepared and analyzed as described *(14)*. The extracts were stored at –70°C until analyzed in <24 h.

Measurements of Respirative and Fermentative Capacities

The cells were pregrown aerobically or semiaerobically with 6% glucose or xylose, harvested by centrifugation, washed with 50 m*M* Sørensen phosphate buffer (pH 5.5), and diluted with the buffer to an OD of 2.0. This suspension was used for the estimation of respirative and fermentative activities and dry weight.

To determine the activities of anaerobically cultivated cells, the cells were pregrown in a fermenter at a constant aeration rate and stirrer speed (300 rpm) with an initial 3% glucose. When the oxygen tension fell to zero, the amount of consumed glucose was added to restore the concentration to 3%, and the fermenter was gassed with nitrogen for 14 h. The activities were determined using standard Warburg manometers. All manometric measurements were made at 28°C, and lasted 1 h.

Analytical Procedures

Concentrations of sugar and ethanol were determined by HPLC with an Aminex Fermentation Monitoring column (Bio-Rad, Hercules, CA) using RI detector (Melz) with methanol as an internal standard. Dry cell weights were determined gravimetrically from samples containing 10 mL of cell suspension. The samples were harvested by centrifugation, washed twice with distilled water, and incubated at 105°C until they maintained a constant weight. In all cases, three or four measurements were made.

RESULTS

Respirative and Fermentative Activity

The rates of O_2 consumption and CO_2 formation were measured in correlation to different sugar concentrations and aeration conditions over a 60-min period. The cells were precultivated aerobically and semiaerobically on a high level (6%) of sugar, and then quickly transferred to Warburg vessels. No significant differences were found for sugar concentrations from 1.5 to 5.2%.

The respiratory quotients of cells pregrown aerobically on glucose or xylose were between 1.1 and 1.2 in every case, the Qo_2 values were between 81 and 84, and the Qco_2 values were between 93.6 and 99. This proves that no fermentation took place as a result of the high sugar concentrations.

The initial RQ value 1.65 of cells precultivated semiaerobically on 6% glucose indicated that the cells were fermenting. However, the RQs decreased, independently of the sugar concentration, within 30 min to values between 1.2 and 1.0, which were comparable to the RQs obtained under aerobic conditions. The Qo_2 remained constant over the period of the test at 120.0–130.0, which was even higher than in cells cultivated aerobically. The Qco_2 decreased from in the mean 189.4 to 156.9. Similar results were obtained with cells that were precultivated on xylose with the exception that the initial value of the RQ was 1.4. The Qco_2 of these cells decreased from 186.0 to 149.6 within 30 min.

In cells that were precultivated for 14 h under anaerobic conditions, the values of the O_2 consumption and the CO_2 production were only 50–60% of that obtained under semiaerobic conditions. The lower values may be owing to an enhanced number of unviable cells, caused by the anaerobic stress or ethanol. The initial RQ was 1.4–1.6 and decreased to 1.0–1.2 within 30 min, independent of the sugar concentration in the assay. Again, the decrease was caused by a reduction in CO_2 formation. The oxygen consumption remained nearly constant for the duration of the assay.

It can be concluded that the respiration activity either does not change or is slightly activated during oxygen limitation. The respiratory chain is not repressed in the absence of oxygen. In contrast, fermentation is rapidly inactivated at high oxygen concentrations. The rates of both O_2 consumption and of CO_2 formation were lower if a sugar concentration of 7.4% was used. This was probably because of substrate inhibition of the uptake *(20,21)* or osmotic effects of the high sugar concentrations.

Induction of Enzyme Activities

Activity of PDH

The specific activities of PDH ranged from 0.1–0.5 U/mg and, therefore, were comparatively higher than in *S. cerevisiae (15,22)*, although the measurements of the activity of the *S. cerevisiae* PDH were achieved after a step of purification. The same activities were found with the carbon sources glucose, xylose, or ethanol, and at all levels of oxygen tension tested. These results show that the PDH genes are expressed constitutively in *P. stipitis*. The differences in the specific activities may be caused by a degradation of the multienzyme complex owing to strong protease activities in the crude extract, even though protease inhibitors and mercaptoethanol were added to prevent protein degradations.

Activities of PDC, ADH, and AldDH

In *S. cerevisiae*, addition of glucose to cells grown in ethanol leads to an induction of the PDC *(23,24)*. We compared the activity of PDC in *P. stipitis* cells grown with 2% ethanol and after the addition of 3% glucose. No effect on the activity of PDC by addition of glucose was found within 4 h, and the activity values remained <0.05 U/mg protein.

We tested the induction of the enzyme activities after an oxygen tension shift from aerobic to anaerobic conditions in cells grown on 3% glucose. Figure 1 shows the typical profiles of the activities of PDC, ADH, and AldDH after such a shift. After a lag, the activity of PDC was enhanced 10-fold, that of ADH around 20-fold, and that of aldehyde dehydrogenase 10-fold of the aerobic activity.

To find out whether the induction is an effect resulting from the low oxygen tension or its difference, we tested the activation of the enzymes after a shift from 80 to 20%. After the shift, aeration was further necessary to keep the oxygen tension constant, as respiration occurred. We found a remarkably rapid activation of the enzymes directly after the shift (Fig. 2). This rapid activation of the three enzymes occurred within the first 30 min. The activation of PDC and ADH lasted up to 4 h. Thereafter, the activities of the enzymes decreased. Inactivation of PDC occurred rapidly, indicating perhaps some type of programmed inactivation. An activation of AldDH was only found after the first half hour. Then its activity decreased and reached its initial value within 8 h. We also investigated the production of ethanol, and we found a small amount of ethanol 2–8 h after the shift with maximum rates of 0.1 g/g dry wt and hour.

Figure 3 shows the activation of the three enzymes after the oxygen tension was shifted from 80 to 0% in cells grown on xylose. An induction of the three enzymes also occurred. However, the induction speed of PDC and ADH was slower than in cells grown on glucose, and the activities that were reached after 8 h were only the half of those of cells grown on glucose. The activation of AldDH was similar to that of cells grown on glucose.

Concentrations of Glycolytic Metabolites

To characterize the induction mechanism, we investigated the changes in the intracellular concentrations of the glycolytic metabolites after an addition of glucose

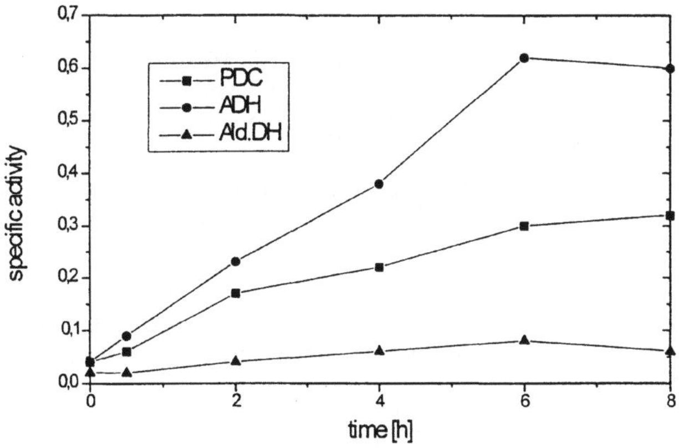

Fig. 3. Specific activities of PDC, ADH, and AldDH of glucose growing cells after a shift of the oxygen tension from 80 to 0%.

to ethanol grown cells, and after an oxygen tension shift with glucose or xylose as carbon source.

Table 1 shows the results from cells grown on ethanol and such cells 2 h after an addition of 3% glucose. Most of the glycolytic metabolites behaved as in *S. cerevisiae* induced with galactose. The concentration of glucose-6-P increased by the same magnitude as in *S. cerevisiae* induced by glucose. However, in *S. cerevisiae*, fermentation is induced after addition of glucose or galactose *(11,25)*.

No significant change in the concentration of glycolytic metabolites, except for pyruvate, was induced by an oxygen tension shift from 80 to 0% in glucose grown cells (Table 2). A threefold increase in the concentration of pyruvate was induced within 30 min by the shift. Thereafter, its concentration decreased to a value about twofold that of the initial level. Similar concentrations of the glycolytic metabolites were measured after a shift of the oxygen tension from 80 to 20%. Glucose-6-phosphate decreased within the first 30 min but after 4 and 8 h, the initial value was nearly reached. The concentration of pyruvate increased as in the shift to 0%. A decrease in the further course of time was also found, but it was not as large as with the shift to anaerobic conditions.

Xylose grown cells differed in their changes of the glycolytic metabolites from cells grown on glucose (Table 3). The concentration of pyruvate marginally decreased in 8 h. Only in the concentrations of glucose-6-phosphate and fructose-6-phosphate was an increase detected. The concentration of pyruvate was found to be much higher in cells grown on glucose or xylose after the shift than in cells grown in ethanol after addition of glucose (Tables 1–3). Moreover, cells grown aerobically with xylose were found to have a higher intracellular concentration of pyruvate than cells grown aerobically with glucose (Tables 2 and 3).

Kinetic Properties of PDC and PDH

We further tested some kinetic properties of PDC and PDH in crude cell extracts. PDC had a Michaelis constant for pyruvate of about $10^{-3}M$, which is nearly in accordance with that of *S. cerevisiae (26)*. The Hill constant 1.4–1.8 was also similar to *S. cerevisiae (26,27)*. The V_{max} in the crude extracts was found to be 0.8–1.1 U/mg

Table 1
Intracellular Concentration of Glycolytic Metabolites (nmol/mg [dry wt]) in Ethanol Grown Cells and of Ethanol Grown Cells 2 h After Addition of Glucose[a]

Growth conditions	G-6-P	F-6-P	FDP	Triose phosphate	3-PG	2-PG	PEP	Pyruvate
Ethanol	1.1	0.5	0.1	0.2	0	2.4	0.4	0.7
Glucose	3.4	0.8	0.2	0.2	0	1.1	0.4	2.3

[a]The results are averages of four experiments. G-6-P: glucose-6-phosphate, F-6-P: fructose-6-phosphate, FDP: fructose-1.6.-di-phosphate, 3-PG: 3-phospho-glycerate, 2-PG: 2-phospho-glycerate, PEP: phospho-enol-pyruvate.

Table 2
Time-Course of the Intracellular Concentration of Glycolytic Metabolites (nmol/mg [dry wt]) of Glucose Growing Cells After a Shift of the Oxygen Tension[a]

Time, h	0	0.5		4		8	
O_2	80	0	20	0	20	0	20
G-6-P	5.7	2.1	5.5	4.2	4.7	5.0	4.6
F-6-P	1.4	0.4	1.3	1.0	1.1	0.6	1.0
FDP	2.0	1.0	2.6	1.4	1.1	1.0	1.0
Triose-P	1.0	1.8	0.6	1.9	0.6	1.8	0.7
3-PG	0.6	1.4	1.3	1.6	0.8	1.1	0.5
2-PG	1.0	1.9	0.4	4.0	0.2	2.8	0.7
PEP	0	2.6	1.5	6.4	0	3.4	0.2
Pyruvate	8.8	23.7	19.2	13.0	15.4	14.2	18.9

[a]The oxygen tension at the start was 80% and then shifted to either 0 or 20%. The concentrations of the listed metabolites at various times are listed under 0 (0% O_2) or 20 (20% O_2). The results are averages of three experiments.

Table 3
Intracellular Concentration of Glycolytic Metabolites (nmol/mg [dry wt]) of Xylose Growing Cells After a Shift of the Oxygen Tension from 80 to 0%[a]

Time, h	0	0.5	4	8
G-6-P	1.6	2.3	2.7	3.8
F-6-P	0.4	0.6	1.8	1.0
FDP	3.5	3.8	3.6	3.5
Triose-P	1.2	1.0	1.5	1.4
3-PG	0	0	0	0
2-PG	0.5	0.2	0.4	0.5
PEP	0.8	0.8	0.8	0.8
Pyruvate	15.2	15.1	9.0	13.7

[a]The results are averages of three experiments.

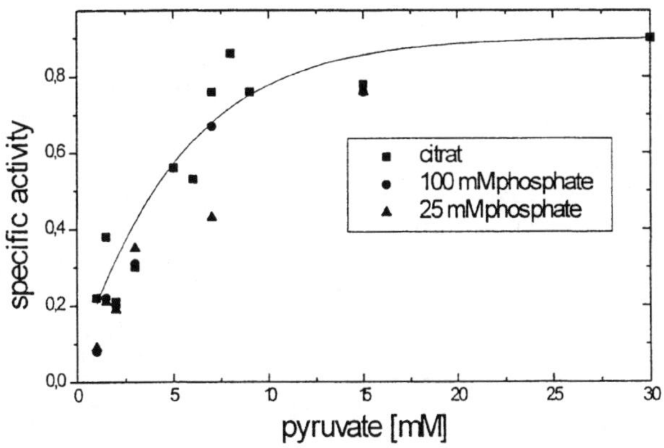

Fig. 4. Kinetic behavior of PDC of *P. stipitis* in phosphate or citrate buffer.

protein. However, no inhibition by phosphate was found. Figure 4 shows the typical behavior of PDC in semiaerobically grown cells.

The Hill constant of PDH with regard to pyruvate was approx 1.0, and the Lineweaver-Burt plot was always linear (results not shown). Therefore, this enzyme was concluded to be a Michaelis-Menten enzyme, as has also been described for *S. cerevisiae (22)*. The Michaelis constant for pyruvate was about $10^{-4}M$, which corresponds to *S. cerevisiae (15)*. NADH was found to inhibit the PDH (results not shown).

DISCUSSION

In *P. stipitis*, the activity of the respiratory chain is not repressed as a result of high concentrations of fermentable sugars or oxygen limitation. Moreover, high concentrations of fermentable sugars do not activate its fermentative pathways. This is in contrast to *S. cerevisiae*, where respiration is repressed and fermentation is activated by both high concentration of glucose and limitation of oxygen *(23,28,29)*.

In *S. cerevisiae* the induction of PDC is well investigated. At least, two different glycolytic signals trigger this induction. The main signal originates from an increase of metabolites of the lower part of glycolysis, including the concentration of pyruvate *(11)*. No induction of PDC was found after addition of glucose to cells grown in ethanol. Nevertheless, the intracellular concentrations of pyruvate behaved similarly to *S. cerevisiae* induced with galactose. Intracellular glucose-6-P was enhanced as in *S. cerevisiae* induced with glucose *(11)*.

PDC was induced in glucose and xylose grown *P. stipitis* cells as a reaction to a shift in the oxygen tension. The concentration of pyruvate increased significantly in glucose grown cells after the shift. However, in xylose grown cells, no increase in the concentration of pyruvate was found under similar conditions. Therefore, we conclude that in *P. stipitis* the main signal for the activation of PDC does not originate from the glycolytic pathway.

The accumulation of pyruvate could act as a minor signal resulting in a stronger induction of PDC in cells grown on glucose. This minor signal perhaps depends on reaching a certain critical concentration of pyruvate. No activation of PDC was found in ethanol grown cells induced with glucose, although the concentration of pyruvate was increased almost three times the initial value. The crucial concentration of pyruvate is estimated to be more than 15 nmol/mg dry wt because this concentration was found in aerobic xylose grown cells without any activation of PDC.

These results demonstrate that the mechanism of the induction of PDC in *P. stipitis* is totally different to that of *S. cerevisiae*.

Boles et al. *(11)* detected in *S. cerevisiae* a minor signal derived from the upper part of glycolysis. We found an increase in glucose-6-phosphate and fructose-6-phosphate in both ethanol grown cells induced with glucose and xylose grown cells induced by an oxygen tension shift. This implies that the upper part of glycolysis does not play a role in the regulation of PDC in *P. stipitis*.

The character of the main signal that triggers the PDC activity remains unclear. It was shown, however, from the oxygen shift experiments, that this signal is tightly coupled with the oxygen tension in the medium. Furthermore, the reaction of the cells to the 80–20% shift demonstrates that the signal detects a difference in the oxygen tension rather than a certain low O_2 concentration. Permanent cultivation at 20% oxygen tension does not lead to permanent ethanol formation *(3)*. However, in this study, a transient induction of PDC, ADH, and AldDH was observed after the shift. The presence of oxygen or respiration activity seems to cause a very quick induction of fermentation. The activities of the three measured enzymes were enhanced within 30 min up five times their initial value, which is a faster activation than was described for glycolytic enzymes of *S. cerevisiae (30)*.

ADH and aldehyde dehydrogenase could be regulated in *P. stipitis* by the same O_2-dependent mechanism of PDC. However, acetaldehyde is also a candidate as an inducer of both enzymes. Acetaldehyde has been shown to induce ADH I and AldDH of *Aspergillus nidulans (31)*. The function of AldDH in anaerobiosis may be to prevent an accumulation of acetaldehyde, which is known to be toxic (for a review, *see 32*). However, at anaerobic conditions, the cells have to tolerate the formation of NADH and NADPH by the AldDH reaction *(33)* and, therefore, possibly the generation of a redox imbalance.

Our data demonstrate that, in *P. stipitis* under aerobic conditions, all pyruvate is metabolized via the PDH reaction because PDC is not active. Under oxygen limitation, PDC is induced and the fermentative pathway becomes available for the cell. The activity of PDH and its affinity for pyruvate are relatively high. Initially, this competitive advantage provides the respiratory pathway with pyruvate. However, oxygen limitation is also connected with an increase of intracellular NADH. This high NADH concentration will now inhibit PDH and, therefore, cause the preferential channeling of pyruvate into the fermentative pathway.

Moreover, Skoog and Hahn-Hägerdal *(10)* demonstrated a strong decrease in the intracellular pyruvate concentration after a lengthy time of oxygen limitation. This may be owing to an inhibition of an early step of glycolysis, perhaps the sugar uptake *(2)*. At these low concentrations of pyruvate, however, the velocity of the PDC reaction is slow because of the allosteric kinetics of this enzyme. Therefore, also the ethanol formation rate of *P. stipitis* is low.

The PDC induction signal functions independently of the activities in the glycolytic pathway. It enables the activation of the enzyme in spite of the low glycolytic flux. On the other hand, after a shift to aerobic conditions, the cells are able to switch quickly to the respiratory mode of metabolism because the PDH and the respiratory chain are constitutively expressed, and the inhibition of PDH by NADH is abolished.

The peculiarities of the fermentation regulation mechanism have to be taken into account when manipulating strains to improve their ethanol production. Limitation of oxygen leads to an activation of PDC and ADH, but also to a significant decrease in the glycolytic flux *(10)*. The mechanism of this phenomenon is unknown. Furthermore, *P. stipitis* needs a certain low aeration rate for growth *(3)*.

Ethanol production under aerobic conditions could circumvent the inactivation of the early stages of glycolysis and the requirement for a regulation of the oxygen transfer rate. Such aerobic ethanol production could be achieved by overexpression of *S. cerevisiae* PDC and ADH in *P. stipitis*. However, the PDC of both organisms differ with regard to their inhibition by phosphate. These differences may be disadvantageous during expression of the *S. cerevisiae* PDC in *P. stipitis*.

Furthermore, the aerobic inactivation of PDC has to be taken into account because the inactivation may lead to futile cycles of production and destruction of the enzyme. Moreover, a greater amount of the substrate may be respired, leading to a decrease in the ethanol yield, owing to the high activity of the PDH.

Our results demonstrated large differences between *S. cerevisiae* and *P. stipitis* in the regulation of fermentation and respiration. The model organism of yeast physiology, *S. cerevisiae*, is not comparable in many cases with the physiology of *P. stipitis*. Therefore, improving strains for ethanol production also needs further investigations into the physiology of *P. stipitis*. This knowledge may open up new ways to produce ethanol from xylose and for the understanding of the physiology of Crabtree-negative yeasts.

ACKNOWLEDGMENTS

We thank A. Carlier for technical assistance and B. Kosier for linguistic advice. This work was supported by the Deutsche Forschungsgemeinschaft.

REFERENCES

1. du Preez, J. C. (1994), *Enzyme Microb. Technol.* **16,** 944–956.
2. Hahn-Hägerdal, B., Jeppsson, H., Skoog, K., and Prior, B. A. (1994), *Enzyme Microb. Technol.* **16,** 933–943.
3. Rizzi, M., Klein, C., Schulze, C., Bui-Thanh, N.-A., and Dellweg, H. (1989), *Biotechnol. Bioeng.* **34,** 509–514.
4. Lighthelm, M. E., Prior, B. A., and du Preez, J. C. (1988), *Appl. Microbiol. Biotechnol.* **28,** 63–68.
5. Skoog, K., Jeppson, H., and Hahn-Hägerdal, B. (1992), *Appl. Biochem. Biotechnol.* **34/35,** 369–375.
6. Bruinenberg, P. M., de Bot, P. H. M., van Dijken, J. P., and Scheffers, W. A. (1984), *Appl. Microbiol. Biotechnol.* **19,** 256–260.
7. Prior, B. A., Kilian, S. G., and du Preez, J. C. (1989), *Process Biochem.* **24,** 21–31.
8. Kurtzman, C. P. (1990), *Antonie van Leeuwenhoek* **57,** 215–222.
9. Alexander, M. A., Yang, V. W., and Jeffries, T. W. (1988), *Appl. Microbiol. Biotechnol.* **29,** 282–288.
10. Skoog, K. and Hahn-Hägerdal, B. (1990), *Appl. Environ. Microbiol.* **56,** 3389–3394.
11. Boles, E., Heinisch, J., and Zimmermann, F. K. (1993), *Yeast* **9,** 761–770.
12. Dellweg, H., Rizzi, M., Methner, H., and Debus, D. (1984), *Biotechnol. Lett.* **6,** 395–400.
13. Girio, F. M., Peito, M. A., and Amaral-Collaco, M. T. (1989), *Appl. Microbiol. Biotechnol.* **32,** 199–204.
14. Ciriacy, M. and Breitenbach, I. (1979), *J. Bacteriol.* **139,** 152–160.
15. Kresze, G.-B. and Ronft, H. (1981), *Eur. J. Biochem.* **119,** 573–579.
16. Ullrich, J. (1970), *Methods Enzymol.* **18,** 109–115.
17. Bradbury, S. L., Clark, J. F., Steinman, C. R., and Jakoby, C. R. (1975), *Methods Enzymol.* **41,** 354–360.

18. Bergmeyer, H. U. (1977), Verlag Chemie. Weinheim.
19. Wieker, H. -J., Johannes, K.-J., and Hess, B. (1970), *FEBS Lett.* **8,** 178–185.
20. Delgenes, J. P., Moletta, R., and Navarro, J. M. (1988), *Appl. Microbiol. Biotechnol.* **29,** 155–161.
21. Grootjen, D. R. J., van der Lans, R. G. J. M., and Luyben, K. C. A. M. (1991), *Enzyme Microb. Technol.* **13,** 648–654.
22. Keha, E. E., Ronft, H., and Kresze, G.-B. (1982), *FEBS Lett.* **145,** 289–292.
23. Polakis, E. S. and Bartley, W. (1965), *Biochem. J.* **97,** 284–297.
24. Schmitt, H. D. and Zimmermann, F. K. (1982), *J. Bacteriol.* **151,** 1146–1152.
25. Sierkstra, L. N., Nouwen, N. P., Verbakel, J. M. A., and Verrips, C. T. (1993), *Yeast* **9,** 787–795.
26. Boiteux, A. and Hess, B. (1970), *FEBS Lett.* **9,** 293–296.
27. Ullrich, J. and Donner, I. (1970), *Hoppe-Seyler's Z. Physiol. Chem.* **351,** 1026–1029.
28. Polakis, E. S., Bartley, W., and Meek, G. A. (1964), *Biochem. J.* **90,** 369–374.
29. Tustanoff, E. R. and Bartley, W. (1962), *Biochem. J.* **84,** 40p,41p.
30. Maitra, P. K. and Lobo, P. K. (1971), *J. Biol. Chem.* **246,** 475–488.
31. Pateman, J. A., Doy, C. H., Olson, J. E., Norris, U., Creaser, E. H., and Hynes, M. (1983), *Proc. Roy. Soc. Lond.* **B217,** 243–264.
32. Jones, R. P. (1989), *Enzyme Microb. Technol.* **11,** 130–153.
33. Steinman, C. R. and Jakoby, W. B. (1968), *J. Biol. Chem.* **243,** 730–734.

0273-2289/96/5758—0213$7.25

Inactivation of an Aldehyde/Alcohol Dehydrogenase Gene from *Clostridium acetobutylicum* ATCC 824

EDWARD M. GREEN AND GEORGE N. BENNETT*

Department of Biochemistry and Cell Biology, Rice University, Houston, TX 77251

ABSTRACT

A nonreplicative plasmid containing an internal *aad* gene fragment has been integrated into the chromosome of *Clostridium acetobutylicum* ATCC 824. Transformation was accomplished by electroporation with relatively high concentrations of methylated plasmid DNA. Southern hybridization experiments revealed that integration occurred by single crossover homologous recombination inactivating the *aad* gene. Integrants were relatively stable after 25 generations. Inactivation of the *aad* gene drastically reduced solvent production. This result suggests that aldehyde/alcohol dehydrogenase (AAD) plays a important role in butanol production.

Index Entries: *Clostridium acetobutylicum*; aldehyde/alcohol dehydrogenase; solvent production; gene inactivation; homologous recombination.

INTRODUCTION

Clostridium acetobutylicum is a gram-positive, spore-forming, saccharolytic bacterium capable of fermenting a wide variety of sugars to acids (acetate and butyrate) and solvents (ethanol, acetone, and butanol). The production of acetone and butanol by *C. acetobutylicum* was a thriving industrial process, but the fermentation can no longer compete with the chemical synthesis of solvents from petroleum. However, renewed interest in using butanol as a fuel extender or chemical feed stock has focused research on the genetic manipulation of *C. acetobutylicum* with the aim of developing a more competitive industrial process.

The biochemical pathways for the conversion of carbohydrates to acids and solvents by *C. acetobutylicum* have been well reviewed *(1)*. Acetyl-CoA, acetoacetyl-CoA, and butyryl-CoA function as key intermediates for ethanol, acetone, and butanol production, respectively. Acetyl-CoA is first reduced to acetaldehyde and then to ethanol in reactions catalyzed by acetaldehyde dehydrogenase (ACDH) and ethanol dehydrogenase (EDH), respectively. Likewise, butyryl-CoA is reduced to butyraldehyde and then butanol in reactions catalyzed by butyraldehyde dehydrogenase (BYDH) and butanol dehydrogenase (BDH). The reassimilation of acetate and butyrate is coupled to the irreversible production of acetoacetate from acetoacetyl-CoA by acetoacetyl-CoA:acetate/butyrate:coenzyme transferase

*Author to whom all correspondence and reprint requests should be addressed.

(CoAT). Acetone and carbon dioxide are produced from the decarboxylation of acetoacetate by acetoacetate decarboxylase (AADC).

Genes encoding AADC *(adc)*, the two subunits of CoAT (*ctf*A and *ctf*B), two BDH isozymes (*bdh*A and *bdh*B) and aldehyde/alcohol dehydrogenase (AAD) *(aad)* have been cloned and sequenced from *C. acetobutylicum* ATCC 824. The *aad* gene resides immediately upstream of the previously cloned *ctf* genes *(2)*, and the three genes form an operon *(3,4)*. The *adc* gene lies downstream of the *ctf* genes in a separate operon *(5)*. The *bdh* genes are linked in adjacent monocistronic operons, but are not in close proximity to other solventogenic genes *(6)*. Complementation studies suggest that BYDH activity resides largely in AAD *(7)*, but it is unclear what role AAD and the BDH isozymes play, if any, in ethanol and butanol formation. We hope to finally resolve the role of AAD in solvent production by inactivating *aad* on the chromosome.

Nonreplicative integrational plasmids have been used to advance genetic studies in gram-positive bacteria *(8,9)*. Such plasmids need a DNA fragment from the host and a genetic marker for which selection can be made. After transfer, plasmids may become established by integrating into the homologous region on the host replicon in a Campbell-like fashion *(10)*. Plasmid integration may result in two direct repeats of the homologous region flanking the plasmid DNA. Insertion can be mutagenic if the homologous DNA fragment on the integrational plasmid lies entirely within a gene or transcription unit *(11)*. Insertion may also result in stable gene amplification if the homologous DNA fragment on the plasmid contains the entire gene *(12)*.

Gene integration involving nonreplicative plasmids is usually limited to bacteria that can only be transformed at a high frequency. Since clostridia are notoriously difficult to transform, the use of this type of integration system has been limited to *Clostridium beijerinckii* (formerly known as *C. acetobutylicum* NCIMB 8052), for which conjugative transfer systems have been developed *(13,14)*. Several integrational plasmids have been established in the chromosome of *C. beijerinckii* by homologous recombination *(15)*.

An efficient and reliable plasmid transformation protocol, involving in vivo methylation, now exists for *C. acetobutylicum* ATCC 824 *(16)*. In addition, most of the genes involved in acid and solvent formation pathways have recently been identified *(17)*. Since cloned genes provide possible sites for integration studies, it should now be possible to exploit integrational plasmid technology fully in this strain. We report on use of a nonreplicative integrational plasmid containing an internal *aad* fragment to inactivate *aad* on the chromosome. This article provides valuable information about gene function and demonstrates the potential of using integrational plasmids to modify C. *acetobutylicum* ATCC 824 genetically.

METHODS

Bacterial Strains and Plasmids

Bacterial strains and plasmids are listed in Table 1.

Growth Conditions and Maintenance

Escherichia coli was grown aerobically at 37°C in Luria-Bertani (LB) medium. *C. acetobutylicum* was grown anaerobically at 37°C in 10-mL tube cultures of Clostridium growth medium (CGM; *18*). Colonies were obtained on agar-solidified rein-

Table 1
Bacterial Strains and Plasmids

Strain/plasmid	Relevant characteristics[a]	Source of gram-positive replicon	Source/reference
Strain			
C. acetobutylicum			
ATCC 824			ATCC
PJC4AAD	*aad*⁻Em[r]		Insertion of pJC4aad into ATCC 824
Escherichia coli			
ER2275	*rec*A⁻*mcr*BC⁻		New England Biolabs
Plasmid			
pAN1	Cm[r]ϕ3T I		*(16)*
pJC4	Em[r] Tc[r]		*(24)*
pCAAD	Ap[r] Em[r] *aad*	*B. subtilis* pIM13	*(3)*
pSYL2	Ap[r] Em[r]	*C. butyricum* pCBU2	*(24)*
pJC4aad[b]	Em[r]		This study

[a]Abbreviations: *aad*, aldehyde/alcohol dehydrogenase; Ap[r], ampicillin resistant; Cm[r], chloramphenicol resistant; Em[r], erythromycin resistant; *mcr*BC⁻, lacking methylcytosine-specific restriction system; *rec*A⁻, homologous recombination abolished; Tc[r], tetracycline resistant; ϕ3T I, ϕ3T methylase.

[b]pJC4aad was constructed by subcloning a 1.3-kb *Bam*HI-*Hind*III *aad* fragment from pCAAD into pJC4.

forced clostridial medium (RCM; Difco, Detroit, MI). LB medium was supplemented, as required, with erythromycin (200 μg/mL), tetracycline (15 μg/mL), and chloramphenicol (32 μg/mL). RCM and CGM media were supplemented with 20 μg/mL and 40 μg/mL of erythromycin, respectively.

DNA Isolation and Manipulation

The manipulation, transformation, and isolation of plasmid DNA from *E. coli* ER2275 were performed using standard procedures *(19)*. Large-scale plasmid isolation was undertaken with a plasmid purification kit (Qiagen, Studio City, CA). The restriction enzymes and T4 DNA ligase were used in accordance with the supplier's instructions (Promega, Madison, WI). Prior to transformation of *C. acetobutylicum*, pJC4aad was methylated in *E. coli* ER2275 (pAN1) by the *Bacillus subtilis* phage ϕ3T methyltransferase, which protects the plasmid DNA from restriction by the clostridial endonuclease *Cac*824I. The plasmid DNA was desalted and concentrated using MICROCON 100 microconcentrators (Amicon, Beverly, MA). Approximately 15 μg of methylated plasmid DNA were used to transform *C. acetobutylicum* by electroporation using a previously described method *(20)*. Chromosomal DNA was prepared from *C. acetobutylicum* using a PUREGENE DNA isolation kit (Gentra Systems, Research Triangle Park, NC). Cells were harvested from a 10-mL CGM tube culture during the exponential phase (A_{600} ~ 0.8).

Enzyme Assays

Cell extracts were prepared from 50 mL of solventogenic *C. acetobutylicum* culture. The cells were harvested by centrifugation (20,000*g* for 20 min at 4°C). The cell pellet was resuspended in 10 mL of 15 m*M* potassium phosphate buffer (pH 7.0) containing 1 m*M* dithiothreitol and 0.1 m*M* $ZnSO_4$. The cells were lysed by two passes through a precooled French pressure cell (SLM model FA-078) at an operating pressure of 10,000 psi, and the cell debris was removed by centrifugation (35,000*g* for 20 min at 4°C). BDH assays were performed in the physiological direction as previously described *(21)* with butyraldehyde at a final assay concentration of 25 m*M*. One unit (U) is defined as the amount of enzyme necessary to oxidize 1 µmol of NADH/min. The protein concentration was determined using a dye reagent in accordance with the manufacturer's instructions (Bio-Rad, Hercules, CA).

Southern Hybridization

Chromosomal DNA was digested to completion with *Sca*I, transferred from an agarose gel to a Hybond-N$^+$ nylon membrane (Amersham, Arlington Heights, IL) by capillary blotting *(22)*, and then probed with a radiolabeled 1.3-kb *Bam*HI-*Hin*dIII *aad* fragment, isolated from pCAAD (Table 1). The gene fragment was labeled with [α-^{32}P]dATP using a random primer DNA labeling system (Gibco BRL, Gaithersburg, MD), and unincorporated radionucleotides were removed by exclusion chromatography on Sephadex G-50. The prehybridization, hybridization, and washing steps were performed in accordance with the membrane manufacturer's instructions, and the radioactivity was detected by autoradiography.

Analytical Methods

Cell growth was monitored at 600 nm using a Beckman DU 64 spectrophotometer with cuvets of 1-cm light path. Samples were diluted in distilled water, if necessary, so that the absorbance at 600 nm ($A_{600\ nm}$) did not exceed 0.8. Fermentation products (acetate, butyrate, ethanol acetone, and butanol) were determined by gas chromatography using a Hewlett-Packard 5890 Series II Gas Chromatograph with a flame ionization detector and reporting integrator. Samples of culture supernatant and a standard solution (10 m*M* ethanol, 20 m*M* acetone, 40 m*M* butan-1-ol, 30 m*M* acetate, and 30 m*M* butyrate) were acidified with an equal volume of 20% (v/v) phosphoric acid containing 12.5 m*M* pentan-3-one (internal standard). Acidified samples (5 µL) were injected onto a glass column (4 ft × 4 mm internal diameter) containing Porapak Q (80–100 mesh; Hewlett-Packard, Wilmington, DE). Separation of sample components was achieved using a temperature program. The column temperature was held at 135°C for 8 min, increased by 30°C/min for 2 min, and then held at 195°C for 10 min. The injector and detector temperatures were 245° and 215°C, respectively. The carrier gas (N_2) flow rate was 25 mL/min.

Measurement of the Recombination Frequency

The integrational strain was grown overnight in CGM with erythromycin. The culture was diluted approx 1000-fold in 50 mL of prewarmed CGM lacking erythromycin and incubated until the A_{600} reached 1.0. The culture was serially diluted in CGM, and plated on nonselective and selective CGM medium. The proportion of antibiotic-resistant cells was determined by dividing the number of colonies found on selective medium by the number of colonies found on nonselective medium. Subculturing steps and colony counts were repeated until the bacterial population had undergone about 25 generations of growth. The segregation fre-

Table 2
Electrotransformation Efficiencies of Replicative and Nonreplicative Plasmid DNA

Plasmid	Insert size, kb	Ability to replicate in *C. acetobutylicum*	DNA concentration, µg	Number of transformants
pAN1	0	–	5	0
pAN1 and pSYL2	0	+	0.05	500
pAN1 and pJC4	0	–	15	0
pAN1 and pJC4aad	1.3	–	15	13

quency was calculated by dividing the proportion of antibiotic-sensitive colonies by the number of generations of growth *(23)*.

RESULTS

Electrotransformation

C. acetobutylicum ATCC 824 was transformed by electroporation with the methylation plasmid (pAN1) and plasmid mixtures containing pAN1 and either pSYL2, pJC4, or pJC4aad. The DNA in these mixtures was methylated, in vivo, by the pAN1-encoded methylase, which protects vector DNA from *Cac*824I restriction *(16)*. The number of transformants observed after electroporation is shown in Table 2. The methylation plasmid (pAN1) did not transform *C. acetobuylicum*. This plasmid lacks a suitable origin of replication and cannot therefore propagate in *C. acetobutylicum*. The plasmid also lacks an erythromycin marker and cannot be maintained in selective media. Plasmid pSYL2 is an efficient shuttle vector *(16)*, and transformation with this vector serves as a positive control. The plasmid contains a gram-positive replicon *(24)*, and the results show that this vector transformed ATCC 824 at a relatively high frequency (1×10^4 colonies/µg DNA). Plasmid pJC4 and its derivative pJC4aad lack suitable replicons that would enable them to replicate in *C. acetobutylicum*. They do, however, contain erythromycin markers, and selective pressure may force them to integrate into the chromosome. Plasmid pJC4 did not transform ATCC 824, whereas pJC4aad transformed at a low frequency (0.9 colony/µg DNA). This result suggests that a small proportion of the incoming pJC4aad molecules integrated into the chromosome by homologous integration.

Southern Hybridization

Three pJC4aad transformants were selected and designated PJC4AAD (a, b, and c). Chromosomal DNA from the integrants and parental strain was digested with *Sca*I and characterized by Southern hybridization (Fig. 1). *Sca*I was chosen because this enzyme has a single site in the backbone of the vector pJC4, but none in the clostridial *aad* insert. If a single copy of pJC4aad integrated into the chromosome, digestion with *Sca*1 should generate two fragments whose combined size equals the combined size of pJC4aad and the *ScaI* fragment on the parental chromosome that contains the aad gene. The *aad* probe hybridized to two *Sca*I fragments (approx 5.5 and 6.3 kb) from the PJC4AAD mutant strains and one *Sca*I fragment (5.6 kb) from the parental strain. Since pJC4aad is 6.2 kb and the *aad* gene resides

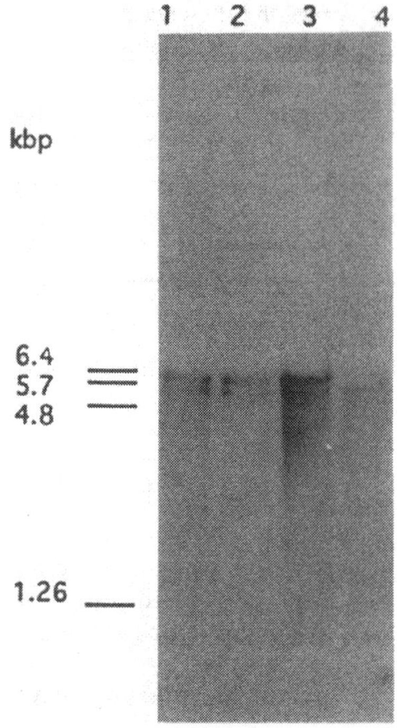

Fig. 1. Hybridization of a 1.3-kb *aad* fragment to *Sca*I-digested chromosomal DNA from three *C. acetobutylicum* PJC4AAD integrants and the parental strain ATCC 824. Lanes 1, 2, and 3 are integrants PJC4AAD(a), PJC4AAD(b), and PJC4AAD(c); lane 4 is ATCC 824. Size markers (kbp) are *Bst*EII-digested λ DNA.

within a 5.6-kb *ScaI* fragment on the chromosome *(3)*, it appears that a single copy of pJC4aad integrated into the *aad* gene on the chromosome in all three mutants tested.

Product Formation

The parental and integrational strains were grown for 48 h in 10-mL CGM tube cultures supplemented, as required, with erythromycin. The culture supernatant was analyzed for acid and solvent production by gas chromatography (Table 3). The PJC4AAD integrants (a, b, and c), previously characterized by Southern hybridization, produced on average 15% more butyrate, 9% more acetate, 38% more ethanol, and 84% less butanol than the parental strain. The integrants did not produce any acetone. The average cell biomass (A_{600}) of the mutant cultures was 11% lower than the parental strain.

Enzyme Assays

Since AAD appears to depend on either nicotinamide adenine dinucleotide (NAD) or reduced NAD (NADH) *(7)*, crude cell-free extracts from wild-type and mutant PJC4AAD(a) cultures were assayed for NADH-dependent BDH. The spe-

Table 3
Production of Acids and Solvents
by *C. acetobutylicum* ATCC 824 Wild Type and PJC4AAD[a]

C. acetobutylicum ATCC 824	$A_{600\,nm}$	Concentration (m*M*) Ethanol	Acetone	Butanol	Acetate	Butyrate
PJC4AAD(a)	4.3	10.2	0	6.9	31.1	56.0
PJC4AAD(b)	4.2	10.4	0	8.1	27.4	46.8
PJC4AAD(c)	4.1	8.8	0	6.7	24	49.2
PJC4AAD average	4.2	9.8	0	7.2	27.5	50.6
Wild-type average	4.7	7.1	14.3	43.9	25.2	44.1

[a]Wild-type average values are calculated from three cultures (results not shown). PJC4AAD average values are calculated from the three mutant strains selected (PJC4AAD a, b and c).

cific activity (U/mg of protein) of BDH from PJC4AAD(a) and wild type was 0.012 and 0.018, respectively. PJC4AAD had 33% less BDH activity than the wild type.

Integrational Stability

Strain PJC4AAD(a) was grown for approx 25 generations in the absence of selective pressure. After plating serial dilutions on selective and nonselective growth medium, the segregation frequency (rate of plasmid loss per generation) was found to be about 3.5×10^{-3}.

DISCUSSION

Southern hybridization results indicate that the nonreplicative plasmid pJC4aad integrated into the homologous *aad* region on the chromosome. Integration, assumed to have taken place by homologous recombination involving a Campbell-like mechanism, resulted in two copies of the homologous region provided. Duplicated sequences are potential substrates for homologous recombination and can result in plasmid loss *(9)*. However, the integrant tested appeared to be relatively stable after 25 generations. The segregation frequency (3.5×10^{-3}) was comparable to values reported for *Clostridium beijerinckii* ($1.3–37 \times 10^{-3}$; *15*), but was one order of magnitude lower than values reported for *Lactococcus lactis* *(9)* and *B. subtilis* *(25)*. The integrants grew slower and reached slightly lower cell densities than the parental strain. Slow integrant cell growth probably increased selection for faster-growing nonintegrants, which contributed to the segregation frequency.

Analysis of the fermentation products revealed that the PJC4AAD integrants did not produce any acetone. The integrants also produced considerably less butanol than the parental strain. These results suggest that integration of pJC4aad into the *aad* gene structurally inactivates *aad* and disrupts transcription of the two genes (*ctf*A and *ctf*B) that code for CoAT, an enzyme involved in acetone production. This is no surprise, since the *ctf* genes reside immediately downstream of *aad* and are located in the same operon *(3,4)*. The three genes appear to be cotranscribed.

C. acetobutylicum AAD displays both aldehyde (ALDH) and alcohol dehydrogenase (ADH) activity *(3)*. Inactivation of the ADH domain on the *aad* gene reduced NADH-dependent BDH activity by 33% and decreased butanol production by 84%. These findings suggest that the ADH function of AAD is primarily involved in the reduction of butyraldehyde to butanol. These results are consistent with findings from a complementation study that suggests that the ALDH function of AAD is primarily involved in the reduction of butyryl-CoA to butyraldehyde. Nair and Papoutsakis *(7)* transformed a mutant strain of *C. acetobutylicum* (M5), deficient in BYDH, with a replicative plasmid containing *aad* and found that the plasmid restored BYDH activity and butanol production. It appears that AAD is primarily responsible for most of the BYDH and BDH activity in the cell. This multifunctional AAD protein must therefore play a significant role in butanol production.

The BDH isozyme genes *bdh*A and *bdh*B encode BDHII and BDHI *(6,26)*, but their role in solvent production is unclear. Both enzymes possess EDH and BDH activity but BDHII has a higher specificity for butyraldehyde than for acetaldehyde *(6)*. The integrants produced 35% more ethanol than the parental strain, probably in an attempt by the cells to get rid of excess reducing power. The enzymes encoded by one or both *bdh* genes may be responsible for the increased ethanol production and remaining BDH activity in the AAD mutant, but their precise role in solvent production cannot be determined until the genes have been inactivated on the chromosome.

Inactivation of the *aad* gene on the chromosome greatly reduced the solvent-producing capability of the cell without drastically affecting cell growth. This result demonstrates that gene integration technology can be used to redirect cellular metabolism in *C. acetobutylicum* ATCC 824. It should be possible either to eliminate acid-producing pathways (acetate or butyrate) by gene inactivation or enhance solvent-producing pathways (acetone and butanol) by gene duplication. Not only does this technique provide valuable information about gene function, but it may be used to redirect a greater proportion of the carbon and electron flow toward solvent production, improving both the product yield and selectivity. We believe gene integration technology moves the acetone–butanol fermentation one step closer to becoming a viable alternative to the chemical synthesis of solvents.

ACKNOWLEDGMENTS

We thank Ramesh Nair, Northwestern University, Evanston, IL for supplying plasmid pCAAD. This research was supported by a grant from the National Science Foundation (BCS 9209627).

REFERENCES

1. Jones, D. T. and Woods, D. R. (1986), *Microbiol. Rev.* **50,** 484–524.
2. Cary, J. W., Petersen, D. J., Papoutsakis, E. T., and Bennett, G. N. (1990), *Appl. Environ. Microbiol.* **56,** 1576–1583.
3. Nair, R. V., Bennett, G. N., and Papoutsakis, E. T. (1994), *J. Bacteriol.* **176,** 871–885.
4. Fischer, R. J., Helms, J., and Durre, P. (1993), *J. Bacteriol.* **175,** 6959–6969.
5. Petersen, D. J., Cary, J. W., Vanderleyden, J., and Bennett, G. N. (1993), *Gene* **123,** 93–97.
6. Walter, K. A., Bennett, G. N., and Papoutsakis, E. T. (1992), *J. Bacteriol.* **174,** 7149–7158.
7. Nair, R. V. and Papoutsakis, E. T. (1994), *J. Bacteriol.* **176,** 5843–5846.
8. Perego, M. (1993), in Bacillus subtilis *and other Gram-Positive Bacteria*, Sonenshein, A. L., Hoch, J. A., and Losick, R., eds., American Society for Microbiology, Washington, DC, pp. 615–624.
9. Leenhouts, K. J., Kok, J., and Venema, G. (1989), *Appl. Environ. Microbiol.* **55,** 394–400.
10. Campbell, A. M. (1962), *Adv. Genet.* **11,** 101–146.
11. Pigot, P. J., Curtis, C. A. M., and de Lencastre, H. (1984), *J. Gen. Microbiol.* **130,** 2123–2136.

12. Young, M. (1984), *J. Gen. Microbiol.* **130,** 1613–1621.
13. Young, M., Staudenbauer, W. L., and Minton, N. P. (1989), *FEMS Microbiol. Rev.* **63,** 301–325.
14. Williams, D. R., Young, D. I., and Young, M. (1990), *J. Gen. Microbiol.* **136,** 819–826.
15. Wilkinson, S. R. and Young, M. (1994), *Microbiology* **140,** 89–95.
16. Mermelstein, L. D. and Papoutsakis, E. T. (1993), *Appl. Environ. Microbiol.* **59,** 1077–1081.
17. Papoutsakis, E. T. and Bennett, G. N. (1993), in *The Clostridia and Biotechnology*, Woods, D. R., ed., Butterworth-Heinemann, Stoneham, MA, pp. 157–199.
18. Roos, J. W., McLaughlin, K., and Papoutsakis, E. T. (1985), *Biotechnol. Bioeng.* **27,** 681–694.
19. Maniatis, T., Fritsch, E. F., and Sambrook, J. (1982), *Molecular Cloning: A Laboratory Manual*, Cold Spring Harbor Laboratory, Cold Spring Harbor Press, NY.
20. Mermelstein, L. D., Welker, N. E., Bennett, G. N., and Papoutsakis, E. T. (1992), *Bio/Technology* **10,** 190–195.
21. Welch, R. W., Rudolph, F. B., and Papoutsakis, E. T. (1989), *Arch. Biochem. Biophys.* **273,** 309–318.
22. Southern, E. (1975), *J. Mol. Biol.* **98,** 503–517.
23. Noirot, Ph., Petit, M.-A., and Ehrlich, S. D. (1987), *J. Mol. Biol.* **196,** 39–48.
24. Lee, S. Y., Bennett, G. N., and Papoutsakis, E. T. (1992), *Biotech. Lett.* **14,** 427–432.
25. Young, M. and Ehrlich, S. D. (1989), *J. Bacteriol.* **171,** 2653–2656.
26. Petersen, D. J., Welch, R. W., Rudolph, F. B., and Bennett, G. B. (1991), *J. Bacteriol.* **173,** 1831–1834.

0273-2289/96/5758—0223$7.25

Manipulation of Microalgal Lipid Production Using Genetic Engineering

TERRI G. DUNAHAY,* ERIC E. JARVIS, SONJA S. DAIS, AND PAUL G. ROESSLER

National Renewable Energy Laboratory, Applied Biological Sciences Branch, 1617 Cole Blvd., Golden, CO 80401

ABSTRACT

Genetic transformation of two species of diatoms has been accomplished by introducing chimeric plasmid vectors containing a bacterial antibiotic resistance gene driven by regulatory sequences from the acetyl-CoA carboxylase (ACCase) gene from the diatom *Cyclotella cryptica*. The recombinant DNA integrated into one or more random sites within the algal genome and the foreign protein was produced by the algal transformants. This is the first report of genetic transformation of any chlorophyll c-containing microalgal strain. We are using this system to introduce additional copies of the ACCase gene into diatoms in an attempt to manipulate lipid accumulation in transformed strains.

Index Entries: Biodiesel; diatoms; genetic transformation; lipid synthesis; microalgae.

INTRODUCTION

Microalgae are single-celled or colonial photosynthetic organisms that are attracting significant interest for industrial applications, including the production of specialty chemicals and nutritional supplements. In many microalgae, excess photosynthate accumulates as triacylglycerol storage lipids, and many species can be induced to accumulate increased amounts of lipids under conditions of nutrient deficiency *(1)*. The lipid, which is chemically similar to common vegetable oils, can be extracted and converted to fatty acid methyl esters for use as a diesel fuel substitute *(2)*. This type of fuel, which can be derived from animal, plant, or algal lipids, is known as "biodiesel." Biodiesel has advantages over fossil-derived diesel in that it is renewable, biodegradable, and produces fewer particulates and less SO_X when burned. Biodiesel from vegetable oil sources, such as soybean and canola, is currently used in a number of European markets, and is being tested in the US. Our goal is to develop microalgae as an economically viable source of biodiesel fuel.

One approach we are taking toward this goal is to use genetic engineering to optimize the lipid production capabilities of microalgae that show potential for biodiesel fuel production. A key step in this process is the development of genetic transformation systems for oleaginous microalgae. Prior to this article, the only

*Author to whom all correspondence and reprint requests should be addressed.

reproducible genetic transformation systems that had been developed for any microalgal species were for the fresh-water green alga *Chlamydomonas reinhardtii* (reviewed in ref. *3*) and a related colonial species, *Volvox carteri (4)*. Efficient nuclear transformation in these strains was accomplished only by using homologous genes as selectable markers; thus, the techniques were not directly transferable to other microalgal species.

Diatoms are a group of microalgae that has shown significant potential for biodiesel production. These organisms are major primary producers in the global ecosystem, producing approx 25% of the world's oxygen. Diatoms are characterized by the presence of chlorophyll c and fucoxanthin in their chloroplasts, and by their intricately patterned, siliceous cell walls *(5)*. We report here on the successful genetic transformation of two species of diatoms. Chimeric plasmid vectors were constructed that utilize regulatory sequences from a diatom gene to drive the expression of a bacterial antibiotic resistance gene. These vectors were introduced into diatoms via microprojectile bombardment, and the transformants were selected based on their acquired resistance to the antibiotic. Transformants were analyzed to demonstrate the presence of the foreign DNA and production of the foreign protein. Progress toward developing recombinant microalgal strains with altered lipid production capabilities will also be discussed.

METHODS

Algal Strains

The centric diatom *Cyclotella cryptica* Reimann, Lewin, and Guillard strain T13L *(6)* was obtained from the Provasoli-Guillard National Center for Culture of Marine Phytoplankton (West Boothbay Harbor, ME). *C. cryptica* strain CYCLO1 and the pennate diatom *Navicula saprophila* NAVIC1 are from the NREL Microalgal Culture Collection *(7)*. Cells were grown axenically in artificial sea-water medium (ASW) *(8)* supplemented with 1.07 m*M* sodium silicate. In liquid culture, *C. cryptica* T13L was grown in 50% ASW, and *N. saprophila* and *C. cryptica* CYCLO1 were grown in 10% ASW. Cultures were grown at 26°C in Erlenmeyer flasks without agitation under a 16:8 h light/dark cycle, with a light level of 50 $\mu E/m^2/s$. For solid media, 10% ASW supplemented with 20 m*M* glucose (ASWG) and 1% agar was used for all strains.

Plasmids

Two plasmids containing selectable markers that were used to transform the diatoms (pACCNPT10 and pACCNPT5.1) are shown in Fig. 1. Details of the construction of these plasmids will be described elsewhere *(9)*. Both plasmids contain the coding region of the neomycin phosphotransferase gene (*npt*II) from *E. coli* transposon Tn5; the neomycin phosphotransferase enzyme (NPTII) confers resistance to aminoglycoside antibiotics, such as G418 (geneticin). The *npt*II coding sequence is flanked by putative promoter and terminator sequences from the acetyl-CoA carboxylase (ACCase) gene (designated *acc1*) from *C. cryptica* T13L *(10)*. Plasmid pACCNPT10 contains 445 bp of upstream and 592 bp of downstream untranslated region (UTR) from *acc1*; this vector also contains 275 bp from the 3' end of the *acc1* coding region between the translational stop signal of the *npt*II gene and the *acc1* 3'-UTR. Plasmid pACCNPT5.1 was derived from pACCNPT10 by removing all

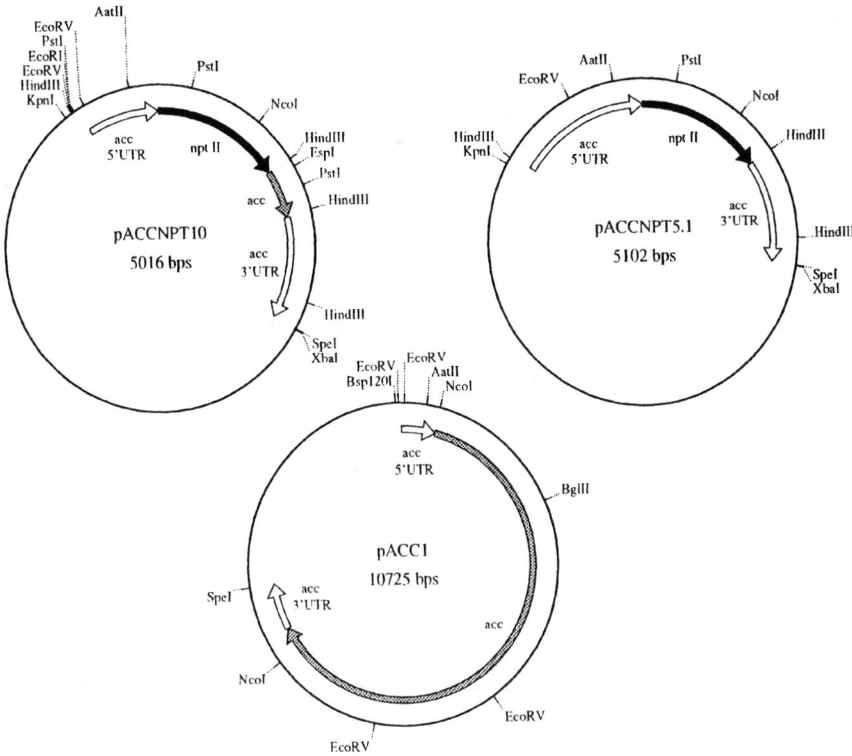

Fig. 1. Maps of the two plasmid vectors, PACCNPT10 and pACCNPT5.1, used as selectable markers for microalgal transformation. Also shown is pACC1, which contains the entire ACCase coding sequence from *C. cryptica* T13L cloned into pBluescript KS+. The *npt*II coding sequence is shown in solid black, *acc1* regulatory regions are in white, and *acc1* coding regions are shaded.

but 13 bp of the *acc1* coding region following the *npt*II gene, and by replacing the 445 bp of 5'-untranslated sequence with a corresponding 816-bp fragment to extend the putative promoter region. Plasmid pACC1 (also shown in Fig. 1) contains the entire coding region of *acc1*, including 445 bp of upstream and 592 bp of downstream sequence. The parental plasmid in all cases was pBluescript KS + (Stratagene, La Jolla, CA). Plasmid pCaMVNEO (not shown) was a gift from Michael Fromm.

Transformation

DNA was introduced into algal cells using the DuPont/Bio-Rad PDS1000-He microprojectile accelerator (Bio-Rad Laboratories, Hercules, CA). In this technique, plasmid DNA is precipitated onto tungsten or gold beads and propelled into the algal cells by a burst of helium pressure *(11)*. Before bombardment, concentrated algal cells were spread onto the center two-thirds of an ASWG agar plate that had been supplemented with 50 μg/mL ampicillin to reduce bacterial contamination. Approximately 3×10^7 cells were used for each bombardment for the *C. cryptica* strains, and 3×10^8 cells for *N. saprophila*. Plasmid DNA was coated onto tungsten particles (0.5-μm m3N particles from Alfa Research Chemicals, Ward Hill, MA), and the cells were bombarded with the particles using a burst pressure of 1100 psi. Ten microliters of the DNA/particle suspension, containing 3 mg tungsten and 0.8–1.0 μg plasmid DNA, were used for each transformation. After bombardment, the plates were placed in the growth room for 2 d to allow the cells to recover and express the foreign gene. To select transformants, cells on each plate were resuspended in 2 mL of 10% ASW and replated onto eight ASWG plates containing G418 (Sigma, St. Louis, MO). The concentration of G418 required to inhibit the growth of the cells completely was determined empirically for each algal strain, and depended on the cell plating density and on the salt concentration of the media used to grow the cells. G418-resistant strains of *C. cryptica* T13L, *C. cryptica* CYCLO1, and *N. saprophila* were selected on 100, 50, and 25 μg/mL G418, respectively. G418-resistant colonies were typically seen within 7–10 d. The putative transformants were picked after 14 d, and tested for continued growth on G418 plates and for the presence of the *npt*II gene and NPTII protein.

For the cotransformation experiments, pACCNPT5.1 and pACC1, in a molar ratio of approx 1:1, were coprecipitated onto tungsten beads and used to transform *C. cryptica* T13L or *N. saprophila*. G418-resistant colonies were screened for the presence of pACC1 sequences using the polymerase chain reaction (PCR).

Analysis of Transformants

DNA was isolated for Southern analysis by the glass bead/vortexing protocol as described previously *(12)*. The DNA was digested with restriction endonucleases, separated on 0.8% agarose gels, and transferred to nylon membranes. The foreign DNA sequences within the algal DNA were detected using the Genius nonradioactive DNA detection system (Boehringer Mannheim Biochemicals, Indianapolis, IN) according to the manufacturer's instructions. The probes were labeled with digoxigenin using PCR *(13)*. To detect *npt*II sequences, the entire *npt*II coding region was used as a probe. *C. cryptica* T13L *acc1* sequences in cotransformants were detected using a probe that recognizes a 312-bp conserved sequence at the 5'-end of the *acc1* coding region.

A rapid assay for the presence of NPTII protein in putative transformants was performed by Western blotting. Cells were scraped from plates using a sterile inoculating loop (approx 10-μL packed cell volume) and placed in 50 μL of water in a microfuge tube. An equal volume of SDS 2X extraction buffer (125 m*M* Tris, 4% SDS, 20% glycerol, 10% β-mercaptoethanol) was added, and the sample was boiled for 5 min. Cell debris was removed by centrifugation, and 10–20 μL samples were separated on 8–16% SDS-polyacrylamide gels (Novex, San Diego, CA). The separated proteins were transferred to a nitrocellulose membrane, and NPTII protein was detected using anti-NPTII primary antibodies (5-Prime→ 3-Prime, Inc., Boulder, CO) and alkaline phosphatase-conjugated goat antirabbit IgG secondary antibodies.

RESULTS

Two chimeric plasmid vectors were constructed that contain the bacterial *npt*II gene flanked by regulatory sequences from *acc1* from the diatom *C. cryptica* T13L (Fig. 1); these vectors were introduced into microalgal cells via microprojectile bombardment. The results from several transformation experiments are summarized in Table 1. Both plasmids were used successfully to mediate the transforma-

Table 1
Transformation of *C. cryptica* and *N. saprophila*

Recipient Strain	Plasmid	Form of input DNA	# Plates treated	# G418-resistant colonies	Ave # transf. per treatment	Transformants per 10^8 cells[a]
T13L	pACCNPT10	Supercoiled	5	25	5.0	16.7
T13L	pACCNPT5.1	Supercoiled	10	19	1.9	6.3
T13L	pACCNPT5.1	Linear[b]	7	16	2.3	7.7
T13L	pACCNPT10	Linear	7	9	1.3	4.3
T13L	Control[c]	—	4	0	0	0
T13L	pCaMVNEO	Supercoiled	2	0	0	0
CYCLOI	pACCNPT10	Supercoiled	2	2	1.0	3.4
CYCLOI	pACCNPT5.1	Supercoiled	2	12	6.0	20.4
CYCLOI	Control	—	2	0	0	0
NAVIC1	pACCNPT5.1	Supercoiled	5	42	8.4	2.8
NAVIC1	pACCNPT10	Supercoiled	1	18	18.0	6.0
NAVICI	Control	—	1	0	0	0

[a]*C. cryptica* plates contained ten times as many cells as the *N. saprophila* plates.
[b]Linearized plasmids were produced by digestion with *Kpn*I, which cuts in the polycloning site of pBluescript KS+ at the 5'-end of the *acc1* promoter region (Fig. 1).
[c]Controls consisted of cells not bombarded with plasmid or cells bombarded with supercoiled pBluescript KS+.

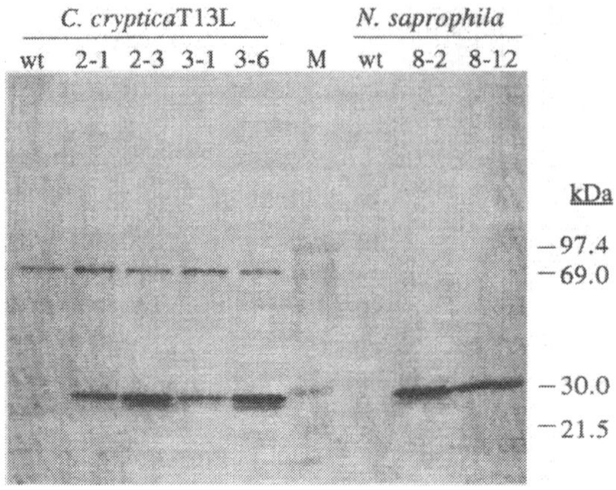

Fig. 2. Western blot showing the presence of the 30-kDa NPTII protein in transformants of *C. cryptica* T13L and *N. saprophila* NAVIC1. The sizes (in kDa) of the mol- wt markers in the 6th lane (M; Rainbow Markers, Amersham, Arlington Heights, IL) are indicated on the right.

tion of two strains of the centric diatom, *C. cryptica*, as well as the pennate diatom *N. saprophila*. The numbers of transformants obtained in each experiment were low, from about 1–18 from each plate treated. Similar numbers of transformants were obtained using either supercoiled or linearized plasmids. No G418-resistant colonies were seen on untreated controls or on plates containing cells bombarded with pBluescript KS+. We also did not obtain any transformants using pCaMVNEO as the transformation vector; this plasmid contains the *npt*II gene driven by regulatory sequences shown to work well in a number of higher plants systems *(14,15)*.

DNA was isolated from wild-type and G418-resistant strains of the algae to test for the presence of the *npt*II gene by Southern blot analysis. DNA fragments that hybridized to the digoxigenin-labeled *npt*II probe were seen in all G418-resistant isolates tested; the probe did not hybridize to any fragments in the DNA from wild-type cells (data not shown). In the *C. cryptica* transformants, the hybridization patterns suggest that the plasmid DNA integrates into one or more random sites in the host cell genome, often in the form of tandem repeats. Similar analyses were also performed on DNA isolated from transformed strains of *N. saprophila* NAVIC1. All G418-resistant isolates of these transformants also contained DNA fragments that hybridized to the *npt*II probe. In contrast to the results with *C. cryptica*, the data suggest that *N. saprophila* may be less likely to integrate the foreign DNA in the form of tandem repeats, and more likely to integrate the plasmid at multiple sites or to insert rearranged copies of the plasmid DNA.

To ensure that the G418-resistant phenotype of the transformants was caused by the expression of a functional NPTII protein, and not by a spontaneous mutation, crude cell extracts were analyzed for the presence of the NPTII protein by Western blotting. A blot containing protein extracts from wild-type and transformed cells of *C. cryptica* T13L and *N. saprophila* is shown in Fig. 2. All G418-resistant isolates produced a 30-kDa protein that is recognized by the anti-NPTII antibody. This protein is not seen in wild-type cells. Of the G418-resistant isolates listed in Table 1, 58 were

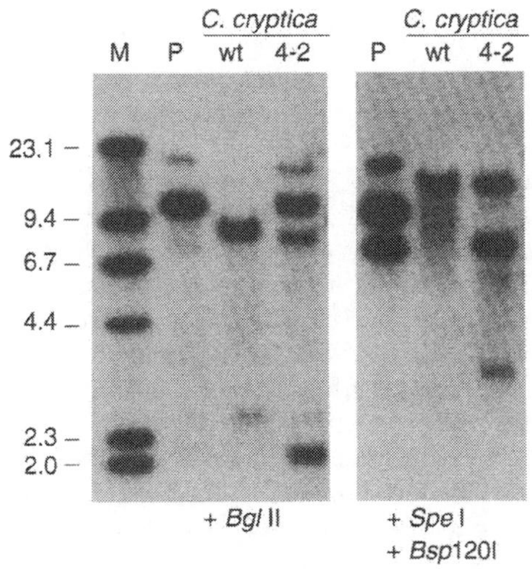

Fig. 3. Southern blot showing the presence of additional *acc1* sequences in an isolate of *C. cryptica* T13L cotransformed with pACCNPT5.1 and pACC1. The DNA was hybridized to a digoxigenin-labeled probe specific for a 312-bp region at the 5'-end of *acc1*. *Bgl*II cuts pACC1 once, within the ACCase coding sequence (*see* Fig. 1). The presence of a band of 10.7-kb in the cotransformant that comigrates with linearized pACC1suggests that the plasmid has integrated in the form of two or more tandem repeats, as was seen for the *npt*II containing plasmids. *Bsp*120I and *Spe*I recognize single sites in pACC1 that flank a 7.9-kb fragment that includes the entire *acc1* coding region. The presence of a band of this size demonstrates that at least one full-length copy of *acc1* from the plasmid is present in the recombinant strain. The first lane (M) contains digoxigenin-labeled λ DNA digested with *Hind*III as size markers (Boehringer Mannheim Biochemicals). Lanes marked "P" contain pACC1 plasmid, digested with the indicated enzymes.

chosen at random and tested for the presence of NPTII protein by Western blotting. All contained the NPTII protein.

As discussed above, one goal of this research is to use the genetic transformation technology to manipulate microalgal lipid biosynthetic pathways. One potential engineering target is the enzyme ACCase, which catalyzes the carboxylation of acetyl-CoA to form malonyl-CoA, the primary building block of fatty acids. This enzyme is believed to be a key regulatory enzyme in lipid production for higher plants *(16)*, and results from our laboratory suggest a role for ACCase in the accumulation of storage lipids in *C. cryptica* in response to nutrient deficiency *(17)*. Therefore, we decided to test whether overexpressing the ACCase gene (*acc1*) in diatoms could affect lipid accumulation. We successfully introduced additional copies of *acc1* into *C. cryptica* T13L and *N. saprophila* by cotransforming these organisms with pACC1, which contains the genomic ACCase gene, and pACCNPT5.1. At least one isolate of each strain has been shown by Southern analysis to contain one or more full-length copies of *acc1* in addition to the endogenous sequences (Fig. 3). Analyses of these recombinant strains for enhanced ACCase protein production and activity, and for their effects on lipid accumulation, are in progress. Preliminary results suggest that ACCase activity may be increased two- to threefold as compared to wild-type cells in several recombinant strains.

DISCUSSION

There are two basic aspects to the development of a genetic transformation system: (1) a method to get DNA into the cell; and (2) a way to detect expression of the foreign gene.

For microalgae, the cell wall is a significant barrier to DNA entry. There has been little success with producing and regenerating microalgal protoplasts because of the diversity and complexity of microalgal cell walls. For this reason, we have been exploring methods to introduce DNA into microalgal cells directly through the cell walls. We recently reported the use of silicon carbide fibers to mediate the entry of DNA into walled cells of the green alga *C. reinhardtii (18),* but preliminary studies suggested that this method would not work to penetrate the cell walls of diatoms, such as *C. cryptica.* During the past few years, microprojectile bombardment (also known as biolistics), has been widely used to introduce DNA into walled cells of higher plants and *C. reinhardtii,* as well as intact cells of some bacteria and fungi. We found that biolistics also works very well to mediate DNA entry into at least two species of diatoms through the siliceous cell wall. Based on results in *C. reinhardtii* and the results reported here, biolistics should work well to introduce DNA into most microalgal species.

To be able to detect the expression of the foreign gene, we chose a selectable marker gene, since this allowed us to screen for rare transformation events in a large population of cells. In *C. reinhardtii,* nuclear transformation required the use of homologous genes as selectable markers. The DNA of these organisms is unusually rich in the nucleotides guanosine and cytosine, which may relate to codon bias and poor expression of foreign genes *(3).* Previous work from our laboratory indicated that DNA from diatoms does not have an elevated G + C content *(12),* so expression of bacterial marker gene by the diatom should not be problematic. We believe that the use of diatom regulatory sequences was critical to obtaining high enough expression of the bacterial gene to allow growth of the transformed cells in the presence of G418. This is supported by our inability to recover *C. cryptica* T13L transformants when pCaMVNEO, which contains the *npt*II gene driven by a strong plant promoter (the 35S RNA promoter from the cauliflower mosaic virus, CaMV), was introduced into the algal cells (Table 1). However, when we cotransformed *C. cryptica* T13L with pACCNPT5.1 and plasmid pDO432 (which contains a bacterial luciferase gene driven by the same CaMV promoter *[19,20]*), a number of transformants selected based on their resistance to G418 also demonstrated significant luciferase activity (unpublished results). Thus, although the microalgae can recognize and utilize foreign promoter sequences, the level of expression may be insufficient to allow the use of the foreign promoters in selectable marker systems.

We have recently begun experiments designed to use genetic engineering to optimize lipid production in microalgae. The preliminary results are encouraging, because it appears that introducing additional copies of the ACCase gene into *C. cryptica* does result in enhanced activity of this enzyme. We are working to characterize these transformants and to identify other key genes in the various algal metabolic pathways that would be likely genetic engineering targets for modifying the lipid production capabilities of microalgae.

ACKNOWLEDGMENTS

We would like to thank Dana Heacox and John Buen for excellent technical support. This work was funded by the Feedstock Development Element of the Biofuels Energy Systems Program of the US Department of Energy.

REFERENCES

1. Roessler, P. G. (1990), *J. Phycol.* **26,** 393–399.
2. Shay, E. G. (1993), *Biomass and Bioenergy* **4,** 227–242.
3. Kindle, K. L. and Sodeinde, O. A. (1994), *J. Appl. Phycol.* **6,** 231–238.
4. Schiedlmeier, B., Schmitt, R., Mueller, W., Kirk, M. M., Gruber, H., Mages, W., and Kirk, D. L. (1994), *Proc. Natl. Acad. Sci. USA* **91,** 5080–5084.
5. Round, F. E., Crawford, R. M., and Mann, D. G. (1990), *The Diatoms*. Cambridge University Press, Cambridge, UK, p. 747.
6. Reimann, B. E. F., Lewin, J. M. C., and Guillard, R. R. L. (1963), *Phycologia* **3,** 75–84.
7. Barclay, W., Johansen, J., Chelf, P., Nagle, N., Roessler, P., and Lemke, P. (1986), Microalgae Culture Collection 1986–1987, Solar Energy Research Institute, Golden, CO, SERI/SP-232-3079.
8. Brown, L. (1982), *Phycologia* **21,** 408–410.
9. Dunahay, T. G., Jarvis, E. E., and Roessler, P. G. (1995), *J. Phycol.* **31,** 1004–1012.
10. Roessler, P. G. and Ohlrogge, J. B. (1993), *J. Biol. Chem.* **268,** 19,254–19,259.
11. Sanford, J. C., Smith, F. D., and Russell, J. A. (1993), *Methods Enzymol.* **217,** 483–509.
12. Jarvis, E. E., Dunahay, T. G., and Brown, L. M. (1992), *J. Phycol.* **28,** 356–362.
13. Lion, T. and Haas, O. A. (1990), *Anal. Biochem.* **188,** 335–337.
14. Fromm, M. E.,Taylor, L. P., and Walbot, V. (1986), *Nature* **319,** 791–793.
15. Gruber, M. Y. and Crosby, W. L. (1993), In *Methods in Plant Molecular Biology and Biotechnology* (Glick, B. R. and Thompson, J. E., eds.), CRC, Boca Raton, FL, pp. 89–119.
16. Post-Beittenmiller, D., Roughan, G., and Ohlrogge, J. B. (1992), *Plant Physiol.* **100,** 923–930.
17. Roessler, P. G. (1988), *Arch. Biochem. Biophys.* **267,** 521–528.
18. Dunahay, T. G. (1993), *BioTechniques* **15,** 452–460.
19. Ow, D. W., Wood, K. V., DeLuca, M., de Wet, J. R., Helinski, D. R., and Howell, S. H. (1986), *Science* **234,** 856–859.
20. Jarvis, E. E. and Brown, L. M. (1991), *Curr. Genet.* **19,** 317–321.

0273-2289/96/5758—0233$7.50

Screening for L-Arabinose Fermenting Yeasts†

BRUCE S. DIEN,*,[1] CLETUS P. KURTZMAN,[2] BADAL C. SAHA,[1] AND RODNEY J. BOTHAST[1]

[1]Fermentation and Biochemistry Research and [2]Microbial Properties Research, National Center for Agricultural Utilization Research, Agricultural Research Service, US Department of Agriculture, 1815 North University Street, Peoria, IL 61604-3999

ABSTRACT

Utilization of pentose sugars (D-xylose and L-arabinose) derived from hemicellulose is essential for the economic conversion of biomass to ethanol. Xylose-fermenting yeasts were discovered in the 1980s, but to date, no yeasts have been found that ferment L-arabinose to ethanol in significant quantities. We have screened 116 different yeasts for the ability to ferment L-arabinose and have found the following species able to ferment the sugar: *Candida auringiensis*, *Candida succiphila*, *Ambrosiozyma monospora*, and *Candida* sp. (YB-2248). Though these yeasts produced ethanol concentrations of 4.1 g/L or less, they are potential candidates for mutational enhancement of L-arabinose fermentation. These yeasts were also found to ferment D-xylose.

Index Entries: L-arabinose; pentose; ethanol; fermentation; yeasts.

INTRODUCTION

Ethanol has the potential to be an economic and environmentally sound alternative to petroleum for use in vehicular transportation. Corn starch is currently the primary feedstock for the US biofuel industry. However, a significant increase in ethanol production will depend on the fermentation of lignocellulosic biomass. Both the cellulosic and hemicellulosic components of the biomass will need to be fermented for the conversion to be economical *(1)*.

Fermentation of many hemicelluloses in agricultural biomass is problematic because it consists primarily of D-xylose and other pentoses that cannot be fermented by the yeast traditionally used to produce ethanol from starch. Much of the past research on hemicellulose fermentation has concentrated on D-xylose, because D-xylose is the most abundant pentose in almost all hemicelluloses *(2,3)*. However, L-

*Author to whom all correspondence and reprint requests should be addressed.

†Names are necessary to report factually on available data. However, the USDA neither guarantees nor warrants the standard of the product, and the use of the name by USDA implies no approval of the product to the exclusion of others that may also be suitable.

arabinose is present in significant amounts in many hemicelluloses. One important example is the fiber that is separated during the wet milling of corn. Other examples include rice hulls, sugar beet residues *(4)*, sugar cane residues *(5)*, and the hemicelluloses of herbaceous species *(2)*.

Several bacteria *(4,6–11)*, three of which are recombinant, and a single filamentous fungus, *Paecilomyces* sp. NF1 *(12,13)*, have been reported to convert L-arabinose to ethanol efficiently. No yeast has been discovered to date that produces more than a trace amount of ethanol from L-arabinose *(2,14,15)*. Before the 1980s, yeasts were also thought to be unable to ferment all pentoses *(16–18)*. Many of the yeasts, which were subsequently discovered to ferment D-xylose, were screened by using microaerophilic culture conditions and directly testing for the presence of ethanol *(19–21)*.

Previous screenings for L-arabinose fermenters have been incomplete in that they have been limited to a few of the better known yeasts *(14)* or to some of the more efficient D-xylose-fermenting species *(2,14,15,22–27)*. We have now carried out an extensive screening of 116 yeast strains, from the ARS Culture Collection (National Center for Agricultural Utilization Research, Peoria, IL), some of which had previously been screened for arabitol production by this laboratory *(28)*. The yeasts were screened for L-arabinose fermentation under microaerophilic culture conditions, and ethanol was detected directly using gas chromatography (GC). The yeast strains screened were chosen on the basis of one or more of the following four sets of criteria:

1. Yeast species that aerobically catabolize D-xylose and L-arabinose and ferment glucose (101 strains).
2. Yeast strains that ferment glucose, aerobically catabolize L-arabinose, and do not catabolize D-xylose (six strains).
3. Yeast strains that are superior D-xylose fermenters (10 strains).
4. Yeast strains isolated from environments rich in L-arabinose (12 strains).

METHODS

Organisms

The yeast screened are listed in Table 1. All strains were obtained from the ARS Culture Collection (NRRL).

Media

Media were sterilized by autoclaving for 15 min. Sugars were autoclaved separately. Yeast cultures were maintained on YM slants (0.3 % [w/v] yeast extract, 0.3% [w/v] malt extract, 0.5% [w/v] peptone, 1.0% [w/v] glucose, and 1.5% [w/v] agar) incubated at 28°C and stored at 4°C. Cultures were transferred at least once to slants before being screened for L-arabinose fermentation.

All precultures and fermentations used yeast peptone (YP) (1% [w/v] yeast extract, 2% [w/v] Bacto peptone) supplemented with either L-arabinose or D-xylose. Precultures were grown on the same sugar as used for the fermentation and, unless stated otherwise, the sugar concentration was 8% (w/v). Medium ingredients were purchased from by Difco (Detroit, MI), and sugars were purchased from Sigma (St. Louis, MO).

Culture Conditions for L-Arabinose Fermentation Screening

Yeasts were screened for L-arabinose fermentation under microaerophilic culture conditions. Yeast strains were transferred from slants to 12-mL sterile polypropylene test tubes (Fisher Scientific, Springfield, NJ) containing 3 mL of medium. The tilted culture tubes were incubated with shaking (220 rpm) at 30°C in a Innova 4000 New Brunswick (Edison, NJ) incubator/shaker. After approx 24 h, 0.5-mL aliquots were removed from the culture tubes and used to inoculate foam-capped (Fisher Scientific) 125-mL Erlenmeyer flasks containing 30 mL of medium. The culture flasks were incubated with shaking as described for 48–72 h before being used for inoculation.

L-Arabinose fermentations were done in 50-mL Erlenmeyer flasks containing 30 mL of medium and 12 mL of inoculum from the above precultures. The flasks were sealed with rubber septa (Sigma). Each rubber septum was pierced with a 20-gage needle, with cotton inserted into the syringe end, to vent CO_2. The cultures were incubated as described for 12 d.

Fermentations were sampled (1-mL vol) every approx 48 h. Samples were centrifuged (8000*g* for 6 min) and the supernatant fluids recovered. Samples were analyzed for ethanol immediately, and then stored at –20°C for later analysis of sugars and sugar alcohols.

After the initial screening was completed, all the available strains that corresponded to the yeast species that fermented L-arabinose were screened. These strains were tested for the ability to ferment D-xylose using the same culture conditions. Yeast species that fermented L-arabinose were also tested to make sure they did not ferment the basal medium. The fermentations were performed as described earlier, except the cells used for inoculating these fermentations were washed three times with YP, and the medium used for the fermentations was not supplemented with sugar. The inoculum was grown on YP supplemented with L-arabinose.

Variation of Culture Conditions for L-Arabinose and D-Xylose Fermentations

In an attempt to increase ethanol yield from L-arabinose and D-xylose, the fermentation temperature was lowered (from 30 to 25°C) and aeration increased. Lower temperature and moderate aeration have been reported to give larger ethanol yields for some xylose fermentations *(29–31)*. Yeast strains were grown in culture tubes as described previously at 25°C, and after approx 24 h, 0.5-mL aliquots were transferred to foam-capped 250-mL Erlenmeyer flasks containing 50 mL of medium. Cultures were incubated with shaking for approx 24 h and 1-mL aliquots were used to inoculate foam-capped 500-mL Erlenmeyer flasks, each containing 100 mL of medium. These cultures were grown with shaking for 24 h. The cells were harvested by centrifugation (6 min at 6000*g*) and resuspend in 4 mL of YP. Two-milliliter aliquots from each suspension were transferred to the final fermentation flasks. The 125-mL Erlenmeyer flasks used for the fermentation each contained 50 mL of YP medium supplemented with either 10% (w/v) L-arabinose or D-xylose and were capped with Silicone Sponge Plug Closures (Bellco Glass). Cultures were incubated at 25°C and shaken at 140 rpm for 12 d. Cultures were sampled once a day as described.

Table 1
Yeasts Screened for L-Arabinose Fermentation

Species	NRRL no.	Species	NRRL no.
Ambrosiozyma monospora	Y-1081	*C. tenuis*	Y-12987
A. monospora	Y-1484	*C. tenuis*	Y-1498
A. monospora	Y-5955	*C. tenuis*	Y-17105
A. monospora	Y-5956	*C. tenuis*	Y-17106
A. monospora	Y-7403	*C. tropicalis*	Y-7661
Candida auringiensis	Y-11848	*C. verstilis*	Y-12819
C. auringiensis	Y-11849	*C. viswanathii*	Y-6726
C. auringiensis	Y-11850	*C. wickerhamii*	Y-2563
C. boidinii	Y-2332	*Citeromyces matritensis*	Y-11797
C. buinensis	Y-11706	*Debaryomyces (Wingea) robertsiae*	Y-6670
C. chilensis	Y-17141	*D. castellii*	Y-7423
C. conglobata	Y-1504	*D. hansenii*	Y-7426
C. dendronema	Y-7781	*D. hansenii*	YB-434
C. diddensiae	Y-2324	*D. nepalensis*	Y-7108
C. diddensiae	Y-7589	*D. nepalensis*	Y-7534
C. entomaea	Y-7785	*D. polymorphus*	Y-2022
C. entomophila	Y-7783	*D. pseudopolymorphus*	YB-4228
C. ernobii	Y-12940	*D. yamadae*	Y-11714
C. guilliermondii	Y-7572	*Geotrichum fermentans*	Y-1492
C. guilliermondii	YB-4760	*Kluyveromyces marxianus*	Y-1195
C. halophila	Y-2483	*K. marxianus*	Y-6373
C. hellenica	Y-17319	*K. marxianus*	Y-7571
C. insectorum	Y-7787	*Pachysolen tannophilus*	Y-2460
C. intermedia	Y-981	*Pichia tannicola*	Y-7499
C. kefyr	Y-1204	*P. acaciae*	Y-7773
C. membranamfaciens	Y-2089	*P. amylophila*	YB-1287
C. mogii	Y-17032	*P. angophorae*	Y-7842
C. naeodendra	Y-10942	*P. angusta*	Y-2214
C. nitratophila	YB-3654	*P. bimundalis*	Y-5343
C. parapsilosis	Y-7659	*P. bovis*	YB-4184
C. parapsilosis	YB-433	*P. burtonii*	Y-1933
C. peltata	Y-6888	*P. burtonii*	Y-2057
C. rhagii	Y-2596	*P. burtonii*	Y-7144
C. shehatae	Y-12856	*P. capsulata*	Y-1842
C. shehatae	Y-12858	*P. capsulata*	Y-2234
C. shehatae	Y-17029	*P. capsulata*	YB-2252
C. silvanorum	Y-7782	*P. ciferrii*	Y-1031
C. silvicola	YB-2251	*P. etchellsii*	Y-7121
C. silvicultrix	Y-7789	*P. etchellsii*	Y-7547
C. sp.	Y-10925	*P. guilliermondii*	Y-2075
C. sp.	YB-2090	*P. hangzkouana*	Y-17346
C. sp.	YB-2248	*P. haplophila*	Y-7860
C. sp.	YB-2249	*P. holstii*	Y-2155
C. sp.	YB-3082	*P. kodamae*	Y-17234
C. steatolytica	Y-7136	*P. methanolica*	Y-11993
C. succiphila	Y-11997	*P. methanolica*	Y-7685
C. succiphila	Y-11998	*P. methylovora*	Y-17250

Species	NRRL no.	Species	NRRL no.
P. mexicana	Y-11818	*P. silvicola*	Y-1679
P. mississippiensis	Y-11748	*P. silvicola*	Y-7005
P. mississippiensis	YB-1294	*P. stipitis*	Y-11545
P. naganishii	Y-7654	*P. stipitis*	Y-7124
P. nakazawae	Y-7903	*P. sydowiorum*	Y-10995
P. philogaea	Y-7813	*P. sydowiorum*	Y-10996
P. philogaea	Y-7814	*P. sydowiorum*	Y-7130
P. pinus	Y-11528	*P. trehalophila*	Y-6781
P. pinus	Y-7928	*Saccaromycopsis capsularis*	Y-17639
P. scolyti	Y-11540	*Saccaromycopsis capsularis*	Y-676
P. scolytii	Y-5512	*Zygoascus hellenicus*	Y-7923

Analytical Methods

Ethanol concentrations were determined by gas–liquid chromatography on a Hewlett Packard 5890A gas chromatograph (Hewlett Packard, Wilmington, DL) with a 80/100 Porapak Q column (6 ft × 1.8 in., Supelco, Bellefonte, PA) using a flame ionization detector. The injection temperature was set at 200°C, the oven temperature at 150°C, and the detector temperature at 250°C. Integration was performed by a Hewlett Packard 3393A Integrator. Sugars (L-arabinose and D-xylose) and sugar alcohols (arabitol, xylitol, and ribitol) were determined by high-pressure liquid chromatography with an HPX-87C column (300 × 7.8 mm; Bio-Rad Laboratories, Hercules, CA) at 85°C. The column was eluted with distilled H_2O at a flow rate of 0.6 mL/min. The detector used was either a Waters R401 or 410 Differential Refractometer (Milford, MA). Detected peaks were separated baseline to baseline.

RESULTS

Of the 116 strains screened for L-arabinose fermentation (Table 1), seven strains produced detectable levels of ethanol (Table 2). Most of the 116 strains screened converted L-arabinose to arabitol, and for some strains, trace amounts of xylitol were also detected (data not shown). The greatest amount of ethanol produced was 3.4 g/L by *Candida* sp. (YB-2248). However, yeasts fermented the L-arabinose slowly, and maximum ethanol concentrations were measured at or near the end of the fermentations, 14 d after inoculation (data not shown). Most yeast strains consumed less than half of the L-arabinose supplied and not all of the *A. monospora* or *C. auringiensis* strains fermented L-arabinose. The yeasts that fermented L-arabinose did not produce any ethanol from YP medium not supplemented with L-arabinose (data not shown).

All 11 strains fermented D-xylose (Table 3). The ethanol concentration measured ranged from 2.1 to 20.4 g/L. As observed for L-arabinose fermentations, the maximum ethanol concentration for most strains was measured on the last day of the fermentation. These yeasts tended to leave less residual D-xylose than L-arabinose (Table 2) and most also produced xylitol as a coproduct (Table 3). In general, the better D-xylose fermenters were the yeasts that also fermented L-arabinose.

The fermentation temperature was lowered from 30 to 25°C, and aeration increased in an attempt to increase the yield of ethanol from L-arabinose. Only the

Table 2
Variation in Ethanol Production by L-Arabinose-Fermenting Yeasts[a]

Species	NRRL no.	Max. ethanol, g/L	Max. arabitol, g/L	% L-Arabinose used
A. monospora	Y-1081	0.8 ± 0.2	0 ± 0	14 ± 6
	Y-5955	1.8 ± 0.7	0 ± 0	19 ± 8
	Y-5956	nd	0 ± 0	18 ± 4
	Y-7403	nd	0 ± 0	13 ± 5
	Y-1484	2.3 ± 0.4	4 ± 0	36 ± 4
C. auringiensis	Y-11849	nd	37 ± 20	75 ± 26
	Y-11850	nd	11 ± 3	35 ± 3
	Y-11848	1.4 ± 0.6	17 ± 1	32 ± 1
C. succiphila	Y-11997	2.1 ± 0.3	8 ± 1	43 ± 2
	Y-11998	2.3 ± 0.4	8 ± 0	32 ± 6
Candida sp.	YB-2248	3.4 ± 0.1	4 ± 0	33 ± 5

[a]All strains were tested in duplicate. nd, not detected.

Table 3
Ethanol Production from D-Xylose Fermentation by L-Arabinose-Fermenting Yeast Species

Species	NRRL no.	Max. ethanol, g/L	Max. xylitol, g/L	% Xylose used
A. monospora	Y-1081	6.2	10	44
	Y-5955	2.1	5	20
	Y-5956	4.7	0	13
	Y-7403	7.6	7	44
	Y-1484	20.4 ± 1.6	31 ± 3	78 ± 8
C. auringiensis	Y-11849	6.6	0	33
	Y-11850	4.1	0	18
	Y-11848	3.6 ± 0.2	0 ± 0	20 ± 4
C. succiphila	Y-11997	5.5	16	71
	Y-11998	17.1 ± 0.8	18 ± 1	66 ± 7
Candida sp.	YB-2248	14.2 ± 1.1	25 ± 0	96 ± 5

[a]For each strain tested in duplicate, the average ± the half difference is shown.

strain that produced the highest concentration of ethanol at 30°C from each species was tested. The results are shown in Table 4. Both *A. monospora* and *C. succiphila* reached approximately twice the ethanol concentration under the new culture conditions compared to the previous fermentations (Table 2). However, *Candida* sp. (YB-2248) produced less ethanol, and *C. auringiensis* produce none. All four yeasts fermented L-arabinose slowly. Two of the strains (*C. auringiensis* and *C. succiphila*) consumed all of the L-arabinose, converting most of it into arabitol. In contrast, *A. monospora* did not produce arabitol as a coproduct.

These yeast strains, along with *Pichia stipitis*, were also tested for fermentation of D-xylose (Table 5) under the same conditions. *P. stipitis*, which is one of the better D-xylose fermenters *(32,33)*, produced 35.0 g/L of ethanol. In comparison, the L-arabinose-fermenting yeast produced approximately a third of this amount

Table 4
Fermentation of L-Arabinose at 25°C with Moderate Aeration[a]

Species	NRRL no.	Max. ethanol, g/L[b]	Max. arabitol, g/L	% L-Arabinose used
A. monospora	Y-1484	4.1 ± 0.3 (7)	0 ± 0	27 ± 1
C. auringiensis	Y-11848	nd	73 ± 3	100 ± 0
C. succiphila	Y-11998	3.9 ± 0.6 (7)	81 ± 1	100 ± 0
Candida sp.	YB-2248	1.7 ± 0.7 (9)	6 ± 0	42 ± 1

[a]All strains tested in duplicate.
[b]Numbers in parentheses are the average number of days before maximum ethanol concentration was measured.

Table 5
Fermentation of D-Xylose at 25°C with Moderate Aeration[a]

Species	NRRL no.	Max. ethanol, g/L[b]	Max. xylitol, g/L	% D-Xylose used
A. monospora	Y-1484	18.2 ± 0.5 (9)	22 ± 3	100 ± 0
C. auringiensis	Y-11848	11.5 ± 1.3 (6.5)	4 ± 3	67 ± 5
C. succiphila	Y-11998	12.5 ± 1.0 (9)	12 ± 1	57 ±13
Candida sp.	Y-2248	12.3 ± 0.3 (9)	20 ± 1	96 ± 0
P. stipitis	Y-7124	35.0 ± 0.4 (3)	0 ± 0	100 ± 0

[a]Replicates were run for all strains.
[b]Numbers in parentheses are the average number of days into the fermetation when the maximum ethanol concentration was measured.

and fermented the sugar more slowly. *A. monospora* produced the most ethanol (18.2 g/L). Although *P. stipitis* produced ribitol as a coproduct (data not shown), the other yeast produced xylitol. Except for *C. auringiensis*, the yeasts produced similar concentrations of ethanol under both aeration conditions.

DISCUSSION

Rarity of L-Arabinose Fermenters

The majority of the 101 yeast species tested were able to convert L-arabinose to arabitol. The results of previous, more limited, screenings also have reported that yeasts and other fungi able to catabolize L-arabinose aerobically will commonly convert it to arabitol when oxygen limited *(2,14,28)*. Only four species, 4% of the yeast species screened, were able to ferment L-arabinose. The rarity of L-arabinose fermenters might arise from a putative redox imbalance in the L-arabinose pathway and/or the low incidence of L-arabinose-rich environments found in nature.

The L-arabinose and D-xylose pathways in yeasts and other fungi share the common intermediate xylitol, but whereas D-xylose is converted directly to xylitol *(18,34,35)*, the conversion of L-arabinose to xylitol requires two additional steps. L-Arabinose is converted to xylitol according to the following sequence: L-arabinose to arabitol to L-xylulose to xylitol *(18,36–38)*. Once xylitol is formed, whether from L-arabinose or D-xylose, it is oxidized to D-xylulose, phosphorylated to D-xylulose-5-phosphate, and enters the pentose phosphate pathway *(18,36,38)*. The

Table 6
Past Results for Fermentation of L-Arabinose by Fungi and Yeast

Species	Ethanol, g/L	References
Candida sp.	0.16	*15*
Candida sp. XF 217	0.68	*15*
C. tropicalis	Trace	*42*
Fusarsium oxysporum	1	*43*
Pachysolen tannophilus	Not reported	*26*
Paecilomyces sp. NF1	13.8	*12*
Schizosaccharomyces pombe	trace	*14*

formation of D-xylulose from L-arabinose compared to D-xylose requires an extra NADPH and NAD^+.

The characteristic formation of xylitol from D-xylose under oxygen-limited conditions and the inability of most yeasts to ferment D-xylose are theorized to be consequences of yeasts' inability to regenerate the NAD^+ used in the oxidation of xylitol to D-xylulose *(34,39,40)*. Yeasts that do ferment D-xylose well have D-xylose reductase activity that is cospecific for NADH and NADPH *(39)*, which allows for the regeneration of NAD^+. Conversion of L-arabinose to D-xylulose requires two oxidation steps, which are thought to use NAD^+ *(18,36,38)*. The inability of yeasts to regenerate NAD^+ might, therefore, hinder the ability of yeast to ferment L-arabinose to ethanol and instead favor arabitol production *(2)*.

Another possible reason for the rarity of L-arabinose fermenters might be the low incidence of L-arabinose-rich environments found in nature. Although L-arabinose is present in many hemicelluloses, the predominant pentose is almost always D-xylose *(2,3)*. Therefore, an environment relatively rich in L-arabinose is likely to be much richer in D-xylose. An exception is the water-soluble hemicellulose fraction of larch wood *(41)*. In this regard, it is significant that one of the L-arabinose fermenters (*Candida* sp. YB-2248) was isolated from larch wood. In fact, all the better D-xylose fermenters were isolated from D-xylose rich environments *(20,23)*.

Efficiency of L-Arabinose Fermentations

Of the yeast screened, *A. monospora* produced the highest concentration of ethanol (4.1 g/L) from L-arabinose. Those yeasts that fermented L-arabinose did so slowly. Previous results reported for yeasts and other fungi fermenting L-arabinose are listed in Table 6. The ethanol concentrations reported here are greater than those previously reported, except for the fungus *Paecilomyces* sp. NF 1.

Aeration was increased and temperature lowered in an attempt to increase final ethanol concentrations. Changing the culture conditions doubled the maximum ethanol concentrations measured for *A. monospora* and *C. succiphila*. However *Candida* sp. (YB-2248) and *C. auringiensis* produced more ethanol at the lower aeration and higher temperature.

L-Arabinose Fermenters Also Ferment D-Xylose

The yeast strains we found that fermented L-arabinose also fermented D-xylose. Only *C. succiphila* has previously been identified as a D-xylose fermenter

(20). The direct correlation between the ability to ferment L-arabinose and D-xylose is probably a consequence of the similarity of the biochemical pathways for fermentation of both of these sugars. A near-perfect correlation has been observed between aerobic catabolism of L-arabinose and D-xylose by yeasts (18,44). Of the 214 yeast species identified by Barnett et al. (1983) (45) that aerobically catabolize L-arabinose, only 13 do not catabolize D-xylose (5). As an aside, the six yeasts present in the ARS Culture Collection that utilize L-arabinose and not D-xylose were included in the screen, and none fermented L-arabinose.

The L-arabinose-fermenting yeasts fermented D-xylose more efficiently, producing on average five times more ethanol from D-xylose than L-arabinose. Like the L-arabinose fermentations, the D-xylose fermentations were also run under different aeration conditions and at 25°C. *P. stipitis* was included in the experiment for comparison. The maximum ethanol concentrations produced by the L-arabinose fermenters were approximately half of that formed by *P. stipitis* and the rates of ethanol formation much slower. The L-arabinose fermenters also produced the coproduct xylitol. *P. stipitis* did not produce xylitol as a coproduct during this fermentation, but has been reported under some culture conditions (29).

The seven strains that have been discovered to ferment L-arabinose do not yet produce enough ethanol to be of commercial interest. However, these strains might be improved by mutagenesis. Previous to the discovery of *Pachysolen tannophilis* and other more efficient D-xylose fermenters, mutagenesis was used to increase the ethanol yield of an unknown *Candida* sp.-fermenting D-xylose fivefold (15). If, instead natural enrichment is used to find a better L-arabinose fermenting yeast, the yeasts should probably be screened first for D-xylose fermentation and second for L-arabinose fermentation. All the yeasts found here to ferment L-arabinose fermented D-xylose better.

ACKNOWLEDGMENTS

The critical review of the manuscript by Robert Hespell is gratefully acknowledged.

REFERENCES

1. Ladisch, M. R. and Svarczkopf, J. A. (1991), *Bioresource Technol.* **36,** 83–95.
2. McMillan, J. D. and Boynton, B. L. (1994), *Appl. Biochem. Biotechnol.* **45/46,** 569–584.
3. Timell, T. E., Wolfrom, M. L., and Tipson, R. S., eds. (1964), *Advances in Carbhydrate Chemistry*, vol. 19, Academic, New York, pp. 247–302.
4. Finn, R. K., Bringer, S., and Sahm, H. (1984), *Appl. Microbiol. Biotechnol.* **19,** 161–166.
5. Meyer, P. S., Du Preez, J. C., and Kilian, S. G. (1992), *System. Appl. Microbiol.* **15,** 161–165.
6. Beall, D. S., Ohta, K., and Ingram, L. O. (1991), *Biotechnol. Bioeng.* **38,** 296–303.
7. Bothast, R. J., Saha, B. C., Flosenzier, A. V., and Ingram, L. O. (1994), *Biotechnol. Lett.* **16,** 401–406.
8. Bringer, S., Durst, E., Sham, H., and Finn, R. K. (1984), *Biotechnol. Bioeng. Symp.* **14,** 269–278.
9. Patel, G. B. (1984), *Appl. Microbiol. Biotechnol.* **20,** 111–117.
10. Tolan, J. S. and Finn, R. K. (1987), *Appl. Environ. Microbiol.* **53,** 2033–2038.
11. Tolan, J. S. and Finn, R. K. (1987), *Appl. Environ. Microbiol.* **53,** 2039–2044.
12. Wu, J. F., Lastick, S. M., and Updegraff, D. M. (1986), *Nature* **321,** 887,888.
13. Singh, A., Kumar, P. K. R., and Schugerl, K. (1992), *Adv. Biochem. Eng. Biotechnol.* **45,** 30–55.
14. Gong, C. S., Claypool, T. A., McCracken, L. D., Maun, C. M., Ueng, P. P., and Tsao, G. T. (1983), *Biotechnol. Bioeng.* **25,** 85–101.
15. Gong, C. S., McCracken, L. D., and Tsao, G. T. (1981), *Biotechnol. Lett.* **3,** 245–250.

16. Slininger,,P. J., Bothast, R. J., van Cauwenberge, J. E., and Kurtzman, C. P. (1982), *Biotechnol. Bioeng.* **24,** 371–384.
17. Du Preez, J. C., Bosch, M., and Prior, B. A. (1986), *Appl. Microbiol. Biotechnol.* **23,** 228–233.
18. Barnett, J. A., Tipson, R. S., and Horton, D., eds. (1976), *Advances in Carbhydrate Chemistry,* vol. 32, Academic, New York, pp. 126–234.
19. Schneider, H., Wang, P. Y., Chan, Y. K., and Maleszka, R. (1981), *Biotechnol. Lett.* **3,** 89–92.
20. Toivola, A., Yarrow, D., van den Bosch, E., van Dijken, J. P., and Scheffers, W. A. (1984), *Appl. Environ. Microbiol.* **47,** 1221–1223.
21. Du Preez, J. C. and van der Walt, J. P. (1983), *Biotechnol. Lett.* **5,** 357–362.
22. Jeffries, T. W. and Sreenath, H. K. (1988), *Biotechnol. Bioeng.* **31,** 502–506.
23. Kurtzman, C. P. (1983), *Adv. Biochem. Eng. Biotechnol.* **27,** 73–84.
24. Du Preez, J. C., Bosch, M., and Prior, B. A. (1986), *Appl. Microbiol. Biotechnol.* **23,** 228–233.
25. Delgenes, J. P., Moletta, R., and Navarro, J. M. (1988), *Appl. Microbiol. Biotechnol.* **29,** 155–161.
26. Dekker, R. F. H. (1982), *Biotechnol. Lett.* **4,** 411–416.
27. Schneider, H., Maleszka, R., Neirinck, L., Veliky, I. A., Wang, P. Y., and Chan, Y. K. (1983), *Adv. Biochem. Eng. Biotechnol.* **27,** 57–71.
28. Saha, B. C. and Bothast, R. J. (1996), *Appl. Microbiol. Biotechnol.* **44,** in press.
29. Slininger, P. J., Bothast, R. J., Ladisch, M. R., Okos, M. R. (1990), *Biotechnol. Bioeng.* **35,** 727–731.
30. Slininger, P. J., Bolen, P. L., and Kurtzman, C. P. (1987), *Enzyme Microb. Technol.* **9,** 5–16.
31. Du Preez, J. C., Bosch, M., and Prior, B. A. (1986), *Enzyme Microb. Technol.* **8,** 360–364.
32. Slininger, P. J., Bothast, R. J., Okos, M. R., and Ladisch, M. R. (1985), *Biotechnol. Lett.* **7,** 431–436.
33. Du Preez, J. C. and Prior, B. A. (1985), *Biotechnol. Lett.* **7,** 241–246.
34. Bruinenberg, P. M., de Bot, P. H. M., van Dijken, J. P., and Scheffers, W. A. (1983), *Eur. J. Appl. Microbiol. Biotechnol.* **18,** 287–292.
35. Chiang, C. and Knight, S. G. (1959), *Biochim. Biophys. ACTA* **35,** 454–463.
36. Magee, R. J. and Kosaric, N. (1985), *Adv. Biochem. Eng. Biotechnol.* **32,** 61–93.
37. Chiang, C. and Knight, S. G. (1961), *Biochim. Biophys. ACTA* 271–278.
38. Witteveen, C. F. B., Busink, R., van de Vondervoort, P., Dijkema, C., Swart, K., and Visser, J. (1989), *J. Gen. Microbiol.* **135,** 2163–2717.
39. Bruinenberg, P. M., de Bot, P. H. M., van Dijken, J. P., and Scheffers, W. A. (1984), *Appl. Microbiol. Biotechnol.* **19,** 256–260.
40. Bruinenberg, P. M. (1986), *Antonie van Leeuwenhoek* **52,** 411–429.
41. Timell, T. E. (1965), in *Advances in Carbohydrate Chemistry,* Wolfrom, M. L. and Tipson, R. S., eds., Academic, New York, pp. 410–484.
42. Karczewska, H. (1959), *Compt. -rend. Lab. Carlsberg* **31,** 251–258.
43. Suihko, M. L. (1983), *Biotechnol. Lett.* **5,** 721–724.
44. Barnett, J. A. (1966), *Nature* **210,** 565–568.
45. Barnett, J. A.. Payne, R. W., and Yarrow, D. (1983), *Yeasts: Characteristics and Identification.* Cambridge University Press, Cambridge, UK.

0273-2289/96/5758—0243$7.50

Microbial Conversion of Synthesis Gas Components to Useful Fuels and Chemicals

G. R. MADHUKAR, B. B. ELMORE,* AND H. K. HUCKABAY

Louisiana Tech University, Department of Chemical Engineering, Ruston, LA 71272-0046

ABSTRACT

Enriched culture techniques have been used to isolate microbial cultures exhibiting growth on synthesis gas components. Three rod-shaped, gram-positive cultures have been isolated from petroleum-contaminated soil, a cow manure–soil mixture, and sheep rumen fluid. Each culture exhibits growth on carbon monoxide as its primary carbon source, producing alcohols and acids in the fermentation medium. Quantities of up to 7.5, 0.58, and 0.25 g/L of acetate, ethanol, and methanol, respectively, have been produced in batch culture with lesser amounts of acetone, butyric, and propionic acid detected.

Index Entries: Synthesis gas; carbon monoxide; unicarbonotrophic growth; anaerobic microorganisms.

INTRODUCTION

The combustion of carbonaceous materials (e.g., oil, coal, natural gas, and lignocellulosics) to produce energy results in the release of tremendous quantities of carbon dioxide—one of the predominant greenhouse gases *(1,2)*. However, these materials may be gasified under controlled conditions to produce synthesis gas. Synthesis gas consists predominantly of carbon monoxide (CO), carbon dioxide (CO_2), and hydrogen (H_2) with lesser amounts of sulfur and nitrogen compounds. Through subsequent processing, synthesis gas may be upgraded catalytically to valuable fuels and chemical feedstocks *(3,4)*.

One approach to upgrading synthesis gas employs microorganisms as biocatalysts. These microbes have a demonstrated ability to utilize synthesis gas components as their primary or sole carbon source and are known as unicarbonotrophs. An attractive feature of such a process is the ability, through gasification, to convert a variety of complex carbon materials to the molecularly simple components of synthesis gas—thus improving the process concerns for the microbial conversion step (e.g., no solids handling, enhanced mass transfer and reaction kinetics).

Several microorganisms have been identified as having the ability to convert synthesis gas components to liquid and gaseous products *(5–7)*. Sulfur gas poisoning (e.g., carbonyl sulfide and hydrogen sulfide), of particular concern when using

*Author to whom all correspondence and reprint requests should be addressed.

chemical catalysts for upgrading synthesis gas, has received attention as a factor in the microbial conversion step. Syngas-converting microbes have demonstrated considerable sulfur tolerance. In some cases, cultures capable of sulfur gas removal may be used in sequential or cocultures, further enhancing the attractiveness of a microbial-conversion step *(8,9)*. The rich variety of microbial flora presents a strong likelihood that new species will be discovered with similar capabilities and that new metabolic products will emerge through continued research *(10)*.

This work presents preliminary results from a microbial screening program. Its purpose is to examine an array of natural inocula for microorganisms with a demonstrated ability to convert synthesis gas components to valuable products. Enriched cultures of three isolates have shown promising synthesis gas conversion rates with a variety of chemicals secreted in the fermentation medium.

MATERIALS AND METHODS

Isolation Medium

To isolate cultures from natural inocula selectively, a basal salts medium was prepared following the modified Hungate method *(11,12)* for working with anaerobic microbial cultures. The medium (constituents shown in Table 1) was selected for its suitability for meeting the nutritional requirements of a variety of anaerobic bacteria. Following medium preparation by boiling under a nitrogen (N_2)/carbon dioxide (CO_2) atmosphere, 50-mL aliquots were transferred to 150-mL (nominal volume) bottles via an automatic pipet flushed with the N_2/CO_2 mixture. The 150-mL bottles, rated for use at pressures near 4–5 atm, served as batch reactors for this stage of work. Once filled, bottles were closed with butyl rubber stoppers and aluminum crimp seals. Prior to inoculation, the growth medium was reduced with cysteine-hydrochloride and/or sodium sulfide (2.5 wt%, respectively) to provide a strictly anaerobic environment. Anticipating the likelihood of methane production, as with many anaerobic cultures growing in such environments, an agent to block methanogenesis chemically was added to initial inoculations. The chemical 2-bromoethanesulfonic acid (BESA) was chosen as a proven inhibitor to methane production *(6)*. It was added in a concentration of 0.01 g/50 mL of medium. Once an absence of methane production was noted, BESA was removed from the medium in subsequent transfers.

Natural Inocula Sources

Raw inocula were obtained from a variety of sources, including sewage sludge, crude petroleum-contaminated soil, manure–soil mixtures, and rumen and intestinal fluids from cows, sheep, and pigs. To optimize microbial viability and activity, contaminated soil samples were collected after several days of rainy weather and warm temperatures. Rumen and intestinal fluids were removed from freshly slaughtered animals at the university meat processing facility. All inocula were immediately transferred to reduced basal salts medium under a nitrogen atmosphere. After mixing with the medium, 5-mL samples were transferred via hypodermic syringe to medium in the 150-mL bottles followed by immediate gassing with a synthesis gas mixture to 10 psig. The mixture chosen for this study consisted of the following mole percentages: 13.2% CO_2, 22.5% H_2, and 64.3% CO. Gassed bottles were placed horizontally on a shaker/incubator at 37°C and 85 rpm.

Table 1
Basal Salts Medium for Isolating Anaerobic Cultures

	Concentration, mg/L, unless specified otherwise
Pfennig minerals	
KH_2PO_4	500
$MgCl_2{\cdot}6H_2O$	330
NaCl	400
NH_4Cl	400
$CaCl_2{\cdot}2H_2O$	50
Pfennig trace metals	
$ZnSO_4{\cdot}7H_2O$	0.1
$MnCl_2{\cdot}4H_2O$	0.03
H_3BO_3	0.3
$CoCl_2{\cdot}6H_2O$	0.2
$CuCl_2{\cdot}H_2O$	0.01
$NiCl_2{\cdot}6H_2O$	0.02
$Na_2MoO_4{\cdot}2H_2O$	0.03
$FeCl_2{\cdot}4H_2O$	1.5
Na_2SeO_3	0.01
B-Vitamins	
Biotin	0.1
Folic acid	0.1
Pyridoxal-HCl	0.05
Lipoic acid	0.3
Riboflavin	0.25
Thiamine-HCl	0.25
Ca-D-pantothenate	0.25
Cyanocobalamin	0.25
P-aminobenzoic acid	0.25
Nicotinic acid	0.25
Supplements	
Yeast extract	0.1%
$Na_2{\cdot}9H_2O$, 2.5% soln (w/w)	2.0 mL
Cysteine-HCl, 2.5% soln (w/w)	2.0 mL
Resazurin, 0.1% solution	0.1 mL

Sampling and Analytical Procedures

Following the addition of synthesis gas, 50 mL of argon were added to each bottle with a gas-tight syringe. The addition of this biologically inactive, insoluble gas allowed the quantification of gas-phase components by the relation:

$$N_{CO} = [(\%CO)/(\%Ar)] * N_{Ar} \quad (1)$$

Similar determinations were made for carbon dioxide and hydrogen.

Gas samples of 150 µL were obtained periodically for analysis. Gas-phase analyses were made using a Hewlett Packard 5890 gas chromatograph equipped with a thermal conductivity detector (TCD) and a CarboPLOT 007 molecular sieve capillary column (25 m × 0.53 mm × 25 µm) using helium as the carrier gas. The injector, oven, and detector temperatures were 50, 80, and 110°C, respectively.

Table 2
Initial Inoculation for Isolating Anaerobic Cultures

Culture ID	Time, d	pH	Co, mmol	CO_3, mmol	Comments
Petroleum– soil inoculum	0	7.0	3.23	0.52	
	3	5.30	3.05	0.76	
	6	5.39	2.93	1.14	
	8	5.34	2.41	1.66	Regassed
	9	—	3.53	1.07	
	13	5.33	3.60	1.28	Regassed
	14	—	2.04	1.86	
	15	5.30	0.84	3.81	
Cow–manure soil inoculum	0	6.0	3.14	0.67	
	3	5.63	2.95	0.78	
	6	5.60	3.05	0.86	
	8	5.46	2.61	1.35	Regassed
	9	—	2.04	1.45	
	13	4.08	3.75	0.83	Regassed
	14	—	3.80	0.74	
	15	3.79	1.38	2.76	

Two-milliliter liquid samples were removed for analysis. Liquid samples were acidified with 3% (w/w) hydrochloride acid (HCl) and mixed with isopropanol (i-P) as an internal standard once i-P was determined not to be a liquid product. Samples were analyzed by gas chromatography using an HP 5890 equipped with a flame ionization detector (FID) and an HP-FFAP capillary column (30 m × 0.53 mm × 1.0 μm) with helium as a carrier gas. The injector and detector were operated at 200°C. The oven was operated with a program initiated at 45°C ramped to 150°C over 5 min.

RESULTS AND DISCUSSION

Initially, raw inocula were added to basal medium at various initial pH values ranging from 5 to 7. Gas and liquid samples were obtained several times weekly. Prior work *(13,14)* indicated that a relatively slow adaptation to growth on synthesis gas would be required over a period of weeks or months. However, a rapid onset of CO depletion occurred with several inocula sources, including the petroleum–soil source, the cow manure–soil source, and to a lesser extent the sheep rumen fluid inoculum. Over a period of 2 mo, no other inocula exhibited significant synthesis gas consumption and were subsequently discarded. Representative data for the petroleum–soil (PS) and cow–soil (CS) inocula are shown in Table 2. The data are reported at total millimoles of gaseous substrate (CO and CO_2) present. The cultures were regassed several times during this experiment. With each bottle of PS inoculum, the pH moderated from its initial range of 5–7 to approx 5.3 within 72 h, and remained relatively constant throughout the 15-d period of measurement. Likewise, all CS inocula initially approached a pH near 5.5. A gradual downward trend in pH was observed to be near 3.8 accompanying higher concentrations of acetate—compared to relative amounts observed with the PS inocula (data not shown). Acetate concentration showed a continued

Table 3
Parent Culture Conditions for pH Study

Culture ID	Age, d	pH	Acetate, g/L	Ethanol, g/L	Methanol, g/L	Comments
Petroleum–soil inoculum PS3-1	7	6.16	3.15	0.287	0.105	Rod-shaped gram +
Cow manuer–soil inoculum CS1-1	7	6.27	4.55	0.322	0.112	Rod-shaped gram +
Sheep rumen Fluid SR4-1	7	6.50	2.91	0.469	—	Rod-shaped gram +

increase in the CS bottles to an ultimate concentration between 1.5 and 2.0 g/L. Within 48 h of inoculation, acetate levels in the PS bottles reached and remained at approx 0.5 g/L. A third inoculum, sheep rumen fluid (SR), exhibited a similar CO consumption and production of liquids, but to a lesser extent than with the PS and CS inocula described earlier in this paragraph.

In all bottles, a variety of liquid products were observed. Although acetate was the predominant product, propionate, butyrate, methanol, and ethanol were also observed. An increase in CO_2 production was noted in all bottles, as well. No significant amounts of methane were produced with any inocula during this initial experiment. Therefore, BESA was removed from subsequent transfers—a decision revealing interesting consequences with the petroleum–soil culture in later transfers.

A second transfer (to triplicate bottles) exhibited a relative increase in the rate of CO consumption for the PS inocula (data not shown). Within 48 h, all bottles were essentially depleted of CO, requiring regassing. At day 5, after two regassings, the PS bottles had accumulated an average of 7.5 g/L acetate and 0.36 g/L ethanol with lesser amounts of methanol, propionate, and butyrate. Identification of trace amounts of acetone were tentatively detected in some bottles containing the PS inoculum. The CS bottles exhibited a slower rate of CO depletion in comparison to its parent inoculum. Acetate accumulated to an average of 4.0 g/L in each of the triplicate bottles in 5 d. The ethanol and methanol concentrations at this time averaged 0.37 and 0.10 g/L, respectively.

pH Effects on CO Consumption and Product Concentrations

The PS, CS, and SR isolates were transferred successively at 5–7 d intervals to enrich the adaptive species. After four successive transfers, an experiment was conducted to examine the initial pH effects on gas consumption and the relative amounts of liquid products obtained in the fermentation broth. Seed inocula for each isolate were selected from bottles that had shown a rapid depletion of CO with time and an accompanying high level of liquid products (when compared to other bottles of the same inoculum source). Table 3 provides the conditions for the source inocula used in this experiment. Each of these cultures was initiated at pH 7.0 and regassed five times over 7 d. Preliminary observations indicate an abundance of rod-shaped, gram-positive bacteria in each culture.

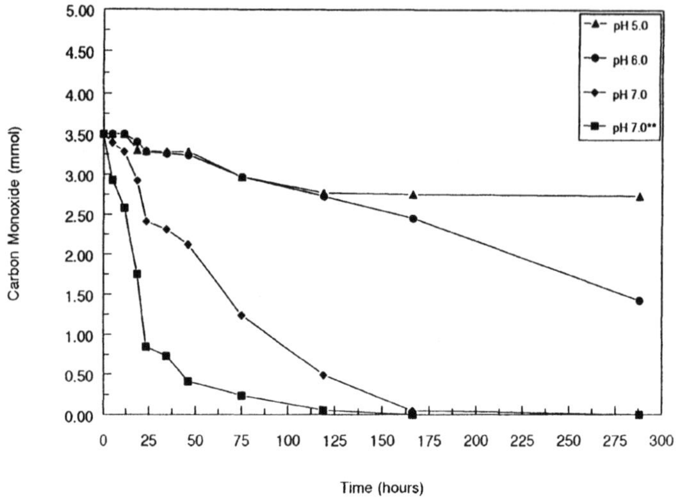

Fig. 1. Petroleum/soil isolate CO consumption at initial pH values. -▲- pH 5.0, -●- pH 6.0, -◆- pH 7.0, -■- pH 7.0 (added reducing agent *[see text]*).

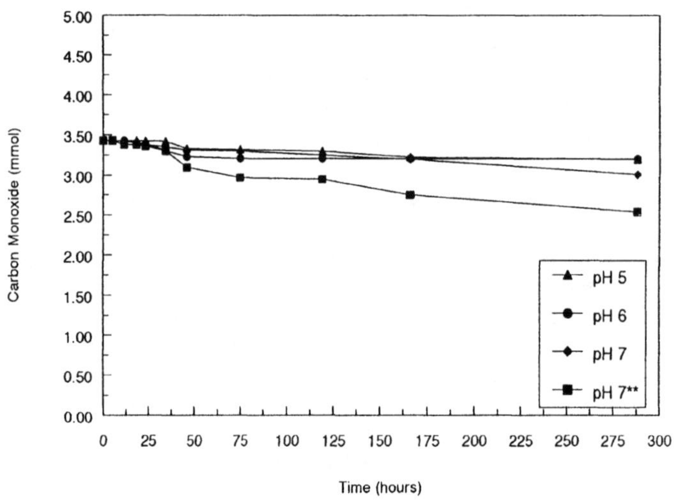

Fig. 2. Cow manure/soil isolate CO consumption at initial pH values. -▲- pH 5.0, -●- pH 6.0, -◆- pH 7.0, -■- pH 7.0 (added reducing agent *[see text]*).

Each bottle contained 50 mL of medium initially at a pH of 5.0, 6.0, or 7.0. All bottles were reduced with 0.5 mL of 2.5% (w/w) cysteine-HCl. In addition, one set of bottles at pH 7.0 received an added amount of sodium sulfide (0.5 mL of 2.5 wt%). This combination of reducing agents provided for a lower redox potential than cysteine-HCl alone—a condition often favoring growth of obligate anaerobes *(15)*.

Results for CO consumption with time are shown in Figs. 1–3. Individual data points are connected for visual clarity. It is clearly seen that, at increasingly lower initial pH values, the rate and extent of CO consumption declined dramatically. The distinction is particularly marked with the cultures PS3-1 and SR4-1. A similar trend

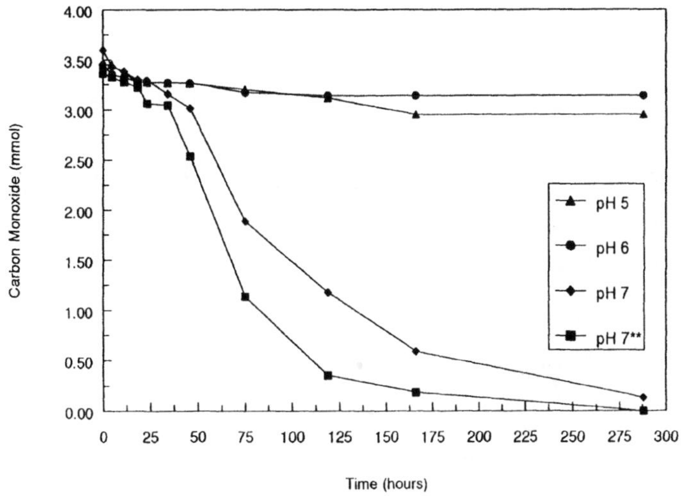

Fig. 3. Sheep rumen isolate CO consumption at initial pH values. -▲- pH 5.0, -●- pH 6.0, -◆- pH 7.0, -■- pH 7.0 (added reducing agent *[see text]*).

was observed for isolate CS1-1, though it generally displayed a much slower rate of CO depletion (with incomplete utilization for any of the bottles over the observed time period). Bottles containing the CS1-1 isolate also produced relatively small amounts of alcohols and acids (data not shown). The PS3-1 and SR4-1 isolates initiated at pH 7.0 consumed essentially all CO over the 12-d observation period.

Data for the three primary liquid products (acetate, ethanol, and methanol) are shown for the PS3-1 and SR4-1 isolates in Figs. 4 and 5. Only data for the bottles containing the added reducing agent are depicted. These bottles produced significantly higher quantities of acetate, methanol, and ethanol than those reduced with only cysteine-HCl.

With the PS3-1 isolate, the acetate concentration rose to 5.3 g/L within 24 h, peaking at 6.0 g/L followed by a gradual decrease in concentration. The rate of methanol production closely followed acetate rising to a concentration of 0.14 g/L before quickly dropping to zero. During the production of acetate and methanol, the ethanol concentration dropped slightly from an initial value of 0.042 g/L and then slowly rose to an ultimate concentration of 0.116 g/L.

Initially, the SR4-1 production rates of all three major liquid products were closely matched. After 24 h, product concentrations peaked at 4.85, 0.581, and 0.25 g/L for acetate, ethanol, and methanol, respectively. Although the methanol concentration dropped to zero (also seen in the PS3-1 isolate), both acetate and ethanol showed decreases and again rose to ultimate concentrations of 6.38 and 0.26 g/L, respectively.

The liquid products for the remaining bottles containing only cysteine-HCl were significantly lower with concentrations reaching only 1.0, 0.15, and 0.05 g/L (average values) for acetate, ethanol, and methanol.

In all bottles, CO_2 was produced with liquid products. However, the two bottles containing PS3-1 at pH 7.0 exhibited consumption of CO_2 with a concomitant production of methane once CO was depleted. Methane production was detected at 166 h, rising to ultimate levels of 2.51 and 2.11 mmol by the completion of the experiment at 288 h.

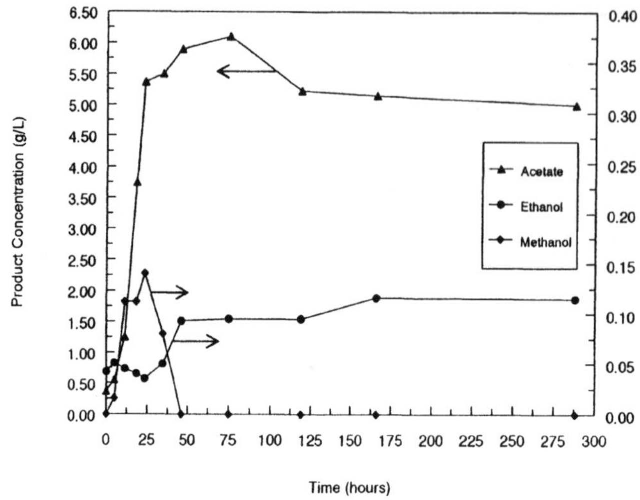

Fig. 4. Petroleum/soil isolate (PS3-1) liquid products at pH 7.0.* -▲- Acetate, -●- ethanol, -◆- methanol (added reducing agent *[see text]*).

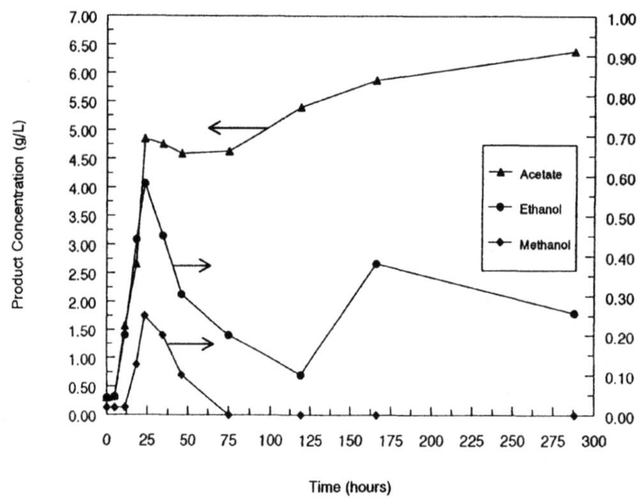

Fig. 5. Sheep rumen isolate (SR4-1) liquid products at pH 7.0.* -▲- Acetate, -●- ethanol, -◆- methanol (added reducing agent *[see text]*).

CONCLUSIONS

Three rod-shaped, gram-positive cultures have been isolated and enriched in basal salts medium—each demonstrating growth on synthesis gas components as their primary carbon source. A variety of liquids were produced (predominantly acetate, ethanol, and methanol). These liquid products serve as valuable fuels or chemical feedstocks. Quantities of up to 7.5, 0.58, and 0.25 g/L of acetate, ethanol, and methanol, respectively, were produced in various batch cultures. Traces of acetone were noted in the petroleum–soil isolate, as well. Future studies will focus on fermentation conditions that maximize the production of selected products. The potential for producing alcohols at the expense of acetate will be investigated. When compared to parallel work conducted by Barik et al. *(6)*, the cultures isolated in this study exhibited similar rates of CO consumption, but at significantly earlier stages of culture transfer (e.g., in just two to three transfers over a couple of weeks). Of particular interest is the strong growth and consumption of CO by the petroleum–soil isolate.

Further characterization is needed to identify the strain(s) present in each culture and to describe the growth characteristics of individual species in the event of identifying a new strain or species. Optimization of the growth medium (e.g., removal of yeast extract–an expensive medium constituent) is necessary to improve process economics.

ACKNOWLEDGMENTS

The authors would like to express thanks to the Louisiana Tech Departments of Chemical Engineering and Chemistry for support and assistance in conducting this research.

REFERENCES

1. den Elzen, M., Janssen, M., Rotmans, J., Swart, R., and de Vries, B. (1992), *Intl. J. Global Energy Issues* **4:4,** 287–301.
2. Garret, C. W. (1992), *Prog. Energy Combustion Science* **18,** 369–407.
3. Alfantanzi, A. M. and Pickles, C. A. (1992), *Materials at High Temperature* **10:1,** 51–59.
4. Machio, G., Koufopanos, C., and Lucchesi, A. (1992), *Biores. Technol.* **42,** 219–231.
5. Barik, S., Corder, R. E., Clausen, E. C., and Gaddy, J. L. (1987), *Energy Progress* **7:3,** 157–160.
6. Barik, S., Prieto, S., Harrison, S. B., Clausen, E. C., and Gaddy, J. L. (1988), *Appl. Biochem. Biotechnol.* **18,** 363–378.
7. Grethlein, A. J. and Jain, M. K. (1992), *TIBTECH* **10,** 418–423.
8. Vega, J. L., Klasson, K. T., Kimmel, D. E., Clausen, E. C., and Gaddy, J. L. (1990), *Appl. Biochem. Biotechnol.* **24/25,** 329–340.
9. Basu, R., Klasson, K. T., Clausen, E. C., and Gaddy, J. L. (1994), *Appl. Biochem. Biotechnol.* **45/ 46,** 787–797.
10. Cheetham, P. S. J. (1987), *Enzyme Microb. Technol.* **9,** 194–213.
11. Bryant, M. P. (1972), *Am. J. Clin. Nutr.* **25,** 1324–1328.
12. Miller, T. L. and Wolin, M. J. (1974), *Appl. Microbiol.* **27,** 985–987.
13. Vega, J. L., Prieto, S., Elmore, B. B., Clausen, E. C., and Gaddy, J. L. (1989), *Appl. Biochem. Biotechnol.* **20/21,** 781–797.
14. Elmore, B. B. (1990), *Biological production of ethanol from coal synthesis gas using* Clostridium ljungdahlii, *strain PETC,* doctoral dissertation.
15. Holland, K. T., Knapp, J. S., and Shoesmith, J. G. (1987), *Anaerobic Bacteria,* Blackie & Son Ltd., Glasgow and London, p. 10.

0273-2289/96/5758—0253$8.25

Saccharification of Native Sugar Cane Bagasse Pith by the Cross-Synergistic Action of Cellulases from *Penicillium* sp. CH-M-001 and *A. terreus* CH-M-013

OSCAR GARCIA-KIRCHNER* AND CARLOS HUITRON

Departmento de Biotecnología, Instituto de Investigaciones Biomédicas. UNAM. A.P. 70228, México, D.F. 04510 México

ABSTRACT

Filter paper, carboxymethylcellulase, β-glucosidase, and xylanase activities were determined and compared to cellulases originating from *Aureobasidium* sp. (Ab), *Penicillium* sp. (Pe), and *Aspergillus terreus* (At). The formation of total reducing sugar was measured as a function of time for the hydrolysis of the native sugar cane bagasse pith using culture filtrates with the same quantity of extracellular protein content from Ab, Pe, At, or from the following mixtures Ab:Pe/Ab:At and Pe:At at different volumetric ratios. The saccharification progress indicated a synergistic effect for the mixture of all the enzymes, the highest being in the 40:60 relation of Pe and At, respectively. The synergistic action has been assigned to a better balance of endo- and exoglucanases in this system and essentially to the addition of β-glucosidase and xylanases from At to Pe cellulases.

Index Entries: Cellulases; enzymatic hydrolysis; synergism; *Aureobasidium* sp; *Penicillium* sp; *Aspergillus terreus;* bagasse pith; mixtures.

INTRODUCTION

Lignocellulosic biomass includes such materials as agricultural and forestry waste, municipal solid waste, waste paper, and wood and herbaceous energy sources. These materials represent one of the most abundant renewable resources on earth in that both the cellulose and hemicellulose fractions of biomass can be converted to simple sugars that can subsequently be fermented.

Cellulose, the major constituent of plant cell walls, is recycled in nature essentially as a microbial energy source. This cellulosic biomass also represents a

*Author to whom all correspondence and reprint requests should be addressed.

Present address: Departmento de Bioprocesos, Unidad Profesional Interdisciplinaria de Biotecnología (UPIBI/IPN). Instituto Politécnico Nacional. Av. Acueducto s/n Barrio La Laguna; Ticomán, Zacatenco. México 07340, D.F.

source potentially inexpensive for the bioproduction of a variety of chemicals *(1)*. There is currently great interest in the degradation of lignocellulose complexes to provide suitable raw materials for food, fuel, or chemical needs. Much of this interest is centered on the hydrolysis of cellulose by cellulase enzymes.

Acid and enzyme processes based on hydrolysis are the two most common methods of converting cellulosics to simple sugar. Various processes based on dilute and concentrated acids are used to hydrolyze the cellulose and hemicellulose to their respective sugars *(2,3)*. Acid hydrolysis reduces the degree of polymerization of cellulose and hemicellulose, and ultimately results in the formation of glucose. Acid treatments proceed with rapid rates of conversion, but suffer either high catalyst consumption in concentrated acid processes or low sugar yields in high-temperature, dilute-acid processes owing to sugar destruction.

Enzymatic hydrolysis is favored over these processes, since this catalyst is noncorrosive environmentally harmless, potentially reusable, and offers an efficiency above 90% *(4)*.

The inherent recalcitrance of the lignocellulosic feedstock to bioconversion requires a pretreatment of the biomass to render the cellulose fraction amenable to enzymatic conversion to glucose, which then can be utilized. Physical, chemical, and biological means, used alone or in combination, have been reported *(5,6)* to be effective pretreatment techniques. Enzymatic processes are based on the use of a pretreatment step to hydrolyze the hemicellulose or dissolve the lignin, thereby increasing the digestibility of the cellulose. Cellulolysis is complex, since it involves interacting enzymes and an insoluble substrate constituted by crystalline microfibrils, recalcitrant to enzymatic attachment.

Thus, the enzymatic hydrolysis of cellulose has been widely investigated *(7,8)*. Efficient hydrolysis is accomplished by the synergistic action of three types of enzymes: endoglucanase (endo-1,4-β-glucose 4-glucanohydrolase, EC 3.2.1.4), cellobiohydrolase (1,4-β-D-glucose cellobiohydrolase, EC 3.2.1.91), and cellobiase (β-D-glucoside glucohydrolase, EC 3.2.1.21). It is important to note that during cellulolysis, the products glucose and cellobiose accumulate and act as end-product inhibitors.

Cellulases from several fungal and bacterial systems have been identified, characterized, and studied for this ability to react synergistically. There also have been several studies *(9–12)* investigating how various cellulases act synergistically to yield higher rates and extents of hydrolysis than would be predicted from the sum of the hydrolytic activities of the individual enzymes when the individual cellulolytic enzymes act alone on a crystalline cellulosic substrate, such as cotton. These components are unable to hydrolyze this substrate to any great extent, but in combination they cooperate to effect its solubilization *(13)*. Because native cellulose is both insoluble and structurally heterogeneous, comparisons of activity between different enzymes or enzyme mixtures are difficult. Most synergism experiments have utilized a constant time of hydrolysis and/or a constant amount of enzyme *(14)*, and a cross-synergism between cellulases from different microbial systems has been shown by several investigators *(15)*.

Studies about synergism in cellulose systems have shown that two broad classifications of cellulase synergism are possible: synergism between endocellulases and exocellulases, and synergism between different exocellulases. Synergism between endoglucanases and exoglucanases can be explained by the action of the EG component on cellulose (which catalyzes the hydrolysis of the glycosidic bonds along the length of the cellulose chain), resulting in the formation of new cellulose chain ends through which the CBH component acts, releasing cellobiose. Wood and McCrae *(16)* have suggested that the formation of a complex of endoglucanase and cellobiohydrolase on the surface of the cellulose chain is essential for synergism.

The process development for the biological conversion of native cellulosic materials requires further investigation dealing with the mechanism of enzyme attachment. Native cellulose in many agricultural byproducts is a miscellaneous material, characterized by a sequence of amorphous and crystalline regions.

Considerable attention has been focused on the enzymatic hydrolysis of cellulosic substrates in recent years; for this reason, and taking into account the abundance of sugar cane bagasse annually available and regarded as an underutilized resource in Mexico (approx 13 million ton/yr, about 35% in this total was bagasse pith), the present study was aimed at reaching cellulose saccharification of native sugar cane bagasse pith, using simultaneously two different cellulase complexes from three filamentous fungi isolated in our laboratory. This way cellulases from three different strains, *Aureobasidium* sp., *Penicillium* sp. and *A. terreus*, mixed in different ratios, were tested under proper operational conditions; these cellulase complexes were selected mainly because of their different predominant activities, which allows the exploration a large variety of enzyme compositions. Therefore, our goal was to examine the saccharifying capacity of the enzymes of the three cellulolytic strains, acting separately or jointly on an agroindustrial byproduct like native sugar cane bagasse pith.

MATERIALS AND METHODS

Microorganisms

Aureobasidium sp. CH-ME-18, *Penicillium* sp. CH-ME-001, and *A. terreus* CH-ME-013 were isolated in our laboratory from soil of a sugar cane field of the state of Morelos, México as previously reported *(17,18)*. The fungi strains were grown at 29 ± 1°C for 5 d on potato dextrose slants and maintained on the same medium at 4°C. Subculturing was carried out at monthly intervals.

Cellulase and Xylanase Production

Four grams of cellulosic material (native sugar cane bagasse pith collected locally in the "Emiliano Zapata" sugar mill, Morelos, México) were placed in a 500-mL Erlenmeyer flask, pouring in it only 180 mL of fermentation medium. All chemicals used in this culture medium were of industrial grade, and tap water was according to the formulation indicated for *Aureobasidium* sp. *(19)*.

For *Penicillium* sp. and *A. terreus*, we developed a modified medium with the same composition of chemicals, but using corn syrup liquor (CSL) as nitrogen source instead of urea *(20)*. Inocula were prepared by harvesting spores from 1-wk-old PDA slants of the three strains in the proportion 3:1 in sterile distilled water containing a small amount of Tween-80.

Fermentation flasks were inoculated to give 10^6–10^7 spores/mL. Incubation was at 29 ± 1°C for *Penicillium*, but for *Aureobasidium* sp. and *A. terreus*, it was at 37°C on a rotary shaker (180 rpm). The pH value was initially adjusted to 4.5 for

Aureobasidium sp. and 5.5 for the other strains. Samples were removed aseptically and were centrifuged to remove solids and to obtain cell-free filtrates.

Enzymatic Assays and Soluble Protein Determination

These were carried out using the supernants after removing the cell growth and residual bagasse pith. The protein content was performed according to Lowry *(21)* in aliquots of the obtained mixture after a ±4°C dialysis overnight. The culture filtrates of *Aureobasidium* sp. and *Penicillium* sp. were concentrated by ultrafiltration to obtain the same protein content with respect to the *A. terreus*.

Total cellulase activity (FPU) was determined on filter paper following the protocol of Mandels et al. *(22)*. The determination of carboxymethylcellulose saccharifying activity (CMCase) was applied according to the method developed by us. Briefly, the assay determining reducing groups on CMC is: 0.5 mL of diluted culture filtrate, 1 mL of CMC at 7.5% suspension in citrates buffer 0.075*M*, pH 4.8, 60 min of incubation at 50°C.

The β-glucosidase activity was determined as in ref. *23* as an aryl-β-glucosidase by mean of the hydrolysis of *p*-nitro-phenyl-β-D-glucoside measuring the amount of 1 μmol of *p*-nitrophenol liberated by the enzyme/min and xylanase activity was determined by quantifying the reducing sugars (as xylose) as in ref. *24*.

All reducing sugar determinations were performed by the 3,5-dinitrosalicylic (DNS) method *(25)*. The enzyme activities were calculated as indicated by IUPAC *(26)*, using the recommendations of Ghose for enzyme solutions for which the undiluted solution does not reach the target glucose-release value. For this reason, the activity is taken as directly and linearly proportional to whatever amount of reducing sugar is released. The filter paper and CMC hydrolyzing activity was defined as 0.185/(enzyme releasing 1 mg glucose) U/mL enzyme solution.

Enzymatic Hydrolysis

The saccharification of nonpretreated sugar cane bagasse pith initially was carried out in 22 × 175 mm test tubes, containing 50 mg of material, suspended in 1.0 mL citrate buffer, 0.075*M*, pH 4.8, and 0.5 mL of the respective filtrate culture. At the same time, during the experiment the combinations mentioned below were made, using the following volumetric proportions: 80:20 (v/v); 60:40 (v/v), and 40:60 (v/v), using the same 0.5 mL final volume. Citrate buffer was used as a control instead of the respective portion of culture filtrate.

The tubes were parafilm-covered and shaken in a reciprocal water bath at 50°C for 60 min, separating a group of test tubes with their content in every 15-min period until the 60 min were completed. The reaction was stopped by adding 3 mL of DNS reagent to determine the reducing sugars, using glucose as standard.

Saccharification of Sugar Cane Bagasse Pith in a Stirred Reactor

This was carried out in a 1000 mL Berzelius beaker containing 20 g of bagasse pith, suspended in 500 mL of citrate buffer, 0.075*M*, pH 4.8. The beakers were shaken in a reciprocal water bath at 50°C. The total volume was 600 mL, which included 100 mL of filtrate, or a mixture of *Penicillium* sp. and *A. terreus* in the proportions of 60:40 (v/v) and 40:60 (v/v). A control was run, containing the respective proportion of the filtrate and the rest, was completed with citrate buffer. A few merthiolate drops were added to each system and the beakers were covered with a rubber top.

The total incubation time at 50°C was 48 h, taking samples of 5 mL every 3 h during the first 12 h to determine reducing sugars; and after every 12 h until the 48-h period was fulfilled. The percentage of saccharification (% SACCH) was calculated according to the formula: % SACCH: = P/S (0.55) where P is the reducing sugar concentration (g/L) in the hydrolysate estimated and S is the concentration (g/L) of bagasse, which is multiplied by 0.55, since bagasse pith contains approx 55% cellulose. The contribution of the pentose reducing groups was not taken into account to calculate the conversion percentage.

RESULTS AND DISCUSSION

The first stage of our work was the production of the cellulolytic and xylanolytic activities, using three different strains of filamentous fungi, previously chosen for the following reasons: *Aureobasidium* sp. CH-M-18 with characteristics of a yeast-like fungus. Maximum cellulase activity produced by this strain from crystalline cellulose was reached in a shorter time than that required by other microorganisms *(24)*. The microorganism grew well at 37°C on different carbon sources. *Penicillium* sp. CH-M-001 and *A. terreus* CH-M-013 were chosen based on their characteristics found in prior investigations *(18)*, which mainly point out different degrees of relations in the cellulolytic and xylanolytic activities obtained from a byproduct like bagasse cane pith; *Penicillium* sp. is characterized by its capacity to degrade filter paper. It is considered a microorganism with great ability to take advantage of the crystalline portion of cellulosic materials and grows at 29°C.

A. terreus is capable of producing a high degrading activity of amorphous regions of cellulosic materials like xylans, since it produces a high CMCase activity from the first 2 d of fermentation, as well as xylanases and β-glucosidase, and it may grow at 37°C just as *Aureobasidium* sp.

Another important aspect to be mentioned is that there have been very few reports on the use of byproducts like sugar cane bagasse pith as starting material for the production of enzymes of industrial interest *(27,28)*. This aspect is important, since agroindustrial byproducts contain large amounts of compounds, such as cellulose and hemicellulose, which could serve as inducers for the production of extracellular enzymes, such as cellulases and xylanases. The polysaccharide content could therefore be hydrolyzed until soluble sugar or other products with different applications are obtained. For these reasons, it would be interesting to develop processes for the use of these byproducts.

In prior reports *(19)*, similar results were presented of simple and economical culture media and the optimal inoculum of *Aureobasidium* sp. for the production of cellulases and xylanases by submerged fermentation of untreated sugar cane bagasse pith. Most of the studies on microbiological enzyme production are performed with analytical-grade salts and distilled water. However, for industrial production, it would be advantageous to the costs involved to simplify the medium and use industrial-grade salts and tap water.

Figure 1 shows the enzyme activities after 5 d of inoculation with culture of *Aureobasidium* sp., *Penicillium* sp., and *A. terreus* under liquid fermentation on bagasse sugar cane pith in a medium formulated with industrial materials. This culture medium was initially designed for *Aureobasidium* sp. (however, the production of cellulolytic enzymes with urea for the other strains was not appropriate).

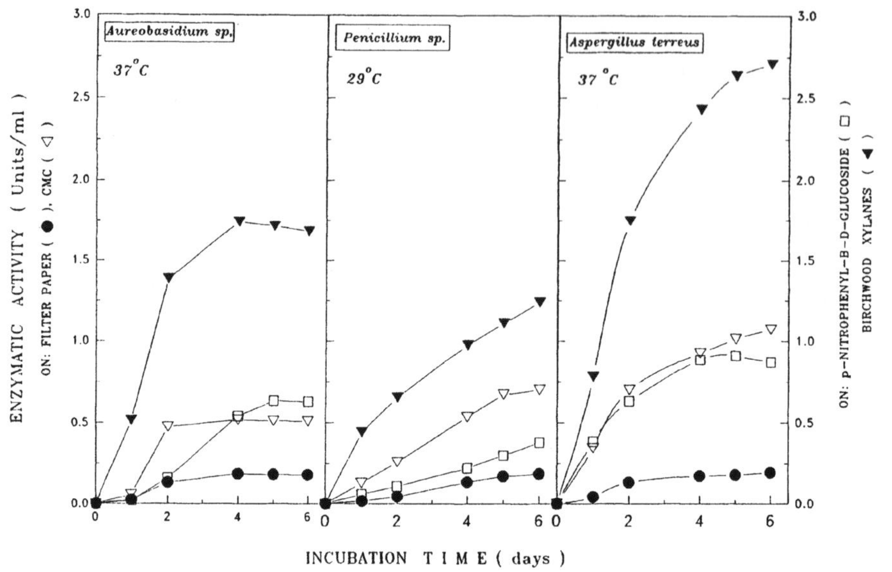

Fig. 1. Production of extracellular cellulolytic activity by fermentation of 2% (wt/vol) native sugar cane bagasse pith by *Aureobasidium* sp., *Pennicillium* sp., and *A. terreus* at the indicated temperature in shaking flask (180 rpm) in a culture medium containing industrial-grade components and tap water. For *Aureobasidium*, the nitrogen source was urea and $(NH_4)_2SO_4$; for the others strains, CSL replaced urea. Enzymatic activity on: (●) filter paper, (△) CMC, (□) *p*-nitrophenyl-β-D-glucoside, (▲) birchwood xylanes.

These fungi were initially tested to produce cellulases with a proteose peptone medium, originally reported by Mandels and Weber *(29)*. Because of economic considerations, we developed a modification that replaces this organic nitrogen source with a cheaper one, the corn syrup liquor (CSL). It was a good replacement, and it was no longer necessary to use urea.

The use of tap water has no negative effects on the production of cellulolytic and xylanolytic activity, or on β-glucosidase activity. Table 1 shows the gross composition of the culture medium mentioned before.

Since our main goal was the utilization of agroindustrial byproducts, such as sugar cane bagasse pith, in a saccharification reaction to produce soluble sugars useful for other purposes, we initially evaluated the capacity of hydrolysis of the cellulases and mixtures obtained by fermentation on a cellulosic material, such as filter paper, which is characterized as having amorphous as well as crystal regions, similar to native lignocellulosic materials. Table 2 shows the results of these considerations.

In these experiments, the existence of cross-synergesis between the cellulases of different origins became apparent. Because of this, we decided to evaluate these systems on native sugar cane bagasse pith, as indicated in Figs. 2, 3, and 4. We conclude from these experiments that the synergistic effect is similar to the one obtained with filter paper on all the combinations carried out.

According to Woodward et al. *(30)*, we calculated the degrees of synergistic effect (DSE). The results obtained are indicated in Table 3. DSE is defined as the

Table 1
Culture Medium Composition Utilized
for the Production of Cellulolytic and Xylanolytic Activities

Mandels Medium	Industrial grade medium, original	Industrial grade medium, modified
Sugar cane bagasse pith	Sugar cane bagasse pith	Sugar cane bagasse pith
KH_2PO_4	KH_2PO_4	KH_2PO_4
$(NH_4)_2SO_4$	$(NH_4)_2SO_4$	$(NH_4)_2SO_4$
$CaCl_2$		
Trace elements		
$FeSO_4 \cdot 7H_2O$		
$MnSO_4 \cdot 7H_2O$		
$ZnCl_2$		
$CoCl_2$		
Urea	Urea	
Proteose peptone		CSL
Distilled water and analytical grade chemicals	Tap water and Industrial-grade chemicals	Tap water and Industrial-grade chemicals
Tween-80		$Mazu_{6000}$
pH 5.5	pH 5.5	pH 5.0

product concentration over time in the hydrolysis of a cellulose mixture divided by the sum of the product concentration of the individual cellulases at the concentration in the mixture. DSEs are comparable to those reported in the literature by other investigators *(10,12).*

Since the combination of *Penicillium* and *A. terreus* was higher in DSE value than the others, mainly with volumetric relations 40:60 (v/v) and 60:40 (v/v), we considered it worthwhile to evaluate them in a stirred reactor. Figure 5 shows the behavior obtained with the individual culture filtrates and with the combinations mentioned. Table 4 shows the saccharification percentage obtained with this combination compared to each one of the culture filtrates individually. A maximum increase of 42% in the extent of saccharification was obtained when culture fluids with these cellulolytic strains were mixed in a 40:60 volumetric proportion higher than that obtained separately with the most active filtrate and gives a 12% greater extent of saccharification than would result from the theoretical sum of the action of each individual cellulase. The observed synergism in reducing sugar formation suggests a cooperative interaction between cellulases, which opens up the cellulose structure for increased degradation; the synergistic effect shown by the cellulases from *Penicillium* sp. and *A. terreus* could be assigned to a better balance of endo- and exoglucanase action, as well as to the high level of β-glucosidase and xylanase activities in the latter strain, acting on the native sugar cane bagasse pith. The synergism between cellulase components also appears to be independent of the ratio in a reaction mixture and dependent only on their concentration. This is important because a practical goal of cellulase research is to define both the optimal types and proportions of the enzymes necessary to gain efficient hydrolysis. The latter is especially difficult to define, in that the ratio of enzymes depends on the changing status of the substrate throughout cellulolysis.

Table 2
Reducing Sugar Concentration at Short Time of Hydrolysis on Filter Paper for Cellulase Mixtures

Mixture of culture filtrates	Volumetric relation V/V	Reducing sugars, mg/mL[a] time, min 0	15	30	45	60
Aureobasidium sp.	100	0	0.30	0.56	0.74	0.90
Penicillium sp.	100	0	0.23	0.44	0.60	0.78
A. terreus	100	0	0.30	0.56	0.74	0.92
Aureobasidium sp.	80:20	0	0.50	0.76	0.96	1.14
+ *Penicillium* sp.	Control (80)	0	0.30	0.46	0.58	0.70
	Control (20)	0	0.12	0.18	0.26	0.30
	Theoretical sum	0	0.42	0.64	0.84	1.00
	60:40	0	0.48	0.74	0.94	1.10
	Control (60)	0	0.24	0.42	0.56	0.64
	Control (40)	0	0.16	0.26	0.32	0.36
	Theoretical sum	0	0.40	0.68	0.88	1.00
Aureobasidium sp.	80:20	0	0.48	0.74	1.00	1.18
+ *A. terreus*	Control (80)	0	0.22	0.44	0.60	0.78
	Control (20)	0	0.10	0.16	0.22	0.26
	Theoretical sum	0	0.32	0.60	0.82	1.04
	60:40	0	0.40	0.72	1.00	1.24
	Control (60)	0	0.20	0.38	0.54	0.68
	Control(40)	0	0.14	0.24	0.32	0.40
	Theoretical sum	0	0.34	0.62	0.86	1.08
Penicillium sp.	80:20	0	0.44	0.70	0.90	0.98
+ *A. terreus*	Control (80)	0	0.24	0.40	0.52	0.54
	Control (20)	0	0.12	0.20	0.30	0.34
	Theoretical sum	0	0.36	0.60	0.82	0.88
	60:40	0	0.48	0.72	0.92	1.04
	Control (60)	0	0.14	0.24	0.32	0.38
	Control (40)	0	0.26	0.38	0.46	0.52
	Theoretical sum	0	0.40	0.62	0.78	0.90
	40:60	0	0.56	0.74	0.86	0.94
	Control (40)	0	0.20	0.26	0.30	0.32
	Control (60)	0	0.16	0.30	0.38	0.44
	Theoretical sum	0	0.36	0.56	0.68	0.76

[a]The values of reducing sugars are the mean by triplicate.

Controls: Samples of filtrates were run at the same time as controls, and its volume was completed with citrates buffer 0.075*M*, pH 4.8.

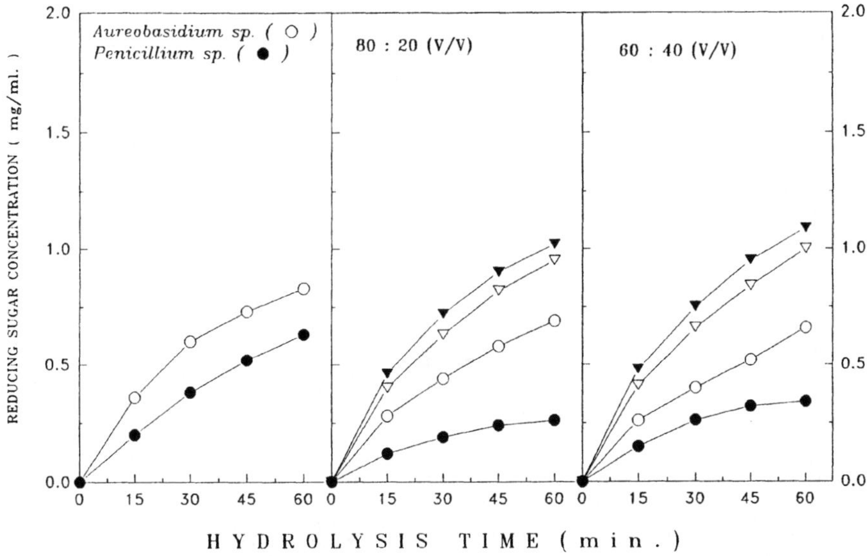

Fig. 2. Saccharification of native sugar cane bagasse pith (50 mg) by culture filtrates of *Aureobasidium* sp. (○) and *Penicillium* sp. (●) alone and with the indicated volumetric combination (▼), pH 4.8, 50°C, and shaking of 120 rpm. Theoretical sum of the action of each filtrate, acting separately (▽).

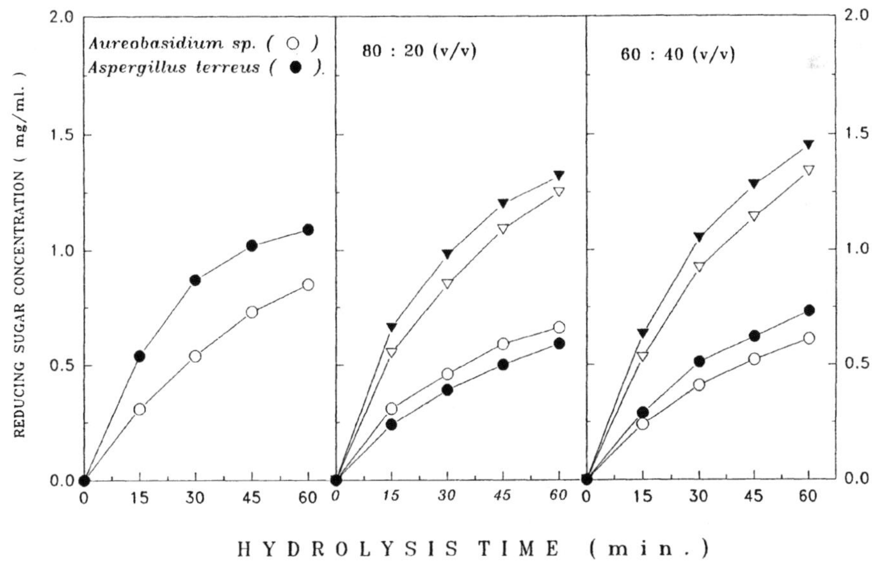

Fig. 3. Saccharification of native sugar cane bagasse pith (50 mg) by culture filtrates of *Aureobasidium* sp. (○) and *A. terreus* (●), alone and with the indicated volumetric combination (▼), pH 4.8, 50°C, and shaking of 120 rpm. Theoretical sum of the action of each filtrate, acting separately (▽).

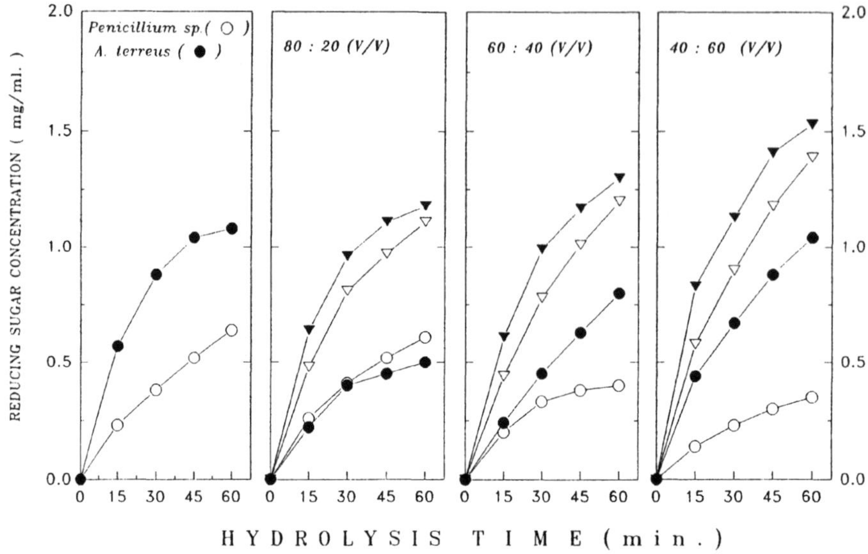

Fig. 4. Saccharification of native sugar cane bagasse pith (50 mg) by culture filtrates of *Penicillium* sp. (○) and *A. terreus* (●), alone and with the indicated volumetric combination (▼), pH 4.8, 50°C, and shaking of 120 rpm. Theoretical sum of the action of each filtrate, acting separately (▽).

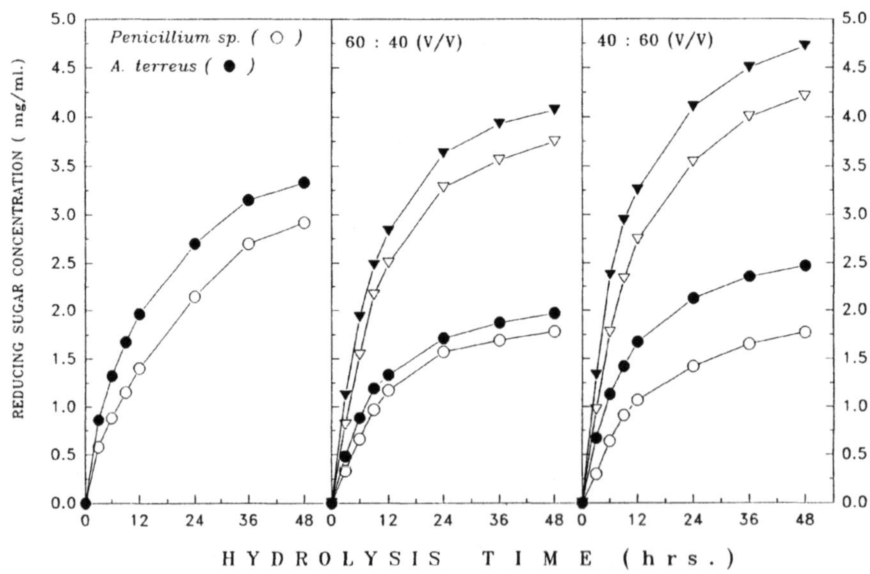

Fig. 5. Saccharification of native sugar cane bagasse pith (20 g) by culture filtrates from *Penicillium* sp. (○) and *A. terreus* (●), alone and with the indicated volumetric combination (▼), pH 4.8, 50°C, in a stirred reactor of 120–150 strokes/min. Theoretical sum of the action of each filtrate, acting separately (▽).

Table 3
Reducing Sugar Concentration at Short Time of Hydrolysis on Sugar Cane Bagasse Pith, and DSE for Cellulase Mixtures

Mixture of culture filtrates	Volumetric Relation, V/V	Time, min									
		Reducing Sugars, mg/mL					DSE[a]				
		0	15	30	45	60	0	15	30	45	60
Aureobasidium sp.	80:20	0	0.46	0.72	0.90	1.02	0	1.15	1.14	1.10	1.07
+ *Penicillium* sp.	60:40	0	0.48	0.75	0.95	1.09	0	1.17	1.14	1.13	1.09
Aureobasidium sp.	40:60	0	0.65	0.98	1.20	1.32	0	1.18	1.15	1.10	1.06
+ *A. terreus*	60:40	0	0.63	1.05	1.28	1.45	0	1.19	1.14	1.12	1.08
Penicillium sp.	80:20	0	0.64	0.96	1.13	1.15	0	1.33	1.19	1.14	1.06
+ *A. terreus*	40:60	0	0.61	0.99	1.17	1.30	0	1.39	1.27	1.16	1.08
	60:40	0	0.83	1.13	1.41	1.53	0	1.43	1.33	1.19	1.10
Aureobasidium sp.	100	0	0.31	0.54	0.73	0.85	NA	NA	NA	NA	NA
Penicillium sp.	100	0	0.23	0.38	0.52	0.64	NA	NA	NA	NA	NA
A. terreus	100	0	0.57	0.88	1.04	1.08	NA	NA	NA	NA	NA

[a]DSE = (Reducing sugar concentration released by the mixture of cellulases/sum of reducing sugar concentration released by each individual cellulase).

Table 4

Percentage of Saccharification of Nonpretreated Sugar Cane Bagasse Pith by Culture Filtrates from *Penicillium* sp., *A. terreus*, and Their Mixtures

Source of culture filtrate[a]	Concentration of protein, mg/mL	Cellulolytic Enzyme Activities,[c] U/mL				% Saccharification of native sugar cane bagasse pith, h				
		FPase	CMCase	β-glucosidase	Xylanase	3	6	12	14	48
Pennicillium sp.	0.50	0.185	0.703	0.380	1.240	3.17	4.80	7.64	11.68	15.94
A. terreus	0.54	0.194	1.073	0.870	2.708	4.69	7.21	10.70	14.74	18.18
Mixtures[b] of *Pennicillium* and *A. terreus*:										
60:40	0.52	—	—	—	—	6.11	10.53	15.45	19.81	22.22
40:60	0.52	—	—	—	—	7.26	12.94	17.74	22.38	25.66

[a]After 6 d of cultivation for both filamentous fungi.

[b]At the volumetric ratios indicated.

[c]The enzymatic activities were determined by methods indicated, and the U/mL of each activity was estimated as recommended by IUPAC *(26)*.

ACKNOWLEDGMENTS

We thank UCOAS/IPN for granting a sabbatical year at the Biomedical Research Institute of the National University of Mexico under the guidance of Carlos Huitrón. We also thank Bárbara Pérez for English translation, and Claudia García and Lilia Ramírez, who typed this manuscript. Isabel Pérez Montfort corrected the English version of the manuscript.

REFERENCES

1. Giovanozzi-Sermanni, G., Bertoni, G., and Poosi, A. (1989), in *Enzyme Systems for Lignocellulose Degradation*, Coughlan M. P., ed., Elsevier, London, pp. 373–382.
2. Brenmen, A. H., Hoagland, W., and Schell, D. J. (1987), *Biotechnol. Bioeng. Symp.* **17**, 53–70.
3. Wright, J. D. and Power, A. J. (1985), *Biotechnol. Bioeng. Symp.* **15**, 511.
4. Cuskey, S. M., Frein, E. M., Montenecourt, B. S., and Eveleigh, D. E. (1984), in *Overproduction of Microbial Metabolites*, Krumphanzyl, V., Sihyta, B., and Vaneck, A. eds., Academic, New York, pp. 405–416.
5. Fan, L. T., Lee, Y. H., and Gharpuray, M. M. (1987), *Adv. Biochem. Eng.* **23**, 157–187.
6. Grohman, K., Wyman, C. E., and Himmel, M. E. (1992), in *Emerging Technologies for Materials and Chemical Biomass*, Rowel, R. M., Schults, T. P., and Narayan, R., eds., ACS Symp. Ser. No. 476, pp. 354–392.
7. Alber, J. P., Beguin, P., and Millet, J. (1988), *Biochemistry and Genetics of Cellulose Degradation*, Academic, New York.
8. Kubick, C. P., Eveleigh, D. E., Esterbauer, U., Steiner, W., and Kubiceck-Pranz, E. M. (1990), *Trichoderma Cellulases; Biochemistry, Genetics, Physiology and Application*, The Royal Chemicals Society, Cambridge, UK.
9. Beldman, A. G. V., Voragen, T. M., Rombouts, F. M., and Pilnik, W. (1988), *Biotechnol. Bioeng.* **31**, 173–178.
10. Wood, T. M., Mc Crae, S. I., and Bhat, K. M. (1989), *Biochem. J.* **260**, 37–43.
11. Woodward, V., Luna, M., and Lee, N. E. (1988), *Biochemistry* **255**, 895–899.
12. Henrissat, B., Driguez, H., Viet, C., and Schulein, M. (1985), *Bio/Technol.* **3**, 722–726.
13. Wood, T. M. and Mc Crae, S. I. (1979), *Adv. Chem. Ser.* **181**, 181–209.
14. Irwin, D. C., Spezio, M., Walker, L. P., and Wilson, D. B. (1993), *Biotechnol. Bioeng.* **42**, 1002–1013.
15. Wood, T. M., McCrae, S. I., and Macfarlane, C. C. (1980), *Biochem. J.* **189**, 51.
16. Wood, T. M. and McCrae, S. I. (1978), *Biochem. J.* **171**, 61.
17. Gilbón, A., Larios, G., and Huitrón, C. (1981), *Tecnol. Aliment.* **16(5)**, 24–30.
18. García-Kirchner, O. (1983), Producción de Celulasas y su Utilización en la Sacarificación de Materiales Celulósicos. Tesis de Licenciatura. Fac. de Química, UNAM, México.
19. Huitrón, C., Saval, S., and Acuña, M. E. (1984), *Ann. NY Acad. Sci.* **434**, 110.
20. García-Kirchner, O. (1985), Memorias del Congreso Nacional de Microbiología. Universidad Autónoma de Puebla-Sociedad Mexicana de Microbiología. Pue., Pue. México.
21. Lowry, O. H., Rosenbrough, M. V., Farr, A. L., and Randall, R. J. (1951), *J. Biol. Chem.* **193**, 256.
22. Mandels, M., Andreotti, R., and Roche, C. (1976), *Biotechnol. Bioeng. Symp.* **6**, 21–33.
23. Larios, G. and Huitrón, C. (1981), *Rev. Tecnol. Aliment (México)* **16(5)**, 24–29.
24. Larios, G., Gilbón, A., Lara, Y., and Huitrón, C. (1987), *Enzyme Eng.* **6**, 353,354.
25. Miller, G. L. (1959), *Anal. Chem.* **31**, 426–428.
26. Ghose, T. K. (1987), *Pure Appl. Chem.* **59**, 257–268.
27. Zetelaki, K. (1976), *Proc. Biochem.* July/August: 11–18.
28. García, M. D. V., Ogawa, T., Shinmyo, L., and Enatsu, T. (1974), *J. Ferment. Technol.* **52**, 378–386.
29. Mandels, M. and Weber, M. (1969), *Adv. Chem. Ser.* **95**, 391–414.
30. Woodward, J., Lima, M., and Lee, N. E. (1988), *Biochem. J.* **255**, 895–899.

0273-2289/96/5758—0267$7.50

Increased Xylose Reductase Activity in the Xylose-Fermenting Yeast *Pichia stipitis* by Overexpression of *XYL1*

KRISTINE M. DAHN,[1,2] BRIAN P. DAVIS,[2,3] PAUL E. PITTMAN,[1] WILLIAM R. KENEALY,[4] AND THOMAS W. JEFFRIES*,[2,3]

[1]*Department of Biochemistry, University of Wisconsin, Madison, WI;* [2]*Institute for Microbial and Biochemical Technology, Forest Products Laboratory, USDA Forest Service, One Gifford Pinchot Drive, Madison, WI 53705;* [3]*University of Wisconsin, Department of Bacteriology, Madison, WI, 53706; and* [4]*Current address: J. Whittier Biologics Inc., Madison, WI*

ABSTRACT

The *Pichia stipitis* xylose reductase gene *(XYL1)* was inserted into an autonomous plasmid that *P. stipitis* maintains in multicopy. The plasmid pXOR with the *XYL1* insert or a control plasmid pJM6 without *XYL1* was introduced into *P. stipitis*. When grown on xylose under aerobic conditions, the strain with pXOR had up to 1.8-fold higher xylose reductase (XOR) activity than the control strain. Oxygen limitation led to higher XOR activity in both experimental and control strains grown on xylose. However, the XOR activities of the two strains grown on xylose were similar under oxygen limitation. When grown on glucose under aerobic or oxygen-limited conditions, the experimental strain had XOR activity up to 10 times higher than that of the control strain. Ethanol production was not improved, but rather it decreased with the introduction of pXOR compared to the control, and this was attributed to nonspecific effects of the plasmid.

Index Entries: *Pichia stipitis*; xylose reductase; gene expression; fermentation; metabolism.

INTRODUCTION

Over the past decade, considerable research has focused on the metabolism of xylose, and strains of microorganisms, which can ferment xylose, have been identified or developed *(1–13)*. D-Xylose is a primary component of biomass hemicellulosic materials, and it is an excellent substrate for fuel production because it can be

*Author to whom all correspondence and reprint requests should be addressed.

fermented to ethanol. In the long-term view, environmentally friendly renewable fuels could displace fossil fuels used in transportation.

Pichia stipitis, Candida shehatae, and *Pachysolen tannophilus* are three yeast species generally accepted as the best wild-type xylose fermenters *(1)*. In *P. stipitis,* oxygen levels greatly influence the fate of xylose. Under fully aerobic culture conditions, where oxygen is in excess, no fermentation occurs *(14)*. Ethanol is produced in culture only during oxygen limitation *(15)*. During oxygen limitation, yeast cells grow more slowly and use less xylose for respiration *(16)*. Skoog and Hahn-Hägerdal *(17)* demonstrated that pyruvate decarboxylase (PDC) activity in *P. stipitis* increases with declining oxygen uptake. Alexander et al. *(18)* showed that in a similar yeast, *C. shehatae,* alcohol dehydrogenase (ADH) activity is elevated when oxygen availability declines. Other enzymes are probably also regulated during the transition from aerobic conditions to oxygen limitation.

Efforts have been made to broaden the substrate range of *Saccharomyces cerevisiae,* a high-productivity glucose fermenter that is incapable of xylose utilization. The *Escherichia coli* gene for xylose isomerase (XOI), an enzyme that directly converts xylose into xylulose, was integrated into *S. cerevisiae,* but the protein was inactive in *S. cerevisiae (4)*. When the *E. coli* gene for XOI was incorporated into *Schizosaccharomyces pombe,* active protein was produced, but xylitol formation and low ethanol productivity were problems *(2)*.

The major enzymes involved in xylose assimilation (xylose reductase [XOR], xylitol dehydrogenase [XDH], and xylulose kinase [XUK]) have been investigated *(19)*. XOR and XDH activities are induced in *P. stipitis, P. tannophilus,* and *C. shehatae* when these yeast are grown on xylose *(18,20)*. The gene for XUK from *S. cerevisiae* has been cloned and introduced back into *S. cerevisiae* on a high-copy-number plasmid, resulting in an increased level of XUK activity *(5)*. The gene for XOR was cloned from *P. stipitis* and introduced into *S. cerevisiae,* resulting in XOR activity *(6,8,10)*. This recombinant *S. cerevisiae* strain could neither grow on nor ferment xylose, but it did produce xylitol. Recently, *P. stipitis* genes for both XOR and XDH were introduced into *S. cerevisiae,* resulting in the expression of active protein products and production of ethanol in low titers *(11,12)*.

P. stipitis does not produce ethanol under fully aerobic conditions, but its ethanol yield improves and approaches theoretical as conditions become oxygen limited *(21)*. However, the growth rate of this yeast is significantly reduced when the culture is oxygen limited *(22)*. To increase ethanol production, it would be useful to develop a *P. stipitis* strain in which relevant fermentative enzymes are expressed on induction by xylose, rather than being oxygen regulated. We have begun to investigate the regulation of the *XYL1* promoter in *P. stipitis,* and we have overexpressed *XYL1* in *P. stipitis.* This strain has elevated levels of XOR activity, and it was analyzed for its ability to ferment xylose and glucose separately.

MATERIALS AND METHODS

Microbial Strains

E. coli DH1 (*recA*1, *endA*1, *gyrA*96, *thi*-1, *hsdR*17, *supE*44, *relA*1) or *E. coli* DH5α (*recA*1 *endA*1 *hsdR*17 *supE*44 *thi*-1 *gyrA*96 *relA*1 Δ*lacU*169[Ø80*lacZ*ΔM15]) were used as bacterial hosts for plasmids. *P. stipitis ura3* strain TJ26 *(23)* was used as the recipient strain in yeast transformations.

Media

The minimal medium used in yeast fermentation analysis contained 6.5 g/L urea, 10% glucose or xylose, and 1.7 g/L yeast nitrogen base without amino acids or ammonium sulfate (Difco, Detroit, MI) at pH 6.6. *E. coli* was cultivated in Luria-Bertani medium. When selection was required for *E. coli*-harboring plasmids, 50 μg/mL ampicillin was added to the medium.

DNA Manipulations and Molecular Biological Enzymes

Yeast genomic DNA was isolated and purified as described by Rose et al. *(24)*, except Novozyme 234 was used instead of Zymolyase 100,000. Bacterial plasmids were purified according to Sambrook et al. *(25)* or by the Wizard DNA purification systems from Promega (Madison, WI). Restriction enzymes were obtained from Promega or New England Biolabs (Beverly, MA), and digestions were performed as recommended by the suppliers. Restriction fragments were purified from low melting agarose by phenol extraction.

Cell Growth

Inoculum (100 mL) was grown in a 250-mL baffled flask at 30°C for 72 h shaken at 150 rpm. The cells were harvested by centrifugation, resuspended in fresh medium, and used to inoculate shake flasks. To obtain oxygen-limited conditions, 100 mL of medium was put in a foam-plugged 125-mL flask. For the aerobic cultures, 75 mL of medium was placed in a 250-mL baffled flasks. Four flasks were used for each condition, so that cells could be harvested at four different time points. Glucose and xylose fermentations of both strains under the different aeration conditions were done side by side in a single incubator/shaker. The fermentations were done at 30°C with continuous shaking at 180 rpm.

Fermentation Analysis

Cells from each condition and sugar source were harvested every day for 4 d at indicated times. A small amount of culture was taken immediately before harvest, and the optical density of the sample was determined using a Beckman DU spectrophotometer at 600 nm.

The cells were collected by centrifugation. The supernatant was passed through a 0.2- or 0.45-μm pore Nalgene filter, and stored at –20°C for later gas chromatographic and high-performance liquid chromatography (HPLC) analysis. The cells were then washed with water and the volume was split in two. One-half was washed with 1*M* sorbitol for later DNA analysis. The other half was washed with 0.1*M* MOPS (pH 6.8) for later enzyme activity analysis. The cells were then quick-frozen in liquid nitrogen and stored at –80°C.

Preparation of Cell Extracts and XOR Activity Assays

Cell pellet (0.5 mL) slurry in 0.1*M* MOPS (pH 6.8) buffer was added to 0.5 g of 425–600 μ glass beads (Sigma, St. Louis, MO) in a borosilicate glass culture tube. Cells were vortexed vigorously for five 1-min bursts, with cooling in crushed ice for 1–2-min in between bursts. Homogenates were centrifuged at 13,800*g* for 20 min at 4°C, and the supernatant solution was transferred to a new tube. Within 4 h of cell breakage, the xylose-dependent oxidation of reduced nicotinamide adenine dinucleotide phosphate (NADPH) was followed at 340 nm, 30°C, according to a modification of the protocol by Chiang and Knight *(26)*. The assay mixture contained 0.1*M* Tris-HCl (pH 7.2), 5–50 µL cell extract, and 100 µL $1.5 \times 10^{-3}M$ NADPH in a total volume of 900 µL. The reaction was started with the addition of 100 µL 0.1*M* xylose.

Protein Assays

Total protein was measured by the bicinchoninic acid assay (Pierce, Rockford, IL) using bovine serum albumin as standard.

Southern Blot Analysis

Southern transfer was performed by capillary blotting *(27)*. DNA hybridizations were done with a Genius 1 kit (Boehringer Mannheim, Indianapolis, IN) as described by the manufacturer. Zeta-Probe GT membrane (Bio-Rad, Hercules, CA) was used to crosslink the nucleic acids. Hybridizations were typically done in 25% formamide at 37°C. The 65°C washing times were shortened from 15 to 5 min.

Transformation

Electrotransformation of *P. stipitis* was performed as previously described *(23)*. Transformation of *E. coli* was done by electroporation with a Gene Pulser together with a Gene Controller apparatus and cuvets from Bio-Rad according to the manufacturer's protocol.

Densitometry

Densitometry was done using a Hewlett-Packard Desk Scanner with the Quantitative Images program (National Institutes of Health, Washington, DC) on a Macintosh II computer.

HPLC

The sugars, ethanol, and byproducts of fermentation were separated by HPLC (Hewlett Packard series 1050, Wilmington, DE) using a Bio-Rad carbohydrate column (HPX 87C, 300 × 7.8 cm) maintained at 85°C. The mobile phase was degassed distilled water at a flow rate of 0.5 mL/min at a pressure of 50–55 bar. The clear filtered sample (980 µL) was mixed with 20 µL of sucrose (500 g/L) as an internal standard before injection. The sugars and ethanol were quantitated by a refractive index (RI) detector.

Gas Chromatography (GC)

Ethanol was separated by GC (Hewlett Packard series 2, model 5890) using Porapack Q column from Alltech (Deerfield, IL) with the initial and final oven temperature at 180°C. The detector temperature was 220°C, and the injector temperature was 200°C. The carrier gas was helium, and a flame ionization detector was used.

RESULTS

pTN131 *(10)* was cleaved with *Bam*HI restriction enzyme, and the approx 5.5-kb DNA fragment containing the *P. stipitis XYL1* gene was isolated and purified. This

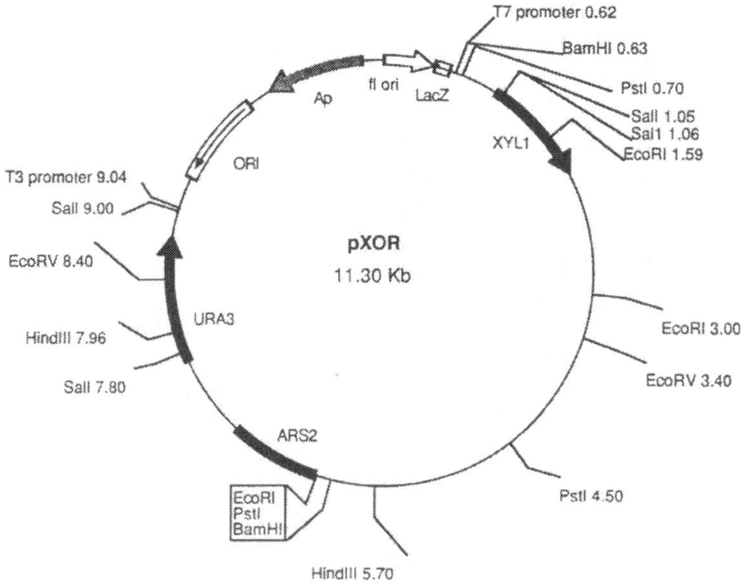

Fig. 1. Physical map of plasmid pXOR containing the *P. stipitis* xylose reductase gene (*XYL1*).

fragment was inserted at the unique *Bam*HI site in pJM6 *(23)*, creating pXOR (Fig. 1). Uncut pXOR was introduced into *P. stipitis* TJ26 (*ura3-1*), and a Ura$^+$ colony harboring pXOR as an autonomous plasmid was isolated (TJ26 [pXOR]). To obtain a control strain, a TJ26 Ura$^+$ colony harboring pJM6 as an autonomous plasmid was isolated (TJ26 [pJM6]).

TJ26 (pXOR) had about 1.8-fold higher XOR activity than TJ26 (pJM6) after 1 d of growth in xylose medium under aerobic conditions (Fig. 2). At the same time point under oxygen-limited conditions, the XOR activities in both the pXOR strain and the control strain were higher, and the activities of the two strains were similar. After 2 d of xylose fermentation (Fig. 3), the XOR activities of both strains at both levels of aeration decreased and were similar.

Introduction of the pXOR plasmid increased XOR activity levels even when glucose was the sole carbon source. After 1 d of growth in glucose medium with either high or low levels of aeration, TJ26 (pXOR) showed up to 10 times higher XOR activity than the control (Fig. 2). The pXOR strain continued to show higher XOR activity on glucose after 2 d of fermentation (Fig. 3).

Following 4 d of fermentation, ethanol production by the pXOR strain was compared to that of the control (Fig. 4). Surprisingly, the ethanol production of the pXOR strain was lower than the pJM6 strain under the conditions tested. In fact, the pJM6 strain consistently produced more ethanol than the pXOR strain on each of the 4 d of fermentation (data not shown). The shake-flask fermentations were repeated, producing similar trends in XOR activity and ethanol production (data not shown).

The cell growth of TJ26 (pJM6) and TJ26 (pXOR) during the shake-flask fermentation was determined spectrophotometrically. Optical density at 600 nm was

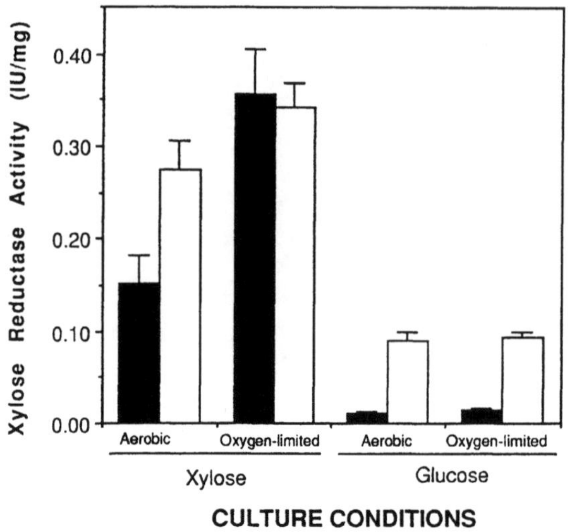

Fig. 2. Xylose reductase activity in the TJ26::pJM6 control strain (■) and in the *P. stipitis* TJ26::pXOR strain (□) after 1 d of growth in minimal medium. The aerobic cultures were harvested at 24 h postinoculation, and the oxygen-limited cultures were harvested after 27 h postinoculation. Error bars show the standard deviation of at least three replicate assays from the same sample, taken from a representative shake-flask experiment. Values are given in IU/mg of total protein in the cell extract.

determined from the shake flasks at the time of harvest, one flask for each condition at each time point (data not shown). Optical density was also monitored daily from the shake flask harvested at the final time point. The growth patterns of the pJM6 strain and the pXOR strain were compared and found to be similar, except the pJM6 strain grew to a higher cell density when grown on xylose under oxygen-limited conditions. In other repetitions, the pJM6 strain grew to a higher cell density under all conditions tested.

We compared carbohydrate utilization by the two strains under aerobic and oxygen-limited conditions (data not shown). Under both conditions, TJ26 (pXOR) used less sugar than TJ26 (pJM6) when grown on xylose. The glucose utilization patterns of the two transformants varied with the plasmid and the aeration conditions employed. No direct correlation between xylose or glucose consumption and ethanol production was seen.

In order to determine the copy number of plasmids pJM6 and pXOR, we harvested cells from one experiment (the same experiment from which all the other results shown were obtained) daily. Total DNA was extracted from the cells grown under the high aeration condition and digested to completion with *Bam*HI restriction enzyme. The DNA fragments were separated by agarose gel electrophoresis and analyzed by Southern blotting. Probing with *P. stipitis URA3* yielded two bands of different sizes, one band from the native genomic copy of the gene and the other band from the plasmid-borne copy of the gene. Copy number was calculated by comparing the intensity of the two bands using densitometry. Between one and eight copies of the plasmid (either pJM6 or pXOR) were found per cell under the conditions used in this experiment.

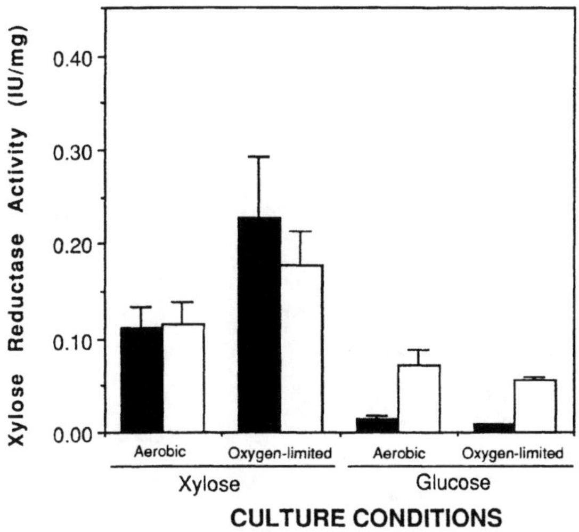

Fig. 3. Xylose reductase activity in the TJ26::pJM6 control strain (■) and in the *P. stipitis* TJ26::pXOR strain (□) after 2 d of growth in minimal medium. The aerobic cultures were harvested at 46.5 h postinoculation, and the oxygen-limited cultures were harvested after 42.5 h postinoculation. Error bars show the standard deviation of at least three replicate assays from the same sample, taken from a representative shake-flask experiment. Values are given in IU/mg of total protein in the cell extract.

DISCUSSION

When the xylose reductase gene *(XYL1)* from *P. stipitis* was introduced into *P. stipitis* TJ26 *(ura3)* on a multicopy plasmid, XOR activity increased when the transformant was grown on glucose as a sole carbon source. Introduction of pXOR also increased XOR activity in aerobically grown cells with xylose as a sole carbon source. These increases can be explained by an increase in the copy number of *XYL1*. Oxygen levels did not affect the XOR activities of either the pXOR or pJM6 strains when grown on glucose. When grown on xylose, both strains had higher XOR activity when limited for oxygen. However, under oxygen limitation, the XOR activities of the two strains grown on xylose were similar. Under oxygen limitation, there are likely other levels of regulation in addition to gene dosage for expression of *XYL1*. It is possible that the XOR activity of *P. stipitis* is regulated not only at the level of transcription, but also at the level of translation and post-translationally by such factors as other gene products and metal ions.

Ethanol production was not increased by overexpression of *XYL1*, suggesting that either XOR is not rate limiting under the conditions employed for the strain tested or that any benefit resulting from overexpression XOR was countered by negative effects owing to the multicopy plasmid. Introduction of pXOR and the resulting increase in XOR activity could alter the metabolism of the *P. stipitis* in an unanticipated manner, resulting in lower ethanol production, slower growth, and decreased carbohydrate utilization. For example, in some *E. coli* strains bearing multicopy plasmids, changes in cell growth were seen and postulated to occur owing to interaction of the host cell machinery with the plasmid and plasmid

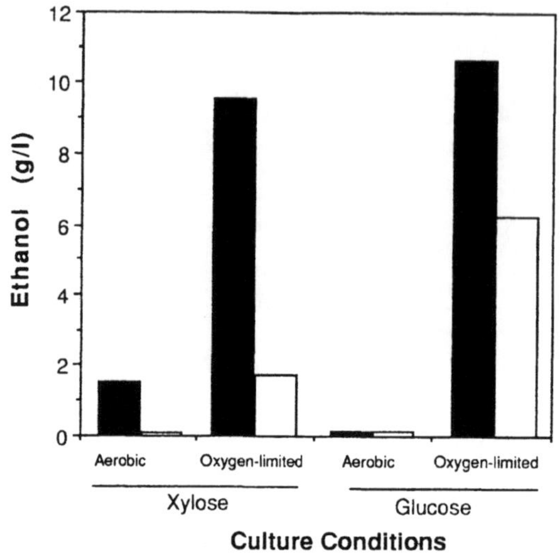

Fig. 4. Quantification of ethanol in culture supernatant after 4 d of a representative fermentation by *P. stipitis* TJ26 strains harboring plasmid pJM6 (■) or pXOR (□). The aerobic culture samples were taken at 92 h postinoculation. The oxygen-limited samples were taken at 95.75 h postinoculation. The ethanol concentrations shown were determined from the shake flask supernatant samples at the time of harvest, one flask for each condition at each time point.

gene products *(28,29)*. pXOR is about twice the size of pJM6, this difference might account in part for the poor fermentative performance of the TJ26 (pXOR) strain.

The *XYL1* DNA fragment used to construct pXOR is large, and different results might be obtained if less of the *XYL1* 3'-flanking region were used. The genetic background of the host strain also determines how gene overexpression affects metabolism. We know from other studies that *P. stipitis* TJ26 is a poor fermentative strain (data not shown). If this deficiency results from decreased fermentative or increased respirative enzyme activity, overexpressing genes for xylose assimilation, such as *XYL1*, can be expected to have little effect.

The copy number of the autonomous plasmid varies from one time point to the next. We assume that the copy number may vary also from one fermentation experiment to the next. The basis for variability is not fully understood, but copy number appears to increase with frequent subculturing (data not shown). Therefore, subculturing may be useful to increase the gene dosage further beyond the relatively low copy numbers (between one and eight) that were achieved in the experiments described here. Alternatively, expression vectors may be incorporated into the host genome for prolonged stability, eliminating the requirement of maintaining selective pressure.

Regulation of *XYL1* expression by glucose was altered in TJ26 (pXOR), resulting in higher XOR activity under conditions when transcription is normally not induced owing to the presence of glucose. This suggests that the plasmid-borne *XYL1* 5'-flanking sequence does not have all the necessary regulatory and control regions for full and regulated expression. Alternatively, the increase in gene dosage may be titrating negative transcriptional regulators of *XYL1*. However, the 5'-flanking sequence could be used in developing strains to carry out simultaneous fermentation of glucose and xylose. In a commercial process, fermentation of both sugars would be desirable.

Oxygen appears to affect XOR activity when TJ26 (pXOR) was grown on xylose, resulting in higher activity than the control strain under relatively aerobic conditions. No such effect was seen when glucose was the carbon source. Normally, ethanol production by *P. stipitis* is low under aerobic conditions, and this is an area for improvement. Therefore, the *XYL1* 5'-flanking sequence may be useful when higher levels of desired enzyme activity are required in the presence of high levels of oxygen.

Other fermentative enzymes need to be studied. Pyruvate is a key fermentative intermediate, and two enzymes involved in the conversion of pyruvate to ethanol are PDC and ADH. In *S. cerevisiae*, increasing PDC and ADH activity does not lead to an increased rate of ethanol production *(30)*. However, ethanol production by *E. coli* is increased when PDC and ADH activities are increased *(31)*. *P. stipitis* could be transformed with these genes under the regulation of the *XYL1* 5'-flanking sequence, thus allowing for increased ethanol production in the presence of xylose. Strains that overexpress two or more genes simultaneously could be developed in an effort to produce a strain that has higher ethanol production.

ACKNOWLEDGMENTS

We thank Tatsuji Seki for his generous donation of plasmid pTN131. Paul E. Pittman was supported by NSF grant no. GER9253952. This investigation was supported in part by National Research Service Award 5 T32 GM08349 from the National Institute of General Medical Sciences, by USAID project no. 10:299, and National Renewable Energy Laboratory subcontract no. AUX-4-11193-02.

REFERENCES

1. Toivola, A., Yarrow, D., van den Bosch, E., van Dijken, J. P., and Scheffers, W. A. (1984), *Appl. Environ. Microbiol.* **47,** 1221–1223.
2. Chan, E.-C., Ueng, P. P., and Chen, L. (1986), *Biotechnol. Lett.* **8(4),** 231–234.
3. Ingram, L. O., Conway, T., Clark, D. P., Sewell, G. W., and Preston, J. F. (1987), *Appl. Environ. Microbiol.* **53,** 2420–2425.
4. Sarthy, A. V., McConaughy, B. L., Lobo, Z., Sundstrum, J. A., Furlong, C. E. and Hall, B. D. (1987), *Appl. Environ. Microbiol.* **53(9)**, 1996–2000.
5. Deng, X. X. and Ho, N. W. Y. (1990), *Appl. Biochem. Biotechnol.* **24/25,** 193–199.
6. Amore, R., Kötter, P., Küster, C., Ciriacy, M., and Hollenberg, C. P. (1991), *Gene* **109,** 89–97.
7. Hallborn, J., Walfridsson, M., Airaksinen, U., Ojamo, H., Hahn-Hägerdal, B., Penttilä, M., and Keränen, S. (1991), *Bio/Technology* **9,** 1090–1095.
8. Ho, N. W. Y., Petros, D., and Deng, X. X. (1991), *Appl. Biochem. Biotechnol.* **28/29,** 369–375.
9. Ohta, K., Beall, D. S., Mejia, J. P., Shanmugam, K. T., and Ingram, L. O. (1991), *Appl. Environ. Microbiol.* **57,** 893–900.
10. Takuma, S., Nakashima, N., Tantirungkij, M., Kinoshit, S., Okada, H., Seki, T., and Yoshida, T. (1991), *Appl. Biochem. Biotechnol.* **28/29,** 327–340.
11. Kötter, P. and Ciriacy, M. (1993), *Appl. Microbiol. Biotechnol.* **38,** 776–783.
12. Tantirungkij, M., Nakashima, N., Seki, T., and Yoshida, T. (1993), *J. Fermentation Bioeng.* **75,** 83–88.
13. Zhang, M., Eddy, C., Deanda, K., Finkelstein, M., and Picataggio, S. (1995), *Science* **267,** 240–243.
14. Bruinenberg, P. M., de Bot, P. H. M., van Dijken, J. P., and Scheffers, W. A. (1984), *Appl. Microbiol. Biotechnol.* **19,** 256–260.
15. Alexander M. A. and Jeffries, T. W. (1989), *Enzyme Microb. Technol.* **12,** 2–19.

16. Grootjen, D. R. J., van der Lans, R. G. J. M., and Luyben, K. Ch. A. M. (1990), *Enzyme Microb. Technol.* **12,** 20–22.
17. Skoog, K. and Hahn-Hägerdal, B. (1990), *Appl. Environ. Microbiol.* **56,** 3389–3394.
18. Alexander, M. A., Yang, V. W., and Jeffries, T. W. (1988), *Appl. Microbiol. Biotechnol.* **29,** 282–288.
19. Lachke, A. H. and Jeffries, T. W. (1986), *Enzyme Microb. Technol* **8,** 353–359.
20. Webb, S. R. and Lee, H. (1990), *Biotechnol. Adv.* **8,** 685–697.
21. Boynton, B. L. and McMillan, J. D. (1994), *Appl. Biochem. Biotechnol.* **45/46,** 509–514.
22. du Preez, J. C., van Driessel, B., and Prior, B. A. (1989), *Arch. Microbiol.* **152,** 143–147.
23. Yang, V. W., Marks, J. A., Davis, B. P., and Jeffries, T. W. (1994), *Appl. Environ. Microbiol.* **60,** 4245–4254.
24. Rose, M. D., Winston, F., and Hieter, P. (1990), *Methods in Yeast Genetics: Laboratory Course Manual*, Cold Spring Harbor Laboratory Press, Cold Spring Harbor, NY.
25. Sambrook, J., Fritsch, E. F., and Maniatis, T. (1989), *Molecular Cloning: A Laboratory Manual*, 2nd ed., Cold Spring Harbor Laboratory Press, Cold Spring Harbor, New York.
26. Chiang, C. and Knight, S. G. (1966), *Methods Enzymol.* **9,** 188–193.
27. Southern, E. M. (1975), *J. Mol. Biol.* **98,** 503–517.
28. Bailey, J. E., Da Silva, N. A., Peretti, S. W., Seo, J.-H., and Srienc, F. (1986), *Ann. NY Acad. Sci.* **469,** 194–211.
29. Seo, J.-H. and Bailey, J. E. (1985), *Biotechnol. Bioeng.* **27,** 1668–1674.
30. Schaaff, I., Heinisch, J., and Zimmermann, F. K. (1989), *Yeast* **5,** 285–290.
31. Ingram, L. O., Conway, T., Clark, D. P., Sewell, G. W., and Preston, J. F. (1987), *Appl. Environ. Microbiol.* **53,** 2420–2425.

0273-2289/96/5758—0277$9.00

The Relationship Between Growth Enhancement and *pet* Expression in *Escherichia coli*

HUGH G. LAWFORD* AND JOYCE D. ROUSSEAU

Department of Biochemistry, University of Toronto, Toronto, Ontario, Canada M5S 1A8

ABSTRACT

The *pet* operon consists of genes coding for enzymes responsible for ethanol production and consists of pyruvate dehydrogenase and alcohol dehydrogenase II from the high-performance ethanologen *Zymomonas mobilis*. This article describes the physiological influence of *pet* expression in *Escherichia coli* B (ATCC 11303) in terms of growth rate and overall concentrations of cell mass and catabolic end products achieved under well-defined cultivation conditions that included constant pH and carbon (energy) limitation. Glucose, mannose, and xylose were used as substrates, because they represent the principal fermentable components of lignocellulosic biomass and because fermentation of these sugars involves different metabolic pathways. Two different types of ethanologenic recombinants were used—a strain in which *pet* expression was via a multicopy plasmid (pLOI297) and a chromosomal integrant, strain KO11. Under the condition of sugar substrate limitation, there was no growth enhancement by *pet* expression with either glucose or mannose. Whereas the host strain produced exclusively lactic acid from glucose and mannose, both recombinants produced mostly ethanol. Both the plasmid-carrying strain and the *pet* integrant exhibited slower growth compared to the host culture with glucose or mannose as fermentation substrate. With mannose, the plasmid recombinant grew appreciably slower than either the host culture or the recombinant KO11. Use of a magnesium-deficient medium produced different results with glucose since substrate and turbidometric measurements proved to be unreliable in terms of estimating overall biomass levels. At pH 6.3, *pet* expression improved overall biomass yield; but at pH 7.0, the cell yields exhibited by the plasmid recombinant and the host strain were the same. *E. coli* B did not grow well on xylose as sole carbon source. With xylose, *pet* expression increased the growth rate, but had no effect on the overall biomass yield. In comparing our observations with the reports of others, it was concluded that the effect of *pet* expression on growth of *E. coli* is dependent on several different biochemical, physiological, genetic, and environmental factors, which largely precludes a statement of generality regarding this phenomenon.

Index Entries: ATP yield; xylose; recombinant *E. coli*; ethanol; bioenergetics; defined medium mannose; growth yield; *pet* plasmid; *E. coli* KO11.

*Author to whom all correspondence and reprint requests should be addressed.

INTRODUCTION

Recombinant DNA technology provides interesting possibilities for analyzing and studying in vivo behavior of complex systems through the introduction of well-defined perturbations. The essence of metabolic engineering is the introduction of a change in metabolic flux at a metabolic junction (branch point in a pathway). It is important that we understand why and how flux distribution is accomplished in nonlinear metabolic systems in order to predict the total physiological response to metabolic perturbations introduced by genetic manipulations.

The facultative gram-negative bacterium *Escherichia coli* is capable of utilizing a broad spectrum of sugar substrates and is generally regarded as being heterofermentative, producing a mixture of metabolic end products *(1–3)*. In addition to strain specificity, certain physicochemical environmental factors are known to result in various degrees of end-product selectivity, but the nature of cellular control strategies that determine selectivity by regulating flow through competing pathways is poorly understood. Various excreted metabolites can be differentiated by the energy gain (G_{ATP}), and it has been proposed that selectivity is determined by energy demand *(4,5)*. For example, acetate production generates extra energy via acetate kinase, and a correlation has been suggested between acetate production and high growth rates *(5)*. Although knowledge of mechanisms by which cells generate and conserve energy has advanced *(6)*, an understanding of how energy fluxes are regulated in relation to the needs of cell synthesis is largely lacking.

Ingram's recombinant *E. coli (7–11)* represents a model with which to examine the physiological influence of a particular metabolic perturbation designed to linearize terminal carbohydrate metabolism and maximize ethanol yield. Anaerobic metabolism is branched at the level of pyruvate through competition for substrate by several different enzymes, including pyruvate dehydrogenase, pyruvate formate lyase, and lactic acid dehydrogenase *(2)*. Pyruvate decarboxylase activity (coded for by the *pdc* gene) is absent from *E. coli (12)*. Overexpression of *Zymomonas* pyruvate decarboxylase and alcohol dehydrogenase II (collectively called the *pet* operon) *(9)* results in a redirection of metabolism from mixed-acid fermentation to the production of ethanol *(9)*. It is postulated that the high affinity of pyruvate decarboxylase for pyruvate permits this enzyme to compete successfully against the other pyruvate catabolic enzymes, principally lactic acid dehydrogenase (pyruvate reductase *[9]*); however, it is known that a high level of *pdc* expression is critical to ensuring high ethanol selectivity *(9–11)*. Furthermore, other potentially competitive enzymes are regulated both by effector modulation and at the level of transcription through induction or repression *(1)*. For example, respiratory NADH oxidase is induced under aerobic conditions and is a potential competitor for the NADH produced by pyruvate decarboxylase *(13)*. Similarly, the amount and the activity of pyruvate dehydrogenase is subject to regulation *(14,15)*. Hence, flux control at the level of the pyruvate "metabolic intersection" is a complicated matter *(13)* and one that is not entirely understood.

During the formative stages of research relating to the creation of the *pet* operon, it was observed that recombinants exceeded the growth of the host on solidified complex medium containing fermentable sugar *(7)*, and enlarged colony morphology was a useful distinguishing phenoptypic characteristic of *pet* transformants. Recombinants also were observed to grow to higher cell densities generally in unbuffered broth media *(7)*, and this was interpreted in terms of the beneficial effect of eliminating the synthesis of the potentially toxic acidic end products and modifying pH during fermentation whereby the pH did not fall as rapidly to a level intolerable for growth. Collectively, these observations led to the concept that *pet* expression has a positive influence on growth yield *(9–11)*. Diaz-Ricci et al. *(16,17)* used well-defined cultivation conditions to explore further the influence of the expression of the *pet* operon on *E. coli*. Using different host cultures (K12 and HB101) from that used by Ingram, batch fermentations were conducted under glucose-limiting conditions, at constant pH. Diaz-Ricci et al. *(16)* confirmed that the growth yield of *pet*-expressing recombinants was improved over that of the wild-type culture. Furthermore, in an investigation of the effect of *pet* expression on changes in the concentration of various intracellular metabolites in *E. coli* using phosphorus-31 nuclear magnetic resonance (NMR), it was concluded that growth enhancement could not be attributable to effects on external pH or ΔpH, but rather to the effect of the switch in pathways from mixed acid to ethanol fermentation *(17)*.

The purpose of this investigation was to examine the physiological influence of the expression of the *pet* operon on the growth of wild-type *E. coli* B (ATCC 11303) under defined environmental conditions. The concept of growth enhancement was examined from both a kinetic and stoichiometric perspective in terms of growth rate and growth yield under carbon (energy) limitation. Two different types of recombinants were used—a strain in which *pet* expression was via a multicopy plasmid (pLOI297) *(10)* and a chromosomally integrated construct, strain KO11 *(18)*. Batch fermentations were conducted under anaerobic conditions with both complex and defined media at constant pH. This investigation employed three different sugars—namely, glucose, mannose, and xylose. In addition to representing the major fermentable components of lignocellulosic biomass, these sugars are not metabolized by the same pathway, and this introduced another variable hitherto unexplored with respect to this phenomenon.

MATERIALS AND METHODS

Organisms

The wild-type, host culture *E. coli* B (ATCC 11303) was obtained from The American Type Culture Collection (Rockland, MD). Recombinant *E. coli* B (ATCC 11303 carrying the *pet* plasmid pLOI297) *(10)* and the chromosomally integrated strain KO11 *(18)* were received from L. O. Ingram (University of Florida, Gainesville, FL). Cultures grown from single colony isolates on selective antibiotic-containing agar medium were stored at –10°C in Luria broth (LB) medium supplemented with glycerol (20 mL/dL) and sodium citrate (1.5 g/dL). Inocula were prepared as previously described *(19)* using complex or defined media buffered with 100 m*M* phosphate (pH 7.0). Batch fermentations were inoculated by transferring approx 100 mL of the overnight flask culture directly to 1400 mL of medium in the stirred-tank bioreactor. The same sugar was used for preculture and fermentation. The initial cell density was monitored spectrophotometrically to give an OD_{550} in the range 0.1–0.2 corresponding to 30–50 mg dry wt cells/L.

Culture Media

The nutrient-rich, complex culture medium LB consists of 5 g Bacto yeast extract (Difco Laboratories, Detroit, MI), 10 g Bacto Tryptone (Difco), and 5 g NaCl/L of distilled water. In experiments with glucose, a modified formulation of this medium (designated as "mLB") as described by Grohmann et al. *(20,21)* was also used, and it contained reduced amounts of yeast extract (2.5 g/L) and tryptone (5 g/L), but no NaCl. The defined minimal medium was comprised of 2.25 g NH_4Cl, 0.1 g $MgSO_4$, 0.7 mg $FeCl_3 \cdot 6H_2O$, 2.72 g KH_2PO_4, 3.48 g K_2HPO_4, 5.0 g NaCl, 12 mg $CaCl_2 \cdot 2H_2O$, 9.9 mg $MnCl_2 \cdot 4H_2O$, 0.05 mg thiamine, and 0.21 g citric acid/L of distilled water. All media were sterilized by autoclaving. Stock sugar solutions were autoclaved separately and added at the concentration specified. When the plasmid-bearing strain was used, filter-sterilized antibiotics (final concentration of 40 mg/L ampicillin and 10 mg/L tetracycline) were added to the autoclaved fermentation media after cooling.

Fermentation Equipment

pH-stat batch fermentations were conducted in a volume of 1500 mL in MultiGen (model F2000) stirred-tank bioreactors fitted with agitation, pH, and temperature control (30°C) (New Brunswick Scientific, Edison, NJ). Unless specified otherwise, the pH was controlled at 6.3 by the addition of 4*N* KOH.

Analytical Procedures

Growth was measured turbidometrically at 550 nm (1-cm lightpath), and culture dry wt was measured by microfiltration. Following vacuum filtration, the filters (0.45-μ pore size) were washed and dried to constant weight under an infrared heat lamp. Compositional analyses of culture media and cell-free spent broths were determined by HPLC using an HPX-H column (Bio-Rad, Hercules, CA) as described previously *(19)*. The final concentrations of metabolic end products in spent fermentation broths was not corrected for the dilution caused by the addition of titrant during fermentation.

Determination of Growth Yield and Percent Carbon Recovery

The mass-based growth yield coefficient with respect to carbon (energy) source (i.e., the sugar added to the medium) was calculated by dividing the maximum cell density (g dry wt cells/L) by the concentration of sugar consumed. When complex media were used, the cell yield was corrected for the contribution of the nonsugar nutrients to the total cell mass (cor.$Y_{x/s}$). This value was experimentally determined to be on average 0.41 and 0.25 g dry wt cells/L for LB and mLB media, respectively. Carbon balances (expressed as percent carbon recovery) were calculated as described previously *(19)*. The carbon content of the *E. coli* cell mass was assumed constant at 47.6% carbon *(19)*.

RESULTS AND DISCUSSION

Under the conditions used in this investigation, the host culture, *E. coli* B (ATCC 11303), and the *pet*-expressing recombinants exhibit the same pattern of growth in LB. Anaerobic growth in LB is characterized by a final cell density of 0.41g dry wt cells/L achieved in about 4–5 h. With the nutritionally leaner mLB medium, the final cell density was only 0.25 g dry wt cells/L (results not shown). Growth in the defined salts medium was entirely dependent on the addition of sugar, since this was the sole source of assimilable carbon. In this study, the mass-based growth yield coefficient with respect to carbon and energy source was used to assess the physiological effect of *pet* expression on growth quantitatively. Hence, to eliminate the contribution made to cell mass by the nonsugar components of the complex media, the corresponding amounts of biomass produced in the absence of added sugar were routinely subtracted from the values determined for the maximum cell density. Overall growth yield values are limited in their informative value by virtue of the fact that they reflect the aggregate of all effects produced during the entire time period of batch growth. In addition, there exists the potential for growth interference by different metabolic end products. Therefore, in an attempt to minimize possible inhibitory effects, either by ethanol or organic acid end products, the concentration of sugar employed was relatively low at about 20 g/L. Also at this relatively low sugar concentration, carbon-limited growth can be assured using LB *(19,22)*. Under a condition of growth limitation by the carbon (energy) source, the growth yield for carbon can be viewed in bioenergetic terms as an indirect estimate of the net ATP gain (G_{ATP}, mol ATP/mol sugar) *(6,9,23)*. However, the inability to account quantitatively for changes with respect to nongrowth associated energy metabolism (i.e., the amount of ATP required for homeostasis or "maintenance") in an environment of batch culture, limits the generality of interpretation of growth yield determinations with respect to bioenergetics *(19)*. For example, a condition producing an increase in the maintenance energy requirement could result in a decreased overall growth yield without affecting the G_{ATP}. Such an "uncoupling effect" in *E. coli* is known to be produced by acetic acid *(24,25)* and by the maintenance of high-copy-number plasmid vectors *(16)*.

Effect of *pet* Expression with Glucose

Anaerobic batch fermentations were conducted in stirred-tank bioreactors in LB with the pH-controlled at 6.3. Figure 1 shows that under a condition of glucose limitation, the growth rate of the wild-type culture *E. coli* B (ATCC 11303) was faster than that of either the plasmid-carrying strain (11303:pLOI297) or the chromosomally integrated strain KO11. Because the final cell density is, in part, a function of the limiting nutrient and because the sugar concentration was not exactly invariable for different batches, the growth yield provides a means of normalizing the effect of differences in sugar concentration. Figure 1B shows that with LB, the growth yield with respect to glucose (cor.$Y_{x/s}$) for both the wild-type culture and the plasmid recombinant is similar (0.093 and 0.091 g dry wt cells/g glu, respectively) whereas for strain KO11, it is reduced by about 10% (0.083 g dry wt cells/L). A similar observation with respect to the lack of growth enhancement by *pet* expression was made when the nutritionally rich LB medium was replaced with a chemically lean, defined medium (Fig 1B). Table 1 shows stoichiometric data with respect to the distribution of metabolic end products associated with these batch fermentations. The high ethanol selectivity for both recombinant strains shows that *pet* expression clearly functions in the manner consistent with the purpose for which it was designed. Metabolism is redirected by the *Zymomonas* pyruvate decarboxylase and alcohol dehydrogenase from the production of lactic acid as the major end product in the host to ethanol in the recombinants (Table 1). However, our observations with respect to the lack of growth enhancement are in conflict with those of Diaz-Ricci et al. *(16)* who reported that, under similar conditions of glucose limitation in LB medium with pH controlled at 7.0, whereas expression of the *pet* plasmid pLOI295 in *E. coli* HB101 had no effect on growth rate, there was a substantial increase in growth yield, from 0.10 to 0.16g dry wt. cells/L. *E. coli* HB101 is a hybrid of strains

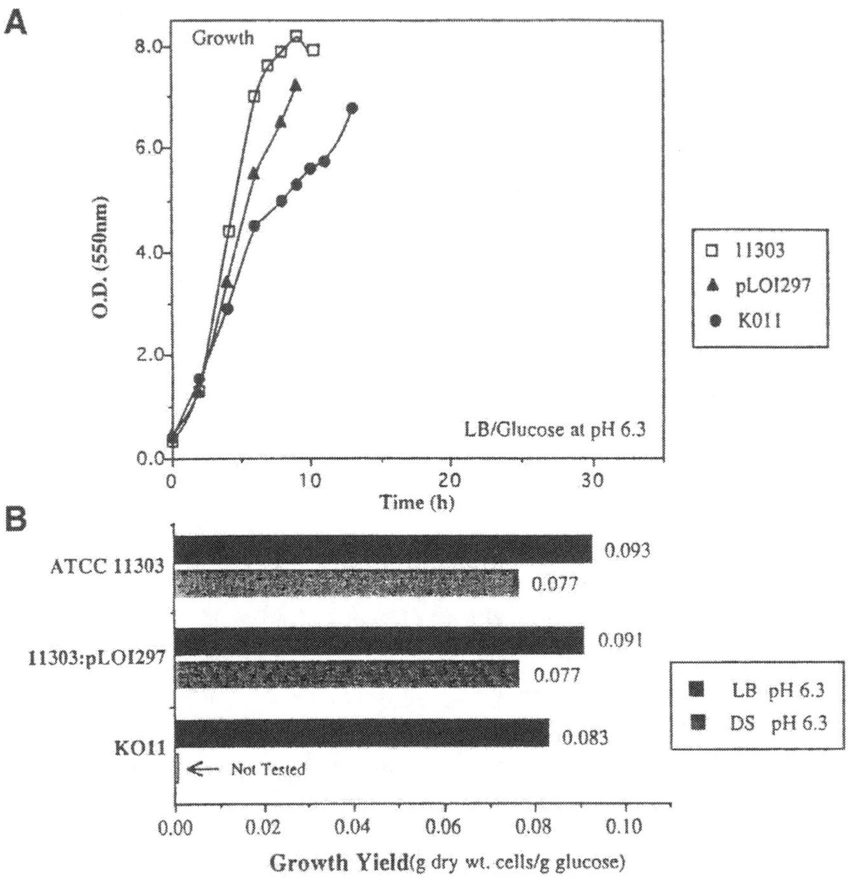

Fig. 1. The effect of *pet* expression on growth of *E. coli* recombinants 11303:pLOI297 and KO11 with glucose as carbon (energy) source. **(A)** Effect on growth rate and **(B)** effect on growth yield coefficient for carbon. The host culture *E. coli* B (ATCC 11303) is included as a control. In the case of the complex media LB, the final biomass concentration was corrected for contribution to growth from nonsugar components of the medium (*see* Materials and Methods).

K12 and Luria strain B *(26)*. In the fermentations reported by Diaz-Ricci et al. *(16)*, *pet* expression affected the end-product distribution, causing a decrease in acetate concentration from 83 to 20 m*M*. Acetic acid is a known byproduct of glucose fermentation by *E. coli*, but the yield of this acid appears to be dependent on both the particular strain and environmental conditions employed *(1–3)*. Furthermore, with *E. coli*, sensitivity to acetic acid toxicity has been shown to be a strain-dependent phenomenon *(27)*. At pH 7.0, 83 m*M* acetate are equivalent to 0.46 m*M* undissociated acid, and whereas this concentration has been shown to have little effect on the growth yield in *E. coli* B *(25)*, it has a pronounced effect of *E. coli* K12 *(28,29)*. In the context of considering the possible effect of acetic acid, it is interesting to note that in another study conducted under similar conditions, expression of the *pet*

Table 1
The Effect of *pet* Expression on Anaerobic Glucose Catabolism by *E. coli* B (ATCC 11303) and Recombinants 11303:pLOI297 and KO11

Genotype medium	Expt.	pH	Glucose, m*M*	Biomass, g/L	EtOH, m*M*	Lactic, m*M*	Acetic, m*M*	Succ., m*M*	Formic, m*M*	% Carbon recovery
ATCC 11303										
LB	138c	6.3	122	2.46	0	226	0	0	0	106
mLB	128b	6.3	118	1.36	0	183	0	0	0	85
mLB	128d	7.0	117	1.80	20	181	0	0	0	88
DS	112b	6.3	100	1.36	40	125	0	13	0	89
11303:pLOI297										
LB	137a	6.3	116	2.32	224	6	0	8	0	117
mLB	123c	6.3	105	1.78	208	0	0	4	0	113
mLB	139a	7.0	107	1.82	178	22	0	7	0	110
DS	112	6.3	95	1.31	172	0	0	0	0	91
KO11										
LB	139c	6.3	112	2.08	210	11	14	0	11	121
mLB	123e	6.3	104	1.60	203	0	11	0	0	113
mLB	140d	7.0	126	1.54	216	21	35	0	0	113
DS	Not tested									

plasmid pLOI308-10 in a double-deletion mutant of *E. coli* K12 lacking both acetyl phosphotransferase and acetate kinase activity, and thereby genetically blocked in its ability to synthesize acetic acid, did not alter either the anaerobic growth rate or the growth yield with respect to glucose *(30)*. Fortunately, the lack of acetic acid production in our fermentations (Table 1) removed this complication in interpreting the results.

In nutritional terms, LB is an excessively rich medium, and much leaner complex media have been shown to be compatible with good fermentation performance by ethanologenic recombinant *E. coli (31,32)*. One such formulation, recommended by Grohmann et al. *(20,21)*, was used in this study (designated as "mLB"). However, the mLB medium is known to be deficient in both phosphate and magnesium *(31)*, and whereas the phosphate deficiency is compensated by the volumetric transfer from the phosphate-buffered preculture during inoculation, the concentration of magnesium (estimated at 2 mg/L) remains below the minimal amount required to support a final cell concentration of about 2 g dry wt cells/L. Hence, because growth in mLB medium was not glucose limited, the glucose growth yield coefficient was not an appropriate comparative parameter. Since others have used a variety of environmental conditions, including the lack of pH control and growth that was not limited by the carbon (energy) source, we considered our observations with mLB pertinent for the purpose of comparison. Figure 2 shows the growth of the host culture and the two recombinants in mLB at a nominal glucose concentration of 20 g/L. With the pH controlled at 6.3, turbidometric data indicate an enhancement effect for *pet* expression in the plasmid-bearing culture, but not for the chromosomal integrant (Fig. 2A). However, direct determination of final biomass concentrations revealed that both ethanologenic recombinant cultures surpassed the lactic acid-producing host strain (Table 1). Lactic acid toxicity is pH-dependent, and by changing the pH control set point from 6.3 to 7.0, the final cell density of the host culture was significantly improved (Fig. 2B and Table 1). Furthermore, in a separate experiment with an elevated glucose concentration (80 g/L), but where the mLB medium was replaced with a magnesium and phosphate-fortified LB medium, at pH 6.3, the host culture achieved a final cell density of 3 g dry wt cells/L and a final lactate concentration in excess of 600 m*M* lactic acid (results not shown). Since mLB is magnesium-deficient, we are led to conclude that magnesium may play a role in preventing the negative effect of lactic acid on cell yield. With the pH controlled at 7.0, the optical density (OD) of the host culture surpasses both recombinants (Fig. 2B). However, the final cell densities of the host culture and plasmid-bearing recombinant are similar at 1.8 and 1.82 g dry wt cells/L, respectively (Table 1). Clearly, the ratio of dry cell mass to OD is not the same for the ethanol and acid-producing cultures, and this difference may reside in the light-scattering properties of an ethanologenic culture grown in a magnesium-deficient medium.

In addition to ethanol, recombinant strain KO11 also synthesizes acetic acid (Table 1). Although the growth-inhibiting effects are reduced at the higher pH, the amount produced by strain KO11 at pH 7.0 is equivalent to 1 m*M* undissociated acid, which has been shown to affect the growth of *E. coli* B *(25)*. Collectively, our observations suggest that, under glucose limitation, *pet* expression in *E. coli* B (ATCC 11303) does not enhance growth either with respect to rate or yield. However, in comparing our observations to those of others, it would appear that the effect is dependent on several biochemical, physiological, genetic, and environmental factors, which largely precludes a statement of generality regarding this phenomenon.

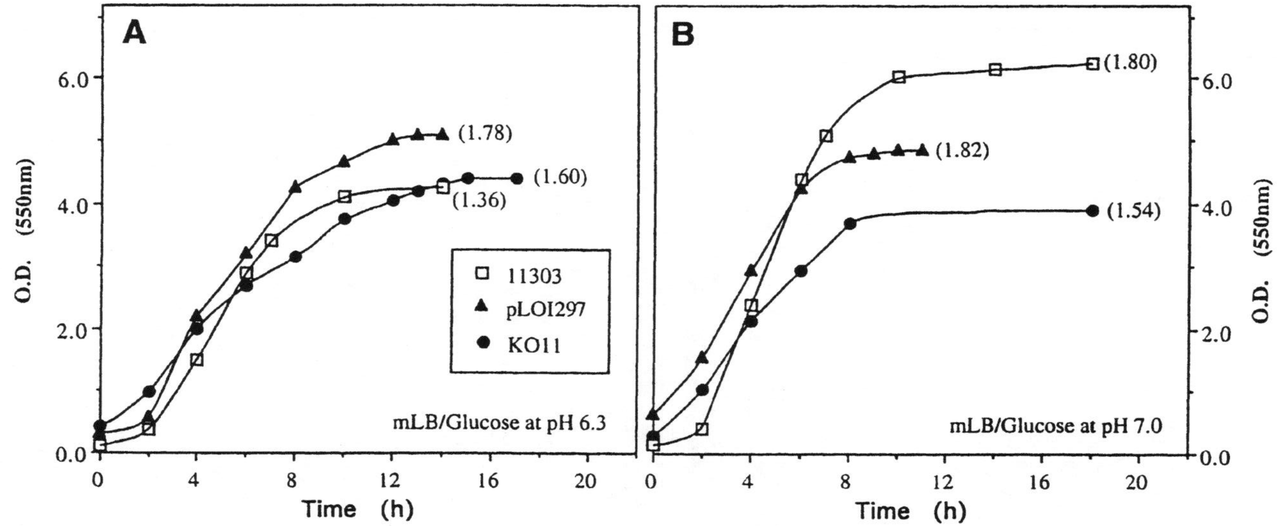

Fig. 2. Anaerobic growth of wild-type *E. coli* B (ATCC 11303) and *pet* recombinants 11303:pLOI297 and KO11 in modified LB medium with glucose at 20 g/L. **(A)** pH controlled at 6.3 and **(B)** pH controlled at 7.0. Maximum biomass and final concentrations of metabolic end products are given in Table 1.

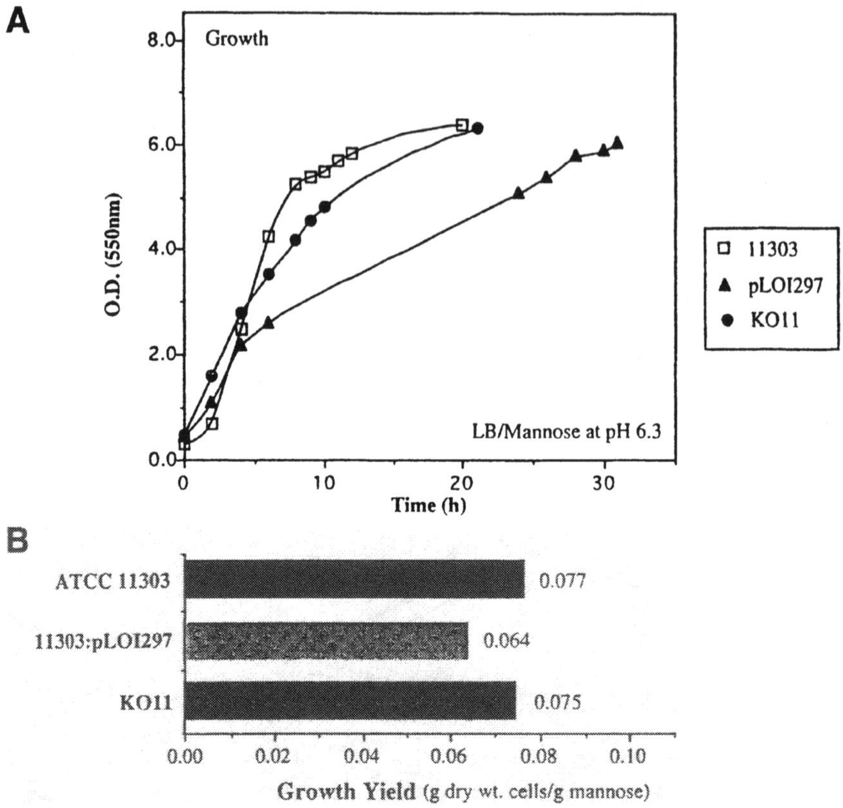

Fig. 3. The effect of *pet* expression on growth of *E. coli* recombinants 11303:pLOI297 and KO11 with mannose as carbon (energy) source. **(A)** Effect on growth rate and **(B)** effect on growth yield coefficient for carbon. The host culture *E. coli* B (ATCC 11303) is included as a control.

Effect of *pet* Expression with Mannose

D-Mannose is an aldohexose and differs from D-glucose only with respect to the configuration of the hydroxyl group at carbon 2. Mannose is a prominent component of the hemicellulose of certain lignocellulosic biomass, for example, softwood *(33)*. Previous studies have examined the physiology of the plasmid-bearing recombinant 11303:pLOI297 with mannose as carbon source *(34)*, and the fermenation performance of both 11303:pLOI297 and KO11 in softwood prehydrolysates has been investigated *(35a)*.

Under a condition of carbon limitation at pH 6.3, the wild-type culture exhibits a slightly reduced growth rate and a marked growth yield relative to glucose as carbon source (Figs. 1 and 3); however, lactic acid is the sole end product of both glucose and mannose catabolism (Tables 1 and 2). The apparent reduction in growth rate may be owing to the short lag that was experienced by the host culture on introduction into the mannose LB medium. The relatively poor growth performance of the plasmid recombinant has been observed previously with mannose as principal carbon source *(34,36)* but no explanation for the reduced growth has

Table 2
The Effect of *pet* Expression on Anaerobic Mannose Catabolism by *E. coli* B (ATCC 11303) and Recombinants 11303:pLOI297 and KO11

Genotype medium	Expt.	pH	Mannose, m*M*	Biomass, g/L	EtOH, m*M*	Lactic, m*M*	Acetic, m*M*	Succ., m*M*	Formic, m*M*	% Carbon recovery
ATCC 11303										
Lo	134b	6.3	124	2.12	0	240	0	0	0	109
11303:pLOI297										
LB	052b	6.3	118	1.78	199	5	0	8	0	103
KO11										
LB	137d	6.3	118	2.01	224	4	14	3	1	116

been forthcoming. Figure 3B shows that the growth yield with respect to mannose is similar for both the host culture at strain KO11 being 0.077 and 0.075 g dry wt cells/L, respectively. The growth yield for the plasmid recombinant is 0.064 g dry wt cells/L (Fig. 3B). Measurements of growth yield have previously been used to infer differences in energetics associated with glucose and xylose metabolism by the plasmid-bearing recombinant *(19)*. A similar approach can be used in analyzing the differences in growth yield elicited by glucose and mannose. Assuming that the growth yield with respect to energy (ATP) is constant when either glucose or mannose serves as the energy source for growth, then the reduction in growth yield exhibited under the condition of mannose-limitation suggests that the net gain in ATP is less for mannose than glucose. Carrying a high-copy-number plasmid places an energetic burden on the cell *(16)* and the associated increase in maintenance energy coefficient would reduce the net energy yield, but the issue of why the magnitude of energy required for plasmid maintenance might be different for glucose and mannose remains problematical. In the context of this study, it is important to note that *pet* expression did not enhance growth when mannose was the limiting carbon and energy source (Fig. 3).

Effect of *pet* Expression with Xylose

Because the recombinant cultures under investigation were specifically designed to convert xylose to ethanol efficiently, the majority of previous work with these cultures has involved xylose either as the sole sugar supplement *(35b–37)* or as a component of either synthetic mixtures *(38)* or biomass prehydrolysates *(35a,39)*. Xylose is a poor growth substrate for *E. coli (12)*. Under a condition of carbon limitation at pH 6.3, *pet* expression results in an improvement in both the growth rate (Fig. 4A) and growth yield (Fig. 4B) of *E. coli* B (ATCC 11303) in LB with xylose as the carbon source. With glucose or mannose as fermentation substrates, *pet* expression alters the homofermentative pattern in *E. coli* B from lactic acid to ethanol production. However, with xylose as substrate, *E. coli* B is heterofermentative. In addition to lactic, acetic, and succinic acids, ethanol is synthesized (Table 3). The xylose-based ethanol yield is 0.22 g/g and this observation has been reported previously *(37)*. Hence, with xylose as substrate, *pet* expression improves the xylose-to-ethanol conversion efficiency. The ethanol yield for the plasmid recombinant and KO11 is 0.45 and 0.47 g/g, respectively (Table 3). At pH 6.3, the concentration of the undissociated (inhibitory) form of acetic acid is 1.86 and 0.76 m*M* for the host culture and KO11, respectively (Table 3). The difference in growth rate between these cultures can be explained in terms of the amounts of acetic acid produced (Fig. 4A). The absence of acetic acid production by the plasmid recombinant gives this culture a distinct growth advantage (Fig. 4A). The xylose-based growth yield coefficient in the host culture was 0.055 g dry wt cells/g xylose (Fig. 4B). In the case of the plasmid recombinant, the growth yield of 0.054 g dry wt cells/g xylose has been previously reported *(19)*. The growth yield varies only slightly for the three cultures (Fig. 4B), and under the conditions of assay, this parameter is considered to be unaffected by *pet* expression.

The Concept of Ethanologenic *E. coli* as a Superior Host for Heterologous Expression

Largely because of the vast amount of research that has been done on *E. coli*, especially in terms of molecular biology and genetics, it has been a popular selec-

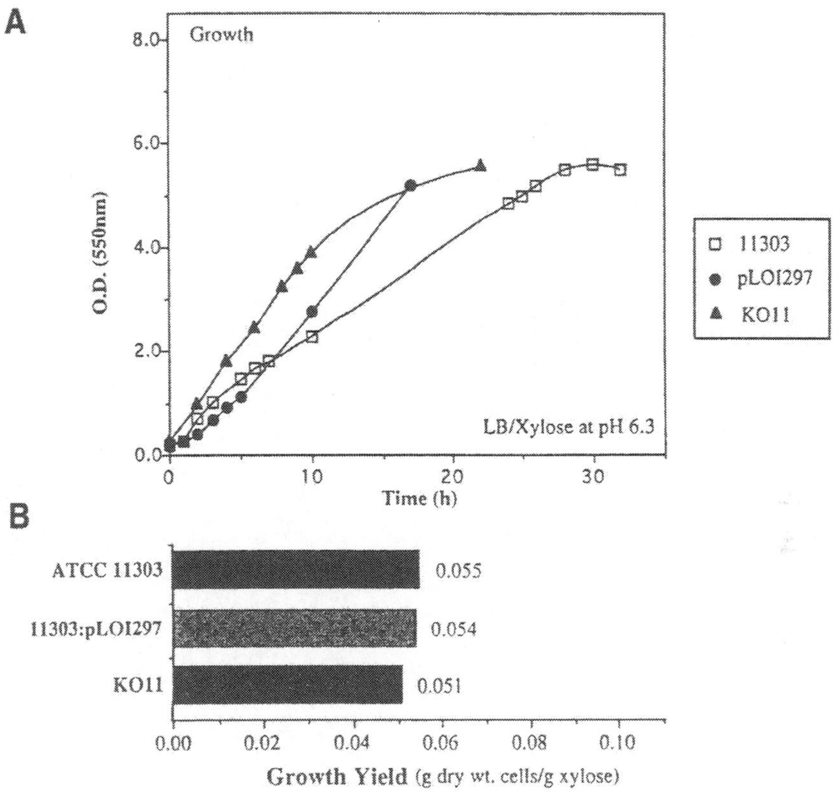

Fig. 4. The effect of *pet* expression on growth of *E. coli* recombinants 11303:pLOI297 and KO11 with xylose as carbon (energy) source. **(A)** Effect on growth rate and **(B)** effect on growth yield coefficient for carbon. The host culture *E. coli* B (ATCC 11303) is included as a control.

tion as host for the expression of heterologous proteins from a variety of origins. To achieve maximum cell yield, recombinants are usually grown under aerobic conditions. In general, aerobic growth of plasmid-bearing cultures is often depressed compared to the host culture, and this has led to studies on optimization of growth of recombinant *E. coli (40–43)*. The excretion of acetic acid (HAc) is cited as the causative agent responsible for both depressed growth yield and protein biosynthesis *(44,45)*. Recombinant cultures are reported to be more HAc-sensitive than wild-type cultures *(46)*.

Based on the claim that *pet* expression in *E. coli* results in growth enhancement, it has been suggested that ethanologenic *E. coli* would be a superior host for the expression of foreign proteins *(9,18)*. However, there are seveal considerations that detract from the general utility of ethanologenic constructs in the biotechnology industry. The inherent instability of plasmid-carrying constructs is a serious consideration *(18,47)*, and the energetic burden imposed by a high-copy-number *pet* plasmid *(16)*, such as pLOI297, not only lessens the potential growth (protein) yield through an increase in energy consumption that is unrelated to growth or

Table 3
The Effect of *pet* Expression on Anaerobic Xylose Catabolism by *E. coli* B (ATCC 11303) and Recombinants 11303:pLOI297 and KO11

Genotype medium	Expt.	pH	Xylose, m*M*	Biomass, g/L	EtOH, m*M*	Lactic, m*M*	Acetic, m*M*	Succ., m*M*	Formic, m*M*	% Carbon recovery
ATCC 11303										
LB	011c	6.3	137	1.53	98	38	67	25	28	106
11303:pLOI297										
LB	030b	6.3	141	1.54	206	0	0	12	0	108
KO11										
LB	136d	6.3	152	1.58	219	3	26	0	18	109

protein biosynthesis, but also increases the sensitivity to HAc toxicity by reducing the amount of energy available for intracellular pH homeostasis. Under aerobic conditions, expression of the *pet* operon is purported to divert >90% of carbon to ethanol production *(9)*. Successful competition by the *Zymomonas* ethanol production enzymes with the respiratory NADH oxidase for NADH and pyruvate dehydrogenase for pyruvate would result in an energetic debilitation, since the G_{ATP} associated with fermentative metabolism is far less than for aerobic catabolism. As a consequence, it could be anticipated that ethanologenic recombinant *E. coli* would be at an energetic disadvantage under aerobic conditions *(48)*. However, it has been shown that the ethanologenicity of recombinant 11303:pLOI297 is greatly reduced under aerobic conditions *(37,48)*, and this observation challenges the claim regarding the ability of the *pet* enzymes to control and monopolize pyruvate catabolism under aerobic conditions. To overcome the limitations imposed by the potential for genetic instability in plasmid-based *pet* constructs, a number of different chromosomal integrants of *E. coli* B (ATCC 11303) have been developed *(18)*. Furthermore, one of the integrants (strain KO12) has a recA mutation, thereby making it a more suitable host for the production of recombinant proteins *(18)*. It has been suggested that "additional genes can be inserted into plasmids to direct the synthesis of recombinant proteins as valuable coproducts with no expected interference in ethanol productivity" *(18)*. This suggestion has not yet been tested, but if proven tenable, the concept of recombinant protein biosynthesis as a byproduct of fuel ethanol production has interesting economic ramifications. From the perspective of the fuel ethanol industry, it is not known what effect the imposed energy demand of protein production would have on ethanol yield and productivity. Strains KO11 and KO12 were selected on the basis of their resistance to high levels of chloramphenicol (600 μg/mL), and the ultimate utility of these ethanologenic *pet* integrants as hosts for heterologous protein expression is predicated on claims relating to their genetic stability *(47)*. However, at this symposium, we reported the results of our investigation into the genetic stability of Ingram's ethanologenic *E. coli* recombinants, and our observations with continuous culture shed new light on the purported stability of both the plasmid construct 11303:pLOI297 and the chromosomal integrant, strain KO11 *(18)*. There appears to be sufficient evidence to discount the suggested utility of these ethanologenic recombinants as candidate host strains for use by the biotechnology industry.

ACKNOWLEDGMENTS

This research was supported by the Natural Sciences and Engineering Research Council of Canada and The University of Toronto. We are greatful to Lonnie Ingram for the gift of the recombinant *E. coli* cultures.

REFERENCES

1. Knappe, J. (1987), in Escherichia coli *and* Salmonella typhimurium, vol. 1, Neidhart, F. C., ed., Academic, New York, pp. 151–155.
2. Tempest, D. W. and Neijssel, O. M. (1987), in Escherichia coli *and* Salmonella typhimurium, vol. 1, Neidhart, F. C., ed., Academic, New York, pp. 797–806.
3. Gottschalk, G. (1986), in *Bacterial Metabolism*, 2nd ed., Springer-Verlag, New York, pp. 208–282.
4. Roos, J. W., McLaughlin, J. K., and Papoutsakis, E. T. (1985), *Biotechnol. Bioeng.* **27**, 681–694.

5. Thauer, R. K., Jungermann, K., and Decker, K. (1977), *Bacteriol. Rev.* **41,** 100–180.
6. Stouthamer, A. H. (1977), in *Microbial Energetics, 27th Symp. Soc. Gen. Microbiol.*, Haddock, B A. and Hamilton, W. A., eds., Cambridge University Press, London, pp. 285–315.
7. Ingram, L. O., Conway, T., Clark, D. P., Sewell, G. W., and Preston, J. F. (1987), *Appl. Environ. Microbiol.* **53,** 2420–2425.
8. Ingram, L. O., Conway, T., and Alterthum, F. (1991), United States Patent 5,000,000.
9. Ingram, L. O., Alterthum, F., Ohta, K., and Beall, D. S. (1990), in *Developments in Industrial Microbiology*, vol. 31, Pierce, G. E., ed., Elsevier Science, New York, pp. 21–30.
10. Alterthum, F. and Ingram, L. O. (1989), *Appl. Environ. Microbiol.* **54,** 397–404.
11. Ingram, L. O. (1991), in *Energy from Biomass and Wastes XIV*, Klass, D. L., ed., Institute of Gas Technology, Chicago, IL, pp. 1105–1126.
12. Neale, A. D., Scopes, R. K., and Kelly, J. M. (1988), *Appl. Microbiol. Biotechnol.* **29,** 162–167.
13. Holmes, W. H. (1986), in *Current Topics in Cellular Regulation*, vol. 28, Academic, New York, pp. 69–105.
14. Nimmo, H. G. (1987), in Escherichia coli *and* Salmonella, vol. 1, Neidhart, F. C., ed., Academic, New York, pp. 156–169.
15. Guest, J. R., Cole, S. T., and Jeyaseelan, K. (1981), *J. Gen. Microbiol.* **127,** 65–79.
16. Diaz-Ricci, J. C., Tsu, M., and Bailey, J. E. (1992), *Biotechnol. Bioeng.* **39,** 59–65.
17. Diaz-Ricci, J. C., Hitzmann, B., and Bailey, J. E. (1991), *Biotechnol. Prog.* **7,** 305–310.
18. Ohta, K., Beall, D. S., Mejia, J. P., Shanmugam, K. T., and Ingram, L. O. (1991), *Appl. Environ. Microbiol.* **57,** 893–900.
19. Lawford, H. G. and Rousseau, J. D. (1995), *Appl. Biochem. Biotechnol.* **51/52,** 179–195.
20. Grohmann, K., Baldwin, E. A., Buslig, B. S., and Ingram, L. O. (1994), *Biotechnol. Lett.* **16,** 281–286.
21. Grohmann, K., Cameron, R. G., and Buslig, B. S. (1995), *Appl. Biochem. Biotechnol.* **51/52,** 423–435.
22. Lawford, H. G. and Rousseau, J. D. (1991), *Appl. Biochem. Biotechnol.* **28/29,** 221–236.
23. Stouthamer, A. H. (1976), in *Yield Studies in Microorganisms*, Meadowfield, Dewbury, UK.
24. Lawford, H. G. and Rousseau, J. D. (1992), *Appl. Biochem. Biotechnol.* **34/35,** 185–204.
25. Lawford, H. G. and Rousseau, J. D. (1993), *Appl. Biochem. Biotechnol.* **39/40,** 301–322.
26. Boyer, H. W. and Rouland-Dussoix, D. J. (1969), *J. Mol. Biol.* **41,** 459–472.
27. Lui, G. W. and Strohl, W. R. (1990), *Appl. Environ. Microbiol.* **56,** 1004–1011.
28. Smirova, G. V. and Oktybr'skii, O. N. (1985), *Microbiology (USSR)* **54,** 205–209.
29. Smirova, G. V., and Oktybr'skii, O. N. (1985), *Microbiology (USSR)* **57,** 446–451.
30. Diaz-Ricci, J. C., Regan, L., and Bailey, J. E. (1991), *Biotechnol. Bioeng.* **38,** 1318–1324.
31. Lawford, H. G. and Rousseau, J. D. (1996), *Appl. Biochem. Biotechnol.* **57/58,** 307–326.
32. Guimaraes, W. V., Dudley, G. L., and Ingram, L. O. (1992), *Biotechnol. Bioeng.* **40,** 41–45.
33. Lawford, H. G., and Rousseau, J. D. (1993), in *Energy from Biomass and Wastes XVI*, Klass, D. L., ed., Institute of Gas Technology, Chicago, IL, pp. 559–597.
34. Lawford, H. G. and Rousseau, J. D. (1993), *Biotechnol. Lett.* **15,** 615–620.
35a. Barbosa, M. F. S., Beck, M. J., Fein, J. E., Potts, D., and Ingram, L. O. (1992), *Appl. Environ. Microbiol.* **58,** 1382–1384.
35b. Ohta, K., Alterhum, F., and Ingram, L. O. (1990), *Appl. Environ. Microbiol.* **56,** 463–465.
36. Beall, D. S., Ohta, K., and Ingram, L. O. (1991), *Biotechnol. Bioeng.* **38,** 296–303.
37. Lawford, H. G. and Rousseau, J. D. (1992), in *Energy from Biomass and Wastes XV*, Klass, D. L., ed., Institute of Gas Technology, Chicago, IL, pp. 583–622.
38. Lawford, H. G. and Rousseau, J. D. (1994), *Appl. Biochem. Biotechnol.* **45/46,** 367–382.
39. Lawford, H. G. and Rousseau, J. D. (1991), *Biotechnol. Lett.* **13,** 191–196.
40. Kracke-Hem, H. A., Rinas, U., Hitzmann, B., and Schügerl, K. (1991), *Curr. Microbiol.* **23,** 71–74.
41. Yang, X.-M. (1992), *J. Biotechnol.* **23,** 271–389.
42. Yee, L. and Blanch, H. W. (1992), *Bio/Technol.* **10,** 1550–1556.
43. Yee, L. and Blanch, H. W. (1993), *Biotechnol. Bioeng.* **41,** 221–230.
44. Landwall, P. and Holme, T. (1977), *J. Gen. Microbiol.* **103,** 353–358.
45. LeVine, S. M., Ardeshir, F., and Ferro-Luzzi Ames, G. (1980), *J. Bacteriol.* **143,** 1081.
46. Koh, B. T., Nakashimada, U., Pfeiffer, M., and Yap, M. G. S. (1992), *Biotechnol. Lett.* **14,** 1115–1118.
47. Lawford, H. G. and Rousseau, J. D. (1996), *Appl. Biochem. Biotechnol.* **57/58,** 293–305.
48. Lawford, H. G. and Rousseau, J. D. (1994), *Appl. Biochem. Biotechnol.* **45/46,** 349–365.

0273-2289/96/5758—0293$8.25

Factors Contributing to the Loss of Ethanologenicity of *Escherichia coli* B Recombinants pLOI297 and KO11

HUGH G. LAWFORD* AND JOYCE D. ROUSSEAU

Department of Biochemistry, University of Toronto, Toronto, Ontario, Canada M5S 1A8

ABSTRACT

To be economic and to be compatible with modern continuous bioconversion systems, it is imperative that the process organism exhibit an extremely high degree of stability. In the case of ethanol production from lignocellulosic biomass, functional stability of the potential process biocatalyst can be assessed in terms of the capacity to sustain high-performance fermentation during the continuous fermentation of biomass-derived sugars.

This investigation employed glucose- or xylose-limited chemostat culture to examine the functional stability of two patented, genetically engineered *E. coli*—namely *E. coli* B (ATCC 11303) carrying the *Zymomonas* genes for pyruvate decarboxylase and alcohol dehydrogenase II on a multicopy plasmid pLOI297 and a chromosomal *pet* integrant of strain 11303, designated as strain KO11. Both recombinants carry markers for antibiotic resistance and have been reported to exhibit genetic stability in the absence of antibiotic selection.

Chemostats were fed with Luria broth (LB) (with 25 g/L sugar) at a dilution rate of 0.14 and 0.07/h when the feed medium was glucose-LB and xylose-LB, respectively. The pH was controlled at 6.3. With glucose, both recombinants exhibited a rapid loss of ethanologenicity even when selection pressure was imposed by the inclusion of antibiotics in the feed medium. With strain KO11, increasing the concentration of chloramphenicol from 40 to 300 mg/L, resulted in a retardation in the rate of loss of ethanologenicity, but it did not prevent it. Under xylose limitation, the plasmid-bearing recombinant appeared to be stabilized by antibiotics, but this did not reflect genetic stability, since the slower-growing revertant was washed out at a dilution rate of 0.07/h. With both recombinants, interpretation of functional stability with xylose was complicated by the inherent ethanologenicity associated with the host culture.

Based on an average cost for large bulk quantities of antibiotics at $55/kg and an amendment level of 40 g/m^3, the estimated economic impact regarding the potential requirement for operational stabilization by antibiotics in a plant operating in batch mode varied from a maximum of 29¢/gal of E95 ethanol for antibiotic amendment of all fermentation media to a minimum of 0.45¢/gal where antibiot-

*Author to whom all correspondence and reprint requests should be addressed.

ics were used exclusively for the preparation of the inocula for every fourth batch fermentation cycle. The high degree of instability observed in these continuous fermentations does not auger well for the proposed potential industrial utility of these patented, genetically engineered constructs for the production of fuel ethanol from biomass and wastes.

Index Entries: Ethanol; xylose; stability; recombinant *E. coli*; strain KO11; antibiotics; *pet* operon.

INTRODUCTION

To be economic, it is imperative that the production of fuel ethanol from lignocellulosic biomass and wastes involve maximal efficiency of the conversion of all sugars to ethanol *(1,2)*. The five-carbon sugar xylose is a major component of hemicellulose *(3)*, which is hydrolyzed during pretreatment of lignocellulosic biomass *(4)*. Considerable research has been directed to the search for organisms capable of high-performance fermentation of biomass prehydrolysates. This search for xylose-fermenting ethanologenic microorganisms has produced several alternatives, including bacteria, yeasts, and fungi (for review, *see* ref. *5*). In addition to natural isolates, several genetically engineered biocatalysts have been constructed for this purpose, and prominent among these have been the patented ethanologenic *Escherichia coli* cultures that carry genes for ethanol production, namely pyruvate decarboxylase *(pdc)* and alcohol dehydrogenase II *(adhB)* from *Zymomonas* mobilis *(6–8)*. In the early stages of developing ethanologenic recombinants, genetic engineering involved insertion of these genes (referred to as the *pet* operon) *(7)* on multicopy plasmids carrying genes responsible for resistance to tetracycline and ampicillin *(8)*. In one particular construct, designated as plasmid pLOI297, expression of the *pet* operon was under the control of the *E. coli* lac promoter *(9)*. The construction of this transformation vector is illustrated in Fig. 1A (for details, *see* ref. *9*). Although the pioneering work was done with *E. coli* K12 *(10–12)*, a subsequent physiological assessment of growth characteristics of several different potential host cultures of *E. coli* identified the wild-type Luria strain B (ATCC 11303) as a "hardy strain and a suitable host for the *pet* plasmid pLOI297" *(13)*. For several years, we have been assessing the fermentation performance characteristics of this patented recombinant *E. coli* 11303:pLOI297 using both synthetic lab media *(14–18)* and biomass prehydrolysates prepared by different thermochemical processes from a variety of biomass/waste feedstocks, including both hardwood (aspen) *(19)* and softwood (pine) *(20)*, newsprint *(21)*, spent sulfite liquors *(22)*, and corn crop residues *(23)*.

Regarding plasmid stability, it has been claimed that "plasmid pLOI297 is stably maintained in ATCC 11303 in the absence of antibiotic selection" *(24)* based on the observation of 98% retention of antibiotic resistance for over 25 generations of growth with glucose in the absence of antibiotic selection *(9)*. Pyruvate decarboxylase is absent in *E. coli (25,26)*, and it is known that in order to function as an efficient ethanologen, there needs to be hyperexpression of the *pet* operon *(9)*. Although the replicon associated with pLOI297 ensures a high number of copies of the plasmid in the host (typically hundreds of copies per cell) *(27)* and consequently promotes high levels of activity of the *Zymomonas* ethanol production enzymes, the genetic instability associated with pUC-based plasmid constructs is also well known *(27,28)*. The potential for instability in recombinant 11303:pLOI297 is exacerbated

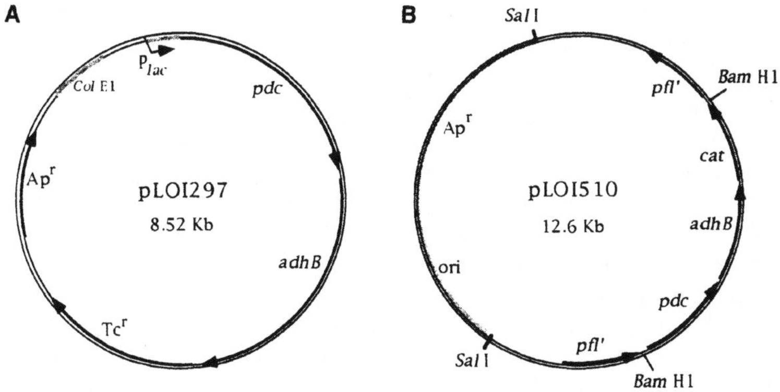

Fig. 1. The construction of vectors used to transform *E. coli* B (ATCC 11303). **(A)** Plasmid pLO12971—derived from the high-copy-number plasmid pUC18 with the *ColEI* replicon; **(B)** plasmid pLOI5102—Transformation via chromosomal integration was accomplished using a circularized 8.6-kbp derivative of pLOI510 containing promotorless *Zymomonas pdc* and *adhB* genes adjacent to *cat* (with a native promoter) inserted within a promotorless *E. coli* pyruvate formate lyase (the 4-kbp portion shown in gray was removed by *Sal* 1 digestion). Nomenclature: replicons *Col*E1 and ori—origins of replication; *pdc*, *Zymomonas* pyruvate decarboxylase; *adh*B, *Zymomonas* alcohol dehydrogenase II; Ap^r, β-lactamase (ampicillin resistance gene); Tc^r, tetracycline resistance gene; *pfl'*, split *E. coli* pyruvate formate lyase; *cat*, chloramphenicol acetyl transferase; P_{lac}, *E. coli* lac promotor; *Sal*1 and *Bam*H1, restriction endonucleases *(9,32)*.

by the fact that the host harbors low-copy-number cryptic plasmids *(29)*. Finally, high-copy-number plasmids are known to impose an energetic burden on the host *(30)*, and because energy is diverted from growth to plasmid maintenance, this effect is often reflected in reduced growth rate and cell yield *(31)*.

Recognizing that "plasmid-bearing recombinants are typically less stable than strains in which the foreign genes have been integrated into the host chromosome" *(32)*, Ingram and his associates at the University of Florida created a series of *pet* integrants *(32)*. One of the plasmids used for the purpose of inserting the *Zymomonas pet* genes into the chromosome of *E. coli* B ATCC 11303 is described in Fig. 1B (for details, *see* ref. 32). Chromosomally integrated strains were constructed by inserting the *Zymomonas pdc* and *adhB* genes within the pyruvate-formate lyase gene *(pfl)* of the host *(32)*. However, it was discovered that single-copy inserts of the *pdc* and *adhB* genes did not result in the high level of activities of *Zymomonas* enzymes that had been achieved in multicopy plasmid-based recombinants *(32)*. The circularized fragment of DNA used to transform *E. coli* also contained the gene for chloramphenicol acetyl transferase *(cat)*, which is responsible for conferring resistance to chloramphenicol (Cm) (Fig. 1). Strain KO11 expresses high levels of both the *Zymomonas* genes and *cat*, and was a spontaneous mutant that was selected for resistance to high levels (600 μg/mL) of chloramphenicol. In addition, strain KO11 carries a mutation in its fumarate reductase gene, which impairs its ability to produce succinate as a fermentation end product *(32)*.

In a recent collaborative, international, interlaboratory study designed to compare, under standardized conditions, the fermentation performance of several different microorganisms in a xylose-rich medium consisting primarily of a dilute acid hydrolysate of corn cobs, it was concluded that the *E. coli* recombinant strain KO11 "is currently the best known pentose-fermenting organism" *(33)*. This conclusion was based on a combination of criteria with a primary focus on ethanol yield and productivity. The issue of stability of this genetic construct was not addressed in this comparative study *(33)*. Likewise, in a recent review of hemicellulose conversion technology, McMillan *(2)* concluded that Ingram's recombinant *E. coli* is currently one of the "the best candidates" for converting hemicellulosic hydrolysates, but there was surprisingly no reference to the issue of stability of this candidate biocatalyst. A recently published technical and economic analysis of large-scale ethanol production from willow based on the use of recombinant *E. coli* KO11 admits to lack of information relating to the genetic stability of KO11 and its ability to sustain a high level of ethanologenicity during prolonged batch cycling and recirculation *(34)*. This study failed to consider the economic impact of the use of antibiotics to provide constant selective pressure as an aid to genetic stability *(34)*. Clearly, there is a deficiency in our knowledge regarding the issue of stability of these genetically engineered ethanologenic biocatalysts, and the purpose of this study was to extend our preliminary assessment of the stability of recombinant *E. coli* 11303:pLOI297 and the chromosomal integrant strain KO11 in which we used serial transfer cultivations with and without antibiotic selection *(35)*. Since the potential utility of these constructs relates principally to their ability to convert various biomass-based sugars to ethanol, we have chosen to assess the functional stability in terms of the ethanologenicity of the chemostat, whereby system stability is viewed as the capacity of the chemostat culture to sustain high fermentation performance with respect to the conversion of sugar to ethanol. Furthermore, continuous culture of recombinant ethanologenic *E. coli* is another area in which there is little available information and continuous fermentation offers significant potential for economic improvement through lowered capital and operating costs.

In this study, we used both glucose and xylose because these two sugars represent the major components of cellulose and hemicellulose, respectively *(3)*. The apparent linear growth pattern exhibited by recombinant *E. coli* in batch fermentations *(9,14,20)* made it difficult to predict an appropriate dilution rate for chemostat culture. Consequently, we have adopted a heuristic approach to the set point with respect to dilution rate. Based on our previous experience with batch fermentations, and knowing that growth with glucose was approximately twofold faster than with xylose *(14)*, the dilution was set at 0.14 and 0.07/h for these sugars, respectively.

MATERIALS AND METHODS

Organisms

The wild-type, host culture, *E. coli* B (ATCC 11303) was obtained from The American Type Culture Collection (Rockland, MD). Recombinant *E. coli* B (ATCC 11303 carrying the *pet* plasmid pLOI297) *(9)* and the chromosomally integrated strain KO11 *(32)* were received from L. O. Ingram (University of Florida, Gainesville, FL). Cultures grown from single-colony isolates on selective antibiotic-containing agar medium were stored at –10°C in luria broth (LB) medium supplemented with glycerol (20 mL/dL) and sodium citrate (1.5 g/dL).

Continuous Culture Conditions

Continuous cultures were conducted in benchtop chemostats (Bioflo model C30, New Brunswick Scientific Co., Edison, NJ) with a working volume (V) of approx 350 mL. The chemostats were fed sterile LB medium *(36)* (containing approx 2.5 wt% sugar) at a constant flowrate (F) by means of peristaltic pump (dilution rate [D] where D = F/V h^{-1}). The temperature was 30°C, the agitation (multiple flat-blade turbines) was constant at approx 150 rpm and the pH was maintained at 6.3 by the addition of 2*N* KOH. In the case of recombinant 11303:pLOI297 selection medium, the LB medium was amended with 40 mg/L ampicillin (Ap) and 10 mg/L tetracycline (Tc). For recombinant KO11, Cm was used either at 40 or 300 mg/L as specified in the figure captions. Inocula were prepared as in phosphate-buffered LB in flask cultures described previously *(37)*.

Analytical Procedures

Growth was estimated turbidometrically at 550 nm. Compositional analyses of fermentation media and cell-free spent media were determined by high-performance liquid chromatography (HPLC) as described previously *(37)*.

Assaying Stability of Recombinant Cultures

Functional stability of recombinant *E. coli* cultures was assessed in practical terms as ethanologenicity of the culture and the degree of ethanol selectivity among metabolic end products. Genetic stability was assessed phenotypically by plating cultures on selective and nonselective agar. The phenotypic characteristics of the recombinant culture were related to antibiotic resistance, colony size, and morphology. Ethanologenic recombinants were recognized on antibiotic-selective media by the formation of distinctive large, yellowish, opaque colonies *(14,38)*.

Determination of Fermentation Parameters

Culture ethanologenicity was expressed in terms of the ethanol yield ($Y_{p/s}$), which was calculated as the mass of ethanol produced per mass of sugar consumed. The theoretical maximum ethanol yield for either glucose or xylose is 0.51 g EtOH/g sugar.

RESULTS AND DISCUSSION

Recombinant *E. coli* 11303:pLOI297

Figure 2 represents the continuous culture of recombinant *E. coli* B 11303:pLOI297 in LB with glucose (about 25 g/L) and is a plot of the trajectories for biomass (estimated by OD), effluent glucose, and the different catabolic end products (ethanol, lactic acid, and acetic acid). Flow of medium from the nutrient reservoir was initiated following overnight batch culture, and a period of time equal to 3 vol changes (21 h) was assumed to be sufficient for steady-state growth to be achieved in the selective medium, which contained both ampicillin and tetracycline. After 24 h of elapsed continuous fermentation, the feed was switched to a nonselective medium (25 g/L glucose in LB minus the antibiotics) (Fig. 2). The increase in lactic acid prior to the change over to the nonselective medium suggests instability even under the selective pressure of the presence of antibiotics in

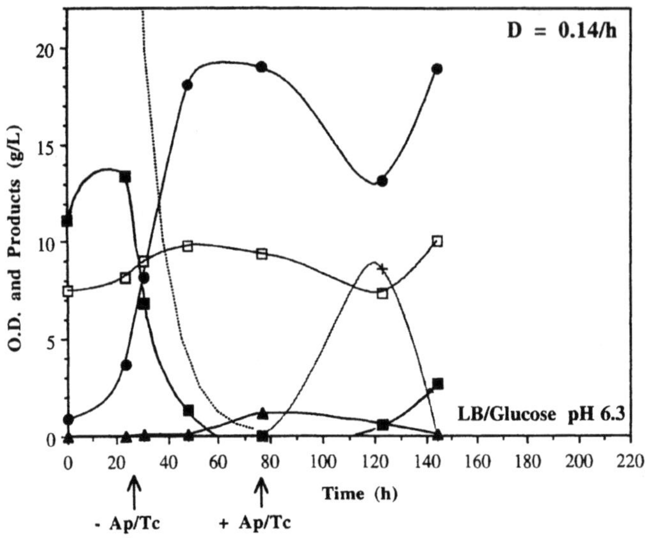

Fig. 2. Continuous culture of recombinant *E. coli* 11303:pLO1297. LB, containing glucose (25 g/L) and antibiotics (40 mg/L ampicillin and 10 mg/L tetracycline), was fed to the chemostat (350 mL constant working volume) at a fixed rate of 49 mL/h (D = 0.14/h). The pH was controlled at 6.3, and the temperature was constant at 30°C. The first arrow (-Ap/Tc) indicates the switch to a medium feed without antibiotics. The dashed line shows the theoretical dilution profile with respect to the concentration of ampicillin(*y*-axis as mg/L) in the chemostat. The second arrow (+Ap/Tc) indicates addition of antibiotics directly to the chemostat and the switch back to the medium containing antibiotics. Symbols: □, OD at 550 nm; +, glucose (g/L); ■, ethanol (g/L); ●, lactate (g/L); ▲, acetate (g/L).

the medium (Fig. 2). The trend to elevated lactic acid with concomitant fall in ethanol following the switch to the nonselective medium indicates a high level of functional instability under these conditions. The ethanol concentration falls to zero within 40 h of the switch to the nonselective medium, and this corresponds to only about eight generations. In fact, the apparent displacement of ethanol from the chemostat precedes that of the antibiotic (Fig. 2).

In this study, we have viewed the chemostat culture from a holistic perspective, and although functional instability almost certainly is a function of the genetic instability of this genetically engineered biocatalyst, we have not attempted to discern the mechanism(s) responsible for such instability. Since it is known that under similar conditions of pH-controlled batch culture, the wild-type host culture (ATCC 11303) exhibits a faster growth rate than recombinant 11303:pLOI297 *(31)*, it is reasonable to assume that the particular environmental conditions of the glucose-limited chemostat (D = 0.14/ h) offer a growth advantage to the revertant and that it rapidly displaces the recombinant culture *(39,40)*. The stoichiometric distribution of fermentation end products in the absence of antibiotics (Fig. 2) is consistent with what would be expected from glucose fermentation by the host culture *(31)*. In this investigation, plating of cultures on selective and nonselective agar was done only for the purpose of qualitative phenotypic screening. Plating of the 60-h chemostat culture revealed that a high proportion of the population

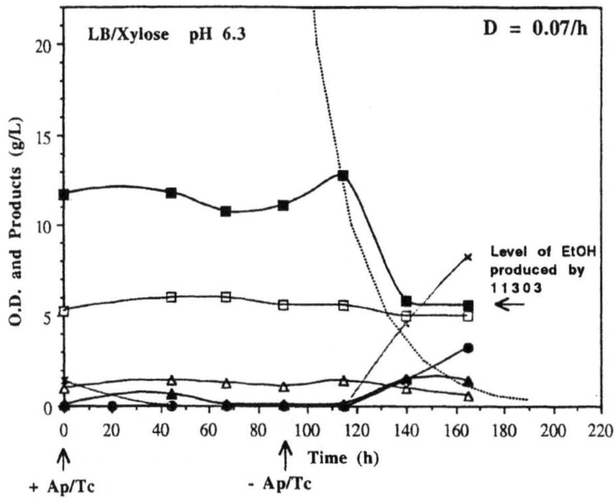

Fig. 3. Continuous culture of recombinant *E. coli* 11303:pLOI297. Conditions were as described in Fig. 2, except that the medium was LB with xylose (25 g/L) and antibiotics (40 mg/L ampicillin and 10 mg/L tetracycline). The dilution rate (D) was 0.07/h. The first arrow (-Ap/Tc) indicates the switch to a medium feed without antibiotics. The dashed line shows the theoretical dilution profile with respect to the concentration of ampicillin (*y*-axis as mg/L) in the chemostat. Symbols: □, OD at 550 nm; X, xylose (g/L); ■, ethanol (g/L); ●, lactate (g/L); ▲, acetate (g/L); △, succinate (g/L).

was not antibiotic-resistant (results not shown). That the culture was not completely devoid of antibiotic-resistant cells was indicated by the lack of "washout" and the partial recovery of ethanol that was observed following reintroduction of antibiotics into the system (Fig. 2).

The first *pet* recombinants of *E. coli* were based on high-copy-number plasmid constructs, and it was reported that the pLOI297 plasmid was "quite stable" in *E. coli* B (ATCC 11303) where plasmid stability was expressed as the percentage of cells retaining antibiotic markers after "25 generations" of growth in LB medium with glucose in the absence of antibiotic selection *(9)*. For recombinant 11303:pLOI297 under these specific assay conditions, it was reported that 98% of the cells retained antibiotic resistance, but the actual time period was not specified nor was the stability determined in terms of retention of culture ethanologenicity *(9)*.

The virtual disappearance of ethanol from the glucose-limited chemostat after only 3 d of operation does not suggest a stable biocatalyst and challenges the earlier claims *(9)* regarding the stability of this construct.

Figure 3 represents an experiment of similar design to that shown in Fig. 2, except that the LB medium contained xylose (about 25 g/L) and the dilution rate was 0.07/h. Over the initial 90-h period, the chemostat was operated under antibiotic selection, and the ethanol concentration was relatively constant at about 12 g/L corresponding to a xylose-to-ethanol conversion efficiency of 95% of the theoretical maximum (Fig. 3). However, when the antibiotics were absent from the feed medium, the decay in ethanol concentration paralleled that of the concentration of antibiotics in the chemostat (Fig. 3). The ethanol concentration decreased to a plateau value of about 6 g/L (Fig. 3), corresponding to an ethanol yield of 0.24 g/g

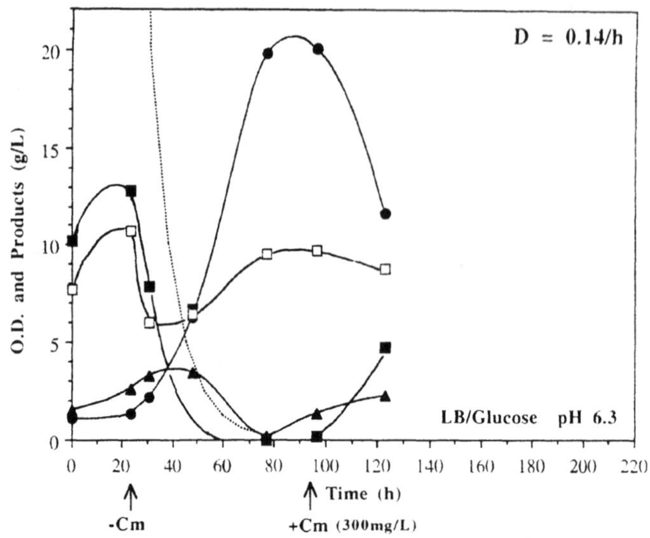

Fig. 4. Continuous culture of recombinant *E. coli* KO11. Conditions were as described in Fig. 2, except that the LB medium contained glucose (25 g/L) and Cm (40 mg/L). The dilution rate (D) was 0.14/h. The arrow at 24 h (-Cm) indicates the switch to a medium feed without Cm. The second arrow at 96 h (+Cm) indicates addition of Cm (final concentration of 300 mg/L) directly to the chemostat and the switch to LB medium containing 300 mg/L Cm. The dashed line shows the theoretical dilution profile with respect to the concentration of Cm (*y*-axis as mg/L) in the chemostat. Symbols: □, OD at 550 nm; ■, ethanol (g/L); ●, lactate (g/L); ▲, acetate (g/L).

xylose that is characteristic of xylose fermentation by the host culture *(8,31)*. Operationally, the xylose-limited chemostat under selective pressure appears to be stable over the 4-d period of assessment. However, this may not reflect genetic stability of the plasmid recombinant for the following reasons. It is known that *E. coli* does not grow well with xylose *(24,31)*, and furthermore, that the recombinant 11303:pLOI297 grows in LB-xylose at a rate that is considerably faster than the host culture *(31)*.Therefore, the *pet*-expressing culture would have an advantage in the chemostat whereby the slower-growing revertant host culture would be washed out *(39,40)*. In a separate experiment (results not shown), we determined that, at this dilution rate with xylose, steady-state growth was not possible with the host culture ATCC 11303. It follows that the apparent stability of the xylose-limited chemostat under selective pressure derives from its capacity for self-regulation with respect to maintenance of the ethanologenic strain.

Recombinant *E. coli* KO11

A similar experimental strategy was employed to assess the functional stability of the chromosomal integrant, strain KO11. Figure 4 represents the continuous culture of the recombinant in LB with glucose (about 25 g/L). Initially the medium contained chloramphenicol (Cm) at 40 mg/L; however, at 21 h, the feed was switched to a nonselective medium in which Cm was omitted. The trajectories for ethanol and lactic acid that are plotted in Fig. 4 are remarkably similar to those of Fig. 2. For purposes of comparison, all the plots have been made with both *x*-axis

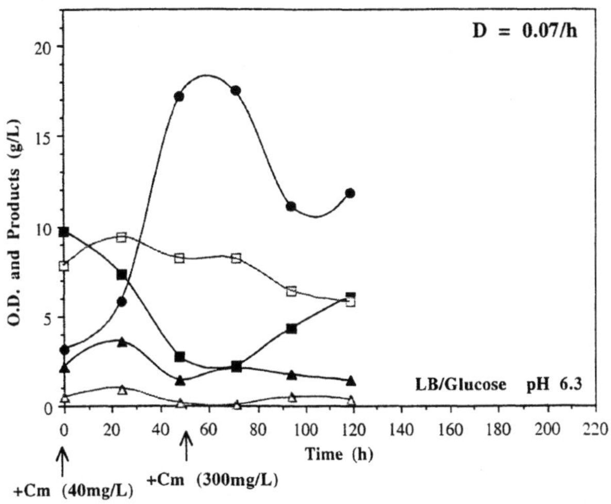

Fig. 5. Continuous culture of recombinant *E. coli* KO11. Conditions were as described in Fig. 4, except that the dilution rate was 0.07/h. Symbols: □, OD at 550 nm; ■, ethanol (g/L); ●, lactate (g/L); ▲, acetate (g/L); △, succinate (g/L).

and *y*-axis of the same scale and magnitude. The initial level of lactic acid is consistent with the known fermentation pattern for strain KO11 *(32)* and was not construed as indicative of instability during the initial period while the chemostat was under selective pressure with Cm in the medium (Fig. 4). The decrease in ethanol parallels the decrease in the concentration of Cm and seems to commence immediately on removal of the Cm from the medium being supplied to the chemostat. However, in comparing Figs. 2 and 4, it can be seen that the rate at which the ethanol decreases, and the lactic acid increases, is not as fast as with the plasmid-bearing recombinant. As was the case with the recombinant 11303:pLOI297, recovery of the system with respect to ethanol production is possible through the reintroduction of Cm (Fig. 4), which demonstrates that KO11 had not been completely eradicated at the time the antibiotic was added (96 h).

Figure 5 shows the effect of dilution rate on the stability of the glucose-limited chemostat with recombinant KO11. In the experiment shown in Fig. 5, the chemostat was operated at a dilution rate of only 0.07/h (compared to Fig. 4 where D = 0.14/h), and the decrease in ethanol along with the concomitant rise in lactic acid indicates functional instability even in the presence of 40 mg Cm/L. The ethanol yield fell to a value of 0.12 g/g by 48 h after flow was started, and this time period corresponds to <5 generations. Recovery was demonstrated by the introduction of 300 mg/L Cm (Fig. 5), suggesting that the system could be operated in a more stable fashion at this higher concentration of Cm. Figure 6 shows that this was not the case since the ethanol concentration began to decline after only 2 d of operation with 300 mg/L Cm in the medium. The higher level of antibiotic retarded the loss of ethanologenicity, but did not prevent it (Fig. 6).

Figure 7 shows the results of a continuous culture of recombinant KO11 in LB medium with xylose (about 25 g/L). The chemostat was operated at a dilution rate of 0.07/h, and for the first 2 d there was Cm (40 mg/L) in the feed medium during which time the ethanol decreased from 13 to 9 g/L (Fig. 7). There was very little

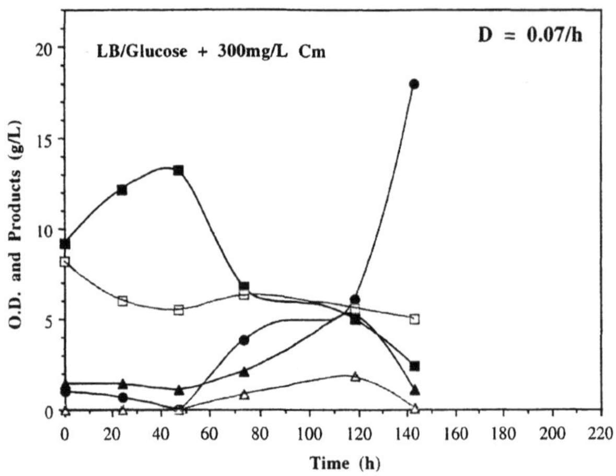

Fig. 6. Continuous culture of recombinant *E. coli* KO11. Conditions were as described in Fig. 5, except that the medium contained 300 mg/L chloramphenicol. Symbols are the same as in Fig. 5.

change in the ethanol concentration following the switch to medium without Cm, although there was a slow progressive increase in the concentration of both acetic acid and succinic acid (Fig. 7). Since strain KO11 is restricted from producing succinate because of a mutation in fumarate reductase (*frd*), the increase in succinate signals the presence of another population different from strain KO11. The pattern seen with xylose-limited KO11 in Fig. 7 is similar to that of the xylose-limited chemostat with plasmid recombinant (Fig. 3) in that the ethanol concentration decreases to a plateau that is defined by the ethanol yield characteristic of xylose fermentation by the host ATCC 11303.

Ethanologenic recombinant strain KO11 is characterized by its resistance to high levels of Cm and was derived from the *pet* integrant, strain KO3, which was selected for resistance to Cm at 40 mg/L *(29)*. Whereas strain KO11 exhibits a characteristic very high degree of ethanol selectivity, strain KO3 produces a stoichiometric pattern of the fermentation end products that is very similar to the host culture *(32)*. The distribution of metabolic end products resulting from xylose fermentation by strain KO3 has not been reported. Strain KO3 is distinguishable from the host culture by virtue of its resistance to Cm *(32)*. The high frequency of mutation for hyperresistance to Cm (10^{-5}) that was observed in selecting strain KO11 from a culture of KO3 *(32)* suggests that the frequency of reversion could be of the same order. Hence, it is conceivable that the integrant KO3 strain could compete with strain KO11 in the chemostat that is being fed medium with 40 mg/L Cm.

These observations with recombinant KO11 challenge the claims of its designers regarding its "stability" *(32)* and call into question the generality of the presumption of stability of ethanologenic chromosomal integrants of other bacteria in which the genetic engineering was modeled after the work with *E. coli*. For example, strain P2 is a recombinant *Klebsiella oxytoca* M5A1 that expresses chromosomally integrated *Zymomonas* genes for ethanol production and is being used as a host strain for the expression of heterologous enzymes for hydrolyzing cellulose and xylan *(41)*. However, a functional stability assessment for these constructs is lacking.

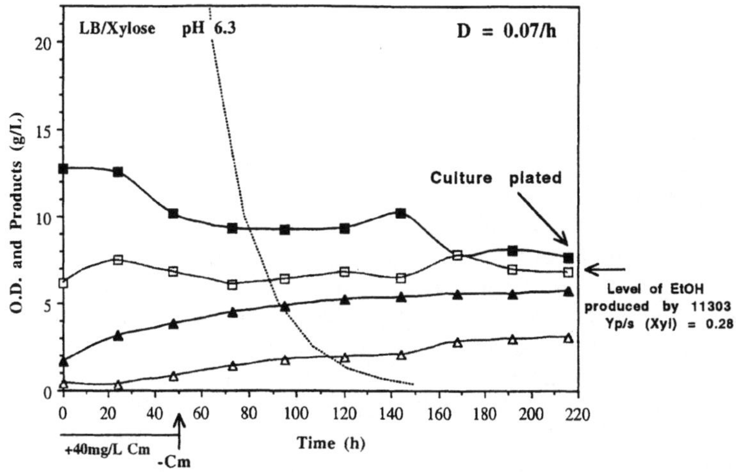

Fig. 7. Continuous culture of recombinant *E. coli* KO11. Conditions were as described in Fig. 4, except that the LB medium contained xylose (25 g/L) and the dilution rate was 0.07/h. The arrow at 50 h (-Cm) indicates the switch to a medium without chloramphenicol. The dashed line shows the theoretical dilution profile with respect to the concentration of Cm (*y*-axis as mg/L) in the chemostat. The arrow (right side) indicates the average level of ethanol produced under these conditions by the host culture *E. coli* B (ATCC 11303). Symbols are the same as in Fig. 5.

Although the stability of strain KO11 *per se* was not reported, the hyper-Cm-resistant integrant from which it was derived, namely strain KO4, was said to be stable after being serially transferred in glucose-LB without antibiotics for more than 68 generations, although the value of "generation time" was not specified *(32)*. Supposedly, strain KO11 differs from KO4 only by virtue of a mutation in the gene coding for *frd*, which is responsible for succinate production. Ohta et al. *(32)* described stability in terms of resistance to high levels of Cm (600 µg/mL) and phenotypic characteristics relating colony morphology on agar plates. According to these criteria of stability, 100% of the cells retained these traits after 68 generations *(32)*. The literature is remarkably silent with respect to the subject of the stability of ethanologenic recombinants, and claimed stability appears to have been tacitly assumed by several investigators who have independently endorsed strain KO11 in terms of its superiority as a xylose fermenter and suggested that it be considered as a candidate biocatalyst for hemicellulose utilization in large-scale fuel ethanol production facilities *(2,33,34)*.

The Economic Impact of a Requirement for Antibiotics on Cost of Fuel Ethanol

von Sivers et al. *(34)* recently published the results of their cost analysis of fuel ethanol production from lignocellulosic biomass based on using recombinant *E. coli* KO11. Their technical and economic analysis focused solely on the fermentation of prehydrolysate produced by steam-treated willow and estimated the total cost of producing fuel-grade E95 ethanol from xylose-rich prehydrolysate batch fermentations to be 48¢/L (equivalent to $1.82/US gal). The cycle time for each batch fermentation was 55 h, and the plant produced 6.9 million liters of E95 ethanol/yr in 141 fermentation cycles. Pertinent to the results of our study with KO11 is the fact that it was assumed in their base case cost estimate that recombinant *E. coli* KO11 was sufficiently stable that only seven inoculations from pure culture would be required/yr. It was estimated that if the frequency with respect to the requirement for reinoculation changed from every 20 cycles to every 5 cycles, there would be an additional cost of 2¢/L of E95 raising the cost from 48 to 50¢/L. However, there was no mention regarding cost of antibiotics. Our results of the present study in combination with our preliminary findings with serial transfer fermentations in the absence of antibiotic suggest not only a more frequent requirement for inoculation, but also a possible involvement of antibiotics in all media. If Cm were to be used at the level of 40 mg/L in all media in the operation described by von Sivers et al. *(34)*, we estimate that the added cost of producing E95 ethanol would be 7.63¢/L (29¢/gal). This estimate is based on a bulk order cost of Cm at $55/kg. von Sivers et al. *(34)* used a pitch rate of 20% (to give an initial cell concentration of 1g dry wt/L), and even if there were a requirement to inoculate only every three cycles *(35)* and if Cm at 40 g/m^3 were used exclusively for the preparation of the inocula, there would still be an added cost of 0.45¢/gal E95. The stability of strain KO11 under the aerobic culture conditions proposed by these investigators remains to be assessed. Since strain KO11 appears to be maintained better at the higher level of 300 g/m^3 Cm, the economic impact would be increased proportionally by a factor of 7.5 times. The estimated economic impact regarding the potential requirement of stabilizing antibiotics is significant when viewed in terms of the cost estimates associated with nutritional supplements. von Sivers et al. *(34)* estimated the cost of using ammonia, phosphoric acid, and magnesium oxide as the sole nutritional supplements to be 10.6¢/gal E95. The results of our investigation being reported at this symposium indicate that corn steep liquor (CSL), at about 8 g/L, can supply all the nutritional requirements of recombinant *E. coli (31)*.The economic impact of CSL supplementation was estimated at 4.2¢/gal E95 ethanol. In the context of our findings with recombinant *E. coli* KO11, it is interesting to note that von Sivers et al. *(34)* made the following concluding remarks in their paper on technical and economic analysis:

> More experiments on both the laboratory and pilot scale are necessary to determnine how cell recirculation will affect the genetic stability of *E. coli* KO11. It is also necessary to establish whether *E. coli* KO11 is able to ferment equally well in a mineral salt medium as in the complex medium used in the laboratory experiments.

ACKNOWLEDGMENTS

This research was supported by The University of Toronto. We are grateful to Lonnie Ingram for the gift of the recombinant *E. coli* culture.

REFERENCES

1. Wyman, C. E. and Hinman, N. D. (1990), *Appl. Biochem. Biotechnol.* **24/25,** 735–753.
2. McMillan, J. D. (1994), in *Bioconversion for Fuels*, Himmel, M. E., Baker, J. O., and Overend, R. P., eds., ACS Symposium Series 566, American Chemical Society, Washington, DC, pp. 411–437.

3. Timell, T. E. (1967), *Wood Science Technol.* **1,** 45–70.
4. Hinman, N. D., Wright, J. D., Hoagland, W., and Wyman, C. E. (1989), *Appl. Biochem. Biotechnol.* **20/21,** 391–401.
5. McMillan, J. D. (1993), Xylose fermentation to ethanol: a review; NREL TP-421-4944; National Renewable Energy Laboratory, Golden, CO.
6. Ingram, L. O., Conway, T., and Alterthum, F. (1991), United States Patent 5,000,000.
7. Ingram, L. O., Alterthum, F., Ohta, K., and Beall, D. S. (1990), in *Developments in Industrial Microbiology*, vol. 31, Elsevier Science, New York, pp. 21–30.
8. Ingram, L. O. (1991), in *Energy from Biomass and Wastes XIV*, Klass, D. L., ed., Institute of Gas Technology, Chicago, IL, pp. 1105–1126.
9. Alterthum, F. and Ingram, L. O. (1989), *Appl. Environ. Microbiol.* **55,** 1543–1948.
10. Ingram, L. O. and Conway, T. (1988), *Appl. Environ. Microbiol.* **54,** 397–404.
11. Brau, B. and Sahm, H. (1986), *Arch. Microbiol.* **144,** 296–301.
12. Neale, A. D., Scopes, R. K., Wettenhall, E. H., and Hoogenraad, N. J. (1987), *J. Bacteriol.* **169,** 1024–1028.
13. Beall, D. S., Ohta, K., and Ingram, L. O. (1991), *Biotechnol. Bioeng.* **38,** 296–303.
14. Lawford, H. G. and Rousseau, J. D. (1991), *Appl. Biochem. Biotechnol.* **28/29,** 221–236.
15. Lawford, H. G. and Rousseau, J. D. (1992), *Appl. Biochem. Biotechnol.* **34/35,** 185–204.
16. Lawford, H. G. and Rousseau, J. D. (1993), *Appl. Biochem. Biotechnol.* **39/40,** 301–322.
17. Lawford, H. G. and Rousseau, J. D. (1993), *Biotechnol. Lett.* **15,** 615–620.
18. Lawford, H.G. and Rousseau, J. D. (1991), in *Energy from Biomass & Wastes XV*, Klass, D. L., ed., Institute of Gas Technology, Chicago, IL, pp. 583–622.
19. Lawford, H. G. and Rousseau, J. D. (1991), *Biotechnot. Lett.* **13,** 191–196.
20. Lawford, H. G. and Rousseau, J. D. (1993), in *Energy from Biomass and Wastes XVI*, Klass, D. L., ed., Institute of Gas Technology, Chicago, IL, pp. 559–597.
21. Lawford, H. G. and Rousseau, J. D. (1993), *Biotechnol. Lett.* **15,** 505–510.
22. Lawford, H. G. and Rousseau, J. D. (1993), *Appl. Biochem. Biotechnol.* **39/40,** 667–685.
23. Lawford, H. G. and Rousseau, J. D. (1992), *Biotechnol. Lett.* **14,** 421–426.
24. Beall, D. S., Ohta, K., and Ingram, L. O. (1991), *Biotechnol. Bioeng.* **38,** 296–303.
25. Neale, A. D., Scopes, R. K., and Kelly, J. M. (1988), *Appl. Microbiol. Biotechnol.* **29,** 162–167.
26. Sanbrook, J., Fritsch, E. F., and Maniatis, T. (1989), *Molecular Cloning: A Laboratory Manual*, 2nd ed., Cold Spring Harbor Laboratory, Cold Spring Harbor, NY.
27. Beall, D. S., Ohta, K., and Ingram, L. O. (1991), *Biotech. Bioeng.* **38,** 296–303.
28. Tolan, J. S. and Finn, R. K. (1987), *Appl. Environ. Microbiol.* **53,** 2033–2038.
29. Ohta, K., Beall, D. S., Mejia, J. P., Shanmugam, K. T., and Ingram, L. O. (1991), *Appl. Environ. Microbiol.* **57,** 893–900.
30. DaSilva, N. A. and Bailey, J. E. (1986), *Biotechnol. Bioeng.* **28,** 741–746.
31. Lawford, H. G. and Rousseau, J. D. (1996), *Appl. Biochem. Biotechnol.* **57/58,** 307–326.
32. Ohta, K., Beall, D. S., Mejia, J. P., Shanmugam, K. T., and Ingram, L. O. (1991), *Appl. Environ. Microbiol.* **57,** 893–900.
33. Hahn-Hägerdal, B., Jeppsson, H., Olsson, L., and Mohagheghi, A. (1994), *Appl. Microbiol. Biotechnol.* **41,** 62–72.
34. von Sivers, M., Zacchi, G., Olsson, L., and Hahn-Hägerdal, B. (1994), *Biotechnol. Prog.* **10,** 555–560.
35. Lawford, H. G. and Rousseau, J. D. (1995), *Biotechnol. Lett.* **17,** 751–756.
36. Luria, S. E. and Delbruck, M. (1943), *Genetics* **28,** 491–511.
37. Lawford, H. G. and Rousseau, J. D. (1995), *Appl. Biochem. Biotechnol.* **51/52,** 179–195.
38. Ingram, L. O., Conway, T., Clark, D. P., Sewell, G. W., and Preston, J. F. (1987), *Appl. Environ. Microbiol.* **53,** 2420–2425.
39. Pirt, J. S. (1975), in *Principles of Microbe and Cell Cultivation*, Wiley, New York, pp. 200–202.
40. Bailey, J. E. and Ollis, D. F. (1977), in *Biochemical Engineering Fundamentals*, McGraw-Hill, New York, pp. 650–653.
41. Wood, B. E. and Ingram, L. O. (1992), *Appl. Environ. Microbiol.* **58,** 2103–2110.

0273-2289/96/5758—0307$10.00

Studies on Nutrient Requirements and Cost-Effective Supplements for Ethanol Production by Recombinant *E. coli*

HUGH G. LAWFORD* AND JOYCE D. ROUSSEAU

Department of Biochemistry, University of Toronto, Toronto, Ontario, Canada M5S 1A8

ABSTRACT

This article describes a systematic study of the nutritional requirements of a patented recombinant ethanologenic *Escherichia coli* (11303:pLOI297) and provides cost-effective formulations that are compatible with the production of fuel ethanol in fermentations of lignocellulosic prehydrolysate characterized by high xylose conversion efficiency. A complex and nutrient-rich laboratory medium, Luria broth (LB), provided the benchmark with respect to fermentation performance standard. Xylose fermentation performance was assessed in terms of the target values for operational process parameters established by the US National Renewable Energy Laboratory (NREL)—final ethanol concentration (25 g/L), xylose-to-ethanol conversion efficiency (90%), and volumetric productivity (0.52 g/L·h). Biomass prehydrolysates that are rich in xylose also contain acetic acid, and in anticipation of a need to reduce acetic acid toxicity, the fermentors were operated with a pH control set-point of 7.0 Growth and fermentation in the minimal defined salts (DS) medium was only about 15% compared to the reference medium. Amendment of the minimal medium containing 6 wt% xylose with both vitamins and amino acids resulted in improved growth, but the volume productivity (0.59 g/L·h) was still only about 54% of that with LB (1.1 g/L·h). Formulations directed at cost reduction through the use of less expensive commercial complex nutritional supplements were within 90% of the NREL process target with respect to yield and provided a productivity at about 80% of the LB medium, but were not economical. Corn steep liquor (CSL) at about 7–8 g/L was shown to be a complete source of nutritional requirements and supported a fermentation performance approaching that of LB. At a cost of CSL of $50/t (dry wt), the economic impact of using this amount CSL as the sole nutritional supplement in a cellulosic ethanol plant was estimated to be about 4¢/gal of ethanol.

Index Entries: Xylose; recombinant *E. coli*; fuel ethanol; corn steep liquor; nutrient requirements.

Abbreviations: LB, Luria broth; sLB, LB supplemented with 0.5 m*M* Mg + 17 m*M* PO_4; mLB, 2.5 g/L YE + 2.5 g/L Tryptone; DS, Defined salts medium; YE, Difco yeast extract (g/L); VYE, Veeprex B430 yeast extract (g/L); Tryp, Bacto-

*Author to whom all correspondence and reprint requests should be addressed.

tryptone (g/L); Pan, pancase S (g/L); TW, Toronto tap water; AP, 1.4 g/L ammonium monohydrogenphosphate; CSL, corn steep liquor (Nacan Products) (mL/L); Vit, Vitamin stock (added 5 mL/L) for 2X Vit—added 10 mL/L; Glut, 0.25 g/L glutamic acid (g/L); AA, amino acid cocktail = 50 mg/L each of aspartic acid, tyrosine, tryptophan, phenylalanine, and histidine; AA^+, 50 mg/L each of arginine, asparagine, aspartic acid, cysteine, histidine, isoleucine, leucine, lysine, methionine, phenylalanine, proline, serine, threonine, tryptophan, tyrosine, and valine.

INTRODUCTION

The economics of the large-scale production of fuel ethanol are very sensitive to the cost of fermentation feedstock, and consequently, less expensive lignocellulosic biomass and waste materials are being targeted as alternatives to cereal or sugar crops *(1,2)*. However, lignocellulosic materials remain recalcitrant to bioconversion because the yeast cultures presently employed in the starch-based fermentation ethanol industry are unable to utilize the five-carbon pentose sugars derived from the hydrolysis of hemicellulose *(3)*. Attention has been focused on D-xylose, because this pentose sugar represents a major constituent of the hemicellulose in the particular biomass species (energy crops) and waste resources presently being considered as potential fermentation feedstocks *(3,4)*.

Current processing technologies employing dilute sulfuric acid are capable of yielding a "prehydrolysis" stream containing a maximum xylose concentration (without concentrating) of about 60 g/L *(5)*. Current research efforts are directed to bioprocesses capable of producing fuel ethanol from hemicellulose hydrolysate at a pentose conversion efficiency of >70%, thereby surpassing that of current bioconversion technologies that propose to use pentose-fermenting yeasts *(6,7)* or thermophilic bacteria, which generally suffer from relatively low tolerance to ethanol *(8)*.

Although the bacterium *Escherichia coli* is capable of utilizing all biomass-derived hexose and pentose sugars, the product of anaerobic metabolism is a mixture of organic acids with ethanol also sometimes appearing as a fermentation end product *(9–14)*. A recombinant that expresses *Zymomonas mobilis* genes for pyruvate decarboxylase and alcohol dehydrogenase II (termed the *pet* operon) *(15)* on a multicopy plasmid designated pLOI297 *(16,17)* has been shown to convert both hexose and pentose sugars to ethanol at near-maximum theoretical efficiency *(14,18–23)*. The physiological characteristics of this plasmid-bearing recombinant (*E. coli* B ATCC 11303:pLOI297) has been extensively investigated *(17–23)*. However, the majority of this research has been conducted using a complex and nutrient-rich medium (LB) composed of laboratory-grade chemicals *(17–20)*. The use of expensive nutrient supplements has a serious negative impact on the economics of ethanol production. The extent to which these complex nutrients are essential for high-performance fermentation is not known.

The objective of this study was twofold. First, to establish the basic nutritional requirements of recombinant *E. coli* (11303:pLOI297) using defined media with LB as the reference for comparative fermentation performance in terms of both ethanol yield and productivity in pH-controlled batch fermentations with xylose as substrate, and, second, by employing inexpensive complex nutritional supplements, to formulate a cost-effective medium compatible with high-performance fermentation in a full-scale commercial fuel ethanol production plant.

Criteria for Assessing Fermentation Performance of the Biocatalyst

Cost reduction is the primary driving force for process improvement and alternative feedstock utilization. Techno-economic sensitivity analyses related to fuel ethanol production from cellulosic biomass have revealed the ordered importance of certain bioconversion process parameters whereby the price of ethanol was most sensitive to yield *(5)*. The efficient utilization of the hemicellulose component of lignocellulosic feedstocks offers an opportunity to reduce significantly the price of ethanol. In terms of the fermentation portion of the overall production process, the ethanol yield is a function of the efficiency with which the biocatalyst converts sugar to ethanol. The price of ethanol was also found to be sensitive to the final concentration of ethanol in the bioreactor, but was less sensitive to the rate of ethanol production *(5)*. Such economic analyses establish the criteria for assessing the performance of the process biocatalyst as ethanol yield, ethanol concentration and productivity. In the case of xylose utilization, the National Renewable Energy Laboratory (NREL) (Golden, CO) has established target values for each of these fermentation process parameters *(4)*. The target for ethanol yield (based solely on xylose) is 0.46 g/g, which is equivalent to a xylose-to-ethanol conversion efficiency of 90% (relative to the theoretical maximum conversion efficiency). The objective in terms of ethanol concentration is 25 g/L and for volumetric productivity it is 0.52 g ethanol/L·h *(4)*.

Elemental Composition of Cell Mass as a Nutritional Indicator

Microorganisms exhibit growth optima with respect to both physical and chemical environmental factors. A prerequisite to growth is that the aqueous environment ("medium") supply the elements of carbon, nitrogen, phosphorous, and sulfur that, in addition to hydrogen and oxygen, are the major components of all biomolecules *(24)*. In addition to sodium, potassium, magnesium, calcium, and iron, certain minerals are required in relatively much smaller amounts and are therefore referred to as "trace elements" *(24)*. These trace elements are primarily the metals such as manganese, copper, cobalt, molybdenum, zinc, and so forth, that act as enzyme cofactors. In addition, there are can be certain other essential elements that, apart from the known "vitamins," are referred to collectively as "growth factors" *(24)*. In general, the elemental composition of the organism provides insights into nutritional requirements. However, it is known that cellular composition can be influenced by the chemical nature of the growth environment. For example, the amount of NaCl in the medium has been shown to influence the sodium level in the cell mass *(25)*. Table 1 provides examples of the elemental composition both with specific reference to *E. coli* and bacteria in general.

In this study, we have focused attention principally on three major nutritional elements, namely nitrogen (N), phosphorus (P), and magnesium (Mg). Based on several literature sources relating to the elemental composition of *E. coli* cell mass *(24–27)*, we have assumed that, on average, the elemental composition with respect to N, P, Mg was constant at about 15, 1.5, and 0.15 wt%, respectively *(24–27)*. Furthermore, it has been assumed that the cellular composition reflects the mass ratio requirements of these various nutritional elements in the culture medium *(25)*. Consequently, it can be inferred that the minimal concentrations (g/L) of N, P, and Mg required to support a yield of 1 g dry wt cells/L, are 0.15, 0.015, and 0.0015, respectively (*see* Table 2).

Table 1
Elemental Composition of Bacteria (%dry wt basis)

Element	*E. coli*,[a]	Bacteria,[b] gram-negative	Bacteria,[c] in general
Carbon (C)	50		ca. 50
Nitrogen (N)	15	12	10
Phosphorus (P)	3.2	1.5	2.0–3.0
Sulfur (S)	1.1	0.33	0.2–1.0
Sodium (Na)	1.3		0.5–1.0
Potassium (K)	1.5	0.75	1.0–4.5
Magnesium (Mg)	0.5	0.15	0.1–0.5
Calcium (Ca)	1.0	0.10	0.01–1.1
Iron (Fe)	0.24	0.015	0.02–0.20
Manganese (Mn)	Trace	0.005	0.001–0.01
Copper (Cu)	Trace	0.001	0.01–0.02
Others	Trace	Trace	Trace

[a]Demain and Solomon *(25)*.
[b]Pirt *(24)*.
[c]Aiba et al. *(11)*.

Selection of Conditions for Assessing Fermentation Performance

E. coli is less ethanol tolerant than yeast or *Zymomonas (28)*, and for this reason, a concentration of 40 g/L D-xylose was selected for routine fermentation performance assessment with different media formulations because the anticipated maximum ethanol concentration would fall below the inhibitory threshold for *E. coli*, thereby avoiding unnecessary complications in interpretation of growth and fermentation kinetics.

The *pet* operon (plasmid pLOI297) contains antibiotic resistance markers for both ampicillin and tetracycline *(17)*. Since a recognized contribution to ethanol yield reduction in fermentations employing plasmid-bearing recombinant *E. coli* relates to the potential for genetic instability *(29,30)*, all media formulations used in this investigation included antibiotics.

Acetic acid toxicity is an obstacle to the fermentation of biomass hydrolysates *(31)*, but since the inhibitory effect on recombinant *E. coli* is minimized at elevated pH *(32)*, a pH value of 7.0 was selected as the control set-point.

The "Reference" Medium

The nutritionally-rich, complex LB medium has been commonly used in studies with recombinant *E. coli (17–20)*. However, LB medium is known not to support carbon-limited growth at sugar concentrations greater than about 30 g/L *(21,33)*. Carbon-limited growth at 40 g/L can be achieved with LB supplemented with mg and phosphate, and for this reason, this amended LB (designated as "sLB") was selected as the reference medium in this study and provided a performance "benchmark" for purposes of comparing fermentation performance of the recombinant in variously formulated media. We have previously described the formulation of a defined minimal medium that is compatible with growth of recombinant *E. coli* using pure xylose as the sole carbon source *(33)* and this medium

Table 2
Composition of Different Fermentation Media
with Respect to the Potentially Limiting Elements–N, P, mg

Designation of medium, formulation/composition	Elements, g/L N	P	Mg
Min. amount required (g/L) per g (dry wt.) biomass[a]	0.15[b]	0.015[c]	0.0015[c]
LB	1.76	0.09	0.006
sLB	1.76	0.62	0.018
mLB	0.56	0.03	0.002
DS	0.59	0.62	0.020
DS + Vit + 0.25 g/L Glut	0.61	0.62	0.020
DS + Vit + 0.25 g/L Glut + AA[+]	0.76	0.62	0.020
TW + AP + 5.0 g/L VYE	0.72	0.35	0.012
TW + 37 mL CSL (15.7 g)	1.04		0.105

The N content of media is based on the determination of "total N."
[a]Based on elemental composition of *E. coli* dry biomass.
[b]Demain and Solomon *(25)* p.108.
[c]Pirt *(24)* pp. 123,124.

(designated as DS) provides a base case for modifications designed not only to examine basic nutritional requirements, but also to improve growth rate and ethanol productivity.

MATERIALS AND METHODS

Organism

The patented recombinant *E. coli* B (ATCC 11303 carrying the *pet* plasmid pLOI297) *(16,17)* was a gift from L. Ingram (University of Florida, Gainesville, FL).

Long-Term Storage/Maintenance of Organism

Plasmid-bearing cultures, grown from single-colony isolates on selective agar medium, were stored at –10°C in LB medium supplemented with antifreeze (glycerol at 20 mL/dL) and sodium citrate (1.5 g/dL). The phenotypic characteristics of the recombinant culture were related to antibiotic (ampicillin and tetracycline) resistance, colony size, and morphology. The recombinant culture was recognized on selective media, containing the antibiotics ampicillin (40 mg/L) and tetracycline (10 mg/L), by the formation of distinctive large, yellowish, opaque colonies *(21,34)*.

Fermentation Equipment

pH-stat STR batch fermentations were conducted in a volume of 1500 mL in MultiGen (model F2000) (New Brunswick Scientific Co., Edison, NJ) stirred-tank bioreactors fitted with agitation, pH, and temperature control (30°C). The central agitator shaft was fitted with three multiblade turbine impellers, and the rotation speed was constant at about 100 rpm. The pH was monitored using a sterilizable combination pH electrode (Ingold) and was controlled at 7.0 by the addition of 4*N*

KOH. On average, about 75 mL of 4*N* KOH were added during the fermentation of 4 wt% xylose in STR pH-stat batch fermentations.

Methods of Preculture and Inoculation Procedures

A 1-mL aliquot of a glycerol/citrate preserved culture was removed from cold storage (freezer) and transferred to about 100 mL of complex medium (LB), containing about 2% xylose and supplemented with antibiotics (40 mg/L ampicillin and 10 mg/L tetracycline), in 125-mL screw-cap flasks and grown overnight at 30°C and low agitation in a water bath shaker. In the case of experiments employing defined media, the inoculum was prepared by transferring an aliquot of a glycerol/citrate preserved culture to DS medium containing about 2 wt% xylose and antibiotics (40 mg/L ampicillin and 10 mg/L tetracycline).

STR batch fermentations were inoculated by transferring approx 100 mL of the overnight flask culture directly to 1400 mL of medium in the bioreactor. The initial cell density was in the OD_{550} range of 0.1–0.2, corresponding to 30–50 mg dry wt cells/L.

For fermentations involving defined (minimal) media, an alternative preculture and inoculation procedure was developed ("plate direct inoculation") during the course of this investigation, which permitted preculture using LB, but avoided the typical centrifugal harvesting/washing procedure, which was found to be detrimental to culture vitality. This procedure also minimized the potential for transfer of nutrients during inoculation. Preculture was accomplished using LB (Ap/Tc) agar plates. A glycerol/citrate culture was used to inoculate several plates using a standard "spreading" technique. Following overnight incubation at 30°C, the culture lawns were scraped off the plates using a sterile inoculation loop, and the plates were rinsed with DS medium. The cell suspensions from several plates were pooled, agitated to achieve homogeneity, and used to inoculate the STR.

Fermentation Media

For comparative purposes, the nutrient-rich, complex culture medium described by Luria and Delbruck *(34)* was used as the benchmark standard. This medium is commonly referred to as "Luria broth" (LB) and consists of 5 g Difco yeast extract (YE), 10 g Difco Tryptone, and 5 g NaCl/L of distilled water *(34)*. Two different modifications of this LB medium were used in this investigation. sLB was fortified with magnesium (0.5 m*M* added as $MgSO_4$) and phosphate (17 m*M*). Modified LB (mLB) contained reduced amounts of YE (2.5 g/L) and Tryptone (2.5 g/L) and no NaCl.

In the case of fermentations with defined media, the DS medium was comprised of 2.25 g NH_4Cl, 0.1g $MgSO_4$, 0.7 mg $FeCl_3{\cdot}6H_2O$, 2.72 g KH_2PO_4, 3.48 g K_2HPO_4, 5.0 g NaCl, 12 mg $CaCl_2{\cdot}2H_2O$, 9.9 mg $MnCl_2{\cdot}4H_2O$, 0.05 mg thiamine, and 0.21 g citric acid/L of distilled water. In certain instances, the DS was fortified with glutamic acid (Glut) at a final concentration of 0.25 g/L. Two different amino acid supplements were used—one contained a mixture of six amino acids (aspartate, glutamate, histidine, phenylalanine, tryptophan, tryrosine—designated as AA) and the other contained a mixture of 17 amino acids (aspartate, arginine, asparagine, cysteine, histidine, glutamate, isoleucine, leucine, methionine, phenylalanine, proline, serine, threonine, tryptophan, tyrosine, and valine—designated as AA^+). The final concentration of each amino acid in the medium was 50 mg/L, except glutamic acid, which was 250 mg/L. In the case of fermentations with semidefined media, the DS medium was supplemented with various complex nutrient supplements, including yeast extracts (YE Bacto-YE from Difco Laboratories, Detroit, MI; VYE Veeprex B430 YE from Champlain Industries Ltd., Mississauga, Canada) and casein hydrolysates (Tryp Bacto-Tryptone from Difco Laboratories, Detroit, MI; Pan–Pancase S from Champlain Industries Ltd., Mississauga, Canada).

Some formulations involved the use of tap water (TW) instead of distilled water and were supplemented with 1.4 g/L ammonium monohydrogenphosphate (AP) or various specified amounts of corn steep liquor (CSL) (Nacan Products Ltd., Collingwood, Canada).

Unless stated otherwise, the xylose concentration in all media was about 40 g/L. All media contained antibiotics (40 mg/L ampicillin and 10 mg/L tetra–cycline). All media were sterilized by autoclaving at 121°C for 30 min. Antibiotics were filter (0.45-μ pore size) sterilized and added to the sterilized medium after cooling.

Analytical Procedures

Growth was measured turbidometrically at 550 nm (1-cm lightpath) and culture dry weight was measured directly by microfiltration, washing and drying the filter to constant weight under an infrared heat lamp. Culture cell mass was estimated using the relationship 1.0 OD_{550} = 0.34 g dry wt cells/L *(14,20,35)*. Carbon balances (% carbon recovery) were calculated by the method previously described *(33)*. Compositional analyses of fermentation media and cell-free spent media were determined by HPLC using an HPX-H column (Bio-Rad Labs, Hercules, CA) as described previously *(33)*. The total nitrogen content of the complex nutrient supplements Bacto-YE, TRP, VYE, Pan, and CSL was calculated based on product specifications provided by the manufacturers. Inorganic elemental analysis of TW and the complex nutrient supplements was performed at the SLOWPOKE Reactor Facility (University of Toronto).

Determination of Growth and Fermentation Parameters

The mass-based growth yield coefficient ($Y_{x/s}$) was calculated by dividing the maximum cell density (dry wt cells/L) by the concentration of xylose consumed. The ethanol (product) yield ($Y_{p/s}$) was calculated as the mass of ethanol produced (final concentration)/mass of xylose added to the medium and was corrected for the dilution caused by the addition of alkali (cor$Y_{p/s}$). However, $Y_{p/s}$ was not corrected for the contribution from fermentable components other than xylose, which accounts for ethanol yield values that are greater than the theoretical maximum of 0.51 g EtOH/g xylose.

The average volumetric rate of xylose utilization (Q_s) was determined by dividing the initial xylose concentration by the total time required to achieve complete depletion of xylose from the medium. The maximum volumetric rate of sugar utilization (max Q_s) was estimated from the maximum slope in plots of sugar concentration vs elapsed fermentation time. The corresponding value of volumetric productivity (Q_p) was calculated by multiplying maxQ_s by cor$Y_{p/s}$. Since the dry wt cell mass was only determined as an end point (i.e., at the time when the culture mass was maximal) and was not determined over the entire time course of the batch fermentation (other than indirectly as OD), the specific rates of xylose utilization and ethanol production were not routinely determined. However, in cases where growth ceased considerably in advance of the complete utilization of the sugar, the linear rate of xylose utilization during the postgrowth phase (maxQ_s) could be divided by the maximum cell mass to yield an estimate of the specific rate of xylose utilization (q_s). Similarly, the specific productivity (q_p) was estimated by multiplying q_s by the overall ethanol yield (cor$Y_{p/s}$).

RESULTS AND DISCUSSION

Basic Nutritional Requirements

The first approach to gaining further insights into the basic nutritional requirements of recombinant *E. coli* was one of "subtraction refinements' of the complex reference medium (sLB). At a xylose concentration of 40 g/L, the maximum cell density under carbon-limiting conditions is 2 g dry wt cells/L *(33)*, and to support this cell density, the medium should contain minimally 0.3 g assimilable N, 0.03 g P, and 0.003 g Mg (Table 2). It should be noted that the value for the N content of the different media shown in Table 2 is based on a determination for "total N" and not "assimilable N." LB contains 5.0 g/L YE and 10 g/L Tryp and one particular reformulation of LB, designated as mLB, that contains 2.5 g/L YE, and 2.5 g/L Tryp has been recommended by others as a useful refinement for fermentations with recombinant ethanologenic *E. coli* cultures *(36,37)*. However, the utility of this formulation is dependent on the concentration of fermentable sugar. Whereas mLB with 40 g/L xylose contains sufficient N to support a cell density of 2 g/L, there is a potential for insufficiency with respect to both P and Mg (Table 2). Figure 1 compares the effect of various subtraction refinements of sLB on the calculated total N content (A), the xylose-based growth yield (B), the ethanol yield (C), and the volumetric productivity (D). The arrow in Fig. 1A indicates the minimum requirement for assimilable N (0.3 g/L), and the arrows in Figs. 1C,D indicate the NREL targets *(4)* for yield and productivity, respectively. Other data pertaining to metabolic end-product distribution and operational parameters in connection with this series of subtraction refinement experiments are summarized in Table 3. For all formulations tested in this subtraction series, the values for both yield and productivity either met or surpassed the NREL goal, with the notable exception of the DS medium, which was included for purposes of comparison to illustrate the scope for improvement with respect to productivity. Since the ethanol yield is routinely based on the amount of sugar added to the medium, there exists the potential for yield inflation owing to the metabolism of nutrients other than added sugar when employing nutritionally rich, complex media such as LB *(37)*. Therefore, the use of the DS provides useful information regarding the contribution to ethanol production by the nonxylose components of the complex media. Our observations lend support to the claim of Grohmann et al. *(37)* regarding ethanol yield inflation by nonsugar components of the medium.

Apart from economic considerations, it is because of the reduction in growth rate and productivity that defined media are rarely used for the production of bulk chemicals *(25)*. When a single source of a complex nutritional supplement (e.g., YE or protein digest) is added to a chemically defined medium, the resulting formulation is termed a "semidefined" ("semisynthetic") medium *(25)*. Generally, the addition of relatively small amounts of organic (complex) nutrients supply growth factors and vitamins that can potentially significantly increase growth compared to

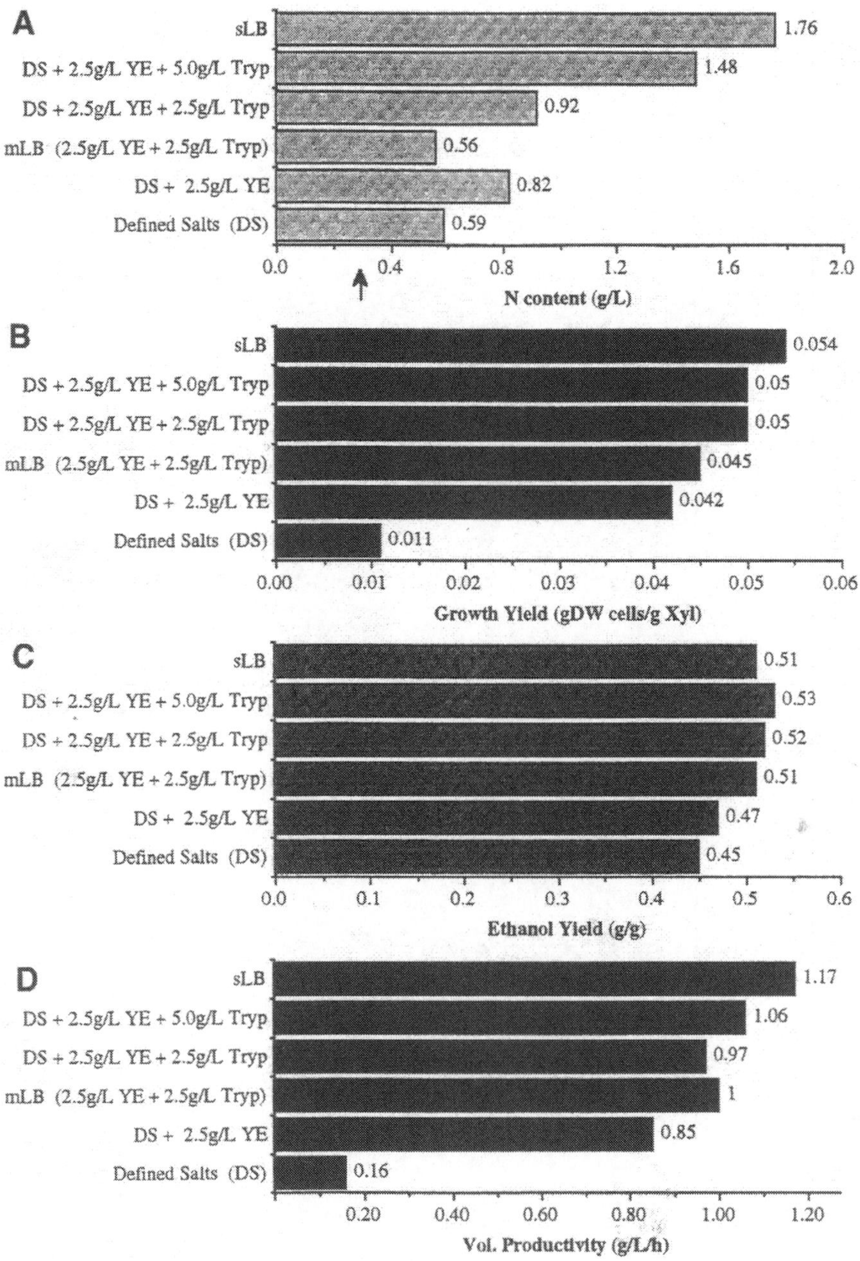

Fig. 1. The effect of subtraction refinements to sLB medium in terms of **(A)** total N content of the medium, **(B)** growth yield coefficient for xylose, **(C)** ethanol yield, and **(D)** volumetric productivity. All media contained about 4 wt% xylose. The arrow in A indicates the assumed minimal N concentration required to support 2 g dry wt cells/L. The arrows in C and D represent the NREL target values for these process parameters. Abbreviations regarding medium formulations are described in the text.

Table 3
Summary of End-Product Distribution and Operational Parameters

Expt. code	Medium, modification	Substrate [Xyl] g/L	Utilization Rate Q_s g/L·h	Utilization Rate max.Q_s g/L·h	Products [Cells] g dry wt/L	Products [EtOH] g/L	Products [Succ] m*M*	Products [Lac] m*M*	Products [Ac] m*M*	Yield cor.$Y_{p/s}$ g/g	pH	% Carbon recovery
B111a	sLB ("benchmark" reference)	42.4	1.93	2.29	2.28	20.52	31.2	0.0	13.2	0.51	7.0	113
B111b	DS + 2.5 YE + 5.0 Tryp	41.8	1.74	2.00	2.08	20.82	24.4	0.0	8.7	0.53	7.0	112
B133a	mLB	42.1	1.28	1.96	1.88	20.06	20.2	0.1	0.8	0.51	7.0	104
B146c	DS + 2.5 YE + 2.5 Tryp	41.9	0.87	2.00	2.11	20.58	22.3	0.1	15.3	0.52	7.0	112
B111c	DS + 2.5 YE	49.3	1.50	1.80	2.06	21.92	31.2	0.0	11.4	0.47	7.0	101
B110a	DS	40.5	0.34	0.36	0.44	17.00	15.9	0.0	18.0	0.45	7.0	92
B134c	DS + Vit + Glut	38.6	0.47	0.66	1.05	18.12	19.6	0.0	17.9	0.50	7.0	106
B135c	DS + Vit + AA	39.8	0.47	0.63	0.98	16.72	18.8	0.0	11.5	0.45	7.0	93
B135d	DS + Vit + AA⁺	42.9	0.76	1.05	1.59	17.52	15.5	14.4	10.3	0.44	7.0	94
B133b	DS + Vit + 2.5 Tryp	42.2	1.03	1.35	1.82	18.97	22.9	3.5	11.9	0.48	6.3	103
B114a	TW + AP + 2.5 VYE	39.9	1.33	1.63	1.45	18.41	33.7	0.0	14.4	0.49	7.0	108
B114b	TW + AP + 1 Pan	41.7	1.30	1.64	1.39	18.00	33.4	0.0	14.2	0.47	7.0	101
B114c	TW + AP + 2.5 VYE + 1 Pan	40.4	1.44	1.66	1.58	17.35	38.9	0.0	15.9	0.46	7.0	104
B121b	TW + 37 mL CSL	36.3	1.45	1.96	1.26	19.56	25.3	0.0	14.3	0.57	7.0	122
B121c	TW + 18 mL CSL	35.9	1.33	1.64	1.10	17.83	46.8	0.0	16.6	0.53	7.0	105
B124a	TW + 18 mL CSL	44.7	1.32	1.57	1.21	18.34	49.6	0.0	16.8	0.44	7.0	100
Fermentations with 6% xylose												
B141a	sLB	62.7	2.02	2.40	2.98	26.71	25.9	0.0	14.5	0.46	7.0	96
B141b	TW + AP + 5.0 VYE	63.9	1.21	1.40	1.70	24.93	51.4	0.0	24.6	0.41	7.0	92
B141c	DS + 2X Vit + 0.25 Glut + AA⁺	62.3	0.67	1.45	0.98	23.88	36.8	0.0	25.0	0.40	7.0	88
B141d	TW + 37 mL CSL	62.2	1.41	1.93	1.25	25.00	50.0	0.0	10.2	0.42	7.0	92

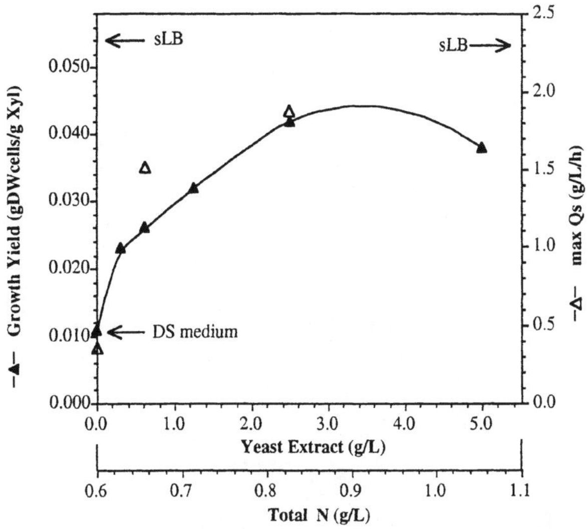

Fig. 2. The effect of incremental supplementation of the basic DS medium with YE. The basic medium contained about 4 wt% xylose and the pHE was controlled at 7.0. Symbols: filled triangles, growth yield coefficient based on xylose; open triangles, maximum rate of xylose utilization—*see* Materials and Methods for methods of determination. The arrows indicate the level of fermentation performance exhibited by recombinant *E. coli* (11303:pLOI297) in reference medium (sLB).

a fully defined medium. Figure 2 shows the effect of adding incremental amounts of Difco YE to the basic DS. In terms of both growth yield and the rate of xylose utilization, the optimal amount of YE appears to be about 2.5 g/L. Since the DS medium satisfies the N requirement (Table 2), the YE can be viewed primarily as a source of vitamins. Table 4 shows the vitamin composition of the DS medium supplemented with 2.5 g/L Difco YE. Based on calculated vitamin content of this medium (DS + 2.5 YE), we formulated a multicomponent synthetic vitamin supplement that, at an amendment rate of 5 mL/L, produces a medium closely approximating the YE-supplemented medium. The vitamin content of these media does not differ substantially from that for a defined *E. coli* medium described by Hernandez and Johnson *(38)* (Table 4). Tryptone is an enzymatic digest of casein, and similar components of complex media mean that such media contain a relatively complete complement of the amino acids. The supply of readily assimilable N in the form of amino acids in the culture medium results in an enhanced rate of growth because the necessity for biosynthesis from inorganic N and carbon metabolites derived from xylose catabolism is circumvented. The amounts of amino acids and vitamins in the complete defined medium used by others *(38,39)* have been based on the composition of bacteria *(27,40)*. In addition to vitamins, YE also supplies organic N, principally in the form of amino acids *(25)*. The content of YE with respect to aspartic and glutamic acids is 5.1 and 6.5 wt%, respectively. Figure 3 shows that supplementing DS with vitamins (5 mL/L stock solution) and 0.25 g/L Glut (DS + Vit + 2.5 Glut) produces a doubling in both the growth yield (Fig. 3B) and productivity (Fig. 3D). Supplementation of this medium with of a mixture containing an additional five different amino acids (DS + Vit + 2.5 Glut + AA) did

Table 4
Vitamin Content of Various Media Formulations

Vitamin/L	sLB	1/2 LB	mLB	DS + 2.5 YE	7.7 g/L CSL[a] 18 mL/L	This study, 5 mL stock/L	Ref.
Choline·Cl					26.9	5.0	
Inositol·$2H_2O$					46.1	7.5	309.0
Niacin (B_6)	1.46	0.728	0.725	0.700	0.640	0.750	0.323
Ca Pantothenate	NS				0.115	0.075	0.107
Pyridoxine·HCl	0.105	0.053	0.057	0.050	0.069	0.075	0.012
Riboflavin	0.096	0.048	0.048	0.048	0.046	0.050	0.103
Thiamine·HCl (B_1)	0.018	0.009	0.009	0.008	0.023	0.010	0.048
Biotin	0.009	0.004	0.004	0.007	0.002	0.005	0.013
Folic acid							0.103

NS, not specified for BYE and Tryp(Difco).
Ref., Hernandez and Johnson *(38)*.
[a]Based on specifications for vitamin content of CSL (Corn Refiners Association Inc., Washington, DC).

not significantly produce an unexpected decrease in growth yield, ethanol yield, and productivity (Fig. 3). Further refinement through the addition of a full complement of amino acids as described in the formulation of Hernandez and Johnson *(38)* (DS + Vit + 2.5 Glut + AA$^+$) produced an improvement in growth (Fig. 3B) and productivity (Fig. 3D). The supply of amino acids in the culture medium bypasses the necessity for biosynthesis frown inorganic N and carbon metabolites derived from xylose catabolism, which results in an enhanced rate of growth.

Further incremental improvement in both growth and productivity was achieved by substitution of the complete amino acid supplement (including the Glut) by 2.5 g/L Tryp (DS + vit + 2.5 Tryp). Although this formulation produced a fermentation performance, in terms of productivity, that surpassed the NREL target value of 0.52 g/L·h, the productivity of 0.64 g/L·h was only about 50% of that achieved with sLB (Fig. 3D) and only 75% of the productivity achieved with basic DS medium that had been amended with 2.5 g/L YE (Fig. 2D). Interestingly, this formulation produced a disproportionate increase in the growth yield (Fig. 3B) relative to productivity (Fig. 3D) suggesting a possible negative impact on specific productivity.

All the formulations involving amendment of the basic DS medium contained sufficient N, P, and Mg to satisfy the requirements for the expected maximum cell yield of 2 g dry wt cells/L (Table 2). In a recently published technical and economical analysis of the production of fuel ethanol from lignocellulosic biomass

Fig. 3. *(Opposite page)* The effect of amendments to defined minimal medium in terms of **(A)** total N content of the medium, **(B)** growth yield coefficient for xylose, **(C)** ethanol yield, and **(D)** volumetric productivity. All media contained about 4 wt% xylose. The arrow in A indicates the assumed minimal N concentration required to support 2 g dry wt cells/L. The arrows in C and D represent the NREL target values for these process parameters. Abbreviations regarding medium formulations are described in the text.

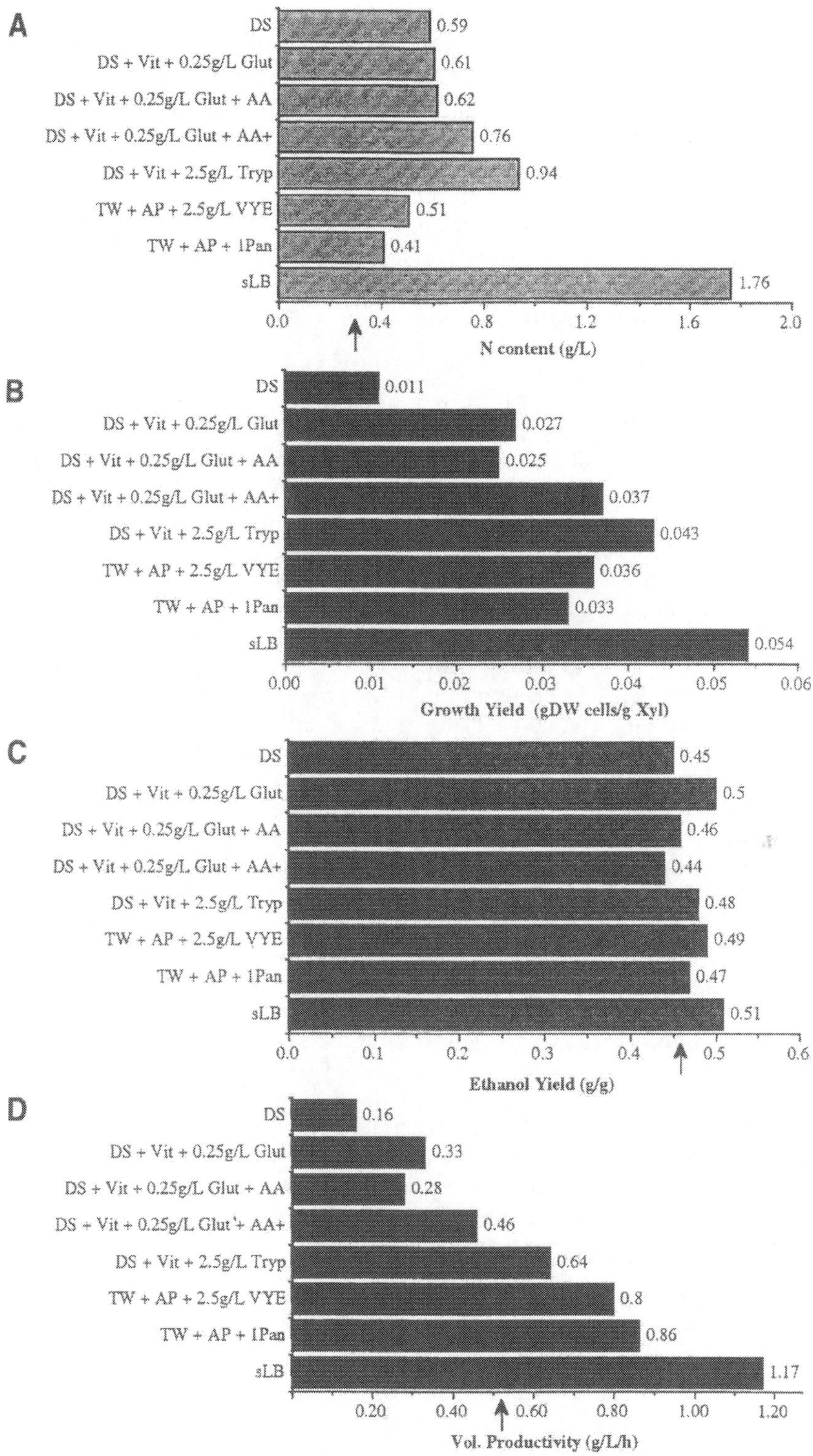
A
DS
DS + Vit + 0.25g/L Glut
DS + Vit + 0.25g/L Glut + AA
DS + Vit + 0.25g/L Glut + AA+
DS + Vit + 2.5g/L Tryp
TW + AP + 2.5g/L VYE
TW + AP + 1Pan
sLB
0.59
0.61
0.62
0.76
0.94
0.51
0.41
1.76
0.0
0.4
0.8
1.2
1.6
2.0
N content (g/L)
B
0.011
0.027
0.025
0.037
0.043
0.036
0.033
0.054
0.00
0.01
0.02
0.03
0.04
0.05
0.06
Growth Yield (gDW cells/g Xyl)
C
0.45
0.5
0.46
0.44
0.48
0.49
0.47
0.51
0.0
0.1
0.2
0.3
0.4
0.5
0.6
Ethanol Yield (g/g)
D
0.16
0.33
0.28
0.46
0.64
0.8
0.86
1.17
0.20
0.40
0.60
0.80
1.00
1.20
Vol. Productivity (g/L/h)

(willow) using recombinant *E. coli* KO11, von Sivers et al. *(41)* suggested that in large-scale fermentations the cell mass would increase from 1 to 3 g dry wt cells/L during each fermentation and that the nutritional requirements would be satisfied by a medium containing 0.4 g N, 0.2 g P, and 0.02 g Mg/L. They did not recommend a specific recipe or source for these elements. These estimates of von Sivers et al. *(41)* are based on measurements of the stoichiometry of these elements in cell mass and therefore are similar to the ratios shown in Table 2.

Cost-Effective Nutrient Supplementation for Large-Scale Fermentations

Utilizing Less-Expensive Sources of Yeast Extract and Tryptone

In the light of nutritional information gained through supplementation of the basic defined minimal medium and subtraction refinements to the chemically rich complex reference medium, the essence of cost reduction involves the substitution of expensive lab-grade ingredients with nutritionally equivalent, less costly, commercially available supplements such that the fermentation performance of the biocatalyst is not compromised as a consequence of the substitution or reformulation. With YE and Tryp costing about $100 and $83/kg, respectively, the cost of the reference LB medium ($1/L) is clearly not economical in terms of the production of bulk chemicals. In a series of flask fermentations with buffered media, it was determined that tap water could be used effectively as a source of trace elements and that 1.4 g/L ammonium phosphate satisfied the requirement for N (results not shown). Two such TW +AP formulations were assessed: one contained 2.5 g/L Veeprex (a commercial YE product costing about $8/kg) and the other contained 1 g/L Pancase S (a commercial casein digest product costing about $27/kg). These formulations met the requirements for N, P, and Mg (Fig. 3A and Table 2) and provided an improved fermentation performance over the formulations involving DS supplementation (Fig. 3). Although these TW formulations surpassed the NREL targets for yield and productivity (Fig. 3C,D), the productivity was only about 70% compared to the "benchmark" value that was established with sLB (Fig. 3D).

Corn Steep Liquor as a Single-Source Nutrient Supplement

CSL is a byproduct of corn wet-milling and is marketed primarily as an animal feed supplement with a current selling price of about $50/t. CSL contains a rich complement of important nutrients (including vitamins—Table 4) to support robust microbial growth and fermentation *(42)*. It was first used as a nutrient source in the 1940s in the development of large-scale penicillin fermentations and continues to be used extensively today in diverse industrial fermentation processes. The process (light) steep water is concentrated by evaporation to about 40–50% solids to produce heavy steep water or CSL *(42)*. The protein content (on average about 40–50%) is estimated from the determination of the total Kjeldahl N (N × 6.25) and, hence, N content quoted in Table 2 does not truly represent the assimilable N content.

The results of a series of pH-controlled batch fermentations designed to determine the optimal amount of CSL as a single-source nutrient supplement are shown in Fig. 4; the relevant data relating to metabolic end product distribution, carbon recovery, and other operational parameters are summarized in Table 3. At

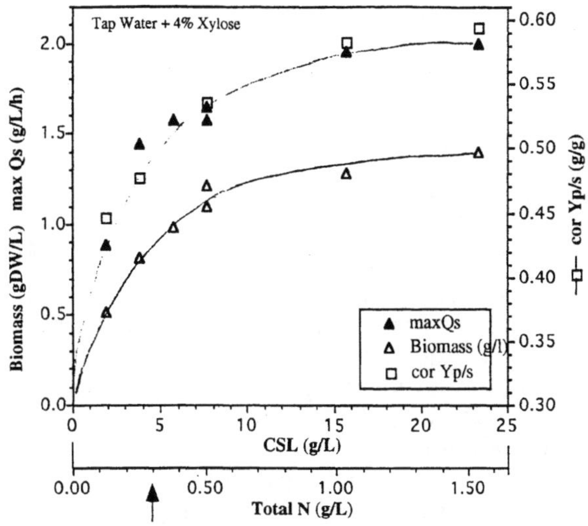

Fig. 4. Comparative fermentation performance achieved by recombinant *E. coli* in a medium consisting of the effect of incremental supplementation of 4 wt% xylose with CSL. The xylose was prepared with tap water. The pH was controlled at 7.0. Symbols: open triangles, maximum cell density (g dry wt cells/L); filled triangles, maximum rate of xylose utilization; open squares, ethanol yield based on xylose consumption and corrected for dilution by alkali added during fermentation. The arrow indicates the assumed minimal N concentration required to support 2 g dry wt cells/L.

4% xylose, there is only a marginal improvement in both growth yield and productivity at supplementation levels >18 mL CSL/L (equivalent to 7.7 g/L) (Fig. 4). Owing to the contribution of nonxylose components of CSL, the ethanol yield at the higher amendments exceeded the theoretical maximum value of 0.51 g/g xylose (Fig. 4). The productivity (estimated from maxQ_s) achieved with optimal levels of CSL corresponded to about 90% achieved with the reference sLB medium.

Based on the N content of *E. coli*, others have estimated that it would require 7 g CSL/L to achieve a cell density of 3 g dry wt cells/L *(43)*. Amartey and Jeffries *(44)* used CSL to replace YE and other nutrients in xylose fermentations by *Pichia stipitis*. Beall et al. *(45)* supplemented a corn hulls acid hydrolysate with 20 mL CSL/L (equivalent to 9 g/L at 45% solids). It should be noted in comparing our results to those of Beall et al. *(45)* that all our media were sterilized by autoclaving, whereas theirs were filter-sterilized. It is not known to what extent the content of vitamins and other growth factors might be affected by the sustained heat treatment of autoclaving. Nevertheless, in general, our findings appear to substantiate Ingram's recommendation regarding the sufficiency of this amount of CSL in supporting good growth and xylose fermentation by recombinant *E. coli*.

Unfortunately, there are disadvantages associated with the use of CSL as a nutritional supplement. Apart from the fact that the composition is not invariant, CSL does not have an indefinite shelf life. The steeping process in corn wet-milling involves addition of SO_2 to the steep water, which can yield substances in the CSL that are potentially inhibitory to bacterial growth. Fortunately, most of the sulfur dioxide is removed during the evaporation of the steep water in the production of heavy steep water (CSL at about 40–50% solids). However, lactic acid bacteria are known to be prevalent in CSL, and they are not totally eradicated during the evaporation stage. CSL contains lactic acid and the concentration of this acid increases with storage even at refrigeration temperatures. Propagation of bacteria can be expected to affect the potential nutritional value of the CSL. It is not known how lactic acid (or other byproducts of LAB metabolism) might affect the fermentation performance of the biocatalyst.

Fermentations with Selected Media Formulations (6% Xylose)

The economic analysis of Hinman et al. *(5)* with respect to the impact of hemicellulose utilization in the conversion of lignocellulosic feedstocks to fuel ethanol was based on 6% xylose, "which seems to be the highest concentration that can reasonably be obtained using dilute-acid pretreatment without a xylose concentrating step." Therefore, in order to test the ability of the recombinant *E. coli* to achieve the target final ethanol concentration of 25 g/L, a series of batch fermentations were conducted with selected media at the upper limit of xylose concentration to be anticipated in dilute-acid prehydrolysates (approx 6 wt%).

Under a condition of xylose-limited growth, a medium containing 60 g/L xylose would be expected to produce a maxium cell yield of about 3 g dry wt cells/L *(33)*. Therefore, for none of these major essential elements to be growth-limiting, the 6% xylose medium should contain (at a minimal level) 0.45 g N/L, 0.045 g P/L, and 0.005 g Mg/L. Table 2 compares the composition of the selected media with respect to N, P, and Mg. Clearly, all the selected media formulations contain an excess of these estimated threshold levels of N, P, and Mg.

The results of this series of batch fermentations are represented in Fig. 5 and Table 3. Only in the case of the reference medium (sLB) was the ethanol yield target value of 0.46 g/g achieved (Fig. 5D). However, the lower than usual value for carbon recovery in these experiments is perhaps a reflection of the underestimation of the ethanol produced in these systems (Table 3). All media formulations resulted in a final ethanol concentration that met the target value of 25 g/L (Fig. 5C). Likewise, with all 6% xylose media formulations, the productivity goal of 0.52 g/L·h was surpassed (Fig. 5D).

In the economic analysis of Grethlein and Dill *(43)*, which was based on a batch process using recombinant *E. coli* in a biomass hydrolysate supplemented with CSL, it was assumed that 40 g ethanol could be derived from 80 g sugars (equivalent to a process yield of 0.50 g/g or 98% conversion efficiency) and that the fermentation would be complete in 48 h (equivalent to a Q_s value of 1.67 or a volumetric productivity of 0.84 g/L·h). In the experiment with the CSL-based medium, the yield (corrected for dilution by alkali) was 0.43 g/g (83.4% conversion efficiency) (Fig. 5B) and the productivity was 0.82 g/L·h (Fig. 5D). Based on the final cell concentration of 1.25 g dry wt cells/L (Table 3), this is equivalent to specific productivity of 0.66 g/g·h. In a recent economic assessment of biomass prehydrolysate fermentation using recombinant *E. coli* KO11, von Sivers et al. *(41)* used a value for specific productivity that was half the value calculated for the fermentation with CSL supplementation.

Economic Impact of Nutrient Supplementation on Cost of Fuel Ethanol from Biomass

With respect to the cost effectiveness of the various media formulations that were examined in this study, nothing can compete with the CSL-based medium. At

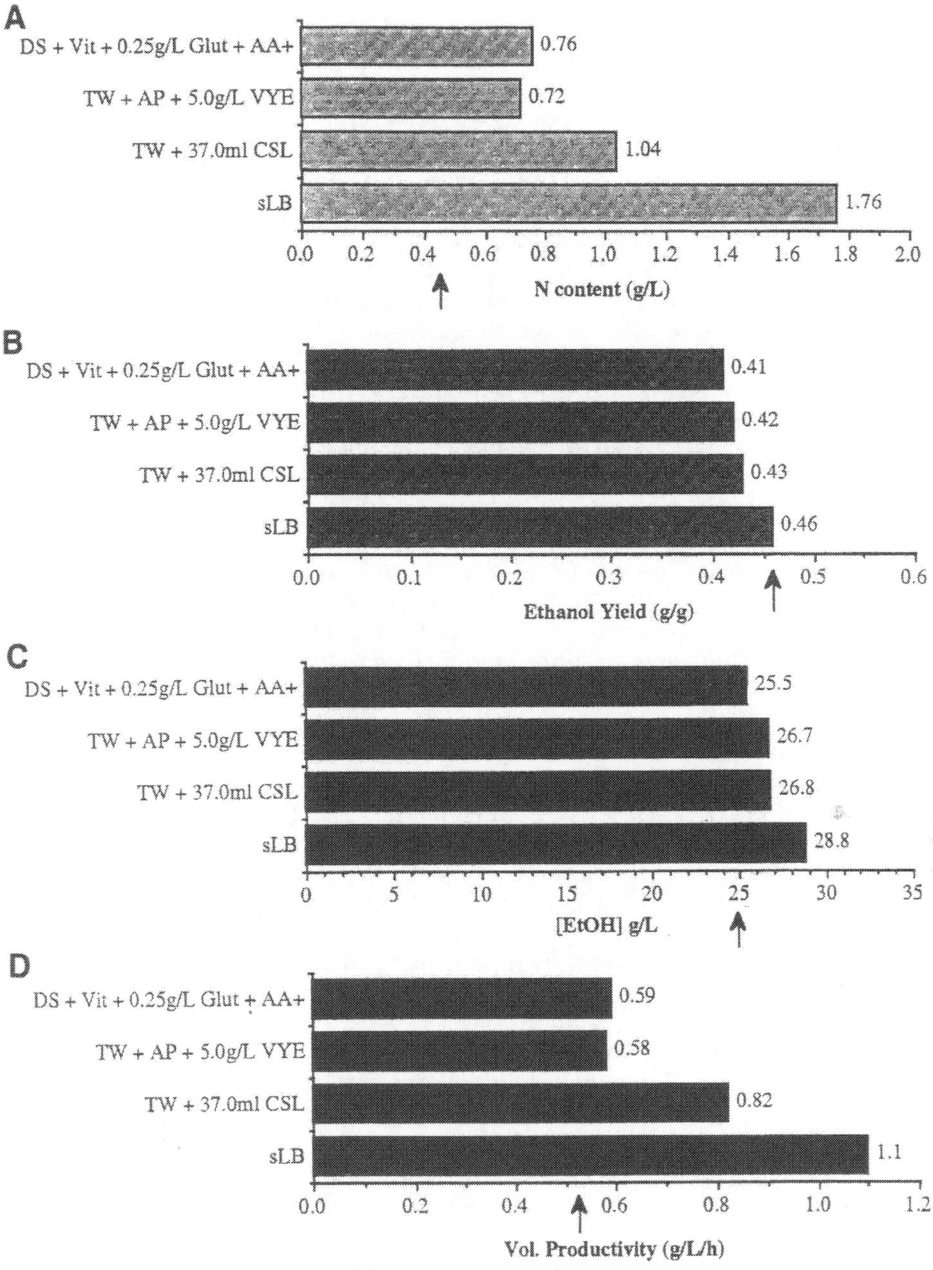

Fig. 5. Fermentation with selected media at 6 wt% xylose in terms of **(A)** total N content of the medium, **(B)** ethanol yield, **(C)** final ethanol concentration, and **(D)** volumetric productivity. The arrow in A indicates the assumed minimal N concentration required to support 3 g dry wt cells/L. The arrows in B, C, and D represent the NREL target values for these process parameters. Abbreviations regarding medium formulations are described in the text.

a cost of about \$1/L, the laboratory reference medium (sLB) is clearly well off the economic scale of feasibility. Even the medium based on VYE (\$8/kg) and ammonium phosphate (fertilizer-grade), at an estimated cost of about \$40/m^3, is well beyond economic feasibility (equivalent to cost of \$4.10/gal denatured ethanol). However, at a cost of \$50/t of CSL (5.5¢/kg) (dry wt basis), the cost of CSL supplementation of prehydrolysate at a rate equivalent to 7.7 g/L is \$0.42/m^3 of fermentation medium. At a level of 4% xylose, this amount of CSL seems saturating in terms of growth, cell concentration, and ethanol productivity. To ensure that CSL was not limiting, we used twice as much CSL in the fermentation at 6% xylose, but it is not known if less CSL would have been sufficient. Based on a xylose loading of 60 kg/m^3 in the prehydrolysate feed to the C_5 fermenter and assuming a xylose-to-ethanol conversion efficiency of 90% of theoretical maximum (equivalent to the NREL yield goal 0.46 g/g) *(4)* and complete product recovery, the cost of using CSL at the lower rate of 7.7 kg/m^3 would be 1.13¢/L or \$0.043/US gal of E95 fuel ethanol. This cost estimate compares favorably with the value of \$0.042/gal estimated by Grethlein and Dill *(43)* based on CSL as a nutritional supplement for a process using enzymatic hydrolysis of 10% lignocellulosic biomass and recombinant *E. coli* KO11 operating at near 100% conversion efficiency.

In an economic analysis of fuel ethanol produced from lignocellulosic biomass (willow) in a process using dilute-acid pretreatment and fermentation by recombinant *E. coli* (KO11), von Sivers et al. *(41)* based the cost of nutrient supplementation solely on ammonia, phosphoric acid, and magnesium oxide, although there has been no experimental confirmation of the sufficiency of this type of supplementation. In their analysis, von Sivers et al. *(41)* assumed an overall sugar-to-ethanol conversion efficiency of 92% (94% efficiency of xylose-to-ethanol conversion by recombinant *E. coli* KO11) with a xylose loading or 42.7 kg/m^3 (also containing 18 kg/m^3 glucose) and estimated the cost of their inorganic supplementation to be \$192,000/yr (based on a plant producing at total of 6.9 million liters of E95 fuel ethanol/yr). This is equivalent to a nutrient supplement cost of 2.78¢/L or \$0.106/US gal of denatured ethanol. von Sivers et al. *(41)* also proposed that the pure culture (inoculum preparation) plant associated with this operation would use prehydrolysate similarly supplemented with these inorganic elements.

Other Relevant Economic Considerations

In this study, the potential nutritional value of the biomass hydrolysate was not taken into account. Similarly, apart from considerations with respect to controlling the pH at 7.0 to minimize the effect of acetic acid, the potential toxicity of the biomass hydrolysate was also ignored. Hence, there has been no regard for the potential cost relating to detoxification. Detoxification costs can be considerable *(4)* and were estimated at 22% of total production cost *(41)*.

In an industrial operation, there is a requirement for make-up water, and apart from the contribution of the biomass hydrolysate, this process water is expected to be a good source of trace elements and other essential minerals. This study examined the nutritional requirements of the biocatalyst within the context of a batch fermentation process, but did not examine the effect of multiple sequential batch fermentations (cycles) in which a portion of the stillage (usually at 8–10% by volume) is used for the purpose of seeding the next batch (back set). This practice has implications in terms of the amounts of supplements required. Furthermore, it is anticipated that in continuous fermentation processes, there is an opportunity for nutrient cost reduction through recycling of cells and stillage.

Finally, this study did not address the issue of the stability of the genetically engineered biocatalyst. Even if antibiotics are required only for the generation of inocula (so-called pure culture operation), the economic impact will depend in a large part on the required frequency of seeding with fresh (pure) culture. In their recent cost analysis based on the use of recombinant *E. coli* KO11, von Sivers et al. *(41)* estimated this requirement to be only three times a year. However, there was no evidence regarding the stability of this organism to validate this assumption, and our recent findings on this subject cast doubt on the presumed stability of this chromosomally integrated recombinant *(30)*. Setting aside any possible regulatory obstacles in terms of employing a genetically engineered biocatalyst, it is estimated the cost of antibiotic ($55/kg) for the purpose of ensuring and maintaining high performance of the biocatalyst is as important a consideration as the cost effectiveness of nutrient supplementation.

ACKNOWLEDGMENTS

This research was part of Phase 1 of a Subcontract AAP-4-11195-03 from the National Renewable Energy Laboratory. We are grateful to Lonnie Ingram for the gift of the recombinant *E. coli* culture.

REFERENCES

1. Wyman, C. E. and Hinman, N. D. (1990), *Appl. Biochem. Biotechnol.* **24/25,** 735–753.
2. Lynd, L. R., Cushman, J. H., Nichols, R. J., and Wyman, C. E. (1991), *Science* **251,** 1318–1323.
3. McMillan, J. D. (1993), Xylose Fermentation to Ethanol: a review; NREL TP-421-4944; National Renewable Energy Laboratory, Golden, CO.
4. McMillan, J. D. (1994), in *Bioconversion for Fuels,* Himmel, M. E., Baker, J. O., and Overend, R. P., eds., ACS Symposium Series 566, American Chemical Society, Washington, DC, pp. 411–437.
5. Hinman, N. D., Wright, J. D., Hoagland, W., and Wyman, C. E. (1989), *Appl. Biochem. Biotechnol.* **20/21,** 391–401.
6. Skoog, K. and Hahn-Hägerdal, B. (1988), *Enzyme Microbiol. Technol.* **10,** 66–88.
7. Prior, B. A., Kilian, S. G., and du Preez, J. C. (1989), *Process Biochem.* **24,** 21–32.
8. Lynd, L. R. (1989), *Adv. Biochem. Eng. Biotechnol.* **38,** 1–52.
9. Knappe, J. C. (1987), in Escherichia coli *and* Salmonella typhimurium—*Cellular and Molecular Biology,* vol. 1, Neidhart, F. C., ed., American Society of Microbiology, Washington, DC, pp. 151–155.
10. Miller, T. L. and Churchill. B. W. (1981), in *Manual of Industrial Microbiology and Biotechnology,* Demain, D. L. and Solomon, N. A., eds., American Society of Microbiology, Washington, DC, pp. 122–136.
11. Aiba, S., Humphrey, A. E., and Millis, N. F. (1973), *Biochemical Engineering,* 2nd ed., Academic, New York, NY, pp. 29,30.
12. Gottschalk, G. (1985), in *Bacterial Metabolism,* Springer-Verlag, New York.
13. Varma, A. and Palsson, B. O. (1994), *Appl. Environ. Microbiol.* **60,** 3724–3731.
14. Lawford, H. G. and Rousseau, J. D. (1992), in *Energy from Biomass and Wastes XV,* Klass, D. L., ed., Institute of Gas Technology, Chicago, IL, pp. 583–622.
15. Ingram, L. O., Alterthum, F., Ohta, K., and Beall, D. S. (1990), in *Developments in Industrial Microbiology,* vol. 31, Elsevier Science, New York, pp. 21–30.
16. Ingram, L. O., Conway, T., and Alterthum, F. (1988), United States patent 5,000,000.
17. Alterthum, F. and Ingram, L. O. (1989), *Appl. Environ. Microbiol.* **54,** 397–404.
18. Ingram, L. O. (1991), in *Energy from Biomass and Wastes XIV,* Klass, D. L., ed., Institute of Gas Technology, Chicago, IL, pp. 1105–1126.

19. Ohta, K., Alterhum, F., and Ingram, L. O. (1990), *Appl. Environ. Microbiol.* **56,** 463–465.
20. Beall, D. S., Ohta, K., and Ingram, L. O. (1991), *Biotechnol. Bioeng.* **38,** 296–303.
21. Lawford, H. G. and Rousseau, J. D. (1991), *Appl. Biochem. Biotechnol.* **28/29,** 221–236.
22. Lawford, H. G. and Rousseau, J. D. (1991), *Biotechnol. Lett.* **13,** 191–196.
23. Lawford, H. G. and Rousseau, J. D. (1993), in *Energy from Biomass and Wastes XVI,* Klass, D. L., ed., Institute of Gas Technology, Chicago, IL, pp. 559–597.
24. Pirt, J. S. (1975), in *Principles of Microbe and Cell Cultivation,* John Wiley, New York, NY.
25. Demain, A. L. and Solomon, N. A. (1981), *Manual of Industrial Microbiology & Biotechnology,* American Society for Microbiology, Washington, DC, p. 108.
26. Stouthamer, A. H. (1976), in *Yield Studies in Microorganisms,* Meadowfield, Dewbury, UK.
27. Luria, S. E. (1960), in *The Bacteria,* vol. 1, Gunsalus, I. C. and Stanier, I. Y., eds., Academic, New York, pp. 1–341.
28. Lawford, H. G. and Rousseau, J. D. (1993), *Biotechnol. Lett.* **15,** 499–505.
29. Ohta, K., Beall, D. S., Meija, J. P., Shanmugam, K. T., and Ingram, L. O. (1991), *Appl. Environ. Microbiol.* **57,** 393–900.
30. Lawford, H. G. and Rousseau, J. D. (1996), *Appl. Biochem. Biotechnol.* **57/58,** 293–305.
31. Lawford, H. G. and Rousseau, J. D. (1992), *Appl. Biochem. Biotechnol.* **34/35,** 185–204.
32. Lawford, H. G. and Rousseau, J. D. (1992), *Appl. Biochem. Biotechnol.* **39/40,** 687–699.
33. Lawford, H. G. and Rousseau, J. D. (1995), *Appl. Biochem. Biotechnol.* **51/52,** 179–195.
34. Luria, S. E. and Delbruck, M. (1943), *Genetics* **28,** 491–511.
35. Ohta, K., Alterthum, F., and Ingram, L. O. (1990), *Appl. Environ. Microbiol.* **56,** 463–465.
36. Grohmann, K., Baldwin, E. A., Buslig, B. S., and Ingram, L. O. (1994), *Biotechnol. Lett.* **16,** 281–286.
37. Grohmann, K., Cameron, R. G., and Buslig, B. S. (1995), *Appl. Biochem. Biotechnol.* **51/52,** 423–435.
38. Hernandez, E. and Johnson, M. J. (1967), *J. Bacteriol.* **94,** 991–995.
39. Lee, S. Y. and Chang, H. N. (1993), *Biotechnol. Lett.* **15,** 971–974.
40. Stokes, J. L. and Gunnes, M. (1946), *J. Bacteriol.* **52,** 195–207.
41. von Sivers, M., Zacchi, G., Olsson, L., and Hahn-Hägerdal, B. (1994), *Biotechnol. Prog.* **10,** 555–560.
42. Anon. (1975), Corn Refiners Association Inc., Washington, DC.
43. Grethlein, H. E. and Dill, T. (1993), "The Cost of Ethanol from Lignocellulosic Biomass—A Comparison of Selected Alternative Processes." April 30, SCA No. 58-1935-2-050, Agricultural Research Service, USDA, Philadelphia, PA.
44. Amartey, S. and Jeffries, T. W. (1994), *Biotechnol. Lett.* **16,** 211–214.
45. Beall, D. S., Ingram, L. O., Ben-Bassat, A., Doran, J. B., Fowler, D. E., Hall, R. G., and Wood, R. E. (1992), *Biotechnol. Lett.* **14,** 857–862.

0273-2289/96/5758—0327$8.00

Coal-Induced Enhancement of Ethanol and Biomass Production

J. Kevin Polman,* Catherine Rae, and Karen M. Delezene-Briggs

Idaho National Engineering Laboratory, Lockheed Idaho Technologies Company, Idaho Falls, ID 83415-2203

ABSTRACT

Inclusion of coal in *Saccharomyces cerevisiae* cultures enhanced ethanol and biomass production (from dextrose) similarly to yeast extract. Coal-induced enhancement was slower in comparison to yeast extract. Ethanol/biomass ratios were twice as high for coal-induced enhancement. Humic acid also enhanced ethanol and biomass production. Coal supplied nitrogen, phosphorus, magnesium, and zinc to cells. Coal facilitated the utilization of cobalt, and reduced the toxicity of aluminum and manganese. Coal-induced enhancement may be a chelation effect; ethylenediaminetetraacetic acid mimicked coal enhancement. Coal also enhanced ethanol and biomass production by *Zygosaccharomyces rouxii*.

Index Entries: Ethanol; coal; chelator; *Saccharomyces*; *Zygosaccharomyces*.

INTRODUCTION

Scientific and engineering investigations, which are directed at improving the efficiency with which ethanol is fermentatively produced, are important because of the huge commercial market represented by ethanol. In 1992, 2.4 billion gallons of ethanol were produced in the United States. Some of the major producers of ethanol are Midwest Grain Products, Archer Daniels Midland, Cargill, Chief Ethanol Fuels, Union Carbide, Quantum Chemical, and Eastman Kodak. The current US presidential administration has supported an increase in the use of ethanol as an oxygenate in reformulated gasoline. This presents a considerable opportunity for the ethanol production industry, which has a projected market growth of at least 10%/yr *(1)*. Citizen Action, a US lobby group, has pressed the administration to encourage this use of ethanol *(2)*. Canada has been involved in a major effort to market gasoline–ethanol blends *(3)*. Problems that occur in using ethanol as a gasoline additive include increased emissions of nitroxides and volatile organic carbon *(4)*.

Raw materials that have been used as fermentable feedstocks for the production of ethanol include sugar cane and sugar beet juices, high-test and blackstrap molasses, sweet sorghum, fruit juices, whey, cereal grains, roots, tubers, cacti, cellu-

*Author to whom all correspondence and reprint requests should be addressed.

losic raw materials, and sulfite waste liquor *(5)*. The microorganisms used for conversion include *Saccharomyces cerevisiae, Saccharomyces ellypsoideus, Saccharomyces carlsbergensis, Saccharomyces fragilis, Schizosaccharomyces pombe, Torula cremoris, Candida pseudotropicalis, Clostridium thermosaccharolyticum, Thermoanaerobacter ethanolicus, Pachysolen tannophilus*, and *Zymomonas mobilis (5)*. Of these, *S. cerevisiae* is probably the most commonly utilized and studied.

Using this organism, theoretical ethanol yield from dextrose is 0.511, actual yield under optimal conditions is 0.485, and observed industrial yield using complex substrates is 0.46 *(5)*. A well-defined, high-productivity/yield growth medium was devised for ethanol-producing yeast, which contains dextrose, nitrogen, sulfur, potassium, chlorine, phosphorus, magnesium, calcium, boron, zinc, aluminum, copper, manganese, iodine, iron, cobalt, pyridoxine, pantothenate, thiamine, inositol, and biotin *(5)*. A cultivation problem that is universal to industrial biotechnological fermentations is toxicity arising from the components that are used to formulate the growth medium *(6)*. These components have been found in carbonaceous feedstocks as well as the feed water, and may be organic or inorganic in nature *(6,7)*. Detoxification of growth medium can be accomplished by ion exchange, calcium hydroxide/sodium sulfite treatment (which results in a separable calcium sulfate sludge that contains inhibitors), molecular sieves, or steam stripping at a cost that can be as high as 22% of the overall production cost *(7)*.

The study that is described herein was directed at determining the effects of coal on ethanol production. We have observed that the inclusion of coal in microbial growth media often enhances growth, although it is rarely used as a primary carbon source (refs. *8–10*, and several unpublished experiments from this laboratory). Other investigators have observed that compounds leached from coal can either inhibit or enhance growth, depending on the microbial species *(11)*. The chemical structure of coal is complex and highly variable; a variety of organic and inorganic moieties are present *(12,13)*. Because coal contains several elements, it can provide cells with a variety of nutrients. The chemical complexity of coal also renders different functions to coal molecules, such as the ability to chelate metals *(13)*. As we will demonstrate in this article, such functions play a role in how coal affects microbial metabolism.

METHODS

Soluble coal, solid coal, humic acid, and yeast extract were used as metabolic enhancers in initial ethanol and biomass production experiments using *S. cerevisiae* ATCC 36026 and *Zygosaccharomyces rouxii* ATCC 28253. The basal medium used in these experiments contained (g/100 mL): dextrose (1), NH_4Cl (0.02), NaCl (0.0001), K_2HPO_4 (0.001), $MgS0_4{\cdot}7H_2O$ (0.001), $CaCl_2{\cdot}2H_2O$ (0.0001), sodium EDTA (0.01), $FeSO_4{\cdot}7H_2O$ (0.0035), and 5 mL/L of medium of a trace elements solution containing 0.1 g/100 mL of each of the following: $ZnSO_4{\cdot}7H_2O$, $MnSO_4{\cdot}H_2O$, $CuSO_4{\cdot}5H_2O$, $Na_2B_4O_7{\cdot}10H_2O$, $CoCl_2{\cdot}6H_2O$, and $Na_2MoO_4{\cdot}2H_2O$ *(9)*. Enhancers that were added included soluble North Dakota Beulah Zap lignite *(14)*, soluble Mississippi Wilcox lignite, granular (–100 mesh) Beulah Zap, soluble humic acid (Aldrich Chemicals), and yeast extract (Difco). All experiments were performed using 25 mL of medium in 50-mL culture flasks (200 rpm, 30°C, aerobic). Soluble coal was prepared by dissolving 2 g of lignite in 1 L of 0.1*N* NaOH overnight, neutralizing the mixture with HCl, and then filtering to remove undissolved coal. Ethanol was measured by gas chromatography (Stabilwax column, Restek, 30 m, 0.25 mm id, FID, 0.1-mL injection). Biomass was determined indirectly by measuring turbidity at A_{660} (absorbance owing to dark coal solutions or particulate coal was considered by determining time zero values for cultures containing these components and using these values to correct and thus standardize measurements). Dextrose was measured by the classic phenol/sulfuric acid assay *(15)*.

Other experiments were designed to determine the mechanism of coal-induced enhancement of biomass production by *S. cerevisiae*. Two growth media were used: 25X optimized media (OM) and 1X OM. 25X OM contained (g/100 mL): dextrose (2), $(NH_4)_2SO_4$ (0.35), KCl (0.06), H_3PO_4 (0.04), $MgSO_4 \cdot 7H_2O$ (0.006), $CaCl_2 \cdot 2H2O$ (0.002), H_3BO_3 (0.0005), $ZnSO_4 \cdot 7H_2O$ (0.0003), $Al_2(SO_4)_3 \cdot 17H_2O$ (0.00015), $CuSO_4 \cdot 5H_2O$ (0.00013), $MnSO_4 \cdot H_2O$ (0.00013), KI (0.00005), $FeSO_4 \cdot 7H_2O$ (0.00005), 0.1% $CoCl_2 \cdot 6H_2O$ (0.00003), pyridoxine (0.0003), pantothenate (0.0002), thiamine (0.00013), inositol (0.00013), and biotin (0.0000005) *(5)*. 1X OM was 25X OM diluted 25 times (except the dextrose, which was still present at 2 g/100 mL). 25X OM was used in experiments that examined (1) replacement of ammonium, dextrose, magnesium, and zinc by coal and (2) detoxification of manganese and aluminum. 1X OM was used in experiments that examined:

1. Replacement of phosphate and cobalt by coal;
2. Effects of EDTA; and
3. Effects of coal on culture pH.

RESULTS AND DISCUSSION

Soluble Beulah Zap lignite and yeast extract enhanced ethanol levels produced by *S. cerevisiae* to approximately the same extent (Fig. 1A,B; Table 1). However, enhancement by coal required a significantly longer period of time to reach maximum level. In addition, both Beulah Zap lignite and yeast extract enhanced biomass production. The level of biomass production caused by yeast extract was much higher than that observed for coal (Fig. 1C,D; Table 1). This resulted in ethanol:biomass ratios that were approximately two times higher for coal compared to yeast extract, despite the fact that product:substrate ratios for both coal and yeast extract were the same (Table 1). In the case of coal, perhaps the dextrose was funneled into the production of another metabolite rather than biomass. Nonetheless, it is an unusual observation that warrants further investigation.

Soluble Mississippi Wilcox lignite also enhanced biomass and ethanol production by *S. cerevisiae* (Fig. 2), suggesting that the enhancement effect may be general with respect to coal type. In the cases of both coal types used here, the amount of soluble coal required to bring about enhancement was very low. Granular Beulah Zap lignite was also capable of enhancing ethanol and biomass production by *S. cerevisiae* implying that the solubilization step may not be necessary; this observation might be cost effective with respect to potential industrial processes (Fig. 3A,B). Humic acid enhanced ethanol and biomass production, as well (Fig. 3C,D), which is not surprising since soluble lignite coal has a humic chemical nature *(16,17)*. Overall, the data indicated that humic acid and granular Beulah Zap lignite provided higher ethanol and biomass levels than soluble Mississippi Wilcox and Beulah Zap lignites (Figs. 1–3).

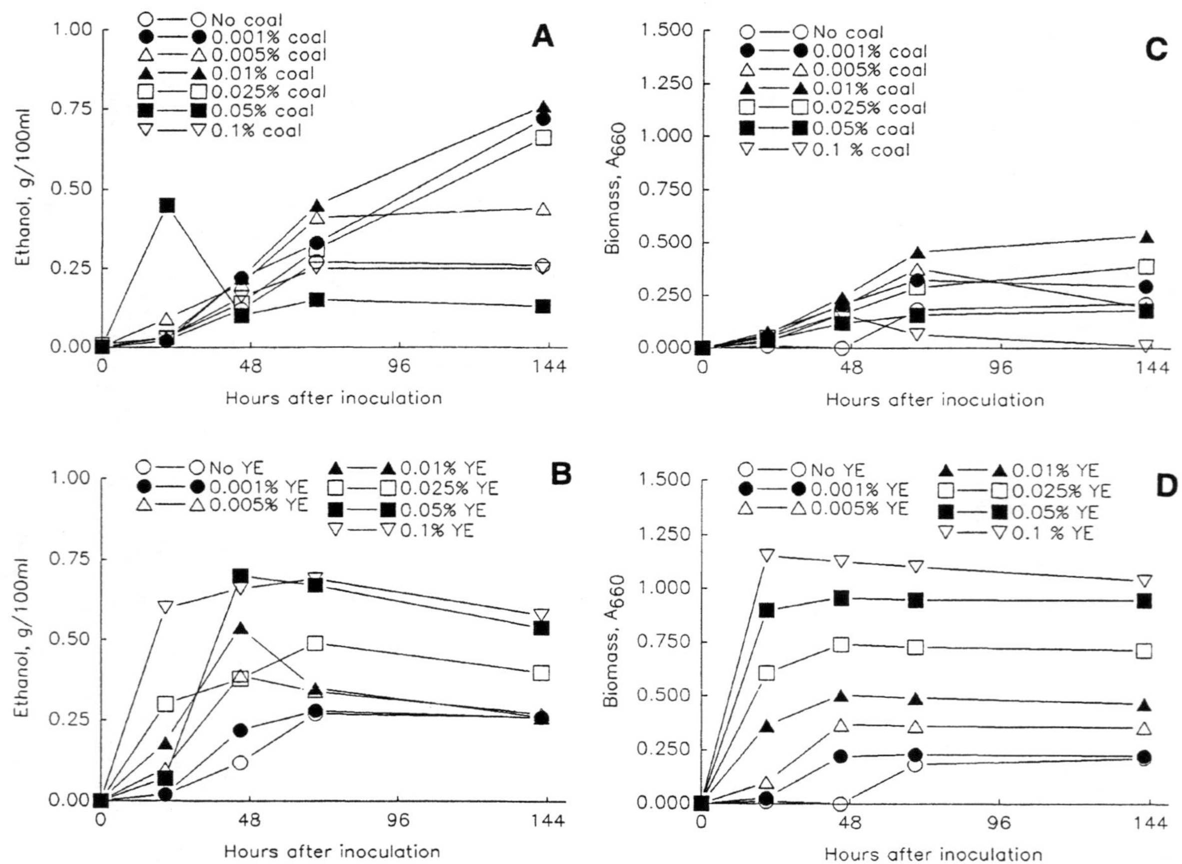

Fig. 1. Enhancement of ethanol and biomass production by *S. cerevisiae* in the presence of soluble Beulah Zap coal and yeast extract.

Table 1
Biomass Production, Ethanol Production, and Dextrose Utilization by *S. cerevisiae* in the Presence of Coal and Yeast Extract

Culture medium[a]	0.001C	0.01C	0.05YE
Time after inoculation (d)	3	6	2
B, biomass (A_{660})	0.144	0.370	0.926
E, ethanol (g/100 mL)	0.32	0.79	0.86
D1, dextrose provided (g/100 mL)	2.384	1.995	2.190
D2, dextrose remaining (g/100 mL)	1.344	0.033	0.046
D1-D2, dextrose utilized (g/100 mL)	1.04	1.962	2.144
E/(D1-D2), product yield (g/g)	0.31	0.40	0.40
E/B, product-to-biomass ratio (g/A_{660})	2.222	2.135	0.929

[a]Media contain dextrose and basal salts, plus 0.001% (0.001C) or 0.01% (0.01C) alkali-soluble Beulah Zap lignite, or 0.05% yeast extract (0.05YE).

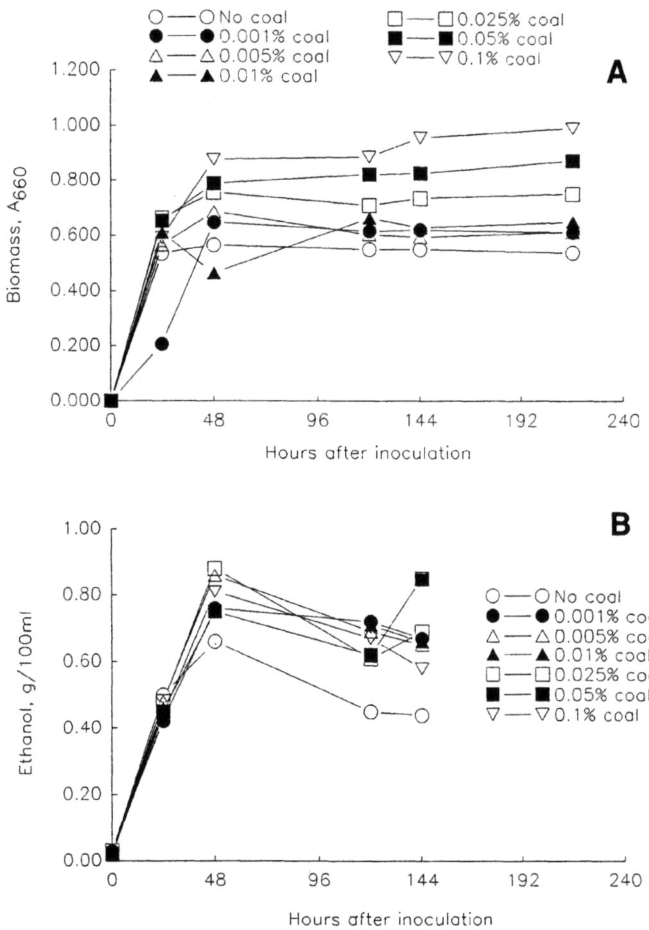

Fig. 2. Enhancement of ethanol and biomass production by *S. cerevisiae* in the presence of soluble Mississippi Wilcox coal.

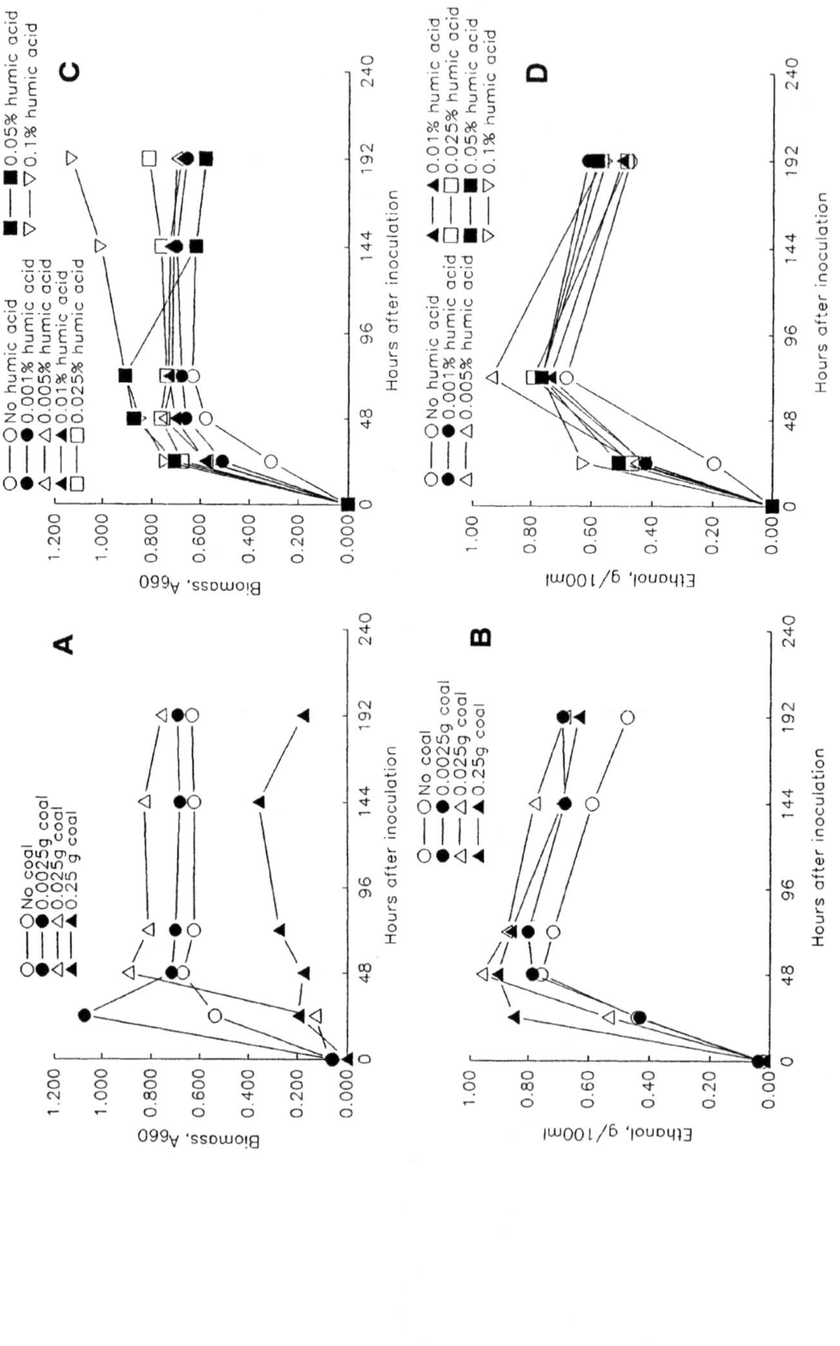

Fig. 3. Enhancement of ethanol and biomass production by *S. cerevisiae* in the presence of granular Beulah Zap coal and humic acid.

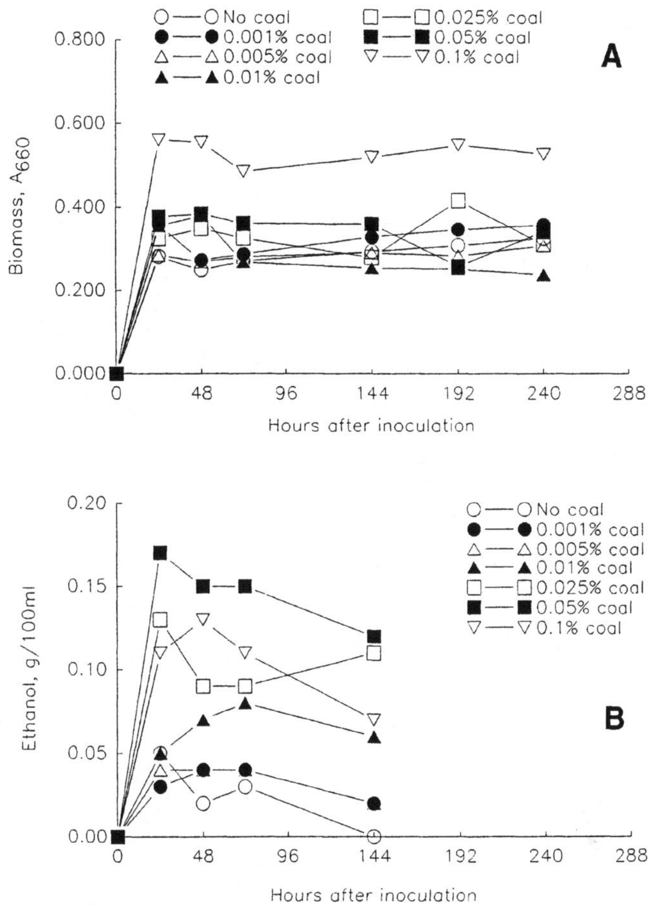

Fig. 4. Enhancement of ethanol and biomass production by *Z. rouxii* in the presence of soluble Beulah Zap coal.

Supporting the hypothesis that coal-induced enhancement of ethanol and biomass production (or metabolic processes in general) may not be limited to one microbial species were the observations of similar enhancements for *Z. rouxii* (Fig. 4). Other results, such as coal-induced enhancement of benzoate degradation by *Pseudomonas chlororaphis,* support this theory *(18).*

Soluble Beulah Zap lignite did not provide carbon to *S. cerevisiae* (Fig. 5A). This is likely due to the recalcitrant nature of carbon skeletons in coal *(10).* Coal did provide some nitrogen (Fig. 5B) and phosphate (Fig. 6A), and was able to replace magnesium (Fig. 5C) and zinc (Fig. 7A) deficiencies completely. Coal appeared either to replace cobalt deficiencies or facilitate a faster uptake of trace cobalt (Fig. 6B).

The presence of soluble coal also seemed to alleviate the toxicity imposed by excessive levels of aluminum and manganese (Fig. 7B,C). We hypothesized that this was due to a chelator effect, since coal is known to bind various metals *(13,19).* This hypothesis was supported by the observation that ethylenediaminetet-

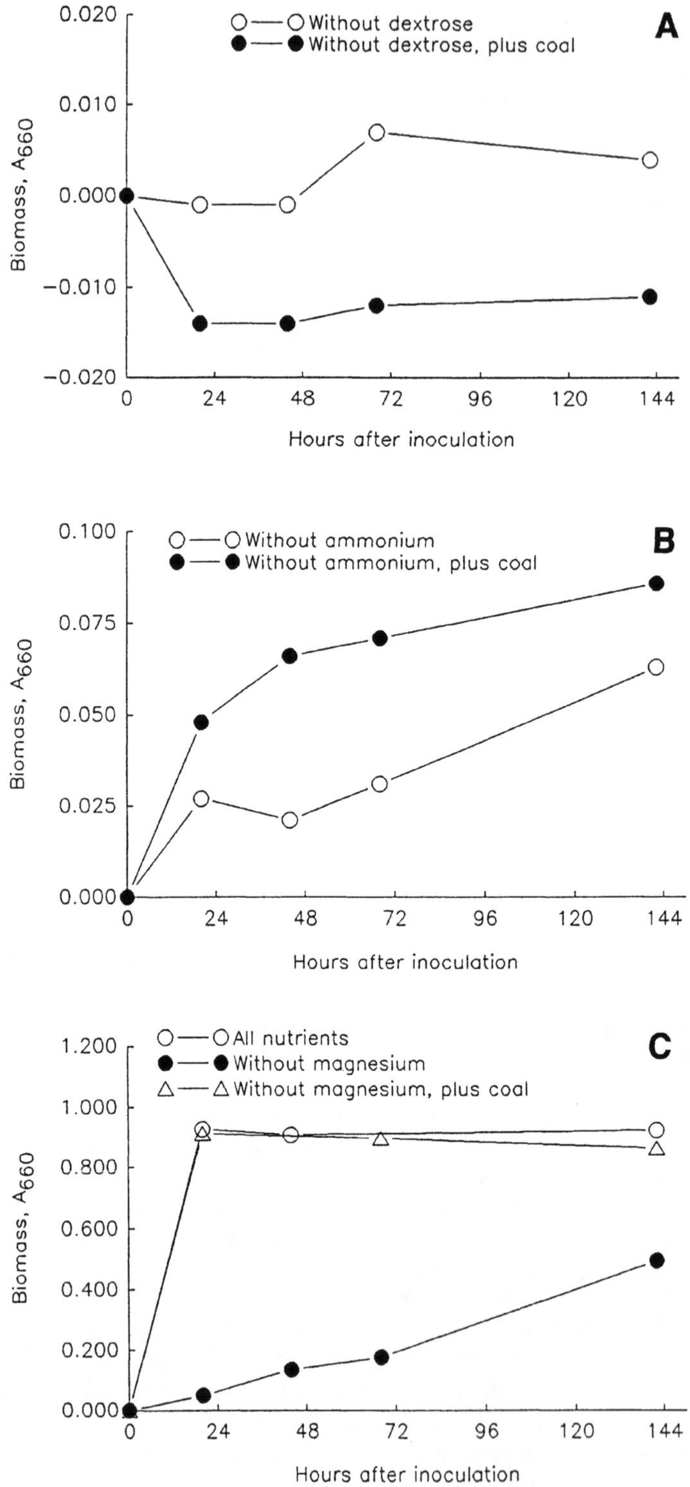

Fig. 5. Replacement of dextrose, ammonium, and magnesium with soluble Beulah Zap coal.

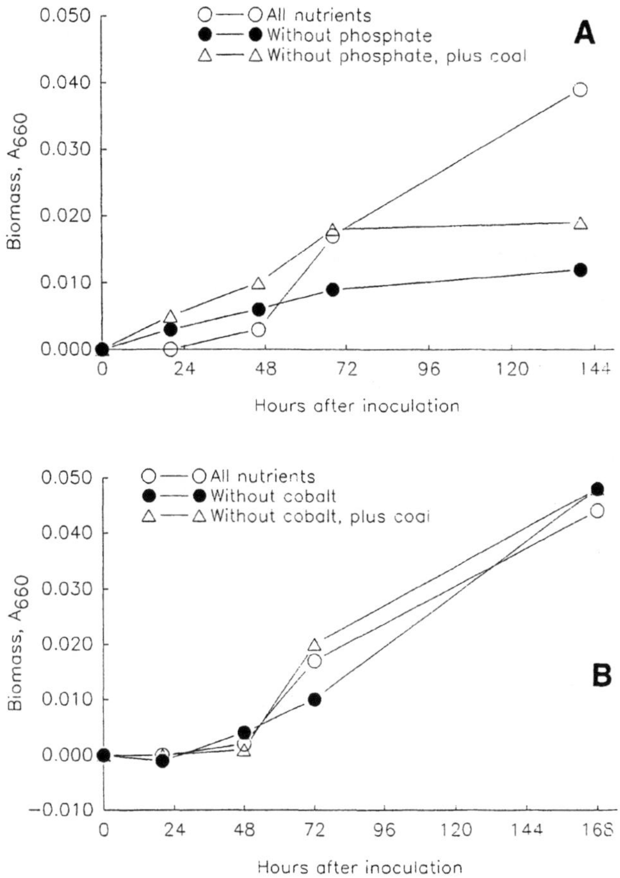

Fig. 6. Replacement of phosphate and cobalt with soluble Beulah Zap coal.

raacetic acid (EDTA) mimicked the enhancing effect that coal provided (Fig. 8B). Coal did not significantly affect growth-related pH changes, indicating that enhancement of growth was not the result of buffering of medium (Fig. 8A).

In conclusion, coal supplies cells with macronutrients and micronutrients and alleviates toxicity associated with certain elements. Since medium toxicity can be a costly problem associated with industrial fermentations *(7)*, a beneficial application of these results is suggested. For reasons mentioned above, the application is likely to be general with respect to bioprocesses, and thus not limited to the production of ethanol. This idea of a general effect is supported by results from many years ago that indicated that humic acids derived from coal were excellent soil fertilizers *(17)*.

ACKNOWLEDGMENT

Funding for the work described herein was provided by the US Department of Energy (Office of Advanced Research and Technology Development; Office of Fossil Energy; contract #DE-AC07-76IDO1570).

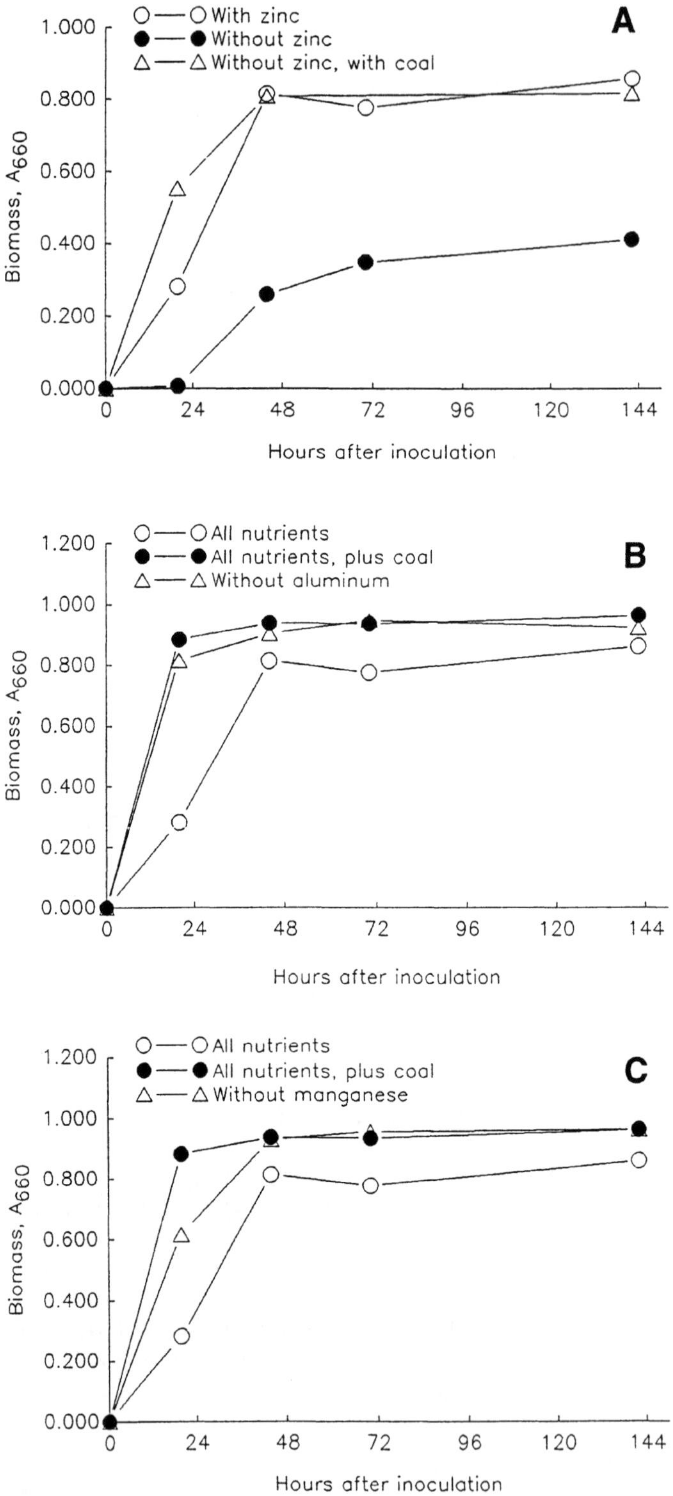

Fig. 7. Replacement of zinc **(A)** and alleviation of aluminum and manganese toxicity **(B,C)** with soluble Beulah Zap coal.

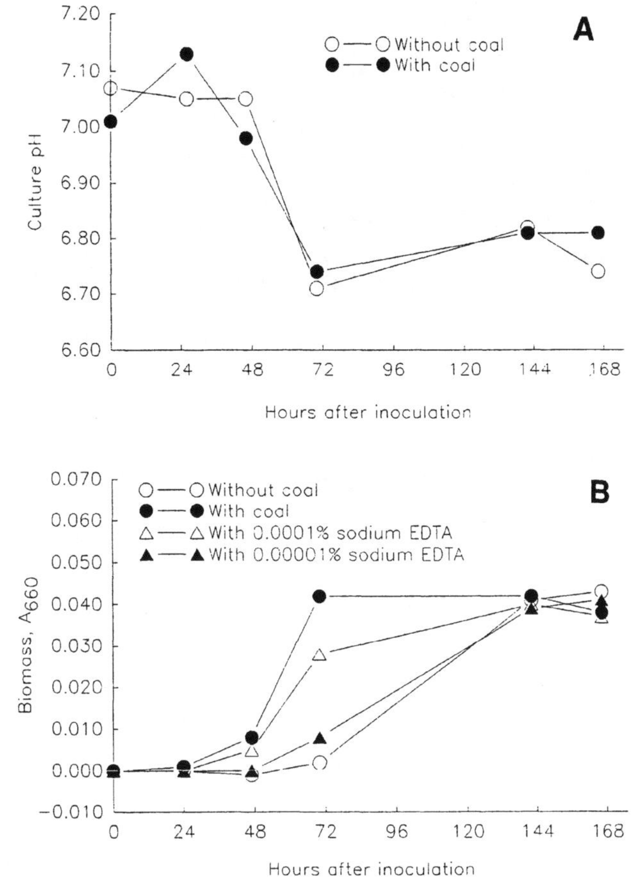

Fig. 8. Effect of soluble Beulah Zap coal on culture pH **(A)** and EDTA mimicry of soluble Beulah Zap coal effect **(B)**.

REFERENCES

1. *Chemical Marketing Reporter* (1994), Dec. 27 issue, p. 7.
2. *The Oil Daily* (1993), Nov. 1 issue, p. 4.
3. *Eco-Log Week* (1993), April 30 issue.
4. *Oil Gas J.* (1993), Oct. 18 issue, p. 19.
5. Maiorella, B. L. (1985), in *Comprehensive Biotechnology*, vol. 3., Moo-Young, M., ed., Pergamon, New York, pp. 861–914.
6. Hodgson, J. (1994), *Bio/technology* **12,** 983–987.
7. von Sivers, M., Zacchi, G., Olsson, L., and Hahn-Hägerdal, B. (1994), *Biotechnol. Prog.* **10,** 555–560.
8. Polman, J. K., Breckenridge, C. R., Dugan, P. R., and Quigley, D. R. (1990), in *Proceedings: First International Symposium on the Biological Processing of Coal*, EPRI, Palo Alto, CA, pp. 4-62–4-72.
9. Polman, J. K., Breckenridge, C. R., Dugan, P. R., and Quigley, D. R. (1991), *Appl. Biochem. Biotechnol.* **28/29,** 487–494.
10. Polman, J. K., Breckenridge, C. R., and Quigley, D. R. (1991), in *Proceedings: Second International Symposium on the Biological Processing of Coal*, EPRI, Palo Alto, CA, pp. P-63–P-77.
11. Olsson, G., Larsson, L., Holst, O., and Karlsson, H. T. (1989), *Fuel* **68,** 1270–1274.

12. Derbyshire, F., Marzec, A., Schulten, H.-R., Wilson, M. A., Davis, A., Tekely, P., et al. (1989), *Fuel* **68,** 1091–1106.
13. Huttinger, K. J. and Michenfelder, A. W. (1987), *Fuel* **66,** 1164,1165.
14. Vorres, K. S. (1990), *Fuel* **4,** 420–426.
15. Daniels, L., Hanson, R. S., and Phillips, J. A. (1994), in *Methods for General and Molecular Bacteriology*, Gerhardt, P., ed., ASM, Washington, DC, pp. 518,519.
16. Polman, J. K. and Quigley, D. R. (1991), *Energy Fuel* **5,** 352,353.
17. *Chem. Eng.* (1962), vol. 69, pp. 68–70.
18. Polman, J. K. (1993), in *Bioprocessing of Fossil Fuels Program: Quarterly Report*, Dugan, P. R., ed., EG&G Idaho, Idaho Falls, ID, pp. 11–20.
19. Quigley, D. R., Breckenridge, C. R., and Dugan, P. R. (1989), *Energy Fuel* 3, 571–574.

0273-2289/96/5758—0339$7.25

Bioconversion of Rice Straw Hemicellulose Hydrolysate for the Production of Xylitol

Effect of pH and Nitrogen Source

INÊS C. C. ROBERTO,*,[1] SILVIO S. SILVA,[1] MARIA G. A. FELIPE,[1] ISMAEL M. DE MANCILHA,[1] AND SUNAO SATO[2]

[1]*Center for Biotechnology, Faculty of Chemical Engineering of Lorena, 12600-000, Lorena, SP, Brazil; and* [2]*Faculty of Pharmaceutical Sciences, University of São Paulo, Brazil*

ABSTRACT

Xylitol production by the yeast *Candida guilliermondii* was evaluated in a rice straw hemicellulose hydrolysate under different conditions of initial pH and nitrogen source. Xylitol production was significantly affected ($p < 0.05$) by the nitrogen source, pH, and the interaction between these factors. The best yield and productivity were observed at initial pH of 5.3 in medium containing ammonium sulfate as nitrogen source. Under these conditions, the xylitol yield factor ($Y_{p/s}$) was 0.68 g/g and volumetric productivity (Q_p) was 0.51 g/L.h.

Index Entries: Rice straw; hemicellulose hydrolysate; xylitol; xylose; *Candida guilliermondii.*

INTRODUCTION

Lignocellulosic materials in the form of agricultural and forest residues accumulate in large amounts all over the world every year. Cereal straw is a major subproduct of agriculture in developing countries, with an annual production of more than 560 million tons in Asian countries *(1)*. In Brazil, these residues are also quite abundant, especially in the southern region where approx 3.85 t of straw are generated per each ha of rice planted. The availability and the possibility of converting these materials into different products by biotechnological processes have called the attention of several researchers to the utilization of the total biomass components. Over the last few years, several studies have been performed aimed at the utilization of the xylose-rich hemicellulosic fraction as a substrate for the production of ethanol *(2,3)*, microbial protein *(4)*, acetic acid *(5)*, and xylitol *(6–8)*.

*Author to whom all correspondence and reprint requests should be addressed.

Xylitol is a xylose pentitol of wide application in the food, pharmaceutical, and dental industry owing to its sweetening power equivalent to sucrose and anticariogenic properties *(9,10)*. Furthermore, xylitol can be used by diabetic patients, since its metabolism is not insulin dependent *(10)*. Despite these many advantages, the use of xylitol has been limited by its high price. This cost is a result of the extensive purification steps needed for the preparation of a pure xylose solution, which is essential for the chemical process *(11)*. The fermentative process of xylitol production has been considered an alternative to the chemical process, since it does not require initial xylose purification *(12)*. Although many yeast species are able to convert xylose to xylitol with high yields in semisynthetic media *(13–16)*, the fermentation conditions may differ from those employed with media containing hemicellulose hydrolysate. This is because of the substrate complexity, which normally includes different compounds that are toxic to the microbial metabolism. By using appropriate environmental conditions, such as initial pH value *(17)* and inoculum concentration *(18)*, negative effects can be minimized.

Research on the utilization of lignocellulosic residues by fermentative processes has been carried out in our laboratory since 1987 *(2,7,8)*. The possibility of converting xylose to xylitol using rice straw hemicellulose hydrolysate obtained by acid hydrolysis has been demonstrated recently *(19)*. In the present article, we report the fermentation results using rice straw hemicellulose hydrolysate under different conditions of initial pH and nitrogen source by the yeast *Candida guilliermondii*.

MATERIALS AND METHODS

Microorganisms

The yeast *C. guilliermondii* FTI 20037, previously selected by Barbosa et al. *(15)*, was used for the study. Cultures were maintained on malt extract agar at 4°C.

Inoculum Preparation

The inoculum was prepared by growing cells in 125-mL Erlenmeyer flasks containing 50 mL medium of the following composition: 20.0 g/L xylose, 3.0 g/L $(NH_4)_2SO_4$, 0.1 g/L $CaCl_2{\cdot}2H_2O$, and 20.0 g/L rice bran. The culture was incubated at 30°C for 24 h under shaking (200 rpm). The cells were separated by centrifugation at 1600*g* for 10 min and used to prepare a suspension in sterile distilled water at the concentration of 3.7 g/L dry wt. This suspension was used to inoculate the fermentation medium to obtain an initial cell concentration of 0.5 g/L.

Preparation and Treatment of Hemicellulose Hydrolysate

The raw material used was rice straw from the region of Lorena, São Paulo, Brazil. Before extraction of the hemicellulose fraction, the straw was dried in the sun and reduced to 100-mm long and 1-mm thick fragments. The hemicellulose hydrolysate was obtained by acid hydrolysis in a 25-L steel reactor (AISI 316) under the following conditions: temperature of 145°C, reaction time of 20 min, and H_2SO_4 (0.7% [w/v]) using a liquid-to-solid ratio of 10:1. After the hydrolysis, the liquid fraction (hemicellulosic) was concentrated at 70°C under vacuum until a xylose concentration of approx 80 g/L was reached. The concentrated hydrolysate (pH 1.37) was treated with NaOH (pellets) to a pH of 10.0 and adjusted to 5.0 by addition of H_2SO_4 (72% [w/w]). After each change in pH, the precipitate was removed by centrifugation at 1000*g* for 15 min.

Medium and Fermentation Conditions

The hydrolysate treated as described above was submitted to pH adjustment to 4.0, 5.0, and 6.0 with NaOH or H_2SO_4 and then supplemented with the same nutrients as used for inoculum culture, except xylose. The effect of the nitrogen source was verifed by keeping fixed the C/N ratio, using ammonium sulfate (3.0 g/L) or urea (1.3 g/L) to provide a concentration of 0.6 g nitrogen/L. The nutrient solutions were sterilized separately by autoclaving at 121°C for 20 min, except the urea solution that was sterilized by filtration through membranes with 0.22-μm pores. All experiments were carried out in 125-mL Erlenmeyer flasks containing 50 mL of the fermentation medium with shaking (200 rpm) at 30°C. The fermentations were monitored by periodic analyses of the samples.

Analytical Methods

The concentrations of xylose, glucose, arabinose, xylitol, and acetic acid were determined by high-performance liquid chromatography (HPLC) using a model HP 1082B Hewlett-Packard (Miami, FL) apparatus, under the following conditions: Aminex (Bio-Rad, Hercules, CA) HPX87H column at 45°C, 0.01*N* sulfuric acid as eluent, flow rate 0.6 mL/min, detector with a 16× refraction index, and 20-μL sample volume.

Furfural and hydroxymethylfurfural were analyzed by HPLC using the apparatus described under the following conditions: RP18 (HP-79916B) column at 25°C, acetonitrile:water (1:8) and 1% acetic acid as eluent, flow rate of 0.8 mL/min, visible UV-light detector at 276 nm, and 20-μL volume sample.

Cell growth was monitored by measuring absorbance at 600 nm using a Micronal model B34211 spectrophotometer. Cell concentration was calculated by a calibration curve (dry wt × absorbance) prepared with cells grown in hydrolysate supplemented with the same nutrients and using the same conditions as those employed for inoculum preparation. Incubation time was 48 h.

Statistical Analysis

The effect of pH, nitrogen source, and the interaction of these two factors was evaluated by a two-way factorial design *(20)*. The analysis of variance was done using the Statgraphics statistical software version 2.7.

RESULTS AND DISCUSSION

Chemical Composition of the Hydrolysate

The chemical composition of the rice straw hemicellulose hydrolysate obtained by acid hydrolysis is given in Table 1. The hydrolysis conditions used resulted in a hydrolysate containing a mixture of xylose, glucose, and arabinose at the proportions of 70, 20, and 10%, respectively. The presence of furfural and hydroxymethylfurfural, products of pentoses and hexoses degradation, respectively, was also detected. These compounds represent a sugar loss, and their pres-

Table 1
Composition of Rice Straw Hemicellulose Hydrolysate
Obtained by Acid Hydrolysis

Components	Concentration, g/L	
	Original hydrolysate	Concentrated hydrolysate
Xylose	16.44	79.3
Glucose	4.40	22.6
Arabinose	2.35	13.4
Furfural	0.41	0.13
Hydroxymethylfurfural	0.096	0.51
Acetic acid	1.40	1.82

ence in the hydrolysates is undesirable owing to their inhibitory effects on microbial metabolism *(21,22)*. In addition to these two compounds, acetic acid resulting from acetylated sugars was also observed. This has a negative effect on different bioconversion processes *(23–26)*.

The hydrolysate evaporation (under vacuum) gave variations on the original component levels (Table 1). Furfural concentration was reduced in about 30%, whereas hydroxymethylfurfural concentration increased proportionally to water removal during evaporation, by about five times, owing to the low volatility of this compound. The levels of these compounds present in rice straw hydrolysate, however, were lower than those reported as inhibitors of the growth of the *Saccharomyces cerevisiae* yeast *(18)*.

Regarding acetic acid levels, only a small variation (1.40–1.82 g/L) was attained during the evaporation process. Higher acetic acid concentrations on the order of 10 g/L have been detected in wood and sugar cane bagasse hemicellulose hydrolysates *(27,28)*. According to these authors, acetic acid was found to be the most important inhibitor to fermentation. The low concentration attained in rice straw values is expected to improve the xylitol yields achieved by *C. guilliermondii*.

Influence of pH and of Nitrogen Source

The results of the fermentation runs using rice straw hemicellulose hydrolysate supplemented with $(NH_4)_2SO_4$ or urea as nitrogen source at different initial pH values are illustrated in Figs. 1–3 (A,B). Independently of the nitrogen source employed at pH 4.5, no sugar consumption and a slight cell growth were observed during the first 15 h of fermentation (Fig. 1A,B). These results suggest the existence of a phase of yeast adaptation as a function of the low initial pH value utilized. The presence of acetic acid in the hydrolysate (Table 1) may have contributed to this behavior, since the toxic effect of acetic acid is a function of the concentration of nondissociated acid and therefore pH dependent *(29)*. These observations were confirmed by the behavior of cell growth in the fermentations carried out at initial pH of 5.3 and 6.0 (Figs. 2 and 3, A,B). Under these conditions, high growth rates and simultaneous utilization of xylose and glucose were observed since the beginning of fermentation. Regarding xylitol production, the higher and similar values were observed in media at pH 5.3 and 6.0, independently of the nitrogen source utilized (Figs. 2 and 3, A,B). However, at pH 4.5 (Fig. 1A,B) the use of urea promoted a 25%

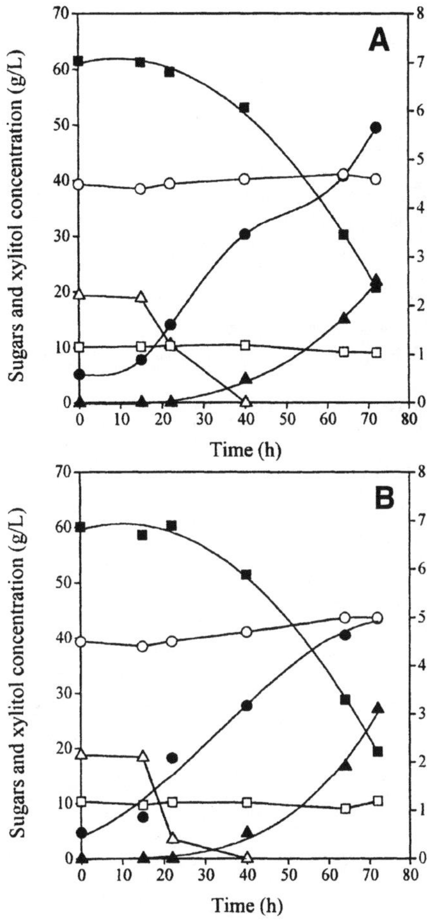

Fig. 1. Fermentation of rice straw hemicellulose hydrolysate by *C. guilliermondii* using $(NH_4)_2SO_4$ **(A)** or urea **(B)** as nitrogen source at pH 4.5. △, glucose; ■, xylose; □, arabinose; ▲, xylitol; ●, cell mass; and ○, pH.

increase in xylitol formation when compared to ammonium sulfate. This behavior can be related to the increase in the pH medium owing to the release of ammonia and CO_2 by urea hydrolysis *(30)*. Furthermore, the consumption of acetic acid by the yeast may also have contributed to this raise in the pH value *(19)*. A similar behavior in terms of pH variation in the presence of urea was observed in *Candida shehatae* cultures on medium containing xylose as a carbon source *(31)*.

With respect to the fermentative parameters (Table 2), it can be seen that the highest xylitol yields were obtained under different conditons of pH and nitrogen source. These yields were 0.68 g/g for the medium containing ammonium sulfate at pH 5.3 and 0.66 g/g in the presence of urea at 4.5. Despite similar yields, higher volumetric productivity was attained when ammonium sulfate was used as nitrogen source. These results reveal the important role of the nitrogen source as a function of pH in the bioconversion of xylose to xylitol using hemicellulose hydrolysates. According to Silva et al. *(16)*, in a semisynthetic medium, the amount of xylitol

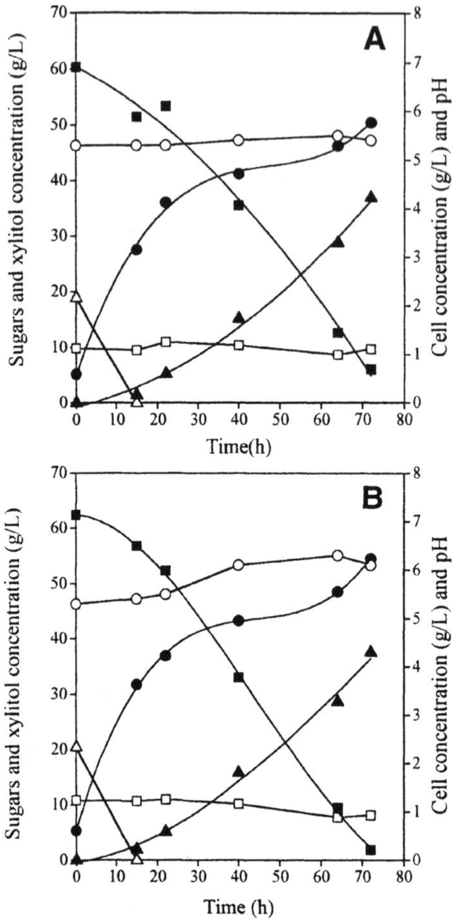

Fig. 2. Fermentation of the rice straw hemicellulose hydrolysate by *C. guilliermondii* using $(NH_4)_2SO_4$ **(A)** or urea **(B)** as nitrogen source at pH 5.3. △, glucose; ■, xylose; □, arabinose; ▲, xylitol; ●, cell mass; and ○, pH.

produced by the yeast *C. guilliermondii* was not affected by the type of nitrogen source utilized. The composition of the fermentation medium may possibly account for these differences.

The results concerning the influence of pH and nitrogen source on xylitol production were also analyzed statistically by analysis of variance of the major effects and their interactions (Table 3). The data revealed a significant effect ($p < 0.05$) of these factors and their interaction in the production of xylitol by *C. guilliermondii* in a rice straw hydrolysate. The effect of the interaction was probably caused by the increase in the pH level when urea was used as nitrogen source at pH 4.5 (Fig. 1 B).

These results demonstrate that the effect of pH cannot be analyzed individually because of the interaction with nitrogen source and possibly with other medium components, such as acetic acid, which is commonly present in hemicellulose hydrolysates.

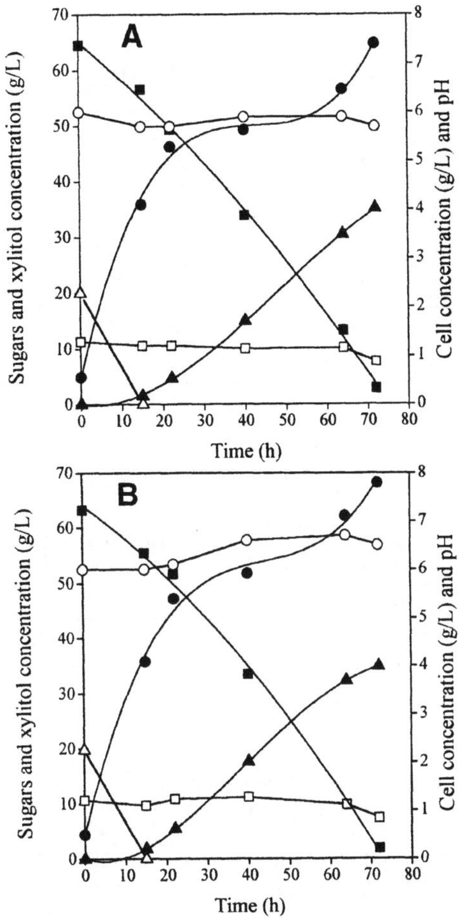

Fig. 3. Fermentation of the rice straw hemicellulose hydrolysate by *C. guilliermondii* using $(NH_4)_2SO_4$ **(A)** or urea **(B)** as nitrogen source at pH 6.0. △, glucose; ■, xylose; □, arabinose; ▲, xylitol; ●, cell mass; and ○, pH.

Table 2
Effect of pH and Nitrogen Source on Xylitol Fermentation of Rice Straw Hemicellulose Hydrolysate by *Candida guilliermondii* after 72 h

Nitrogen source[a]		Ammonium sulfate			Urea	
Initial pH	4.5	5.3	6.0	4.5	5.3	6.0
Xylitol, g/L	21.86	36.92	35.16	27.11	37.57	34.84
Q_p, g/L·h	0.32	0.51	0.49	0.38	0.49	0.48
$Y_{p/s}$, g/g	0.54	0.68	0.57	0.66	0.62	0.57
$Y_{x/s}$, g/g	0.08	0.07	0.08	0.07	0.07	0.09
Efficiency, %	60	74	62	72	64	62

[a]Qp = xylitol volumetric productivity; Y*p/s*, g xylitol produced/g xylose consumed; $Y_{x/s}$, g cells produced/g (xylose + glucose) consumed; efficiency was calculated assuming a theoretical maximum of 0.917 g/g *(1)*.

Table 3
Analysis of Variance for Xylitol Production
by *C. guilliermondii* as a Function of pH and Nitrogen Source (NS)

Source of variation	Degrees of freedom	Sum of squares	Means square	Calculated F	Level of significance
Main effects	3	572.25	190.75	72.83	0.0000
NS	1	15.60	15.60	5.96	0.0311[a]
pH	2	556.54	278.32	106.27	0.0000[a]
Interactions	2	26.52	13.26	5.06	0.0254
NS × pH	2	26.52	13.26	5.06	0.0254[a]
Residue	12	31.43	2.62		
Total (corr.)	17	630.19			

[a]Significant at the 5% level of probability.

ACKNOWLEDGMENTS

The authors acknowledge financial assistance from CNPq/Brazil and Heizir F. de Castro for the helpful revision of this article.

REFERENCES

1. Srnivasan, M. C., Ramarao, R., Deshmukh, S. S., and Sahasrabudhe, N. A. (1983), *Enzyme Microb. Technol.* **5(4),** 269–272.
2. Roberto, I. C., Lacis, L., Barbosa, M. F. S., and Mancilha, I. M. (1991), *Process Biochem.* **26,** 15–21.
3. Lawford, H. G. and Rousseau, J. D. (1993), *Biotechnol. Lett.* **15(5),** 505–510.
4. Meyer, P. S., du Preez, J. C., and Kilian, S. G. (1992), Syst. Appl. Microbiol. **15(1),** 161–165.
5. Brownell, J. E. and Nakas, J. P. (1991), *J. Ind. Microbiol.* **7,** 1–6.
6. Strehaino, P. and Dupuy, M. L. (1990), French Patent 2,641,545.
7. Roberto, I. C., Felipe, M. G. A., Lacis, L., Silva, S. S., and Mancilha, I. M. (1991), *Bioresource Technol.* **36(3),** 271–275.
8. Felipe, M. G. A., Mancilha, I. M., Vitolo, M., Roberto, I. C., and Silva, S. S. (1993), *Arquivos de Biologia e Tecnologia* **36(1),** 103–114.
9. Bar, A. (1986), in *Alternative Sweeteners*, O'Brien-Nabors, L. and Gelardi, R., eds., Marcel Dekker, New York, pp. 185–216.
10. Pepper, T. and Olinger, P. M. (1988), *Food Technol.* **42(10),** 98–106.
11. Melaja, J. and Hämäläinen, L. (1977), US Patent 4,008,285.
12. Heikkilä, H., Nurmi, J., Rahkila, L., and Töyrylä, M. (1992), US Patent 5,081,026.
13. Nolleau, V., Preziosi-Belloy, L., Delgenes, J. P., and Navarro, J. M. (1993), *Curr. Microbiol.* **27,** 191–197.
14. Meyrial, V., Delgenes, J. P., Moletta, R., and Navarro, J. M. (1991), *Biotechnol. Lett.* **13(4),** 281–286.
15. Barbosa, M. F. S., Medeiros, M. B., Mancilha, I. M., Schneider, H., and Lee, H. (1988), *J. Ind. Microbiol.* **3,** 241–251.
16. Silva, S. S., Mancilha, I. M., Queiroz, M. A., Felipe, M. G. A., Roberto, I. C., and Vitolo, M. (1994), *J. Basic Microbiol.* **34(3),** 205–208.
17. van Zyl, C., Prior, B. A., and du Preez, J. C. (1991), *Enzyme Microb. Technol.* **13,** 82–86.
18. Chung, I. S. and Lee, Y. Y. (1985), *Biotechnol. Bioeng.* **27,** 308–315.
19. Roberto, I. C., Mancilha, I. M., Souza, C. A., Felipe, M. G. A., Sato, S., and Castro, H. F. (1994), *Biotechnol. Lett.* **16(11),** 1211–1216.
20. Box, G. E. P., Hunter W. G., and Hunter, J. S. (1978), in *Statistics for Experimenters*, Wiley, New York, pp. 228–244.

21. Tran, A. V. and Chambers, R. P. (1985), *Biotechnol. Lett.* **7(11),** 841–846.
22. Weigert, B., Klein, C., Rizzi, M., Lauterbach, C., and Dellweg, H. (1988), *Biotechnol. Lett.* **10(12),** 895–900.
23. Ferrari, M. D., Neirotti, E., Albornoz, C., and Saucedo, E. (1992), *Biotechnol. Bioeng.* **40,** 753–759.
24. Bjorling, T. and Lindman, B. (1989), *Enzyme Microb. Technol.* **11(4),** 240–246.
25. Meyer, P. S., du Preez, J. C., and Kilian, S. G. (1992), *J. Ind. Microbiol.* **9,** 109–113.
26. Felipe, M. G. A., Vieira, D. C., Vitolo, M., Silva, S. S., Roberto, I. C., and Mancilha, I. M. (1995), *J. Basic Microbiol.* **35,** 171–177.
27. Perego, P., Converti, A., Palazzi, E., Del Borghi, M. D., and Ferraiolo, G. (1990), *J. Ind. Microbiol.* **6,** 157–164.
28. van Zyl, C., Prior, B. A., and du Preez, J. C. (1988), *Appl. Biochem. Biotechnol.* **17,** 357–369.
29. Lawford, H. G. and Rousseau, J. D. (1993), *Appl. Biochem. Biotechnol.* **39/40,** 301–322.
30. Phaff, H. J., Miller, M. W., and Mark, E. W. (1978), in *The Life of Yeasts*, 2nd ed., Harvard University Press, Cambridge, MA, pp. 168,169.
31. Jeffries, T. W. (1985), *Biotechnol. Bioeng. Symp.* **15,** 149–166.

0273-2289/96/5758—0349$8.00

Production of Cellulases and Xylanases by *Trichoderma viride* and Biological Processing of Lignocellulose and Recycled Paper Fibers

U. VIESTURS*,[1,2], M. LEITE,[2] A. TREIMANIS,[2]
T. EREMEEVA,[2] A. APSÍTE,[2] M. EISIMONTE,[2] AND P. JANSONS[1]

[1]*Latvian Biogas Association; and* [2]*Latvian State Institute of Wood Chemistry, 27 Dzérbenes Str., LV-1006 Riga, Latvia*

ABSTRACT

The production of cellulases and xylanases by *Trichoderma viride* L-333 was studied. Significant amounts of extracellular enzymes were obtained when submerged fermentation was performed in the fed-batch regime on cellulose-and xylan-based media with addition of sophorose and Tween-80. The culture filtrate of *T. viride* was used for enzymatic treatment of recycling newspaper (TMP) fibers. To elucidate the mechanism of the enzymatic hydrolysis, the fractionation and separate processing of tree different fiber size fraction were performed. The biological processing of recycling fiber caused an increase in brightness and dewatering properties. The most significant improvement was observed with unfractionated and long fiber fractions at a ratio of cellulase to endoglucanase to xylanase of 1:10:100.

Index Entries: Cellulase; hemicellulase; fed-batch fermentation; enzyme activities; enzyme adsorption; recycled waste paper; freeness; drainage; brightness.

INTRODUCTION

Cellulases and xylanases are the main enzymes used for recycling fiber modification. *Trichoderma* spp. are well known as producers of cellulolytic enzymes. Several fermentation techniques have been applied to enhance mycelial growth and extracellular enzyme production *(1–6)*.

Recycled fibers have lower freeness and strength compared to virgin fibers. The treatment of the secondary fibers with cellulases and hemicellulases can result in the improvement of the fiber quality. It has been reported that the modification of the secondary fiber with a mixture of cellulases and hemicellulases can reverse the decrease in drainage that occurs when the secondary fiber is recycled *(7–9)*. Improved dewatering of pulp was presumed to derive from the "peeling" effect, which removes small fiber fibrils that have high affinity for water *(8–10)*. It has been found that the increase in drainage occurs owing to the hydrolysis of the

*Author to whom all correspondence and reprint requests should be addressed.

fines and an increase in the long fiber content *(7,11)*. Jackson and coauthors *(12)* found that the fines were subjected to both flocculation and hydrolysis by enzymes, and fibrils were removed or fragmented from the surface of large fibers. Visualization of enzyme binding to the secondary fibers by the silver-enhanced colloidal gold affinity labeling technique confirmed the previous finding reported by the authors *(12)* that fines are the preferred fiber fraction for cellulase attack *(13)*. Prasad and coauthors *(11)* found that xylanase was essential to increase brightness of recycled fiber and to reduce the ink area. In a joint effort, a European research consortium of microbiologists and biochemists, as well as wood and fiber technologists *(14)*, found that the action of endoglucanase (EG) was necessary for an improvement in recycled paper drainage. The effect did not appear to be owing to a selective hydrolysis of fines, but was a consequence of the hydrolysis of amorphous cellulose on the surface of the fibers *(14)*. In spite of the fact that this study included various fractions of virgin and recycled fibers and all possible enzymes of the cellulase and hemicellulase system, some unresolved questions on the mechanisms of the enzymatic treatment still remain, because of the large variation in the enzyme sources, activities and dosages, secondary fiber origin and composition, and so forth. The purpose of the present study was to test the culture filtrate of *Trichoderma viride* produced in the fed-batch fermentation regime, and containing a mixture of cellulolytic and hemicellulolytic enzymes. To elucidate the mechanism of the enzymatic hydrolysis of the recycled paper fibers, the study included the fractionation and separate processing of the three fiber-size fractions of the recycled newspaper made of spruce thermomechanical pulp (TMP).

EXPERIMENT

Materials and Methods

Enzyme source

The enzyme source was an unpurified filtrate of *T. viride* L-333 (Institute of Microbiology and Biotechnology, University of Latvia). The medium composition was similar to that used by Mandels et al. *(1)*, with the following alterations: instead of urea, the concentration of $(NH_4)_2SO_4$ was increased from 1.4 g/L to 3.0 g/L and the concentration of KH_2PO_4 from 1.0 g/L to 2.0 g/L (to ensure the buffering capacity). Proteose peptone was replaced by 0.1% yeast extract. The carbon sources were carboxymethylcellulase sodium salt (Sigma, St. Louis, MO) and pretreated or untreated lignocellulosic (LC) substrate (wheat straw, wheat bran, waste newspaper). EG formation was induced by sophorose (500 μg/mL) *(2)*. The fermentations were carried out in a laboratory fermentor FU-8 *(4)*. The use of a special stirring system for mycelial cultures designed by the authors allowed the increase in the LC substrate concentration in the initial nutrient medium from 2–8% dry matter. Tween-80 (0.02%) and the physiological regulation of the metabolic activity of the cell by acetate pulses in *T. viride* fed-batch culture were used to increase the production and excretion of extracellular cellulases and xylanases.

Enzyme Assays

1. Cellulase activities: Filter paper (FPA) and EG (EC 3.2.14) activities were measured against Whatman No. 1 filter paper and carboxymethylcellulase (CMC), respectively, according to Mandels and coauthors *(1)*, and expressed as international units (IU/mL).

2. Xylanase (1.4-β-xylanase, EC 3.2.1.8) was assayed by the method of Bailey and coauthors *(15)*.
3. Reducing sugars were determined by the 3.5-dinitrosalicylic acid (DNS) method *(16)*.
4. Substrate: Unprinted and 6-mo-old printed samples of the Latvian newspaper "Diena," manufactured from spruce TMP was used as a raw material in this study. The newsprint waste paper was torn into small pieces by hand and blended until a uniformity of the raw material and 76° Schopper Riegler (SR) were reached. Three types of fiber fractions were prepared using 17-, 240-, and 250-mesh screens. Type I (17-mesh screen) contained long fibers only. Type II (240-mesh screen) contained both long fibers and fines. Type III (250-mesh screen) consisted only of fines. All the fractions were used for the enzyme treatment.

Enzyme Treatment

Enzyme reactions were performed at pH 4.8, 50°C, and 3.0% consistency for 120 min. After the incubation, the residual enzyme activity adsorbed onto the fibers was destroyed by boiling the pulp for 5 min to prevent further enzyme action. The enzyme dosage was calculated according to FPA, EG, and xylanase activities based on the oven dry fiber.

Fiber Characterization

The influence of the enzyme treatment on brightness of the recycling paper (ISO 3688), freeness (°SR, ISO standard 5 267/1-1979), and dewatering properties (WRV) *(21)* was determined. Lignin was assayed as Klasson lignin according to the standard procedure (Tappi T222m).

High-Performance Size-Exclusion Chromatography (HP SEC) Conditions

HP SEC was used for the analyses of enzymatic hydrolysates and the low-molecular part of the fibers soluble in 5% NaOH. The HP SEC analyses were performed on a size-exclusion chromatograph (Laboratory Instruments, Prague, Czech Republic), equipped with refractometric and spectrophotometric detectors and a Rheodyne Model 7125 fixed-loop (100 μL) injector. Prepacked stainless-steel columns (250 × 8 mm ID) containing Separon S HEMA 1000 E10 100 (for the analysis of enzymatic hydrolysates) and Separon S HEMA 1000 (for the analysis of 5% NaOH soluble part of fiber) were used. Water was used as the eluent in both cases at a flow rate of 0.5 mL/min at ambient temperature.

RESULTS

Production of Enzymes

Technological parameters of the cultivation were studied with *T. viride* to elucidate the most effective way to promote cellulase and xylanase production. Significant amounts of cellulase (2.06–3.1 FPA IU/mL) and EG (20.6–33.0 IU/mL) were obtained on cellulose-based media, when submerged fermentation was performed in the fed-batch regime. Xylanase was produced efficiently on both cellulose- and xylan-based media, although the best yield (up to 116 IU/mL) was observed with birch xylan (1% dry wt) as inducer. The EG was slightly promoted from 19.3 to 23.0 IU/mL by sophorose. Tween-80 was used to favor extracellular enzyme excretion from the cell.

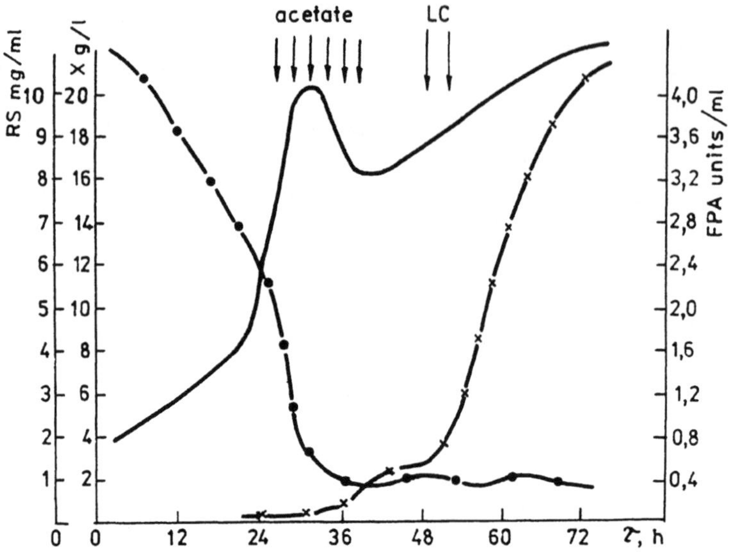

Fig. 1. The growth of *Trichoderma* L-333 biomass (—), cellulase (FPA) synthesis (–x–), and reducing substances (RS) consumption (–●–) on alkali-treated wheat straw (LC). Arrows indicate acetate and LC substrate feeding. The LC substrate feeding allowed to increase the LC substrate concentration from 2–8% dry wt.

The main technological advantage used in this study follows from our previous investigations. It was found that it is possible to promote the exoenzyme synthesis by affecting the general metabolic activity of the cell with energy-rich, cell-energy supplying substrates, such as ethanol, lactose, acetate, or glucose *(17–19)*. The data obtained on the cultivation of *T. viride* L-333 on alkali-treated wheat straw, unprinted newspaper, and so forth, confirmed our previous findings. The acetate pulses at the moment when the biomass and enzyme synthesis begins to decrease and restores both the synthesis of cellulase and cell growth (Fig. 1). Acetate and ethanol are preferable auxiliary sources with weakened carbon limitations, probably because these nonfermentable substrates enter more easily into the Krebs cycle *(18)*. A gradual addition of the LC substrate has a double effect: the intensification of the general metabolism by the soluble carbohydrate pulse and the intensification of the cellulase synthesis at the expense of an increased substrate concentration in the culture media. Owing to the media rheology and mass-transfer problems, it is not advisable to add high LC substrate concentrations at the beginning of the process. The gradual increase in the cellulose concentration and the feeding by small pulses of energy-rich carbon source (acetate) gave the best synthesis of extracellular xylanases and cellulases. The influence of the addition of sorphose, Tween-80, variations of carbon sources, pH profiling, and so on, was comparatively low. Comparative data characterizing enzyme activities in the various fed-batch cultivation processes of *T. viride* L-333 are presented in Table 1.

Treatment of Recycled Fibers

As can be seen from Table 1, the concentration of extracellular enzymes in the *T. viride* culture filtrate was rather different for various fermentation processes. In

Table 1
Effect of Various Carbon Sources and Energy-Rich Carbon Substrate Pulses on the Enzyme Activities During the Fed-Batch Cultivation of *T. viride* L-333

		Enzyme activities, IU/mL			Protein
Substrate	Addition	FPA	EG	Xylanase	% dry wt
CMC	Lactose	2.10	28.4	34.2	22.2
LC untreated	Glucose	1.12	28.3	33.1	19.9
LC alkali Treated, birch xylon, sophorose	LC alkali Treated + acetate	3.40	36.8	116.1	22.1
Unprinted newspaper (TMP), Tween-80	Acetate	1.6	20.1	86.5	20.04

[a]LC, lignocellulosic (wheat straw); CMC, carboxymethylcellulose.

some experiments, the culture filtrate was supplemented with a commercial xylanase preparation, Pulpzyme HB (Novo Nordisk A/S, Bagsvaerd, Denmark). In the range of the enzyme dosages under study, the most significant improvement in WRV and brightness values of the recycled fibers was observed at a comparatively low cellulase content (FPA 0.3 IU/g cellulose dry wt). The optimal ratio of FPA to EG to xylanase was estimated as 1:10:100. The treatment of the printed newspaper "Diena" with such a culture filtrate gave an increase in brightness from 46–52% ISO, and an increase in the WRV values from 109.3–123.0% during a 2-h treatment at 50°C with mixing. It was observed that most of the enzymes were adsorbed rapidly onto fibers within the first 10 min of the contact at 50°C (Fig. 2), which resulted in a rapid increase in the WRV values. The increase in the contact time during a 2-h treatment caused a slight loss in the WRV and °SR values.

Fiber Fractions

To elucidate the mechanism of the enzyme action, the recycled fibers were fractionated into three fractions according to the fiber length and used to separate enzyme treatments. The fractions were subjected to pentosan and Klason lignin analysis. A remarkable difference was established in the carbohydrate and lignin content of the different fractions. As compared to the other two fractions—Type I and Type II—the fines fraction (Type III) probably consisted of more hemicellulose, from the primary cell wall and the outer layer of the secondary cell wall. This was confirmed by the Klason lignin content, which was higher in the Type III fraction (37.2%) than in the longer fiber fractions (24.0 and 24.7% for Type II, respectively). The adsorption of enzymes (Fig. 2) and the hydrolysis rate, according to the release of reducing sugar (Fig. 3), were two times higher for the fines fraction (Type III). At the same time, only the long fiber fraction (Type I) showed an increase in WRV and °SR values (Fig.4), while a rapid decrease in WRV and °SR values for the fines fraction and a slight decrease in those of the Type II fraction were observed (Fig. 5). The HP SEC data confirmed a remarkable difference between the fiber fractions (Table 2). The solubility in 5% alkali and absorbance at 313 and 405 nm for the fines fraction was twice as high compared to the same characteristics for Type I and Type II fibers. The changes in absorbance during the course of experiment, estimated

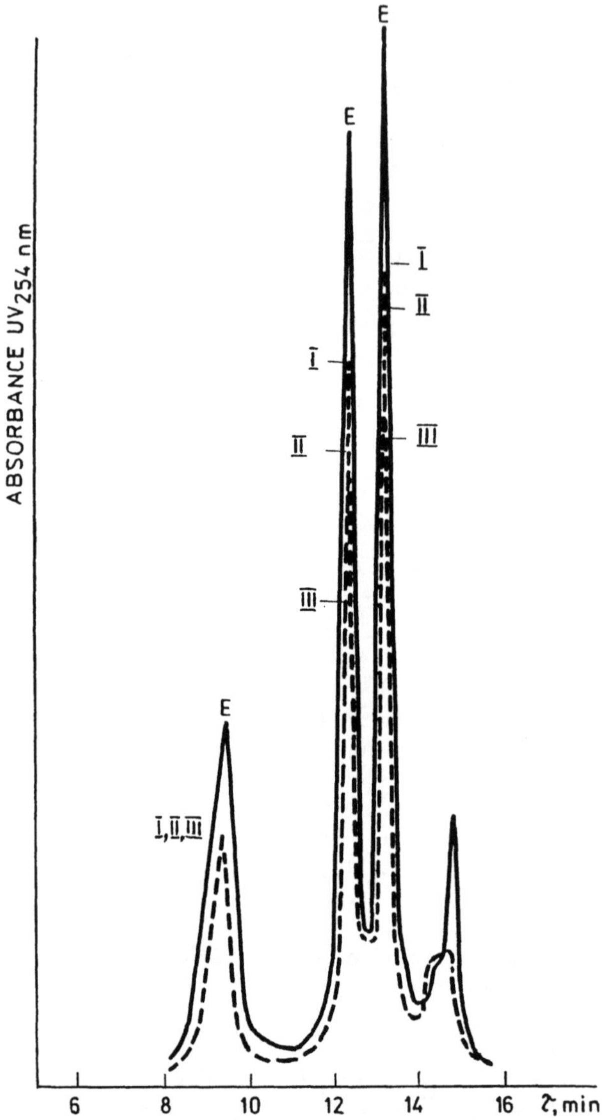

Fig. 2. SEC-chromatograms of the initial enzyme solution, containing *T. viride* culture filtrate, enriched with commercial xylanases (– E) and enzymatic hydrolysates after 10 min of processing of the fiber fractions (– –). I, maxima of the absorption peak for the long fiber fraction (Type I); II, absorption peak for a fraction consisting of both long and short fibers (Type II); and III, absorption peak for fines fraction (Type III).

according to the alkali soluble part of fibers, indicated a dependence of the changes on the type of fiber fraction (Table 2). The absorbance at different wavelengths remained practically unchanged for the long fiber fraction (Type I), but sharply increased for fines (Type III).

An increase in absorption at 313 nm (particularly for the fines fraction) indicated that the carbonyl group content calculated on the mass unit was increased.

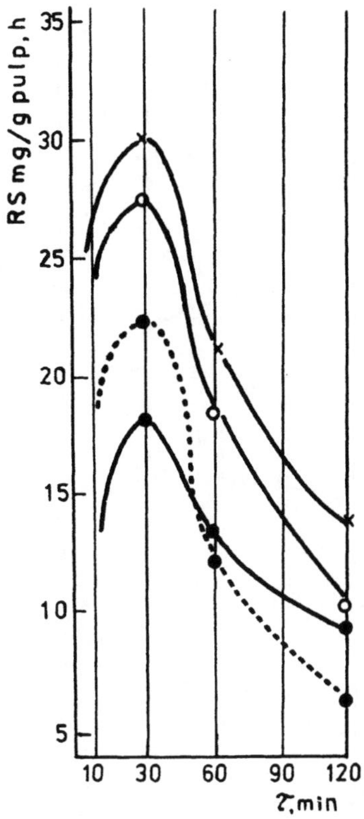

Fig. 3. Rate of hydrolysis as a function of the treatment time for different types of fiber fractions. (-•-) Type I, (-○-) Type II, (-x-) Type III, (—) unfractionated.

An increase at 405 nm, which usually refers to quinone-type substances, confirmed that an oxidation process probably took place in the fines fraction of fibers.

DISCUSSION

The cellulase complex in *T. viride* consists mainly of three enzymes. In processes of submerged cultivation in the Mandels-Andreotti media, these enzyme concentrations are usually as follows: 1.0 IU/mL FPA, 10 IU/mL endoglucanase, and 1.0 IU/mL β-glucosidase *(1)*. A doubling of the cellulase yield is possible either by increasing the cellulose *(20,21)* and nitrogen concentration or controlling pH during the growth *(22)* (or both). The addition of different concentrations of Tween-80 has a similar effect on FPA activities *(3)*. There are reports on the possibility of increasing the cellulase yield by the induction of cellulase synthesis by several soluble inducers, such as sophorose, cellobiose, and lactose *(5,23)*, and of reducing the adsorption action using soluble-substituted derivatives of cellulose, such as carboxymethylcellulose *(24)*. The synthesis of cellulolytic enzymes as regulated by the substrate metabolism has received less attention. The literature reports on the possibility of inducing the cellulase synthesis by affecting the general metabolic activity of the cells by lactose *(5)* or glucose *(25)*. Various organisms, includ-

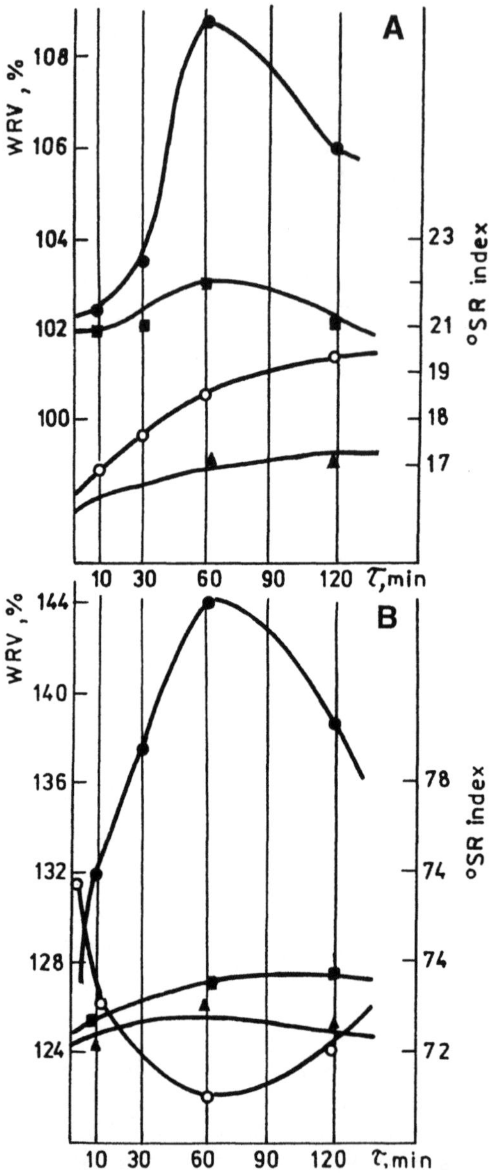

Fig. 4. Effect of enzymetic treatment on the long fiber (Type I) fraction **(A)** and unfractionated fiber **(B)** characteristics. WRV values: (–●–) enzyme treated, (–■–) control. Fiber freeness in °SR: (–○–) enzyme treated, (–▲–) control.

ing, *T. viride*, have been reported as sources for high cellulase and xylanase activities on cellulose and xylan-based media *(3)*. In our previous investigations, a relationship among the cell-energy-supplying substrate, the ATP intracellular concentration, and the exoenzyme activity was observed *(19)*. It was established that within the pulse of the additional carbon substrate, the respiration oscillation was followed by a rapid increase in the ATP intracellular concentration and a slightly slower

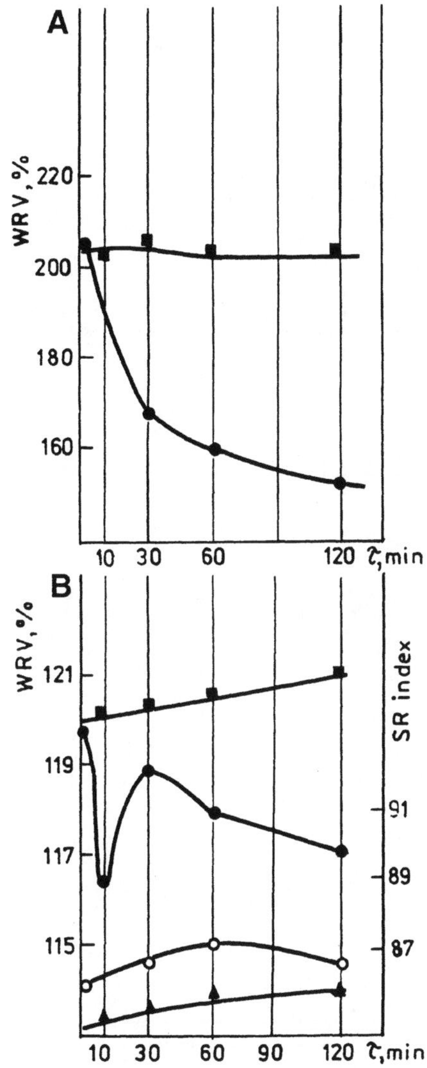

Fig. 5. Effect of enzymatic treatment on the fines (Type III) fraction **(A)** and fraction consisting of both long and short fibers (Type II) **(B)** WRV values: (–•–) enzyme treated, (–■–) control, fiber freeness in °SR: (–○–) enzyme treated, (–▲–) control.

increase in the exoenzyme activity *(19)*. A similar approach used in this study gave high activities of cellulase, EG, and xylanase during fed-batch cultivation of *T. viride* L-333. Our data on the cultivation of *T. viride* L-333 indicate that energy-rich carbon substrate feeding in the cellulose based fed-batch culture of *T. viride* could be an additional energy source and an inducer of extracellular enzyme synthesis to obtain high yields of both cellulase and xylanase.

The effects of enzymatic treatment on pulp characteristics has been studied by several authors. Stork and coauthors *(14)* attributed the shifting of the fibrous material composition from the long fiber fraction to the fines fraction to EG. It

Table 2
HP SEC Characteristics for Different Types of Fiber Fractions

Type of fibers	Time, min	Absorbance coefficient[a]							
		S[b]5,%		290 nm		313 nm		405 nm	
		Buffer	Enzyme	Buffer	Enzyme	Buffer	Enzyme	Buffer	Enzyme
I	0	2.3	2.5	1.5	1.5	1.4	1.4	1.2	1.4
	30	2.3	2.3	1.5	1.6	1.4	1.6	1.3	1.5
	60	2.3	2.1	1.5	1.5	1.4	1.4	1.5	—
	120	2.3	2.1	1.8	1.5	1.7	1.5	1.7	—
II	0	2.4	2.6	1.5	2.0	1.3	1.4	1.2	1.7
	30	2.4	2.6	1.5	2.2	1.2	1.5	1.2	1.8
	60	2.4	2.6	1.5	2.1	1.1	1.5	1.2	1.9
	120	2.4	2.9	1.5	2.1	1.0	1.6	1.1	2.0
III	0	3.1	3.6	1.7	2.5	2.6	3.0	2.3	3.1
	30	3.1	3.2	1.8	2.7	2.6	3.3	2.7	3.2
	60	3.1	3.5	1.9	2.8	2.4	2.9	2.6	3.5
	120	3.1	4.1	2.0	3.1	2.2	3.8	2.8	3.7
Unfractionated	0	2.5	2.5	1.6	1.8	1.5	1.5	1.4	1.5
	30	2.6	2.7	1.6	0	1.6	1.6	1.4	1.7
	60	2.5	2.8	1.6	2.1	1.5	1.7	1.4	1.7
	120	2.4	2.7	2.7	2.4	1.4	1.7	1.2	1.8

[a]The ratio of absorbance to mass unit of soluble part.
[b]Solubility in 5% NaOH.

was found that the presence of the EG activity is a prerequisite for improvement of the WRV values of recycled fibers *(14)*. EG hydrolyzes amorphous cellulose randomly at its glucosid linkages into fragments. Celobiohydrolase (CBH) attacks the nonreducing end of both crystalline and amorphous cellulose macromolecula, releasing cellobiose *(26)*. When the activities of CBH and xylanase were presented, the EG effect was increased *(14)*.

The effect of EG and CBH on the cellulose microfibril structure has been investigated by transmission electron microscopy *(27)*. It was determinate that the EG action evokes the splitting of the microfibril structure into submicrofibrils. This leads to the appearance of whisker-like submicrofibrils at the cellulose microfibril periphery. According to Sprey and Bochem *(27)*, after the surface erosion initiated by EG, microfibrils were transformed into amorphous cellulose by CBH. This step occurred concomitant with an increase in the surface area in the submicrofibril part and an increase in water uptake into the intercrystalline structure *(27)*. Noe and coauthors *(28)* investigated the action of the crude xylanase complex on pulp, and concluded that the enzyme action can be summarized in terms of external and internal fibrillation. Our data confirmed the opinion that the fines fraction is preferable for the enzyme action, particularly when the enzymatic complex enriched in xylanases is used. The hydrolysis of the fines seems to be concomitant with an increase in the recycled newspaper brightness. The improvement of WRV and °SR values can be attributed to the hydrolysis of the long fiber fraction rather then the fines fraction, and it could be assumed that the improvement of drainage occurs owing to the EG activity, improved by its synergistical action with xylanase.

ACKNOWLEDGMENTS

Thanks are owed to our colleague M. Skute for skillful experimental work. This research was financed from the Latvian budget, grant No.772 (Biotechnology).

REFERENCES

1. Mandels, M., Andreotti, R., and Roche, C. (1976), *Biotechnol. Bioeng. Symp.* **6,** 21.
2. Nisizawa, T., Suzuki, H., Nakayama, M., and Nisizawa, K. (1971), *J. Biochem.* **70,** 375.
3. Tangnu, S. H., Blanch, H. W., and Wilke, C. R. (1981), *Biotechnol. Bioeng.* **23,** 1873.
4. Zeltina, M. P., Leite, M. P., Vanags, J. J., Apine, A. J., and Viesturs, U. E. (1987), *Acta Biotechnol.* **2,** 157.
5. Farkaš, V., Kerns, G., Liskovs, M., and Bauer, Š. (1986), *Folia Microbiol.* **31,** 277.
6. Acebal, C., Castillion, M. P., Estrada, P., Mata, I., Aquado, J., and Romero, D. (1988), *Acta Biotechnol.* **6,** 487.
7. Bhat, G. R., Heitmann, J. A., and Joyce, T. W. (1991), *Tappi J.* **74,** 151.
8. Pommier, J. C., Fuentess, J. I., and Goma, G. (1989), *Tappi J.* **72,** 187.
9. Pommier, J. C., Goma, G., and Fuentess, J. L. (1990), *Tappi J.* **73,** 197.
10. Lee, S. B., and Kim, I. H. (1983), *Biotechnol. Bioeng.* **23,** 33.
11. Prasad, D. I., Heitmann, J. A., and Joyce, T. W. (1992), *Prog. Paper Recycling* **1(3),** 21.
12. Jackson, L. S., Heitmann, J. A., and Joyce, T. W. (1993), *Tappi J.* **76,** 147.
13. Jackson, L. S., Heitmann, J. A., and Joyce, T. W. (1994), *Prog. Paper Recycling* **(2),** 32.
14. Stork, G., Pereira, H., Wood, T. M., Dusterhoft, E. M., Toft, A., and Puls, J. (1995), *Tappi J.* **78,** 79.
15. Bailey, G. L., Biely, P., and Pountanen, K. (1992), *J. Biotechnol.* **23,** 257.
16. Miller, G. L. (1959), *Anal. Chem.* **31,** 426.
17. Shvinka, J., Viesturs, U., Baburins, L., and Kalnenieks, U. (1992), *Chem. Biochem. Eng.* **6,** 35.

18. Leite, M., Apine, A., Zeltina, M., and Shvinka, J. (1989), *Acta Biotechnol.* **9,** 421.
19. Viesturs, U. E., Leite, M., Shvinka, J. E., Zeltina, M. O., and Apine, A. J. (1989), *Proceedings of the Soviet–Finnish seminar,* Sept. 26–29, 1988, at Riga, Latvia, Maltseva, O., ed., Scientific Center of Biological Research AS USSR, Puschino.
20. Andreotti, R. E., Mandels, M., and Roche, C. (1977), *Proceedings of the Bioconversion Symposium.* Ghose, T. K., ed., II T, Delhi, India, p. 249.
21. Sternberg, D. (1976), *Biotechnol. Bioeng.* **18,** 75.
22. Wilke, C. R. and Blanch, H. W. (1978), Report No. LB1-8658 December, 1978, Lawrence Berkeley Laboratory, Berkeley, CA.
23. Wang, C.-H., Hsen, T.-H., and Huang, C.-M. (1988), *J. Biotechnol.* **9,** 47.
24. Halliwell, G. and Loveladay, J. (1981), *J. Gen. Microbiol.* **126,** 211.
25. Hesse, G. (1983), *Zeitung Allgemeine Microbiol.* **23,** 93.
26. Fan, L. T. and Lee, Y. H. (1963), *Biotechnol. Bioeng.* **25,** 2707.
27. Sprey, B. and Bochem, H. P. (1992), *FEMS Microbiol. Lett.* **97,** 113.
28. Noe, P., Chevalier, J., Mora, F., and J. Comtat (1986), *J. Wood Chem. Technol.* **6,** 167.

0273-2289/96/5758—0361$6.50

Factors Affecting Polyhydroxybutyrate Biosynthesis in the Marine Photosynthetic Bacterium *Rhodopseudomonas* sp. Strain W-1S

WASIMUL Q. CHOWDHURY, KENJI IDEHARA, ISAMU MAEDA, FUSAKO UMEDA, KIYOHITO YAGI,* YOSHIHARU MIURA, AND TADASHI MIZOGUCHI

Faculty of Pharmaceutical Sciences, Osaka University, 1-6 Yamada-oka, Suita, Osaka 565 Japan

ABSTRACT

A marine photosynthetic bacterium, *Rhodopseudomonas* sp. strain W-1S, accumulated polyhydroxybutyrate (PHB) to 56% of the dry cell weight under microaerobic and photoheterotrophic conditions. The addition of NaCl and phosphate was essential for the accumulation. Acetate was the best carbon source for PHB accumulation, whereas adding malate, succinate, or sugars that did not support the growth in this strain resulted in no accumulation. The strain accumulated PHB to about 50% of the dry cell weight at the logarithmic phase of growth. These results suggested that the accumulation would be growth associated under microaerobic and photoheterotrophic conditions. Vitamins, especially biotin, exerted an inhibitory effect on PHB accumulation and a stimulatory effect on pigment production. There might be regulation of carbon flux to PHB or pigments as storage materials.

Index Entries: Marine; polyhydroxybutyrate; photosynthetic bacteria; hydrogen; *Rhodopseudomonas*.

INTRODUCTION

Many bacteria accumulate polyhydroxybutyrate (PHB) as a carbon and energy storage compound, or use it as a sink for reducing equivalents *(1)*. PHB and probably most other polyhydroxyalkanoates (PHA) are completely biodegradable to CO_2 and H_2O. Since PHA are thermoplastic and can be manufactured on a large scale, they have attracted much attention as possible substitutes for plastics produced from petrochemicals *(2,3)*. Recent studies have shown that the intracellular polymers are produced also by some halobacteria *(4)*. Members of the genus *Halobacterium* belong to archaebacteria, a separate branch of cell evolution *(5)*. Halobacteria are known to require a high concentration of NaCl in the medium in order to survive and grow *(6)*, and their cells lyse if exposed to distilled water *(7)*.

*Author to whom all correspondence and reprint requests should be addressed.

However, there is little information on the regulation of storage material accumulation in halophilic photosynthetic bacteria. The marine photosynthetic bacterium *Rhodopseudomonas* sp. strain W-1S, which was isolated from marine samples in our laboratory *(8)*, was found to be able to evolve hydrogen under light from fermentative products of microalgae, such as acetic acid and ethanol. The efficiency of energy conversion in the total system with the alga and bacterium was also evaluated, and a high molar yield of hydrogen was achieved by this alga–bacterial combination in an alternating light–dark cycle *(8)*. In recent studies, we found that this strain can also accumulate a considerable amount of PHB as intracellular granules. Under anaerobic conditions, there might be competition between PHB accumulation and hydrogen evolution for reducing equivalents. This article describes the effects of some environmental factors on PHB accumulation in *Rhodopseudomonas* sp. strain W-1S.

MATERIALS AND METHODS

Microorganism

The marine photosynthetic bacterium *Rhodopseudomonas* sp. strain W-1S was used for all the experiments. This strain was newly isolated from marine samples collected near the coast in the Kinki region of Japan *(8)*.

Cultivation

Cells were inoculated into a 2-L culture flask containing 1 L of L- broth consisting of 0.5% Bacto-peptone, 0.3% yeast extract, and 2.5% NaCl. The cells were cultivated aerobically at 30°C with vigorous shaking.

PHB Accumulation

After cultivation for 24 h, cells were harvested by centrifugation, washed aseptically with sterilized distilled water, and resuspended into 1-L culture bottles containing 500 mL of modified Okamoto medium (MOM, pH 8.0), consisting of 30 g NaCl, 0.2 g $CaCl_2 \cdot 2H_2O$, 0.25 g $MgSO_4 \cdot 7H_2O$, 20 mg $FeSO_4 \cdot 7H_2O$, 40.8 mg KH_2PO_4, 495 mg K_2HPO_4, 0.5 mg thiamine, 0.5 mg nicotinic acid, 0.3 mg *p*-aminobenzoic acid (PABA), 0.05 mg biotin, 2.86 mg H_3BO_3, 1.81 mg $MnCl_2 \cdot 4H_2O$, 0.22 mg $ZnSO_4 \cdot 7H_2O$, 0.08 mg $CuSO_4 \cdot 5H_2O$, 0.021 mg Na_2MoO_4, 0.01 mg $CoCl_2 \cdot 6H_2O$, and 50 mg EDTA disodium salt in 1 L, with the addition of 5 m*M* NH_4Cl as a nitrogen source, and sodium acetate, sodium succinate, sodium malate, and sodium pyruvate each at a final concentration of 0.1% as carbon sources. Cells were then incubated microaerobically with a gentle shaking by a magnetic stirrer at 30°C, and illuminated under fluorescent and incandescent lamps at 125 μmol photon/m^2 s for 48 h. After incubation for 48 h, cells were harvested by centrifugation, washed twice with sterilized distilled water, and lyophilized with a freeze dryer (Model VR16, Hitachi, Japan) overnight. After lyophilization, the total cell dry weight was measured.

Analysis of PHB Content

Pigments were extracted from lyophilized cells by acetone extraction with reflux for 5 h at 70°C. After careful removal of the extracts from the cell debris, the remaining cell debris was again suspended with chloroform, and PHB was

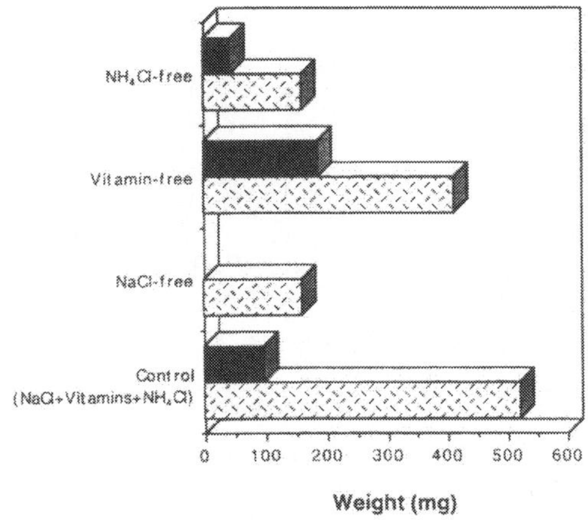

Fig. 1. Effects of NaCl, vitamins, and NH_4Cl on accumulation of PHB in *Rhodopseudomonas* sp. strain W-1S. The control medium contained 3% NaCl, biotin, nicotinic acid, PABA, thiamine, and 5 m*M* NH_4Cl. NaCl, vitamins, or NH4Cl was omitted from the medium, and cells were incubated as described in Materials and Methods. Total cell dry weight (▨) and polymer weight (■) were measured after incubation for 48 h.

extracted for 6 h at 70°C. The cell debris was then removed by passing the suspension through a cellulose filter, and the chloroform solution was concentrated using a rotary vacuum evaporator (Tokyo Rikakikai Co., Ltd., Japan). Finally, the precipitated polymer was separated and the weight of the total PHB content was measured.

RESULTS

The effects of NaCl, vitamins, and NH_4Cl on PHB accumulation in *Rhodopseudomonas* sp. strain W-1S were examined using 0.1% each of acetate, pyruvate, succinate, and malate as substrates. Figure 1 shows the results. In the absence of NaCl, the dry weight of cells after incubation for 48 h was much lower than that of the control, and no PHB accumulation occurred. Cells incubated in the absence of vitamins accumulated twice as much PHB as those incubated in the presence of four vitamins: biotin, nicotinic acid, PABA, and thiamine. The content of PHB under the vitamin-free conditions was 45% (PHB weight/total cell dry weight). Since removal of NH_4Cl decreased both PHB and the total cell dry weight, 5 m*M* NH_4Cl were added in the subsequent experiments.

Figure 2 shows the effect of the NaCl concentration on PHB accumulation. Cells were incubated for 48 h under different NaCl concentrations without adding vitamins. The amount of PHB increased with the NaCl concentration up to 5%, but no PHB accumulation was observed at 10%. The optimum NaCl concentration for growth was 1%. Addition of NaCl at 5% resulted in the best PHB accumulation, but this amount had an inhibitory effect on cell growth. NaCl was therefore added at 3% in the subsequent experiments.

Table 1 shows the effects of vitamins on the PHB content. Each vitamin was omitted in turn to ascertain the inhibitory effect of individual vitamins on PHB accumulation. Removal of thiamine did not affect the PHB content, whereas remov-

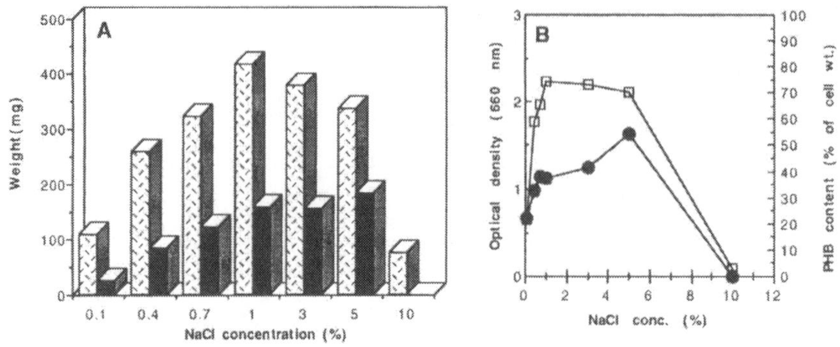

Fig. 2. Effects of NaCl concentration on accumulation of PHB in *Rhodopseudomonas* sp. strain W-1S. Cells were incubated in the presence of various concentrations of NaCl. The medium contained 5 m*M* NH_4Cl and no vitamin. **(A)** Total cell dry weight (⬚) and polymer weight (■). **(B)** Percentage of PHB in total cell dry weight (●) and OD660 (□).

Table 1
Effects of Vitamins on PHB Accumulation in *Rhodopseudomonas* sp. Strain W-1S[a]

Condition	Intitial cell conc., $OK_{660\ nm}$	Final cell conc., $OD_{660\ nm}$	Total cell dry wt, mg	PHB content, mg	PHB ratio, %, PHB/cell dry wt
Control all vitimins	0.3	2.57	521	99	18.9
Vitamin-free	0.3	2.50	411	186	45.2
Biotin-free	0.3	2.32	451	132	29.4
Nicotinic acid-free	0.3	2.09	418	30	7.1
PABA-free	0.3	2.66	561	69	12.4
Thiamine-free	0.3	2.52	504	104	20.7

[a]Cells were grown at 30°C under constant illumination for 48 h in MOM containing 5 m*M* NH_4Cl, and sodium succinate, sodium acetate, sodium pyruvate, and sodium malate each at a final concentration of 0.1%.

ing nicotinic acid and PABA decreased it. However, biotin removal resulted in significantly more PHB accumulation than that obtained under the control conditions in which all four vitamins were added. These results indicate that only biotin had an inhibitory effect on the PHB accumulation in this strain.

Thus far, four carbon sources were used as substrates for PHB accumulation. Next, the effect of each carbon source on the accumulation was examined individually (Table 2). Acetate was shown to be the best substrate, with the PHB content reaching 56% of the cell dry weight. Pyruvate was less effective, giving a PHB content of 12.6%. The addition of malate and succinate, which are good substrates for photohydrogen evolution in this strain, resulted in no PHB accumulation. Cells incubated with glucose, fructose, sucrose, and gluconate also did not accumulate PHB (data not shown). Figure 3 shows the effects of phosphate on PHB accumulation. Cells were incubated for 48 h with 0.5% acetate in the presence and absence of

Table 2
Effects of Carbon Source on PHB Accumulation
in *Rhodopseudomonas* sp. Strain W-1S[a]

Carbon source	Initial cell conc., $OD_{660\ nm}$	Final cell conc., $OD_{660\ nm}$	Total cell dry wt, mg	Total PHB content, mg	PHB % of cell dry wt
Acetate	0.24	1.58	211	118	56.1
Malate	0.24	0.26	50	None	—
Succinate	0.24	0.31	53	None	—
Pyruvate	0.24	0.84	135	17	12.6

[a]Cells were grown at 30°C under constant illumination with each carbon source (final concentration, 0.5%) for 48 h in MOM containing 5 m*M* NH_4Cl and without vitamins.

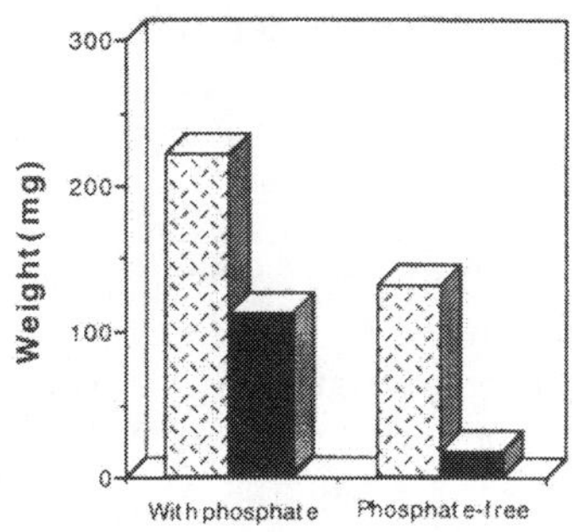

Fig. 3. Effects of phosphate on accumulation of PHB in *Rhodopseudomonas* sp. strain W-1S. Cells were incubated in the presence or absence of phosphate, and the total cell dry weight (⁘) and polymer weight (■) were measured after incubation for 48 h. The medium contained 5 m*M* NH_4Cl, 3% NaCl, and no vitamin.

phosphate. Since the removal of phosphate significantly decreased both PHB and the total cell dry weights as well as NaCl, phosphate was shown to be essential for PHB accumulation in this strain.

The relationship between growth phase and PHB accumulation was examined under microaerobic and photoheterotrophic conditions in MOM. The strain accumulated PHB to about 50% of the dry cell weight at both logarithmic and stationary phases of growth.

DISCUSSION

These results demonstrate that the newly isolated marine photosynthetic bacterium *Rhodopseudomonas* sp. strain W-1S is able to accumulate a considerable amount of PHB under a polymer-enhancing medium in which NaCl is provided. No PHB accumulation was observed in the absence of NaCl in the MOM. We also found that under NaCl-free conditions, the cell growth decreased significantly as compared to that in the presence of NaCl. NaCl thus seems to be an essential component for polymer accumulation as well as for growth in this strain. Phosphate was also shown to be necessary for growth and polymer accumulation, although it has been reported that removal of phosphate greatly enhanced the polymer accumulation in other bacteria *(9,10)*. Usually, cells accumulate more polymer under conditions in which the growth is suppressed by the removal of substrate other than carbon source. On the contrary, the strain W-1S accumulated much polymer in a growth-associated manner under photoheterotrophic conditions. Acetate was shown to be the best carbon source for PHB accumulation. Cells could not accumulate polymer using substrates that did not support photoheterotrophic growth, such as succinate, malate, and sugars. Sugars are, however, good substrates for PHB accumulation in other halobacteria *(9)*.

Removal of vitamins greatly stimulated PHB accumulation. The presence of four vitamins—thiamine, PABA, nicotinic acid, and biotin—might stimulate the production of pigments, such as carotenoid, because the cells displayed a very deep pink color in the presence of vitamins, but were very light pink in their absence. It appears that the carbon flux to PHB or pigments is regulated by certain vitamins under microaerobic and photoheterotrophic conditions in this strain. Under light anaerobic conditions, this bacterium exhibited nitrogenase-dependent hydrogen evolution from organic compounds. Both hydrogen evolution and PHB accumulation could be considered to be electron sinks in the microorganism. Inhibition of nitrogenase greatly enhanced the PHB accumulation under light anaerobic conditions (unpublished data). Therefore, reducing equivalent competition would appear to exist between PHB accumulation and hydrogen evolution.

We have previously proposed a solar energy conversion system based on the microalgal fixation of carbon dioxide *(8)*. The system consists of photosynthetic starch accumulation in microalga, algal fermentation to produce organic compounds, and conversion of the organic compounds to hydrogen by photosynthetic bacteria.

In this study, we have shown that it is possible to control the metabolism in a photosynthetic bacterium. Depending on the purpose, organic compounds from microalga could be used for the production of biodegradable polymer, as well as a source of clean energy—hydrogen.

REFERENCES

1. Anderson, A. J. and Dawes, E. A. (1990), *Microbiol. Rev.* **54,** 450–472.
2. Brandl, H., Gross, R. A., Knee., E. J., Jr., Lenz, R. W., and Fuller, R. C. (1989), *Int. J. Biol. Macromol.* **11,** 49–55.
3. Brandl, H., Gross,. E. J., Lenz., R. W., and Fuller, R. C. (1988), *Appl . Environ. Microbiol.* **54,** 1977–1982.
4. Castillo, F., Rodriguez, R. F., Ramos. J. G., and Berraquero, R. F. (1986), *Appl. Environ. Microbiol.* **51,** 214–216.
5. Woese, C. R. and Fox, G. (1977), *Proc. Natl. Acad. Sci.* USA. **74,** 5088–5090.
6. Larsen, H. (1962), *Halophilism*, vol 4, Academic, New York, pp. 297–342.
7. Abram, D. and Gibbons, N. E. (1961), *Can. J. Microbiol.* **7,** 741–750.
8. Miura,Y., Saitoh, C., Matsuoka, S., and Miyamoto, K. (1992), *Biosci. Biotechnol. Biochem.* **56,** 751–754.
9. Lillo, J. G. and Valera, F. R. (1990), *Appl. Environ. Microbiol.* **56,** 2517–2521.
10. Maness, P.-C. and Weaver, P. F. (1994), *Appl. Biochem. Biotechnol.* **45/46,** 395–406.

0273-2289/96/5758—0367$6.75

Acquisition of the Ability to Grow Under Autotrophic Conditions by Heterotrophic Bacteria Through the Introduction of DNA Fragments from Hydrogen-Oxidizing Bacteria

FUSAKO UMEDA, KAZUHIRO YANO, NORITSUGU GOHDA, KIKUNO FUJIHARA, ISAMU MAEDA, KIYOHITO YAGI,* YOSHIHARU MIURA, AND TADASHI MIZOGUCHI

Faculty of Pharmaceutical Sciences, Osaka University, 1-6 Yamada-oka, Suita, Osaka 565 Japan

ABSTRACT

Pseudomonas oxalaticus OX1 carrying recombinant plasmid pFUS and 29 gram-negative bacteria were used as donor and recipients, respectively, for conjugations. The plasmid pFUS contains genes necessary for chemolithoautotrophic growth in a hydrogen-oxidizing bacterium, *Alcaligenes hydrogenophilus*. By the conjugal transfer of plasmid pFUS, 12 strains acquired the ability to grow in a mineral salts medium under a gas mixture containing H_2, O_2, and CO_2. *Pseudomonas cepacia* carrying pFUS had hydrogenase and ribulose-1,5-bisphosphate carboxylase activities, and could grow at a rate almost equivalent to the growth rates of ordinary hydrogen-oxidizing bacteria. *Pseudomonas acidophila* carrying plasmid pFUS expressed indigenous antibiotic secretion under chemolithoautotrophic conditions. Plasmid pFUS could be stably maintained in these strains in the absence of selection pressure created by the addition of an antibiotic or chemolithoautotrophic conditions. This study shows that introduction of DNA fragments responsible for chemolithoautotrophic growth into a producer of valuable substances could be a promising technique for CO_2 recycling.

Index Entries: Hydrogen-oxidizing bacteria; *Alcaligenes hydrogenophilus*; in vivo cloning; chemolithoautotrophy; carbon dioxide.

INTRODUCTION

The most important greenhouse gas, carbon dioxide, is emitted when fossil fuels are burned. The rate at which carbon dioxide has accumulated in the atmosphere closely parallels the increased use of energy, and efforts to stabilize emis-

*Author to whom all correspondence and reprint requests should be addressed.

sions of carbon dioxide have now assumed great urgency. Recently developed technology enables carbon dioxide to be recovered from the gases emitted by steam power plants *(1)*. In light of this, we have been trying to develop means of recycling recovered carbon dioxide as useful materials, such as antibiotics and biodegradable polymer, using microorganisms. Hydrogen-oxidizing bacteria can grow in a mineral salts medium with carbon dioxide as a sole carbon source and hydrogen as an energy source. These characteristics could be used for carbon dioxide recycling to produce many valuable substances. *Alcaligenes hydrogenophilus* is a gram-negative, facultative, hydrogen-oxidizing bacterium *(2)* harboring a megaplasmid, pHG21-a, which has genes encoding carbon fixation (Cbb), hydrogen oxidation (Hox), and formate oxidation (Fox) *(3)*. Recently, we succeeded in cloning the DNA region of the plasmid responsible for carbon fixation and hydrogen oxidation in an IncP1 R-plasmid, R68.45, and the recombinant plasmid pFU7 was obtained *(4)*. Further, a σ^{54}-like chromosome gene that acted in the expression of *hox* was inserted into pFU7, and the recombinant plasmid pFUS was obtained *(5)*. Since the R-plasmid is self-transmissible and has a broad host range *(6)*, the recombinant plasmid can be transferred to a wide variety of bacteria by conjugation. In this study, we attempted to confer chemolithoautotrophic growth (Aut) ability to bacteria which can secrete antibiotics and to confirm their production from carbon dioxide.

MATERIALS AND METHODS

Bacterial Strains and Plasmids

pFUS has three resistance markers derived from R68.45: resistance against ampicillin (Ap), kanamycin (Km), and tetracycline (Tc). pFUS was kept not in *A. hydrogenophilus*, but in *Pseudomonas oxalaticus* OX1 because the transfer gene of R68.45 was highly expressed in strain OX1 *(7)*. *P. oxalaticus* strain OX1-MH212 (an auxotrophic mutant of strain OX1) carrying pFUS was used as the donor. *Bacillus subtilis* IFO3513 was used as the test bacterium for antibiotics secreted from conjugants containing pFUS.

Conjugation

Conjugations were done by filter mating at 30°C. L-broth overnight cultures of donor and recipient were mated on a membrane filter in an L-plate overnight. Bacteria were then suspended in 4 mL of water, and the bacterial suspension was spread on selective medium plates at 0.1 mL/plate using an appropriate dilution. Transconjugants were selected by chemolithoautotrophic growth and antibiotic resistance markers. Chemolithoautotrophic growth was performed in a mineral salts medium under a gas mixture containing H_2, O_2, and CO_2 at a ratio of 8:1:1 (v:v:v) *(8)*.

Selection

Selection by antibiotic resistance markers was carried out under heterotrophic conditions. Glucose, fructose, or malate was used as the substrate, which was added at a final concentration of 0.2% (w/v). Antibiotics were used at the following final concentrations (µg/mL): Ap, 100; Km, 100; Tc, 10. The antibiotic and substrate used for each recipient strain are given in the Tables 1, 2, and 3. Transconjugants were cultivated at 37°C for *Escherichia coli* strains and 30°C for other strains.

Table 1
Strains Acquiring Chemolithoautotrophic Growth Ability by Introduction of the Recombinant Plasmid pFUS

Strain	Autotrophic conditions, CFU/mL	Antibiotic resistance, CFU/mL[a]
P. acidophila IFO 13774	2.8×10^5	1.0×10^8 (G/Tc)
Pseudomonas aeruginosa IFO 13561	1.5×10^6	n.t.[b]
P. cepacia IFO 14595	4.0×10^4	1.8×10^7 (G/Km)
P. cepacia IFO 15124	2.0×10^4	n.t.
P. dacunhae IAM 1152	1.5×10^6	3.6×10^6 (M/Tc)
Pseudomonas fluorescens ATCC 39502	3.2×10^2	1.1×10^4 (M/Km)
P. mesoacidophila IFO 13884	1.3×10^3	n.t.
Pseudomonas oleovorans ATCC 13474	2.8×10^5	3.1×10^7 (M/Tc)
Pseudomonas putida TN100	2.5×10^3	2.7×10^8 (F/Km)
Pseudomonas stutzeri ATCC 31147	1.2×10^3	n.t.
Rhodopseudomonas rubrum NCIB8255	8.4×10^2	1.4×10^4 (M/Km)
Rhodopseudomonas spheroides NCIB8253	1.2×10^2	1.4×10^7 (M/Km)

[a]The carbon source for heterotrophic cultivation and the antibiotic used are indicated as follows: malate (M), glucose (G), fructose (F), kanamycin (Km), tetracyclin (Tc).
[b]n. t., not tested.

Table 2
Strains Acquiring Organotrophic Growth Ability by Introduction of the Recombinant Plasmid pFUS

Strain	Formate oxidation, CFU/mL	Antibiotic resistance, CFU/mL[a]
P. aeruginosa IFO3898	4.5×10^6	2.6×10^6 (G/Km)
P. fluorescens NCIB8286	1.4×10^6	2.9×10^6 (F/Tc)
P. oleovorans IFO13583	3.7×10^4	0.2×10^5 (M/Km)
P. putida ATCC15175	1.0×10^6	1.0×10^9 (M/Km)
P. putida IFO3738	6.6×10^2	5.4×10^6 (G/Km)
Pseudomonas sp. TA8	6.4×10^5	2.2×10^7 (F/Km)

[a]Abbreviations are the same as in Table 1.

Inhibition of Bacterial Growth

The Aut^+ transconjugants obtained in this study were cultivated under chemolithoautotrophic conditions. Cells were eliminated from the culture broth by centrifugation. The culture filtrate was adjusted to pH 7.0 and concentrated by evaporation. The mineral salts medium was used as the standard. Cell suspension of *B. subtilis* was inoculated into L-plate, and cups were put on the plate. Samples were poured gently into the cups, and the plate was incubated at 30°C overnight.

Table 3
Strains Acquiring Antibiotic Resistance
by Introduction of the Recombinant Plasmid pFUS

Strain	Antibiotic resistance,[a] CFU/mL
E. coli IFO 3210	4.3×10^8 (G/Tc)
E. coli IFO 3552	8.3×10^8 (G/Tc)
E. coli IFO 13500	1.4×10^7 (G/Ap)
E. coli IFO 13502	1.7×10^7 (G/Ap)
Pseudomonas chlororaphis IFO 3521	2.2×10^8 (G/Km)
P. fluorescens ATCC 13475	1.4×10^7 (M/Km)
P. fluorescens IFO 3925	1.3×10^8 (G/Km)
P. fluorescens IFO 13658	5.5×10^7 (G/Km)
Pseudomonas mendocina IFO 14162	6.8×10^5 (G/Km)
Pseudomonas reptilivora IFO 3461	2.4×10^8 (G/Km)
Pseudomonas sp. IFO 13302	1.2×10^6 (M/Km)

[a]Ap, ampicillin; other abbreviations are the same as in Table 1.

Assays

Activities of hydrogenases and ribulose-1,5-bisphosphate carboxylase were measured as described previously *(4)*. Protein was measured by the method of Lowry et al. *(9)*.

RESULTS

Conjugations were done as described in Materials and Methods using *P. oxalaticus* strain OX1-MH212 carrying pFUS as the donor. Table 1 shows strains acquiring Aut ability by conjugal transfer of pFUS. In the first eight strains listed in Table 1, the transfer of pFUS was determined by Aut and the antibiotic resistance marker, whereas in the remaining four strains, the transfer was determined only by Aut, because they indigenously possessed resistance to Ap, Km, and Tc. Strains in Table 1 formed colonies on the selection plate consisting of mineral salts and antibiotic under a mixture H_2, O_2, and CO_2. Aut ability and antibiotic resistance were derived from inserted DNA in pFUS and the vector plasmid, R68.45, respectively. Comparing the Aut^+ and antibiotic-resistant transconjugants, *Pseudomonas dacunhae* IAM1152 exhibited almost the same number of transconjugants under both selections, but the other five strains had several orders of magnitude fewer transconjugants in the selection by Aut than that by the antibiotic-resistant marker. Although 29 strains in all acquired antibiotic resistance, only 12 could express chemolithoautotrophy. Since pFUS encodes formate-dependent organoautotrophic growth (Fox) ability as well as that of Hox, we next examined whether the remaining 17 strains carrying pFUS had organoautotrophic growth ability. Cells were inoculated on plates containing formate and antibiotic, and cultivated aerobically for 1 wk at 30°C. Table 2 shows that six strains carrying pFUS could grow on formate as both an energy and carbon source. The remaining 11 strains, as shown in Table 3, only expressed antibiotic resistance in the region of R.68.45.

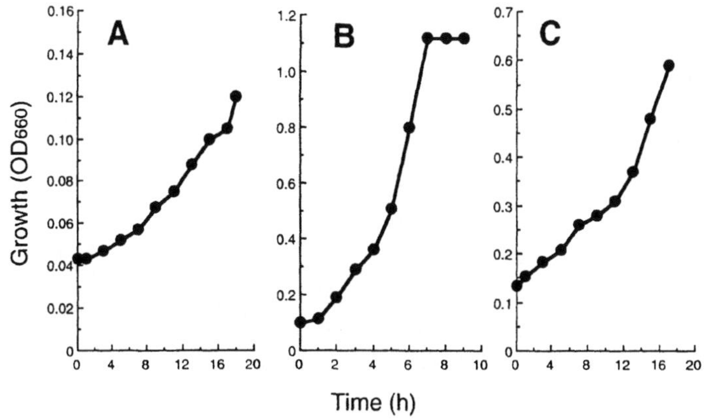

Fig. 1. Chemolithoautotrophic growth of *Pseudomonas* strains carrying pFUS. Recombinant *P. acidophila* **(A)**, *P. cepacia* **(B)**, and *P. mesoacidophila* **(C)**, were each inoculated into test tubes containing 6 mL of mineral salts medium and cultivated at 30°C under a mixture of H_2, O_2, and CO_2 at a ratio of 8:1:1 (v:v:v). The OD at 660 nm was measured at the times indicated.

Out of the 12 strains listed in Table 1, we selected 3, *Pseudomonas acidophila, Pseudomonas cepacia,* and *Pseudomonas mesoacidophila*, which produce the antibiotics G-6302, Altericidins, and SB-72310, respectively, for further experiments. The three strains were inoculated into liquid medium containing only mineral salts and cultivated at 30°C under an H_2, O_2, and CO_2 gas mixture. There was no growth of their parent strains under these conditions in the absence of pFUS (data not shown). As shown in Fig. 1, the Aut characteristics of the three strains were also confirmed in liquid culture. Table 4 shows the growth rates and activities of hydrogenases and ribulose-1,5-*bis*-phosphate carboxylase, which are key enzymes of H_2 oxidation and CO_2 fixation, respectively. The growth and activities varied among the three strains. Notably, *P. cepacia* carrying pFUS grew fastest at a rate almost equivalent to the growth rates of ordinary hydrogen-oxidizing bacteria, whereas *P. acidophila* containing pFUS grew very slowly and had only very slight enzyme activity for Aut ability. The stability of pFUS was examined in the three strains by checking the Aut ability after heterotrophic cultivation. The strains were cultivated overnight in L-broth and inoculated onto L-plates after appropriate dilution. One hundred colonies were picked, transferred onto a mineral-salts-medium plate, and cultivated for several days under a mixture of H_2, O_2, and CO_2. All tested colonies of three strains were confirmed to be grown under chemolithoautotrophic conditions (data not shown). Therefore, pFUS could be stably maintained in these strains in the absence of selection pressure, such as the addition of antibiotic or chemolithoautotrophic conditions.

The ability of the three strains containing pFUS to secrete antibiotics was examined in the culture filtrate obtained from chemolithoautotrophic culture as described in Materials and Methods. Contrary to the Aut ability, the culture filtrate from *P. acidophila* showed strong antibiotic activity toward the test bacterium, *B. subtilis* (Fig. 2). The filtrate from *P. cepacia* did not cause an inhibition zone (data not shown).

Table 4
Chemolithoautotrophic Growth and Activities of Hydrogenases and Ribulose-1,5-Bisphosphate Carboxylase in Recombinant Strains

Strain harboring pFUS	Doubling time, h	RuBPCase, U/g protein[a]	Hydrogenase, U/mg protein SH[b]	MH[c]
P. acidophila IFO13774	11.2	7.8	n.d.[d]	0.03
P. cepacia IFO14595	4.4	14.0	0.46	1.60
P. mesoacidophila IFO13884	8.6	19.0	0.36	1.41

Pseudomonas strains carrying pFUS were cultivated under chemolithoautotrophic conditions.
[a]Ribulose-1,5-bis-phosphate carboxylase.
[b]Soluble hydrogenase.
[c]Membrane-bound hydrogenase.
[d]Not detected.

DISCUSSION

Hydrogen-oxidizing bacteria belonging to diverse taxonomic groups have been reported, including both gram-positive and gram-negative bacteria. It is of interest to elucidate how many taxonomic groups containing hydrogen-oxidizing bacteria can express *aut* genes from *A. hydrogenophilus.* In this study, we chose gram-negative, aerobic, heterotrophic bacteria as the recipients for the following reasons: *A. hydrogenophilus* is gram-negative and the R68.45 used as cloning vector has a broad host range among gram-negative bacteria; the donor and recipient must grow on a nutrient broth like L-broth during conjugation under aerobic conditions, and the Aut$^+$ transconjugant of each recipient has to be cultivated under an atmosphere of 10% O_2. *Pseudomonas*, *Rhodopseudomonas*, and *Escherichia* strains were used as the recipients. *P. oxalaticus* OX1-MH212 carrying pFUS was used as the donor, and the appearance of Aut$^+$ transconjugants was determined. Conjugants were divided into three groups:

1. Those expressing Aut ability;
2. Those expressing only Fox ability; and
3. Those expressing only antibiotic resistance.

In Group 1, a comparison of the number of colonies selected by Aut and by antibiotic resistance shows that *aut* genes might be deleted from pFUS during conjugation. However, once pFUS was successfully transferred into the recipient, it was shown to be stably maintained even in heterotrophic conditions. Six strains belonging to Group 2 did not have Aut ability, but did possess Fox ability. Hydrogen-oxidizing bacteria can oxidize formate by formate dehydrogenase to form CO_2, and assimilate it via the Calvin cycle *(10)*. These strains did not grow under chemolithoautotrophic conditions, probably because *hox* genes in pFUS could not be expressed or deleted during the conjugation. Since there were 11 strains belonging to Group 3, an extensive analysis to determine the reason why Aut could not be expressed will have to be made.

It was confirmed that metabolically engineered *P. acidophila* and *P. mesoacidophila* secreted antibiotics during chemolithoautotrophic growth. There was an inverse relationship between the growth rate and the amount of antibiotic. Since rapidly growing cells cannot afford to secrete a secondary metabolite, a two-stage system in which growth and production are separated might be effective.

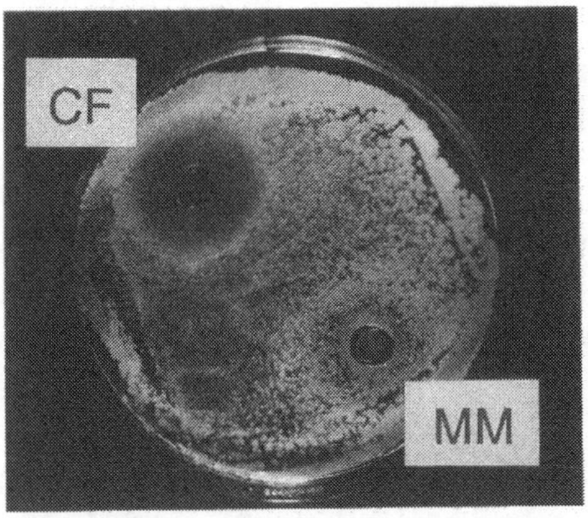

Fig. 2. Growth inhibition of *B. subtilis* IFO3513 by culture filtrate from *P. acidophila* IFO13774 carrying pFUS grown under chemolithoautotrophic conditions. The culture filtrate from *P. acidophila* IFO13774 carrying pFUS (CF) and mineral salts medium (MM) were used as the sample and control, respectively.

In this study, we demonstrated the possibility of constructing a novel system to recycle CO_2. Recently, many attempts have been made to produce valuable products from CO_2 using hydrogen-oxidizing bacteria *(11,12)*. The characteristic feature of this work is that a producer of valuable products can be changed to a chemolithoautotroph by the introduction of the DNA fragment responsible for Aut ability, thereby enabling many substances to be selected as targets for production.

ACKNOWLEDGMENT

This work was supported by the Research Institute of Innovative Technology for the Earth (RITE) Grant.

REFERENCES

1. Kiyohara, M. (1993), *Energy and Resorces* **14,** 91–97.
2. Ohi, K., Takada, N., Komemushi, S., Okazaki, M., and Miura, Y. (1979), *J. Gen. Appl. Microbiol.* **25,** 53–58.
3. Friedrich, B., Friedrich, C. G., Meyer, M., and Schlegel, H. (1984), *J. Bacteriol.* **158,** 331.
4. Umeda, F., Tanaka, N., Kimura, N., Nishie, H., Yagi, K., and Miura, Y. (1991), *J. Ferment. Bioeng.* **71,** 379–383.
5. Umeda, F., Seko, T., Yano, K., Goda, N., Yagi, K., Miura, Y., and Mizoguchi, T. (1995), *J. Ferment. Bioeng.* **79,** 59–61.
6. Haas, D. and Holloway, B. W. (1976), *Mol. Gen. Genet.* **144,** 243–251.
7. Umeda, F., Urushihara, M., Tsujino, Y., Yagi, K., and Miura, Y. (1989), *J. Ferment. Bioeng.* **68,** 207–209.
8. Yagi, K., Min, H., Urushihara, M., Manabe, Y., Umeda, F., and Miura, Y. (1986), *Biochem. Biophys. Res. Commun.* **137,** 114–119.
9. Lowry, O. H., Rosebrough, N. J., Farr, A. L., and Randall, R. J. (1951), *J. Biol. Chem.* **193,** 265–275.
10. Friedrich, C. G., Bowien, B., and Friedrich, B. (1979), *J. Gen. Microbiol.* **115,** 185–192.
11. Hayase, N., Ishiyama, A., and Niwano, M. (1994), *Appl. Environ. Microbiol.* **60,** 3336–3342.
12. Ishizaki, A. and Tanaka, K. (1991), *J. Ferment. Bioeng.* **71,** 254–257.

0273-2289/96/5758—0375$6.75

Production, Characterization, and Application of the Cellulase-Free Xylanase from *Aspergillus niger*

QU YINBO,*[,1] GAO PEIJI,[1]
WANG DONG,[1] ZHAO XIN,[1] AND ZHANG XIAO[2]

[1]*Institute of Microbiology, Shandong University, Jinan 250100, P. R. China; and* [2]*Chemical Engineering College, Nanjing Forestry University, Nanjing 210037, P. R. China*

ABSTRACT

The effects of carbon source on xylanase and cellulase production were studied. The extract of steam-exploded corn stover was found to be the best raw material for producing cellulase-free xylanase. The xylanolytic enzymes of *Aspergillus niger* An-76 were purified by chromatography, and the properties of the four purified components were analyzed. When the enzyme was used to treat the birch Kraft pulp, and followed by a subsequent CEH bleaching, a brightness of 6.8% SBD more than that of the untreated one using the same chlorine dosage, or a saving of nearly 50% of chlorine consumption with the same brightness, was realized.

Index Entries: Xylanase; cellulase-free; *Aspergillus niger*; pulp bleaching.

INTRODUCTION

Xylanolytic enzymes play an important role in the bioconversion of lignocellulosic materials. For example, the energy used for lignin degradation during white rot decay is obtained from hydrolysis of xylan *(1)*. In 1986, Viikari et al. found that when Kraft pulp was pretreated with xylanase, a higher brightness was obtained after bleaching *(2)*. Since then, many reports dealing with the pretreatment of pulp with xylanases have been published. These results show that the xylanase treatment enables significant reduction of chlorine usage, resulting in less production of toxic chlorinated compounds from pulp mills. The xylanase preparations should be free from cellulase activities, since the cellulase would destroy the strength properties of the pulp *(3,4)*.

It has been reported that a large number of mesophilic and thermophilic fungi, e.g., *Aspergillus* strains *(5–7)*, *Trichoderma* strains *(3,6)*, *Chaetomium cellulolyticum (1)*, and *Thermomyces lanuginosus (8)*, can produce xylanases. Most of them produce xylanase and cellulase simultaneously when they were grown on plant residues, but the synthesis of the two enzymes is regulated under a separate control system *(5)*.

*Author to whom all correspondence and reprint requests should be addressed.

In this article, we report the production of a high level of cellulase-free xylanase by using the extract of steam-exploded corn stover, and report on the characterization and application of the cellulase-free xylanase from *Aspergillus niger* An-76.

MATERIALS AND METHODS

Microorganisms

Six filamentous fungi, two bacteria, and one yeast were obtained from the stock cultures in our laboratory *(5)*. The white rot basidiomycetes, *Phanerochate chrysosporium* ME 448, was kindly offered by the Wood Research Institute, Kyoto University, Japan.

Substrates

The steam explosion of corn stover was carried out in pilot-scale equipment with a working volume of 1 m^3. The corn stover was heated to 180°C by steam under a pressure of 1 MPa for 20 min, and exploded by releasing the pressure instantaneously. The exploded corn stover was extracted by hot water (about 75°C) with a water ratio of 1:10. About 15 g/L of total sugar were obtained in the extract. More than 85% of it was pentoses, and the ratio of monose to olignose was 1:9.

Medium and Cultivation

The basic mineral medium of Mandels and Reese *(9)* was used. Twenty grams per liter of carbon source and 10 g/L of wheat bran were added to the medium. After inoculation, the 300-mL flasks containing 80 mL medium were shaken on an orbital shaker at 200 rpm and 28°C for 3–4 d (*P. chrysosporium* was at 39°C for 6 d). The cultured medium was centrifuged at 5000*g* for 10 min, and the clear supernatant was used for enzyme assay and purification.

Enzyme Assay

Endo-1,4-β-xylanase activity was determined by estimating the xylose liberated from 1% of beechwood xylan (Sigma, St. Louis, MO) suspended in acetate buffer (0.05*M*, pH 5.0) after 20 min of reaction at 50°C. CMCase activity and filter paper activity (FPA) were assayed as recommended by IUPAC *(10)*, and the reducing sugar was determined by the DNS method using glucose as standard sugar. One unit of enzyme activity was defined as the amount (μmol) of reducing sugar released per minute.

Chromatography

The crude enzyme in the clear supernatant was precipitated with ammonium sulfate at 80% saturation, and the precipitate was collected by centrifugation at 5000*g* for 20 min. Crude enzyme precipitate (about 2.5 g of dry wt) was dissolved in 20 mL of 0.05*M* phosphate buffer (pH 6.8). The enzyme solution was chromatographed on Sephadex G-100 column (26 × 1000 mm), DEAE-Biogel A, and CM-Biogel A column (26 × 200 mm). Eluted fractions were analyzed for xylanase, CMCase, and FPase activities.

Table 1
Xylanase and Cellulase Production in the Corn Stover and Wheat-Bran-Based Medium by Various Microorganisms[a]

Strains	Xylanase, IU/mL	CMCase, IU/mL	FPase, IU/mL
Bacillus subtilus 921	8.4	12.0	n.d.
Xanthomonas sp.	7.2	8.4	n.d.
Endomyces 9304	8.6	4.3	n.d.
A. niger An-76	38.6	11.3	0.6
A. niger L22	32.4	24.2	1.2
Chaetomium globosum SM27	4.0	8.2	0.6
Neurospora crassa 1606	4.1	11.2	0.5
Trichoderma harzianum 912	8.2	10.3	0.4
Lentinus lepidens	10.2	4.7	n.d.
P. chrysosporium ME446	14.3	10.2	0.7

n.d., not detectable.
[a]All data in the tables (except some of data in Table 5) are means of three tests.

Gel electrophoresis on a 10% polyacrylamide gel using the discontinuous buffer system and thin-layer gel isoelectric focusing (LKB-2117 Multiphor) were used to analyze enzyme purity, molecular weight, and pI (stained by Coomassie blue). Bovine serum albumin (mol wt 68,000), ovalbumin (mol wt 45,000), and chymotrypsinogen (mol wt 25,000) were used as the standard proteins (Bio-Rad, Hercules, CA).

Treatment of Pulp by Xylanase

The treatment was carried out in plastic bags. Thirty grams of Kraft pulp were added to each bag with various amounts of enzyme. The pulp consistency was 5%, temperature was 45°C, and the pH was adjusted to 4.8 by HAc-NaAc buffer. Incubation time was 3 or 4 h.

After enzyme treatment, the pulp was washed to neutrality and bleached by the CEH three-stage bleaching. The pulp and paper were tested according to the standard methods.

RESULTS AND DISCUSSION

High Level Cellulase-Free Xylanase Production Using the Extract of Steam-Exploded Corn Stover

The effects of carbon source on xylanase and cellulase productions were tested by adding three different carbon sources to the basic mineral medium. As shown in Table 1, the yields of both xylanase and cellulase were high in the corn stover–wheat bran-based media. When xylose was selected as the carbon source, the production of both enzymes was repressed (Table 2). The best results of cellulase-free xylanase production were obtained by using the extract of steam-exploded corn stover as the carbon source (Table 3). In this case, the xylooligosaccharide acted not only as a carbon source, but also as an inducer for xylanase production. Small

Table 2
Xylanase and Cellulase Production
in the Xylose-Based Medium by Various Microorganisms

Strains	Xylanase, IU/mL	CMCase, IU/mL	FPase, IU/mL
B. subtilus 921	n.d.	n.d.	n.d.
Xanthomonas sp.	n.d.	n.d.	n.d.
Endomyces 9304	n.d.	n.d.	n.d.
A. niger An-76	3.2	n.d.	n.d.
A. niger L22	0.3	1.2	n.d.
C. globosum SM27	0.2	n.d.	n.d.
N. crassa 1606	0.4	n.d.	n.d.
T. harzianum 912	0.8	n.d.	n.d.
L. lepidens	1.2	n.d.	n.d.
P. chrysosporium ME446	0.6	n.d.	n.d.

n.d, not detectable.

Table 3
Xylanase and Cellulase Production in the Extract of Steam-Exploded
Corn Stover-Based Medium by Various Microorganisms

Strains	Xylanase, IU/mL	CMCase, IU/mL	FPase, IU/mL
B. subtilus 921	4.2	0.2	n.d.
Xanthomonas sp.	3.1	0.3	n.d.
Endomyces 9304	6.3	0.2	n.d.
A. niger An-76	27.3	0.1	0.02
A. niger L22	18.6	0.1	0.01
C. globosum SM27	7.4	0.3	n.d.
N. crassa 1606	12.1	0.2	n.d.
T. harzianum 912	14.3	0.1	n.d.
L. lepidens	8.2	0.2	n.d.
P. chrysosporium ME446	7.3	0.4	0.04

n.d., not detectable.

amounts of glucose existed in the extract (<10% of total sugar), and the production of cellulase was repressed. Corn stover is a cheap agricultural residue, and the cost of its pretreatment (steam-explosion process) is also relatively inexpensive. The extract of steam-exploded corn stover was used for the further studies. *A. niger* An-76 was selected as the xylanase producer according to its xylanase productivity.

Various nitrogen compounds were added to the culture media with the same concentration as the nitrogen content of ammonium sulfate in the basic mineral medium. The xylanase activities after 2, 3, or 4 d of cultivation were determined, and the effects of nitrogen source are shown in Table 4. Except in the medium containing urea, the final pH values of all broth were 3.6–3.8 after 4 d of incubation. According to the data in Table 4, $NaNO_3$ was chosen as the optimal nitrogen source for enzyme production.

Table 4
Effects of Various Nitrogen Sources on the Production of Xylanase by *A. niger* An-76 inthe Extract of Steam-Exploded Corn Stover-Based Medium

	Xylanase, IU/mL		
Nitrogen source	48 h	72 h	96 h
$(NH_4)_2SO_4$	20.5	40.3	46.2
$(NH_4)_2HPO_4$	24.3	36.2	38.2
NH_4NO_3	10.2	30.1	36.5
$NaNO_3$	8.4	34.2	48.0
Beef extract	16.8	28.4	30.1
Peptone	14.6	28.6	48.4
Yeast extract	12.0	24.3	38.8
Urea	10.2	18.3	22.0

Purification and Characterization of Xylanolytic Enzymes from *A. niger* An-76

The procedure used for the fractionation of the xylanolytic enzymes from *A. niger* An-76 is shown in Fig. 1. Four peaks appeared in the eluted fractions from Sephadex G100 column by eluting with 10 mmol/L phosphate buffer, pH 6.8. Two of them (fraction I and fraction II) showed β-xylosidase and endoxylanase activities. Fraction I was concentrated and applied to a DEAE-Biogel A column equilibrated with 0.05*M* phosphate buffer, pH 6.8. After a nonadsorbed protein peak was washed out with the same buffer, the column was eluted using a 0–0.4 mol/L NaCl linear gradient. A β-xylanase component was obtained. The fraction II showing endoxylanase activity was also applied to a DEAE-Biogel A column and eluted as was done with the fraction I. Three endoxylanase components were obtained. The first peak (fraction II-1) was further purified by CM-Biogel A column equilibrated with 0.05*M* acetate buffer, pH 3.8, and eluted with a 0–0.3 mol/L NaCl linear gradient. The endoxylanase I was obtained.

All four components of the xylanolytic enzymes appeared as a single band on SDS polyacrylamide gel electrophoresis. Some of their chemical and physical properties are shown in Table 5.

It was reported that the cellulase and xylanase were usually produced at the same time in the strains of *Trichoderma reesei (11)*. It is difficult to produce cellulase-free xylanase using these strains. The tests of substrate specificity in this study demonstrated that the three endoxylanase components from *A. niger* An-76 showed no CMCase and FPase activity, and the β-xylosidase could not hydrolyze cellobiose. This suggests that the xylanases from *A. niger* An-76 are specific for xylan. This property is very useful for the production of cellulase-free xylanase.

Application of the Cellulase-Free Xylanase as a Pulp Bleaching Booster

The cellulase-free xylanase was used in the treatment of wheat straw pulp before bleaching at 50°C for 3 or 4 h with different enzyme dosages (IU/g pulp). The results show that the more xylanase that was used in the process, the more xylose

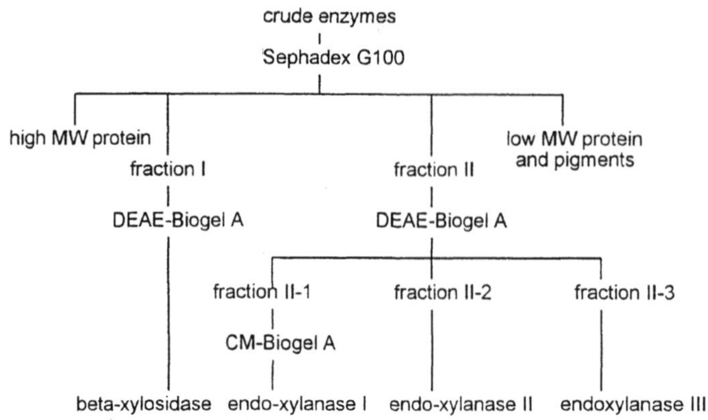

Fig. 1. The procedure used for the purification of xylanolytic enzymes from *A. niger* An-76.

Table 5
Characterization of Four Xylanolytic Enzymes from *A. niger* An-76

	Molecular weight, kDa	pI	Carbohydrate content, %	Temp. optimum, °C	pH optimum
β-xylosidase	147	5.9	4.3	45–55	3.5–4.0
Endoxylanase I	23	4.1	12.5	45–55	5.0
Endoxylanase II	22	3.9	9.4	45–55	5.0
Endoxylanase III	41	4.6	7.8	45–55	3.0

Table 6
The Results of Wheat-Straw Pulp Treatment
Using the Cellulase-Free Xylanase from *A. niger* An-76

	Incubation time, 3 h			Incubation time, 4 h	
Xylanase dosage, IU/g	50	100	200	100	200
Reducing sugar, mg/mL	5.4	5.9	6.7	8.0	8.4
Xylose released, mg/mL	3.0	4.4	5.5	6.0	8.6
Xylose/reducing sugar, %	70	76	82	75	81
κ reduction[a]	2.0	2.8	3.0	5.2	6.0

[a]After bleaching.

that was released from the pulp, which implied that more hemicellulose was hydrolyzed (Table 6). As the incubation time increased, the ratio of xylose to reducing sugar did not increase, but the reduction of κ number increased, which suggests an improvement for the bleaching process and the quality of pulp.

When the birch Kraft pulp was pretreated with the cellulase-free xylanase from *A. niger* An-76 (xylanase dosage 230 IU/g pulp), and followed by a subsequent CEH bleaching, a brightness of 6.8% SBD more than that of the untreated one using the same chlorine dosage, or a saving of nearly 50% of chlorine consumption with the same brightness, could be realized (Table 7). The yield decrease

Table 7
Comparison of the Results of CEH Bleaching on Birch Kraft Pulp with or Without Xylanase Pretreatment

Xylanase dosage, IU/g	Total chlorine dosage, %	Brightness SBD, %	κ Number[a]	Yield[b], %
0	10	69.1	3.7	93.35
0	7	66.3	5.0	94.65
0	5	62.0	5.8	95.01
600	5	70.5	3.3	88.26
320	5	69.4	3.5	88.35
230	5	68.8	3.6	92.95
120	5	67.0	3.6	92.30
60	5	64.6	3.8	93.70

[a]After alkaline extraction.
[b]After bleaching.

was slight, and no adverse effect on the strength properties of pulp was found. The application of this cellulase-free xylanase from *A. niger* An-76 in the pulp-paper industry could be promising.

REFERENCES

1. Higuchi, T. (1989), in *Bioprocess Engineering*, Ghose, T. K., ed., Ellis Horwood, Chichester, UK, pp. 39–58.
2. Viikari, L., Kantelinen, A., Poutanen, K., and Ranua, M. (1986), 3rd Int. Conference on Biotechnology in the Pulp and Paper Industry, Stockholm, June 16–19, pp. 67–69.
3. Viikari, L., Kantelinen, A., Siika-aho, M., Tenkanen, M., Buchert, J., Bailey, M., Pere, J., and Linko, M. (1993), in *Trichoderma reesei Cellulase and Other Hydrolases*, Suomium, P. and Reinikaineu, T., eds., Helsinki, Finland, pp. 255–262.
4. Kautelium, A., Hortling, B., Sundquist, J., Linko, M., and Viikari, L. (1993), *Holzforschung* **47**, 318–324.
5. Chen, H. Z., Gao, P., and Wang, Z. (1990), *Acta Microbiologica Sinica* **30**, 351–357.
6. Beldman, G., Voragen, A. G. J., Rombouts, F. M., Searle-van Leeuwen, M. F., and Pilnik, W. (1988), *Biotechnol. Bioeng.* **31**, 160–167.
7. Zeng, Y. C. and Zhang, S. Z. (1993), *Acta Microbiologica Sinica* **33**, 427–433.
8. Gomes, J., Sinner, M., and Steiner, W. (1993), *Enzyme Microb. Technol.* **15**, 677–682.
9. Mandels, M. and Reese, E. T. (1957), *J. Bacteriol.* **73**, 269–278.
10. Ghose, T. K. (1987), *Pure Appl. Chem.* **59**, 2557–2568.
11. Gomes, I., Gomes, J., Steiner, W., and Esterbauer, H. (1992), *Appl. Microbiol. Biotechnol.* **36**, 701–707.

0273-2289/96/5758—0383$6.50

Fermentation of Orange Peel Hydrolysates by Ethanologenic *Escherichia coli*

Effects of Nutritional Supplements†

KAREL GROHMANN,*,[1]
RANDALL G. CAMERON,[1] AND BÉLA S. BUSLIG[2]

[1]*USDA Citrus and Subtropical Products Laboratory, 600 Avenue S, NW, Winter Haven, FL 33881;*** *and [2]Florida Department of Citrus, Winter Haven, FL*

ABSTRACT

Orange peel, an abundant byproduct of the citrus processing industry, is converted to a mixture of glucose, galacturonic acid, fructose, arabinose, galactose, and xylose by hydrolysis with mixed pectinase and cellulase enzymes. All these sugars can be fermented to ethanol or ethanol and acetic acid by the recombinant bacterium *Escherichia coli* KO11. The fermentation efficiency is improved by the addition of yeast extract, tryptone, mixed amino acids, corn steep liquor, or by proteolytic digestion of endogenous proteins. Batch fermentations of supplemented peel hydrolysate containing 111 g/L of initial total sugars produced 35–38 g/L of ethanol in 48–72 h and a 75–85% yield.

Index Entries: Recombinant *E. coli* B; ethanol production; citrus byproducts.

INTRODUCTION

Orange peel and membranes are abundant byproducts of the citrus processing industry *(1,2)*. Although these byproducts are currently dried in the US and sold as cattle feed *(3)*, they do not command a very high price and could be used for the production of fuel ethanol. More than one million dry tons of these byproducts accumulate in two main processing countries, the US and Brazil *(4)*, with additional supplies being potentially available in other countries where citrus production and processing are undergoing expansion.

Previous studies from our laboratory *(5,6)* reported that *Escherichia coli* KO11 *(7–11)* efficiently ferments all neutral sugars and galacturonic acid in orange peel hydrolysates to ethanol and smaller amounts of acetic and lactic acids in the pH range of 5.8–6.5 and at 34–37°C, when these hydrolysates are supplemented with

*Author to whom all correspondence and reprint requests should be addressed.
**South Atlantic Area, Agricultural Research Service, US Department of Agriculture.
†Mention of a trademark or proprietary product is for identification only and does not imply a guarantee or warranty of the product by the US Department of Agriculture. All programs and services of the US Department of Agriculture are offered on a nondiscriminatory basis without regard to race, color, national origin, religion, sex, age, marital status, or handicap.

1.25 g/L of yeast extract and 2.5 g/L tryptone. Supplementation with these complex nutrients increases the cost of fermentation. Therefore, we have undertaken a more detailed study of nutritional requirements of this fermentation with the aim to develop more practical methods for improvement of the nutritional content of these hydrolysates.

MATERIALS AND METHODS

Bacterial Strains and Media

A recombinant strain of *E. coli* B (ATCC11303), strain KO11 was a gift from Lonnie O'Neal Ingram, University of Florida, Gainesville, FL. Cultures were maintained as frozen stock (–75°C) in 40% glycerol.

Culture Medium

The medium for fermentation experiments consisted of filtered Valencia orange peel hydrolysate prepared by hydrolysis of orange processing byproducts using a mixture of commercial cellulase, pectinase, and β–glucosidase enzymes as described previously *(5,6,12,13)*. The filtrate was partially neutralized by addition of calcium carbonate to pH 5.5, and the precipitated solids were removed by filtration through a 1.2-µm glass fiber filter. The medium and nutritional supplements were filter-sterilized (0.45-µm membrane filter), and chloramphenicol was added to a final concentration of 60 mg/L. The supplementation of the media is described below in the text (*see* Table 1). Filter-sterilized proteinase preparations were added to some fermentations at concentrations recommended in a previous publication *(11)* or at concentrations described later in this article. The proteinase preparations were added immediately after inoculation.

Fermentation Experiments

Fermentations were carried out in pH-controlled, stirred fermenters at pH 6.0, as described previously *(6)* at a temperature of 35°C. Fermentations were inoculated to an initial cell density of 0.11 g dry cell wt/L.

Analyses

Growth was monitored by filtration of culture aliquots through 0.45-µm membrane filters and drying the filters to constant weight at 70°C.

Concentrations of ethanol, acetic, lactic, and galacturonic acids were measured by ion-moderated partition chromatography and refractive index detection as described previously *(5,12,13)*. Neutral sugars were determined by a separate ion-moderated partition chromatography using a Pb^{2+} column *(5,12,13)*. Values for ethanol and acid yields were corrected for changes in volume caused by addition of base (2*N* KOH) and smaller amounts of acid (2*N* HCl) used for pH control and for the amount of ethanol introduced with the inoculum.

Materials and Reagents

Streptomyces griseus proteinase (Type XIV, P5147, approximate specific activity 4 U/mg solid) and *Bacillus polymyxa* proteinase (Type XV, P5647, approximate SA 0.4 U/mg solid) were purchased from Sigma (St. Louis, MO). Vitamin assay

Table 1
Fermentation Summary

		Fermentation time									
		24 h			48 h			72 h			
Supplement	Concentration, g/L	Ethanol[a] concentration, g/L	Ethanol[b] yield, % theor.	Volumetric productivity, g/L/h	Ethanol concentration, g/L	Ethanol yield, % theor.	Volumetric productivity, g/L/h	Ethanol concentration, g/L	Ethanol yield, % theor.	Volumetric productivity, g/L/h	Final[c] cell mass, g/L
None		20.9	43.4	1.00	31.7	68.5	0.72	35.0	72.1	0.54	4.2
Tryptone + yeast extract	2.5 + 1.25	25.2	54.9	1.05	36.4	78.5	0.79	38.1	75.1	0.53	6.2
Tryptone	2.5	27.4	59.9	1.13	36.9	84.7	0.80	36.7	84.9	0.53	4.2
Yeast extract	1.25	25.6	55.7	1.05	35.3	79.8	0.75	35.7	81.3	0.51	4.2
Ammonia	10	20.4	39.4	0.70	24.8	60.8	0.54	—	Not done	—	6.6
Phosphate	10	13.0	24.7	0.49	20.9	43.6	0.42	33.5	72.8	0.46	6.4
Phosphate and ammonia	5 + 5	24.2	50.0	0.89	28.4	70.1	0.63	—	Not done	—	6.3
Casamino acids (vitamin free)	1	23.4	50.5	0.93	35.0	78.3	0.72	35.5	81.6	0.50	4.1
Amino acid mixtures:											
#1[d]	0.1 (ea.)	23.3	46.1	0.92	33.3	68.8	0.69	—	Not done	—	4.1
#2[e]	0.1	19.4	44.6	0.75	34.1	69.7	0.69	—	Not done	—	3.9
#3[f]	0.1	24.7	50.3	1.00	34.2	71.2	0.71	—	Not done	—	3.9
#4[g]	0.1	19.6	48.4	0.75	32.5	66.2	0.67	—	Not done	—	3.9

[a]Not corrected for dilution. [b]Calculated from the initial monosaccharide content. [c]Dry basis. [d]Cys, Met. [e]His, Leu, Ileu,Val, Lys. [f]Phe, Tyr, Trp, Thr, Pro. [g]Glu, Ser, Ala, Asp, Arg.

casamino acids were manufactured by Difco (Detroit, MI). Corn steep liquor was a kind gift of Bioenergy International (Gainesville, FL). It assayed at 56% of dry total solids.

RESULTS AND DISCUSSION

The fermentation of orange peel hydrolysates was investigated at 35°C and pH 6.0, since previous studies *(6)* indicated that these conditions are optimal for this fermentation. Filtered peel hydrolysates diluted to 80% strength before the fermentation had the following final concentrations in wt% (standard deviations) of individual monomeric sugars: glucose 4.22 (0.39), fructose 2.39 (0.23), arabinose 0.89 (0.16), galactose 0.58 (0.04), and galacturonic acid 3.02 (0.25), respectively. Average concentration of total neutral sugars was 8.08 and of total sugars 11.1 wt%, respectively. Dilution of peel hydrolysates by addition of inoculum, nutritional supplements, and water decreased the initial concentration of sugars to a level allowing efficient fermentation of all neutral sugars and formation of relatively concentrated ethanol solutions *(6)*. Formation of additional monomeric sugars during fermentation was not observed. Average results of ethanol production at 24, 48, and 72 h in different media and final dry cell concentration are compiled in Table 1. A comparison of results in the first two rows of Table 1 indicates that unsupplemented peel hydrolysate exhibits only marginal levels of nutritional deficiency, because only the rate of ethanol formation is slower when compared to fermentation of supplemented peel hydrolysate. Supplementation with yeast extract or tryptone alone indicates that either supplement improves fermentability of peel hydrolysate to the same extent as richer media utilized previously *(6)*. Addition of phosphate and ammonia either alone or in combination did not have a stimulatory effect on this fermentation, while an addition of vitamin-free casamino acids appeared to be as effective as supplementation with richer media. We also investigated supplementation with pools of amino acids designed to test for auxotrophic requirements *(14)*. The results in Table 1 do not indicate significant stimulation by any particular mixture of amino acids or auxotrophic requirement for any particular amino acid by *E. coli* KO11.

Since supplementation with mixed amino acids provided the most effective improvement in fermentation performance, we investigated supplementation with different concentrations of casamino acids. The results (not shown) indicate that approx 1 g/L of casamino acids provides adequate stimulation of fermentation of peel hydrolysates by *E. coli* KO11.

Supplementation with a corn steep liquor can provide a cheap source of amino acids and other nutrients *(15)*. The results obtained at different levels of supplementation with corn steep liquor are summarized in Fig. 1. Optimal level of supplementation appears to be at 1.5% (v/v) of corn steep liquor or 0.86% of corn steep liquor solids, because at higher concentrations, the corn steep liquor seemed to inhibit the fermentation.

Another avenue for supplementation of orange peel hydrolysates with amino acids can be provided by proteolysis of endogenous proteins *(11)*. We evaluated two protease preparations used successfully for stimulation of cheese whey fermentation by *E. coli* KO11 *(11)*. The results summarized in Fig. 2 indicate that proteinase from *S. griseus* provided efficient supplementation at 20–100 U/L, whereas proteinase from *B. polymyxa* appears to be somewhat inhibitory to these fermentations.

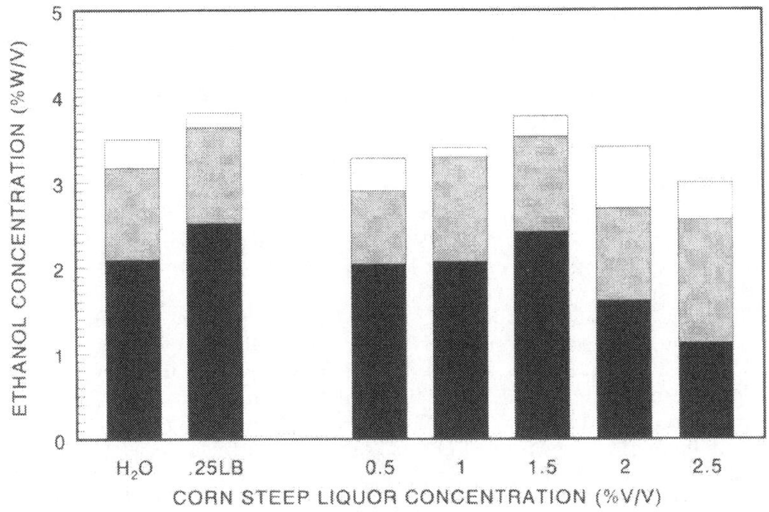

Fig. 1. Effects of supplementation of orange peel hydrolysate with corn steep liquor on ethanol production by *E. coli* KO11. ■, 24 h; ▒, 48 h; ▨, 72 h; H_2O, no supplement; 0.25LB, 2.5 g/L tryptone and 1.25 g/L yeast extract.

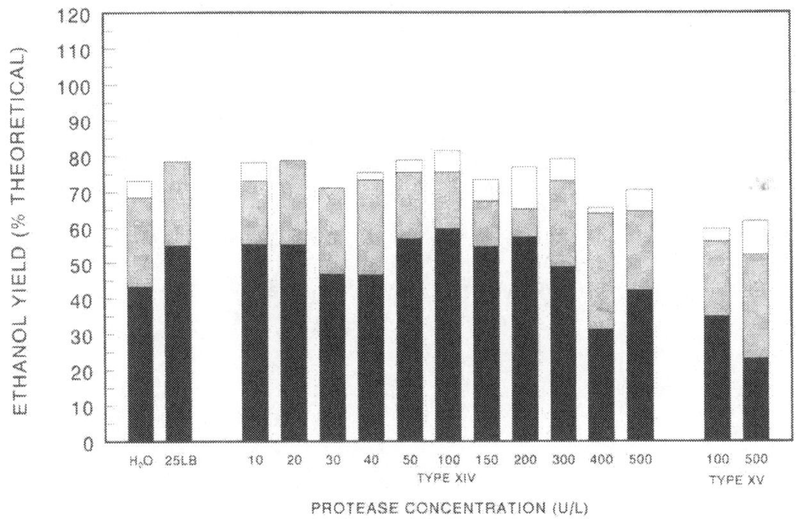

Fig. 2. Effects of supplementation of orange peel hydrolysate with proteases from *S. griseus* (Type XIV) and *B. polymyxa* (Type XV). The bars are coded as in Fig. 1.

CONCLUSIONS

The fermentation of orange peel hydrolysates by *E. coli* KO11 is stimulated by supplementation with low amounts of amino acids and peptides. The stimulation can be readily accomplished by supplementation with mixtures of amino acids, peptides, corn steep liquor, or by proteolysis of endogenous proteins in orange peel hydrolysates.

ACKNOWLEDGMENTS

The authors wish to express appreciation to Lonnie O'Neal Ingram for a kind gift of recombinant *E. coli* KO11. Technical assistance by Veronica Cook and Ricky Lam is also sincerely appreciated.

REFERENCES

1. Agricultural Research Service (1962), *Chemistry and Technology of Citrus, Citrus Products and Byproducts*, Agriculture Handbook No. 98, US Department of Agriculture, Washington, DC.
2. Kesterson, J. W. and Braddock, R. J. (1976), *By-Products and Specialty Products of Florida Citrus.* Bulletin #784, University of Florida, Gainesville, FL.
3. Florida Citrus Processors Association (1995), Statistical Summary 1993–1994 Season. Florida Citrus Processors Association, Winter Haven, FL.
4. Grohmann, K. and Bothast, R. J. (1994), in *Enzymatic Conversion of Biomass for Fuel Production*, Himmel, M. E. and Baker, J. O., eds., American Chemical Society, Washington, DC, pp. 372–390.
5. Grohmann, K., Baldwin, E. A., and Buslig, B. S. (1994), *Appl. Biochem. Biotechnol.* **45/46,** 315–327.
6. Grohmann, K., Cameron, R. G., and Buslig, B. S. (1995), *Appl. Biochem. Biotechnol.* **51/52,** 423–435.
7. Ohta, K., Beall, D. S., Mejia, J. P., Shanmugam, K. T., and Ingram, L. O. (1991), *Appl. Environ. Microbiol.* **57,** 893–900.
8. Barbosa, M. F. S., Beck, M. J., Fein, J. E., Potts, D., and Ingram, L. O. (1992), *Appl. Environ. Microbiol.* **58,** 1382–1384.
9. Beall, D. S., Ingram, L. O., Ben-Bassat, A., Doran, J. B., Fowler, D. E., Hall, R. G., and Wood, B. E. (1992), *Biotechnol. Lett.* **14,** 857–862.
10. Burchhardt, G. and Ingram, L. O. (1992), *Appl. Environ. Microbiol.* **58,** 1128–1133.
11. Guimaraes, W. V., Dudey, G. L., and Ingram, L. O. (1992), *Biotechnol. Bioeng.* **40,** 41–45.
12. Grohmann, K. and Baldwin, E. A. (1992), *Biotechnol. Lett.* **14,** 1169–1174.
13. Grohmann, K., Baldwin, E. A., Buslig, B. S., and Ingram, L. O. (1994), *Biotechnol. Lett.* **16,** 281–286.
14. Roth, R. J. (1970), in *Methods in Enzymology*, vol. 17A, Tabor, H. and Tabor, C. W., eds., Academic, New York, pp. 3–35.
15. Amartey, S. and Jeffries, T. W. (1994), *Biotechnol. Lett.* **16,** 211–214.

0273-2289/96/5758—0389$7.25

Cloning and Expression of Full-Length *Trichoderma reesei* Cellobiohydrolase I cDNAs in *Escherichia coli*

ROBERT A. LAYMON, WILLIAM S. ADNEY, ALI MOHAGHEGHI, MICHAEL E. HIMMEL, AND STEVEN R. THOMAS*

Applied Biological Sciences Branch, Alternative Fuels Division, National Renewable Energy Laboratory, 1617 Cole Blvd., Golden, CO 80401

ABSTRACT

The process of converting lignocellulosic biomass to ethanol via fermentation depends on developing economic sources of cellulases. *Trichoderma reesei* cellobiohydrolase (CBH) I is a key enzyme in the fungal cellulase system; however, specific process application requirements make modification of the enzyme by site-directed mutagenesis (SDM) an attractive goal. To undertake SDM investigations, an efficient, cellulase-free host is required. To test the potential of *Escherichia coli* as a host, *T. reesei* CBH I cDNA was expressed in *E. coli* strain GI 724 as a C-terminal fusion to thermostable thioredoxin protein. Full-length expression of CBH I was subsequently verified by molecular weight, Western blot analysis, and activity on soluble substrates.

Index entries: Cellobiohydrolase I (CBH I); *T. reesei;* cellulases; fusion proteins.

INTRODUCTION

During the simultaneous saccharification and fermentation (SSF) process, ethanol is produced by concurrent saccharification and yeast or bacterial fermentation that utilizes the glucose produced by hydrolyzing the cellulosic content of biomass. The cost of cellulase enzymes and the efficiency of the saccharification process are of great interest not only for separate hydrolysis and fermentation (SHF) applications *(1)*, but also for SSF processes *(2,3)*.

Cellulase enzymes act synergistically on cellulose to produce cellobiose and glucose by cleaving the β-*[1,4]*-D-glucosidic linkages in the polymer chain. CBH I (EC 3.2.1.91; β-*[1,4]*-D-glucan cellobiohydrolase) is a glycoprotein with a polypeptide chain of about 52,214 Dalton and recently has been shown to cleave cellobiose exclusively from the reducing terminus of cellulose chains *(4)*. The importance of this component enzyme of the cellulase system is reflected by the observation

*Author to whom all correspondence and reprint requests should be addressed.

that CBH I and CBH II compose approx 80% of the *Trichoderma reesei* cellulase complex, with CBH I representing approx 75% of the cellobiohydrolase activity *(5)*. β-*(1,4)*-D-Exoglucanases probably perform most of the hydrolytic events during the degradation of cellulose *(6,7)*. Indeed, the major purpose of the endoglucanases seems to be the production of new reducing and nonreducing ends at which cellobiohydrolases can engage. CBH I activity is absolutely crucial to the efficient and complete saccharification of crystalline cellulose by *T. reesei* or synthetic cellulase preparations *(8,9)*.

The first report of successful cloning of *T. reesei* CBH I in yeast was in 1984 from Shoemaker *(10)*. Shoemaker used a yeast-expression plasmid constructed with a phosphoglycerokinase promotor to introduce a functional CBH I gene into *Saccharomyces cerevisiae*. The recombinant (r)CBH I produced in these studies exhibited a specific activity on amorphous cellulose of about 50% of the native enzyme and was heavily *N*-glycosylated. These sugars could be removed by treatment with endoglycosidase H or heat *(10)*. In 1987, van Arsdell et al. *(11)* and Penttilä *(12)* also independently reported the expression of *T. reesei* CBH I in *S. cerevisiae*. The yeast-produced rCBH I from van Arsdell's studies was about 60% as active as the native fungal enzyme (substrate not specified). Consistent with earlier work *(10)*, the rCBH I recovered from both the intracellular and extracellular fractions of yeast reported by Penttilä was a highly glycosylated enzyme (>200 kDa) that showed some activity against phosphoric acid-swollen cellulose *(13)*. Barnett and Shoemaker also noted the overglycosylation of *T. reesei* cellulases expressed in another filamentous fungus, *Aspergillus* sp. *(14)*. Overglycosylated cellulases, however, are undesirable for protein engineering studies, because they may be expected to display nonnative activity characteristics by virtue of their altered physical size and ionic properties *(15)*. This study provides a progress report on our efforts to express CBH I in *E. coli* using a thioredoxin fusion protein and, in doing so, examine this approach as a means of providing active unglycosylated rCBH I for subsequent enzyme engineering studies.

MATERIALS AND METHODS

Isolation of mRNA

Total RNA was isolated from 10 mL of frozen *T. reesei* RUT-C30 (ATCC 56765) cells that were grown at 28°C according to Tangnu et al. *(16)* with 1% w/v Solka-floc as the carbon source and lactose (0.1 wt%) induction. The cells were ground to a fine powder using a mortar and pestle under liquid nitrogen. The powder was then added to denaturing solution (Stratagene, La Jolla, CA: mRNA Isolation Kit) without β-mercaptoethanol. The slurry was then extracted two times with 10 m*M* Tris, pH 8.0, buffer containing 1 m*M* EDTA (TE)-equilibrated phenol and extracted one time with chloroform to remove proteins. (All solutions and glassware are RNase-free, and the techniques used follow commonly practiced RNA extraction and isolation procedures.) The aqueous phase was transferred to a 50-mL conical tube and the RNA precipitated with 8*M* LiCl to a final concentration of 2*M*. The RNA was precipitated overnight at 4°C, then centrifuged (12,000*g*, 10 min, 4°C), and the pellet resuspended in 5 mL diethylpyrocarbonate (DEPC)-treated water. The RNA was transferred to a new conical tube, and 3*M* sodium acetate (pH 5.5) was added to a final concentration of 0.3*M* followed by the addition of 3 vol of 100% ethanol (EtOH) to precipitate the RNA. The EtOH precipitate was stored for 1 h at –80°C, and the RNA was recovered by centrifugation (as above) and resuspended in 10 mL of room temperature elution buffer (Stratagene, La Jolla, CA).

Construction of a *T. reesei* cDNA Library

A cDNA library was constructed using 10 μg of poly(A)$^+$ mRNA from *T. reesei* as described in the ZAP express cDNA synthesis protocol. Original clones are λ phage clones, which were screened by plaque hybridization on lawns of XL1-Blue MRF′ *E. coli* host cells (Stratagene, ZAP Express cDNA Synthesis Kit).

Isolation of Full-Length CBH I Clones from the *T. reesei* cDNA Library

Oligonucleotide primers (Macromolecular Resources, Fort Collins, CO) specific for the *T. reesei* CBH I coding sequence (5'>ATG TAT CGG AAG TTG GCC GTC ATC<3'; 5'>TTT ACA GGC ACT GAG AGT AGT AAG<3'; *17*) were used to synthesize a digoxigenin (DIG)-labeled polymerase chain reaction (PCR) probe specific for the CBH I gene template (i.e., DNA extracted from the *T. reesei* cDNA library). Labeling with DIG and PCR amplification was performed according to the method described by Perkin Elmer (Norwalk, CT; GeneAmp PCR Reagent Kit with AmpliTaq DNA Polymerase), except that the dTTP was partially replaced with DIG-11-dUTP (Boehringer Mannheim, Indianapolis, IN). Plaque hybridization was performed using the Genius System (Boehringer Mannheim, Genius System User's Guide for Filter Hybridization, Version 2. 0, and the Genius Nonradioactive Nucleic Acid Labeling and Detection Kit). Plasmid (phagemid) DNA from selected clones was recovered using the in vivo excision protocol of the ExAssist/XLOLR system (Stratagene).

Construction of the CBH I Expression Vector

The expression vector pTrxFus (Invitrogen, San Diego, CA) was used to clone and express the rCBH I enzyme in the *E. coli* strain GI 724 as a C-terminal fusion to the thermostable thioredoxin protein. Purification of rCBH I consisted of the isolation of inclusion bodies, followed by their solubilization, renaturation *(18)*, and enterokinase digestion (Invitrogen: ThioFusion Expression System, Version 1.1). Restriction enzymes were obtained from New England Biolabs (Beverly, MA) or Boehringer Mannheim (Indianapolis, IN), and used according to manufacturer's specifications.

Isolation and Purification of Inclusion Bodies (IB) from pTF-CBHI

Inclusion bodies (IB) were produced from an 8-mL culture of pTF-CBH I cells induced after 4 h of incubation. The cells were centrifuged, suspended in 4 mL of TE buffer, and lysed by successive sonication and freeze/thaw cycles. The IBs were separated by centrifugation, and the pellet washed three times with TE buffer. IBs were solubilized overnight at 4°C by the addition of 8*M* urea in TE buffer with 10 m*M* dithiothreitol (DTT). The recombinant CBH I was then renatured by slowly removing the urea by diafiltration against 50 m*M* Tris, pH 8.0, buffer containing 5 m*M* $CaCl_2$ using a 500-mL Amicon stirred cell equipped with a 10,000 mol-wt cutoff (PM10) membrane for a period of 2 d at 4°C.

CBH I Enzyme Assay

The activity of the recombinant enzyme was determined by hydrolyzing the fluorogenic substrate analogs 4-methlyumbelliferyl-β-D-lactopyranoside (MUL) and 4-methlyumbelliferyl-β-D-cellobioside (MUC) (Sigma, St. Louis, MO) at 50°C. Enzyme activity on MUL/MUC was determined by adding a 0.1-mL enzyme aliquot to 0.9 mL of 50 m*M* acetate, pH 5.0, buffer containing 0.5 m*M* MUL/MUC solution. Successive, timed 50-µL samples were removed over a period of 60 min, and the reaction stopped by the addition of 0. 1 mL of 150 m*M* glycine-NaOH, pH 10.3 in a Falcon 96-well, flat-bottom microtiter plate. Fluorescence of 4-methylumbelliferone (MU) was calculated from a series of standards measured with a fluorometer equipped with a 360-nm excitation filter and a 460-nm emission filter (Model 7620; Cambridge Technologies, Cambridge, MA). One unit of activity was defined as that amount of enzyme that releases 1.0 µmol of MU/min at 50°C in the described assay conditions.

RESULTS

T. reesei Poly(A)+ mRNA Isolation and cDNA Library Construction

T. reesei poly (A)+ mRNA was used to construct a cDNA library as described in Materials and Methods. The library was amplified and screened using a digoxigenin-labeled probe. Of the approx 50,000 plaques/plate, an average of 20 positive clones were observed. Positive clones were isolated after a second screening, and phagemids were then excised (Stratagene). DNA from four of the positive clones was purified and cut with several restriction enzymes to verify identity of the clones. Of these clones, three gave the expected restriction digestion patterns as determined from previously published DNA sequence data *(17)*.

Construction of the Thioredoxin-CBH I Expression System

The strategy for constructing the CBH I expression vector began with sequencing both the 5' and 3' ends of the cDNA (Sequenase Version 2.0 Sequencing Kit, United States Biochemical, Cleveland, OH) of two positive phagemid clones (pB210 and pB291). It was determined, based on the aforementioned DNA sequence data, that these cDNA clones contained the entire CBH I coding sequence, including the signal peptide and 5'-upstream sequences (data not shown). In Fig. 1, the 6.4-kb phagemid, pB210, is shown and includes restriction sites found in the insert DNA and the location of the two primers (P1, P2) used in a subsequent PCR reaction employed in the construction of the expression vector. The sequences of P1 and P2 were as follows:

XmaI
P1: 5'>gag aga ccc ggg CAG TGG CCT GCA TCT C<3'

XbaI
P2: 5'>aga gag tct aga CAA TGG AGA GGC TGT TAC CCG<3'

The upper-case nucleotides in these two oligonucleotides represent those bases that are homologous to the CBH I clone and are used to prime the 270-bp PCR reaction. The lower-case nucleotides were added to facilitate cloning of the PCR product.

This 270-bp fragment was designed to include a unique *Xma*I restriction site at the 5' end of the fragment and a unique *Xba*I site at the 3' end of the DNA fragment,

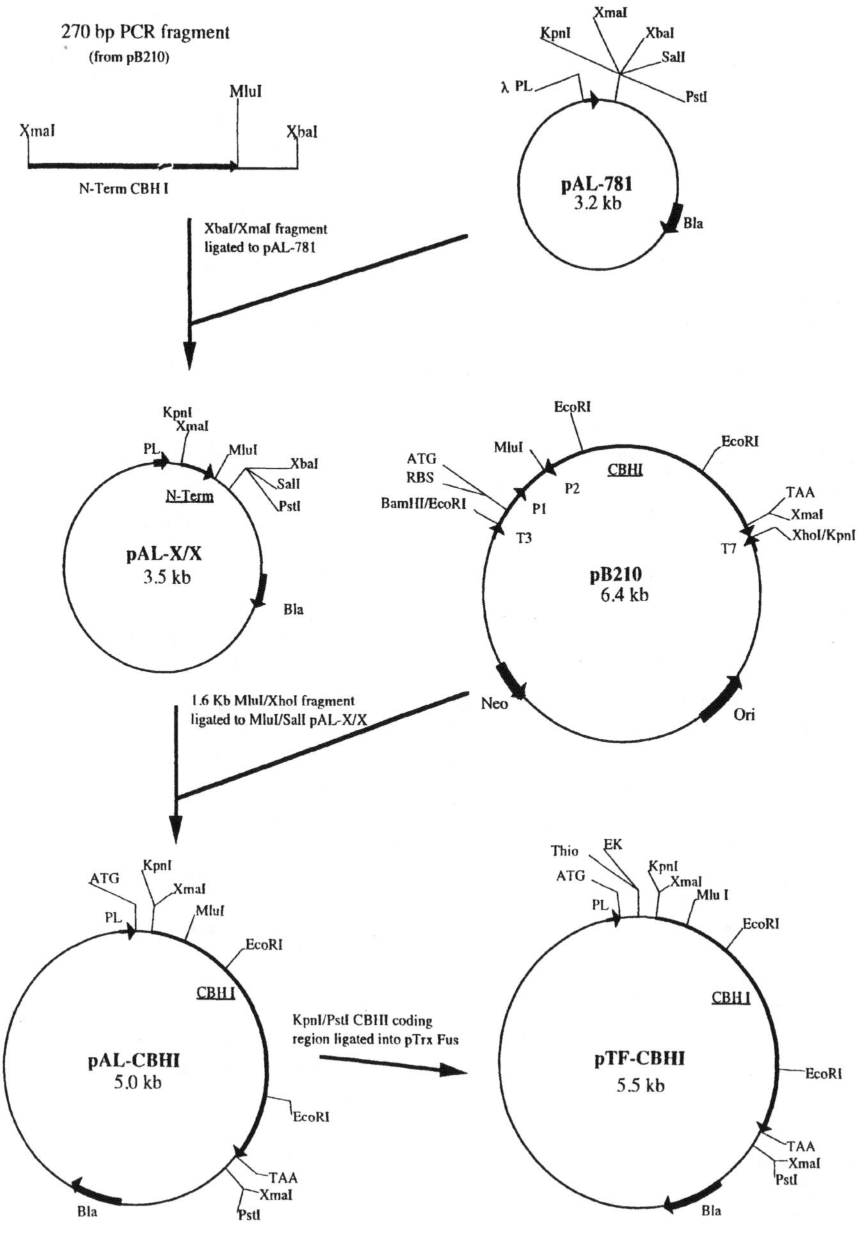

Fig. 1. The 270-bp DNA *Xma*I/*Xba*I fragment was synthesized by PCR using pB210 as a template and ligated into pAL-781, the parent of the pTrxFus expression vector. This small fragment contains the nucleotide coding region for the N-terminal 77 amino acids of CBH I and is inserted in frame with the thioredoxin/enterokinase coding sequence of the vector. The remaining 1.6-kb CBH I coding region was removed from pB210 using *Mlu*I and *Xho*I, and ligated into pAL-X/X cut with *Mlu*I and *Sal*I. The entire coding sequence was then removed from pAL-CBHI with *Kpn*I and *Pst*I and ligated into the same cut pTrxFus vector. The phagemid pB210, isolated from a *T. reesei* λ ZAP library, contains a cDNA encoding CBH I. P1 and P2 represent primers used for the PCR reaction. RBS, ATG, and TAA are the ribosome binding site, initiation codon, and termination codon, respectively. T3 and T7 represent the T3 and T7 promoters, respectively.

which were later used to facilitate construction of expression vectors. This 270-bp PCR fragment encodes the mature N-terminal 77 amino acids of CBH I and was cloned into the *XmaI/XbaI* sites in pAL-781, the parent plasmid of pTrxFus (Invitrogen) (*see* Fig. 1).

The phagemid clone, pB210, was cleaved with *MluI* and *XhoI* to liberate a 1.6-kb fragment of the *cbh1* cDNA (Fig. 1). This fragment was ligated into pAL-X/X cut with *MluI* and *SalI* to produce pAL-CBHI. The entire CBH I gene was then excised from pAL-CBH I with *KpnI* and *PstI* and ligated into pTrxFus (Invitrogen). This construct, pTF-CBHI, was then used to transform *E. coli* GI 724.

Expression of the CBH I Gene Product in *E. coli*

After inducing the cells with tryptophan and a shift in temperature from 30 to 37°C, hourly samples were taken and the cells prepared for sodium dodecyl sulfate-polyacrylamide gel electrophoresis (SDS-PAGE) by boiling in 2X sample buffer. As shown in Fig. 2A, 4 h after induction, the rCBH I fusion protein migrates as a prominent band with a calculated mol wt of 68 kDa. Following treatment with enterokinase and analysis using SDS-PAGE, the size of rCBH I was reduced by approx 12 kDa (data not shown).

In Fig. 2B, the SDS-PAGE gel was transferred to a nitrocellulose membrane and probed with an MAb to native *T. reesei* CBH I protein *(19)*. The time points t = 0 and t = 1 h showed no detectable amount of rCBH I, whereas the samples taken at 2 and 4 h after induction reacted strongly with the CBH I antibody. Figure 2B shows anti-CBH I reactive species of approx 68, 58, and 35 kDa. It is likely that the 58- and 35-kDa mol wt forms represent in vivo protease digestion products of rCBH I.

Phase-contrast microscopy indicated that a large fraction of induced cells contained inclusion bodies. This insoluble protein was purified, solubilized, and renatured according to methods described by Marston and Hartley *(18)*.

Activity Levels of the Recombinant *T. reesei* CBH I Protein

Activity of the rCBH I at 50°C on soluble substrates, MUL and MUC, was confirmed using the methods described in the section on CBH I Enzyme Assay. The renatured rCBH I recovered from purified inclusion bodies, both with and without enterokinase treatment, was found to release 1.1 and 4.3 μmol MU/min/mg crude IB protein, from MUL and MUC, respectively. After consideration for the approx 10% rCBH I content of IB protein, these values lie in the same order of magnitude as specific activities found in this study for MUL and MUC hydrolysis using purified native CBH I, i. e., 18.0 and 1.84 μmol MU/min/mg protein, respectively.

DISCUSSION

Several full-length cDNAs encoding the *T. reesei* CBH I protein have been expressed in *E. coli*. The expression system depends on a construct, pTrxFus, which utilizes a tryptophan-inducible λ P_L promoter to drive expression of a thioredoxin-CBH I fusion protein. Unfortunately, the expression product is mostly localized in insoluble inclusion bodies, which must be purified and solubilized before either enterokinase cleavage or refolding. This result is like a problem identified by Teeri when attempting to express CBH I as either an *E. coli* β-galactosidase or λCro peptide-fusion protein in *E. coli (20)*. Their CBH I-fusion proteins, however, showed activity only on a soluble substrate (4-methylumbelliferyl-β-D-lactoside). The poor

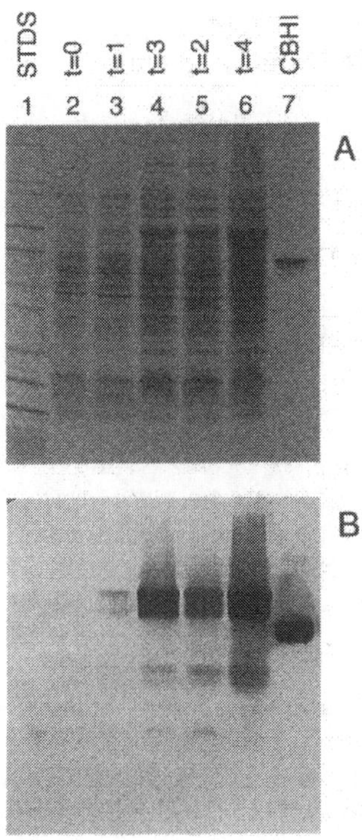

Fig. 2. **(A)** SDS-PAGE of CBH I overexpressed in *E. coli* transformed with pTF-CBHI. Lane 1 was loaded with Novex (San Diego, CA) Mark 12 mol-wt standards: (200, 116, 97, 66, 55.4, 36.5, 31, 21.5, 14.4, and 6 kDa). Cells were grown to early log phase, and then induced with tryptophan at a final concentration of 0.1 mg/mL. After induction, samples were taken every hour, for 4 h (Lanes 2–6) to determine the optimal induction time. The samples were lysed in standard lysis buffer (Invitrogen), boiled for 5 min, and centrifuged before loading. Lane 7 is a purified CBH I-positive control (56 kDa). Times (t) are shown in hours since induction. The gel was stained with Coomassie blue. The overexpressed protein migrates at approx 68 kDa on a 4–20% gradient gel. **(B)** Western blot from a duplicate SDS-PAGE gel loaded as the samples were loaded on the Coomassie-stained gel shown above. The probe is a mouse MAb prepared against purified CBH I.

survivability of nonfused CBH I and other fungal cellulases in *E. coli* cytosol has been documented and is attributed to proteolytic digestion of unprotected, soluble foreign proteins *(20)*. In the present study, the N-terminal thioredoxin was removed from a fraction of the fusion protein sample by digestion with enterokinase (data not shown). Catalytically active enzyme was obtained in reasonable yields following enterokinase digestion; moreover, enzyme recovered from inclusion bodies without subsequent enterokinase treatment was also catalytically active.

Considering that the apparent molecular weight of the *E. coli* derived rCBH I before protease treatment (68 kDa) includes the 12-kDa fusion peptide and that the native CBH I peptide (mol wt = 52.2 kDa *[21]*) is *O*-linked to 6 kDa of carbohydrate content *(21)*, the estimate of rCBH I molecular weight by SDS-PAGE is consistent with the nonglycosylated fusion protein. Whether the differences in migration on SDS-PAGE observed for the native and recombinant enzymes are caused by differ-

Table 1
Comparison of Native and Recombinant CBH I Proteins

	Native CBH I[a]	rCBH I
Molecular weight (SDS-PAGE shown in Fig. 2; kDa)	56	68
Molecular weight of CBH I polypeptide (inferred from this study; kDa)	50	56
Molecular weight of CBH I from DNA sequence (*21*); kDa	52. 2	52. 2
Glycosylation *(21)*	6	None

[a]Purified according to the procedure of Shoemaker et al. *(22)*.

ences in glycosylation patterns or in incomplete/incorrect refolding and renaturation that accompany the purification of the rCBH I protein is unclear. However, because yeast-derived rCBH I treated with endoglycosidase H *(10)* and rCBH I expressed in *E. coli (21)* retain much of their specific activity on soluble substrates, the refolding steps in the present study are suspected as the major culprit leading to the altered electrophoretic behavior of enterokinase-treated, thioredoxin-fused rCBH I. The characteristics of the purified, native, and *E. coli*-derived CBH I product are in Table 1.

Future experiments should evaluate methods designed to fractionate differently folded forms of rCBH I (perhaps using a thio-cellobioside affinity chromatography), so that the specific activities of these recombinant species, instead of the mixture, can be determined. Because of the uncertainties concerning the tertiary structure of rCBH I produced as a fusion protein in *E. coli* (demonstrated by SDS-PAGE behavior), this expression system is probably not ideal for testing CBH I site-directed mutagenesis concepts.

ACKNOWLEDGMENTS

This work was funded by the Biochemical Conversion Element of the US Department of Energy Biofuels Energy Systems Program.

REFERENCES

1. Wright, J. D., Power, A. J., and Douglas, L. J. (1986), *Biotechnol. Bioeng. Symp.* **17,** 285–302.
2. Wright, J. D., Wyman, C. E., and Grohmann, K. (1988), *Applied Biochem. Biotechnol.* **17,** 75–90.
3. Wilke, C. R., Yan, R. D., and von Stockar, U. (1976), *Biotechnol. Bioeng. Symp.* **6,** 55.
4. Teeri, T. T., Koivula, A., Reinikainen, T., Ruohonen, L., and Srisodsuk, M. (1994), Hydrolysis of crystalline cellulose by native and engineered *Trichoderma reesei* Cellulases, Proceedings of the Symposium on Enzymic Degradation of Insoluble Polysaccharides, The 1994 Annual American Chemical Society Meeting, San Diego, CA.
5. Enari, T.-M. (1983), in *Microbial Enzymes and Biotechnology*, Fogarty, W. M., ed., Applied Science, London, pp. 183–223.
6. Eveleigh, D. E. (1987), *Phil. Trans. Royal Soc. Lond.* **A321,** 435–447.
7. Fägerstam, L. G. and Pettersson, L. G. (1980), *FEBS Lett.* **119,** 97.
8. Irwin, D. C., Spezio, M., Walker, L. P., and Wilson, D. B. (1993), *Biotechnol. Bioeng.* **42,** 1002–1013.
9. Thomas, S. R., Laymon, R. A., Chou, Y.-C., Tucker, M. P., Vinzant, T. B., Adney, W. S., Baker, J. O., Nieves, R. A., Mielenz, J. R., and Himmel, M. E. (1995), in *Advances in the Bioconversion of Lignocellulosics*, Saddler, J. N., and Penner, M. N., eds., ACS Series 618, American Chemical Society, Washington, DC, in press.

10. Shoemaker, S. P. (1984), in *The Cellulase System of* Trichoderma reesei: Trichoderma *Strain Improvement and Expression of* Trichoderma *Cellulases in Yeast*, Online, Pinner, UK, pp. 593–600.
11. van Arsdell, J. N., Kwok, S., Schweickart, V. L., Ladner, M. B., Gelfand, D. H., and Innis, M. A. (1987), *Bio/Technology* **5,** 60–64.
12. Penttilä, M. E. (1987), *Construction and Characterization of Cellulolytic Yeast*, Publications No. 39, VTT, Biotechnical Laboratory, Espoo, Finland.
13. Penttilä, M. E., Andre, L., Lehtovaara, P., Bailey, M., Teeri, T. T., and Knowles, J. K. C. (1988), *Gene* **63,** 103–112.
14. Barnett, C. and Shoemaker, S. P. (1987), in *FEMS Symposium, Biochemistry and Genetics of Cellulose Degradation*, Paris, France, pp. 2–18.
15. Teeri, T. T., Jones, A., Kraulis, P., Rouvinen, J., Penttila, M., Harkki, A., Nevalainen, H., Vanhanen, S., Saloheimo, M., and Knowles, J. K. C. (1990), in Trichoderma reesei *Cellulases*, Kubicek, C. P., Eveleigh, D. E., Esterbauer, H., Steiner, W., and Kubicek-Pranz, E. M., eds., Royal Society of Chemistry, London, UK, pp. 156–167.
16. Tangnu, S. K., Blanch, H. W., and Wilke, C. R. (1981), *Biotechnol. Bioeng.* **23,** 1837–1849.
17. Shoemaker, S., Schweikart, V., Ladner, M., Gelfand, D., Kwok, S., Myambo, K., and Innis, M. (1983), *Bio/Technology* **1,** 691–696.
18. Marston, F. A. O. and Hartley, D. L. (1990), *Methods Enzymol.* **182,** 264–276.
19. Nieves, R. A., Himmel, M. E., and Ellis, R. P. (1990), *Appl. Biochem. Biotechnol.* **24/25,** 397–406.
20. Teeri, T. T. (1987), Publications No. 38, VTT, Biotechnical Laboratory, Espoo, Finland.
21. Salovuori, I. (1987), Construction and Characterization of Cellulolytic Yeast, Publications No. 34, VTT, Biotechnical Laboratory, Espoo, Finland.
22. Shoemaker, S., Watt, K., Tsitovsky, G., and Cox, R. (1983), *Bio/Technology* **1,** 687–689.

0273-2289/96/5758—0399$8.00

Ethanol Production from Cellulose by Two Lignocellulolytic Soil Fungi

LUCIA R. DURRANT

Faculdade de Engenharia de Alimentos, Universidade Estadual de Campinas (UNICAMP), 13081-970 Campinas-SP, Brazil

ABSTRACT

The present work examines the production of ethanol via direct fermentation of pure celluloses and lignocellulosic wastes by two soil fungi isolated under anaerobic conditions. The strains were cultured on a defined medium containing filter paper slurry as the carbon source under anaerobic, microaerophilic, and aerobic conditions. After complete degradation of the cellulose, lignocellulases and fermentation products were determined. Highest activities for *Trichocladium canadense* (strain Q10) and the basidiomycete strain (strain H2), were obtained when cultures were incubated under microaerophilic conditions and air, respectively. Laccase activity was present in the culture supernatants of both strains, but peroxidase was only produced by strain H2. Ethanol was the major nongaseous fermentation product. Highest conversion of available cellulose to ethanol was obtained with strain Q10 (90–96%), under microaerophilic conditions. Ethanol production decreased when microcrystalline cellulose and lignocellulosic substrates were used.

Index Entries: Cellulolytic fungi; lignocellulose degradation; microaerophilic fungi; fermenting fungi.

INTRODUCTION

The presence of anaerobic cellulose-degrading fungi in the rumen of various animals and in the alimentary tracts of other herbivorous nonruminants is well documented *(1,2)*. They have been isolated from the rumen and stomach of herbivores belonging to a wide range of mammalian families *(3–7)*. These zoosporic fungi have also been isolated from the saliva, and fresh and dry feces from a wide range of animals. All species studied to date utilize cellulose and hemicelluloses for growth, and produce the enzymes necessary for solubilizing both amorphous and highly ordered celluloses present in plant fibers *(8–12)*. Furthermore, they also contribute to lignocellulose digestion in the rumen of herbivorous mammals because they are able to colonize and degrade the lignin-containing tissues of plant cell walls *(11,13–16)*.

Attempts have been made to isolate obligately anaerobic zoosporic fungi from a variety of other anaerobic environments, such as lake and river sediments and sewage treatment plants, but despite an extensive search, these fungi have not been detected in samples from these environments *(17)*. The possibility of the presence of anaerobic fungi in habitats other than the rumen, where they would participate in the degradation and fermentation of lignocellulosic material, is an interesting one.

To determine the presence of free-living anaerobic fungi that participate in the degradation of cellulose, an isolation procedure was developed *(18)*. Two strains of cellulose-fermenting fungi were isolated, under anaerobic conditions, from soil samples collected from poorly drained areas, such as marshes and swamps. The intent of this work was to investigate the presence of lignocellulose-degrading enzymes in culture fluids of these two fungi. The production of ethanol via direct fermentation of pure celluloses and lignocellulosic wastes by these soil fungi isolated under anaerobic conditions was also examined.

MATERIALS AND METHODS

Microorganisms

Trichocladium canadense strain Q10 and strain H2 (a basidiomycete species), isolated under anaerobic conditions, as previously described *(18)*, were used throughout these experiments.

Media and Culture Conditions

A defined liquid medium containing per liter: 1.0 g KH_2PO_4, 0.54 g NH_4Cl, 0.3 g $CaCl_2$, 1.0 g urea, 1.0 g L-cysteine hydrochloride, 0.1 mL resazurin solution (0.1% [w/v]), 1.0 mL/L of a mineral solution (containing 0.22 g/L ammonium tartrate, 0.66 g/L manganese sulfate, 0.15 g/L iron III sulfate, 0.10 g/L cobalt sulfate, 0.10 g/L zinc sulfate, 6.4 mg/L copper sulfate, and 10 mg/L aluminum potassium sulfate), 0.5 mL/L vitamin solution (2.0 mL biotin, 2.0 mL folic acid, 5.0 mL thiamin-HCl, 5.0 mL riboflavin, 10.0 mL pyridoxine-HCl, 0.10 mL cyanocobalamine, 5.0 mL nicotinic acid, 5.0 mL DL-calcium panthothenate, and 5.0 mL thioctic acid), plus cellulose was used throughout these experiments. The cellulose used was filter paper Whatman No. 1, as a slurry prepared by wet ball milling *(19)*. Media for anaerobic cultures were prereduced *(20)* and stored in an atmosphere of 5% CO_2, 10% H_2, and 85% N_2 in an anaerobic chamber (Coy Laboratory Products, Ann Arbor, MI), until the resazurin indicator became colorless.

Enzyme Preparation and Assays

The production of enzymes capable of degrading cellulose, xylan, and lignin was investigated in cultures incubated anaerobically, microaerophilically, and in air. The defined liquid medium, as just described, containing 0.25% (dry w/v) ball-milled filter paper slurry was used to culture the two fungal isolates. Medium for anaerobic cultures was prereduced, placed in an anaerobic chamber, inoculated, and incubated for 4–5 wk. Medium for microaerophilic cultures was prereduced, prior to inoculation, which was carried out in air with cells cultured on plates of cellulose agar medium under microaerophilic conditions (achieved by using BBL CampyPak Plus gas generator envelopes for the production of a reduced oxygen atmosphere [BBL Microbiology Systems, Becton Dickinson,Cockeysville, MD]). Flasks containing these cultures (800 mL medium in 1-L flasks were used for all three conditions) were incubated for 3 wk in tightly closed jars containing the BBL gas generator envelopes. For aerobic cultures, aerated medium was inoculated in air, and flasks were stoppered with loosely fitting aluminum lids and shaken every other day of the incubation period (up to 3 wk). After growth and complete disappearance of the cellulose present in the medium, the supernatant fluids were collected, filtered, and concentrated by ultrafiltration using a Novacell stirred cell

(Filtron Technology, Clinton, MA) equipped with a 10-kDa mol-wt cutoff membrane. Avicelase, carboxymethylcellulase (CMCase), and xylanase activities were investigated by measuring the production of reducing sugar from either avicel, carboxymethylcellulose (CMC; sodium salt, low viscosity; Sigma), or oat spelt xylan (Sigma), by means of the dinitrosalicylic acid-reducing sugar method *(21)*. Activities were determined using 1% of each substrate in 0.05*M* acetate buffer, pH 5.0, and a final volume and reaction time of 0.3 mL and 1 h for avicelase and 0.6 mL and 30 min for CMCase and xylanase, carried out at 50°C. β-glucosidase activity was assayed using 0.9 mL of 5 m*M* *p*-nitrophenyl-β-D-glucopyranoside in 0.05*M* acetate buffer, pH 5.0, and a final volume and reaction time of 1.0 mL and 20 min at 50°C. The *p*-nitrophenol produced by the addition of 2.0 mL 1*M* $NaCO_3$ was measured spectrophotometrically at 410 nm. One unit of enzyme activity was defined as the number of micromoles of reducing sugar or micromoles of *p*-nitrophenol produced/mL of enzyme. Protein was determined by the method of Bradford *(22)*, with the Bio-Rad protein assay kit (Bio-Rad, Hercules, CA) and bovine serum albumin as the protein standard.

The detection of enzymes involved in lignin degradation was carried out using syringaldazine for the detection of laccase *(23)*, and syringaldazine plus hydrogen peroxide for the detection of peroxidases *(24)*. The presence of peroxidases was also determined using *o*-dianisidine to replace syringaldazine.

Growth on Lignin-Derived Compounds

Lignin sulfonic acid sodium salt, at final concentrations of 0.5% (w/v) and lignin organosolv 2-hydroxyethylether (0.2% [w/v], final concentration), were used to test for the possible utilization of lignin-derived compounds by the isolates. Test tubes containing the lignin source prepared in 10 mL of the defined liquid media were inoculated and incubated under anaerobic, microaerophilic, and aerobic conditions, as described. After growth, the supernatants were collected and subjected to high-performance liquid chromatography (HPLC) analysis, using a fast acid column (Bio-Rad) for the determination of aromatic compounds. The elution buffer used was 15% acetonitrile plus 85% 5 m*M* H_2SO_4.

Fermentation Products

The two strains were inoculated on 20 mL of the defined liquid media containing the following carbon sources: filter paper slurry (0.25% [w/v]), microcrystalline cellulose (0.25% [w/v]), coarse pine sawdust (1% [w/v]), coarse maple sawdust (1% [w/v]), and fine pine sawdust (1% [w/v]). Incubation was carried out under microaerophilic conditions, except for filter paper slurry, which was under the three oxygenation conditions described. For ethanol yield calculations, the amount of cellulose in the lignocellulosic substrates used here was considered as 50%. The conversions of cellulose to ethanol were calculated based on the assumption that 1 g of cellulose on hydrolysis would yield 1.1 g of reducing sugars *(25,26)*.

Chemicals and Cellulosic Substrates

All chemicals used were reagent grade. Avicel (microcrystalline cellulose type PG-105) was a gift of FMC Corp. (Philadelphia, PA). CMC (sodium salt, low viscosity) and oat spelt xylan were obtained from Sigma. Syringaldazine, lignin sulfonic acid-sodium salt, and lignin organosolv 2-hydroxyethylether, were purchased from Aldrich (Milwaukee, WI).

Table 1
Cellulase and Xylanase Activities in Cultures of Strains H2 and Q10

Growth conditions[b]	Enzyme activities, μmol/mL/min[a]			
	CMCase	Avicelase	β-glucosidase	Xylanase
Strain H2				
Anaerobic	1.20	0.038	0.069	1.65
Microaerophilic	1.50	0.047	0.078	2.74
Aerobic[b]	2.11	0.067	0.050	3.13
Strain Q10				
Anaerobic	0.91	0.017	0.23	1.72
Microaerophilic	1.61	0.013	0.38	2.46
Aerobic[c]	0.80	np	0.10	0.89

[a]The SD of each of the values is <2%. np, not present.
[b]Growth substrate: 10 mL of a 0.2% (dry w/v) ball-milled filter paper medium.
[c]Nonagitated cultures.

Table 2
Detection of Enzymes Involved in Lignin Degradation
After Growth of Strains H2 and Q10 Under Microaerophilic Conditions

Substrates[b]	Enzyme activities, μmol/mL/min[a]		
	Laccase	Peroxidases	
	Syringaldazine	Syringadazine plus H_2O_2	*o*-Dianisidine plus H_2O_2
Strain H2	2.77 (0.07)	4.15 (0.09)	10.20 (0.51)
Strain Q10	6.46 (0.12)	np	np

[a]Numbers in parentheses are standard deviations. np, not present.
[b]Growth substrate: 10 mL of a 0.25% (dry w/v) ball-milled filter paper medium.

RESULTS

Enzyme Activities

CMCase, β-glucosidase, avicelase, and xylanase activities present in the culture supernatants of the two strains grown under the three oxygenation conditions described, are shown in Table 1. Strain Q10 exhibited highest activities after growth under microaerophilic conditions. Strain H2 showed highest levels of enzyme activities after growth under aerobic conditions. As shown in Table 2, laccase was detected in culture supernatants of strains H2 and Q10 by the reaction using syringaldazine as the substrate. Strain Q10, however, produced negative results with syringaldazine plus hydrogen peroxide. Peroxidase was detected in strain H2 after reaction with either syringaldazine plus hydrogen peroxide or *o*-dianisidine plus hydrogen peroxide.

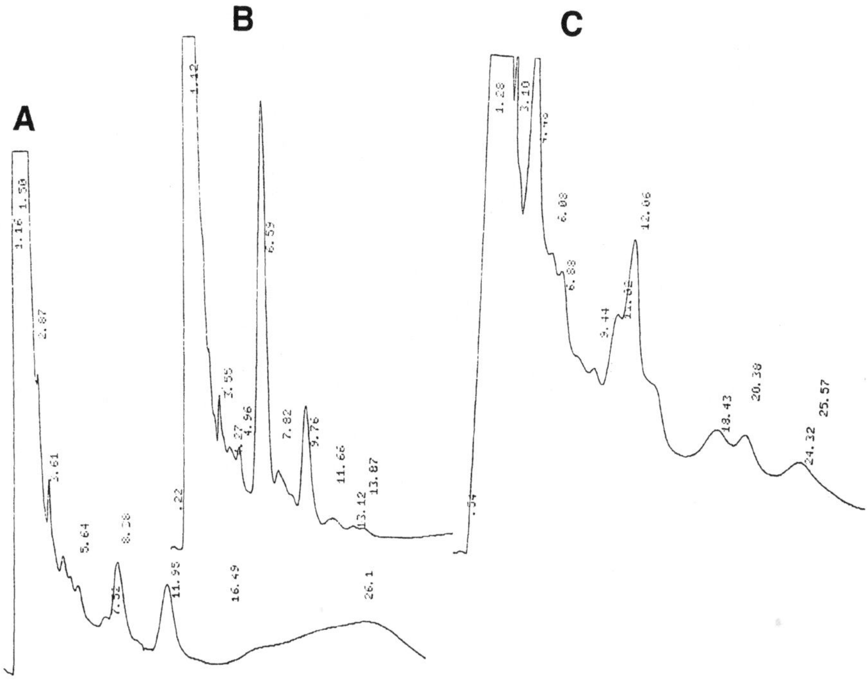

Fig. 1. HPLC chromatograms of supernatant samples of strains H2 and Q10, after growth on lignosulfonate 0.2%, as the sole carbon source, for 3 wk under microaerofilic conditions. Chromatography was carried out using a fast acid column. Aromatic compounds were eluted with a solution containing acetonitrile (15%) and 5 m*M* H_2SO_4 (85%). **(A)** Uninoculated control; **(B)** strain Q10 culture fluid; **(C)** strain H2 culture fluid.

Growth on Lignin-Derived Compounds

In a medium containing lignin sulfonic acid, moderate to good growth of both strains was observed under microaerophilic and aerobic conditions. However, no growth was observed under anaerobic conditions. Only strain H2 was able to use lignin organosolv as growth substrate. The HPLC chromatograms of supernatant samples indicate that the substrates that supported growth were apparently degraded or modified, inasmuch as the composition of the growth medium changed during growth. Moreover, the changes that occurred in the lignin-related substrates were strain specific (Figs. 1 and 2).

Fermentation Products

Both isolates produced ethanol as the major nongaseous fermentation product. Acetate was also produced. Table 3 shows the ethanol produced by strains H2 and Q10 after growth on ball-milled filter paper slurry as the carbon source, and Table 4 shows the ethanol yields obtained after growth of both strains under microaerophilic conditions on lignocellulosic wastes.

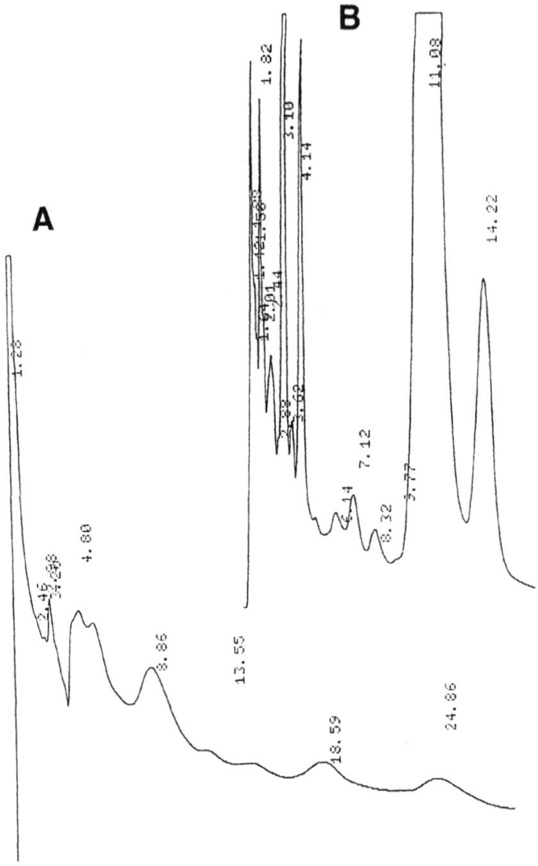

Fig. 2. HPLC chromatograms of supernatant sample of strain H2, after growth on lignin organosolv as the sole carbon source, for 3 wk under microaerofilic conditions. Chromatography was carried out using a fast acid column. Aromatic compounds were eluted with a solution containing acetonitrile (15%) and 5 m*M* H_2SO_4 (85%). **(A)** Uninoculated control; **(B)** strain H2 culture fluid.

Discussion

Both strains grew and degraded cellulose under anaerobic conditions. Strain H2 exhibited best growth and lignocellulases production under aerobic conditions. The results obtained with strain Q10 suggest that it prefers low oxygen concentrations for growth and degradation of lignocellulosic materials, and that it may be a facultative microaerophilic fungus. In this respect, it should be noted that active aeration did not improve growth and degradation of cellulose by strain Q10, and in cultures incubated in air without agitation, noticeable growth occurred only near the bottom of tubes. Both strains were able to grow in lignin sulfonic acid, but only strain H2 was able to grow on the more complex lignin-related substrates used here (lignin organosolv). The results obtained with the syringaldazine and *o*-dianisidine tests indicate that a different enzyme system is utilized by each strain for growth on lignin-related compounds. Strain H2 produces laccase and peroxidases, enzymes reported to be involved in lignin degradation *(27)*. According to the detection tests

Table 3
Ethanol Produced by Strains H2 and Q10
Grown on Ball-Milled Filter Paper Slurry

Growth conditions[a]	Ethanol Yields, %[b]	
	Strain H2	Strain Q10
Anaerobic	36.5 (3.05)	96.0 (4.65)
Microaerophilic	8.0 (3.42)	90.5 (5.83)
Aerobic	81.0 (4.86)	65.6 (2.49)

[a]Growth substrate: 10 ml of a 0.25% (dry w/v) ball-milled filter paper medium.

[b]Values represent % of theoretical yield. Numbers in parentheses are standard deviations.

Table 4
Ethanol Produced by Strains H2 and Q10
Grown on Lignocellulosic Materials

Growth substrate[a]	Ethanol yields, %[b]	
	Strain H2	Strain Q10
Coarse pine sawdust	8.4 (0.79)	20.2 (1.38)
Coarse maple sawdust	8.0 (0.76)	37.6 (1.57)
Fine pine sawdust	67.2 (5.71)	6.2 (0.46)

[a]Growth conditions: Incubation carried out under microaerophilic conditions.

[b]Values represent % of theoretical yield. Numbers in parentheses are standard deviations.

used here, strain Q10 does not produce peroxidases. The lack of peroxidases production by strain Q10 is probably the reason for its incapability to grow on the lignin organosolv substrate. The enzyme systems utilized by these strains to grow on lignocellulosic substrates still need further investigation.

The microbial production of chemicals by lignocellulose fermentation has generated considerable research interest. Since the commercially available celluloses are unlikely to be used as substrates for industrial fermentations, fuel production from biomass in the form of agricultural and forest residues is attractive as an alternative renewable energy resource. Fungi having the capacity to convert carbohydrates to ethanol have been increasingly studied in recent years, and their technological use is being explored *(28,29)*. The production of ethanol and other useful chemical feedstocks from lignocellulosic materials by direct fermentation is, however, more complex than isolated forms of cellulose or hemicellulose, owing to the presence of lignin, the crystallinity of cellulose, and also the presence of a complex mixture of sugars. The two fungal strains used here were able to ferment cellulose and lignocellulosic substrates to ethanol. When lignocellulosic materials were used the ethanol yields were lower than that obtained with filter paper slurry, but strain H2 was able to produce good yields of ethanol when the particle size of the material was reduced, as in the case of the fine pine sawdust. The two strains are also able to grow on xylan, cellobiose, xylose, glucose, and other monossacharides *(18)* as sole carbon sources. The fact that they are able to grow on lignin-related compounds and also produce enzymes involved in the degradation of lignin when growing on cellulose is an advantage. The main disadvantage is the slow conversion rate when compared to ethanol production by yeast. However, more detailed information on certain growth parameters, such as substrate concentration, aeration conditions, temperature, and pH, may lead to improvement in the conversion rates. For instance, in the case of strain H2, a combination in aeration conditions during growth may offer an easy way to speed up the process and is now under investigation.

ACKNOWLEDGMENTS

The author thanks FAEP (Foundation for the Support of Research and Teaching at UNICAMP), FAPESP (Foundation for the Support of Research in Sao Paulo State), and CNPq (National Coucil for Scientific and Technological Development) for financial support.

REFERENCES

1. Orpin, C. G. (1975), *J. Gen. Microbiol.* **91,** 249–262.
2. Orpin, C. G. (1977), *J. Gen. Microbiol.* **99,** 107–117.
3. Bauchop, T. (1989), *Biosystems* **23,** 53–64.
4. Kostiukovskii, V. A., Okunev, O. N., and Tarakanov, B. V. (1990), *Mikrobiologiia* **59,** 1067–1074.
5. Phillips, M. W. and Gordon, G. L. (1988), *Biosystems* **21,** 377.
6. Teunissen, M .J., Op den Camp, H. J., Orpin, C. G., Huis in't Veld, J. H., and Vogels, G. D. (1991), *J. Gen. Microbiol.* **137,** 1401.
7. Teunissen, M. J., Smith, A. A. M., Op den Camp, H. J., Orpin, C. G., Huis in't Veld, J. H., and Vogels, G. D. (1991), *Arch. Microbiol.* **156,** 290–296.
8. Barichievich, E. M. and Calza, R. E. (1990), *Appl. Environ. Microbiol.* **56,** 43–48.
9. Lowe, S. E., Theodorou, M. K. and Trincy, A. P. (1987), *Appl. Environ. Microbiol.* **53,** 1216–1223.
10. Orpin, C. G. (1988), *Biosystems* **21,** 365–370.
11. Teunissen, M. J. and Op den Camp, H. J. (1993), *Antonie Van Leeuwenhoek* **63,** 63–76.
12. Wilson, C. A. and Wood, T. M. (1992), *Enzyme Microb. Technol.* **14,** 258–264.
13. Akin, D. E. (1987), *Anim. Feed Sci. Technol.* **16,** 273–285.
14. Akin, D. E. and Borneman, W. S. (1990), *J. Dairy Sci.* **73,** 3023–3032.
15. Akin, D. E., Borneman, W. S., and Windham, W. R. (1988), *Biosystems* **21,** 385–391.
16. Akin, D. E. and Rigsby, L. L. (1987), *Appl. Environ. Microbiol.* **53,** 1987–1995.
17. Wubah, D. A., Akin, D. E., and Borneman, W. S. (1993), Critical Rev. Microbiol. **19,** 99–115.
18. Durrant, L. R., Canale-Parola, E., and Leschine, S. B. (1995), in *The Significance and Regulation of Soil Biodiversity,* Collins, H. P., Robertson, G. P., and Klug, M. J., eds., Kluwe Academic, The Netherlands.
19. Leschine, S. B. and Canale-Parola, E. (1983), *Appl. Environ. Microbiol.* **46,** 728–737.
20. Hungate, R. E. (1969), *Methods in Microbiology,* vol. 3B, Norris, J. R. and Ribbons, D. W., eds., Academic, New York.
21. Miller, G. L. R., Blum, R., Glennon, W. E., and Burton, A. L. (1960), *Anal. Biochem.* **2,** 127–132.
22. Bradford, M. M. (1976), *Anal. Biochem.* **72,** 248–254.
23. Kawai, S., Umezawa, T., and Higushi, T. (1988), *Arch. Biochem. Biophys.* **262(1),** 99–110.
24. Harkin, J. M. and Obst, Jr. (1973), *Experimentia* **29,** 381–508.
25. Christakopoulos, P., Macris, B., and Kekos, D. (1989), *Enzyme Microb. Technol.* **11,** 236–239.
26. Christakopoulos, P., Koullas, D. P., Kekos, D., Koukios, E. G., and Macris, B. (1991), *Enzyme Microb. Technol.* **13,** 272–274.
27. Morohoshi, N. (1991), *Enzymes in Biomass Conversion,* Leatham, G. F. and Himmel, M. E., eds., ACS Symposium Series 460. ACS, Washington, DC.
28. Singh, A., Kumar, P. K., and Schuger, K. (1992), *Biochem. Int.* **27,** 831–839.
29. Singh, A., Kumar, P. K. and Schuger, K. (1992), *Biotechnol. Appl. Biochem.* **16,** 296–302.

0273-2289/96/5758—0407$6.50

Hexokinase Production from *S. cerevisiae*

Culture Conditions

JOSÉ ABRAHÃO-NETO,
PATRÍCIA INFANTI, AND MICHELE VITOLO*

Faculdade de Ciências Farmacêuticas, University of São Paulo, PO Box 66083, 05389-970, São Paulo, SP, Brasil

ABSTRACT

The effects of pH (4.0, 4.5, or 5.0), temperature (*T*) (30, 35, or 40°C) and dissolved oxygen (DO) (0.2, 2.0, 4.0, or 6.0 mg O_2/L) on hexokinase and invertase formation by yeast were studied. The highest enzyme activities were attained at pH 4.0, DO = 4.0 mg O_2/L, and T = 35 or 40°C.

Index Entries: Hexokinase; invertase; *S. cerevisiae*; baker's yeast.

INTRODUCTION

Hexokinase (EC 2.7.1.1), a key enzyme in carbohydrate metabolism, catalyzes the conversion of glucose into glucose-6-phosphate (G-6-P). The G-6-P, in turn, is an intermediate for several metabolic pathways, including glycolysis. Hexokinase (HK) has an important role in the control of the catabolic repression *(1,2)*, as well as on the glucose uptake mechanism through the cytoplasmic membrane *(3,4)*. Furthermore, in yeasts, the HK is found to be predominantly solubilized in the cytoplasm *(5)*.

In addition to its importance in biochemical studies, HK is used in analytical methods for measurement of glucose, fructose, mannose, ATP, and creatin-kinase activity *(6)*. Furthermore, since HK allows for the measurement of glucose concentrations in the presence of fructose, it constitutes an important tool in the wine and fruit juice industries for the detection of illegal sugar additions in the final products *(7)*.

This article deals with the effect of pH, temperature, and dissolved oxygen on the HK production by *Saccharomyces cerevisiae*. In addition, the effect of these culture conditions on invertase activity of intact cells is also reported.

MATERIALS AND METHODS

Inoculum Preparation

S. cerevisiae (isolated from pressed yeast cake) was maintained on slant tubes containing 23.0 g/L nutrient-agar (Difco, Detroit, MI) and 1.0 g/L glucose. The cells were transferred to test tubes containing 2.5 mL of growth medium

*Author to whom all correspondence and reprint requests should be addressed.

(GM) (10.0 g/L glucose, 5.0 g/L peptone, and 3.0 g/L yeast extract, pH 4.5) and incubated for 48 h at 33°C. Following this procedure, one tube was used to inoculate 50 mL culture medium (CM) in a 250-mL Erlenmeyer flask, followed by incubation at 30°C for 22 h in a rotary shaker (100 rpm) (Super-Hohm, Piracicaba, SP). The composition of CM was: 3.0 g/L yeast extract, 5.0 g/L peptone, 2.0 g/L glucose, 15.0 g/L sucrose, 2.4 g/L $Na_2HPO_4 \cdot 12H_2O$, 0.075 g/L $MgSO_4 \cdot 7H_2O$, and 5.1 g/L $(NH_4)_2SO_4$. All media were sterilized at 120°C for 20 min.

Batch Fermentation

A volume of 0.45 L of inoculum (0.70 g dry cell/L) was poured into a 5-L NBS-MF 200 bench fermenter (coupled with NBS dissolved oxygen controller, DO-81) containing 2.55 L of CM. The culture was then carried out at a temperature of 30, 35, or 40°C; impeller speed = 400 rpm, 0.1 mg/L dimethylpolysiloxane (addition dropwise when needed), pH 4.0, 4.5, or 5.0, and dissolved oxygen (DO) 0.2, 2.0, 4.0, or 6.0 mg O_2/L. Every hour, an aliquot of 20.0 mL of the fermenting medium was taken for analysis.

In all tests, the pH of the medium during fermentation was maintained at the desired value by automatic addition of 1*M* NaOH and 0.5*M* H_2SO_4.

Measurement of Glucose and Cell Concentrations

Five milliliters of fermenting medium were filtered through a Millipore membrane (pore diameter = 0.45 μ). The cell concentration, expressed as g dry cell/L, was measured by drying the cell cake (105°C for 2 h). The glucose concentration in the filtrate was measured by an enzymatic method using the system glucose oxidase (GOD)/peroxidase (POD) according to Bergmeyer *(6)*.

Measurement of Enzymes Activities

Invertase

The invertase activity determinations (always in duplicate) were carried out at 37°C in a mixture of 1.5 mL 0.010*M* acetate-acetic acid buffer (pH 4.6), 2.5 mL 0.3*M* sucrose solution, and 0.5 mL cell suspension. After 3 min, the hydrolysis was stopped by adding 1.0 mL of the Somogyi reagent *(8)* quickly followed by immersion in a boiling water bath for 10 min. The reducing sugar concentration (RS) was then measured as described previously *(9)*.

One invertase unit (U) was defined as the amount of enzyme catalyzing the formation of 1 g of RS/min at the assay conditions. Specific invertase activity (IA) was expressed as U/g dry cell.

The cell suspension was prepared as follows: 5 mL of the fermenting medium was centrifuged (3000*g*/15 min); the sediment was washed with distilled water, centrifuged (3000*g*/15 min), and suspended in distilled water in order to obtain a known suspension volume.

Hexokinase

A volume of 10 mL of fermenting medium was centrifuged (5000*g*/10 min), the sediment was washed with 3.0 mL of TRIS-HCI buffer (50 m*M*, pH 7.5), centrifuged (5000*g*/10 min), and resuspended in the same buffer (3.0 mL) containing: 2 m*M* $MgCl_2$, 0.2 m*M* EDTA, 2 m*M* aminocaproic acid, 2 m*M* DTT, and 1 m*M* PMSF. The cells were disrupted by submission to a vortex (PHOENIX AT56) in the pres-

Table 1
Variation of the Specific HK Activity, Specific IA, and tg for Batch Cultures of *S. cerevisiae* Carried Out at Different DO Concentrations, T = 35°C and pH = 4.5

		HK, U/mg dry cell			IA, U/g dry cell		
O_2, mg/L	tg, h	0 h	2.5 h	5.0 h	0 h	2.5 h	5.0 h
0.2	2.31	0.062	0.076	0.238	0.504	0.244	0.608
2.0	1.47	0.072	0.520	0.540	0.576	0.200	1.22
4.0	1.39	0.075	0.360	0.480	0.610	0.530	2.70
6.0	1.58	0.080	0.230	0.280	0.553	0.152	1.02

ence of 3.0-mL glass beads (0.5 mm diameter). Cell debris and glass beads were removed by centrifugation (5000*g*/10 min), and the supernatant was collected.

The HK activity was measured in the supernatant by spectrophotometric determination at 30°C of reduced NADP according to the method described by Bergmeyer *(6)*.

One HK unit (U) was defined as the amount of enzyme catalyzing the reduction of 1 µmol of NADP/min at the assay conditions. Specific HK activity was expressed as U/mg dry cell. The coefficient of variation (the ratio between the standard deviation and the HK activity mean value) was about ±5%.

RESULTS AND DISCUSSION

As can be seen from Table 1, dissolved oxygen (DO) clearly affects the specific HK activity of *S. cerevisiae* in the DO interval studied (from 0.2 to 6.0 mg O_2/L). HK of about 0.54 and 0.48 U/mg dry cell were attained at DO 2.0 and 4.0 mg O_2/L, respectively. Furthermore, a diminution of approx 50% of HK was observed at the upper and lower DO limits. This can be explained by taking into account that the pair HK/O_2 can be related to the glucose uptake and conversion into energy by yeast. Probably, the glucose/energy conversion should be optimized at 2.0 mg O_2/L ≤ DO ≤ 4.0 mg O_2/L by the fact that cellular growth, which was considered in terms of decreased generation time (tg), was markedly stimulated in that DO interval. The tg was calculated as proposed by Borzani *(10)*.

Since sucrose (a nonfermentable sugar) was the main carbon source present in the culture medium, the availability of glucose for the yeast depends on sucrose hydrolysis by invertase, an enzyme located at the yeast cell wall *(11)*. The invertase (EC 3.2.1.26) represents commercial interest owing to its generalized use for inverted syrup production and in sensors for continuous sucrose determination *(12,13)*. Thus, the invertase production was also considered.

From Table 1 it can be seen that the highest specific invertase activity (IA) (2.70 U/g dry cell) and the lowest tg (1.39 h) occurred at DO = 4.0 mg O_2/L. From the same table, a decreasing of IA is also observed around 55% at DO equal to 2.0 or 6.0 mg O_2/L and an average variation of tg of about 9%, as DO ranged from 2.0 to 6.0 mg O_2/L.

Table 2 indicates that at 35°C, the highest IA (2.70 U/g dry cell) was attained at pH 4.5. However, at the same temperature, the specific invertase activity diminished around 50 and 82% at pH 4.0 and 5.0, respectively. At first, the effect of pH on

Table 2
Variation of the Specific HK Activity, Specific IA, and tg for Batch Cultures of *S. cerevisiae* Carried Out at DO = 4.0 mg O_2/L and Different Temperature and pH

		HK, U/mg dry cell			IA, U/g dry cell			
pH	T, °C	0 h	2.5 h	5.0 h	0 h	2.5 h	5.0 h	tg, h
4.0	35	0.289	0.619	0.754	0.482	0.789	1.36	1.58
	40	0.370	0.560	0.780	0.488	0.207	0.703	1.82
4.5	30	0.260	0.180	0.410	0.336	0.160	0.982	1.73
	35	0.075	0.360	0.480	0.610	0.530	2.70	1.39
	40	0.120	0.300	0.500	0.476	0.380	0.658	2.04
5.0	35	0.149	0.185	0.549	0.325	0.220	0.486	2.10

IA could occur either intracellularly or extracellularly. As the pH was varied from 4.0 to 5.0, a change in the intracellular pH would be unlikely because *S. cerevisiae* has an effective internal buffering capability *(14)*. It would be acceptable therefore to presume that the pH had affected the enzyme activity extracellularly. Thereby, the invertase could have its tertiary and/or quaternary structures modified during budding leading to an inadequate insertion into the cell wall *(15)*.

Figure 1 shows that the reducing sugar concentration (RS) in the medium during yeast growth depends on the DO. Therefore, at the fermenting time between 0 and 2 h, the lowest increase in IA occurred in the test carried out at DO = 4.0 mg O_2/L, at which the average RS was about 6.0 g/L. It was also during this time when the highest final IA was attained (Table 1). Such behavior of IA may be the result of a glucose repression/derepression mechanism *(16)*. On the other hand, the HK activity was not affected by RS concentration in the culture medium (Tables 1 and 2, Fig. 1).

It is interesting to note from Fig. 1 that the RS in the medium increases up to t = 2 h, but decreases afterward. The accumulation of RS could be the result of sucrose hydrolysis by the invertase present in the inoculum cells and to the low cell concentration at this phase of the growth.

Table 2 indicates that at 35 and 40°C, the highest HK values (0.754 and 0.780 U/mg dry cell) were attained at pH 4.0. This result gives rise to increased glucose uptake by yeast at the level of the cytoplasmic membrane as observed by Bisson and Fraenkel *(4)*. The increased glucose uptake could also be observed from Fig. 2, because at pH 4.0 the glucose was completely consumed at t = 3 h, whereas at pH 5.0, this occurred after t = 4 h. Although the concentration of glucose was found to depend on invertase activity of the yeast, the formations of HK and invertase are uncoupled events. This observation is illustrated in Table 2 (pH 4.0 and 35 or 40°C), in which it is seen that high HK values occurred at the low values of invertase activity. Moreover, the same table indicates that the temperature influenced the invertase more sharply than the HK. Tests are being carried out to explain such results.

From the data presented, we have established suitable culture conditions for HK production by *S. cerevisiae*, which were 35 or 40°C, pH 4.0, and 4.0 mg O_2/L.

ACKNOWLEDGMENT

We are grateful to Mauro S. S. Candido for technical assistance.

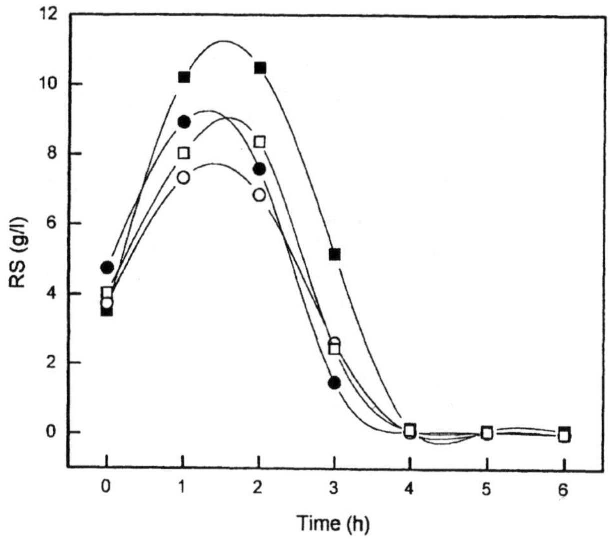

Fig. 1. Variation against time of RS concentration for batch cultures of *S. cerevisiae* carried out at pH 4.5, 35°C, and under the following DO (mg O_2/L): 0.2 (■), 2.0 (●), 4.0 (○), and 6.0 (□).

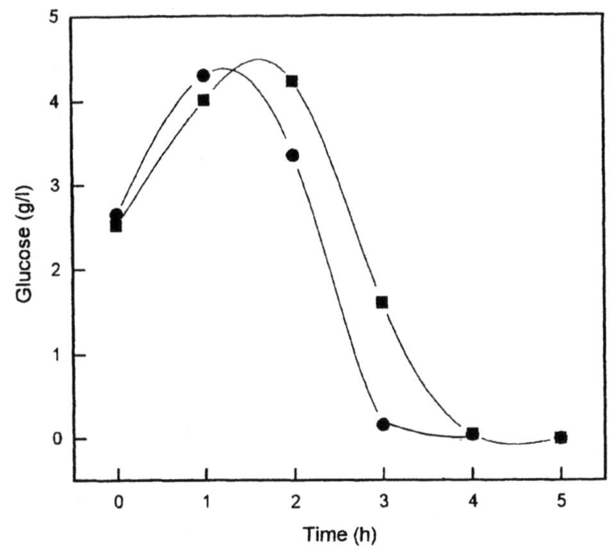

Fig. 2. Variation against time of the glucose concentration for batch cultures of *S. cerevisiae* carried out at 35 °C, DO = 4.0 mg O_2/L and at following pH: 4.0 (●) and 5.0 (■).

REFERENCES

1. Entian, K. D. and Zimmermann, F. K. (1980), *Mol. Gen. Genet.* **177,** 345–350.
2. Baley, R. B. and Woodword, A. (1984), *Mol. Gen. Genet.* **193,** 507–512.
3. Takeda, M. (1981), *Biochem. J.* **198,** 281–287.
4. Bisson, L. F. and Fraenkel, D. G. (1983), *Proc. Natl. Acad. Sci. USA* **80,** 1730–1734.

5. Kovák, L., Nelson, B. D., and Ernster, L. (1986), *Biochem. Biophys. Res. Commun.* **134(1),** 285–291.
6. Bergmeyer, H. U. (1984), in *Methods of Enzymatic Analysis*, vol. 2, 3rd ed., Bergmeyer, H. V., Bergmeyer, J., and Grasl, M., eds., Verlag Chemie, Weinheim, pp. 222,223.
7. Whitaker, J. R. (1991), in *Food Enzymology*, Elsevier Applied Science, New York.
8. Somogyi, M. (1952), *J. Biol. Chem.* **195,** 19–23.
9. Vitolo, M. and Borzani, W. (1983), *Anal. Biochem.* **130,** 469,470.
10. Borzani, W. (1975), in *Biochemical Engineering*, Edgard Blucher, São Paulo, Brasil.
11. Trumbly, R. J. (1992), *Mol. Microbiol.* **6,** 15–21.
12. Layman, P. L. (1986), *Chem. Eng. News* **15,** 11–14.
13. Barlikova, A., Svorc, J., and Miertus, S. (1991), *Anal. Chim. Acta* **247,** 83–87.
14. Weitzel, G., Pilatus, U., and Rensing, L. (1987), *Exp. Cell. Res.* **170,** 64–79.
15. Reddy, A. V., Marcoll, R., and Maley, F. (1990), *Biochemistry* **29(10),** 2482–2487.
16. Patkar, A. and Seo, J. H. (1992), *Biotechnol. Bioeng.* **40,** 103–109.

0273-2289/96/5758—0413$7.25

Culture of the Astaxanthinogenic Yeast *Phaffia rhodozyma* in Low-Cost Media

JOSÉ D. FONTANA,*[,1] MANOEL F. GUIMARÃES,[1] NATALIA T. MARTINS,[2] CLÁUDIO A. FONTANA,[3] AND MADALENA BARON[1]

[1]*Biomass Chemo/Biotechnology Laboratory, Department of Biochemistry, Federal University of Paraná (LQBB-UFPR), PO Box 19046, 81.531-990 Curitiba-PR-Brazil;* [2]*Pescados do Brazil (Pescobras), Curitiba-PR-Brazil; and* [3]*Medicine/Pontifical Catholic University of Paraná (M-PUC-PR), Curitiba-PR-Brazil*

ABSTRACT

Growth of the yeast *Phaffia rhodozyma* was carried out in a simplified medium based on less expensive nutrient sources, such as diluted sugar cane juice, urea, and sodium phosphate. The usual content of the astaxanthin, an oxygenated pink carotenoid useful for fish flesh staining, was improved along with with good cell yields (respective values of >1300 μg/g cells and >5 g cells/L were observed). Yeast invertase and urease must therefore play an important role in the implementation of low-cost culture media.

Index Entries: *Phaffia rhodozyma;* astaxanthin; sugar cane juice; urea; carotenoids.

INTRODUCTION

Sales in the increasing world market for the oxygenated carotenoids astaxanthin and canthaxanthin were $130 million US dollars in 1992. The bioproduced and synthetic counterparts of these pigments are expected to be 67 and 33%, respectively, of the $455 million dollars market projected for 2000. Their current prices average from $1500–$4000/kg for a 100% product formulation *(1)*. The main use of astaxanthin is in farming of selected species, such as salmon or trout. Pink staining of salmonid flesh is a natural process owing to their feeding on colored marine live sources, such as krill and pigmented alga, since they are not able to perform the *de novo* biosynthesis of the pigments *(2)*. Similar or improved patterns of flesh staining may be achieved using feed supplemented with synthetic astaxanthin or whole *Phaffia rhodozyma* yeast cells. Markets for both synthetic and natural carotenoid pigments are now firmly established; in both cases, the question of isomeric forms of the highly oxygenated carotenoids should be considered *(3)*. *P. rhodozyma*, a single genus and single species yeast, was described only two decades ago *(4)*. This basidiomycetous yeast displays two striking features, namely, an intense dark-

*Author to whom all correspondence and reprint requests should be addressed.

orange color of mature cultures, and, as opposed to other pigmented yeasts, the biochemical ability to ferment several mono- and disaccharides. The major pigment produced by *Phaffia* is 3,3'-dihydroxy β-carotene-4,4'-dione or astaxanthin *(5)*. An upper limit of 15 g/L was recorded for glucose as an efficient carbon source, if supplemented with yeast extract and salts, but the resulting astaxanthin concentration of 421 μg/g dry cell yeast could be further improved to a maximum of 652 μg/g when the monosaccharide was replaced by reagent-grade cellobiose. Other disaccharide sources, such as reagent-grade maltose and sucrose, resulted in intermediate values of astaxanthin productivity. A mildly acidic medium and temperatures below 25°C were reported as other optimal culture conditions *(6)*. *P. rhodozyma* whole yeast cells and their main carotenoid pigment should be biotechnologically more advanced than the fish farming basic processes, since recent reports indicate outstanding biological functions for astaxanthin, which serves as a provitamin A, active oxygen species scavenger and/or quencher, tumorogenesis supressor, and in in vitro antibody enhanced production *(7)*. Furthermore, singlet molecular oxygen quenching was proposed to be 500–1000 times more efficient with astaxanthin than with tocopherol *(8)*. The antioxidant role of carotenoids during the aging of *Phaffia* cells was recently documented *(9)*. Another example for the use of astaxanthin from whole or broken *Phaffia* yeast cells is as a poultry feed, since both egg production and yolk coloration are increased *(10)*. Aquaculture is an active and growing enterprise in developed and developing countries *(11,12)*, farming of channel catfish in Georgia, is one example. Exportation of salmon by Chile is one of its outstanding income sources. Paraná State (South of Brazil) has recently set up factories for processing of tilapia. There is an urgent need for novelties and technology improvement in Brazilian fish farming, since highly priced fish, such as salmonids, are being introduced. The flesh pigmentation pattern, being gastronomically attractive, remains an important requirement of fish and seafood consumers, and is directly related to their commercial value. *Phaffia* cells, in addition to astaxanthin, can provide carbohydrates, protein, and glyceride lipids as useful (co)nutrients for the purpose of fish farming. Considering that natural astaxanthin is getting an increased volume in the market of carotenoids *(1)*, and *Phaffia* is one of the best sources of this compound *(5)*, we started investigations on less expensive and readily available carbon and nitrogen sources for growth of *P. rhodozyma* and astaxanthin production.

MATERIALS AND METHODS

Yeast Source and Culture Maintenance

P. rhodozyma (reference strain 24202) was purchased from ATCC (Rockville, MD) and was maintained on agar slants containing 2 g/L glucose and 1 g/L yeast extract at 26°C.

Small-Scale Culture Growth and Routine Analytical Procedures

The preferential carbon source adopted was freshly collected sugar cane juice, previously clarified by centrifugation in order to remove cell debris and solid impurities arising from milling. The clarified juice was then diluted 10 times to ensure an appropriate sugar concentration. The routine supplements were 1 urea and 1.25 g/L sodium phosphate, pH = 6.5. Culture media components (sugars and their complements) were separately autoclaved at 1 atm for 20 min. The comparative cultures of yeast (multifactorial approach) were grown in Erlenmeyer flasks containing 25 mL of medium at a 5:1 air:medium ratio and 100 rpm of agitation for 5 d at 25–26°C (26–27°C in the first trials). Flasks were loosely plugged with cotton wool in order to allow free gas exchange. Inoculation from 48 to 72 h solid medium precultures resulted in an initial value of $A_{650\ nm}$ = 0.05–0.075.

The same wavelength was used to measure growth by turbidimetry. Cell-free supernatant analyses were carried out for residual sugars by high-performance liquid chromatography (HPLC) in a SugarPak column (Waters do Brazil, Sao Paulo), using water at 85°C as eluant and a differential refractometer as detector. Reducing sugars were measured by the dinitrosalicylate method *(13)*. Samples of lyophilized cells (up to 200 mg) were swollen with dimethyl sulfuxide (DMSO) (1 h) and extracted with acetone. Residues were reswollen in DMSO and extracted with chloroform-methanol (1:1). Both extracts were combined and astaxanthin spectrophotometrically quantitated at 478 nm. A purified (silicic acid column) preparation of the pigment from the commercial product Caryophill Pink-8% (Hoffmann-La Roche, Basel, Switzerland) was used for calibration.

Scaled-Up Cultures

Phaffia cultures were scaled up to 2.5 L using a BioFlo IIc fermenter (New Brunswick, Edison, NJ) at 200 rpm, aeration of 0.5–0.75 working vol/min, and 25°C. Antifoam (Sigma, St. Louis, MO; 2 mL/run) was added, and sampling carried out at 12–24-h intervals.

Spectral and Chromatographic Analyses

The spectroscopic scan for pink pigments in acetone was obtained using either a Varian DMS-80 or a Hewlett Packard 8542 spectrophotometer. The composition of the whole and purified (silicic acid column) lipid extracts was assessed using chromatoplates of silica gel 60 (Merck, Darmstadt, Germany) developed with hexane:acetone (7:3). Plates were inspected with long-wave UV and developed at 110°C with 0.1% *p*-anisaldehyde in methanol containing 5% sulfuric acid or with saturated chloroform solution of antimony trichloride *(13)*. The tentative identification of carotenoids was carried out according a previously established protocol *(14)*. The β-carotene standard (all trans) was purchased from Sigma.

RESULTS AND DISCUSSION

Culture of *P. rhodozyma* on pure and expensive carbon and nitrogen sources, such as glucose or pure disaccharides supplemented with yeast/malt extracts and/or peptone, produced yeast cell and astaxanthin yields in the range of 3–4 g/L and 500–650 μg/g dry yeast cell, respectively *(6)*. In order to launch a strategy of double and simultaneous replacement of C and N sources, we first determined 1:10 diluted sugar cane juice to be the C source (mean content of 175 g/L sucrose in the fresh juice) *(15)*. Because of the huge production of sugar cane in Brazil (250 million tons in the 1994/1995 year crop; 13 billion liters of resulting ethanol) (Amorim, H. V., personal communication), sugar cane juice may be considered the most inexpensive C source. Paraná State, in addition to its extensive cane plantations also has one of the Brazilian petrochemical plants that produces urea as a traditional fertilizer. Urea was thus selected as the N source. Despite the physiological importance of

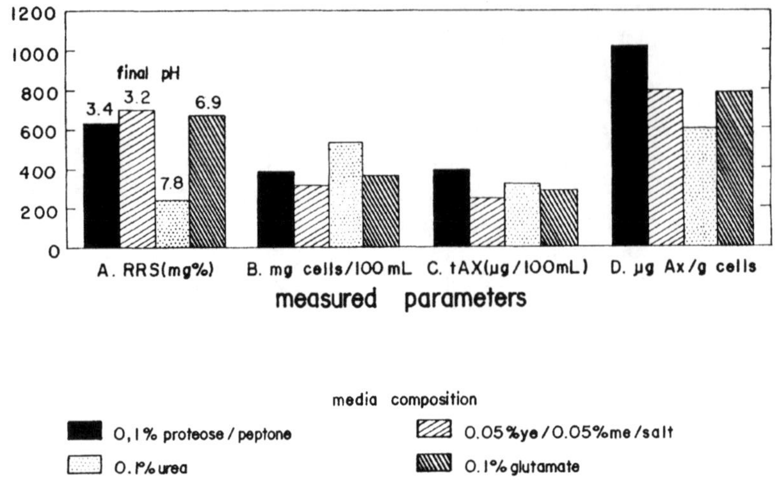

Fig. 1. Effect of nitrogen sources on *P. rhodozyma* growth and pigmention when using sugar cane juice as C source. The four column blocks refer to residual reducing sugars (RRS), cell growth, total astaxanthin (tAx), and astaxanthin productivity as measured in the indicated units.

phosphate, its addition as the third and last ingredient of the medium was to control the alkaline pH imbalance resulting from hydrolysis of urea. Significantly, a slightly acidic condition has been reported as favorable for *Phaffia* growth *(6)*.

Using yeast/malt extracts or proteose/peptone as complex reference sources of N, the results obtained with an initial multifactorial trial are summarized in Fig. 1. If the comparison is restricted to the more successful experiments, the main following parameters for the 25-mL cultures (measured parameters extrapolated to 1/10 L volume) were found. The highest efficiency of sugar consumption and yeast cell yield was obtained in an experiment using 1 g/L urea supplement: 2.4 g/L of residual reducing sugar and 536 mg cells/100 mL medium. Experiments with 1 g/L proteose/peptone and with 0.33 g/L of ammonium phosphate reinforcing the yeast/malt extracts yielded 392 and 249 µg of total astaxanthin/cells from 100 mL of culture, respectively, as compared to 323 µg in the urea-based experiment. The proteolyzed N supply resulted in the deepest colored cells. However, cell yields (g/100 mL media) in these complex nitrogen supplies were only 384 and 312 mg, respectively, and more residual reducing sugar was left over (6.3 and 7.0 g/L, respectively). The corresponding data for the urea-based run were thus superior. The only other result worthy of note was with 1 g/L sodium glutamate as N source, which yielded 364 mg cells and 286 µg of pigment/100 mL medium, but with a less efficient use of sugars (6.7 g/L residual). A remarkable acidification of the stationary-phase media (5 d) resulted from complex N sources (pH 3.4–3.2), whereas urea alone led to a mild alkalinization, which, as further confirmed, could be avoided by the inclusion of diluted phosphate buffer. Since the degree of medium acidity was previously shown to affect more the qualitative carotenoid profile than the content of astaxanthin *(6)*, phosphate salt was maintained as an appropriate cosupplement to urea, except for data in Fig. 2, where a specific focus was pursued as explained below. For this set of experiments, a fine control was set up for some operational

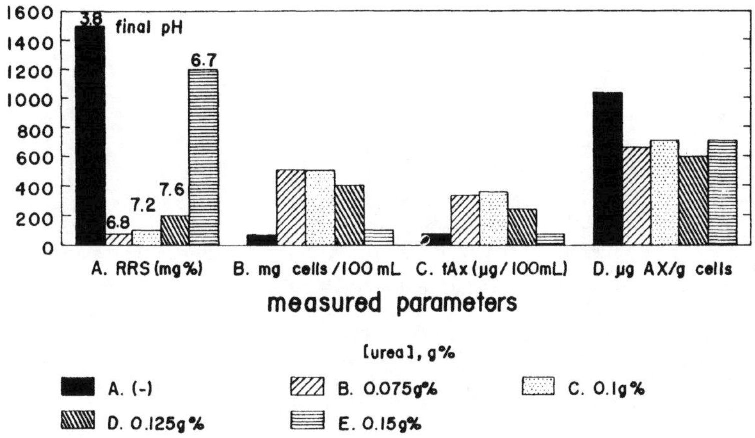

Fig. 2. Effect of urea concentration on *P. rhodozyma* growth and pigmentation when using sugar cane juice as C source (measured parameters as in Fig. 1).

parameters, namely the incubation temperature was lowered by 1° (that is, from 26–27°C to 25–26°C), owing to the high thermo-sensitiviy of *Phaffia;* and inoculum started from 48-h grown cells, which ensured a slightly higher initial optical density ($A_{650\,nm}$ = 0.06–0.07). With the goal of optimizing cane juice and urea pair, the effect of the C and N sources ratio was then determined (results in Fig. 2). The extremities of the dose–response curve, i.e., no urea addition and 1.5 g/L, led to poor cell yields as a result of drastically diminished sugar uptake. Urea supplementation in the 0.75–1.0 g/L range resulted in the highest production of cells (512 and 507 mg/100 mL medium, respectively) and similar astaxanthin production (339 and 360 μg), the pigment production being slightly superior in the second case (709 μg/g cells). The actual increase in astaxanthin could be related to the operational changes, but no net increase in cell biomass was observed. Addition of urea to sugar cane juice resulted in a pH shift from 3.8 (control; no urea addition) to the neutral range, as desired. Accordingly, the optimum urea concentration range also ensured a maximum of C source consumption (<1 g/L of residual reducing sugars). The apparent high productivity of astaxanthin in the urea-free control (first bar; last block of columns; Fig. 2) is not to be considered since the cell yield (only a threefold increase in the inoculation absorbance) is insignificant as compared to the most successful experiments. With the establishment of the beneficial effect of urea, we decided to introduce a small buffering effect through the phosphate input itself. The choice was the sodium salt (1.25 g/L, pH adjusted to 6.5). In the the first trial, maintaining 1 g/L urea as N source, the addition of phosphate resulted in a pH decrease of 1 U with a concomitant increase in cell yield from 404 to 536 mg/100 mL medium and total astaxanthin from 360 to 720 μg (as a reference proteose/peptonle resulted in 260 mg cells and 488 μg of pigment) (Fig. 3). Doubling the sugar cane concentration (i.e., using an 1:5 dilution) and maintaining the C/N ratio (i.e., also doubling urea concentration) did not result in a higher astaxanthin production, probably because of the inhibitory effect of higher urea inputs (as previously shown in Fig. 2). Proteose/peptone (0.5 g/L) allowed the best pigment production, but the corresponding cell yield was less than half that obtained from urea/phosphate. An enhancement

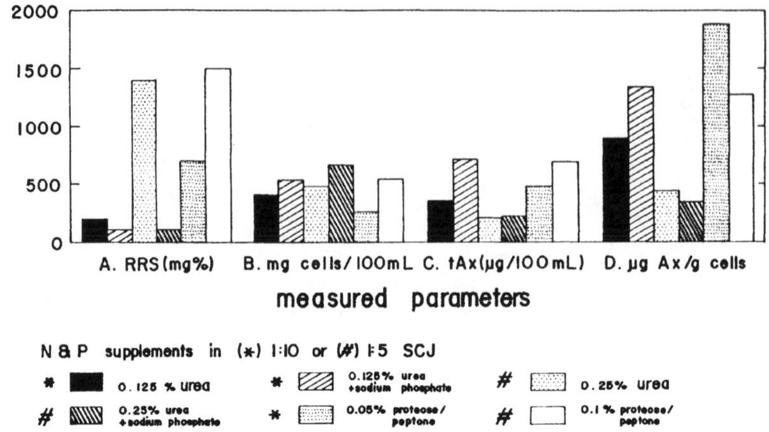

Fig. 3. Effect of phosphate addition on *P. rhodozyma* growth and pigmentation when using sugar cane juice and urea as C and N sources. Buffering pH 6.5 salt at a final concentration of 0.125 g%; (*) and (#), 1:10 and 1:5 dilutions of fresh sugar cane juice; the last two columns of each block.

of this yield (but not in astaxanthin production; last bar; last block of columns, Fig. 3) could be achieved at the expense of more concentrated C and N sources (1:5 diluted juice and 1 g/L of N supply as proteosepeptone), implying a higher cost. As a final attempt dealing with the nature of the N source, multifactorial assessment was carried out with a selected mix of free amino acids and native or proteolyzed leather and soya meal. Complete data will be reported elsewhere, but some of the main conclusions are presented below. The best cell yield arose from a urea supplement with phosphate (699 mg cells), the highest absolute pigment production being obtained with enzyme hydrolyzed soya meal (1089 µg/100 mL culture), and the optimum astaxanthin productivity was found with a mixture of valine, threonine, methionine, and leucine (>2000 µg pigment/g dry cells). The culture of *P. rhodozyma* and moderate productivity of astaxanthin was also feasible using corn, potato, and cassava starches, as well as sugar cane bagasse. The native polysaccharides were previously depolymerized with commercial enzymes (NOVO's disaccharidogenic Thermamyl and Celluclast), and the C input was maintained at 20 g/L along with urea/phosphate supplementation. Cell growth showed no variation, and astaxanthin reached a 550–750 µg/g cell range. The best pigment concentration (ca. 800 µg/g dry cells) resulted from a cassava starch hydrolyzate, a value that was surpassed, however, using, 20 g/L brown sugar utilized as a reference for the fresh sugar cane juice at the same concentration. Sugar cane molasses, the dark mother liquor after sucrose crystallization from concentrated sugar cane juice, also supported yeast growth and astaxanthin production.

P. rhodozyma, grown in sugar cane juice/urea/Na phosphate, was scaled up to 2.5 L in a New Brunswick IIc fermenter. Similar results were obtained to those of shake flasks (as an example, 12 g of dry yeast cells displaying a carroty orange hue). Thin-layer chromatography (TLC) indicated a coincident spot of $R_f = 0.24$ for both astaxanthin standard and the resulting reddish purified pigment. Sugar consumption was monitored by HPLC in the time course of the experiment. Following inoculation (even with washed cells), sucrose was quickly hydrolyzed into its

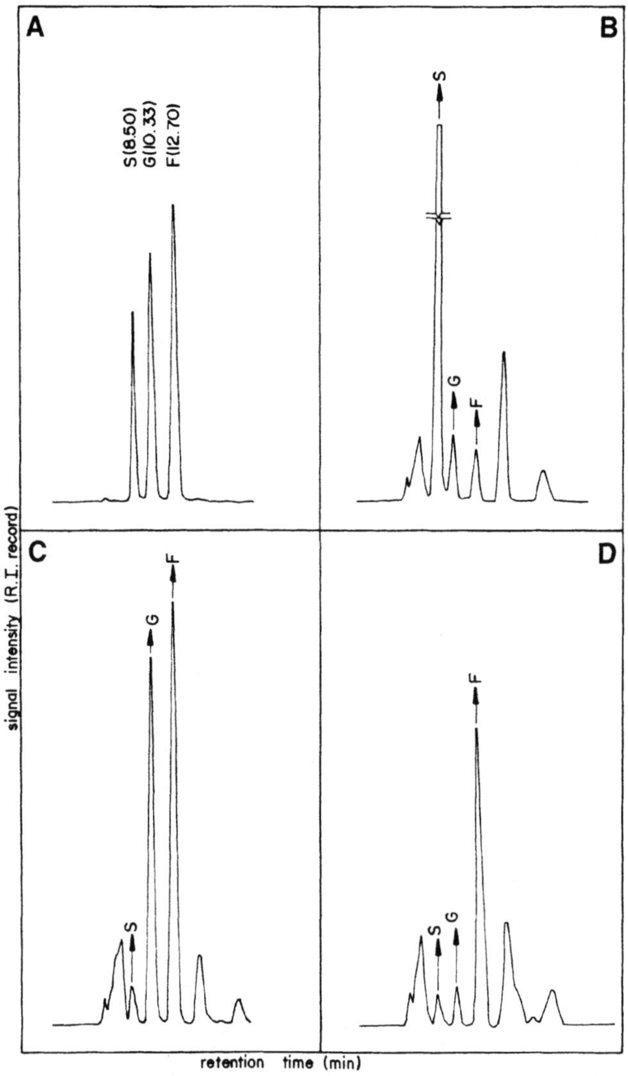

Fig. 4. HPLC analysis for sugar cane juice saccharide composition in this time course of a scaled-up culture of *P. rhodozyma*. The retention times are indicated only for the sugar standard **(A)**, fermenter sampling just after inoculation **(B)**, at 37 h **(C)**, and at 72 h **(D)**. S, sucrose; G, glucose; F, fructose.

halves (Fig. 4B), glucose being preferentially consumed (Fig. 4C), and some fructose remaining at the end of fermentation. Measurements of invertase and urease in the washed cells and in the cell-free supernatant indicated that most, if not all, of these enzymes are associated with the cells. Mild ultrasonication did not release the enzymes into the supernatant.

In order to confirm astaxanthin as the pink to orange carotenoid pigment being produced by *P. rhodozynma*, a spectroscopic analysis was carried out using acetone as solvent (Fig. 5). The pure astaxanthin standard prepared from

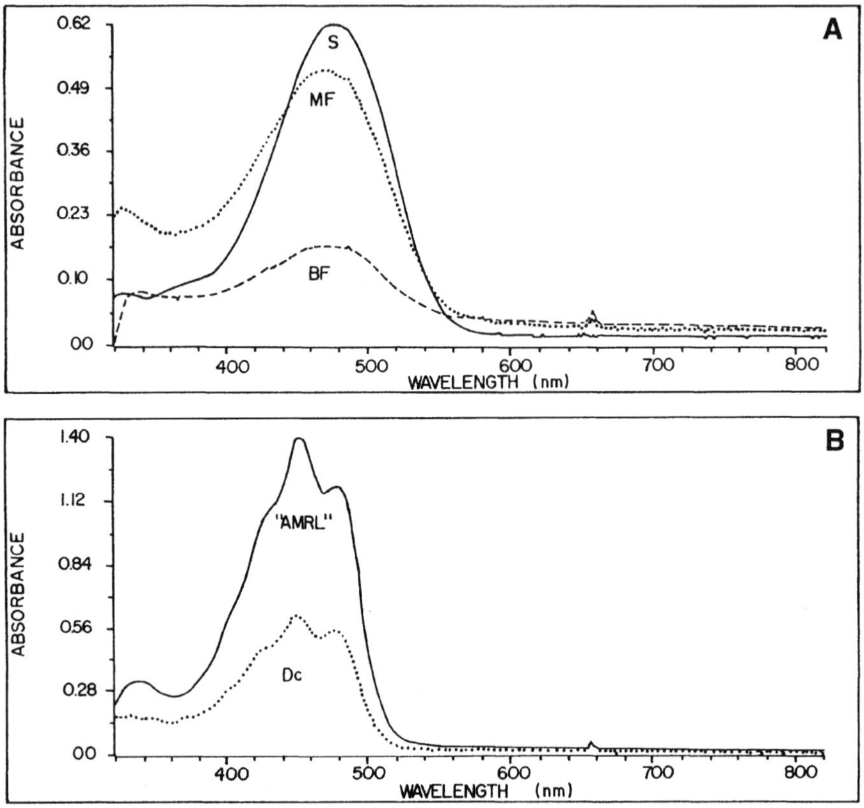

Fig. 5. Spectroscopic analysis of pigments. **(A)** Astaxanthin standard (S) and *P. rhodozyma* pigmented crude lipids from agitated flasks (MF; multifactorial) and scaled-up cultures (BF, BioFlo IIc fermenter). **(B)** Carrot *(Daucus carota)* chloroform extract (Dc); crude carotenoid extract from a carotenogenic bacterium ("AMRL" isolate).

Caryophill Pink, as expected, gave maximum absorbance at 482 nm (Fig. 5A). Absorbance peaks, for two different samples of *Phaffia* pigments obtained either in a multifactorial experiment (MF) or in the fermenter (BF), showed a small displacement toward a lower wavelength (ca. 472 nm) possibly owing to interference of the predominant contamination with triglycerides (as verified by TLC developed with the generic chromogenic reagent *p*-anisaldehyde; results not shown). The spectroscopic runs for carrot chloroform extract (Dc) and for a crude carotenoid extract from a yet uncharacterized carotenogenic bacterium (isolate "AMRL") are shown in Fig. 5A, for comparison purposes. Pigment quantitation *(14)* (Fig. 6A) on *Phaffia* cells from scaled-up cultures indicated 3R, 3′R-astaxanthin as the major component (63%) followed by the β-carotene (12%) (Fig. 6B) and the less oxygenated carotenoids (25%).

CONCLUSIONS

Inexpensive sugar cane juice, because of its high content of sucrose, proved to be a suitable carbon source for the growth of the yeast *P. rhodozyma* and for

A

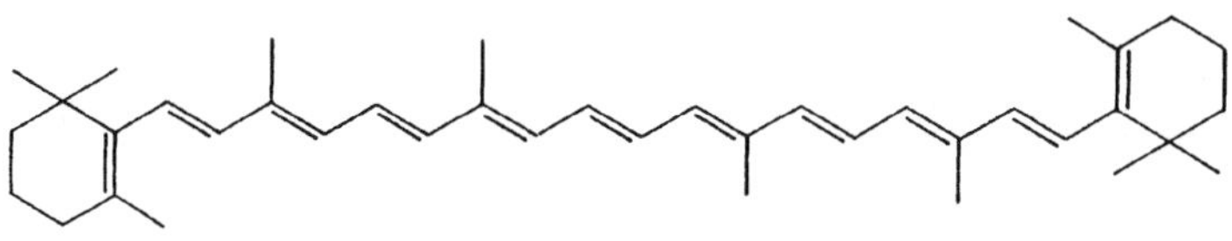

B

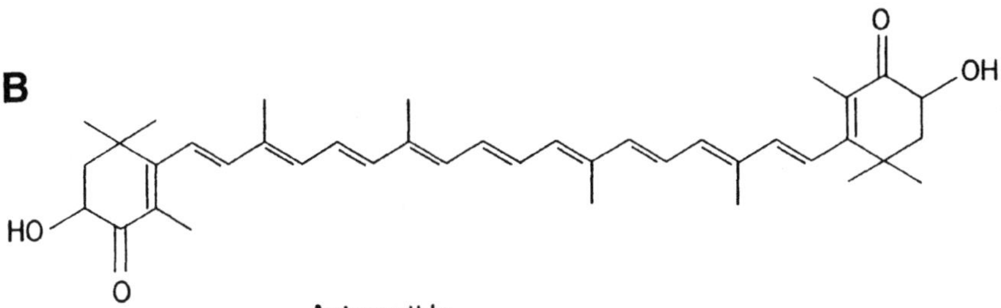

Fig. 6. Carotenoid structures: **(A)** β-carotene and **(B)** astaxanthin.

astaxanthin production. The contribution of the minor components of sugar cane juice to yields of yeast cells and to the carotenoid pigment remains to be established. Urea, a simple and also less expensive source of nitrogen, manifested itself as a productive supplement. The sugar cane juice and urea formulation was further improved by addition of sodium phosphate. Astaxanthin was also produced from disaccharide-rich C supplies, that is, crude maltose or cellobiose as the respective components from depolymerized starches and lignocellulose (bagasse).

ACKNOWLEDGMENTS

The authors thank CNPq and CAPES (funding agencies from Brazil), FUNPAR-UFPR (travel allowance), Braz H. Oliveira (comparative spectroscopy of carotenoids), J. M. G. Moura (Pescobras), and M. de Lima (drawings).

REFERENCES

1. Dean, K. L. (1992), *Ind. Bioprocessing* **14(5)**, 4,5.
2. Storebakken, T. and No, H. K. (1992), *Aquaculture* **100**, 209–229.
3. Pfander, H. (1992), *Methods Enzymol.* **213 (Part A)**, p. 7.
4. Miller, M. W., Yoneyama, M., and Soneda, M. (1976), *Int. J. Syst. Bacteriol.* **26**, 73.
5. Andrews, A. G., Phaff, H. J., and Starr, M. P. (1976), *Phytochemistry* **15**, 1003–1007.
6. Johnson, E. A. and Lewis, M. J. (1979), *J. Gen. Microbiol.* **115**, 173–183.
7. Yokoyama, A., Izunida, H., and Miki, W. (1994), *Biosci. Biotech. Biochem.* **58(10)**, 1842–1844.
8. Miki, W. (1991), *Pure Appl. Chem.* **63**, 141–146.
9. Schroeder, W. A. and Johnson, E. A. (1993), *J. Gen. Microbiol.* **139**, 907–912.
10. Dike, A, O., Lettner, F., and Zollitsch, W. (1992), *Arch. Gefluegelkd* **56(5)**, 205–209.

11. Tacon, A. G. J. (1994), *Production of Feeds for Aquatic Organisms in the Tropics with Particular Reference to Latin America and the Caribbean* (FAO/FIDI), VII Brazilian Aquaculture Symposium, Piracicaba-SP, p. 36.
12. Meyers, S. (1986), *Feedstuffs* **29,** 22,23.
13. Miller, G. L. (1959), *Anal. Chem.* **31(3),** 426–428.
14. Czeczuga, B., Ferraro, L. I., and Baron, M. (1992), *Bol. Botanica* **(Sao Paulo), 13,** 23–29.
15. Amorim, H. V. and Zago, E. A. (1988), *Annual Report in Alcoholic Fermentation Research* 9, ESALQ-USP, Sao Paulo, p. 95.

0273-2289/96/5758—0423$7.00

Scientific Note

Effect of Culture Conditions on Xylitol Production by *Candida guilliermondii* FTI 20037

MARTIN J. PFEIFER,[2] SILVIO S. SILVA*,[1] MARIA G. A. FELIPE,[1] INÊS C. ROBERTO,[1] AND ISMAEL M. MANCILHA[1]

[1] *Faculdade de Engenharia Quìmica de Lorena-Centro de Biotecnologia e Quìmica-Rodovia Itajubá-Lorena, Lorena S.P. Brazil; and* [2] *Hochschule für Technik-Fachhochschule Mannheim-Fachbereich Biotechnologie, Mannheim, Germany*

Index Entries: *Candida guilliermondii;* fermentation; sugar cane bagasse; D-xylose, xylitol.

INTRODUCTION

Lignocellulosic materials like sugar cane bagasse include about 35% of hemicellulose *(1,2)*. Hemicelluloses consist of polymeric substances, such as xylans and glucomannans, which differ from cellulose in having shorter molecular chains, a homo- or heteropolymeric backbone structure, and branch molecules, like acetic acid and a variety of pentoses and hexoses *(3)*. The use of D-xylose, the main component of xylan, to obtain chemical products has been a challenge in wood chemistry for the last 10 years.

Much biotechnological research has already been carried out in order to achieve higher ethanol yields using hemicellulosic biomass *(4,5)*. However, these investigations have been disregarding the fact that xylitol, an economically more interesting product, can be obtained from D-xylose.

Xylitol is a five-carbon sugar alcohol that presents characteristics, such as sweetening power, comparable to sucrose and anticariogenic properties *(6)*. Furthermore, the direct human metabolism of xylitol is insulin-independent. Xylitol is therefore well suited as a substitute of sucrose in cases of diabetes *(7,8)*.

Current technologies in xylitol production are based on catalytic hydrogenation, with Raney-Nickel-catalysts *(9)*, and therefore demand a very pure xylose solution. The combination of dilute-acid hydrolysis that can be carried out efficiently at relatively low cost and a biotechnological hydrogenation process has significant advantages, since it does not require pure xylose solutions. Initially, xylose was regarded to be nonfermentable by yeasts *(10)*, but later xylose-fermenting yeasts were discovered *(5,11,12)*. The screening of various yeast strains by

*Author to whom all correspondence and reprint requests should be addressed.

Barbosa et al. *(13)* proved that the *Candida guilliermondii* FTI 20037 yeast has the ability to convert xylose to xylitol with high efficiency and the maximum theoretical yield of xylitol from D-xylose is 0.917 g/g *(13)*.

Even though the exact mechanism of xylitol formation in *C. guilliermondii* is not fully understood, fermentation parameters of the D-xylose-to-xylitol bioconversion need to be optimized in order to reach maximal xylitol productivity under microaerobic conditions.

This investigation was carried out to determine the effects of age of inoculum and carbon source in the growth medium for xylitol production. These are important factors that should help the development of a more efficient fermentation process.

MATERIALS AND METHODS

Microorganism

C. guilliermondii FTI 20037 was obtained from the Faculdade de Engenharia Quìmica de Lorena, Lorena, São Paulo, Brazil, and maintained at 4°C on malt extract agar slants.

Inocula and Culture Media

The growth inoculum and fermentation media had the following basic composition:

12 g/L yeast extract (Oxoid), 40–50 g/L of xylose, 10 g/L glucose, 2 g/L $(NH_4)_2SO_4$, 0.1 g/L $CaCl_2$.

The inoculum was prepared from one loopfull of cells from stock culture inoculated in 50 mL of inoculum medium of the above mentioned composition.

For the age of inoculum experiment, cultures were grown on mixed substrate medium (MS): glucose (20%) plus xylose (80%) with a total of 50 g/L of these sugars for 15, 24, 48, and 72 h.

For the carbon source experiment, cultures were grown on xylose, glucose, and MS medium. These media were also used in determining the specific growth rates.

For the fermentation of the sugar cane bagasse hemicellulosic hydrolysate, cultures were grown in two steps: first, in synthetic MS medium and then in a medium containing hydrolysate, which was prepared according to the method of Felipe et al. *(14)*.

The inoculum was grown in wool-stoppered 125-mL Erlenmeyer flasks at 200/min in an orbital shaker at 30°C for 24 h, unless otherwise stated.

Fermentation Conditions

The initial cell concentration in all fermentations was 1 g dry wt cells/L was used. The fermentations were carried out under the same conditions as the inoculations. The fermentation in the bioreactor was conducted in a 1-L fermenter (Multigen—New Brunswick Scientific Incorp, Edison, NJ) under the following conditions: 30°C, 300/min, using a flat-blade turbine and an oxygen supply of 0.8 vvm (volume of air per volume of medium per minute).

Analytical Methods

Xylose, glucose and xylitol were analyzed with a Shimadzu high-performance liquid chromatograph (HPLC, Kyoto, Japan), using a Bio-Rad (Hercules, Ca) Aminex

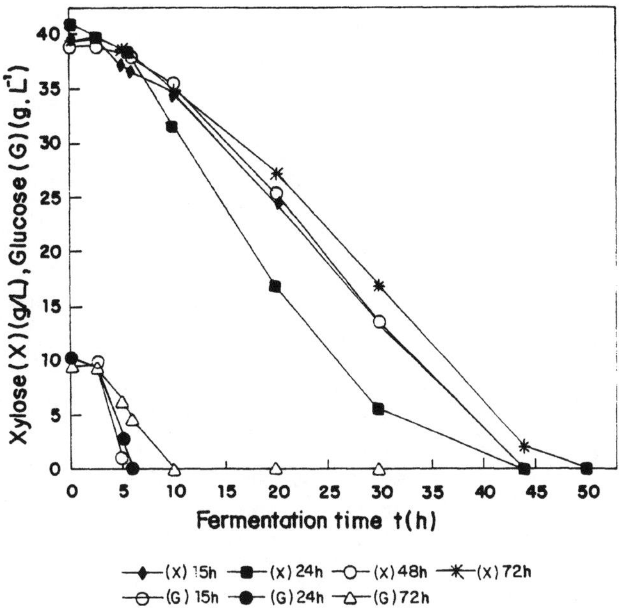

Fig. 1. Xylose (X) and glucose (G) consumption during batch fermentation with *C. guilliermondii* cultures of different ages.

HPX-87H column at 45°C and 0.02*N* H_2SO_4 as the eluent at a flowrate of 0.6 mL/min. Cell concentration was estimated by measuring absorbance at 600 nm. The relationship between absorbance and dry cell wt (g/L) was given by a standard curve (1 OD unit = 2.06 g dry wt cells/L).

RESULTS AND DISCUSSION

Effect of Age of the Inoculum on Xylitol Production

Substrates-to-product conversions can be characterized by the metabolic activity of the cells. In general, culture age is strongly related to this parameter. The results presented in Figs. 1 and 2 and Table 1 show to what extent age of inoculum influenced the fermentation of xylose to xylitol. According to Table 1, during the experiment, no substantial differences occurred in sugar (xylose and glucose) consumption or final xylitol concentration. The best results were achieved with a 24-h culture owing to more rapid fermentation, which ended after 30 h (Figs.1 and 2; Table 1). Probably, the 24-h culture used up most rapidly the NH_4^+ added to the medium, resulting in the liberation of H^+ in the medium. This may be related to the close relationship between growth and fermentation in this bioprocess.

An inoculum growth of 15 h appeared to be too short for effective fermentation (in the range of 48–72 h of growth) in comparison to 24 h of growth. A culture picked after that time still has not terminated all the metabolic changes necessary for the consumption of yeast extract and the fermentation of xylose.

The 72-h cultures in particular represented a stage of growth (Figs. 1 and 2; Table 1) in which the organisms do not have the same metabolic activity as in a

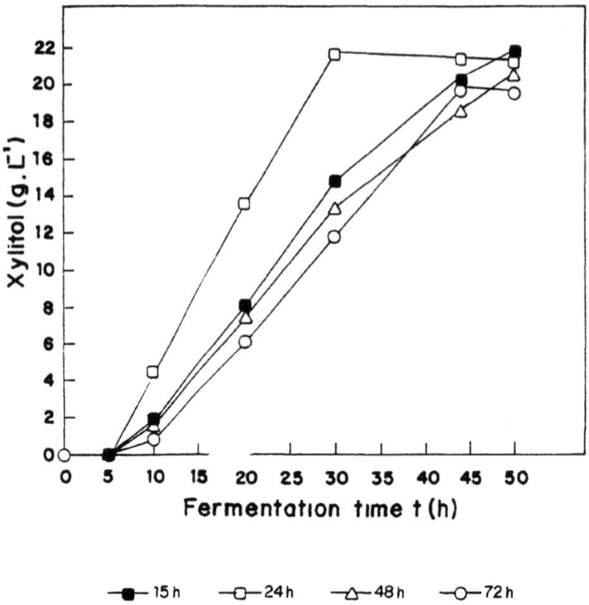

Fig. 2, Xylitol production during batch fermentation with *C. guilliermondii* FTI 20037 cultures of different ages.

Table 1
Effect of the Age of Inoculum on Xylitol Production

Age of inoculum h	'Final h	$Y_{P/S}$ $g \cdot g^{-1}$	δ $g \cdot L^{-1} \cdot h^{-1}$	η %	Final biomass $g \cdot L^{-1}$	ΔXOH $g \cdot L^{-1}$
15	44	0.56	0.50	61.0	17.53	21.92
24	30	0.53	0.66	57.7	18.56	21.50
48	44	0.53	0.49	58.2	18.30	20.60
72	50	0.50	0.39	55.2	17.01	19.80

'Final = time needed to consume at least 90% xylose, $Y_{P/S}$ = xylitol produced/xylose consumed, η = efficiency of fermentation, δ = volumetric productivity, ΔXOH = xylitol produced.

young culture. This may be the reason why the latter culture seemed to form xylitol later and was less productive, and consumed the substrate more slowly than the 24-h culture. Our observations agree with those of Sreenath et al. *(15)*, who reported that the metabolic activity of the cells changed appreciably as the cells aged.

Effect of Carbon Source in the Growth Inoculum on the Fermentation Performance

The choice of carbon source for the growth medium can be an important factor influencing the efficiency of a subsequent fermentation. Furthermore, mixed monosaccharide substrates can cause problems of biphase kinetics as reported by

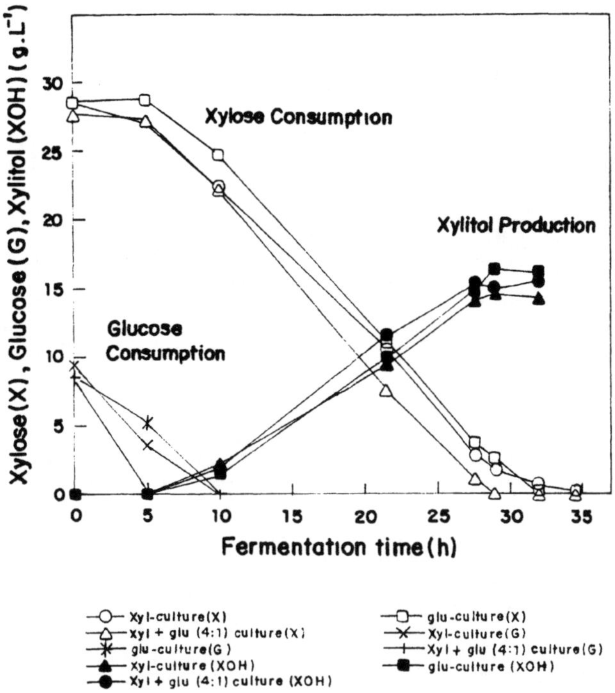

Fig. 3. Xylose (X) and glucose (G) consumption and xylitol (XOH) production during batch fermentation with *C. guilliermondii* cultures grown on xylose (xyl), glucose (glu), and mixed substrate (xyl + glu) medium.

Hsiao et al. *(16)*. According to Figs. 3 and 4 and Table 2, the results showed that the carbon source of the growth inoculum has an effect on the metabolism of *C. guilliermondii*. According to Fig. 3, the initial rate of xylose consumption was fastest in the medium with the xylose grown inoculum, followed by the MS medium grown inoculum and the glucose grown inoculum, which showed no initial xylose consumption at all. This suggests a catabolite repression *(17)* or catabolite inactivation *(18)* mechanism, similar to those reported by others *(16,19,20)*. Furthermore, the xylose consumption rate increases strongly when glucose enters low concentration ranges. This fact suggests that xylose reductase, the main enzyme in xylose metabolism, is reversibly inactivated by the presence of glucose at higher concentrations. This behavior has also been verified for *Pachysolen tannophilus* by Lee *(20)*.

As can be seen in Fig. 3, xylitol formation was hardly affected, since all the final concentrations were in the same range, althrough the glucose grown inoculum seemed to produce the highest final xylitol concentration. This is perhaps correlated to the problem of cofactor regeneration in a xylose grown inoculum, which uses xylitol for regeneration of NADPH under anaerobic conditions. Similar results have been obtained by Jeffries *(21)*, who found that the addition of glucose at low concentrations enhanced the product yield in xylose-to-ethanol fermentations. On the other hand, MS grown inoculum seemed to exhibit the highest xylitol productivity of 0.53 g/L/h, as shown in Table 2.

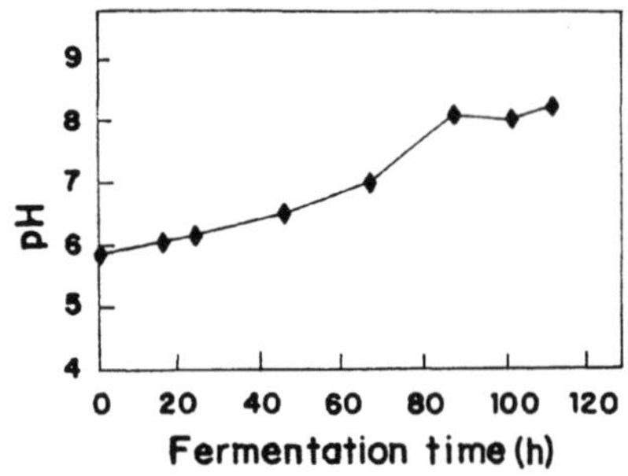

Fig. 4. Variation of pH during batch fermentation of sugar cane bagasse hemicellulosic hydrolysate by *C. guilliermondii* FTI 20037.

Table 2
Effect of Carbon Source in the Growth Inoculumon the Fermentation Performance

Composition of growth medium	[a]Final h	$Y_{P/S}$ $g \cdot g^{-1}$	δ $g \cdot L^{-1} \cdot h^{-1}$	η %	Final biomass $g \cdot L^{-1}$	ΔXOH $g \cdot L^{-1}$
Xylose	34	0.51	0.43	56.3	17.97	14.6
Glucose	32	0.57	0.51	62.9	18.49	16.4
MS medium	29	0.55	0.53	60.5	18.24	15.3

[a]Final = time needed to consume at least 90% xylose, $Y_{P/S}$ = xylitol produced/xylose consumed, η = efficiency of fermentation, δ = volumetric productivity, ΔXOH = xylitol produced.

The use of the carbon source (glucose) in the fermentation medium occurred without any adaptation time (Fig. 3), since all of the metabolic changes had already occurred during the growth phase in the growth inoculum of preliminary experiments. During this adaptation phase, no growth occurred but when the synthesis of all the necessary enzymes was completed, the growth seemed to enter the exponential phase directly.

The maximum specific growth rate (μ_{max}) was 0.04/h and 0.11/h in xylose and glucose medium, respectively. The different types of curves where perhaps owing to the rapid consumption of carbon source in the case of glucose and the slow, but constant consumption of xylose owing to the need for $NADPH_2$ as the reducer *(13)*.

Fermentation of Sugar Cane Bagasse Hemicellulosic Hydrolysate with *C. guilliermondii* FTI 20037

Fermentation in a complex medium, such as a hydrolysate, is always critical, since the hydrolysate still contains various substances that interfere with microbial metabolism. It is difficult to say in advance whether the organism will accept the mixture of nutrients, salts, and toxic substances, which are difficult to remove before conducting a fermentation. According to Figs. 4 and 5, the *C. guilliermondii* yeast was able to grow on a hydrolysate (initial pH: 5.86) still containing a variety of unknown substances and inhibitors, such as furfural, hydroxymethylfurfural, and acetic acid. The yeast fermented xylose at an uptake rate of 0.61 g/L/h, with productivity of 0.38 g/L/h and fermentation efficiency of 68.54%.

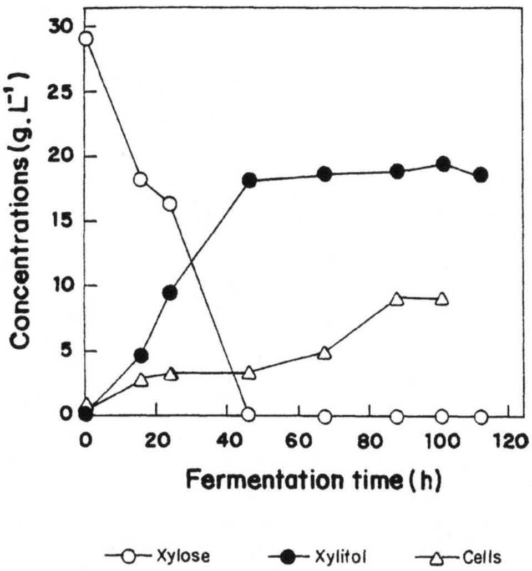

Fig. 5. Xylose consumption, xylitol production, and cell concentrations during batch fermentation of sugar cane bagasse hemicellulosic hydrolysate by *C. guilliermondii* FTI 20037.

The pH increased constantly, suggesting that the organism used the acetic acid present as a carbon source as well. According to the literature *(22,23)*, acetic acid at high concentrations strongly interferes with the energy metabolism of yeasts by reducing the H^+ gradient across the mitochondrial cell membrane used for ATP generation.

As shown by the results (Fig. 5), the hydrolysate did not inhibit the growth of *C. guilliermondii* significantly. Furthermore, the fermentation efficiency in this process was higher than in the foregoing fermentations. Probably, the organism uses carbon sources other than the sugars for growth and confirms the high adaptation ability of *C. guilliermondii* FTI 20037 to different media.

CONCLUSION

The demand for application of xylitol in food and pharmaceutical industries has created a strong market for low-cost xylitol production processes. The biotechnological approach for xylitol production appears to be efficient, and a high xylitol concentration can be obtained under controlled physiological conditions. The choice of a carbon source in the growth inoculum had only minor effects on the fermentation performance. Age of inoculum affected xylitol productivity considerably, but the final xylitol concentration was not influenced.

C. guilliermondii FTI 20037 yeast used in this bioprocess demonstrated high xylitol conversion rates, confirming its potential in sugar cane bagasse hemicellulosic hydrolysate fermentation.

Physiological parameters, such as age of inoculum and carbon source in the growth medium for xylitol formation by *Candida guilliermondii* FTI 20037, were investigated. Age of inoculum seemed to influence only the productivity and not the final xylitol concentration. The highest xylitol productivity (0.66 g/L/h) was reached using a 24-h culture. The specific growth rates (μ) in the different media ranged from 0.04/h for xylose to 0.11/h for glucose, with intermediate values obtained for mixtures of these sugars. Cultivation of the inoculum using different carbon sources did not significantly influence fermentation performance. Batch fermentation was carried out in an 1-L fermenter using a sugar cane bagasse hydrolysate. The yeast was able to ferment this medium, producing 18.40 g/L xylitol from 29.50 g/L of xylose, at a production rate of 0.38 g/L/h. This lower value, compared to that of the synthetic medium, may be attributed to the various toxic substances that interfere with microbial metabolism (e.g., acetic acid).

REFERENCES

1. Dekker, R. F. H. (1982) 12th Int. Union of Biochem. Con., Perth, **15-21** August pp. 1–4.
2. Tsao, G. T., Ladisch, M. R., Voloch, M., and Bienkowski, P. (1982), *Process Biochem.* **17(5),** 34–38.
3. Fengel, D. and Wegener, G., eds. (1989), *Wood—Chemistry, Ultrastructure, Reactions.* De Gruyter, Berlin-New York , pp. 106–107.
4. Jeffries, T. W. (1981), *Biotechnol. Letts.* **3,** 213–218.
5. Du Preez, J. C. (1983), *Biotechnol. Letts.* **5(5),** 357–362.
6. Mäkinen, K. K. and Isokangas, P. (1988), *Prog. in Food and Nutr. Sci.* **12,** 73–109.
7. Lang, K. (1971), *Klin. Wochenschr.* **49,** 233–245.
8. Manz, U., Vanninen, E., and Voirol, F. (1973), Food R. A. Symp. Sugar and Sugar Replacem., London Oct. 10, 1973.
9. Wisniak, J., Hershkowitz, M., and Stein, S. (1974), *Ind. Eng. Chem. Prod. Res. Dev.* **3(4),** 232–236.
10. Barnett, J. A. (1976), *Adv. Carbohydr. Chem. Biochem.* **32,** 125–234.
11. Schneider, H., Wang, P. Y., Chan, V. K., and Maleszka, R., (1981), *Biotechnol. Letts.* **3(2),** 89–92.
12. Toivola, A., Yarrow, D., Van den Bosch, E., Van Dijken, J. P. and Scheffers, W. A. (1984), *Appl. and Environ. Microbiol.* **47,** 1221–1223.
13. Barbosa, M. F. S., Medeiros, M. B., Mancilha, I. M., Schneider, H., and Lee, H. (1988), *J. Ind. Microbiol.* **3,** 241-251.
14. Felipe, M. G. A., Mancilha, I. M., Vitolo, M., Roberto, I. C., Silva, S. S., and Rosa, S. A. M. (1988), *Arquivos de Biol. e Tecnol.* **36(1),** 103–114 .
15. Sreenath, H. K., Chapman, T. W., and Jeffries, T. W. (1986), *Appl. Microbial Biotechnol.* **24,** 294–299.
16. Hsiao, H-Y., Chiang, L. C., Ueng, P. P., and Tsao, G. T. (1982), *Appl. Environ. Microbiol.* **43,** 840–845.
17. Magasanik, B. (1961), *Cold Spring Harbour Symp. Quant. Biol.* **26,** 244–256.
18. Holzer, H. (1976), *Trends in Biochem. Sci.* **1,** 178–181.
19. Webb, S. R. and Lee, H. (1990), *Biotech. Adv.* **8,** 685–697.
20. Lee, H. (1992), *Microbiol. Letts.* **92,** 1–4.
21. Jeffries, T. W. (1985), *Biotechnol. and Bioeng. Symp.* **15,** 149–166.
22. Van Zyl, C., Prior, B. A., and Preez du, J. C. (1988), *Enzyme and Microb. Technol.* **13,** 82–86.
23. Ferrari, M. D., Neirotti, E., Albornoz, C., and Saucedo, E. (1992), *Biotechnol. Bioeng.* **40,** 753–759.

Session 3
Bioprocessing Research

George P. Philippidis[1] and David F. Ollis[2]

[1]*National Renewable Energy Laboratory, Golden, CO;*
and [2]North Carolina State University, Raleigh, NC

Bioprocessing research is crucial to the production of chemicals and fuels from biomass, because it assesses not only the scientific merits of the biochemical process, but also its engineering soundness and technoeconomic feasibility. Hence, the session on bioprocessing research covered a variety of interesting topics from metabolic and thermodynamic studies and material balances to mass transfer and fermentation investigations for the production of both chemicals and enzymes.

In the area of cell metabolism, the fermentation of xylose to ethanol was used as a model system. A metabolic competition model was developed for *Saccharomyces cerevisiae* to depict the spontaneous oscillations observed in the concentration of various fermentation parameters during the organism's diauxic utilization of glucose and xylose. This simple cybernetic model can serve as a tool in efforts to enhance the uptake of 5-carbon sugars by several yeast species. Besides modeling, the availability of thermodynamic data is also important to the success of bioprocess engineering. Using experimental measurements and theoretical calculations, the equilibria of a variety of biochemical reactions over a wide range of temperature, pH, and ionic strength were determined. This information can then be used to predict the feasibility of biochemical reactions and their product yields. With respect to product yield calculations, another presentation emphasized the need for detailed material balances in order to achieve process optimization. Based on developed material balance templates, the true yield of ethanol from pretreated cellulosic biomass was determined.

In the area of fermentation studies, the production of cellulase enzyme and succinic acid were investigated. Using NaOH-pretreated wheat straw, the effect of environmental conditions on the solid-state cultivation of *Trichoderma reesei* was measured for the synthesis of cellulase and β-glucosidase. On the same subject, the rheological, mass transfer, and biological kinetics of cellulase production were outlined in another presentation, based on shear measurements of the suspension of *T. reesei* and the development of a biological kinetic model. Finally, the bacterial production of succinic acid by an *Anaerobiospirillum* species was presented. Using sodium carbonate for pH control and bubbling carbon dioxide through the fermentation broth, high yields of succinic acid were achieved.

0273-2289/96/5758—0433$8.00

Production of Cellulase in Solid-State Fermentation with *Trichoderma reesei* MCG 80 on Wheat Straw

PARMINDER S. CHAHAL,[1]
DEVINDER S. CHAHAL,*[,1] AND GEORGE B. B. LÊ[2]

[1]Institut Armand-Frappier, Université du Québec, 531, Boulevard des Prairies, Laval, Québec, Canada H7N 4Z3; and [2]Ministry of Natural Resources, Charlesbourg, Québec, Canada

ABSTRACT

It is an accepted fact that ethanol production from lignocellulosic materials is not economical as yet because of the high cost of cellulase production. To reduce the cost of cellulase production, lignocellulosic material (wheat straw [WS]), a comparatively much cheaper substrate, was used instead of costly substrates (pure cellulose or lactose). A pan bioreactor was developed for solid-state fermentation (SSF) that required a small capital investment. High yields of complete cellulase system were obtained compared to that in the liquid-state fermentation (LSF) from WS, when treated with 4.25% NaOH at 121°C for 1 h and mixed with Mandels' medium. A complete cellulase system is defined as one in which the ratio of β-glucosidase activity to filter paper activity in the enzyme solution is close to 1.0. The cellulase system derived from SSF using the pan bioreactor gave more than 85% hydrolysis of delignified WS. The prototype pan bioreactor requires further improvements so that optimum quantity of substrate can be fermented to obtain high yields of complete cellulase system per unit space. The SSF process provides a means for the production of complete cellulase system for the economical bioconversion of renewable biomass into ethanol.

Index Entries: Cellulase; hydrolytic potential; lignocelluloses; solid-state fermentation; *Trichoderma reesei* MCG 80.

INTRODUCTION

Economical production of ethanol from lignocellulosic materials is hindered because of the high cost of cellulase production *(1,2)*. One effective approach to reduce the cost of enzyme production is to replace pure cellulose or lactose with relatively cheeper substrates, such as lignocellulosic materials. There have been reports of successful attempts to produce cellulase on lignocellulosics *(1–4)*. The development of technology with minimum capital investment is another approach

*Author to whom all correspondence and reprint requests should be addressed.

to reduce the cost of cellulase production. In our earlier work *(4)*, we were able to produce cellulase in a solid-state fermentation (SSF) process that required relatively inexpensive equipment compared to the conventional fermenter used for liquid-state fermentation (LSF) process. SSF is a process by which insoluble substrate is fermented with sufficient moisture, but without free water. In LSF, the insoluble substrate is fermented in a slurry of 1–5% suspension *(4)*.

For complete hydrolysis of cellulose to glucose, cellulase systems must contain the following enzymes:

1. Endo-1,4-β-glucanase (1,4-β-D-glucan glucanohydrolase, EC 3.2.1.4);
2. Exo-1,4-β-glucanase (1,4-β-D-glucan cellobiohydrolase, EC 3.2.1.91); and
3. β-glucosidase (EC 3.2.1.21, β-D-glucoside glucohydrolase or cellobiase).

Since the hydrolysis of cellulose is owing to the synergistic action of endo- and exoglucanases, we refer to their synergistic activity as filter paper activity (FPA) in this article. The cellulase systems of hypercellulase mutants of *Trichoderma reesei* are generally deficient in β-glucosidase *(3,5–7)*. During the hydrolysis of cellulose with this type of enzyme system, cellobiose accumulates, which inhibits the action of cellulases *(8–10)*. From the hydrolysis results reported by various researchers, we can deduce that a cellulase system having a ratio of β-glucosidase activity (βGA) to FPA close to 1.0 is necessary to obtain the highest rate of hydrolysis and highest glucose content in the hydrolysate *(3,5–7)*.

In this article, we present our efforts to produce a cellulase system using *T. reesei* MCG 80 which has a βGA-to-FPA ratio close to 1.0 by varying the medium composition and pretreatment parameters of wheat straw (WS) for better exposure of cellulose to growing microorganism.

MATERIALS AND METHODS

Microorganism

T. reesei MCG 80 (NRRL#12368) was obtained from J. L. Swezey, ARS Patent Culture Collection, USDA (Peoria, IL). Its culture was maintained on delignified WS agar medium in Petri plates *(4)*.

Substrate and Nutrients

WS, ground to 20-mesh size, was pretreated with various concentrations of sodium hydroxide at 121°C for 1 h. Nutrients described by Mandels and Weber *(11)* were used in the culture medium, but proteose peptone was replaced by yeast extract, potassium phosphate concentration was variable and is described where necessary in the text, and the urea was omitted from the culture medium. The quantity of nutrients required was proportional to the total carbohydrate content (70%) of the WS. WS fermentation medium for SSF was prepared as explained by Chahal *(4)*. Four grams of WS pretreated and mixed with Mandels' medium was dispensed in each Erlenmeyer flask of 500-mL capacity. Twenty grams of WS were used in the pan bioreactor.

Enzyme Assay

The enzyme from SSF was extracted by mixing the whole content of the flask (4 g WS) in a total of 100 mL water and then shaking for 1 h at 200 rpm at room temperature and then centrifuged at 10,000 rpm (11,000*g*) for 30 min. FPA was expressed in international units (IU = μmol of glucose released/min) from 50 mg filter paper after 60 min of incubation at 50°C according to the method described by Mandels et al. *(12)*, which was approved by the International Union of Pure and Applied Chemistry (IUPAC) *(13)*. The activity of β-glucosidase was measured in IU using 0.5% salicin solution for 30 min of incubation at 50°C. The ratio of enzyme activities was defined as β-glucosidase activity divided by FPA.

Hydrolysis and Analysis of Sugars

The hydrolysis of delignified WS was done in 5% (w/v) concentration with enzyme loading of 20 IU FPA/g delignified WS at 45°C in 0.05*M* citrate buffer at pH of 4.8. The quantities of glucose, cellobiose, xylose, and arabinose in the hydrolysate of delignified WS were determined by Beckman 344 HPLC equipped with Altex 156 Refractive Index detector and Spherogel 7.5% carbohydrate column with a flow rate of 0.5 mL/min in the mobile phase of deionized and degassed water at 90°C.

Delignification of WS

The WS was delignified by the method described by Toyama and Ogawa *(14)*.

Experimental Design

An orthogonal composite design was applied to explore the response surfaces of activities of filter paper and β-glucosidase *(15–18)*. This design requires that the distance from the center point to each of the star points be such that all effects and interactions estimated in the second-order model are orthogonal to one another. This distance, *d*, is 1.414 for two factors, and the model has the following form:

$$y = a_0 + a_1X_1 + a_2X_2 + a_{11}X_1^2 + a_{22}X_2^2 + a_{12}X_1X_2 \quad (1)$$

where y is the response variable, a_is are the coefficients of the equation, and X_i is the normalized value (dimensionless) of factor i, normalized according to the following equation:

$$X_i = \{V_i - [(H_i + L_i/2)]/[(H_i - L_i/2)]\} \quad (2)$$

where, H_i = upper limit of factor i, L_i = lower limit of factor i, and V_i = value of operating variable i.

Linear regression analysis was applied to fit the data to experimental values and the coefficients, a_js, were determined at 95% confidence level.

RESULTS AND DISCUSSION

Effect of NaOH Concentration for the Pretreatment of WS and KH_2PO_4 Concentration in the Medium for Growth

Sodium hydroxide was used to pretreat the WS to expose the cellulose for mycelial growth. When too little NaOH is used, not much cellulose is exposed. Whereas, when NaOH is used in a very high concentration, there is a possibility that certain chemicals are produced during the heating cycle of the pretreatment process that are toxic to mycelial growth or high concentrations of hemicelluloses are released that inhibit the cellulase production. Therefore, there was a need to

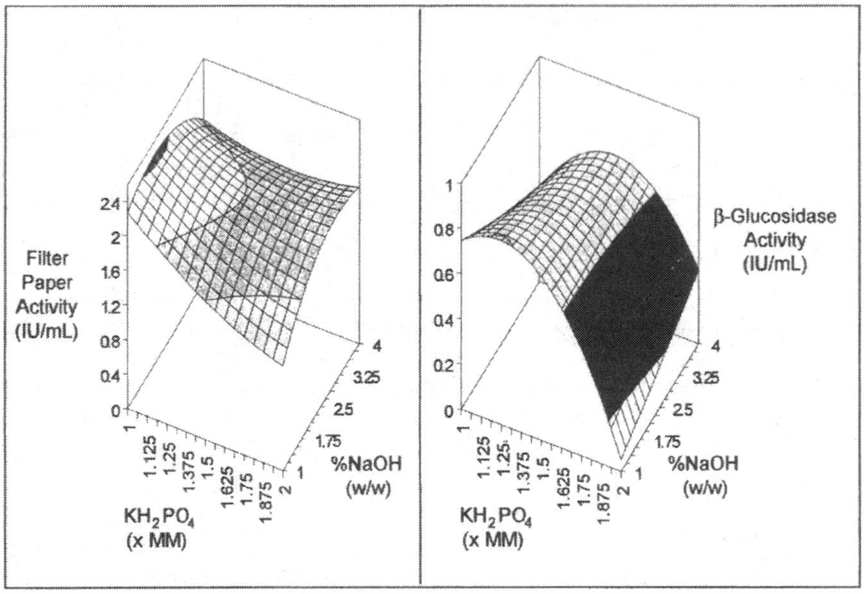

Fig. 1. FPA and βGA as a function of NaOH concentration required for the pretreatment of WS and the potassium phosphate concentration required in growth medium in LSF process.

find the best NaOH concentration required for the pretreatment of WS for the maximum production of cellulase. The potassium phosphate in the medium is a source of phosphorus for the growth of *T. reesei*, and it also acts as a buffer, preventing a fast drop in pH. The pH below 4.0 is not favorable for the biosynthesis of β-glucosidase. Therefore, there was a need for a medium that had a proper buffering capacity.

Experiments were designed with two factors having normalized *–d* and *+d* values representing 1.0 and 4.0% (w/w) for NaOH concentration for the pretreatment of WS, and 1.0 and 2.0 times the concentration of KH_2PO_4 required in Mandels' medium (×MM), respectively. MM contains 2 g KH_2PO_4 for 10 g carbohydrates.

Figure 1 shows the filter paper and βGAs activities in the LSF as functions of concentrations of NaOH (%w/w) and KH_2PO_4 (×MM). The increase in concentration of NaOH used for pretreatment of WS resulted in an increase in FPA. The highest FPA value, 2.4 IU/mL, was achieved when phosphate was in low concentration and the NaOH concentration used for the pretreatment of WS was about 2% (w/w). The increase in phosphate concentration decreased the FPA regardless of the concentration of the NaOH used for the pretreatment of WS. There was a sharp drop in βGA when phosphate concentration in the medium was high regardless of NaOH concentration required for the pretreatment of WS. The highest βGA, 0.8 IU/mL, was obtained at low concentration of NaOH (around 1%) used for the pretreatment with 1.25 x MM phosphate concentration in the growth medium. The best ratio of the activities obtained was 0.4.

When the fermentation was carried out in the SSF, the activities of both the enzymes increased considerably (Fig. 2). The FPA of 6 IU/mL was obtained for low-phosphate concentrations and high concentrations of NaOH for the pretreatment

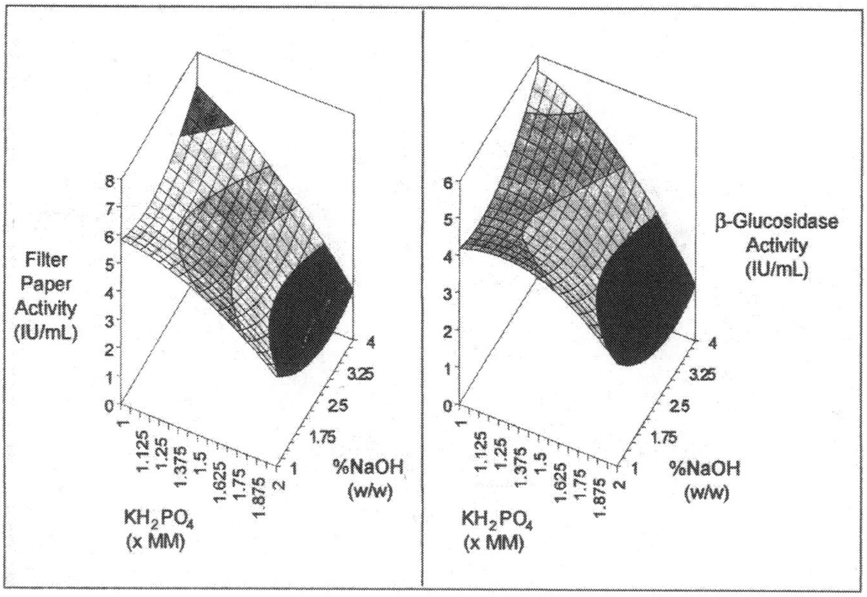

Fig. 2. FPA and βGA as a function of NaOH concentration required for the pretreatment of WS and the potassium phosphate concentration required in growth medium in SSF process.

of WS. The case was similar for βGA production. A value of about 5 IU/mL was achieved for low-phosphate concentration and high concentration of NaOH for pretreatment of WS. The ratio of activities in this case was about 0.8, significantly higher than that in LSF. The higher yields and ratio of activities obtained suggested further exploration of the SSF technology.

Effect of Water Content and NaOH Concentration in the Pretreatment of WS

It was necessary to determine the best water content to WS ratio in the pretreatment of the WS along with NaOH concentrations, because it was assumed that with higher water content in the pretreatment process, a better exposure to cellulose could be achieved with proper sodium hydroxide concentration. Therefore, it was decided to do another orthogonal composite design around the center point of 4.25% (w/w) of NaOH and 6-mL volume of NaOH/4 g WS with low and high value equivalents of 2.5 and 6.0 for % NaOH and low and high value equivalents of 4 and 8 for volume of NaOH solution (mL) required for the pretreatment. At the time of fermentation, the final water content in WS culture medium would be 76% after adding nutrient solution and inoculum. Figure 3 shows the FPA and βGA responses for % NaOH and volume of NaOH solution used for the pretreatment of 4 g WS. There was a maximum obtained in FPA (higher than 5 IU/mL) for % NaOH between 4.25 and 6.31, and water content between 4 and 6 mL/4 g WS. The βGA values were high for high NaOH concentrations regardless of the volume of NaOH solution used for the pretreatment. Although high βGA could

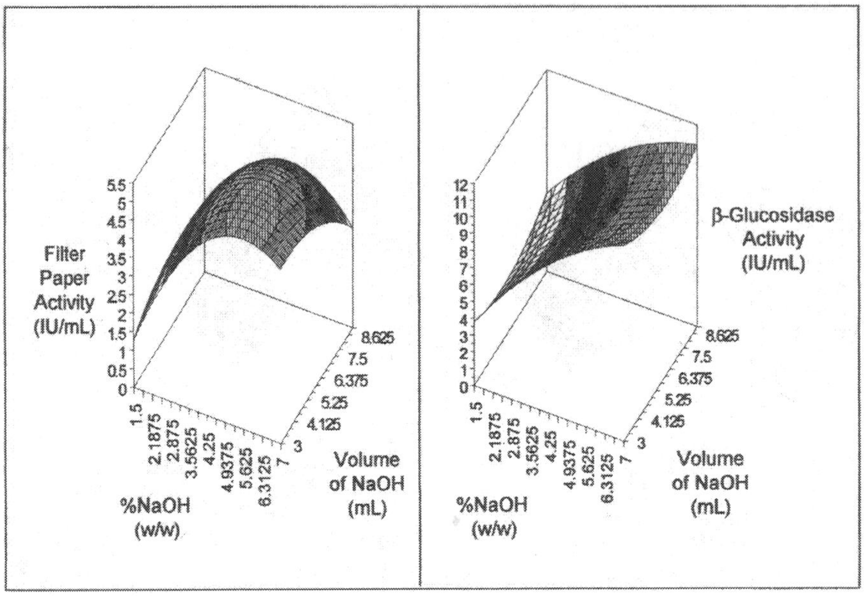

Fig. 3. FPA and βGA as a function of NaOH concentration and the volume of NaOH solution required for the pretreatment of WS in SSF process.

have been obtained by pretreating WS with higher NaOH concentration, it would not affect the overall hydrolysis of the cellulose, because the ratio of activities higher than 1.0 in the cellulase system does not increase the hydrolytic potential of the cellulase system *(3,5–7)*.

Development of Pan Bioreactor

Now there was a need to develop some means to grow *T. reesei* in such a way that a larger scale SSF could be tested. Therefore, a pan bioreactor was developed, the shape and the dimensions of which are shown in Fig. 4. In this bioreactor, a screen (60 mesh) was placed in the middle of the pan. A bed of pretreated WS mixed with medium was spread on the top of this screen. Forced aeration may be supplied from the bottom port leading to a cross with four slits on each side for uniform distribution of air under the screen (insert in Fig. 4). The upper port is for the exit of forced air or natural convectional air movement.

One set of experiments in flasks and two sets of experiments in pan bioreactors (one with forced aeration) were arranged by selecting the following best conditions (although some of the βGA was sacrificed by minimizing the NaOH concentration required for the pretreatment): the pretreatment used 4.25% NaOH (w/w) with 6 mL water/4 g WS, at 121°C for 1 h, the phosphate concentration was 1 × MM (= 2.0 g KH_2PO_4/10 g carbohydrates), the final moisture content of the WS medium was 76% at the time of fermentation, and the loading of WS in the Erlenmeyer flask and the pan bioreactor was 4 and 20 g, respectively. One set of the pan bioreactors was forced aerated. The results of LSF and SSF processes are summarized in Table 1. The best way to compare the activities of enzyme was to report the yields of activities, such as FPA and βGA/g WS supplied.

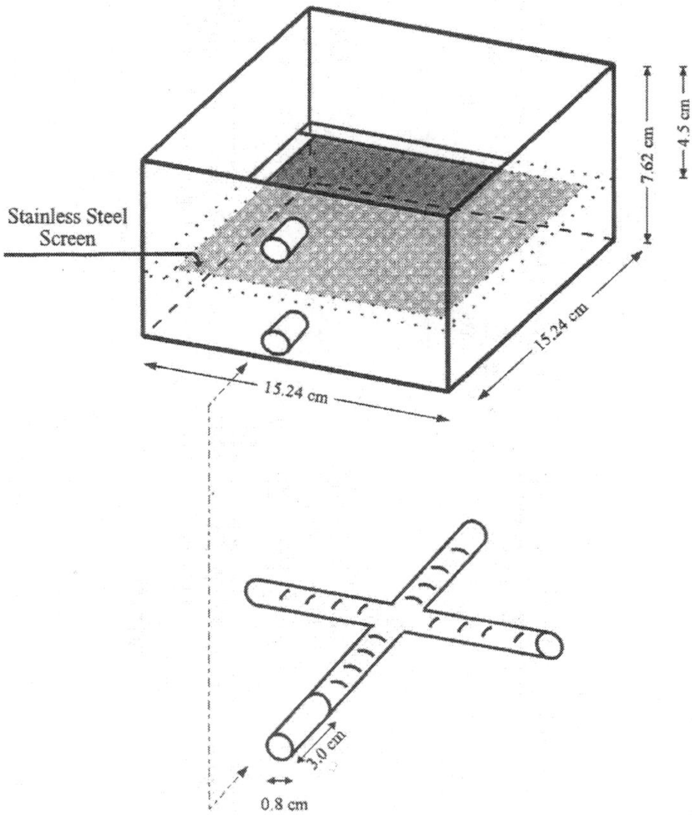

Fig. 4. Pan Bioreactor for the SSF of WS.

The FPA and βGA/g WS and their ratio in LSF in flasks were lower than that of SSF in flasks. In SSF the FPA/g WS in flask and pan bioreactor was the same after 21 d of fermentation. However, the βGA/g WS was slightly higher in the flask than in the pan bioreactor (Table 1). The values in the forced aerated pans were lower than the unforced aerated pan, probably because of the drying of the WS, which restricted the growth of the microorganism in certain regions of the pan bioreactor. It was decided to use humidified air in the future experiments. The ratios of the activities were close to 1.0 in all of the cases of SSF.

Hydrolytic Potential of the Cellulase System Produced in SSF

Figure 5 shows the hydrolysis of 5% delignified WS when loaded with the enzyme extracted from the SSF in the pan bioreactor at a concentration of 20 IU FPA/g delignified WS. A total of 4.33 g sugars/100 mL of hydrolysate were produced in 94 h of hydrolysis at 45°C with shaking at the rate of 200 rpm. The hydrolysate of 5 g delignified WS contained 4.33 g total sugars (3.2 g glucose, 0.8 g xylose, 0.08 g arabinose, and 0.25 g cellobiose) in 100 mL of solution. This represents the hydrolytic potential of over 85% with this cellulase system. Because of high βGA, very little accumulation of cellobiose (0.25 g/100 mL) in the hydrolysate was recorded with this cellulase system (Fig. 5). The accumulation of cellobiose in

Table 1

Cellulase Enzyme Yields in LSF and SSF in Erlenmeyer Flasks and Pan Bioreactors

Type of bioreactors	Amount of WS/ bioreactor, g	FPA[a]		βGA[a]		FPA, IU/g WS[b]	βGA, IU/g WS[b]	Ratio, β-GA/FPA
		IU/mL	Total, IU	IU/mL	Total, IU			
LSF								
Erlenmeyer flask	4	2.0	400	0.8	160	100	40	0.40
SSF								
Erlenmeyer flask	4	6.55	655	7.72	772	164	194	1.18
Pan bioreactor	20	6.53	3266	7.11	3555	163	178	1.09
Pan Bioreactor (Forced Aeration)	20	5.82	2912	5.61	2807	146	141	0.97

[a]Enzyme extraction: In Erlenmeyer flasks: 200 mL total water/4 g of WS in LSF process. In Erlenmeyer flasks: 100 mL total water/4 g of WS in SSF process. In Pan bioreactors: 500 mL total water/20 g of WS in SSF process. For high concentration of enzyme per unit volume, the volume of water required for extraction of enzyme should be reduced accordingly.

[b]WS contains 70% carbohydrate (30–35% cellulose, and 35–40% hemicelluloses).

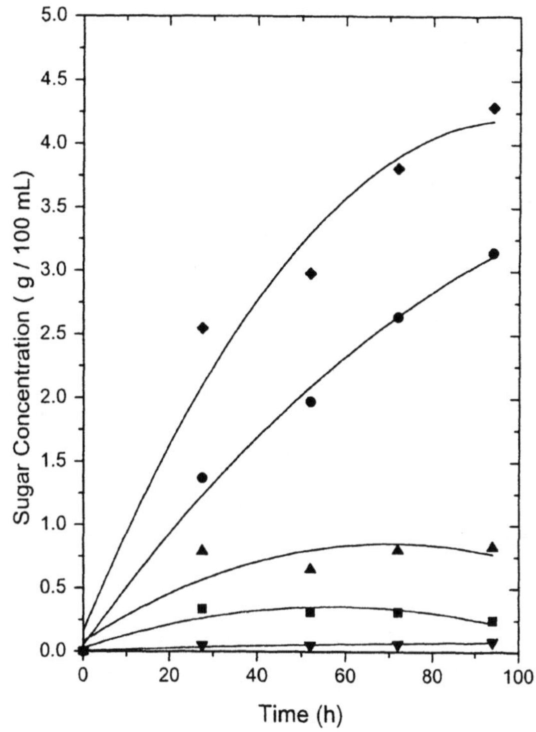

Fig. 5. Hydrolysis of delignified wheat straw by the enzyme system produced by 21 d of solid state fermentation in pan bioreactor. ■, Cellobiose; ●, glucose; ▲, xylose; ▼, arabinose; and ◆, total sugars.

the hydrolysate of cellulose is known to inhibit the cellulase action on cellulose *(8–10)*. The content of residual lignin in the delignified WS was not estimated; therefore, the hydrolytic potential of the cellulase system produced in SSF could be considered more than the reported 85% level.

Advantages of SSF Over LSF

The cellulase system produced in the conventional LSF is normally deficient in the βGA, and thus the ratio is usually <0.5. Therefore, it is necessary to add β-glucosidase from other sources to complement the cellulase system to obtain higher conversions of cellulose into sugars *(3,5–7)*. The LSF cellulase system is often very dilute, so there is a need for ultrafiltration to concentrate the enzyme so that reasonable loadings of the enzyme to substrate can be achieved, especially when the goal is to obtain reasonable glucose concentrations for subsequent ethanol fermentation. Furthermore, to achieve a large amount of enzyme, a large expensive fermenter is required to produce cellulase system in LSF.

In this study, SSF was achieved with minimum water (76% moisture content), thus requiring a very small fermenter (pan bioreactor). The cellulase system can be extracted by a plate and frame filtering unit by adding just enough water as required to achieve the desired enzyme concentration per unit volume. The cellulase system produced by SSF usually contains high βGA. Therefore, the enzyme ratio close to 1.0 is easily obtained. The overall equipment cost for SSF is relatively less than that for LSF.

CONCLUSIONS

The following conclusions can be made as a result of this study:

1. A low-cost lignocellulosic substrate, WS, was successfully used instead of costly substrate (e.g., pure cellulose or lactose) to reduce the cost of cellulase production.
2. The SSF process appeared promising for obtaining high yields of a complete cellulase system and may be superior to the one obtained by LSF process.
3. To reduce the cost of enzyme production still further, a prototype biorector has been developed for SSF, which requires little capital investment. This prototype pan bioreactor requires some further modifications so that an optimum quantity of substrate, lignocellulosic materials, can be fermented to obtain high yields of complete cellulase system.
4. The SSF process provides a means for the production of complete cellulase system for the economical bioconversion of renewable biomass into ethanol.

ACKNOWLEDGMENTS

Financial help to D. S. Chahal for the development of SSF process by the Ministry of Natural Resources, Government of Québec, Canada is gratefully acknowledged.

REFERENCES

1. Chahal, D. S., Chahal, P. S., André, G., and Ishaque, M. (1987), in *Sixth Canadian Bioenergy R & D Seminar*, Stiassny, Z. Z., ed., Elsevier Applied Science, New York, pp. 306–310.
2. Chahal, D. S. (1982), in *Enzymatic Hydrolysis of Cellulose: State-of-the-Art*, Division of Energy Research and Development, National Research Council Canada, Ottawa, Canada, pp. 74–95.
3. Chahal, D. S., Mcguire, S., Pikor, H., and Noble, G. (1982), *Biomass* **2,** 127–138.
4. Chahal, D. S. (1985), *Appl. Environ. Microbiol.* **49,** 205–210.
5. Stockton, B. C., Mitchell, D. J., Grohmann, K., and Himmel, M. E. (1991), *Biotechnol. Lett.* **13,** 57–62.
6. King, K. W. and Vessal, M. I. (1969), *Adv. Chem. Ser.* **95,** 7–25.
7. Wood, T. M. (1981), *Biochem. J.* **121,** 353–362.
8. Ghose, T. K. and Das, K. (1971), *Adv. Biochem. Eng.* **1,** 55–62.
9. Gritzali, M. and Brown, R. D. (1979), *Adv. Chem. Ser.* **181,** 237–260.
10. Ryu, D. and Mandels, M. (1980), *Enzyme Microb. Technol.* **2,** 91–102.
11. Mandels, M. and Weber, J. (1969), *Adv. Chem. Ser.* **95,** 391–414.
12. Mandels, M., Andreotti, R., and Roche, C. (1976), *Biotechnol. Bioeng. Symp.* **6,** 21–33.
13. Ghose, T. K. (1987), *Pure and Appl. Chem.* **59,** 257–268.
14. Toyama, N. and Ogawa, K. (1972), *Fermentation Technology Today*. Society of Fermentation Technology, Japan, pp. 743–757.
15. Box, G. E. P., Hunter, W. G. and Hunter, J. S. (1978), *Statistics for Experimenters, An Introduction to Design, Data Analysis, and Model Building*, Wiley, New York, pp. 510–539.
16. Box, G. E. P. and Hunter, J. S. (1957), *Ann. Math. Stat.* **28,** 1–25.
17. Schmidt, S. R. and Launsby, R. G. (1992), *Understanding Industrial Designed Experiments*, 3rd ed. Air Academy Press, Colorado Springs, CO, pp. 7.1–7.17.
18. Cochran, W. C. and Cox, G. M. (1968), *Experimental Designs*, 2nd ed., Wiley, New York, pp. 335–375.

0273-2289/96/5758—0443$9.25

Detailed Material Balance and Ethanol Yield Calculations for the Biomass-to-Ethanol Conversion Process

CHRISTOS HATZIS,*,[1] CYNTHIA RILEY,[2] AND GEORGE P. PHILIPPIDIS[1]

[1]Bioprocess Development Branch, and [2]Analysis and Project Management Branch, Alternative Fuels Division, National Renewable Energy Laboratory (NREL), 1617 Cole Boulevard, Golden, CO 80401

ABSTRACT

Applying material balance calculations to the evaluation and optimization of lignocellulosic biomass conversion processes is fundamentally important. The lack of a general framework for material balance calculations and inconsistent compositional analysis data have made it difficult to compare results from different research groups. Material balance templates have been developed to follow accurately the distribution of carbon in lignocellulosic substrates through the pretreatment and simultaneous saccharification and fermentation (SSF) processes, and provide information on overall carbon recovery, recovery of individual sugars, and solubilization of biomass components. Based on material balance considerations, we developed equations that allow us to compute overall ethanol yields for biochemical conversion of biomass correctly.

Index Entries: Carbon balance; dilute-acid pretreatment; simultaneous saccharification and fermentation; enzymatic conversion; overall yield.

Nomenclature: C-mol, amount (mass) of a substance containing 1 mol of the element carbon (g); $[C_5]$, concentration of pentose sugars in hydrolysate (g/L); $[C_6]$, concentration of hexose sugars in hydrolysate (g/L); d_L, density of liquor (SSF or hydrolysate) (g/L); $[E]$, ethanol concentration (g/L); f_s, fraction of insoluble solids (g insoluble solids/g slurry); L_p, mass-loss factor of insoluble solids from pretreatment (g pretreated solids/g raw solids); m, mass of whole slurry (g); m_L, mass of liquid phase (g); m_s, mass of insoluble solids (g); Q_{C5}, mass fraction of pentose sugars in raw substrate (g/g dry mass); Q_{C6}, mass fraction of hexose sugars in raw substrate (g/g dry mass); Y_E, ethanol yield (g/100 g hexose sugar). Subscripts: L, liquid-phase composition or mass; *max*, maximal potential yield; 0, raw substrate or beginning of SSF (t = 0); p, substrate or solids compositions after pretreatment; s, solids composition or mass.

*Author to whom all correspondence and reprint requests should be addressed.

THE BIOMASS-TO-ETHANOL TECHNOLOGY

Cellulosic biomass, the most abundant natural polymer on earth, is a renewable resource that can be transformed efficiently into commodity and specialty chemicals using biotechnology *(1)*. In the United States, biomass feedstocks include energy crops, agricultural and forestry residues, pulp and paper industry waste streams, and municipal solid waste. Ethanol-from-biomass production, in particular, is seen as a high-potential technology that offers several advantages to the country and the environment. By replacing gasoline with ethanol as transportation fuel, urban air pollution and global warming may be ameliorated, the US economy will become less dependent on foreign oil, and jobs will be created in the agricultural sector.

Cellulosic biomass typically consists of cellulose (40–45% [w/w]), hemicellulose (15–20% [w/w]), and lignin (25–30% [w/w]), but its composition varies considerably among different feedstocks. Although rich in carbohydrates (cellulose and hemicellulose), biomass is an insoluble substrate with a complex structure: cellulose fibers are embedded in a sheath of hemicellulose and lignin, and held together by hydrogen and van der Waals bonds. Nevertheless, despite its compact structure, cellulose can be converted first to fermentable sugars (primarily glucose) and eventually to ethanol or other chemicals using cellulose-hydrolyzing cellulase enzymes and ethanol-producing microorganisms.

There are several ways to increase the digestibility of cellulose before it is exposed to enzymes: mechanical, physical, or chemical pretreatment. The use of dilute sulfuric acid (0.5–1% [w/w]) and high temperatures (160–200°C) has significantly improved the accessibility of cellulose to cellulase *(2)*. During the process, a significant fraction of hemicellulose is hydrolyzed to its components: the pentoses xylose and arabinose and the hexoses mannose, galactose, and some glucose. At the same time, part of the lignin is also solubilized to phenolic compounds, along with small amounts of cellulose, which yields glucose. Clearly, the selected pretreatment conditions will determine the efficiency of the pretreatment process. The release of readily fermentable pentose monomers (primarily xylose) is highly desirable, as opposed to the unfermentable xylose oligomers. Measuring the concentration of each component is essential to selecting the optimal conditions. Similarly, the pretreatment conditions will determine the degree of lignin solubilization and recondensation, which directly affects the digestibility of the remaining solids that are enriched in cellulose.

Although sugar release is desirable, free sugars are vulnerable to degradation under the high-temperature acidic conditions of pretreatment; xylose degrades to furfural and glucose to hydroxymethyl furfural (HMF). Those compounds, along with lignin-derived phenolics and hemicellulose- derived acetic acid, are known inhibitors of cell metabolism and may have a detrimental effect on the fermentability (simultaneous saccharification and fermentation [SSF] performance) of the pretreated biomass. In addition, their formation represents a waste of biomass carbon that needs to be accounted for when ethanol yields are calculated.

Following the pretreatment, the cellulosic content of biomass is converted to ethanol using the SSF process. The SSF combines the enzymatic hydrolysis of polymeric cellulose to monomeric glucose with the fermentative conversion of glucose to ethanol in the same environment. The hydrolytic step is catalyzed by the synergistic action of endoglucanases, exoglucanases, and β-glucosidases *(3)*. Endos and

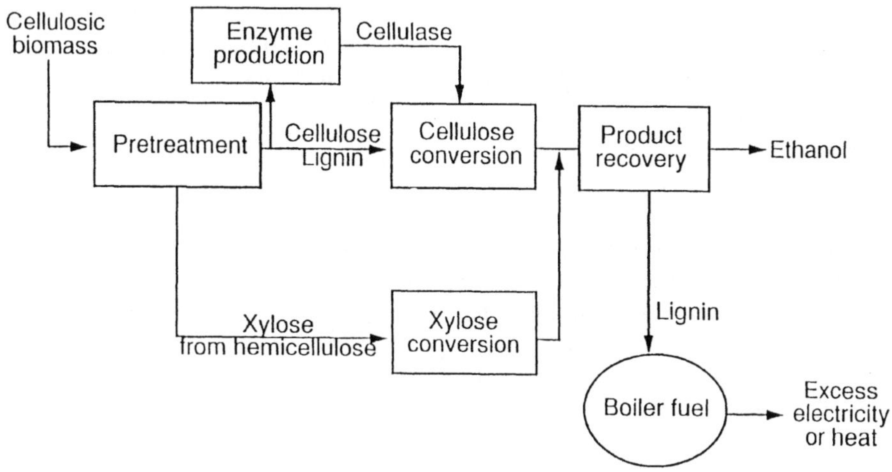

Fig. 1. Simplified schematic of the NREL biomass-to-ethanol conversion process.

exos (collectively referred to as cellulase) randomly attack cellulose chains to produce (primarily) cellobiose, as well as oligosaccharides and glucose. The dimeric and oligomeric sugars are then hydrolyzed to glucose by β-glucosidases. Finally, fermentative microorganisms convert glucose to ethanol, which is recovered through downstream processing. Depending on the SSF conditions and the characteristics of the microorganism, other metabolic byproducts, such as glycerol and organic acids, may also be synthesized. To the extent that they cannot be recovered as high-value products to generate revenue, those byproducts constitute a diversion of carbon to low-value biogass from anaerobic digestion, and their formation should be minimized.

Pretreatment and SSF are at the heart of the bioconversion process (Fig. 1): collectively, they account for the largest portion of the ethanol production cost, almost 45% *(4)*. The efficiency of the pretreatment determines the extent of xylose recovery and the quality of the resulting xylose-rich liquor (hydrolysate) with regard to sugar and inhibitor concentration. On the other hand, the efficiency of the SSF process is equally important, since it determines the yield and productivity of ethanol from solubilized sugars and insoluble cellulose. Because ethanol is the major value-added product of the process, measuring the loss of biomass carbon to other products and calculating ethanol yield based on the actually utilized raw material are imperative for a technically sound process design. Only by calculating the combined efficiency of these two sequential operations (pretreatment and SSF), can we determine the overall feasibility and economic viability of the biomass-to-ethanol conversion technology with high accuracy and reliability.

To date, this reality has been largely ignored because of the complexity of the biomass conversion system. As a result, high SSF yields are erroneously considered equivalent to high overall biomass yields and analytical errors remain unchecked because of the complete lack of carbon balance calculations. This article presents, for the first time, a concise analysis of the accounting aspects and engineering calculations of biomass conversion by (1) formulating a comprehensive carbon balance spreadsheet for the pretreatment and SSF operations to determine carbon distribution, and (2) developing mathematical formulas to calculate product yields for the pretreatment, SSF, and overall process. Such calculations are a necessary precondition for any credible scale-up, optimization, and commercialization study.

EXPERIMENTAL PROCEDURES

Substrate

Yellow poplar wood *(Liriodendron tulipifera)* sawdust, milled to pass through a 4-mm screen, was pretreated with dilute sulfuric acid (0.40% [w/w]) at 150–190°C for 15 min/stage according to a two-stage, reverse-flow percolation process *(5)*. The pretreated biomass was employed in the SSF process at a concentration of 12% solids (8% cellulose) on a dry-wt basis. The composition of raw and pretreated biomass is given in Table 1.

Enzyme

Cellulolytic enzyme, synthesized by a *Trichoderma reesei* strain (CPN International, Jupiter, FL), was employed at a concentration of 25 IFPU/g cellulose. The cellulase and β-glucosidase volumetric activities of the enzyme preparation were 88 and 230 IFPU/mL, respectively, when measured according to the IUPAC methods *(6)*.

SSF

The SSF runs were performed with 12% of pretreated sawdust in yeast-extract, peptone (YP) medium (10 g/L yeast extract, 20 g/L peptone) of initial pH 5.0 at 38°C in 250-mL flasks equipped with stoppers and carbon dioxide traps to simulate anaerobic conditions. In all cases, agitation was maintained at 150 rpm. The pretreatment liquor was partially detoxified by an overliming treatment *(7)* before use. Inocula were prepared in the same media supplemented with 50 g/L glucose under aerobic conditions. A preadaptation step of the inoculum was carried out in 20% (v/v) of the hydrolysate before the SSF run. The fermentative microorganism was *Saccharomyces cerevisiae* D_5A, a strain developed at NREL from baker's yeast *(8)*. A 5% (v/w) inoculum was used to initiate each SSF run.

Chemical Analysis

The biomass was analyzed to determine its glucan, mannan, galactan, xylan, arabinan, lignin, and ash content, according to National Renewable Energy Laboratory (NREL) standard procedures. Fermentation samples were filtered through 0.45-µm nylon filters, and analyzed by high-performance liquid chromatography (HPLC) for the presence of cellobiose, monomeric, and total sugars, and metabolites (ethanol, organic acids, and glycerol). Total sugars were determined after a secondary hydrolysis of the pretreatment or fermentation liquors with 4% (w/w) sulfuric acid for 1 h at 121°C. The HPLC unit 1090LC (Hewlett-Packard, Avondale, PA) was equipped with a refractive index detector. All compounds except for sugars were analyzed using a Bio-Rad Aminex HPX-87H column (Bio-Rad, Hercules, CA), operated at 65°C using 0.2-µm filtered 0.01*N* sulfuric acid as eluent at a flow rate of 0.6 mL/min. Sugars were analyzed using a Bio-Rad Aminex HPX-87C column for glucose, xylose, and arabinose, or a Bio-Rad

Table 1
Pretreatment Carbon Balance

TWO-STAGE SAWDUST PERCOLATION (2x12")

Process:	Percolation
Catalyst:	Sulfuric Acid
Feedstock:	Yellow Poplar
Run:	INHPOP16

CONDITIONS

Temperature (°C):	150-190
Time (min):	15
Concentration (wt%):	0.4015

TOTAL MASS BALANCE

	Raw Feed	*Unwashed Pretreated*
Wet Solids Mass (g):	383.8	580.7
% Dry Solids (%):	59.8	22.9
Dry Solids Mass (g):	229.4	133.0
Liquor (mL):	3370.0	3370.0
Solids Solubilized (%):		42.0

Researcher:	R. Torget
Run Date:	20-Mar-95
CAT Reports:	95-004
	95-005

Carbon Balance: Pretreatment

Component	Unpretreated		Pretreated							
	Dry Biomass		*In Solids*			*In Liquor*			*Total*	**Recovery**
	(% dry weight)	(C-mole)	(% dry weight)	(C-mole)	(% C in Feed)	(g/L)	(C-mole)	(% C in Feed)	(C-mole)	(% C in Feed)
Glucose	46.55	3.556	65.61	2.906	*81.7*	4.43	0.497	*14.0*	3.416	96.0
Mannose	3.66	0.280	0.09	0.004	*1.4*	2.11	0.236	*84.5*	0.240	85.9
Galactose	0.49	0.037	0.00	0.000	*0.0*	0.35	0.039	*104.9*	0.039	104.9
Xylose	17.59	1.344	0.74	0.033	*2.4*	11.30	1.269	*94.4*	1.353	100.7
Arabinose	0.89	0.068	0.00	0.000	*0.0*	0.56	0.062	*91.6*	0.062	91.6
Acetic Acid	4.70	0.359	0.00	0.000	*0.0*	2.76	0.309	*86.1*	0.309	86.1
Formic Acid										
Levulinic Acid										
Lignin	26.65	2.709	31.33	1.846	*68.1*	6.04	0.901	*33.3*	2.748	101.4
Furfural						0.30	0.052	*3.7*		
HMF						0.08	0.013	*0.3*		
Total	*91.83*	***8.354***	*91.11*	*4.789*	*57.3*		*3.379*	*40.5*	***8.168***	97.8

C - RECOVERY:	**97.8%**

Aminex HPX-87P column if cellobiose, mannose, and galactose were also to be determined. Both columns were operated at 85°C using 0.2-μm filtered and degassed HPLC-grade water at a flow rate of 0.6 mL/min. The HPLC-based measurements of glucose and ethanol concentration were confirmed using a YSI glucose analyzer (YSI, Yellow Springs, OH) and a gas chromatograph (Model 5890, Hewlett-Packard), respectively.

MATERIAL BALANCE CALCULATIONS

A necessary first step in assessing the feasibility and potential of new technologies is a techno-economic evaluation of various process scenarios. Such evaluations usually identify the critical parameters of the process and dictate minimal performance criteria for its feasibility. The objective of laboratory bench-scale or demonstration-scale work is then to provide experimental evidence for whether or not the assumed process performance is attainable. Therefore, the reliability of the experimental process data is critical for the unbiased assessment of the technology.

In the biomass-to-ethanol process, the pretreatment of the lignocellulosic feedstock before enzymatic hydrolysis, and the subsequent enzymatic hydrolysis and fermentation of the resulting sugars to ethanol are the essential parts of the process. Technoeconomic evaluation of the NREL biomass- to-ethanol process has identified the SSF step as being the most critical for overall process performance *(4)*. Effective pretreatment, however, can have a large impact on the economics by improving the effectiveness of the enzymatic hydrolysis. Therefore, laboratory evaluations of the biomass- to-ethanol process need to consider pretreatment and SSF steps concurrently.

The efficacy of the various pretreatment methods is typically evaluated based on the following criteria:

1. The yield or recovery of the main sugars in soluble form;
2. The rate and extent of the pretreated substrate's enzymatic digestibility; and
3. The rate and degree of the solubilized and insoluble sugars fermentability through SSF for ethanol production.

The last criterion is the most direct measure of the conversion process performance, and is designed to assess both the digestibility of the solids and the fermentability of the prehydrolysate (pretreatment liquor). This is typically limited by the presence of inhibitors, such as phenolic fragments from solubilized lignin, acetic acid from hemicellulose, and the sugar degradation products furfural and hydroxymethylfurfural (HMF). The concentration of inhibitors increases with the pretreatment severity and the degree of solubilization of the cellulosic, hemicellulosic, and lignaceous components of the feedstock.

The need for a general framework under which pretreatment and SSF experimental data for different feedstocks can be consistently evaluated is therefore of primary importance, as stated in the pretreatment literature *(10,11)*. A comprehensive framework should include:

1. Material balance evaluation;
2. Sugar recovery information; and
3. SSF and overall process ethanol yields.

Material balance calculations are essential for verifying the integrity and consistency of the compositional analysis data both after pretreatment and after SSF, especially in light of reported difficulties associated with the accurate determination of the composition of pretreated lignocellulosic feedstocks and SSF residues *(9,12,13)*. The information on sugar yields and recoveries can then be readily obtained from the material balance calculations around the pretreatment process. Similarly, the SSF and overall process ethanol yields can be calculated from SSF and pretreatment mass balance information, and the produced ethanol.

In an effort to adopt a consistent evaluation procedure to facilitate comparison of results between different groups throughout NREL's biomass-to-ethanol project, a set of EXCEL templates was developed to perform material balance calculations around the pretreatment, SSF, and overall conversion processes. These templates are described next. In addition, a set of formulas developed to calculate ethanol yields correctly for the overall conversion process involving pretreatment and SSF is described in the following section.

Pretreatment Carbon Balance

The material balance for the pretreatment process was followed both in terms of total mass and for the individual components, in terms of carbon. This choice for the individual component balances was made (1) to avoid complications with the water balance resulting from the various hydration and dehydration reactions, and (2) to illustrate clearly changes in the distribution of carbon in the liquid and solid phases that occur between the various carbon-containing substances during the course of the pretreatment process. All major components of the biomass hemicellulosic, cellulosic, and lignaceous factions were considered for individual component balances. These include:

1. The hexose sugars, glucose, mannose, and galactose;
2. The pentose sugars, xylose and arabinose;
3. Lignin;
4. Acetic acid, which results from hydrolysis of the acetyl groups of hemicellulose *(14)*;
5. The sugar degradation products furfural and HMF *(15,16)*; and
6. 4-oxopentanoic or levulinic acid and formic acid, the former resulting from degradation of HMF *(17)*, and the latter mainly from HMF *(17)* and to a lesser degree from furfural *(18)*.

The template for the pretreatment carbon balance is shown in Table 1. The type of pretreatment, pretreatment equipment, catalyst, feedstock, and pretreatment conditions are listed on the top of the template for easy reference. The first part of the spreadsheet consists of the total mass balance calculations. The information needed here is the dry solids loading and liquor volume (catalyst solution) used and the corresponding amounts recovered at the end of the pretreatment. From this data, the hydrolyzed biomass fraction (an important design parameter) is calculated.* In addition to providing this information, the accuracy of the total mass balance is critical, since it affects the balance calculations on the individual

*The percent solubilized solids is calculated as $(1-m_{s/p}/m_{s,0}) \times 100$, where $m_{s/p}$ and $m_{s,0}$ denote the dry solids mass after and before pretreatment, respectively.

Table 2
Molecular and C-Mol Weights

Component	C	H	O	N	OMe	Formula weight, g	C-mol weight, g
Glucose	6	12	6			180.16	30.03
Mannose	6	12	6			180.16	30.03
Galactose	6	12	6			180.16	30.03
Xylose	5	10	5			150.13	30.03
Arabinose	5	10	5			150.13	30.03
Glycerol	3	8	3			92.10	30.70
Acetic acid	2	4	2			60.05	30.03
Lactic acid	3	6	3			90.08	30.03
Succinic acid	4	6	4			118.09	29.52
Lignin	9	8.05	2.70		1.41	203.09	22.57
Furfural	5	4	2			96.09	19.22
HMF	6	6	3			126.11	21.02
Formic acid	1	2	2			46.03	46.03
Levulinic acid	5	8	3			116.12	23.22
Cellulose	6	10	5			162.14	27.02
Cellobiose	12	22	11			342.30	28.53
Cell mass (D5A)	3.874	6.657	2.081	0.752		97.07	25.06
Ethanol	2	6	1			46.07	23.03
Carbon dioxide	1		2			44.01	44.01

components. Therefore, extra attention needs to be given to obtaining the needed mass and moisture content of the untreated and pretreated solids.

The second and main part of the spreadsheet consists of the balance calculations on individual components. Compositional analysis data for the raw solids, pretreated solids, and pretreatment liquor, together with the total mass balance information, are used in these calculations. The solids composition is typically reported on a dry-wt basis, with the sugar concentrations reported as monomeric equivalents. For the particular sample shown in Table 1, the analyzed components make up 91.8% of the raw dry biomass, after applying weight adjustments for anhydro-sugars and acetyl groups,** ash is 0.6%, and the remaining 7.6% is most likely uronic acids and extractives. Total soluble sugar concentrations, including both monomeric and oligomeric forms, are measured in the liquid streams in grams per liter. To use a common basis for the calculations, the compositions of the solids and liquid streams were converted to carbon moles (C-mol), using information from the total mass balance and the C-mol weights of the components,† which are summarized in Table 2. A C-mol is the amount of a substance containing

**These adjustments are needed to account for gain in mass owing to water addition during hydrolysis and are 0.9, 0.88, and 0.7 g polymer/g monomer for hexoses, pentoses, and acetic acid, respectively.

†The number of C-mol of a component x in the solids and liquor is calculated as $Q_x m_s/(CMW)_x$ and $[C_x] V_L / (CMW)_x$, respectively, where (CMW) denotes the C-mol weight of component x (*see* Abbreviations for definition of the remaining symbols).

1 mol of the element carbon *(27)*, i.e., it equals the molecular mass divided by the number of carbon atoms in the molecule. A compositional formula for hardwood (beech) lignin *(19)* was used to calculate the molecular weight of yellow poplar lignin. A slightly different formula has been reported for softwood lignin *(19)*, resulting in a 7% lower molecular weight.

An advantage of using C-mol as the basis for the calculations is that the C-mol ratio of the various components within a stream reflects the distribution of carbon within that stream. In the example of Table 1, more than 60% of the carbon in the pretreated solids is in the form of glucan, and about 38% of the carbon comes from lignin. Furthermore, the fraction of each component that becomes solubilized as a result of the pretreatment is calculated as the fraction of the total C-mol of that component appearing in the liquid stream. For example, the data of Table 1 show that 14–18% of the glucan, 94% of the xylan, and 33% of the initial lignin are solubilized. Furthermore, based on the summary carbon balance information,[††] 57% of the initial carbon remains in the solids after pretreatment, which agrees well with the 42% solubilization that is calculated from the total mass balance data (*see above*, discussion on total mass balance calculations).

For the nonreacting components, the total recovery is calculated as the fraction of the initial carbon recovered in the solids and liquid streams after pretreatment. An exception is glucose and xylose recovery, for which the carbon associated with HMF and furfural, respectively, is added to the amounts of glucose and xylose recovered in the pretreatment solids and liquor to calculate overall recoveries. We therefore assume that all HMF comes from glucose dehydration and similarly that all furfural comes from xylose. This assumption is not entirely correct, however, since all hexoses have been reported to produce HMF when treated in dilute sulfuric acid and all pentoses to produce furfural, among other products, in acidic media *(20)*. However, the reactivity of individual sugars in dilute acid varies according to the approximate order xylose > arabinose > mannose > glucose ≈ galactose *(21)*, and different sugars give different yields of HMF and furfural, with arabinose giving much lower furfural yield than xylose and mannose and galactose significantly lower yields of HMF than either glucose or fructose *(20)*. Therefore, the above assumption is rather justifiable. Similarly, since levulinic acid and formic acid are degradation products of HMF, the carbon associated with these components is being added to the recovery of glucose. Finally, the overall carbon recovery for the pretreatment process is calculated as the percentage of the total initial carbon recovered in all components in both the liquid and solids streams.

The individual component and overall carbon recoveries shown in Table 1 are typical of the percolation process *(5)*. Larger losses on a percentage basis are observed with the minor sugars and are mainly attributed to analytical error. The recoveries from batch dilute sulfuric acid pretreatment are comparable to those reported here, except for xylose, for which losses of 10–25% are usual. Such losses are mainly caused by the formation of polymeric humic compounds that result from furfural degradation *(22)*, which can be analyzed as lignin and therefore upset the lignin balance as well.

[††]Calculation of the "Total" in Table 1 involves different calculations depending on the column. For the "% dry weight" columns, it is calculated as explained in the above footnote.[**] For the "C-mol" columns, it is a straight sum, and for the "% C in feed" columns, it is the ratio of the total C-mol in the solids or liquid streams over the total C-mol in the raw substrate.

SSF Carbon Balance

As with the pretreatment balance, the SSF material balance was followed in terms of carbon. During SSF, the polymeric sugars are hydrolyzed by cellulolytic enzymes and released in the liquor as soluble sugars, which are then fermented to ethanol by the ethanologenic organism. The conversion of the major sugars, both soluble and polymeric, and the production of ethanol and major byproducts are followed through the SSF process. Lignin is also included in the balance for consistency verification purposes, since it is expected to be neither produced nor consumed during SSF. The SSF balance spreadsheet is shown in Table 3. Various concentrations are converted to the common basis of C-mol/kg slurry using the mass fraction of insoluble solids before and after SSF.[§] Because this information is critical to the accuracy of the overall balance, it needs to be determined as accurately as possible.

In addition to the pretreated solids and liquor, the compositions of which are known (*see* Table 1), other streams, including the yeast inoculum, growth media, and enzyme solution, contribute to the overall carbon entering the SSF. The SSF solids residue composition and the concentrations of residual sugars, ethanol, glycerol, and the main organic acid byproducts in the SSF liquor also need be measured at the end of the SSF. Carbon dioxide evolution is typically measured using head-space CO_2-analyzer measurements or mass spectrometry *(23)*. At a smaller scale, where it is often impractical to perform such measurements accurately, CO_2 production can be estimated stoichiometrically based on the amounts of ethanol and byproducts formed during fermentation. For yeast, the anaerobic fermentation pathway produces equimolar amounts of ethanol and CO_2, and 2 mol of CO_2 are produced/mol succinate. Carbon dioxide does not accompany acetate or glycerol formation *(24)*. Measuring cell mass concentration in the presence of solids exhibits well-known challenges; several techniques have been developed and applied with varying degrees of success *(25)*. One promising technique currently under development uses two-color fluorescence staining to detect differentially yeast cells in the presence of solids *(26)*. Although colony forming unit (CFU) measurements is a very time- intensive and often inaccurate technique for solids suspensions, this technique is currently used to determine cell mass in SSF. A calibration curve of CFU vs dry cell mass is used to convert to the appropriate concentration units.

The balance on the individual components consists of three major parts:

1. Carbon in solids and liquid streams entering SSF;
2. Carbon in solids, liquid, and gaseous streams leaving SSF; and
3. Sugar conversions and product formation balance.

Unlike the pretreatment balance, mass closure on individual components cannot be determined for the SSF because of the biotransformations that occur during this process. However, the yields of the fermentation products based on the consumed sugars can be determined, providing useful process information and, at the same time, a material balance for product formation. Because the distribution of product yields is characteristic of a given organism for fixed growth conditions, they can be used to detect contamination in SSF and quantify its impact on ethanol

[§]The compositions of a component x in the solids and liquid streams are converted to C-mol/g slurry as $Q_x f_s$ / $(CMW)_x$ and $[C_x]$ $(1 - f_s)$ / d_L $(CMW)_x$, respectively (*see* † and Abbreviations for definition of symbols).

Table 3
SSF Carbon Balance

SSF CARBON BALANCE

Enzyme: 25 FPU/g cellu (CPN)
Organism: Yeast D5A
Feedstock: Yellow Poplar
Pretreatment: Percolation
Run: INHPOP16
Sample: 48 hrs

Researcher : T.K. Hayward
Run Date : 5-Apr-95
CAT Reports : 95-0010
95-0011

SOLIDS BALANCE

	In	*Out*
Lignin (%):	32.28	51.97
Insoluble Solids (%):	12.00	7.10

PERFORMANCE PARAMETERS

Cellulose Conversion:	63.6%
Overall C6-Sugar Conversion:	60.8%
Overall C5-Sugar Conversion:	21.2%
Ethanol Process Yield (% theor):	57.4%
Ethanol Metabolic Yield (% theor):	94.4%

Carbon Balance: SSF

Component	**Carbon In**							**Carbon Out**							**Conversion**	**Yield**
	In Solids			***In Liquor***			***Total***	***In Solids***			***In Liquor***			***Total***	*100 (In-Out)/In*	100 x g /
	(% dry wt)	(C-mole/Kg Sln)	(% Total In)	(g/L)	(C-mole/Kg Sln)	(% Total In)	(C-mole/Kg Sln)	(% dry wt)	(C-mole/Kg Sln)	(% Total Out)	(g/L)	(C-mole/Kg Sln)	(% Total Out)	(C-mole/Kg Sln)	(%)	g C6 cons
Cellobiose				1.00	0.031		0.031				0.43	0.014		0.014		
Glucose	75.04	2.999	*94.3*	6.20	0.182	*5.7*	3.181	46.17	1.092	*91.5*	3.28	0.101	*8.5*	1.194	62.5	
Galactose	0.00	0.000	*0.0*	0.39	0.011	*100.0*	0.011	0.00	0.000	*0.0*	0.36	0.011	*100.0*	0.011	1.3	
Mannose	0.65	0.026	*30.8*	1.99	0.058	*69.2*	0.084	1.08	0.026	*32.8*	1.69	0.052	*67.2*	0.078	7.6	
Xylose	0.86	0.034	*38.8*	1.85	0.054	*61.2*	0.089	0.62	0.015	*19.2*	1.99	0.062	*80.8*	0.076	13.8	
Arabinose	0.04	0.002	*8.1*	0.62	0.018	*91.9*	0.020	0.05	0.001	*13.2*	0.25	0.008	*86.8*	0.009	54.3	
Lignin	32.28	1.717	*82.7*	9.22	0.359	*17.3*	2.076	51.97	1.636	*79.0*	10.57	0.435	*21.0*	2.071	0.2	
Ethanol				1.10	0.042		0.042				32.40	1.307		1.307		48.3
Cell Mass				0.31	0.011		0.011				1.94	0.072		0.072		2.5
Carbon Dioxide												0.638		0.638		46.5
Glycerol				0.32	0.009		0.009				1.37	0.041		0.041		1.6
Acetic Acid				2.72	0.080		0.080				2.70	0.084		0.084		0.2
Lactic Acid				1.18	0.034		0.034				1.36	0.042		0.042		0.4
Succinic Acid				0.44	0.013		0.013				0.76	0.024		0.024		0.5
Total	*101.19*	*4.778*	*84.1*		0.904	*15.9*	***5.681***	*95.08*	*2.770*	*48.9*		2.891	*51.1*	***5.661***		*100.0*

C - RECOVERY:	***99.6%***

process yields. These calculations are shown in the last column of Table 3 The important SSF process parameters, including conversions of glucan, total hexose and pentose sugars, and ethanol yields based on total (process yield) and consumed (metabolic yield) hexose sugars, are summarized on the top section of the spreadsheet. Finally, overall carbon closure is determined as the percentage of total carbon in the SSF recovered in the solids residue and the fermentation products.

ETHANOL YIELD CALCULATIONS

The ethanol yields of the biomass-to-ethanol technology should be based on the raw biomass fed to the pretreatment reactor because the cost of feedstock is a significant component of the production cost. Hence, to evaluate the overall performance of the biomass conversion process, ethanol yield calculations need to consider the efficiency of the pretreatment and SSF operations, not only the fermentation step. Any losses of fermentable carbon either in the form of sugar degradation products or as part of unused hydrolysate should be accounted for, because in reality they decrease the ethanol yield from the purchased biomass. Based on this rationale, we first present yield calculations for the SSF process, and then formulate overall yield expressions.

SSF Ethanol Yields

In high solids fermentation processes, such as SSF, failing to account for the amount of suspended insoluble solids can bias mass balance closure and the determined yields. In an effort to address these issues comprehensively expressions for the ethanol yield from raw and pretreated substrates were developed based on the following assumptions:

1. All hexoses (glucose, mannose, galactose), which are the products of hydrolysis of the cellulosic and hemicellulosic components of lignocellulosic substrates, can be fully utilized by the organism;
2. The yields for the overall process are based on the total hexose content of the raw substrate, as fed to the pretreatment process; and
3. A significant amount of insoluble solids is present in the SSF media, and therefore, yield calculations need to correct for the reduction in the liquid volume.

The ethanol yield based on the total hexose content is therefore defined as:

$$Y_E = \text{g EtOH produced / g C6 sugars in raw substrate} \qquad (1)$$

The maximal theoretical (stoichiometric) yield of ethanol from hexoses is 0.511 g ethanol / g C6 sugar.[§§] The ethanol yield can be expressed as a percent of the stoichiometric yield as $(Y_E / 0.511) \times 100$. This yield is always <100% as part of the sugars is converted to cell mass and other byproducts by the organism.

The mass fraction of the insoluble solids in the SSF medium is denoted by f_s, where $f_s = m_s / m$ in g insoluble solids/g slurry. As slurry, we define the liquid–solid mixture of pretreated biomass exiting the pretreatment process and entering the

[§§]It is worth pointing out that, because the Gibbs free energy for the conversion of glucose to ethanol is negative, the stoichiometric yield is also thermodynamically achievable.

SSF process. The mass fraction of the liquid phase, which includes any dissolved solids, is therefore $1 - f_s$. The total solids fraction consists of insoluble solids, dissolved solids, and cell mass. A typical dry-wt determination of an SSF sample provides a measure of the total solids fraction. The fraction of insoluble solids f_s can be determined with a standard dry-wt determination procedure, if the SSF residue is thoroughly washed before drying; washing is essential to removing any dissolved solids from the sample. The contribution from the cell mass, expressed as dry cell weight, to the total insoluble solids in an SSF system needs to be quantified with alternative methods such as CFUs.

Some cellulosic materials, such as office paper, can be converted to ethanol efficiently without pretreatment. The vast majority of biomass species, however, require pretreatment to varying degrees of severity. Based on Eq. (1), the following calculations provide the appropriate SSF ethanol yield formulas for both feedstock categories.

SSF of Raw Substrate

The amount of ethanol present in the SSF flask or reactor at any time is equal to the residual ethanol concentration in the liquid phase of the SSF suspension multiplied by the volume of the liquid phase, i.e., $[E]\, m_L/d_L$ or $[E]\, m\,(1 - f_s)/d_L$. The total amount of hexoses available for ethanol production at the beginning ($t = 0$) of the SSF is $f_{s,0}\, m_0\, Q_{C6,0}$. Therefore, the ethanol yield from raw (unpretreated) substrate is:

$$Y_E^{SSF} = [E]m(1 - f_s) \,/\, f_{s,0}\, m_0\, d_L\, Q_{C6,0} \qquad (2)$$

in (g EtOH) / (g hexoses in raw substrate). Typically, the density of the liquid phase is taken to be equal to that of water.

If all solids are assumed to have been completely solubilized at the end of the SSF (i.e., $f_s = 0$) and there is no mass loss (e.g., via evaporation) during the course of the fermentation (i.e., $m = m_0$), Eq. (2) simplifies to the following form:

$$Y_E^{SSF} = [E] \,/\, (f_{s,0}\, Q_{C6,0} \times 1000) \qquad (3)$$

Any mass loss during SSF will have a significant effect on yield calculations. For example, a 10% mass loss from an SSF residue with 6% (w/w) insoluble solids, which is a typical case, will result in a 20% overestimation of the yield, if Eq. (3) is used instead of Eq. (2).

SSF of Pretreated Substrate

Equation (2) can also be used to calculate the SSF ethanol yield from pretreated substrate. If only washed solids are used in the fermentation, Eq. (2) applies unmodified, but the term $Q_{C6,0}$ now refers to the washed pretreated biomass composition. Although not realistic from a process design standpoint, washed solids are often used at bench scale to eliminate the inhibitory effects of acetic acid, HMF, furfural, and phenolics on cell metabolism and SSF performance.

If the pretreatment hydrolysate is also used in the SSF, the yield calculation will have to account for the hexose sugars present in the hydrolysate. In this case, the ethanol yield for the SSF conversion of unwashed pretreated solids is:

$$Y_E^{SSF,Pr} = [E]\, m\,(1 - f_s) \,/\, [f_{s,0}\, m_0\, d_L\, Q_{C6,0} + [C_6]_0\, m_0\,(1 - f_{s,0})] \qquad (4)$$

in (g EtOH produced)/(g hexoses in pretreated slurry), where the slurry is the sum of solid residue and hydrolysate.

Overall Ethanol Yields from Pretreated Substrate

The ethanol yield in the SSF process defines the efficiency of that operation. To determine the carbon utilization efficiency of the overall process (pretreatment and SSF), we need to base all calculations on the raw biomass that enters the pretreatment reactor and thus take into account any carbon losses that occur during pretreatment.

SSF of Washed Pretreated Solids

Pretreatment of the substrate effectively solubilizes part of the insoluble solids (*see* Table 1). Therefore, we need to account for the carbon lost in those sugars in the ethanol yield calculation from washed solids. The amount of ethanol produced is found in the same way as in the previous case, but the total amount of hexoses available for ethanol production at the beginning of the SSF is $(f_s / L_f)\, m_0\, Q_{C6}$, where L_f is a mass-loss factor expressing the insoluble solids fraction recovered after the pretreatment process (i.e., $1 - L_f$ is the insoluble solids fraction that becomes solubilized by pretreatment). Thus, the ethanol yield from the pretreated solids, after accounting for solubilization, is:

$$Y_E^{Pr,w} = [E]\, m\, (1 - f_s) \,/ [\, (f_{s,0}/L_f) m_0\, d_L\, Q_{C6,0}] \tag{5}$$

in (g EtOH produced)/(g hexoses in raw biomass).

SSF of Unwashed Pretreated Solids and Hydrolysate

Except for sugar degradation products, the pentosans and hexosans that become solubilized by the pretreatment process appear as soluble sugars in the pretreatment liquor (hydrolysate). To minimize the cost of ethanol production, all available sugars (just the hexose sugars for the present analysis), including those in the hydrolysate, should be converted to ethanol. To evaluate the fermentability of those sugars, SSF experiments are routinely set up using the unwashed pretreated solids and the corresponding hydrolysate to make up the SSF suspensions. Because all yields are based on the hexose content of the raw substrate according to Eq. (1), the ethanol yield for unwashed pretreated substrate is again given by Eq. (5). If the hydrolysate does not contain significant amounts of sugar degradation products, the ethanol yield will be higher, because more sugars are made available for fermentation.

When the SSFs are run with unwashed substrate and pretreatment hydrolysate, only a portion of the generated hydrolysate can be utilized. Figure 2 illustrates a typical case. Assuming that 20% of the insoluble solids are solubilized during pretreatment (i.e., $L_f = 0.8$), 125 g of raw substrate need to be pretreated to generate 100 g of pretreated solids in the SSF. If the pretreatment is carried out at a 5% (w/w) solids loading, 2375 g of prehydrolysate will also be generated. If the SSF is run at a 10% (w/w) solids level, only 900 g of the hydrolysate are needed/100 g of pretreated solids. Therefore, only 38% of the generated hydrolysate is actually used in the fermentation test. This is because dilute-acid batch pretreatment experiments are usually carried out at lower solids levels (5% [w/w]) than SSF (10% [w/w]) to achieve good mixing with small-scale pretreatment equipment. Consequently, even when pretreatment hydrolysate is added to the unwashed pretreated solids for SSF evaluation, the yields calculated according to Eq. (5) underestimate the potential ethanol yield for the overall pretreatment-SSF process, because part of

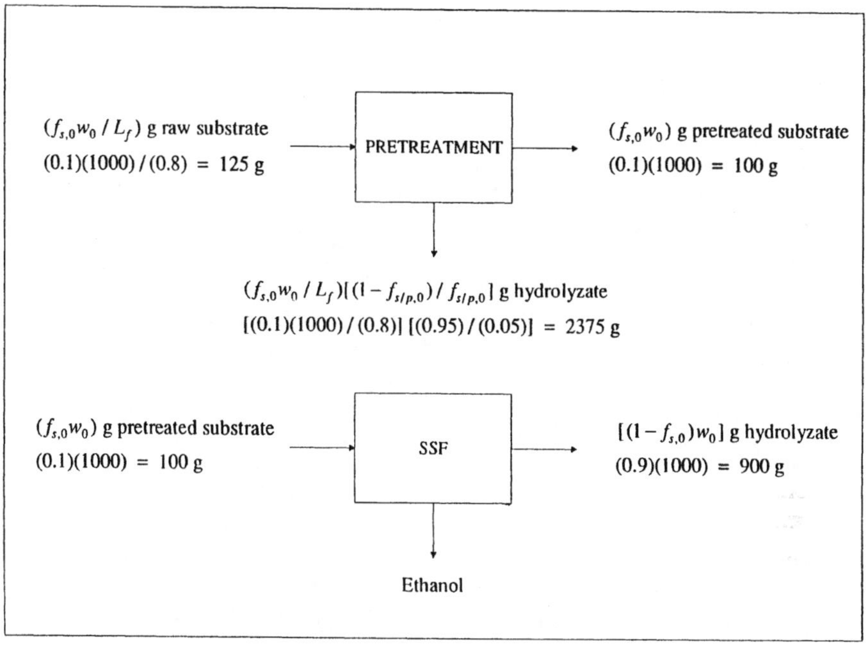

Fig. 2. Mass balance around a typical laboratory-scale pretreatment and SSF processes. Only part of the pretreatment liquor is used in the SSF.

the sugar content of the raw biomass is not made available for fermentation. Ways to account for these effects are presented in the following sections.

Maximal Potential SSF Yield for Pretreated Biomass

If the hydrolysate from a particular pretreatment does not inhibit the organism, the SSF ethanol yield determined from the conversion of the unwashed solids and hydrolysate will approximate the maximal overall process yield (i.e., the yield from converting raw substrate to ethanol). However, if the hydrolysate is inhibitory, the above SSF yield will considerably underestimate the maximal yield that can be potentially attained by an appropriately adapted organism. Therefore, a maximal potential yield can be defined based on the observed yield from the washed residue and the assumption that the organism can utilize all the free hexoses in the hydrolysate, which amount to $[C_6]_0\, m_{L,0}/d_L$ at the maximal theoretical yield of 0.511 g/g. Depending on the assumed degree of utilization of the generated hydrolysate, several expressions can then be formulated to calculate the maximal potential yield.

Use enough hydrolysate to make up SSF suspension: As mentioned in the previous section, if the pretreatment process is run at a lower solids level than the SSF process, only part of the hydrolysate can be used to set up the SSF. In this case, the maximal potential yield is defined as:

$$Y_{E,max}^{Pr} = [\,[E]m(1-f_s) + 0.511[C_6]_0 m_0 (1-f_{s,0})\,]/[(f_{s,0}/L_f) m_0 d_L Q_{C6,0}] \qquad (6a)$$

because only $m_0\,(1-f_{s,0})$ g liquor can be utilized in this case (assuming that no moisture is associated with the remaining $m_0 f_{s,0}$ g of insoluble solids). Based on the assumptions that led from Eq. (2) to Eq. (3), the following approximate expression c an be obtained:

$$Y_{E,max}^{Pr} = [\,[E] + 0.511\,[C_6]_0\,(1 - f_{s,0})\,]/[(f_{s,0}/L_f)Q_{C6,0} \times 1000] \tag{6b}$$

This case is representative of the typical bench-scale evaluation of pretreated substrates.

Use all hydrolysates generated per given amount of pretreated solids: The maximal yields calculated by the above formulas are not appropriate for processes in which all the sugars in the pretreatment hydrolysate are expected to be utilized, as for example in the NREL base-case design where the whole pretreated slurry is fed to the fermentation tanks without an intervening separation step *(4)*. If all the hydrolysate generated by pretreating a given amount of solids is used in the SSF, the maximal potential overall yield is given by an equation similar to Eq. (6a). More specifically, if $m_{s/p,0}$ g of dry solids are pretreated generating $m_{L/p}$ g of hydrolysate with $[C_6]_0\, m_{L/p}\, /\, d_L$ g of free hexose sugars, the maximal amount of ethanol that can be produced is $0.511\,[C_6]_0\, m_{L/p}\, /\, d_L$ g and the yield is now given by:

$$Y_{E,max}^{Pr} = [\,[E]m(1 - f_s) + 0.511\,[C_6]_0\, m_{L/p}]/[(f_{s,0}\,/\,L_f)\, m_0\, d_L\, Q_{C6,0}] \tag{7}$$

If, however, only a portion of the pretreated solids equal to $f_{s,0}\, m_0$ g is used in the SSF, the hydrolysate that corresponds to the amount of the pretreated solids used is $f_{s,0}\, m_0\, /\, m_{s/p,0})\, m_{L/p}$, and in this case, the maximum potential yield is given by:

$$Y_{E,max}^{Pr} = [\,[E]m(1 - f_s) + 0.511\,[C_6]_0\,(f_{s,0}\, m_0/m_{s/p})m_{L/p}]/[(f_{s,0}/L_f)m_0 d_L Q_{C6,0}] \tag{8a}$$

An approximate equation similar to Eq. (6b) can also be derived for this case:

$$Y_{E,max}^{Pr} = [\,[E](1 - f_s) + 0.511\,[C_6]_0\,(f_{s,0}/f_{sp})]/[(f_{s,0}/L_f)Q_{C6,0} \times 1000] \tag{8b}$$

where $f_{s/p}$ is the fraction of insoluble solids after pretreatment.

Simultaneous Saccharification and Cofermentation Ethanol Yields

All the above formulas for SSF yield calculations can be generalized to apply to a cofermentation process in which hexose and pentose sugars are converted to ethanol. In this case, the ethanol yield is still given by Eq. (1), but is now based on total available sugars. As the stoichiometric yield for ethanol production from pentose sugars is again 0.511 g/g, the remaining formulas are still valid, if the correct compositions are used. More specifically, Q_{C6} needs to be replaced by $(Q_{C5} + Q_{C6})$ and $[C_6]$ by $([C_5] + [C_6])$ in all the above formulas.

CONCLUSIONS

A comprehensive framework for evaluating process data and performing the engineering calculations for the biomass-to-ethanol conversion process has been developed. The need for such an analysis scheme has long been recognized *(10,11)*, and it is becoming more urgent as the technology nears commercialization.

The developed framework emphasizes the need to consider the overall process performance when evaluating laboratory data, which are typically obtained from studies on individual unit operations instead of on the integrated process. In addition to providing a consistent platform for conversion and yield calculations, the presented framework is also valuable in directing experimental efforts more effectively by pointing out critical measurements and potential inconsistencies in the analytical data. Although a base-case process, which involves a pretreatment and an SSF step for conversion of hexose sugars, was considered in the present analysis, the results can be readily extended to other process configurations, e.g., an SSF process with recycle or cofermentation processes with simultaneous pentose and hexose sugar conversion.

ACKNOWLEDGMENTS

This work was funded by the Biochemical Conversion Element of the Biofuels Program of the US Department of Energy. We would like to thank Robert Torget for generating the pretereated materials, Tammy Kay Hayward for carrying out the SSF experiments, and Tina Ehrman for performing the numerous analyses.

REFERENCES

1. Philippidis, G. P. and Wyman, C. E. (1992), in *Recent Advances in Biotechnology*, Vardar-Sukan, F. and Sukan, S. S., eds., Kluwer Academic, Dordrecht, The Netherlands. pp. 405–411.
2. Grohmann, K., Torget, R., and Himmel, M. (1985), *Biotechnol. Bioeng. Symp.* **15,** 59–80.
3. Philippidis, G. P. (1994), in *Enzymatic Conversion of Biomass for Fuels Production*, Himmel, M., Baker, J. O., and Overand, R. P., eds., American Chemical Society, Washington, DC, pp. 188–217.
4. Hinman, N. D., Schell. D. J., Riely, C. J., Bergeron, P., and Wyman, C. E. (1992), *Appl. Biochem. Biotechnol.* **34/35,** 639–649.
5. Torget, R., Hatzis, C., Hayward, T. K., Hsu, T., and Philippidis, G. P. (1996), *Appl. Biochem. Biotech.* **57/58,** 85–101.
6. Ghose, T. K. (1987), *Pure Appl. Chem.* **59,** 257–268.
7. McMillan, J. D. (1994), in *Enzymatic Conversion of Biomass for Fuels Production*, Himmel, M., Baker, J. O., and Overand, R. P., eds., American Chemical Society, Washington, DC, pp. 411–437.
8. Spindler, D. D., Wyman, C. E., Grohman, K., and Philippidis, G. P. (1992), *Biotechnol. Lett.* **14,** 403–407.
9. Ehrman, C. I. and Himmel, M. E. (1994), *Biotechnol. Technique* **8,** 99–104.
10. Chum. H. L., Douglas, L. J., Feinberg, D. A., and Schroeder, H. A. (1984), *Evaluation of Pretreatments of Biomass for Enzymatic Hydrolysis of Cellulose.* Solar Energy Research Institute, SERI/TR-231-2183.
11. Dale, B. E. (1985), *Ann. Rep. Fermentation Proc.* **8,** 299–323.
12. Chum, H. L. and Gellerstedt, G. (1991), Modern Methods of Analysis of Wood, Annual Plants and Lignins. Proc. IEA Pre-Symposium, New Orleans, LA.
13. Hsu, T. and Nguyen, Q. (1995), *Biotechnol. Techniques* **9,** 25–28.
14. Fan, L. T., Lee, Y.-H., and Gharpuray, M. M. (1982), *Adv. Biochem. Eng.* **23,** 157–187.
15. Saeman, J. F. (1945), *Ind. Eng. Chem.* **37,** 43–52.
16. Dunlop, A. P. (1948), *Ind. Eng. Chem.* **40,** 204–209.
17. McKibbins, S. W., Harris, J. F., Saeman, J. F., and Neill, W. K. (1962), *Forest Prod. J.* **12,** 17–23.
18. Williams, D. L. and Dunlop, A. P. (1948), *Ind. Eng. Chem.* **40,** 239–241.
19. Sarkanen, K. V. and Ludwig, C. H. (1971), *Lignins: Occurrence, Formation, Structure and Reactions*. Wiley-Interscience, New York.
20. Feather, M. S. and Harris, J. F. (1973), *Adv. Carbohydr. Chem. Biochem.* **28,** 161–224.
21. Harris, J. F. (1975), *Appl. Polymers Symp.* **28,** 131–144.
22. Root, D. F., Saeman, J. F., Harris, J. F., and Neill, W. K. (1959), *Forest Prod. J.* **9,** 158–165.
23. Kadam, K. L., Hayward, T. K., and Philippidis, G. P. (1995), *Solar Eng.* **1,** 339–347.
24. Gancedo, C. and Serrano, R. (1989), in *The Yeast*, vol. 3, 2nd ed., Rose, A. H. and Harrison, J. S., eds., Academic, London, pp. 205–259.
25. Kennedy, M. J., Thakur, M. S., Wang, D. I. C., and Stephanopoulos, G. (1992), *Biotechnol. Prog.* **8,** 375–381.
26. Combs, N. and Hatzis, C. (1996), *Appl. Biochem. Biotech.* **57/58,** 649–657.
27. Roels, J. A. (1983), *Energetics and Kinetics in Biotechnology*, Elsevier Biomedical, Amsterdam, The Netherlands.

0273-2289/96/5758—0461$7.50

Production of Cellulase Systems by Selected Mutants of *Trichoderma reesei* in Solid-State Fermentation and Their Hydrolytic Potentials

VICTOR A. AWAFO,[1,2] DEVINDER S. CHAHAL,*[,2] BEN K. SIMPSON,[2] AND GEORGE B. B. LÊ[3]

[1]Centre de Recherche en Microbiologie Appliquée, Institut Armand-Frappier, Université du Québec, 531, Boulevard des Prairies, Laval, Quebec, Canada H7N 4Z3; [2]Department of Food Science and Agricultural Chemistry, MacDonald College, McGill University, Quebec, Canada; and [3]Ministry of Natural Resources, Quebec, Canada

ABSTRACT

Three mutants of *Trichoderma reesei* were grown in solid-state fermentation (SSF) in flasks and in a pan bioreactor. Mutant strain MCG 80 proved to be best at producing an optimal cellulase system using lignocellulosic material (wheat straw [WS]) as substrate. This preparation exhibited a β-glucosidase activity (βGA) to FPA (FPA) ratio of about 1.0, which is indicative of a high potential for hydrolysis of cellulose. The yields of cellulase systems and the ratio of βGA to FPA produced in flasks were comparable to that of the pan bioreactor. The cellulase system of *T. reesei* MCG 80 having a ratio of βGA to FPA close to 1.0 gave the most complete (88–95%) hydrolysis of 5% delignified wheat straw (DWS). On the other hand, the cellulase system of cocultures of *T. reesei* QMY-1 and *Aspergillus phoenicis* failed to produce high hydrolytic yields in spite of having a very high ratio of βGA to FPA (3.04). This failure was owing to the fact that coculture contained the relatively poor-quality cellulase system of the dominant organism, *A. phoenicis*. The resulting fermented WS can be used, as a source of enzyme (unextracted), for hydrolysis of wheat straw, and it gives increased yields of reducing sugars compared to analogous extracted enzyme preparations. The hydrolytic potential of two commercial enzymes tested were considerably lower than those of the cellulase systems produced on WS. It is evident that a complete cellulase system having a βGA-to-FPA ratio close to 1.0 and high hydrolytic potential can be produced on lignocellulosic feedstocks in SSF.

Index Entries: Cellulase; hydrolytic potential; solid-state fermentation (SSF); *Trichoderma reesei*; wheat straw.

*Author to whom all correspondence and reprint requests should be addressed.

INTRODUCTION

In the hydrolysis of lignocellulosic feedstocks, the cellulase system is part of a complex lignocellulolytic enzyme system that hydrolyzes glucosidic bonds of crystalline cellulose to glucose. The lignocellulolytic enzyme system comprises ligninases, hemicellulases, pectinases, endo-1,4-β-glucanases, exo-1,4-β-glucanases (cellobiohydrolase), and β-glucosidase. The latter three types of enzymes are collectively referred to as the cellulase system. Production of an efficient and inexpensive cellulase system is at the cutting edge in the effort to establish a biotechnological route to complete transformation of cellulose into glucose. A high-quality and complete cellulase system is now realizable using solid-state fermentation (SSF) techniques *(1,2)*. SSF is a preferred technique in cellulase production because of the several advantages it has over the submerged or liquid state fermentation cultivation method *(3,4)*.

The levels of cellulases produced by several cellulolytic microorganisms are comparable with the parent strain, *Trichoderma reesei* QM6a *(5,6)*. Now, several *T. reesei* mutants have been developed for the production of effective and affordable cellulase systems. The driving force has been to obtain high filter paper activity (FPA)/unit volume and/unit of substrate. The term FPA used in this article represents the synergistic action of endo-and exoglucanases. However, the efficiency of hydrolysis of cellulose by a cellulase system is also governed by the quantities of both the FPA and the β-glucosidase activity (βGA). More importantly, a critical factor is the ratio of βGA to FPA, which should be close to 1.0 *(7,8)*. This ratio is particularly important in preventing catabolic repression of the endoglucanase and cellobiohydrolase by ensuring that cellobiose does not accumulate during the course of the reaction. Therefore, a cellulase system having a βGA-to-FPA ratio close to 1.0 is called a "complete" cellulase system, since it establishes a threshold enzyme ratio for the elimination of accumulated cellobiose during hydrolysis of cellobiose. It has been estimated that in cellulose hydrolysis (10–400 g/L), a cellobiose concentration of 0.01–0.1 g/L almost completely inhibits cellobiohydrolase *(9)*.

Although the literature abounds with filter paper and β-glucosidase production values *(10,11)* and a few cellulase quality assessment studies *(12,13)*, very little work is available on the hydrolytic potential of cellulase systems produced by different mutants of *T. reesei* using varied conditions and practical substrates. Hydrolytic potential is defined as the extent to which a given substrate can be hydrolyzed by a set loading of enzyme over a defined period of time and reaction conditions.

In the present article, we report the biochemical characteristics and hydrolytic potentials of cellulase systems of selected mutants of *T. reesei* when cultured individually and also in cocultures with *Aspergillus phoenicis* (a βGA producer) in SSF in flasks and in a specially designed pan bioreactor *(14)*.

MATERIALS AND METHODS

Microorganisms

T. reesei mutants QMY-1 (NRRL 18760), QM 9414 (NRRL 6165), and MCG 80 (NRRL 12368) were cultivated individually and in cocultures with *A. phoenicis* (NBL 1956) in Mandels' medium *(15)*. The genealogy of the *T. reesei* mutants is as follows: QMY-1 is a direct descendant of QM 9414 obtained in our laboratory by D. S. Chahal, whereas MCG 80 is a direct descendant of *T. reesei* Rut. C-30. *A.*

phoenicis was selected for its ability to produce high levels of β-glucosidase *(16,17)*. *T. reesei* MCG 80, QM 9414, and *A. phoenicis* were obtained from J. L. Swezey (ARS Patent Culture Collection, USDA, Peoria, IL). All the test organisms were maintained on agar in Petri plates with Mandels' medium having delignified wheat straw (WS) as carbon source.

Substrate and Enzyme Production

WS has the following approximate chemical composition: cellulose (39%), hemicelluloses (36%), lignin (10%), and silica (6%) *(18)*.

Ground WS (20 mesh) was treated with 4% NaOH (w/w) to a moisture content of 70% on dry-wt basis and led overnight. The treated substrate without washing was autoclaved at 121°C for 30 min. Twenty grams (dry-wt basis) of the treated and autoclaved WS were used for each pan bioreactor, whereas each Erlenrmeyer flask contained 4 g (dry-wt basis) of WS. The nutrient elements of Mandels' medium were calculated by weight of the polysaccharide (cellulose and hemicelluloses) content of WS and added to the substrate, and the final pH was adjusted to 6.5. Five and 25 mL of 36-h grown mycelium of the mutants were inoculated on the surface of each flask and pan, respectively. The moisture content of the substrate after inoculation was about 75% (dry-wt basis). The cultures were held in a humidified incubator (Norco 4100), and samples were harvested after 20 d of cultivation.

Enzyme Extraction and Analysis

Water was added to the fermented substrate to a final volume of 100 mL. The substrate was agitated in a rotary shaker at 200 rpm for 30 min and then centrifuged at 11,000*g* for 30 min. The supernatant was used for enzyme analyses and cellulose hydrolysis. FPA and βGA were determined by the method of Mandels et al. *(19)*, approved by the International Union of Pure and Applied Chemistry (IUPAC Protocol) *(20)*.

Hydrolysis of Cellulose

To examine the hydrolytic potential of cellulase systems, hydrolysis was performed on delignified wheat straw (DWS) using cellulase systems obtained from WS without urea in the cultivation medium. WS was delignified by the method of Toyama and Ogawa *(21)*. The extracted cellulase system was supplied at the rate of 20 IU FPA/g DWS. Crude unextracted enzyme (CUE) as found in the resulting fermented WS was also used for the hydrolysis of DWS. The CUE is that portion of the resulting fermented WS in SSF that was not subjected to extraction. The quantities of CUE to supply 20 IU FPA/g DWS were calculated based on the FPA of the extracted enzyme/g fermented substrate. The required enzyme units were added to 5 and 10% DWS in citrate buffer (0.1*M*, pH 4.8) in Erlenmeyer flasks and incubated at 45°C for 60 h with agitation at 200 rpm.

Analysis of Hydrolysates

Samples were quantitatively analyzed for dominating sugars, such as glucose, cellobiose, xylose, and arabinose, with a Beckman 344 HPLC system, degassed deionized water as mobile phase, an Aminex HPX-87P cohlmn heated to 85°C, and an Altex 156 Refractive Index Detector. In the hydrolysate, glucose and cellobiose (a dimer of glucose) are products of cellulose, whereas xylose and arabinose (pentoses) are from the residual hemicelluloses in the DWS.

RESULTS AND DISCUSSION

Comparison of Cellulase Systems Produced from *T. reesei* Mutants and Their Cocultures with *A. phoenicis*

Economic production of fuel ethanol from cellulose calls for the need to compare different cellulase-producing microorganisms in terms of their ability to produce sufficient quantities of the complete cellulase system to hydrolyze cellulose to glucose efficiently. *T. reesei* occupies a dominant role in the continued search for an efficient cellulase system. This study assesses different mutants evolving from *T. reesei* QM6a (the original strain) for the efficiencies of their cellulase systems.

Table 1 shows the comparison of characteristics of cellulase systems produced from three mutants of *T. reesei*: QM 9414, QMY-1, and MCG 80 alone and in cocultures with *A. phoenicis* in SSF for 20 d with Mandels' medium with or without urea. Urea in the medium not only supplies nitrogen, but also offsets drastic pH drop during growth in SSF. When the *T. reesei* mutants were cultivated individually in Mandels' medium without urea, the MCG 80 appeared to have an edge in both FPA and βGA over all the mutants and had a ratio of βGA to FPA of 1.0. The QMY-1, which is a good producer of cellulase system in SSF with a βGA-to-FPA ratio close to 1.0 *(2)*, failed to achieve this ratio when urea was eliminated from the medium. The QM 9414 also produced low yields of cellulase system with a low ratio of βGA to FPA in the medium without urea. The addition of urea, however, improved the cellulase system of *T. reesei* QMY-1 in the pan bioreactor, and it was comparable to that of MCG 80, except that the ratio of βGA to FPA was a little low (0.71).

Coculturing the *T. reesei* mutants with *A. phoenicis* in SSF did not improve cellulase system production, but instead had a detrimental effect on the FPA yields. The apparent growth of the cocultures indicated that *A. phoenicis* outgrew the *T. reesei* mutants in SSF. *A. phoenicis* alone produced low FPA and high βGA. Therefore, its domination in cocultures resulted in the production of low FPA and high βGA. Thus, a ratio of βGA to FPA of more than 3.0 was obtained.

Commercial enzymes from Genencor and Iogen were also analyzed in our laboratory. These two enzyme sources were added in our study to compare their hydrolytic potentials with those obtained from our cellulase systems produced in SSF. These commercial enzymes, when tested in our laboratory, showed very high FPA and βGA/mL. However, the ratio of βGA to FPA for the Genencor cellulase was 0.32, whereas that of Iogen cellulase was 0.6 (Table 2), which are low with respect to hydrolytic potential and from the point of view of our definition of a "complete" cellulase system.

Hydrolytic Potential of Different Cellulase Systems

For this experiment, the cellulase system of MCG 80 was selected for hydrolysis of DWS, because it had the desired ratio of βGA to FPA of 1.0. The other cellulase systems were selected because of the following qualities: the cellulase system of QMY-1 produced on medium without urea had a low ratio of βGA to

Table 1
Comparison of Cellulase Systems Produced with Three Mutants of *T. Reesei*

Mutant bioreactor[a]	FPA IU/mL	FPA IU/g subt.	FPA IU/g cellulose	βGA, IU/mL	βGA/FPA
QM 9414, no urea in flask	3.75	94	235	1.11	0.30
QMY-1, no urea in flask	3.58	90	224	1.33	0.37
MCG 80, no urea in flask	5.33	133	333	5.33	1.00
QM 9414 + AP, no urea in flask	1.01	25	63	3.78	3.70
QMY-1 + AP, no urea in flask	1.08	27	68	2.80	2.60
MCG 80 + AP, no urea in flask	1.02	26	64	4.10	4.00
A. phoenicis (AP)	1.02	26	64	3.24	3.17
QMY-1 + 0.1% urea on pan	9.55[b]	129	322	6.80	0.71
QMY-1, no urea in pan	6.35[b]	86	214	2.15	0.34
MCG 80, no urea in pan	9.25[b]	125	312	9.55	1.03
QMY-1 + AP, no urea in pan	2.04	28	69	6.96	3.41
Genencor[c]	354.57	–	–	114.71	0.32
Iogen[c]	329.26	–	–	197.46	0.6

[a]Cultural conditions: WS was pretreated with 4% (w/w) NaOH and fermented in Mandels' medium.

[b]The enzyme activity (IU/mL) is high, because in pan bioreactor, the enzymes were extracted fiom 10 g (dry-wt basis) in 100 mL H_2O, whereas it was 4 g (dry-wt basis) from the flask.

[c]Enzymes supplied by Trichromatic Techno-Chem, Québec, Canada, for evaluation.

FPA and that of cocultures of QMY-1 and *A. phoenicis* had a very high ratio of βGA to FPA (3.41). The commercial enzymes were tested because they showed low ratios of βGA to FPA (0.32 and 0.60), and it is assumed that these enzymes were produced by submerged (liquid) state fermentation.

The data presented in Table 2 clearly indicate that the cellulase system of MCG 80 gave the highest hydrolytic potential (88–95%), whereas other enzyme systems gave low hydrolytic potentials (varying from 33.5–51.8%). The low hydrolytic potential of QMY-1 reported here is probably owing to the low ratio of βGA to FPA (0.34) produced in a medium without urea. In an earlier study, when urea was added to the cultivation medium of QMY-1, the ratio of βGA to FPA was close to 1.0, and over 90% hydrolytic potential was recorded *(1)*.

Table 2
Comparison of Hydrolytic Potential of Different Cellulase Systems

Cultrual condition	Type of fermentation	FPA, IU/mL	βGA, IU/mL	Ratio βGA/FPA	Hydrolytic potential, %[b] Extracted enzyme	Unextracted enzyme, CUE
T. reesei QMY-1 without urea	SSF in pan bioreactor	6.34	2.15	0.34	51.8	69.1
T. reesei MCG 80 without urea	SSF in pan bioreactor	9.25	9.55	1.03	88.0	95.3
T. reesei QMY-I + AP without urea	SSF in pan bioreactor	2.04	6.96	3.41	33.5	49.8
Genencor	Not known	354.60	114.71	0.32	44.2	nd[c]
Iogen	Not known	329.26	197.46	0.60	41.3	nd[c]

[a]As given in Table 1.
[b]Hydrolytic potential = 100 (Rs/Wt), where Rs is total reducing sugars and Wt is weight of DWS. Hydrolysis conditions: 5 g DWS, 20 IU/g DWS load based on FPA, and the rest of the conditions are as explained in Materials and Methods.
[c]nd = Not determined.

It is worth noting that the CUE gave higher hydrolytic potential in all the three cases (Table 2). The higher hydrolytic potential could be owing to the fact that the use of CUE could have been an additional source of cellulose and cellulase. The use of fermented WS as such (CUE) has two major advantages: (1) It would eliminate the need to extract enzyme, and (2) unutilized cellulose in the production phase will have a second chance for conversion into glucose. Both factors contribute to reducing the cost of ethanol production from lignocellulosic feedstocks.

The composition of hydrolysates obtained with these cellulase systems indicated that in the case of MCG 80 when the ratio of βGA to FPA was 1.0, the cellobiose concentration was low and therefore did not inhibit the hydrolysis (Figs. 1–3). The hydrolytic efficiency of CUE of *T. ressei* MCG 80 went up as high as 95%, even though the cellobiose concentration in the hydrolysate was as high as 0.8 g/L in 5% DWS digestions (Fig. 1). This calls to question the earlier suggestion of Klyosov *(9)* that cellobiose concentrations of 0.01–0.1 g/L completely inhibit cellobiohydrolase in cellulose hydrolysis concentration of 10–400 g/L. At 5% DWS, CUE from MCG 80 produced less cellobiose in the hydrolysate compared to its analogous extracted enzyme (Fig. 1). This was probably a result of availability of adsorbed enzyme in the CUE during hydrolysis. The cellobiose content was higher in the hydrolysates of the QMY-1 samples compared to the MCG 80 samples (Figs. 1 and 2). The plausible explanation is that MCG 80 had higher βGA than QMY-1 (Table 2), since β-glucosidase has a critical role in preventing cellobiose accumulation during hydrolysis of cellulose.

In the case of the hydrolysate from the cellulase system of coculture of QMY-1 with *A. phoenicis* (Fig. 3), at a ratio of βGA to FPA >1.0, no cellobiose in the hydrolysate was noticed. Even with no inhibition of cellulases because of the absence of cellobiose, the hydrolytic potential was still very low, although the quantity of FPA loading was 20 IU/g in all cases. Low hydrolytic potentials in the cellulase systems

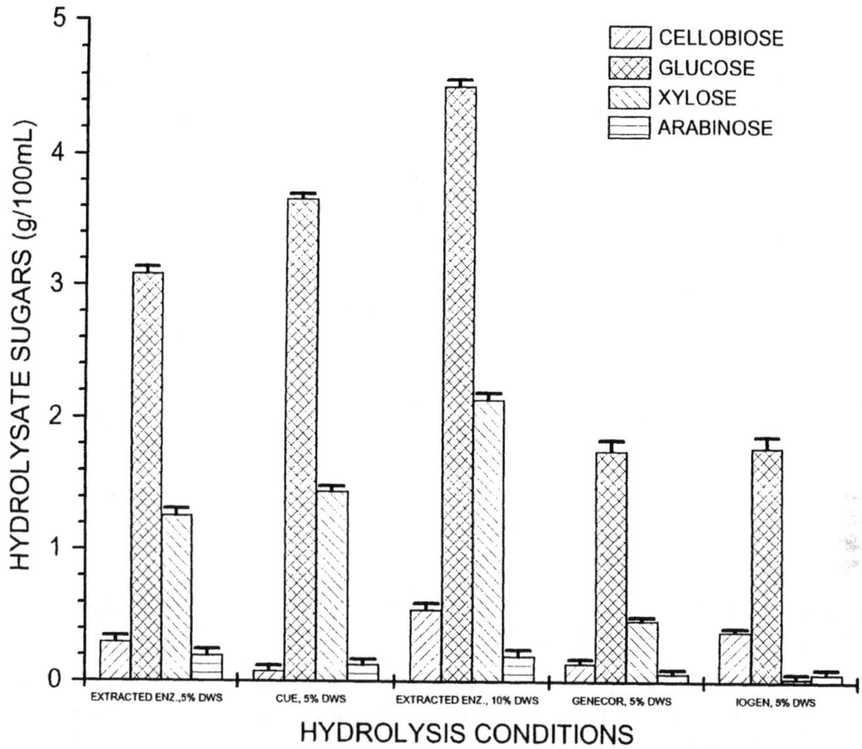

Fig. 1. Hydrolysis of DWS with cellulase system from *T. reesei* MCG 80 and commercial enzymes; Genencor and Iogen after 60 h of incubation.

of cocultures and commercial enzymes could be attributed to the fact of the poor-quality of cellulase systems produced under certain cultural conditions as reported earlier *(2)*. It may, therefore, be noted that equal enzyme loadings from different sources could result in different hydrolytic potentials (Table 2), suggesting the presence of a quality attribute not apparent when enzyme loadings are based on FPA alone.

High sugar concentration in the hydrolysate will also contribute to lower the overall cost of ethanol production. Thus, an efficient cellulase system should have the capability to hydrolyze a high concentration of cellulose. When the substrate concentration for hydrolysis was increased from 5 to 10%, there was an increase on total sugar concentration in the hydrolysate in all the cases (Table 3). Further examination of the data indicated, however, that there was a decrease in percentage of hydrolysis with an increase in substrate concentration. This decrease was lowest (13%) in the case of QMY-1 compared to that of other cellulase systems (about 25%). It is evident from the data (Table 3) that some cellulase systems, like that of QMY-1, can tolerate higher substrate concentration than others (MCG 80) at the same enzyme loading of 20 IU/g.

In fact, when the cellulase system of QMY-1 had the ratio of βGA to FPA of about 1.0, it hydrolyzed 10% DWS to the extent of 89–100% *(1)*. The cellulase system of the coculture was the least tolerant to the high substrate concentration. This

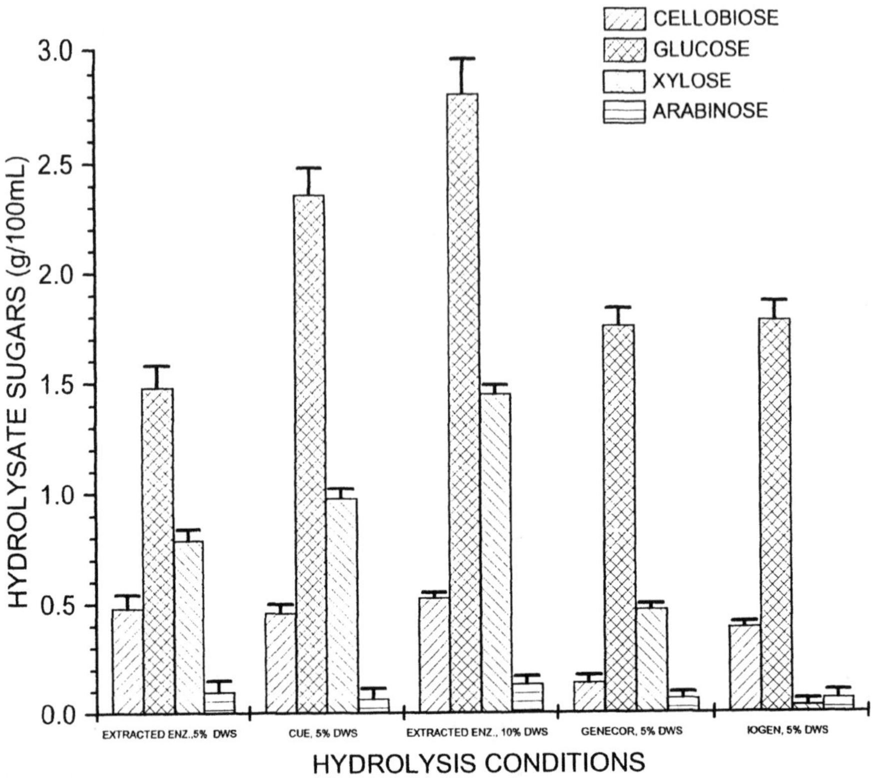

Fig. 2. Hydrolysis of DWS with cellulase system from *T. reesei* QMY-1 after 60 h of incubation. (Results of commercial enzymes from Fig. 1 are also represented here for comparison.)

again might be owing to the poor quality of the cellulase system as already explained. It appears that a cellulase system should also have high tolerance to substrate concentration for a viable hydrolysis process.

CONCLUSION

It is evident from the study that a complete cellulase system having a ratio of βGA to FPA close to 1.0 and high hydrolytic potential can be produced with suitable mutants of *T. reesei* in a pan bioreactor in SSF. Such a complete cellulase system does not require an external source of β-glucosidase for cellulose hydrolysis. Coculturing of *T. reesei* mutants with *A. phoenicis* increased yields of βGA in the cellulase system, but proved to be detrimental to the FPA production in SSF.

ACKNOWLEDGMENTS

Financial support to Devinder S. Chahal for the development of SSF processes by the Ministry of Natural Resources, Government of Quebec, Canada, is gratefully acknowledged.

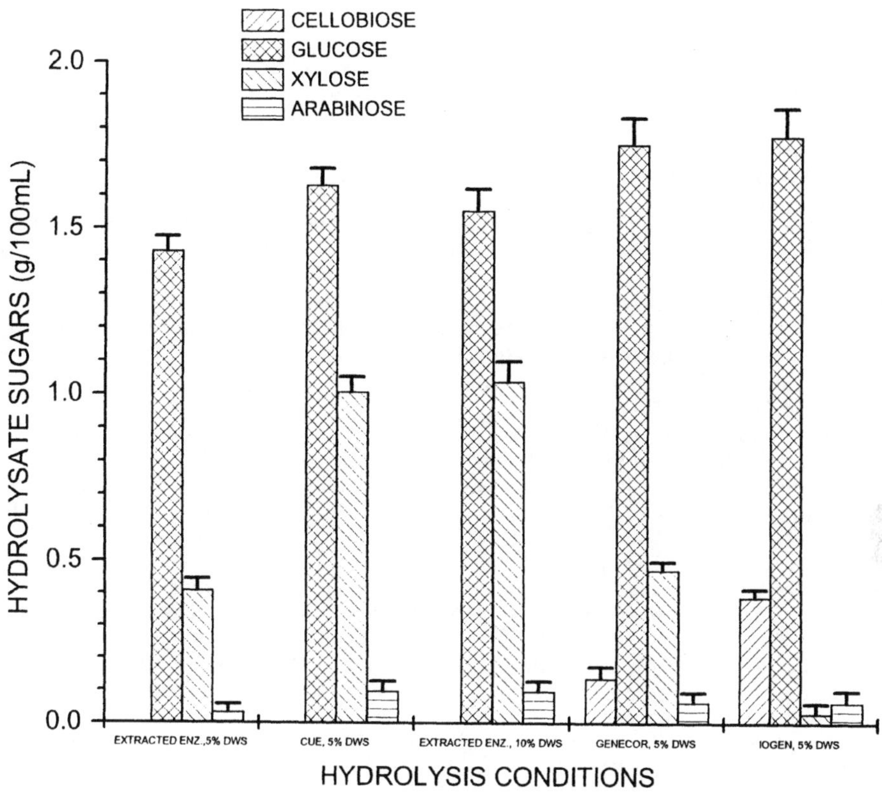

Fig. 3. Hydrolysis of DWS with cellulase system from coculture of *T. reesei* QMY-1 and *A. phoenicis* after 60 h of incubation. (Results of commercial enzymes from Fig. 1 are also represented here for comparison.)

Table 3
Effect of Different Cellulase Systems and Substrate Concentration on the Hydrolysis of DWS

Cellulase system[a]	5% DWS Total sugars,[b] g/100 mL	5% DWS Hydrolysis, %	10% DWS Total sugars,[b] g/100 mL	10% DWS Hydrolysis, %	Decrease in hydrolysis with increase in substrate conc., %
T. reesei MCG 80	4.84	88	7.4	67	24
T. reesei QMY-1	2.85	51.8	4.9	45	13
T. reesei QMY-1 + *A. phoenicis*	1.87	34	2.7	25	25

[a]Extracted cellulase was used for hydrolysis.
[b]Total sugars = (cellobiose + glucose + xylose + arabinose).

REFERENCES

1. Chahal, D. S. (1985), *Appl. Environ. Microbiol.* **49,** 205–210.
2. Chahal, D. S. (1991), *ACS Symposium Series* **460,** 111–122.
3. Hesseltine, C. W. (1972), *Biotechnol.* Bioeng. **14,** 517–532.
4. Mudgett, R. E. (1986), in *Manual of Industrial Microbiology and Biotechnology*, Demain, A. L. and Solomon, N. A., eds., ASM, Washington, DC, pp. 66–83.
5. Saddler, J. N. (1986), *Microbiol. Sci.* **3,** 84–87.
6. Gomes, J., Esterbauer, H., Gomes, I., and Steiner, W. (1989). *Lett. Appl. Microbiol.* **8,** 67–70.
7. Stockton, B. C., Mitchell, D. J., Grohmann, K., and Himmel, M. E. (1991), *Biotech. Lett.* **13,** 57–62.
8. Chahal, D. S., McGuire, S., Pikor, H., and Noble, G. (1982), *Biomass* **2,** 127–138.
9. Klyosov, A. A. (1980), in *Biochemistry and Genetics of Cellulose Degradation*, Aubert, J. P., Beguin, P., and Millet, J., eds., Academic, NY, pp. 87–99.
10. Esterbauer, H., Steiner, W., Labudova, I., Hermann, A., and Hayn, M. (1991), *Bioresource Technol.* **36,** 51–65.
11. Panda, T., Bisaria, V. S., and Ghose, T. K. (1983), *Biotech. Lett.* **5(11),** 767–772.
12. Morisset, W. M. L. and Khan, A. W. (1984), *Biotech. Lett.* **6(6),** 375–378.
13. Breuil, C., Chan, M., Gilbert, M., and Saddler, J. N. (1992), *Bioresource Technol.* **39,** 139–142.
14. Chahal, P. S., Chahal, D. S., and Le, G. B. B. (1995), *Appl. Biochem. Biotechnol.* **57/58,** 433–442.
15. Mandels, M. and Weber, J. (1969), *Adv. Chem. Ser.* **95,** 391–414.
16. Sternberg, D., Vijayakumar, P., and Reese, E. T. (1977), *Can. J. Microbiol.* **23,** 139–147.
17. Srivastava, S. K., Ramachandran, K. B., and Gopalkrishnan, K. S. (1981), *Biotech. Lett.* **3,** 477–480.
18. Staniforth, A. R. (1979), *Cereal Straw.* Oxford University Press, Oxford, UK, p. 27.
19. Mandels, M., Andreotti, R., and Roche, C. (1976), *Biotechnol. Bioeng. Symposium* **No. 6,** 21–23.
20. Ghose, T. K. (1987), *Pure and Appl. Chem.* **59(2),** 257–268.
21. Toyama, N. and Ogawa, K. (1972), *Fermentation Technology Today.* Society of Fermentation Technology, Japan, pp. 743–757.

0273-2289/96/5758—0471$7.50

Recovery and Purification of Lactic Acid from Fermentation Broth by Adsorption

ROQUE L. EVANGELISTA[1] AND ZIVKO L. NIKOLOV*,[1,2]

[1]*Department of Food Science and Human Nutrition; and* [2]*Department of Agricultural and Biosystems Engineering, Iowa State University, Ames, IA 50011*

ABSTRACT

Riedel-de-Haen VI-15, Dowex MWA-1 and Amberlite IRA-35 were employed for lactic acid recovery using model fermentation broth. The broth was first acidified by using a cation exchanger before sorption of lactic acid onto the basic sorbents. Lactic acid was completely recovered from the VI-15 column after 7 bed volumes (BV) of methanol, whereas only 64% was recovered from MWA-1 after 4.5 BV and 18% from IRA-35 after 5 BV. The 5% NH_4OH eluted all lactic acid from the MWA-1 column in 1.5 BV with a maximum effluent concentration of 113 mg/mL. The simple recovery scheme employed was not sufficient to produce heat-stable lactic acid. Other broth components were also adsorbed by the basic sorbents and could not be removed during the rinse step, but eluted with lactic acid during the desorption step.

Index Entries: Lactic acid; recovery; sorption; ion exchange.

INTRODUCTION

As the world's crude oil resources diminish and the prices of petroleum products continue to increase, the production of chemicals by biological processes is becoming more competitive. High-purity and heat-stable lactic acid, which was mainly available from synthetic manufacture, is now being produced by fermentation. About 82% of the estimated 36×10^6 kg/yr of lactic acid produced worldwide is used by the food industry. The rest is being used in pharmaceutical, cosmetics, textile, and leather industries. These markets are projected to grow at 2–4%/yr *(1)*. The use of polylactic acid (PLA) for biodegradable plastics and controlled-release drugs and pesticides can transform lactic acid into a commodity chemical *(2)*. The commercial success of PLA, however, hinges on the cost of producing heat-stable lactic acid. Therefore, one of the major challenges in lactic acid production is to reduce the cost of acid recovery and purification.

Adsorption is a process suitable for recovering substances produced in dilute concentrations and in complex solutions, such as fermentation broth. Synthetic polymeric sorbents with varying degrees of basicity have been used in the recovery of carboxylic acids from fermentation broth *(3–10)*. The polymeric sorbents have also been used to extract lactic acid and control pH during fermentation *(10–13)*.

*Author to whom all correspondence and reprint requests should be addressed.

Batch fermentation of lactic acid is usually conducted at pHs between 5 and 7 to minimize product inhibition by the free acid *(14)*. This is done by adding a base to neutralize the acid as it is produced. The lactic acid in the broth is therefore produced as a salt of the base used. The adsorption of lactate can be achieved by using a strong-base ion exchanger. Recovering lactate from the sorbent, however, requires a stronger desorbent. For example, if NaOH is used, the product is sodium lactate. If the desired product is the undissociated (free) acid form, further processing is necessary. To recover 1 mol of lactic acid, 1 mol each of base and acid will be consumed. One mol of waste salt will also be generated, which creates a disposal problem. Weak-base adsorbents, on the other hand, adsorb only free lactic acid; therefore, they are not effective at pHs above the pK_a of lactic acid (3.86) where mostly lactates are present. Acidifying the broth with mineral acid is not an option because it introduces competing acids *(15)*. The advantage of weak-base sorbents over strong-base sorbents is that the free lactic acid adsorbed is easily recovered by using alcohols (e.g., methanol and ethanol), acetone, and methyl acetate *(4,11,13, 16,17)*, or by temperature-swing desorption using water as eluant *(9)*. A free acid product is produced after evaporating the eluant. Concentrating lactic acid is also cheaper because the low-boiling solvents require less energy to evaporate than water. Furthermore, the solvents can also be recycled, thus significantly reducing waste generation.

The main objective of this research was to investigate the use of weak-base polymeric sorbents in the primary recovery and/or purification of lactic acid from fermentation broth. In the preliminary evaluation of selected basic sorbents using pure lactic acid *(18)*, it was recommended that when weak-base (pyridine and imidazole functional group) and moderate-base sorbents (tertiary amine functional group) were to be used in lactic acid recovery, the broth pH must be below the pK_a of lactic acid for an efficient utilization of the sorption capacity and to minimize leaching of lactates. Of the two weak-base sorbents, Riedel-de-Haen VI-15 was selected over Reillex 425 for further evaluation because of its higher sorption capacity even at low (<70 mg/mL) lactic acid concentrations. The capacity of VI-15 is comparable to that of the moderate-base sorbents evaluated. Dowex MWA-1 and Amberlite IRA-35 were included for comparison. In this article, the results of lactic acid recovery from model fermentation broth will be presented.

MATERIALS AND METHODS

Resin Preparation

The three basic polymeric sorbents selected for this study (Table 1) were conditioned following the procedures outlined in the *Rohm and Haas Ion Exchange Resin Laboratory Guide (21)*. The hydrated sorbents were transferred into a 1-L (35 × 6 cm) column. The sorbent bed was first backwashed with deionized water to remove the fine particles. This was followed by washing (downflow) of the sorbent with 3 BV of 5% HCl, 5 BV of deionized water, 3 BV of 4% NaOH, and another 5 BV of deionized water. The flow rate was adjusted to allow at least 30 min of contact time between the sorbent and HCl or NaOH solutions. These steps were repeated one more time, and the final water rinse was performed until the effluent pH was <8. The cation exchange resin was also conditioned in a similar manner to the basic sorbents, except that the order of NaOH and HCl rinse was reversed to obtain a

Table 1
Properties of Selected Basic Adsorbents

	Resin/manufacturer		
	VI-15	MWA-1	IRA-35
Properties	Riedel-de-Haen, Seelze, Germany	Dow Chemical, Midland, MI	Rohm and Haas, Philadelphia, PA
Type	Gel	Macroporous	Macroporous
Matrix	Methylene-bis-acrylamide	Styrene-divinylbenzene	Acrylic-divinylbenzene
Functional group	Imidazole	3°-amine	3°-amine
pK_a	6.9[a]	8.8[b]	11.1[c]
Capacity for chloride	7.9[d]	3.8[d]	4.9[c]
Capacity for lactic acid (mg/g)	3.1[d]	4.0[d]	4.4[c]
Association constant for lactic acid, K (g/g)	2200[d]	8400[d]	28,160[c]

[a]Weast (1987) *(19)*.
[b]Kuo et al. (1987) *(20)*.
[c]Tung (1993) *(15)*.
[d]Evangelista et al. (1994) *(18)*.

sorbent in H^+ form. The moisture contents of the conditioned sorbents were determined by using the oven method.

Broth Preparation

The model lactic acid broth used in this study contained 10 g yeast extract (Difco Laboratories, Detroit, MI), 100 g lactic acid (Certified ACS, Fisher Scientific, Pittsburgh, PA), 10 g glucose (Sigma, St. Louis, MO), 0.6 g $MgSO_4 \cdot 7H_2O$, 0.5 g K_2HPO_4, and 0.03 $MnSO_4$/L. The pH of the broth was adjusted to 4.5 by using ammonium hydroxide. This starting broth pH was chosen because the final pH of the broth from batch lactic acid fermentation would be around this value.

Analytical Methods

Lactic acid and glucose concentrations were analyzed by using high-performance liquid chromatography (HPLC) (Maxima 820, Waters, Milford, MA) equipped with a refractive index detector and Bio-Rad Aminex HPX-87H column (300 × 7.8 mm) (Bio-Rad, Hercules, CA). The column temperature was maintained at 65°C by using a column oven. The mobile phase was 6 m*M* H_2SO_4 at a flow rate of 0.8 mL/min.

Lactic Acid Recovery Scheme

Lactic acid was recovered from fermentation broth following the scheme shown in Fig. 1. The acidification step was carried out by using a weak-acid cation exchanger (Duolite C-464, Rohm and Haas Co., Philadelphia, PA) prepared in H^+ form. The lactic acid from the acidified broth was then adsorbed on the basic sorbents. The exhausted sorbent was rinsed with water to wash the unbound components of the broth from the bed. The adsorbed lactic acid was eluted from the

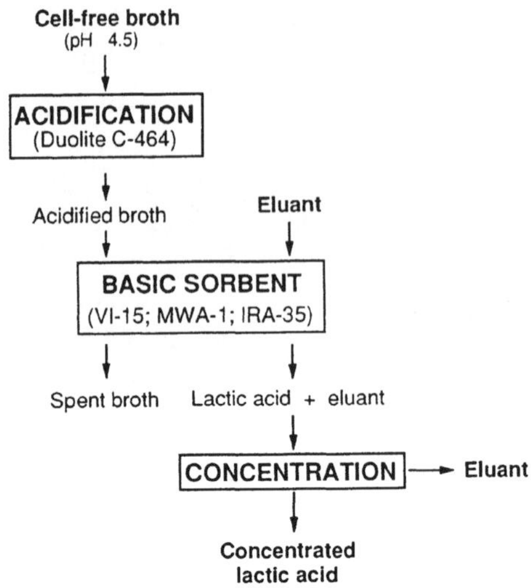

Fig. 1. Scheme for lactic acid recovery and purification.

sorbents by using HPLC-grade methanol (Fisher Scientific) or 5% NH_4OH, depending on the sorbent used.

Broth Acidification

The broth was passed through the Duolite C-464 column in H^+ form at a flow rate of 3 BV/h. Fractions were collected and analyzed for lactic acid and glucose using HPLC. The pH of each fraction was also measured. Fractions containing acidified lactic acid with pH $\leq$ 3.0 were pooled and set aside for sorption on basic sorbents. The column was regenerated by using sulfuric acid. Several cycles were run to produce enough acidified broth for sorption in the basic sorbents.

Sorption in Weak-Base Sorbents

A jacketed 30 × 1 cm column (Kontes Scientific Instrument, Vineland, NJ) equipped with adjustable plungers was charged with conditioned sorbent to a bed height of at least 10 cm. The air trapped in the column was removed by adding water to the column and stirring the sorbent. The plunger was then lowered to the top of the bed. The acidified broth was introduced to the column upflow at 3 BV/h. The column was rinsed with water to remove the excess broth and other unbound broth components. This was followed by elution of bound lactic acid with methanol or 5% NH_4OH. Fractions were collected and analyzed for lactic acid, glucose, and pH. Methanol was evaporated from the eluate by using a rotary evaporator. Eluants from NH_4OH desorption were not concentrated. The column was regenerated by using 3 BV 4% NaOH and rinsed with 5 BV water.

RESULTS AND DISCUSSION

Broth Acidification

The Duolite C-464 column decreased the pH of the broth to as low as 2.1 (Fig. 2). Fractions with pH $\leq$ 3 were pooled, producing 0.4 BV of acidified broth. The

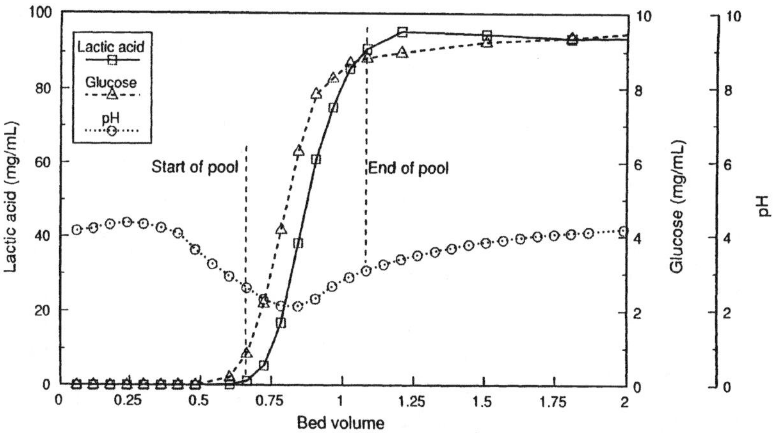

Fig. 2. Effluent profile of model broth on Duolite C-464 column.

acidified broth had a pH of 2.9 and contained 58.2 mg lactic/mL and 6.8 mg glucose/mL. Despite the dilution, the concentration of lactic acid was well above the minimum equilibrium lactic acid concentration (about 10 mg/mL) needed to attain maximum sorption for all three basic sorbents *(15,18)*.

Duolite C-464 also removed colored compounds from the broth, as manifested by the marked color reduction in the broth (from amber to faint yellow). The HPLC chromatograms, however, showed no significant decrease in the number of broth components. The faint yellow color in the pooled acidified broth came from the fractions approaching the cut-off point. The cation exchanger was regenerated back to H^+ form by using sulfuric acid followed by a water rinse.

Recovery of Lactic Acid from Acidified Broth by Using Basic Sorbents

Riedel-de-Haen VI-15

The resin bed required 6 BV of acidified broth to reach saturation (Fig. 3 and Table 2). At this point, the BV increased by 30% owing to the swelling of the resin. No glucose was detected in the effluent after 4 BV of water-rinse effluent. About 55% of the total lactic acid in the column came out with the rinse water and 71 mg lactic acid/mL resin remained adsorbed in the column. In practice, the rinse effluent can be recycled to the adsorption column because the pH is still low—even lower than the pH of the acidified broth. The slightly lower pH in the rinse effluent indicated that, aside from the lactic acid sorbed in the pores, some adsorbed lactic acid was also eluted.

The lactic acid was completely desorbed by using 6.8 BV of methanol. This is about twice the amount of methanol needed to desorb lactic acid from the weaker basic sorbent poly(4-vinylpyridine) *(16)*. The highest lactic acid concentration of the effluent was 31 mg/mL, which would be the expected concentration of lactic acid during desorption in a countercurrent operation. The effluent near the end of the first BV of the methanol eluant was cloudy. Because the micropores are accessible to smaller compounds, but not to lactic acid, the cloudiness may have been caused by salts (like sulfates) or other compounds that are insoluble in methanol. The

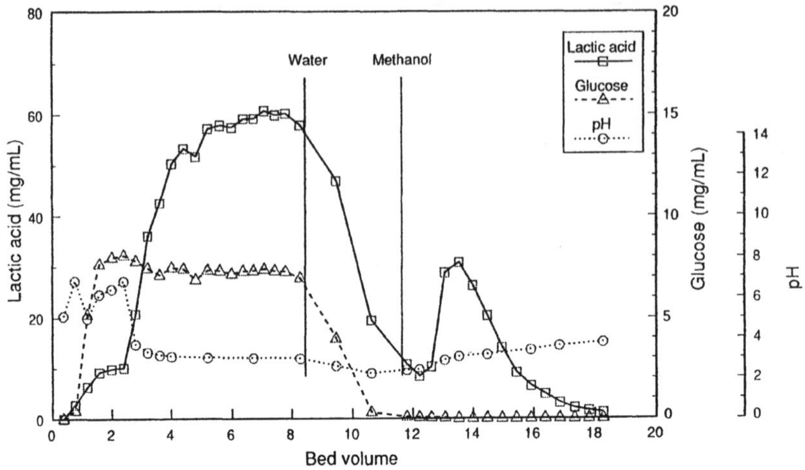

Fig. 3. Effluent profile for VI-15 column.

Table 2
Summary of Lactic Acid Recovery Using Weak-Base Sorbents

	Resin		
Stage of the process	VI-15	MWA-1	IRA-35
Sorption			
Volume of broth to reach saturation, BV	6.0	3.5	3.2
Total lactic acid in the column, mg	1960	2517	1336
Rinse			
Volume when no glucose was detected in the rinse effluent, BV	4	3	3
Fraction of total lactic acid in rinse effluent, %	55	51	48
Lactic acid adsorbed by the resin, mg/mL resin	71	72	75
Desorption			
Methanol			
Adsorbed lactic acid recovered, %	100	64	18
Volume of eluant used, BV	6.8	4.5	5
Highest lactic acid concentration, mg/mL	31	21	5
NH_4OH			
Adsorbed lactic acid recovered, %	nd[a]	100	nd
Volume of eluant, BV	nd	1.5	nd
Highest lactic acid concentration, mg/mL	nd	113	nd

[a]nd, no data.

shrinking of the sorbent as it came in contact with methanol facilitated the exclusion of the liquid from the sorbent, including that in the micropores.

The cloudy fractions contained 29% of the total lactic acid recovered from the column. These fractions were not included in the pool that was subsequently concentrated. The concentrated lactic acid was yellowish, but not turbid. The HPLC chromatogram of the lactic acid product showed that, aside from glucose (0.05%),

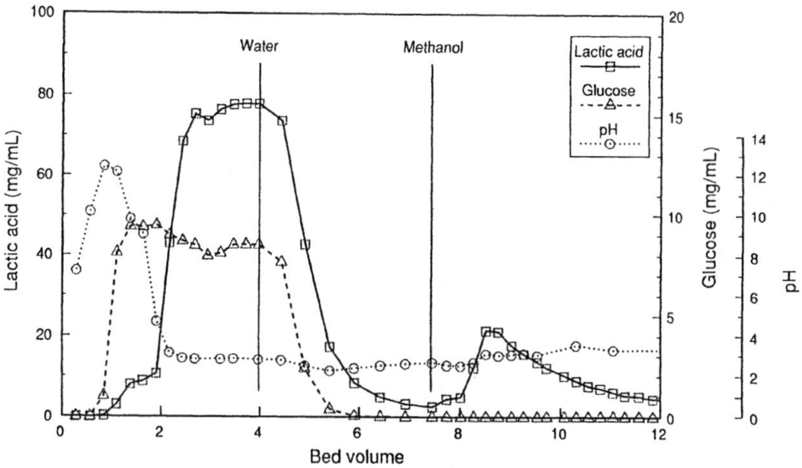

Fig. 4. Effluent profile for MWA-1 column with methanol as desorbent.

small amounts of other impurities were also present. The presence of readily carbonizable residues was verified by the positive result of the sulfuric acid test (formation of brown layer at the interface of sulfuric acid and lactic acid).

The column was regenerated with 1*M* NaOH to remove broth components not desorbed by methanol. The rinse effluent was yellowish, indicating that other compounds were also adsorbed by the resin and could not be eluted readily with methanol. A 30% decrease in column capacity for lactic acid was observed after a 10-cycle run without NaOH rinse. The sorbent, therefore, required periodic regeneration with a strong base to remove these strongly bound broth components and to restore its capacity.

The shrinking of the sorbent during desorption also led to compaction of the sorbent bed, causing a significant flow reduction through the column. The sorbent had to be loosened by stirring the bed and backwashing with water at a high flow rate to prevent compaction of the bed during the next loading cycle. The backwashing step also removed the methanol remaining in the column.

The excessive swelling and shrinking of VI-15 resin would cause attrition of the sorbent and shorten its service life. Swelling of the resin also increases the internal void volume, thus resulting in increased sorption of other broth components because of pore filling. These impurities, if not removed during the rinse step, may be trapped inside or may elute slowly with lactic acid as the sorbent shrinks during desorption with methanol.

Dowex MWA-1

The resin reached saturation after 3.5 BV of acidified broth (Fig. 4 and Table 2), which was about half the volume needed to saturate the VI-15 column. The shorter saturation time for MWA-1 was a combined effect of the matrix structure and the basicity of the functional group. The macroporous structure in MWA-1 allows for a much faster diffusion of lactic acid into the sorbent than in the microporous structure of VI-15. MWA-1 also becomes saturated at a much lower equilibrium concentration because of its higher basicity (pK_a) and, consequently, higher affinity

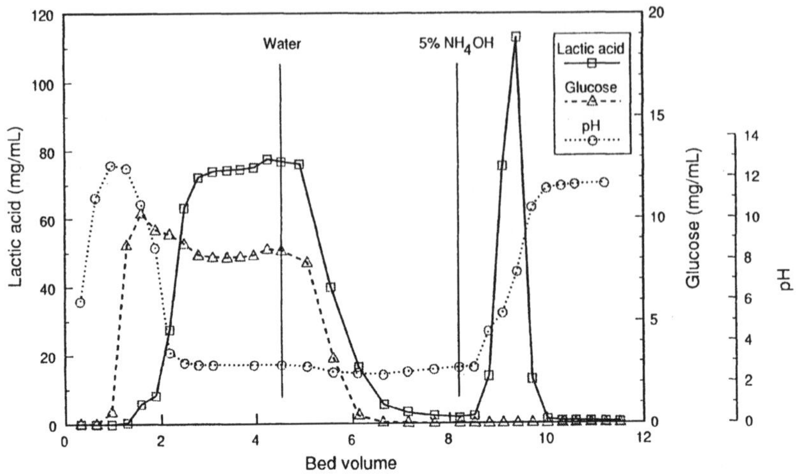

Fig. 5. Effluent profile fore MWA-1 column with NH_4OH as desorbent.

(K) for lactic acid than that of VI-15 (Table 1). The rinse effluent was free of glucose after 3 BV, 1 BV less than was needed by VI-15. Fifty-one percent of the total lactic acid sorbed in the column came out with the rinse water and 72 mg lactic acid/mL resin remained adsorbed by the resin. Methanol was not effective in desorbing lactic acid from the MWA-1 column despite the higher temperature (50°C) used. Only 64% of the adsorbed lactic acid was recovered after 4.5 BV. The highest concentration of lactic acid in the eluate was only 21 mg/mL. The cloudy fractions, which contained 12% of the total lactic acid recovered, were not included in the pool that was concentrated. The concentrated lactic acid was cloudy and yellowish in color. The HPLC chromatogram of the product showed that no glucose was present, but other broth components were detected. The sulfuric acid test for readily carbonizable residues also came out positive.

When 5% NH_4OH was used as desorbent, 100% recovery was achieved in 1.5 BV of eluant (Fig. 5 and Table 2). The highest concentration of lactic acid in the effluent was 113 mg/mL, about 50% greater than the lactic acid concentration in the feed. However, NH_4OH also desorbed other broth components adsorbed by the resin, resulting in a product with more impurities than the one desorbed with methanol. The sulfuric acid test for readily carbonizable residues also turned out positive, despite the absence of glucose.

Amberlite IRA-35

The IRA-35 column required the same amounts of acidified broth (3.2 BV) to reach saturation and the same amount of rinse water to get rid of glucose as in MWA-1 (Fig. 6 and Table 2). About 48% of the lactic acid in the column was washed out with the rinse water, and 75 mg lactic acid/mL resin remained adsorbed by the sorbent. Methanol eluted only 18% of the adsorbed lactic acid after 5 BV. The maximum lactic acid concentration in the effluent was 5 mg/mL. The HPLC chromatogram of the eluted lactic acid revealed that no glucose was present, but other broth components were also eluted. The sulfuric acid test for readily carbonizable residue was also positive despite the dilute concentration of lactic acid used in the test.

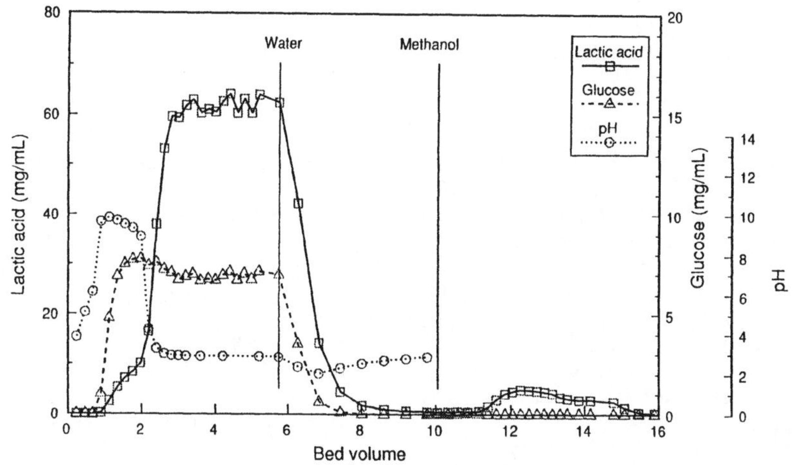

Fig. 6. Effluent profile for IRA-35 column with methanol as desorbent.

The much lower recovery of lactic acid from IRA-35 by methanol was not surprising since IRA-35 has a higher pK_a and higher association constant (K) for lactic acid than MWA-1 (Table 1). This disparity in their basicity is owing to the difference in the environment around tertiary amine groups in each sorbent. The styrene ring and the quaternary ammonium group in MWA-1 have a base-weakening effect on the amine group. In contrast, the aliphatic backbone and the presence of carboxylates in IRA-35 increase the basicity of the amine group *(15)*.

CONCLUSIONS

Duolite C-464 was able to lower the broth pH to an acceptable level, but it generated only 0.4 BV of acidified broth. The amount of acidified broth may be increased by using a strong cation exchanger. This method of broth acidification, however, still produces waste salt during regeneration of sorbent back to H^+ form.

Among the three basic sorbents employed in this study, VI-15 was the most attractive sorbent for lactic acid recovery. Its working capacity compared very well with that of the tertiary amines, and the adsorbed lactic acid can be easily eluted by using methanol. A more physically stable resin, however, is desirable to avoid problems related to excessive swelling and shrinking.

The recovery scheme employed was not sufficient to produce heat-stable lactic acid. Other broth components were also adsorbed by the basic sorbents and could not be removed during the rinse step, but eluted with lactic acid during the desorption step. Therefore, broth pretreatment or polishing steps would be necessary to improve the purity of the product.

ACKNOWLEDGMENTS

The authors thank Rod R. Fisher of Cargill, Inc., for his helpful comments and suggestions. This project was supported by the Iowa Corn Promotion Board and is a contribution of the research program of the ISU Center for Crops Utilization Research. This is journal paper No. J-16374 of the Iowa Agriculture and Home Economics Experiment Station, Ames, IA, Project 3017.

REFERENCES

1. HRA, Inc. (1991), *Lactic Acid and Its Derivatives: A Global Review.* HRA, Inc., Prairie Village, KS.
2. Lipinsky, E. S. and Sinclair, R. G. (1986), *Chem. Eng. Prog.* **82,** 26–32.
3. Yasuda, S., Niwa, T., and Kurohara, T. (1984), Japanese Patent 59-118731.
4. Kiel, K. H., Greiner, U., Engelhardt, F., Kuhlein, K., Heb, G., Reinhold, K., and Schlingmann, M. (1985), German Patent DE 3328093.
5. Obara, H. (1988), Japanese Patent 63188632.
6. Collin, H. and Buresch, E. (1990), European Patent 0377430.
7. Kulprathipanja, S. and Oroskar, A. R. (1991), US Patent 5,068,418.
8. Rossiter, G. J. (1991), in *Preparative and Process-Scale Liquid Chromatography,* G. Subramanian, G., ed., Ellis Horwood, West Sussex, UK.
9. Ernst, E. E. and McQuigg, D. W. (1992), Paper presented at the AIChE Annual National Meeting, November, Miami Beach, FL.
10. King, C. J. and Tung, L. (1992), US Patent 5,132,456.
11. Yates, R. A. (1981), US Patent 4,282,323.
12. Srivastava, A., Roychoudhury, P. K., and Sahai, V. (1992), *Biotechnol. Bioeng.* **39,** 607–613.
13. Tsao, G. T., Lee, S. J., Tsai, G., Seo, J., McQuigg, D. W., Vorhies, S., and Iyer, G. (1993), International Patent Appl. PCT/US92/07738.
14. Vickroy, T. B. (1985), in *Comprehensive Biotechnology: The Principles, Applications and Regulations of Biotechnology in Industry, Agriculture and Medicine,* Moo-Young, M., ed., Pergamon, New York.
15. Tung, L. (1993), Report No. LBL-34669, Lawrence Berkeley Laboratory, University of California, Berkely, CA.
16. Kabawata, N., Yoshida, J., and Tanigawa, Y. (1981), *Ind. Eng. Chem. Prod. Res. Dev.* **20,** 386–390.
17. Frierman, M., Kuo, Y., Joshi, D., Garcia, A. A., and King, C. J. (1987), *Sep. Purif. Methods.* **16,** 91–102.
18. Evangelista, R. L., Nikolov, Z. L., and Mangold, A. J. (1994), *Appl. Biochem. Biotech.* **45/46,** 131–144.
19. Weast, R. C. (1987), *Handbook of Chemistry and Physics,* CRC, Boca Raton, FL, p. D-99.
20. Kuo, Y., Munson, C. L., Rixey, W. G., Garcia, A. A., Frierman, M., and King, C. J., (1987), *Sep. Purif. Methods.* **16,** 31–64.
21. Rohm and Haas (1988), *Rohm and Haas Ion Exchange Resin Laboratory Guide,* Philadelphia, PA.

0273-2289/96/5758—0481$6.75

Production of L-Malic Acid from Fumaric Acid by Resting Cells of *Brevibacterium sp.*

C. S. GONG,*[,1] NINGJUN CAO,[2] YAN SUN,[3] AND G. T. TSAO[1]

[1]*Laboratory of Renewable Resources Engineering,* [2]*Food Science Department, Purdue University, West Lafayette, IN 47907; and* [3]*The Store Institute of Grain Storage Technology, Beijing, China*

ABSTRACT

The ability of the resting cells of *Brevibacterium sp.* B2D, a flocculent strain, to convert fumaric acid to L-malic acid was investigated under different fermentation conditions. The optimal temperature for the bioconversion is approx 36°C, and the optimal pH is around 6.4. The bacteria do not require special cultural conditions for the development of fumarase. In the presence of relatively high concentrations of fumarate (60 g/L), the bacterial cells exhibited a high rate of L-malic acid production with high yield without the coproduction of unwanted byproducts like succinic acid. A yield of 89.8% of L-malic acid from fumaric acid and an average specific productivity of 0.41 g/g/h was obtained. The presence of surfactant, even at low concentrations like 0.02%, resulted in over a twofold increase in specific malic acid production rate.

Index Entries: Fumaric acid; L-malic acid; *Brevibacterium sp.* B2D; fumarase; effect of surfactant.

INTRODUCTION

L-Malic acid, a naturally occurring four-carbon dicarboxylic acid, is a constituent of almost all plant and animal tissues, and is produced as a normal intermediate in basic metabolism. It is a major acid component in many fruits. Apple juice may contain a total acid concentration of up to 7% L-malic acid *(1)*. Malic acid is commlonly used as a food and beverage ingredient. It has also been used for the treatment of hyperammonemia and liver dysfunction, and as a component for amino acid infusion. Recently, malic acid has been considered as a raw material for the biodegradable polymers *(2)*.

L-Malic acid can be obtained from its natural source. However, this is impractical owing to the small quantities in which it occurs. Large-scale production of malic acid has been achieved through the chemical hydration of chemically deriveld fumaric or maleic acids. It can also be produced through enzymatic hydration of fumaric acid mediated by fumarase (fumarate hydratase, EC 4.2.1.2). Unlike malic

*Author to whom all correspondence and reprint requests should be addressed.

acid produced naturally, chemically produced malic acid has an equal proportion of the racemic D- and L-forms.

L-Malic acid is often produced extracellularly by microorganisms during their normal metabolism. Some mycelial fungi, such as those belonging to the genus *Aspergillus*, can produce copious quantities of L-malic acid under controlled environmental conditions from simple sugars *(3,4)*. L-Malic acid can also be produced from fumaric acid by fumarase-containing microbial cells. *Brevibacterium ammoniagenes (5)* and *Brevibacterium flavum (6)* are two well-known microorganisms, owing to their high fumarase activities, that have been used in immobilized forms for large-scale production of L-malic acid resulting from their high fumarase activity.

In this work, we investigated the production of L-malic acid from fumaric acid using the resting cells of a flocculent strain of *Brevibacterium*. We examined the optimal conditions, the substrate tolerance, and the effect of surfactants on the bacterium to achieve the optimal conversion.

MATERIALS AND METHODS

Organism and Medium

Brevibacterium sp. strain B2D was used. It was isolated from a laboratory culture with respect to its ability to flocculate during growth. The medium for cell growth contained the following: 3 g yeast extract (Difco, Detroit, MI), 3 g malt extract (Difco), 5 g peptone (Difco), 2 g glucose, and 1 L distilled water. Sterilization was accomplished by autoclaving at 15 lb/in.^{2} for 15 min.

Inoculation and Fermentation Conditions

Bacterial cells were grown aerobically in Erlenmeyer flasks containing the above medium at 25°C on a rotary shaker at 150 rpm for 72 h. Cells were collected aseptically by centrifugation using a clinical centrifuge at 500 rpm. After washing the cell pellets with sterile water, they were pelleted again and transferred to a fermentation substrate. Fermentation was conducted under aerobic conditions using 50-mL Erlenmeyer flasks that contained 10 mL of fermentation medium covered with a milk filter paper followed by aluminum foil. The fermentation medium contained yeast extract-malt extract–peptone (YMP) with ammonium fumarate.

Analytical Methods

Dry bacterial cell mass was determined by using a calibration curve made from the relationship between the packed-cell volume and their dry cell weight. Fumaric acid, L-malic acid, succinic acid, and other organic acids were determined and quantified by high-performance liquid chromatography (HPLC) (Hitachi Instrument, L-6200A) using a Bio-Rad (Hercules, CA) Aminex HPX-87H ion-exclusion column (300 × 7.8 mm) with a refractive index detector (Hitachi Instrument, L-3350 RI). The column was eluted with dilute sulfuric acid (0.005*M*) at a column temperature of 80°C and a flow rate of 0.8 mL/min over a 13-min period. Figure 1A shows the liquid chromatogram of organic acids of interest.

RESULTS AND DISCUSSION

Brevibacterium sp. B2D showed the ability to convert fumaric acid to L-malic acid when the resting cells were incubated with fumarate. At an initial substrate

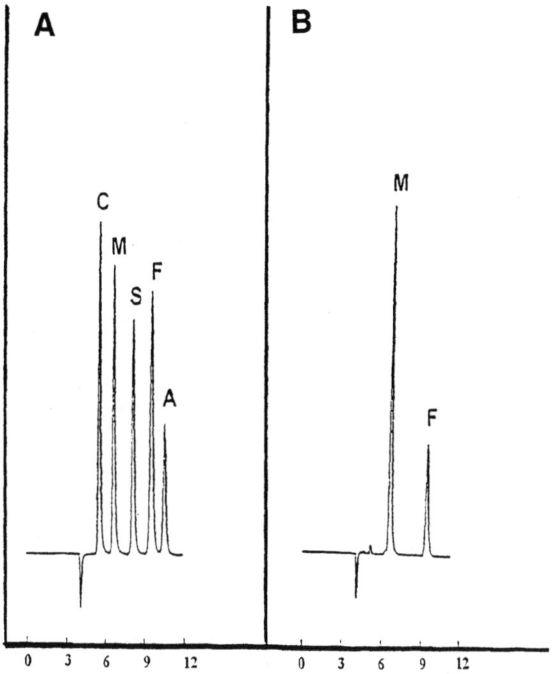

Fig. 1. Liquid chromatograms. **(A)** Retention times for the compounds of interest. C, citric acid (5.98 min); M, L-malic acid (7.19 min); S, succinic acid (8.85 min); F, fumaric acid (10.4 min); A, acetic acid (11.52). **(B)** Bioconversion of fumaric acid to L-malic acid by *Brevibacterium sp.* B2D. Initial fumaric acid concentration, 60 g/L; temperature, 34°C; initial cell density, 2.4 g/L; pH 6.8; incubation time, 48 h. F, fumaric acid (14.4 g/L); M, L-malic acid (47.3 g/L).

concentration of 60 g/L and an initial cell density of 2.4 g/L, fumaric acid was consumed linearly with the production of L-malic acid as the conversion product without apparent lag period (Fig. 2). The results also showed that over 76% (45.6 g/L) of fumaric acid was consumed after 48 h of incubation with an average specific malic acid productivity of 0.41 g/g/h and a concentration of L-malic acid at 47.3 g/L. Cell mass did not increase, and the fumarase activity was maintained for over 100 h of experiments. L-Malic acid yield based on the amount of fumaric acid consumed was 1.037, corresponding to an equivalent of 89.8% of the theoretical value, i.e., 1 g of fumaric acid is converted to produce 1.155 g of L-malic acid. This is based on the following reaction as mediated by fumarase:

$$\text{fumaric acid} + H_2O \rightarrow \text{L-malic acid} \qquad (1)$$

No detectable succinic acid was produced, as evidenced by the liquid chromatographic profile shown in Fig. 1B. Succimic acid is the common undesirable byproduct produced by other L-malic acid-producing bacteria, especially when they are immobilized *(5–7)*.

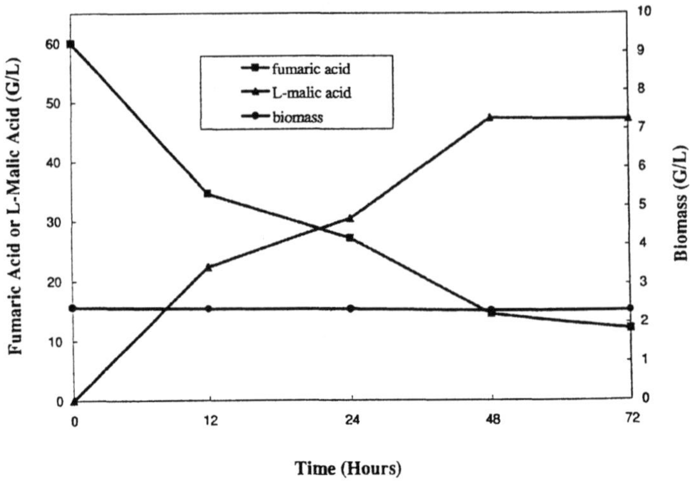

Fig. 2. Bioconversion of fumaric acid to L-malic acid by *Brevibacterium sp.* B2D. ■, Fumaric acid; ▲, L-malic acid; ●, cell density; temperature, 34°C, pH 6.8.

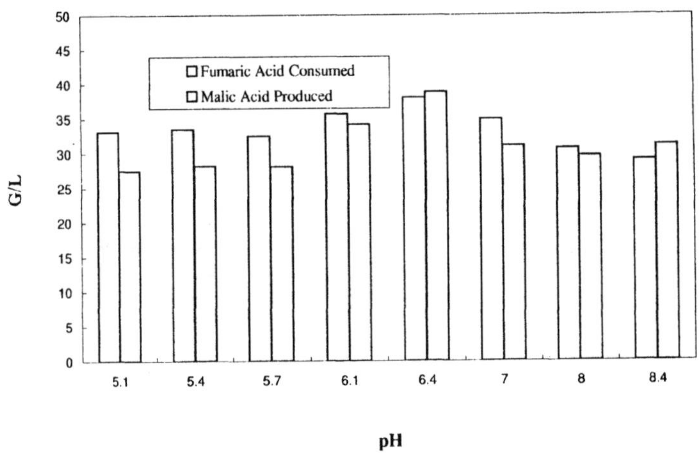

Fig. 3. Consumption of fumaric acid and the production of L-malic acid by *Brevibacterium sp.* B2D at different initial pHs. Initial cell density, 2.4 g/L; initial fumaric acid concentration, 60 g/L; temperature, 34°C; incubation time, 48 h.

Effect of pH and Temperature

The effect of pH was studied in the range of 5.1–8.4. After 36 h of incubation at 34°C with an initial substrate concentration of 50 g/L, over 50% of the fumaric acid was converted to L-malic acid with an effective yield of 88% at pH 6.4. L-malic acid production increased with increasing pH value and reached its optimal at 6.4. At a higher pH value, the rate of L-malic acid production declined slightly (Fig. 3). Similarly, incubation temperature had no significant effect over a range of 25–40°C. The initial conversion rate of fumaric acid to L-malic acid is relatively constant over the temperature range of 32–38°C with the specific L-malic acid produc-

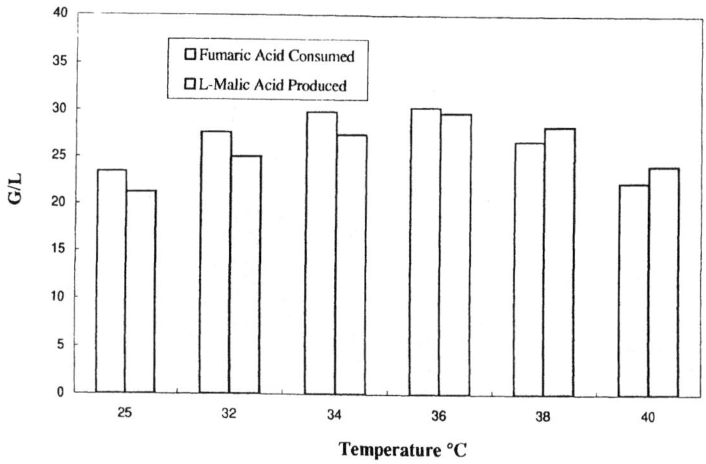

Fig. 4. Consumption of fumaric acid and the production of L-malic acid by *Brevibacterium sp.* B2D at different temperature. Initial cell density, 2.4 g/L; initial furmaric acid concentration, 60 g/L; initial pH 6.8; incubation time, 48 h.

tivity (g malic acid/g cell/h) in the range of 0.4–0.45. The specific productivity was slightly lower at 25 and 40°C (Fig. 4).

Effect of Substrate Concentration

The effect of initial substrate concentration was examined in the range of 20–100 g/L to determine the optimal condition for L-malic acid production. An increase in initial substrate concentration not only led to the increase in accumulation of L-malic acid, but also led to the increase in the average specific L-malic acid productivity (qp). The effects of initial substrate concentration on various parameters are shown in Table 1. The highest value in qp was obtained when the highest S_0 (100 g/L) was used, resulting in a product yield of approx 1, corresponding to 86% of the theoretical. The lower yield of L-malic acid from low substrate concentration is probably the result of the consumption of substrate or product during fermentation.

Effect of Surfactants

Surfactants are known to enhance the rates of conversion of fumaric acid to L-malic acid. In an immobilized system using *Saccharomyces cerevisiae,* such surfactants as Triton X-100 increased the specific malic acid productivity from 0.34 (0.046) to 0.45 mmol/h/g cell (0.06 g/g/h), an increase of 32% *(8)*. Similarly, in immobilized *B. flavum,* the cells were pretreated with surfactant to increase the productivity *(6)*. The most dramatic example of the effect of surfactant is the ability of a strain of *S. cerevisiae* amplified for fumarase to convert fumaric acid to malic acid. In the presence of surfactants, such as anionic deoxycholate, the rate of conversion escalated from zero to as high as 50 mmol/g cell/h or 6.75 g/g/h *(9)*.

We examined the effect of some surfactants on the conversion rate of fumaric acid to malic acid. Triton X- 100, Nonidet p-40, and deoxycholate exhibited a strong enhancement effect on the specific malic acid productivity even at a low

Table 1
Effect of Substrate Concentration[a]

S_o, (g/L)	Fumaric acid used, g/L	L-malic acid produced, g/L	Y(p/s), g/g	qp, g/g/h[b]	QP, g/L/h[c]
20	17.4	14.2	0.810	0.123	0.296
40	33.6	31.4	0.934	0.273	0.654
50	40.2	39.6	0.985	0.344	0.825
60	45.6	47.3	1.037	0.410	0.985
80	48.0	48.0	1.000	0.417	1.000
100	54.3	53.9	0.993	0.468	1.123

[a]Initial cell density, 2.4 g/L; temperature, 34°C; pH 6.8; incubation time, 48 h.
[b]qp, Average specific productivity of L-malic acid (g malic acid/g cells/h).
[c]QP, Average volumetric productivity of L-malic acid (g malic acid/L/h).

Table 2
The Effect of Surfactant on Bioconversion[a]

Surfactant, %	qp, g/g/h[b]
Control	0.67
Triton X-100	
0.02	1.31
0.05	1.34
0.10	1.40
Nonidet p-40	
0.02	1.38
0.05	1.39
0.10	1.39
Deoxycholate	
0.02	1.27
0.05	1.21
0.10	1.21
Tween-80	
0.1	0.85
0.2	0.85
SLS	
0.1	0.00

[a]Initial cell density, 2.4 g/L; temperature, 25°C; pH 6.8; incubation time, 48 h; fumaric acid concentration, 60 g/L.
[b]qp, Average specific productivity of L-malic acid (g malic acid/g cells/h).

concentration of 0.02% (Table 2). A twofold increase in specific productivity was obtained when the aforementioned surfactants were present. At higher concentrations, the effect was similar to that at lower concentrations. Tween-80 exhibited slight enhancement of bioconversion, whereas sodium lauryl sulfate (SLS) exhibited a detrimental effect. Fumarase activity in the present of SLS was not detected nor can the activity be restored after the removal of SLS, indicating the denaturation of fumarase by SLS.

CONCLUSION

From the data presented, *Brevibacterium* sp. B2D is characterized by its strong fumarase activity and has the potential for the large-scale conversion of fumaric acid to L-malic acid. The specific activity is generally higher than those reported for immobilized-cell systems. There is no special pretreatment needed to carry out conversion. Furthermore, the resting cell system exhibited relatively high yield, and in particular, no byproduct, such as succinic acid was detected. The bacterial culture grows well and can be collected by simple centrifugation, since this strain of bacterium flocculates easily. Compared to the chemical method, the biological conversion produces only L-isomer that can be used in food systems for human consumption. With its flocculating characteristics, this bacterial strain can easily be used without immobilization by recycling through centrifugation or by using cell-settling recycling techniques that can be easily adapted for commercial operation.

ACKNOWLEDGMENTS

This research was supported in part by a grant from National Science Foundation. We thank Xuezhi Yu and Ninghua Yao for their technical assistance.

REFERENCES

1. Buck, R. E. and Mottern, H. H. (1947), *Ind. Eng. Chem.* **39**, 1087–1089.
2. Kamenetsky, M., Kharas, G. B., Watson, K., Caywood, G. A., and Kitchell, J. P. (1993), *Poly. Materials* **69**, 333–337.
3. Bercovitz, A., Peleg, Y., Battat, E., Rokem, J. S., and Goldberg, I. (1990), *Appl. Environ. Microbiol.* **56**, 1594–1597.
4. Battat, E., Peleg, Y., Bercovitz, A., Rokem, J. S., and Goldberg, I. (1991), *Biotechnol. Bioeng.* **37**, 1108–1116.
5. Yamamoto, K., Tosa, T., Yamashita, K., and Chibata, I. (1976), *Eur. J. Appl. Microbiol.* **3**, 169–183.
6. Takata, I., Yamamoto, K., Tosa, T., and Chibata, I. (1980), *Enzyme Microb. Technol.* **2**, 30–36.
7. Peleg, Y., Stieglitz, B., and Goldberg, I. (1988), *Appl. Microbiol. Biotechnol.* **28**, 69–75.
8. Figueiredo, Z. M. B. and Carvalho, L. B., Jr. (1991), *Appl. Biochem. Biotechnol.* **30**, 217–224.
9. Neufeld, R. J., Peleg, Y., Rokem, J. S., Pines, O., and Goldberg, I. (1991), *Enzyme Microb. Technol.* **13**, 991–996.

0273-2289/96/5758—0489$8.25

Solid-State Production of Ethanol from Sorghum

LINDA L. HENK* AND JAMES C. LINDEN

Department of Microbiology and Department of Chemical and Bioresource Engineering, Colorado State University, Fort Collins, CO 80523

ABSTRACT

The main goal of this research is to study the solid-state fermentation of sorghum-sudangrass, Grazex II (F_1 hybrid of *Sorghum vulgare X Sorghum sudanese*), to ethanol. Our research focuses on using a modified method of ensiling to produce ethanol directly in the silo. Thirty-eight liters of ethanol/metric ton (L/MT) on a wet-weight basis were produced from sorghum receiving cellulase compared to 23.4 L/MT for sorghum not receiving cellulase additives. Based on total free sugar content, 101 and 84% of theoretical yield are achieved for cellulase-amended and nonamended sorghum, respectively.

Index Entries: Sorghum; sorghum-sudangrass; ethanol; lactic acid; solid-state fermentation.

INTRODUCTION

In the US, environmental regulations, such as the Clean Air Act of 1990, provide incentives for the use of oxygenated fuels in automobiles *(1)*. Ethanol or methyl tertiary butyl ether (MTBE) boost the oxygen content in gasoline and reduce carbon monoxide emissions. One principal advantage for using ethanol is that the fuel is produced from domestically grown renewable resources *(1–4)*. Atmospheric levels of carbon dioxide, a greenhouse gas, can be decreased by replacing fossil fuels with renewable biofuels *(4,5)*.

Typically, grains, such as barley or corn, are utilized for fuel ethanol production; however, biomass, such as sugar cane, corn stover, wood residues, and waste products, can also serve as substrates *(1,3,4)*. Sweet sorghum has been extensively studied as a substrate for fuel alcohol production for several years *(6–9)*. Sweet sorghum is one of the most promising crops that can be grown for biomass in temperate climates *(6,7)*. It is a genetically diversified and drought-tolerant plant that produces high yields of fermentable sugars and biomass, both of which can be used for ethanol production *(6–9)*. Sweet sorghum has demonstrated an ability to grow in widely diverse temperate zones in both irrigated and nonirrigated environments *(6,7)*. Gross green weight averaged 89.8 $Mg \cdot Ha^{-1}$ for irrigated sor-

*Author to whom all correspondence and reprint requests should be addressed.

ghum compared to 65.0 Mg·Ha^{-1} for sorghum grown with natural rainfall *(7)*. In this study, total sugars, comprised of sucrose, glucose, and fructose, averaged about 6 Mg·Ha^{-1}. Sugar concentrations for sucrose, glucose, and fructose averaged 5.1, 1.8, and 2.4% on a fresh wet-wt basis (dry matter content of 24%) *(7)*.

However, in order for sweet sorghum to be used successfully for ethanol production, three issues need to be addressed:

1. Carbohydrate storage;
2. Accessibility of the lignocellulosic fraction to enzymatic hydrolysis of hemicellulose and cellulose; and
3. A more economical means of producing ethanol from sweet sorghum.

Conversion economics is the key to increased use of biofuel in the open market *(10)*. For biofuels to become economical, means of lowering production costs must be found. The purpose of our research is to investigate the traditional farming method of ensiling with regard to storing the carbohydrates in sweet sorghum on a long-term basis, improving accessibility of structural carbohydrates to hydrolytic enzymes, and solid-state fermentation to ethanol in the silo.

Carbohydrate Storage

Seasonal availability and storability of sweet sorghum are important factors in the use of this renewable biomass. Sugar extraction and storability are two serious problems that have limited the use of sweet sorghum as a substrate for ethanol production *(11)*. Traditionally, juice containing 10–15% sugar has been extracted or pressed from the sweet sorghum pulp. The juice is then either fermented directly to alcohol or evaporated to molasses for storage. Typically, press efficiencies are low, averaging 50–65%. Direct fermentation of the juice to ethanol is a seasonal process, accomplished for only a short time after harvest. In addition, sweet sorghum bagasse containing cellulose and hemicellulose is utilized for fodder, not as a biofuel substrate.

Countercurrent extraction technology is used throughout the sugar industry *(12,13)*. Sugar beet cossetts are moved upward in extractors by means of augers. Water is added at the higher end of the extractor, moving in a countercurrent manner, downward with the force of gravity.

Countercurrent extraction has been used to recover sugars and other components in sweet sorghum silage *(12,13)*. Extracts contain a high solute concentration, depending on the liquid:solid ratio used during the extraction. Temperature also affects product recovery. Over 90% of the sugars and organic acid in sweet sorghum silage were extracted using a 2:1 liquid:solid ratio at 70°C *(12)*.

Because sweet sorghum is harvested seasonally at a high moisture level (70–80%), rapid deterioration of readily available carbohydrates can occur unless the material is stabilized by some process that allows storage of the material on a long-term basis. Sweet sorghum deteriorates after harvest, even when bundled and stored dry at ambient temperatures *(11)*. Solid-state fermentation of sweet sorghum to ethanol has been investigated *(14,15)*, as have agronomic systems for harvesting and processing sweet sorghum for fuel alcohol production *(6,8,16)*. Ensiling has also been investigated as an economical method of long-term storage of potential fermentable carbohydrates in sweet sorghum *(17,18)*.

Ensiling is a traditional agricultural practice that has been used for several centuries *(19,20)*. An acidic environment improves the stability of the silage and allows any remaining fermentable carbohydrates to be stored successfully for a long period of time *(19,20–22)*.

Additives have been used to improve the quality of silage. Additives can be divided into four main categories: stimulants, selective inhibitors, total inhibitors, and bacterial inoculants *(23)*. Amending silage with hydrolytic enzymes to release more fermentable sugars can be considered a stimulant additive. Stable silage results from the addition of cellulolytic and hemicellulolytic enzymes and lactic acid bacteria to forage *(17,18,24–27)*.

Our work involves the utilization of hemicellulases and cellulases as additives at an activity level of 5 International Units (IU) of activity/g of dry sorghum *(28)*. Previous work shows that 5 IU/g dry wt is optimal for the hydrolysis of sweet sorghum cellulose *in situ* *(18)*. The pH and temperature profiles within the silo provide compatible conditions for fungal cellulase enzymes to hydrolyze hemicellulose and cellulose to release more fermentable sugars *(18)*.

Acid additives, such as formic and sulfuric acids, increase the concentration of soluble sugars and suppress microbial activity *(23–25,29,30)*. The purpose of adding acid is to lower the pH of the freshly chopped material rapidly to a level of 4.0, thus preserving the soluble carbohydrates in the fresh material. Formic acid is a selective inhibitor of certain bacterial populations, particularly clostridial *(23,24)*. The addition of 0.3% formic acid reduces the initial production of lactic acid. More yeasts and fungi are found in formic acid-treated silage than in silages not receiving formic acid *(20,23)*. Increased levels of ethanol have been detected in silages receiving formic acid as an additive *(20)*.

Ethanol is often produced during ensiling fermentations. Naturally occurring yeast and bacteria capable of a heterolactic fermentation are responsible for the presence of ethanol in silage *(20)*. Ethanol has been detected in unamended sweet sorghum silage at a level of approx 8 g/kg dry matter and 5.8 g/kg in cellulase-amended sorghum *(18)*.

Accessibility of the Lignocellulosic Fraction to Enzymatic Hydrolysis

In order for sorghum lignocellulose to serve as a substrate for fuel alcohol production, accessibility of cellulose to enzymatic hydrolysis must be improved through some form of pretreatment that removes lignin and hemicellulose. Our data show that ensiling is a form of dilute-acid hydrolysis *(17,18,29)*. Ensiling improved the reactivity of the lignocellulosic fibers to enzymatic hydrolysis *(17,29)*.

The National Renewable Energy Laboratory in Golden, CO, is currently using an enzymatic technique known as simultaneous saccharification and fermentation (SSF) to produce ethanol from other forms of lignocellulose, i.e., wood, grasses, and municipal solid waste *(10,31)*. SSF incorporates the enzymatic hydrolysis of cellulose to glucose and the fermentation of glucose to ethanol in one reaction vessel, hence the term simultaneous saccharification and fermentation. Our previous work subjected sweet sorghum with or without cellulase additives to SSF *(21)*. Fresh sweet sorghum produced more ethanol than did material that had been ensiled for 60 d, presumably because easily fermentable sugars were so abundant in the fresh material. In addition, product inhibition of the cellulase enzyme system occurred during SSF owing to the high levels of cellobiose and

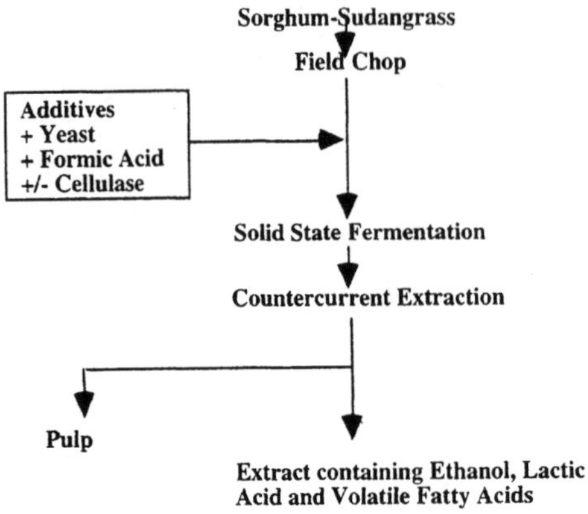

Fig. 1. Flowchart of solid-state fermentation of sorghum-sudangrass to ethanol.

glucose in the cellulase-amended silage. However, it was shown that ensiling enhanced the enzymatic hydrolysis of extracted sweet sorghum silage *(21)*.

Silo Fermentation

The main goal of this research is to study the feasibility of solid-state fermentation of sorghum-sudangrass to ethanol. It may be possible to utilize the silo as a bioreactor to produce ethanol *in situ*, thus lowering equipment costs incurred during SSF. By adding yeasts directly to green-chopped sorghum-sudangrass, more of the sugars that are available initially are converted to ethanol. The addition of cellulase will provide a continuous supply of glucose for ethanol production because of continued *in situ* cellulose hydrolysis. Acid-tolerant yeasts should remain metabolically active throughout the ensiling period. Heat generation resulting from the ethanolic fermentation corresponds to the slow rate of fermentation and may be dissipated. Lactic acid and anaerobic conditions in the silo preserve the sorghum until it is processed by countercurrent extraction to recover soluble components of the fermented sorghum-sudangrass *(17,18)*. A flowchart depicting the process is shown in Fig. 1.

Some supporting objectives used to evaluate ethanol formation from sorghum-sudangrass silage in the silo include:

1. Feasibility studies involving *in situ* ethanol production using sorghum amended with formic acid, cellulase, and additional yeast.
2. Evaluation of ethanol separation technology using countercurrent technology.
3. Evaluation of the process economics based on the first two objectives.

MATERIALS AND METHODS

Cultivation of Sorghum

Two varieties of sorghum were grown at the Colorado State University Agricultural Research, Development and Education Center (ARDEC). A variety of sorghum-sudan grass, Grazex II (F_1 hybrid of *Sorghum vulgare X Sorghum sudanese*) (Sharp Seeds, Greeley, CO) was identified as a variety that grows well in northeastern Colorado. The second variety, sweet sorghum (*Sorghum bicolor* [L.] Moench var.

M 81 E) did not reach maturity before the occurrence of a killing frost on September 14, 1993. In addition, M 81 E had a tendency to lodge, thus making harvest with a field chopper difficult. As a result, M 81 E was not used in any further work. M 81 E was used in our earlier studies regarding the SSF of sweet sorghum to ethanol *(21)*.

Laboratory Silo Setup

GrazexII sorghum was green-chopped with a field chopper and transported to laboratory facilities located in ARDEC for application of silage amendments and silo setup.

The following silage conditions were used:

1. Control, no additives (-C, -Y, -F).
2. Formic acid (75 m*M*), **plus** yeast (15 g of Montrachet wine yeast from Red Star/17 kg) **plus** 5 IU of cellulase activity/g dry wt of sorghum (+C, +Y, +F).
3. Formic acid **plus** yeast **minus** cellulase (-C, +Y, +F).

Silos consisted of 5-gal plastic containers with snap-on lids fitted with O-rings that seal the rims. Sorghum (~17 kg, fresh wt at 68% moisture content) was thoroughly mixed with 67 mL of formic acid (85% diluted to 60%, v/v), 210 mL of cellulase (Genencor, South San Francisco, CA), and 15 g of lyophilized Montrachet wine yeast (Universal Foods, Milwaukee, WI). This yeast is an all-purpose wine yeast that is acid-tolerant *(32)*.

The sorghum mixture was packed tightly into the containers and sealed. Weights were recorded for each silo. Silos were set up in triplicate for each sample date (19, 60, 88, 184, and 231 d of ensiling). The silos are stored under ambient conditions in the Food Engineering Laboratory at ARDEC.

Countercurrent Extraction of Sorghum-Sudangrass

A countercurrent diffuser is located in the Food Engineering Laboratory at ARDEC. This equipment has been used for extraction of sweet sorghum silage *(12,13)*. When silos are harvested, approximately one-third of the material is frozen immediately for the analysis. The remaining two-thirds are subjected to countercurrent extraction as described.

Extraction of ~12 kg of silage takes approx 3 h. Subsets of silage weighing 600 g each are continuously fed into the countercurrent extractor over the period of 10-min intervals. Water is delivered at a rate of 2 to 1 for the mass of sorghum on a wet-wt basis (1200 mL water delivered to 600 g sorghum silage in 10 min). This ratio was optimized for the extraction of sugars *(12,13)*. After the first nine subsets of 600 g silage are delivered to the diffuser, liquid samples are taken every 20 min and are taken for 90–120 min thereafter. These liquid samples are immediately frozen. In addition, a sample of pulp (extracted silage) is also frozen for further analysis. Extractions are conducted under ambient conditions to avoid the possible volatilization of ethanol. No experiments were conducted to quantify the amount of volatilization. Rather, extraction efficiencies were determined by comparing the amount of ethanol present in countercurrent extracts to blender extracts.

Analyses

Moisture Content

Moisture content of solids was determined by conventional oven dry wt (80°C, vacuum, overnight).

Ethanol

Glucose, fructose, xylose, lactic acid, acetic acid, and ethanol are evaluated by HPLC using a Bio-Rad (Hercules, CA) HPX-87H cation-exchange column. The solvent used is 0.008*N* H_2SO_4 at a flow rate of 0.6 mL/min. The column temperature is maintained at 65°C. Peaks are detected by refractive index *(17,18)*.

Lactic Acid

Countercurrent extracts were analyzed for lactic acid, volatile fatty acids, and sugars by HPLC *(17,18)* and enzymatically using a Lactate Diagnostic Kit (Sigma Chemical, St. Louis, MO).

Sugars

Countercurrent extracts and blender extracts were analyzed for sugars by HPLC as follows *(17,18)*. To prepare the blender extracts, 10 g of wet sorghum were added to 90 mL of deionized water, and a final weight was recorded. The mixture was kept on ice until blended at high speed in five short bursts of 10–15 s with periods of cooling (no blending for approx 1 min). This procedure was done in order to avoid heating of the blender blades, which may cause volatilization of ethanol from the solids. Prior to filtering, the blended extracts were returned to the ice bath and allowed to cool.

Both countercurrent extracts and blender extracts were prepared for analysis by first filtering the liquid:solid mixture through Whatman 541 filter papers lined with glass wool. Aliquots of filtrate (1.5 mL) were centrifuged at 18,000*g* in an Eppendorf microcentrifuge, and the supernatant was filtered through a 0.45-μ nylon membrane filter (Gelman Sciences, Ann Arbor, MI). Monosaccharides are analyzed using a Bio-Rad HPX-87P column with degassed, filtered high-purity water (Baxter, McGraw Park, IL) as the solvent at a flow rate of 0.6 mL/min. The column is maintained at 85°C, and sugars are detected by refractive index.

RESULTS AND DISCUSSION

Ethanol Production

Adding yeast, formic acid, and cellulase as silage amendments results in significant ethanol concentrations from 76 to 103 kg/MT on a dry-wt basis during the ensiling fermentation (Fig. 2). The addition of cellulase provides a continuous supply of glucose for ethanol production because of continued *in situ* cellulose hydrolysis. Acid-tolerant yeasts remain metabolically active throughout the ensiling period. As seen in Fig. 2, ethanol levels increase throughout the 231-d fermentation period, thus proving that the ethanolic material is stable over a long period of time.

Data regarding ethanol yields are shown in Table 1 for the day 231 fermentation of sorghum-sudangrass. Theoretical ethanol yield, based on soluble sucrose, glucose, and fructose present at day 0 is obtained from sorghum treated with cellulase, yeast, and formic acid (+C, +Y, +F) because of cellulose hydrolysis *in situ*. Without cellulase, but with formic acid as an additive (-C, +Y, +F), an ethanol yield that is 84% of theoretical is obtained. Total sugar available for fermentation in the unfermented sorghum-sudangrass equals 63.5 kg/MT of wet sorghum. The amount of sugar consumed to support yeast growth was not considered in the calculation of ethanol yield.

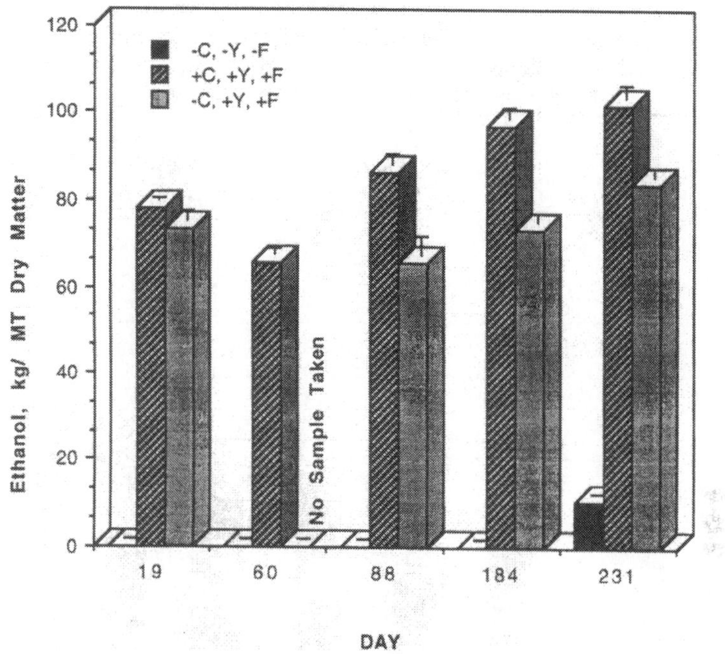

Fig. 2. Ethanol content of sorghum-sudangrass silage as determined by blender extraction. Sorghum-sudangrass control is designated as -C,-Y,-F. Forage receiving yeast, 75 m*M* of formic acid, and cellulase (5 IU/ g dry matter) is designated as +C,+Y,+F, whereas forage receiving yeast and formic acid, but no cellulase is designated as -C,+Y,+F.

The solid-state fermentation (231 d) of sorghum-sudangrass receiving both yeast and formic acid additives can be compared to previous fermentation studies using sweet sorghum *(21)*. The solid-state fermentation produces 64% of the ethanol found in SSF of sweet sorghum ensiled for 60 d (102 g ethanol/kg dry wt compared to 159 g/kg dry matter for SSF).

Lactic Acid Production

Lactic acid is produced during an ensiling fermentation and is responsible for regulating the acidity of the silage *(19,20,33)*. Over 61,000 silage samples preserved with formic acid (4.6 $g \cdot kg^{-1}$ wet wt) were analyzed for pH, lactic acid, and other silage components *(33)*. The potential for continuous lactic acid production exists when there is a minimum of 10 $g \cdot L^{-1}$ reducing sugar present in the silage. The average pH of these samples was 3.87, and the average lactic acid concentration was 57 $g \cdot kg^{-1}$ dry matter.

In our work, 3.0 $g \cdot kg^{-1}$ wet wt of 60% formic acid were added to freshly field chopped sorghum-sudangrass. The pH of sorghum-sudangrass prior to the addition of formic acid was 5.3; after formic acid addition, the pH value was 4.6. In addition to ethanol, lactic acid was produced during the solid-state fermentation of sorghum-sudangrass. More lactic acid was produced in the control than in the forage receiving formic acid, yeast, and with or without cellulase (Fig. 3). L-lactic

Table 1
Ethanol Yield as Determined by Dividing Ethanol Obtained from Blender Extraction by the Total Sugars Available for Fermentation in Sorghum-Sudangrass[a]

Trial	Ethanol, kg/MT, wet wt	Total sugars, kg/MT, wet wt	Yield based on total sugars present at d 0	Percent of theoretical yield based on 0.510
-C,-Y,-F	0.0	63.5	0.000	0%
+C,+Y,+F	32.7	63.5	0.515	101%
-C,+Y,+F	27.1	63.5	0.427	84%

[a]The control is denoted by -C,-Y,-F. Sorghum-sudangrass receiving yeast, formic acid, and cellulase additives is +C,+Y,+F, whereas forage receiving formic acid and yeast additives but no cellulase is -C,+Y,+F.

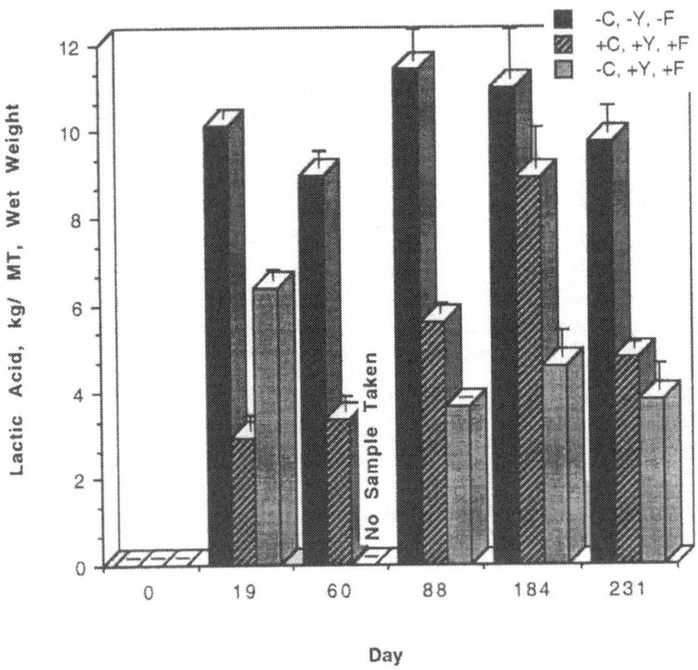

Fig. 3. Lactic acid concentrations in countercurrent extracts of fermented sorghum-sudangrass. Extracts were analyzed for lactic acid by HPLC using a Bio-Rad HPX-87H cation-exchange column. Sorghum-sudangrass control is designated as -C,-Y,-F. Forage receiving yeast, 75 m*M* of formic acid, and cellulase (5 IU/g dry matter) is designated as +C,+Y,+F, whereas forage receiving yeast and formic acid, but no cellulase, is designated as -C,+Y,+F.

acid concentrations are determined enzymatically. The concentrations and distribution of lactic acid at d 231 are given in Table 2. Currently, there is industrial interest in both D- and L-lactic acid for the production of poly-lactic acid *(34,35)*. In addition, the solid-state fermentation of sugar cane bagasse for the production of lactic acid has been investigated *(36)*.

Table 2
Distribution of Lactic Acid After D 231 of Fermentation in Countercurrent Extracts[a]

Trial	L-Lactic acid, kg/MT, wet wt	Total lactic acid, kg/MT, wet wt	L-Lactic acid, %
-C,-Y,-F	4.40	9.72	45.3
+C,+Y,+F	2.78	4.78	57.9
-C,+Y,+F	2.56	3.82	66.6

[a]The control is denoted by -C,-Y,-F. Sorghum-sudangrass receiving yeast, formic acid and cellulase additives is +C,+Y,+F, whereas forage receiving formic acid and yeast additives but no cellulase is -C,+Y,+F.

Sugars

Initially, 63.5 g sucrose/kg wet wt were present in the freshly chopped sorghum. Sucrose was hydrolyzed rapidly to glucose and fructose once sorghum was packed into the silos. These sugars were consumed in all three series throughout the fermentation. In the control series, 49.8% of the sugars were consumed in the first 19 d, as is seen in a typical lactic acid fermentation. Using the value of 11 g glucose/kg wet wt as the glucose remaining in the sorghum after 231 d of ensiling, 17.7% of the sugar remains in the control. The remainder of the glucose was consumed by the lactic acid bacteria over the 231-d fermentation period.

In both series receiving yeast and formic acid, the sugar content drops below levels seen in the control (Fig. 4). Total sugars are the summation of the concentrations of sucrose, glucose, fructose, and xylose at each respective time-point. Yeast consumes fermentable sugars present in the sorghum more rapidly and completely than the lactic acid bacteria utilize the sugars in the control series. This effect is seen at d 19 and remains constant until d 184. At this point, the glucose levels in the sorghum receiving cellulase surpass those seen in the control, indicating that the cellulase enzymes remain active through the 231 d of fermentation. Fructose is not detected in any of the three series after d 19. An increase in the xylose content in the series receiving cellulase may be owing to the continued action of hemicellulases, present in the Genencor cellulase, throughout the 231-d fermentation period (data not shown).

Countercurrent Extraction Efficiencies

Countercurrent technology provides an efficient way to extract sugars from sweet sorghum silage *(12,13)*. Countercurrent extraction also extracts byproducts (lactic acid, acetic acid, and ethanol). In our preliminary work, ethanol was extracted under ambient conditions and a solid:liquid ratio of 2:1. Extraction efficiencies are calculated by dividing the amounts obtained from countercurrent extraction by the amounts obtained from blender extraction (Table 3). A complete extraction would have an extraction efficiency of 1.00. It is assumed that blender extraction is complete and that the quantities found from blender extraction are representative of the total amount found in the solid substrate. Extraction efficiencies for L-lactic acid from sorghum-sudangrass fermented for 184 d are 0.41 and 0.50 for material with and without cellulase, respectively. Extraction efficiencies for

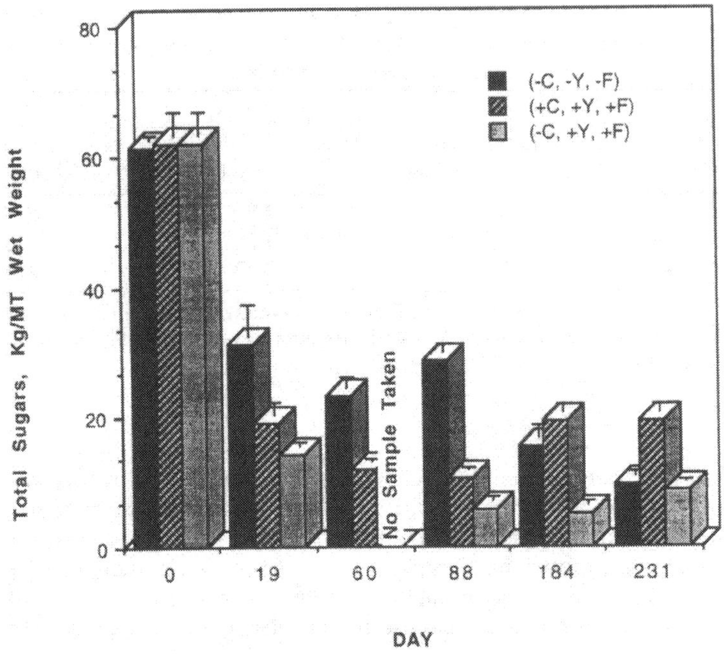

Fig. 4. Changes in sugar content with time suring the solid-state fermentation of sorghum-sudangrass to ethanol. Extracts were analyzed for sugars by HPLC using a Bio-Rad HPX-87P column. Sorghum-sudangrass control is designated as -C,-Y,-F. Forage receiving yeast, 75 m*M* of formic acid, and cellulase (5 IU/g dry matter) is designated as +C,+Y,+F, whereas forage receiving yeast and formic acid, but no cellulase, is designated as -C,+Y,+F.

Table 3
Extraction Efficiencies for the Extraction of Ethanol from Sorghum-Sudangrass Receiving Additives[a]

Day	-C,-Y,-F	+C,+Y,+F	-C,+Y,+F	Extraction temperature, °C
19	(No	0.69	0.75	19.5
60	ethanol	0.65		16.5
88	produced	0.66		16.3
184	until d 231)	0.76	0.72	16.5
231	0.80	0.45	0.63	20.0

[a]The control is denoted by -C,-Y,-F. Sorghum-sudangrass receiving yeast, formic acid, and cellulase additives is +C,+Y,+F, whereas forage receiving formic acid and yeast additives, but no cellulase is -C,+Y,+F. Extraction efficiencies are not available for d 60 and 88 for sorghum-sudangrass without cellulase.

ethanol and lactic acid were lower than those previously achieved at 70°C *(12)*. Average recovery efficiencies were 0.64 for ethanol (Table 3) and 0.45 for lactic acid. By varying temperature and liquid:solid ratios, ethanol and lactic acid should be extracted more efficiently.

Table 4
Estimated Product Value of Ethanol and Lactic Acid from the Solid-State Fermentation of Sweet Sorghum[a]

Treatment	Estimated product value for ethanol, $/MT, wet	Estimated product value for lactic acid $/MT, wet	Estimated cost of production, $/MT, wet	Estimated net product value for both ethanol and lactic acid, $/MT wet
-C,-Y,-F	$2.89	$53.48	$27.56	$28.81
+C,+Y,+F	29.74	24.92	32.48	22.18
-C,+Y,+F	24.67	21.56	30.88	15.35

[a]The control is denoted by -C,-Y,-F. Sorghum receiving yeast, formic acid, and cellulase additives is +C,+Y,+F, whereas forage receiving formic acid and yeast additives, but no cellulase, is -C,+Y,+F.

Economic Evaluation

From the economic analysis of the sorghum-sudangrass control, product values of $27.16, $1.61, and $2.88 could be attributed to lactic acid, ethanol, and wet pulp, respectively, for each metric ton of untreated sorghum-sudangrass. Sorghum-sudangrass has an estimated value of $27.56/wet MT. The (-C, +Y, +F) treatment produced less lactic acid and more ethanol than the control. This series has product values of $10.64, $12.27, and $2.88 for lactic acid, ethanol, and pulp, respectively.

Formic acid additive costs represent $1.60/wet ton at the levels used in the preliminary study. For the (+C, +Y, +F) treatment, product values of $13.44, $14.80, and $2.88 for lactic acid, ethanol, and pulp, respectively, could be reasonably expected. In addition to the raw material costs of $27.56/wet MT of silage, $1.73 for yeast *(32)*, and $1.60 for formic acid *(37)*, enzyme costs currently make the latter option unreasonable at the 5 IU/g dry matter treatment level. However, recent advancements in the solid-state production of fungal cellulases on extracted sweet sorghum pulp in our laboratory *(38)* could reduce the cellulase costs from $375.00/wet ton for commercial enzymes to $1.60/wet ton. After incorporating a value of $1.60 for cellulase, estimated product values, based on yield values for sorghum-sudangrass, are obtained by the solid-state fermentation of sweet sorghum shown in Table 4 are obtained.

Although initially the lactic acid in sweet sorghum ensiled without additives appears to be more valuable than the ethanol and lactic acid in sweet sorghum receiving additives, the cost of product recovery of lactic acid from countercurrent extracts has not been evaluated. Ethanol is recovered by distillation in this process. A complete evaluation of each unit operation in the process will determine the economic potential of our proposed system.

CONCLUSION

Ensiling is an economical means of storing and producing fermentable carbohydrates. We have shown that the silo also serves as a bioreactor for alcohol fuel production. The addition of yeasts and formic acid encourages an ethanolic fermentation within the silo. Ethanol production is 80% complete by d 19 of the fermentation. Since this was the first time-point taken in this study, it may be possible to process ethanol prior to d 19. Ethanolic sorghum silage is stable over a period of at least 230 d, thus potentially producing a low-cost feedstock for continuous ethanol production on a yearly basis.

Yeast additives compete with epiphytic (naturally occurring) microorganisms for the available sugars initially found in sorghum, thus using sucrose, glucose, and fructose for ethanol production. Yeasts are spoilage organisms in silage *(39–44)*. Spoilage occurs because ethanol is not a preservative in silage. When this silage is exposed to aerobic conditions, deterioration occurs in just a few days. However, sorghum-sudangrass receiving yeast and formic acid remained stable over 231 d of storage as long as anaerobic conditions were maintained (Fig. 2). Inhibition studies of acid byproducts and ethanol would provide useful insights into the metabolic activity of the yeasts present during the fermentation.

Separation and recovery of ethanol and lactic acid did not fall within the scope of this research. Countercurrent technology is widely used in the sugar industry; therefore, a broad knowledge base is available. Future work should consider product recovery from the countercurrent liquid extracts. Recovery of lactic acid from these extracts could involve distillation to remove ethanol and volatile fatty acids followed by ion-exchange *(45)* or liquid membrane extraction *(46,47)* to remove lactic acid from the distillery bottoms.

We have addressed the feasibility of *in situ* ethanol production using sorghum-sudangrass as the substrate. Using the silo as a bioreactor in the solid-state fermentation of sorghum as well as for the subsequent storage of the ethanolic material prior to processing is inexpensive. We would expect to double ethanol and lactic acid yields using sweet sorghum instead of sorghum-sudangrass. The utilization of sweet sorghum as an ethanol feedstock will provide renewable biomass equivalent to conventional substrates, such as corn and barley. Sweet sorghum would complement, rather than compete with corn as a substrate for fuel alcohol production, thus adding to the diversity of renewable resources that can be produced for biofuels. Lactic acid may become a valuable byproduct in this system.

ACKNOWLEDGMENTS

This work was sponsored by the National Renewable Energy Laboratory Subcontract No. XC-1-11142-1 under the title of Evaluation of Feedstock Pretreatability and Saccharification/Fermentation Suitability as a Function of Harvesting and Storage. Partial support was received from the Colorado Institute for Research in Biotechnology in the form of a graduate fellowship and from the Colorado State University Agricultural Experiment Station Project No. 622.

REFERENCES

1. Josephson, J. (1991), *Environ. Sci. Technol.* **26(4),** 660–663.
2. Valenti, M. (1992), *Mech. Eng.* December, 23,24.
3. Lynd, L. R., Cushman, J. H., Nichols, R. J., and Wyman, C. E. (1991), *Science* **251,** 1318–1323.
4. Walker, J. N., DeShayer, J. A., and Barfield, B. J. (1992), *Agricultural Eng.* November, 24–26.
5. Sipilä, K., Johansson, A., and Saviharju, K. (1993), *Bioresource Technology* **43,** 7–12.
6. Belletti, A., Petrini, C., Minguzzi, A., Landini, V., Piazza, C., and Salamini, F. (1991), *Maydica* **36,** 283–291.
7. Smith, G. A. and Buxton, D. R. (1993), *Bioresource Technology* **43,** 71–75.

8. Lueschen, W. E., Putnam, D. H., Kanne, B. K., and Hoverstad, T. R. (1991), *J. Prod. Agric.* **4(4),** 619–625.
9. Worley, J. W., Vaughan, D. H., and Cundiff, J. S. (1992), *Bioresource Technol.* **40,** 263–273.
10. Hinman, N. D., Schell, D. J., Riley, C. J., Bergeron, P. W., and Walter, P. J. (1992), *Appl. Biochem. and Biotechnol.* **34/35,** 639–649.
11. Coble, C. G., Shimulevich, I., and Egg, R. P. (1983), *American Society of Agricultural Engineers Paper No. 83-3563.*
12. Noah, K. S. and Linden, J. C. (1989), *Trans. ASAE* **32(4),** 1419–1425.
13. Noah, K. S. and Linden, J. C. (1989), *Trans. ASAE* **32(4),** 1426–1430.
14. Kargi, F., Curme, J. A., and Sheehan, J. J. (1985), *Biotechnol. Bioeng.* **27,** 34–40.
15. Gibbons, W. R., Westby, C. A., and Dobbs, T. L. (1986), *Appl. and Environ. Microbiol.* **51(1),** 115–122.
16. Worley, J. W. and Cundiff, J. S. (1991),*Trans. of the ASAE* **34(2),** 539–547.
17. Linden, J. C., Henk, L. L., Murphy, V. G., Smith, D. H., Gabrielson, B. C., Tengerdy, R. P., and Czako, L. (1987), *Biotechnol. Bioeng.* **30,** 860–867.
18. Henk, L. L. and Linden, J. C. (1992), *Enzyme Microbiol. Technol.* **14,** 923–930.
19. Barnett, A. J. G. (1954), *Silage Fermentation.* Butterworths Publications Ltd., London.
20. McDonald, P. (1981), *The Biochemistry of Silage.* John Wiley, Chichester.
21. Henk, L. L. and Linden, J. C. (1994), in *Enzymatic Conversion of Biomass for Fuels Production.* Himmel, M. E., Baker, J. O., Overend, R. P., eds. ACS Symposium Series 566, pp. 391–410.
22. Beck, T. (1978), in *Fermentation of Silage—A Review,* McCullough, M. E., ed., National Feed Ingredients Assoc., Iowa, pp. 63–115.
23. Lindgren, S., Lingvall, P., Kaspersson, A., de Kartzow, A., and Rydberg, E. (1983), *Swedish J. Agricultural Res.* **13,** 91–100.
24. Narasimhalu, P., Halliday, L. J., Sanderson, J. B., Kunelius, H. T., and Winter, K. A. (1992), *Can. J. Anim. Sci.* **72,** 431–434.
25. Vanhatalo, A., Varvikko, T., and Aronen, I. (1992), *Agricultural Sci. Finl.* **1,** 163–175.
26. Tengerdy, R. P., Weinberg, Z. G., Szakacs, G., Wu, M., Linden, J. C., Henk, L. L., and Johnson, D. E. (1991), *J. Sci. Food Agriculture* **55,** 215–228.
27. Weinberg, Z. G., Ashbell, A., Azrieli, A., and Brukental, I. (1993), *Grassland Forage Sci.* **48,** 70–78.
28. Ghose, T. K. (1987), *Pure Appl. Chem.* **59(2),** 257–268.
29. Morrison, I. M. (1979), *J. Agricultural Sci. Camb.* **93,** 581–586.
30. Martinsson, K. (1991), *Swedish J. Agricultural Res.* **21,** 121–130.
31. Mohagheghi, A,, Tucker, M., Grohman, K., and Wyman, C. (1992), *Appl. Biochem. Biotechnol.* **33,** 67–81.
32. Personal communication with RedStar Sales Representative, Universal Foods Corporation, Milwaukee, WI, 53201, January 6, 1995.
33. Moisio, T. and Heikonen, M. (1994), *Anim. Feed Sci. Technol.* **47,** 107–124.
34. Mikos, A. G., Thorsen, A. J., Czerwonka, L. A., Bao, Y., Langer, R., Winslow, D. N., and Vacanti, J. P. (1994), *Polymer* **35(5),** 1068–1078.
35. Soccol, C. R., Marin, B., Raimbault, M., and Lebeault, J.-M. (1994), *Appl. Microbiol. Biotechnol.* **41,** 286–290.
36. Xavier, S. and Lonsane, B. K. (1994), *Appl. Microbiol. Biotechnol.* **41,** 291–295.
37. *Chemical Marketing Reporter* for the week ending December 9, 1994, December 12.
38. Castillo, M. R., Gutierrez-Correa, M., Linden, J. C., and Tengerdy, R. P. (1994), *Biotechnol. Lett.* **16 (9),** 967–972.
39. Selmer-Olsen, I., Henderson, A. R., Robertson, S., and McGinn, R. (1993), *Grass and Forage Sci.* **48,** 55–63.
40. Muck, R. E., Pitt, R. E., and Leibensperger, R. Y. (1991), *Grass and Forage Sci.* **46,** 283–299.
41. Muck, R. E. and O'Kiely, P. (1992), *J. Sci. Food Agric.* **59,** 145–149.
42. Pitt, R. E., Muck, R. E., and Pickering, N. B. (1991), *Grass and Forage Sci.* **46,** 301–312.
43. O'Kiely, P. and Muck, R. E. (1992), *J. Sci. Food Agriculture* **59,** 139–144.
44. Moon, N. (1983), *J. Appl. Bacteriol.* **55,** 453–460.
45. Vaccari, G., González-Varay, R., A., Campi, A. L., Dosi, E., Brigidi, P., and Matteuzzi, D. (1993), *Appl. Microbiol. Biotechnol.* **40,** 23–27.
46. Lazarova, Z. and Peeva, L. (1994), *Biotechnol. Bioeng.* **43,** 907–912.
47. Lazarova, Z. and Peeva, L. (1994), *J. Biotechnol.* **32,** 75–82.

0273-2289/96/5758—0503$8.25

Continuous and Simultaneous Fermentation and Recovery of Lactic Acid in a Biparticle Fluidized-Bed Bioreactor

ERIC N. KAUFMAN,* STANLEY P. COOPER, MARIA K. BUDNER, AND GERALD R. RICHARDSON

Bioprocessing Research and Development Center, Chemical Technology Division, Oak Ridge National Laboratory, Oak Ridge, TN 37831-6226

ABSTRACT

A continuous biparticle fluidized-bed reactor (BFBR) is developed for the simultaneous fermentation and recovery of lactic acid. In this processing scheme, bacteria are immobilized in gelatin beads and are fluidized in a columnar reactor. Solid particles (weak-base resin IRA-35) with sorbent capacity for the product are introduced at the top of the reactor and fall countercurrently to the biocatalyst, effecting *in situ* removal of the inhibitory lactic acid while also controlling reactor pH at optimal levels. One-week-long fermentation trials using immobilized *Lactobacillus delbreuckii* with sorbent addition demonstrated a volumetric productivity (6.9 g/L·h) at least 16-fold higher than that of a free-cell batch fermentation with base pH control and identical biomass concentration and medium composition. Regeneration of the loaded sorbent from the BFBR has effected a 35-fold concentration of lactic acid compared with original levels in the fermentation broth (70 vs 2 g/L). Lactic acid concentrations as high as 610 g/L have been observed when the loading solution contained 50 g/L lactic acid. Rich medium formulations did not seem to increase BFBR performance. The benefits of this reactor system, as opposed to conventional batch fermentation, are discussed in terms of productivity and process economics.

Index Entries: Lactic acid fermentation; resin; adsorbent; biparticle fluidized bed; continuous.

INTRODUCTION

Lactic acid is a specialty chemical used in the food industry for the manufacture of cheeses, pickles, and yogurt, and also as a preservative *(1)*. It may also be used as feed in plastics production and in the synthesis of other organic acids,

*Author to whom all correspondence and reprint requests should be addressed.
†"The submitted manuscript has been authored by a contractor of the US government under contract DE-AC05-84OR21400. Accordingly, the US government retains a nonexclusive, royalty-free license to publish or reproduce the published form of this contribution, or allow others to do so, for US government purposes."

acrylic acids, acetaldehyde, and ethanol *(2)*, with a current price of $1.15/lb for technical (88%) or food-grade lactic acid *(3)*. Annual US consumption of lactic acid totals over 30 million pounds *(4)*. Biological production of lactic acid accounts for approx 50% of the world's production *(5)*. Biological production is complicated primarily owing to economic considerations arising from product inhibition and the required downstream processing of dilute aqueous product streams. It is estimated that lactic acid purification accounts for 50% of production cost *(6)*. The standard method of biological lactic acid production is the anaerobic fermentation of glucose or sucrose by *Lactobacillus* in batch reactors. The conventional process requires that the base be added to the reactor to control pH *(4,7)*, and/or that calcium carbonate be added to buffer the reactor and to precipitate the lactate *(8,9)*. These processes produce the lactate salt, which must be reacidified (usually by sulfuric acid *[8–10]*), and which yields sulfates, further adding to process chemical costs and waste streams. The lactic acid productivity of these conventional processes are reported in the open literature to be ~1.6 g/L·h *(8)*.

It is postulated that it is the protonated form of lactic acid that is inhibitory to the fermentation *(5,11–13)*. Based on the kinetic modeling of *L. delbreuckii* by Yeh et al. *(11)*, Kaufman et al. *(13)* demonstrated how pH control alone is insufficient to ensure maximal productivity, owing to the small but inhibitory amount of protonated lactic acid that exists even with pH control to 6.0. For instance, according to the kinetic constants of Yeh et al., the production rate of lactic acid is reduced by 50% when the broth concentration of lactic acid reaches 40 g/L at pH 6.0. At a broth concentration of 90 g/L, the fermentation is operating at only 10% of its maximum production rate. It is evident that bioreactors employing *in situ* product removal have the potential to realize great improvements in reactor productivity compared to conventional batch production of lactic acid with pH control alone.

Improvements in volumetric productivity for the lactic acid fermentation have been reported by increasing biomass loading and by reducing product inhibition. High cell densities in reactors have been achieved by the use of cell aggregates *(14)*, growth of biofilm on activated carbon *(15)*, cell immobilization in gelatin beads (for example, *16–19*), cell growth in hollow fibers *(20)*, and cell recycle using hollow fiber membranes *(21,22)*. Product inhibition has been reduced by moving toward a plug flow reactor system through the use of staged CSTR reactors *(23,24)*, such that the product concentration is reduced in the initial stages and inhibition becomes relevant in a smaller portion of the reactor volume. Removal of inhibitory product has been achieved using both liquid extractants *(10,19,25–27)* and solid adsorbents *(25,28)*, either in a product-stripping side stream *(29,30)* or by adding directly to the constantly stirred tank reactor (CSTR) *(31)*. Product removal using ultrafiltration *(22)* and electrodialysis have also been investigated (*see* ref. *27* for a review).

Combining the benefits of cell immobilization and *in situ* product separation, Davison and Thompson *(32)* demonstrated the simultaneous fermentation and separation of lactic acid in a BFBR. In this process, *L. delbreuckii* was immobilized in alginate beads and fluidized by the upflowing liquid media in a tubular reactor. Such fluidized beds have been shown to increase the productivity of fermentations for a variety of processes *(33,34)*. In the demonstration by Davison and Thompson *(32)*, the polyvinyl pyridine resin Reillex 425, which possesses affinity for the lactic acid, was added batchwise to the top of the reactor, fell through the biocatalyst bed, and was found to moderate reactor pH, adsorb the lactic acid product, and increase lactic acid production nearly fourfold over a controlled fluidized-bed reactor without resin addition or pH control. In this proof-of-concept experiment, the resin was added batchwise for a short period of time (7 h) and was added at a constant schedule. Although resin addition moderated the decrease in reactor pH, it did not control the pH at the optimal level. Recently, Kaufman et al. *(35)* screened a series of resins for use in a lactic acid BFBR and demonstrated the potential economic benefits of BFBR as opposed to batch production *(13)*. Two-day-long fermentations in the BFBR were conducted with manual resin addition and were compared to similar runs without pH control or resin addition *(13)*. A 12-fold increase in productivity was reported in BFBR as opposed to immobilized cell fermentations without pH control. In this article, we describe the further demonstration of the BFBR by conducting longer term fermentation trials and providing more relevant control runs in which pH control is utilized in free and immobilized cell reactors with the same media and cell concentrations as the BFBR. This enables the assessment of the impact of *in situ* product removal and cell immobilization on reactor productivity. We also compare the use of rich vs lean media recipes in the BFBR.

METHODS

L. delbreuckii NRRL B445, now being reported as *Lactobacillus rhamnosus (22)* and *Lactobacillus casei* subsp. *rhamnosus (27)*, was used as the biocatalyst in all experiments. Medium for both large-scale biomass growth and cell passaging contained 10 g/L Sheftone T, 5 g/L KH_2PO_4, 5 g/L KH_2PO_4, 0.5 g/L $(NH_4)_2SO_4$, 0.3 g/L $MgSO_4$, 0.002 g/L $FeSO_4$, and 0.10 g/L NaCl. Sheftone T (Sheffield Products, Norwich, NY) is a low-cost alternative nitrogen source to yeast extract. Sheftone T has a 9.8% total nitrogen content with 1.3% amino nitrogen compared to 10.9 and 6.0% for the total and amino nitrogen content of yeast extract. Glucose concentrations varied between 10 and 100 g/L, depending on the expected duration of growth in the given vessel. The media were sterilized by autoclaving and had an initial pH of 6.0. For biomass harvesting in bead production, inocula were grown in Fernbach flasks at 40°C for 2–4 d, after which the cells were used to inoculate a New Brunswick 75-L fermentation system. The system was allowed to ferment for a period of about 3 d, after which the cells were harvested and concentrated using a continuous centrifuge. The resulting paste was washed with sterile water and recentrifuged before adding to the gelling solution below. This step decreased gel viscosity and improved the production of the beads. The bead production method has been described in detail elsewhere *(36)*. Briefly, the bacteria were immobilized into uniform gel beads, 0.7 mm in diameter, using 1% low-viscosity alginic acid (A-2158, Sigma, St. Louis, MO) and stabilized in a 0.2*M* $CaCl_2$ solution. To find the average number of cells per volume of beads, approx 1 mL of beads was placed in 3 mL of water to determine the volume of beads by displacement. Sodium citrate (0.5 g) was then added to dissolve the beads. Cells were counted microscopically using a Petroff-Hausser counting chamber. Cell counts were typically 4×10^{10} cells/mL of bead, or 0.067 g dry wt/mL.

The reactor utilized in the fluidized-bed fermentations has been described previously *(13)* and is shown schematically in Fig. 1. Briefly, the water-jacketed reactor tapered from an inner diameter of 1/2 in. at its base to 1 in. at 23 cm above the liquid entrance. This tapered region at the bottom allowed for efficient disengagement of the denser adsorbent from the biocatalyst beads. A reservoir was added at the reactor base for collecting adsorbent particles. The working volume of

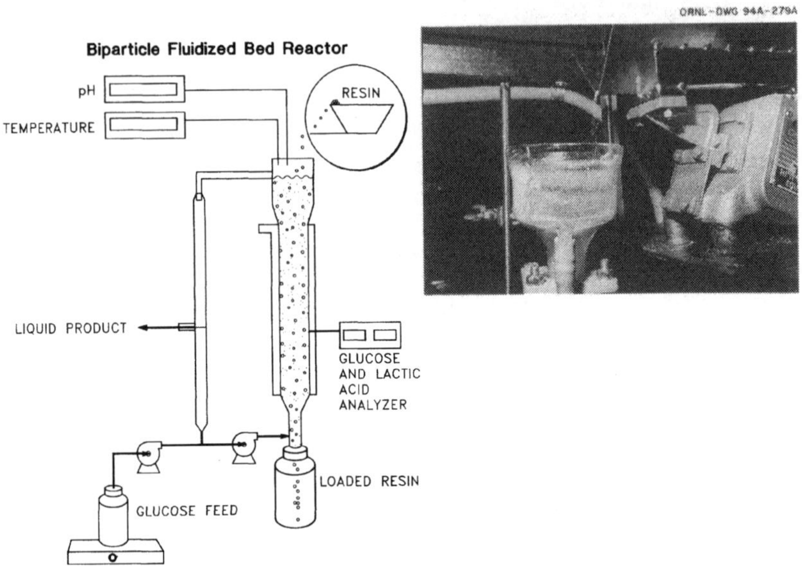

Fig. 1. Schematic of the BFBR. Inset: Resin delivery system in the BFBR. A pH controller located at the reactor apex actuated a vibratory hopper that delivered resin into the BFBR at a controlled rate to affect pH control and *in situ* product removal.

the reactor was 0.7 L, of which 0.2 L was occupied by the biocatalyst. However, surge and resin collection flasks made the total volume of the reactor 2.6 L. A peristaltic pump fluidized the biocatalyst by the upflow of recirculating media at a rate of 50 mL/min. A nitrogen sparge was added to the reactor base with a flow rate of 15 mL/min to improve resin/biocatalyst disengagement. In the envisioned larger scale use of the BFBR, the glucose will be converted in a single pass, and medium recycle will not take place. However, owing to the prohibitively large volume of medium that would be needed to operate this laboratory-scale reactor in a once past medium mode (72 L/d) and the low conversion rate achieved in a single pass, we chose to operate the present reactor in a recycle mode with a small "make-up" feed. The medium that was originally loaded into the BFBR with the biocatalyst beads and recycled during the course of a given experiment consisted of 5 g/L glucose, 10 g/L Sheftone T, 0.5 g/L $(NH_4)_2SO_4$, 0.3 g/L $MgSO_4$, 0.1 g/L $CaCl_2$, 0.002 g/L $FeSO_4$, and 0.1 g/L NaCl. Prior to re-entering the reactor, the media were replenished with glucose via a peristaltic pump drawing from a reservoir containing 250 g/L glucose, 5 g/L Sheftone T, 0.5 g/L $(NH_4)_2SO_4$, 0.3 g/L $MgSO_4$, 0.1 g/L $CaCl_2$, 0.002 g/L $FeSO_4$, and 0.1 g/L NaCl. The glucose level in the reactor was controlled by manipulating the rate of this pump (when the pump was on, the flow rate was typically <0.25 mL/min). The reactor was monitored for pH and temperature by probes placed at the reactor apex. Glucose and lactic acid levels were monitored through the automated sampling allowed by a model 2700 Dual Channel Biochemistry Analyzer (Yellow Springs Instrument Co., Yellow Springs, OH). Bead cell counts as well as the bed height were monitored on a daily basis.

In one set of reactor trials, the columnar reactor with immobilized cells was operated in the "biparticle mode" with continuous, automated addition of sorbent particles added to maintain the reactor pH at 6.0 while simultaneously removing lactic acid. For this operation, a pH probe at the reactor apex was attached to a pH controller (Chem Cadet, Cole Parmer, Niles, IL), which, in turn, actuated a vibrating tray/hopper delivery system (*see* Fig. 1 inset). This tray and hopper were custom designed and manufactured by Eriez Magnetics (Erie, PA) to meet the particularly challenging specifications of low volumetric feed of a self-agglomerating material. When acutated, the hopper delivered resin at a rate of 2.2 dry g/min. The selection of a resin sorbent for this application has been discussed previously *(13,35)*. The resin sorbent used in these studies was Amberlite IRA-35 (Rohm & Haas, Philadelphia, PA), a macroreticular weak-base anion exchanger with tertiary amine functionality in an acrylic-divinylbenzene matrix. This resin possessed a capacity of ~0.2 g/g under the conditions described. The resin was dried under a heat lamp prior to use in the hopper system in order to convey the material in a controlled, reproducible manner. The resin quickly hydrated on addition to the reactor, and this drying process did not affect the resin's capacity. The residence time of a given resin particle under reactor conditions was typically 30 min. As described previously *(13)*, resin removed from the BFBR was regenerated by successive washings with 2*M* sulfuric acid. The volume of eluent and its concentration were recorded on each wash, and the total lactic acid bound to the resin was thus calculated. The resin regenerated in this manner was for the purpose of quickly tabulating the total amount of lactic acid produced and not necessarily for achieving the highest eluent concentration possible. Regeneration experiments designed to achieve high eluent concentration are described in the Results section. In order to assess how the selection of media components would affect lactic acid production in the BFBR, one fermentation trial was conducted with a "rich" media formulation as opposed to the rather lean mixture listed previously. Reactor operation was as described above, except that the reactor media contained 10 g/L pancreatic digest of casein, 10 g/L glucose, 5 g/L yeast extract, 3 g/L tryptose, 1 g/L sodium acetate, 0.2 g/L L-cysteine, 1 mL/L Tween-80, 0.6 g/L $MgSO_4$, 0.2 g/L $MnSO_4$, 0.3 g/L $FeSO_4$, and 0.1 g/L $CaCl_2$. Since the reactor was operated with rich media for a period of about 24 h, no make-up glucose feed was used in this trial.

The columnar reactor was also operated with pH control, but with no *in situ* product removal in order to assess better how product removal can augment productivity in addition to the augmentation afforded by pH control alone. This reactor run utilized the same biocatalyst and normal or "baseline" media recipe used in the BFBR experiment. The pH controller, however, rather than actuating the resin delivery system, actuated a peristaltic pump, which delivered 4*M* ammonium hydroxide.

A control fermentation was performed in a batch, 75-L free-cell reactor (New Brunswick Scientific, Edison, NJ), with pH control via the addition of 4*M* ammonium hydroxide. With the, exception of 100 g/L of glucose, the media were identical to that loaded into the BFBR. In order to operate this reactor with a biocatalyst density similar to that used in the immobilized cell reactors, the batch reactor was operated for a week, and cells were harvested and reintroduced into fresh media in the 75-L reactor. In this manner, the initial cell concentration for the batch reactor run was 5×10^9 cells/mL. Liquid samples were removed daily to allow for cell counts as well as glucose and lactic acid measurements.

Table 1
Media Formulations Used in Screening Experiments[a]

Component, g/L	#1	#2	#3	#4	Baseline
Glucose	20	20	20	20	20
Sheftone T	10	10	10	10	10
$(NH_4)_2SO_4$	0.5	0	0.5	0	0.5
$MgSO_4$	0.3	0.2	0.3	0.3	0.3
$CaCl_2$	0.1	0.1	0.1	0.1	0.1
Peptone	0	10	0	10	0
Sodium acetate	0	5	0	5	0
$MnSO_4$	0	0.1	0.1	0	0
$FeSO_4$	0	0.002	0.002	0	0.002
NaCl	0	0.1	0.1	0	0.1

[a]Four different media formulations, varied according to their salt, protein, and buffer content, were assayed along with the baseline media formulation used in the BFBR. Concentrations of the various components are given in grams/L.

An experiment was also performed with immobilized biocatalyst in serum bottles with various media recipes in order to screen these recipes for use in the BFBR. In this experiment, beads were prepared as described, and 8 g of beads were placed in duplicate 125-mL serum bottles containing 75 mL of the medium formulations shown in Table 1. Flasks were incubated in a rotary shaker at 40°C and 80 rpm, and were assayed daily for glucose and lactic acid concentrations.

RESULTS

Plots of broth lactic acid concentration and cumulative lactic acid produced for the three fermentation runs in different reactor arrangements with the baseline media recipe are shown in Fig. 2 (A and B). The data in Fig. 2B have been normalized to reflect the lactic acid that would be produced in a reactor with an active volume of 0.7 L in order to compare the productivities of various sized reactors. In the batch free-cell bioreactor, the high initial cell loading of 5×10^9 cells/mL plateaued at 1×10^{10} cells/mL after 1 d. The lactic acid concentrations in the broth approached a level of 80 g/L after nearly 10 d of operation. The reactor's productivity was 0.55 g/L·h after 66 h, 0.30 at 162 h, and 0.23 after 234 h of fermentation.

The immobilized-cell reactor with chemical pH control and no lactic acid removal was operated for a period of 62 h. During this time, a lactic acid concentration of 15 g/L was reached, with a reactor productivity of 1.30 g/L·h. The fact that the broth concentration was lower in this case than the free-cell reactor is attributable to the low ratio of active reactor volume to total reactor volume. This allows dilution of the lactic acid produced in the active volume. The cell density remained constant at 4×10^{10} cells/mL biocatalyst, and there were 200 mL of biocatalyst in the reactor; thus, the immobilized- and free-cell reactors had nearly the same amount of biomass per working reactor volume (7×10^{12} and 8×10^{12} cells/700 mL working volume for the immobilized and free-cell reactors, respectively).

The BFBR was operated with continuous resin addition for a period of 1 wk. During this time, the broth lactic acid concentration was kept below 2 g/L by adsorption onto the resin. This simultaneous fermentation and separation minimized the

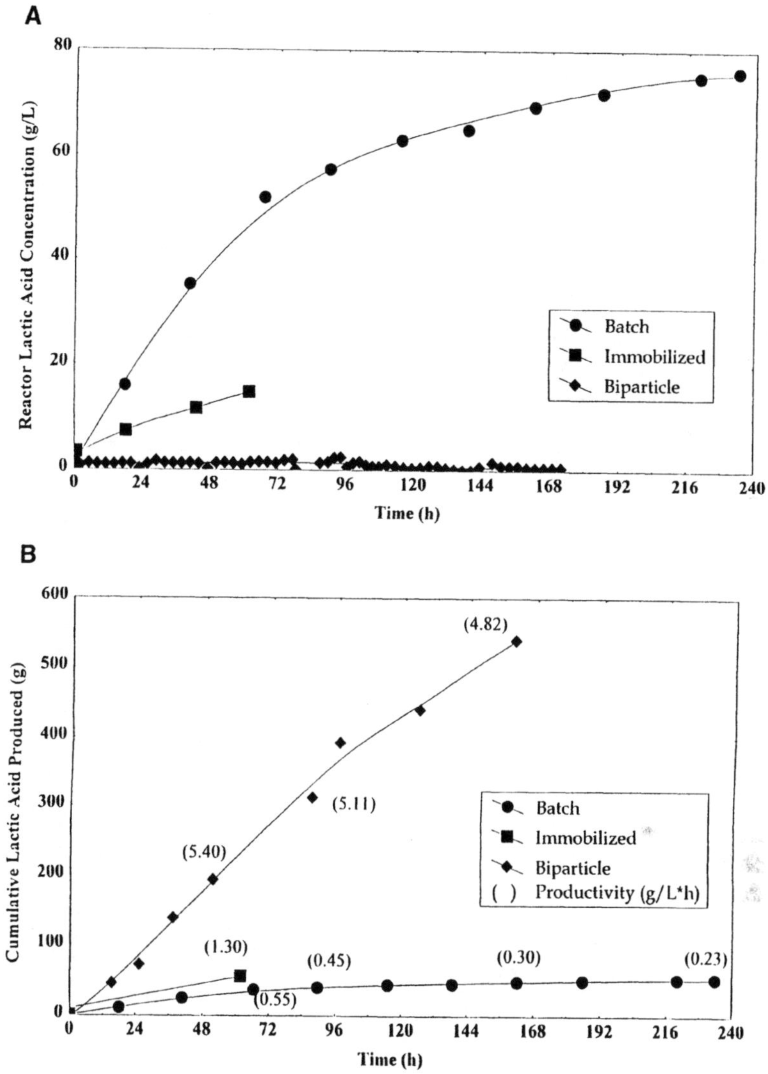

Fig. 2. Fermentation trials in free-cell batch, immobilized-cell fluidized-bed, and BFBRs for the production of lactic acid. **(A)** Broth concentrations of lactic acid as a function of time. The BFBR maintains reactor pH and low-broth concentrations to decrease product inhibition. **(B)** Cumulative lactic acid produced as a function of time. The BFBR outperforms the free-cell fermentation by a factor of 16. Reactor productivities (g/L·h), are shown in parentheses at various times during the fermentation.

product inhibition that occurred in the above fermentation schemes even when the pH of the reactor was controlled to pH 6.0. The cell density in the biocatalyst remained constant during the reactor run at 4×10^{10} cells/mL biocatalyst, and there were 200 mL of biocatalyst in the reactor. Typically, 19 dry g/h of resin were added to the reactor to maintain the pH at 6.0. On regeneration of the resin with 2*M*

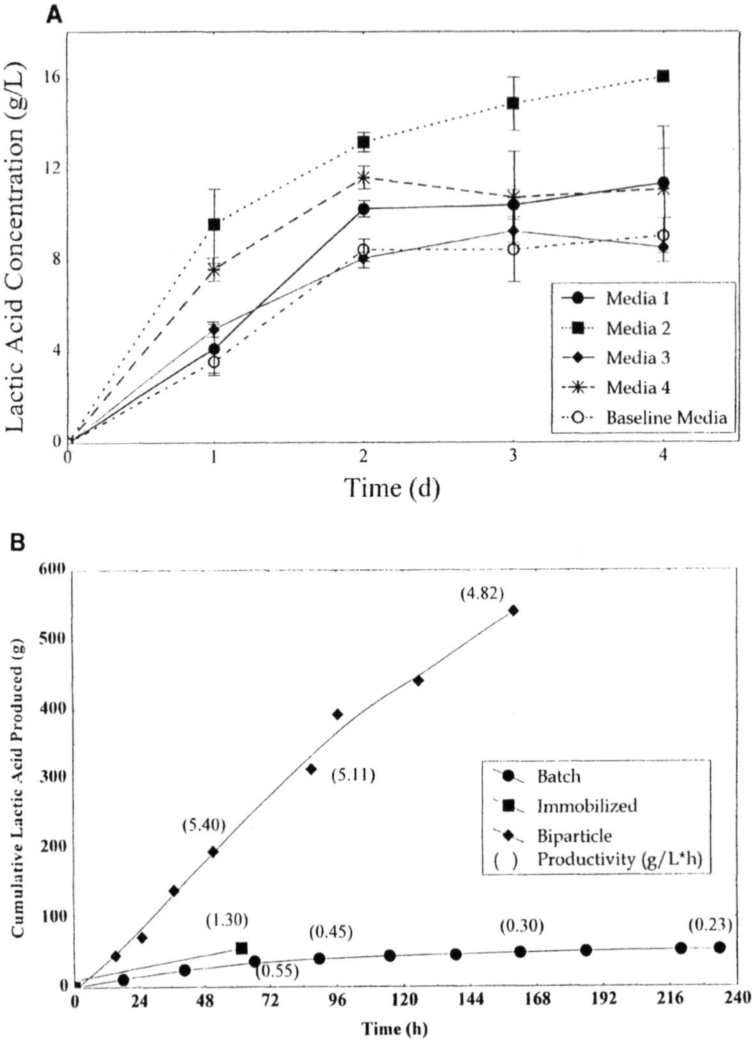

Fig. 3. Lactic acid production by biocatalyst beads in serum bottles as a function of media composition. In conditions in which the pH is not controlled and the inhibitory product is not removed, media formulations with increased buffer and amino acid content yield improved lactic acid production when compared to our baseline medium.

sulfuric acid, product concentrations as high as 70 g/L were achieved, comparable to those obtained in batch fermentations. When the cumulative lactic acid produced as a function of time is studied for this reactor, it is seen that the BFBR far outperforms the conventional bioprocessing schemes. The BFBR exhibited nearly constant volumetric productivity with 5.4 g/L·h after 51 h, and 4.82 after 160 h of operation.

As seen in Fig. 3, "rich" media formulations containing additional buffers and protein extracts resulted in improved lactic acid production when immobilized biocatalyst was incubated in serum bottles without pH control or *in situ* product

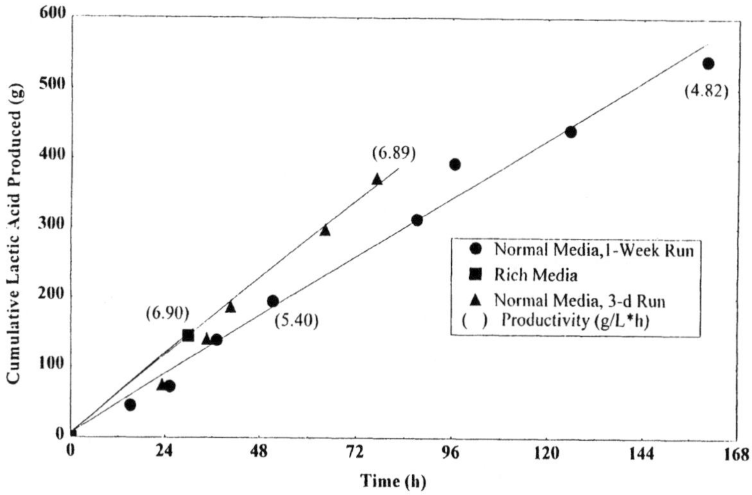

Fig. 4. Cumulative lactic acid produced as a function of time in the BFBR when operated with rich, as opposed to baseline, media formulations. Media recipes are described in the text. Although rich media recipes containing additional buffers and amino acids augment production in a system that is strongly inhibited, they do not seem to impact production in a BFBR in which the pH is controlled and the inhibitory product is continuously removed.

removal. One reason for this improvement is obviously the increased buffering capacity of the media, which allows the biocatalyst to exist at the optimal pH for a longer period of time. It also appears that constituents, such as peptone and tryptose, serve to protect the biocatalyst from the inhibitory effects of lactic acid, which continue to persist even when the pH of the reactor is controlled. Figure 4 demonstrates the cumulative lactic acid produced and the volumetric productivity achieved when the BFBR was operated on rich, as opposed to normal, medium formulations. The three fermentation trials were all performed with identical biocatalyst concentrations. Although the productivity of the BFBR for the 30-h run on rich medium (6.9 g/L·h) was higher than that achieved in the week-long run with normal medium (4.8 g/L·h), this augmentation is believed to be insignificant, since additional fermentation trials with normal medium also exhibited productivities of 6.9 g/L·h (*see* Fig. 4).

Although the BFBR does offer several advantages over batch fermentation in terms of product separation and purification, more research on the selection of an optimal adsorbent and adsorbent regeneration processes is needed in order to realize the full potential of the BFBR system. Owing to the *in situ* product removal, the BFBR does not require separation of the fermentation broth from the biomass as required by batch fermentations. Using a packed column of IRA-35, we have previously demonstrated a 68-fold increase in product-to-substrate ratio over the inlet solution, and a 1734-fold increase over the ratio found in the column effluent *(13)*. In the current study, we have attempted to minimize the volume of eluent in order to achieve higher product concentrations.

To simulate regeneration of resin that is loaded with BFBR fermentation broth, 220 g of IRA-35, as supplied by the manufacturer, were loaded into a 1-in. id packed column and contacted with a single pass (1 L at 5 mL/min) of medium at 40°C.

The baseline media as described was "spiked" with 5 g/L glucose and 5 g/L lactic acid to approximate the concentrations that may be present in the broth from a BFBR. The column was washed with a single volume of water at 40°C followed by 40 mL (at 5 mL/min) of 2*M* H_2SO_4 at 90°C. These 40 mL represent 80% of the column void volume. Eluent was collected in 2-mL fractions and reached a peak concentration of 70 g/L lactic acid, a 14-fold concentration over the loading solution. This single pass of sulfuric acid eluted 75% of the lactic acid loaded onto the column. In order to demonstrate more fully the resin's ability to concentrate lactic acid, this above experiment was repeated using media that contained 10 g/L glucose and 50 g/L lactic acid. On regeneration, lactic acid concentrations reached as high as 610 g/L (~50% [w/v]), a 12-fold concentration of the loading solution. This experiment yielded 72% of the lactic acid originally loaded onto the column. Clearly more research is needed in the area of resin selection and regeneration. Resins with high capacity and specificity are needed in order to decrease the volume of resin required in the reactor at a given time. Improved chromatographic-like techniques need to be developed in order to effect increased product concentrations to decrease downstream processing costs.

DISCUSSION

This study has demonstrated that the BFBR is a viable long-term, continuous method of lactic acid production with several advantages over conventional batch fermentation. The BFBR exhibited volumetric productivities at least 16-fold higher than that of a free-cell reactor with identical biomass concentration and medium composition. On resin regeneration, the BFBR yielded identical product concentrations as the free-cell reactor and has the potential to achieve much higher concentrations with reduced processing cost. Rich media formulations are not needed in the BFBR system, for the cells are already protected from the inhibitory product owing to *in situ* product removal. The BFBR offers considerable economic advantage over batch processing owing to the increased reactor productivity and continuous nature of the process.

The BFBR was operated continuously for a period of 1 wk and demonstrated a productivity that was 16-fold higher than that of a free-cell, pH-controlled reactor with identical media composition and biomass concentration operated for the same period of time. The BFBR exhibited a productivity that was fourfold higher than a pH-controlled, immobilized-cell reactor, again with identical biomass concentration and media composition, but without *in situ* product removal. Although the immobilized-cell reactor appears to have a higher productivity than the free-cell reactor, we believe that this is owing to the "buffering capacity" afforded by the dead volume of the immobilized-cell reactor. The columnar reactor had a total volume of 2.6 L with only 0.2 L of biocatalyst. This dead volume served to dilute the lactic acid produced (15 g/L as opposed to 80 g/L for the free-cell reactor) and minimize product inhibition in this manner. It appears that immobilization of the biocatalyst does not have an adverse effect on activity, in that the biocatalysts in the immobilized-cell reactor are at least as productive as their counterparts in the free-cell reactor. Yield measurements were difficult to obtain in the immobilized-cell systems owing to the aseptic production of the beads and eventual reactor contamination owing to the open configuration of the resin addition system. Product yields on glucose for *L. delbreuckii* have been reported between 74 and 90% *(16,30,37)*. The batch free-cell fermentation reported here exhibited a product yield on glucose of 85%.

Rich media formulations may be considered to assuage product inhibition that exists even when the pH of the reactor is controlled. These rich media contribute significant additional chemical cost to the process, but may pay for themselves in terms of increased productivity. Although rich media were demonstrated to cause increased lactic acid production in serum bottles with no pH control or product removal, these constituents did not increase productivity in the BFBR, presumably because product inhibition was already alleviated owing to the *in situ* product removal afforded by the resin. Since the low-cost medium recipes provide competitive reactor productivities, the BFBR is seen to possess a potential economic advantage in terms of decreased chemical costs.

Since product separation and purification accounts for 50% of lactic acid production costs, the output product concentration of a given fermentation process has tremendous impact on the economic viability of the process. In this article, we demonstrated that the product stream from the BFBR is as concentrated as that from the free-cell batch fermentation (70 g/L). We have also demonstrated that concentrations as high as 610 g/L can be obtained, depending on how the resin is loaded and regenerated. We believe that continued efforts in resin investigation *(6,25,28,30, 35,38–40)* for increased capacity, specificity, and regenerability will lead to great strides in reduced production cost.

Although formal economic assessment of the BFBR process (as opposed to conventional batch fermentation of lactic acid) has yet to be performed, we have previously published a rough comparison of the two processes *(13)*. In this assessment, we demonstrated how increased productivity owing to cell immobilization and *in situ* product removal, increased on-line production time because of the continuous nature of the process, decreased capital costs, and equal raw material, operating, and overhead costs should render the BFBR an attractive alternative to conventional batch fermentation. For a given lactic acid output, the BFBR will be smaller, have increased on-line time, and have equal or decreased chemical costs compared to batch fermentations. Improved resin regeneration will provide further economic advantage. Formal collaboration with a lactic acid producing industrial partner should enable a more complete economic evaluation of this technology.

CONCLUSIONS

A BFBR for the fermentation and simultaneous separation of lactic acid has been demonstrated at the laboratory scale. This reactor has demonstrated a 16-fold increase in volumetric productivity over control experiments in a free-cell, batch reactor. This increased productivity is achieved by providing high cell density and *in situ* removal of the inhibitory product. Rich media formulations are not needed to achieve higher reactor productivity owing to the decreased product inhibition with low broth lactic acid concentrations. Initial economic assessment of this bioprocessing scheme has revealed the potential for cost reduction owing to decreased reactor size, increased on-line production time, and equal chemical, operating, utility, and overhead costs. Further improvements to resin regeneration are expected to reveal additional economic advantages of the BFBR over conventional batch fermentation of lactic acid. The BFBR is a generalized processing scheme that may be applied to any fermentation. It will have the greatest impact on rapid fermentations that yield a strongly inhibitory product.

ACKNOWLEDGMENTS

Research was supported by the Advanced Industrial Concepts Division—Biological and Chemical Technologies Research Program, US Department of Energy under contract DE-AC05-84OR21400 with Martin Marietta Energy Systems, Inc. The authors thank Brian H. Davison for his valuable technical discussions and assistance.

REFERENCES

1. Atkinson, B. and Mavituna, F. (1991), *Biochemical Engineering and Biotechnology Handbook*, 2nd ed. Stockton, New York.
2. Lipinsky, E. S. and Sinclair, R. G. (1986),*Chem. Eng. Prog.* **82,** 26–32.
3. *Chemical Marketing Reporter* (1994), March 14, pp. 28–36.
4. Millis, J. (1993), *Am. Chem. Soc.* **38,** 297–299.
5. Ohleyer, E., Blanch, H. W., and Wilke, C. R. (1985), *Appl. Biochem. Biotechnol.* **11,** 317–331.
6. Evangelista, R. L., Mangold, A. J., and Nikolov, Z. L. (1994), *Appl. Biochem. Biotechnol.* **45,** 131–144.
7. Mercier, P., Yerushalmi, L., Rouleau, D., and Dochain, D. (1992), *J. Chem. Tech. Biotechnol.* **55,** 111–121.
8. Buchta, K. (1983), in *Biotechnology: A Comprehensive Treatise Biomass, Microorganisms for Special Applications, Microbial Products I, Energy from Renwable Resources,* Dellweg, H., ed., Verlag Chemie, Weinheim, Germany, 410–417.
9. Mark, H. F., Othmer, D. F., Overberger, C. G., and Seaborg, G. T., eds. (1981), *Kirk-Othmer Encyclopedia of Chemical Technology,* 3rd ed., vol. 13, Wiley, New York.
10. King, C. J. (1992), *Chem. Tech.* **22,** 285–291.
11. Yeh, P. L. H., Bajpai, R. K., and Iannotti, E. L. (1991), *J. Fermentation Bioeng.* **71,** 75–77.
12. Kuhn, R., Peretti, S., and Ollis, D. (1993), *Appl. Biochem. Biotechnol.* **39/40,** 401–413.
13. Kaufman, E. N., Cooper, S. P., Clement, S. L., and Little, M. H. (1995), *Appl. Biochem. Biotechnol.* **51,** 605–620.
14. Melzoch, K. and Konopaskova, L. (1993), *Biotechnol. Lett.* **15,** 517–520.
15. Andrews, G. F. and Fonta, J. P. (1989), *Appl. Biochem. Biotechnol.* **20,** 375–390.
16. Stenroos, S. L., Linko, Y. Y., and Linko, P. (1982), *Biotechnol. Lett.* **4,** 159–164.
17. Guoqiang, D., Kaul, R., and Mattiasson, B. (1991), *Appl. Microbiol. Biotechnol.* **36,** 309–314.
18. Hang, Y. D., Hamamci, H., and Woodams, E. E. (1989), *Biotechnol. Lett.* **11,** 119,120.
19. Yabannavar, V. M. and Wang, D. I. C. (1991), *Biotechnol. Bioeng.* **37,** 1095–1100.
20. Vick Roy, T. B., Blanch, H. W., and Wilke, C. R. (1982), *Biotechnol. Lett.* **4,** 483–488.
21. Shi, Z., Shimizu, K., Iijima, S., Morisue, T., and Kobayashi, T. (1990), *J. Fermentation Bioeng.* **70,** 415–419.
22. Xavier, A. M. R. B., Goncalves, L. M. D., Moreira, J. L., and Carrondo, M. J. T. (1995), *Biotechnol. Bioeng.* **45,** 320–327.
23. Kulozik, U., Hammelehle, B., Pfeifer, J., and Kessler, H. G. (1992), *J. Biotechnol.* **22,** 107–116.
24. Venkatesh, K. V., Okos, M. R., and Wankat, P. C. (1993), *Process Biochem.* **28,** 231–241.
25. Seevaratnam, S., Holst, J. O., Hjorleifsdottir, S., and Mattiasson, B. (1991), *Bioprocess Eng.* **6,** 35–41.
26. Martin, M. S., Pazos, C., and Coca, J. (1992), *J. Chem. Tech. Biotechnol.* **54,** 1–6.
27. Siebold, M., Frieling, P. V., Joppien, R., Rindfleisch, D., Schugerl, K., and Roper, H. (1995), *Process Biochem.* **30,** 81–95.
28. Garcia, A. A. (1991), *Biotechnol. Prog.* **7,** 33–42.
29. Galliot, F. P., Gleason, C., Wilson, J. J., and Zwarick, J. (1990), *Biotechnol. Prog.* **6,** 370–375.
30. Srivastava, A., Roychoudhury, P. K., and Sahai, V. (1992), *Biotechnol. Bioeng.* **39,** 607–613.
31. Davison, B. H. and Scott, C. D. (1992), *Biotechnol. Bioeng.* **39,** 365–368.
32. Davison, B. H. and Thompson, J. E. (1992), *Appl. Biochem. Biotechnol.* **34,** 431–439.
33. Godia, F., Casas, C., and Sola, C. (1987), *Process Biochem.* **22,** 43–48.
34. Davison, B. H. and Scott, C. D. (1988), *Appl. Biochem. Biotechnol.* **18,** 19–34.

35. Kaufman, E. N., Cooper, S. P., and Davison, B. H. (1994), *Appl. Biochem. Biotechnol.* **45,** 545–554.
36. Scott, C. D. (1987), *Ann. NY Acad. Sci.* **501,** 487–493.
37. Major, N. C. and Bull, A. T. (1989), *Biotechnol. Bioeng.* **34,** 592–599.
38. Ernst, E. E. and McQuigg, D. W. (1992) in AICHE 1992 National Meeting, Miami Beach, FL.
39. Nakagawa, M., Nakamura, I., and Kobayashi, T. (1975), *J. Fermentation Technol.* **53,** 127–134.
40. Yang, C. W., Lu, Z., and Tsao, G. T. (1995), *Appl. Biochem. Biotechnol.* **51,** 57–71.

0273-2289/96/5758—0517$7.00

Evaluation of Nitrogen Supplements for Bioconversion of Municipal Solid Waste to Lactic Acid

SHENG D. ZHOU,[1] THOMAS A. MCCASKEY,*,[1] AND JACQUELINE BRODER[2]

[1]*Department of Animal and Dairy Sciences, Auburn University, AL 36849; and* [2]*Biotechnical Research Department, Tennessee Valley Authority, Muscle Shoals, AL 35660*

ABSTRACT

Previous studies conducted by the authors showed that the ingredients of Elliker broth could be used as fermentation adjuncts for the bioconversion of municipal solid waste (MSW) acid hydrolysate to lactic acid. Five kinds of inexpensive organic nitrogen supplements were evaluated for their potential to replace more expensive nitrogen sources derived from tryptone and yeast extract, which are the nitrogenous components of Elliker broth. At a soluble nitrogen concentration of 0.28%, which is equivalent to that supplied by tryptone and yeast extract, soybean meal and cottonseed meal were the best alternative nitrogen sources and compared favorably to tryptone and yeast extract in terms of lactic acid production. Soybean meal nitrogen of 0.21% was the minimum nitrogen requirement for maximum production of lactic acid. When soybean meal was used as the nitrogen source, the addition of 1.91% phosphate as monobasic potassium phosphate significantly improved lactic acid production, but no improvement occurred below 1.91%. Ascorbic acid and sodium acetate are also components of Elliker broth, and the omission of these from the fermentation substrate did not affect lactic acid production. When 0.21% soluble soybean meal nitrogen and 0.4% sodium chloride were supplied to double-sugar MSW (DSMSW) hydrolysate, the amount of carbohydrate used, the percent of carbohydrate converted, the amount of lactic acid produced, and the percent yield after 3 d of fermentation were 78.3 mg/mL, 91%, 68.7 mg/mL, and 88%, respectively.

Index Entries: Municipal solid waste; fermentation; nitrogen supplements; MSW bioconversion; lactic acid production.

INTRODUCTION

The manufacture of 100% biodegradable polylactic plastics (PLP), requires large quantities of lactic acid *(1,2)*, and lactic acid production via fermentation has been the main source of lactic acid *(3)*. However, to be an economical source of lactic acid, the fermentation should employ a microorganism that produces high

*Author to whom all correspondence and reprint requests should be addressed.

concentrations of lactic acid from inexpensive substrates. Previous studies have demonstrated the ability of *Lactobacillis pentosus* B-227 to ferment acid-hydrolysed municipal solid waste (MSW) to lactic acid, and optimum fermentation conditions have been reported *(4)*. These conditions include nutrient supplementation of MSW substrate with the ingredients of Elliker broth (tryptone, yeast extract, sodium chloride, sodium acetate, and ascorbic acid), an initial fermentation substrate pH of 7.6, substrate buffering with 5% calcium carbonate, 1% culture inoculation, and fermentation under static conditions at 32°C *(4)*. Tryptone and yeast extract are the most expensive components of the MSW fermentation substrate. In order to produce lactic acid on an industrial scale from MSW hydrolysate, a more economical nitrogen source is required to replace tryptone and yeast extract. A previous study indicated that cottonseed meal, soybean meal, and several other nitrogen sources have been used successfully in bioconversion processes *(5)*.

MATERIALS AND METHODS

Microorganism

The lactic acid producing bacterium *L. pentosus* B-227 was provided by the USDA Northern Regional Research Laboratory (Peoria, IL). The culture was serially transferred twice in Elliker broth (Difco, Detroit, MI) before it was used to inoculate the MSW fermentation substrate.

MSW Hydrolysate Substrate

MSW was "classified" in a commercial facility to recover recyclables to produce an organic fraction for hydrolysis. This organic fraction was comprised predominantly of papers, yard wastes, and food wastes. The classified MSW (organic fraction) was processed at the Tennessee Valley Authority's hydrolysis facility. Dilute sulfuric acid (2%) was mixed with the MSW and heated at 180°C for 20–40 min. These process conditions converted the cellulose and hemicellulose in the MSW to the monomeric sugars, glucose, mannose, xylose, and galactose, in that order of predominance. The hydrolyzed MSW was pumped through a filtration system to separate the hydrolysate from the solids. The hydrolysate was treated with calcium hydroxide as described in previous studies *(4,6)*. After treatment, the hydrolysate was used as the fermentation substrate.

Nitrogen Supplements

The nitrogen sources shown in Table 1 were dried at 60°C for 2 d and ground through a 2-mm mesh screen before they were evaluated as nitrogen adjuncts to the MSW substrate.

Fermentation Substrate

The total and water-soluble nitrogen contents of each nitrogen supplement were determined, and an equivalent amount of dry supplement was added to the MSW hydrolysate to obtain 0.28% soluble nitrogen. This percent is equivalent to the nitrogen content supplied by tryptone and yeast extract in Elliker broth. The nitrogen-supplemented MSW was then filtered through four layers of cheesecloth. The sugar content was increased by the addition of sugars rather than by concentrating the sugars in the hydrolysate to simulate hydrolysate with a higher

Table 1
Nitrogen Sources Evaluated and Approximate Costs

Nitrogen source	Price $/kg approx	Soluble N % of DM	Amt. for ESN,[a] g/L
Soybean meal[b]			
41% protein	0.19	1.62	173.0
Cottonseed meal			
40% protein	0.15	0.93	301.1
Feather meal			
75% protein	0.17	1.02	274.5
Fish meal			
57% protein	0.42	1.99	140.7
Meat and bone meal			
58% protein	0.21	3.03	92.4
Tryptone			
74% protein	70.00	11.75	23.8
Yeast extract			
61% protein	99.00	9.70	28.9

[a]Equivalent soluble nitrogen (ESN) is based on 0.28% soluble nitrogen in Elliker broth.
[b]Crude protein determined as %N × 6.25. Tryptone and yeast extract were obtained from Difco Laboratories, and all other nitrogen supplements were animal feed ingredients obtained from local feed companies. Approximate price was calculated from quantity discounts.

content of fermentable sugar. The following sugars (mg/mL), glucose (16.4), mannose (14.0), xylose (6.5), and galactose (4.4), were added to the MSW filtrate to double the concentrations of sugar monomers naturally in the hydrolyzed MSW, henceforth designated as double-sugar MSW (DSMSW). Sodium chloride, sodium acetate, and ascorbic acid were added to the substrate at the same concentrations as in Elliker broth. The substrate was sterilized at 121°C for 15 min and buffered with 5% (w/v) sterile calcium carbonate *(4)*.

Fermentation Conditions

The fermentation was conducted in 250-mL Erlenmeyer flasks containing 100 mL substrate. The flasks were inoculated with 1% (v/v) of *L. pentosus* B-227 broth culture, which had been serially transferred twice in Elliker broth. The fermentation was conducted under static conditions for 3 d at 32°C, and each treatment was replicated three times.

Evaluation of the Effects of Nutrients

The effects of different nitrogen supplements on lactic acid production were evaluated in DSMSW substrate. The nitrogen supplement was added to give a soluble nitrogen content of 0.28% in the fermentation substrate, which is equivalent to the nitrogen content supplied by tryptone and yeast extract in Elliker broth.

Based on the fermentation conditions described, soybean meal was used to make 0.28, 0.21, and 0.14% water-soluble nitrogen in the DSMSW substrate to determine the minimum nitrogen supplement required for maximum production of lactic acid. Calcium carbonate was used as a fermentation buffer and may have caused phosphate deficiency in the fermentation substrate owing to the formation of insoluble calcium phosphate. To determine the effect of phosphate addition on lactic acid production, monobasic potassium phosphate was added to give phosphate concentrations of 0.64, 1.28, and 1.91% in the DSMSW substrate, supplemented with 0.21% soluble soybean meal nitrogen. The concentration of monobasic potassium phosphate used was based on the 1.80% solubility of $Ca(H_2PO_4)_2$ *(7)*.

Elliker broth, which is a good substrate for lactic acid production, also contains 0.05% ascorbic acid and 0.15% sodium acetate. To determine if the adjuncts are beneficial for the bioconversion of DSMSW to lactic acid, the addition or omission of the adjuncts on lactic acid production from DSMSW with 0.21% soluble soybean nitrogen was determined. The adjuncts were added to DSMSW at the same levels as in Elliker broth.

Analyses

Total nitrogen was determined with 1 g of dry supplement by the semimicro-Kjeldahl method *(8)*. Water-soluble nitrogen was extracted from 5 g of dry supplement soaked in 100 mL of distilled water overnight at 25°C, and then filtered through four layers of cheesecloth. The filtrate was then used to determine water-soluble nitrogen by the semimicro-Kjeldahl method *(9)*. Water-soluble phosphate was determined on the filtrate by the vanadomolybdophosphoric acid colorimetric method *(8)*. Additional filtrate was centrifuged and filtered through 0.45-µm membrane filter, and carbohydrate concentration of the filtrate was determined by high-performance liquid chromatography (HPLC) *(4)*.

During fermentation, lactic acid and carbohydrate concentrations of the fermentation substrate were determined as follows: about 15 mL of fermentation substrate were collected, centrifuged at 12,000*g* for 15 min, and the supernatant was filtered through a 0.45-µm filter. Lactic acid and carbohydrate concentrations in the filtrate were determined by HPLC *(4)*. The data were analyzed by least-squares test of variance using the SAS GLM procedures *(10)*.

RESULTS AND DISCUSSION

Nitrogen and Phosphate Contents

Seven organic nitrogen sources, including tryptone and yeast extract, were evaluated as nitrogen adjuncts for the bioconversion of MSW to lactic acid (Table 1). Tryptone and yeast extract are highly refined sources of nitrogen that are used principally as nitrogen adjuncts for laboratory microbiological media. Because of the high cost of tryptone and yeast extract, less costly nitrogen supplements derived from animal protein feeds were evaluated as possible nitrogen substitutes (Table 1). An advantage of using the animal feed protein supplements is their low cost, but a disadvantage (because they are not refined) is their lower content of soluble nitrogen. For example, data in Table 1 indicate that 173.0 g of 41% protein soybean meal would be required to achieve 0.28% soluble nitrogen, which is the amount contained in Elliker broth. If tryptone were used, 23.8 g or about one-seventh of the amount would be required. However, because technical grade tryptone and yeast extract are valued at over 300 times the cost of soybean meal, the latter nitrogen source is more economical, provided that the nitrogen of soybean meal can substitute for the nitrogen of tryptone and yeast extract for lactic acid production.

Table 2
Nitrogen and Phosphate Contents of Nitrogen Supplements

Nitrogen supplement	TKN, %	WSN, % of TKN	WSP, mg/100 g	WEC, mg/mL
Soybean meal	6.6	24.5	26	<1
Cottonseed meal	6.4	14.6	16	<1
Feather meal	12.0	8.5	6	<1
Fish meal	8.9	22.3	177	<1
Meat and bone meal	9.3	32.5	117	<1
Tryptone	11.8	100	303	<1
Yeast extract	9.7	100	146	<1

TKN, total Kjeldahl nitrogen.
WSN, water-soluble nitrogen as a percentage of TKN.
WSP, water-soluble phosphate.
WEC, water-extractable carbohydrate as hexose monomers.

The nitrogen and water-soluble nitrogen contents of the supplements are shown in Table 2. Feather meal had the highest nitrogen content (12.0%), which was slightly higher than tryptone (11.8%) and yeast extract (9.7%). Soybean meal (6.6%), cottonseed meal (6.4%), fish meal (8.9%), and meat and bone meal (9.3%) all contained less nitrogen than tryptone and yeast extract. However, nitrogen must be soluble to be utilized by microorganisms. Meat and bone meal contained the most water-extractable nitrogen (32.5%), followed by soybean meal (24.5%), fish meal (22.3%), cottonseed meal (14.6%), and feather meal (8.5%). Water-soluble phosphate content (mg/100 g), was highest in fish meal (177) and meat and bone meal (117), and least in soybean meal (26), cottonseed meal (16), and feather meal (6).

Equivalent Available Nitrogen Supplements

DSMSW substrates supplemented with equivalent amounts of soluble nitrogen (0.28% [w/v]) from the various sources supported growth of *L. pentosus* B-227, and lactic acid was produced in concentrations ranging from 17.9 to 65.3 mg/mL (Table 3). Based on the carbohydrate used, carbohydrate conversion rate, lactic acid produced, and yield, soybean meal, followed by cottonseed meal, was the best nitrogen supplement, compared to the control supplement of tryptone and yeast extract. After a 3-d fermentation, 65.3 mg/mL of lactic acid (yield 85%), was produced from 95 mg/mL of carbohydrate (conversion 81%) when soybean meal supplement was used. When cottonseed meal was used as the supplement, 61.1 mg/mL of lactic acid (yield 79%) was produced from 89 mg/mL of carbohydrate (conversion 88%). These lactic acid concentrations were not significantly ($p < 0.05$) different from that of the control supplement of tryptone and yeast extract, which produced 67.8 mg/mL of lactic acid (yield 83%) from 94 mg/mL of carbohydrate (conversion 87%). When feather meal, fish meal, and meat and bone meal were used as nitrogen supplements, lactic acid produced and carbohydrate used were significantly ($p < 0.05$) lower than for the tryptone and yeast extract supplements (Table 3).

The data in Table 4 indicate that different levels of soybean meal nitrogen supplement affect lactic acid production. A 0.28% soluble nitrogen content produced the highest level of lactic acid (65.3 mg/mL), with an 85% yield. When

Table 3
Effects of Nitrogen Supplements on Lactic Acid Production from DSMSW Hydrolysate by *L. pentosus* B-227

Nitrogen supplements	Carbohydrate[a] Used, mg/mL	Carbohydrate[a] Conversion,[b] %	Lactic acid Produced, mg/mL	Lactic acid Yield,[c] %
Soybean meal	77.1[A]	81	65.3[A]	85
Cottonseed meal	77.8[A]	88	65.1[A]	79
Feather meal	63.9[B]	75	46.0[B]	85
Fish meal	54.8[C]	65	40.5[C]	74
Meat and bone meal	27.5[D]	33	17.9[D]	65
Tryptone and yeast extract[d]	81.5[E]	87	67.8[A]	83

[a]Initial substrate carbohydrate concentrations were 95, 89, 85, 85, 84, and 94 mg/mL for soybean meal, cottonseed meal, feather meal, fish meal, meat and bone meal, and tryptone and yeast extract, respectively.

[b]Conversion = carbohydrate used/initial carbohydrate × 100.

[c]Yield = lactic acid produced/carbohydrate used × 100.

[d]Nitrogen supplementation is at a rate of 20 g/L tryptone and 5 g/L yeast extract.

[A-E]Different superscript capital letters in a data column are significantly different ($p < 0.05$).

Table 4
Effects of the Levels of Soybean Meal Supplement (SBM) on Lactic Acid Production from DSMSW Hydrolysate by *L. pentosus* B-227

SBM supplement level, %	Carbohydrate[a] Used, mg/mL	Carbohydrate[a] Conversion, %	Lactic acid Produced, mg/mL	Lactic acid Yield, %
0.28	77.1[A]	81	65.3[A]	85
0.21	70.8[B]	74	61.1[A]	86
0.14	60.8[C]	64	39.3[B]	65

[a]Initial carbohydrate concentration was 95 mg/mL.

[A-C]Different superscript capital letters in a data column are significantly different ($p < 0.05$).

the soluble nitrogen level was decreased from 0.28 to 0.14%, lactic acid production decreased from 65.3 to 39.3 mg/mL, carbohydrate used decreased from 77.1 to 60.8 mg/mL, and carbohydrate conversion also decreased from 81 to 64%. However, carbohydrate conversion, lactic acid produced, and yield did not differ significantly ($p < 0.05$) between the 0.28 and 0.21% soluble nitrogen levels supplied by the soybean meal. Therefore, the 0.21% soluble soybean meal nitrogen supplement was used for further studies.

Phosphate Supplement

The addition of phosphate as monobasic potassium phosphate to the DSMSW fermentation substrate showed some benefit on the production of lactic acid from MSW (Table 5). When phosphate addition increased from 0.64 to 1.91%,

Table 5
Effects of Phosphate Addition on Lactic Acid Production from DSMSW Hydrolysate by *L. pentosus* B-227

	Carbohydrate[a]		Lactic acid	
Phosphate addition, %	Used, mg/mL	Conversion, %	Produced, mg/mL	Yield, %
0	82.7^{A}	87	67.4^{B}	81
0.64	80.1^{A}	84	67.2^{B}	84
1.28	79.9^{A}	84	$71.1^{A,B}$	89
1.91	81.4^{A}	85	75.4^{A}	93

[a]Initial carbohydrate concentration was 95 mg/mL.

^{A,B}Different superscript capital letters in a data column are significantly different ($p < 0.05$).

Table 6
Effects of Ascorbic Acid and Sodium Acetate Supplementation on Lactic Acid Production from DSMSW Hydrolysate by *L. pentosus* B-227

	Carbohydrate[a]		Lactic acid	
Substrate	Used, mg/mL	Conversion, %	Produced, mg/mL	Yield, %
Control[b]	77.2^{A}	90	69.4^{A}	90
Without ascorbic acid[c]	78.5^{A}	91	69.7^{A}	89
Without ascorbic acid and NaAC[d]	78.3^{A}	91	68.7^{A}	88

[a]Initial carbohydrate concentration was 86 mg/mL.

[b]DSMSW with 0.21% soybean meal nitrogen, 0.05% ascorbic acid, and 0.15% sodium acetate.

[c]Same as control without 0.05% ascorbic acid.

[d]Same as control without 0.05% ascorbic acid and 0.15% sodium acetate.

^{A}Different superscript capital letters in a data column are significantly different ($p < 0.05$).

lactic acid production significantly increased from 67.2 to 75.4 mg/mL, and yield increased from 84 to 93%. However, phosphate addition had no effect on carbohydrate used or its conversion, which were the same for all treatments. There were two possible reasons why the phosphorus addition did not improve the utilization of carbohydrate or its conversion to lactic acid. First, the organic phosphate in soybean meal might be adequate for bacterial growth and lactic acid production. Second, most of the inorganic phosphorus added to the fermentation substrate might have combined with the calcium from the calcium carbonate buffer and precipitated as the calcium salt, making the phosphate unavailable.

Ascorbic Acid and Sodium Acetate Supplements

The omission of ascorbic acid or both ascorbic acid and sodium acetate from the fermentation substrate supplements had no effect on carbohydrate used, carbohydrate conversion, lactic acid produced, or yield of lactic acid (Table 6). After a 3-d fermentation, 69.4 mg/mL of lactic acid (90% yield) were produced in the control, which was not different ($p < 0.05$) from 69.7 mg/mL (yield 89%) for the omission of ascorbic acid and 68.7 mg/mL (yield 88%) for the omission of both ascorbic acid and sodium acetate. Although ascorbic acid and sodium acetate are ingredients in Elliker broth recommended for the cultivation of lactic acid bacteria, these compounds did not improve lactic acid production from DSMSW hydrolysate and, therefore, need not be used as nutrient adjuncts for DSMSW hydrolysate.

CONCLUSION

Soybean meal and cottonseed meal were suitable substitutes for tryptone and yeast extract as nitrogen supplements for lactic acid production from MSW. A minimum soluble soybean meal nitrogen content of 0.21% was required for maximum production of lactic acid. When soybean meal was used as the nitrogen supplement, the addition of 0.64% phosphate as monobasic potassium phosphate did not improve lactic acid production, but the addition of 1.91% phosphate gave a slight improvement. The omission of ascorbic acid, or both ascorbic acid and sodium acetate, from the fermentation substrate did not affect lactic acid production.

ACKNOWLEDGMENTS

This study was supported in part by the Alabama Agricultural Experiment Station, Auburn University, and the Tennessee Valley Authority, Muscle Shoals, AL.

REFERENCES

1. Lipinsky, E. S. and Sinclair, R. G. (1986), *Chem. Eng. Prog.* **8,** 26.
2. US Department of Energy (1992), *Industrial Innovations for Tomorrow,* National Renewable Energy Laboratory, Golden, CO, pp. 1,2.
3. Vickroy, T. B. (1985), *Comprehensive Biotechnology,* vol. 3, *The Practice of Biotechnology: Commodity Products,* Moo-Young, M., ed., Pergamon, Oxford, Oxfordshire, UK, pp. 761–776.
4. McCaskey, T. A., Zhou, S. D., Britt, S. N., and Strickland, R. (1994), *Appl. Biochem. Biotechnol.* **45/46,** 555–568.
5. Vuyst, L. D. and Vardamme, E. J. (1993), *Appl. Microbiol. Biotechnol.* **40,** 17–22.
6. Broder, J. D. and Barrier, J. W. (1993), *Municipal Solid Waste and Waste Cellulosic Conversion to Fuels and Chemicals,* April 1990–September 1992, Final Report, vol. I: Summary Report, Tennessee Valley Authority, Muscle Shoals, AL.
7. Weast, R. C. and Selby, S. M. (1967/68), *Handbook of Chemistry and Physics,* 48th ed. Chemical Rubber Co., Cleveland, OH, p. B-164.
8. Association of Official Analytical Chemists (1984), Official Methods of Analysis, AOAC, Washington, DC, p. 154.
9. American Public Health Association (1985), *Standard Methods for the Examination of Water and Wastewater,* 16th ed., Greenberg, A. E., Trussell, R. R., and Clesceri, L. S., eds., American Public Association, Washington, DC.
10. SAS Institute (1992), *SAS Language and Procedures: Usage, Version 6,* 1st ed., SAS Institute, Inc., Cary, NC.

0273-2289/96/5758—0525$7.50

Dynamics of the Immobilization of Anaerobic Meshopilic Bacteria on a Plastic Support

MÓNICA MERAZ,*[,1] ALFONSO GONZÁLEZ-BARRERA,[1] JOSÉ ALVAREZ-RAMIREZ,[2] AND OSCAR MONROY[1]

[1]*Department of Biotechnology and* [2]*Department of Process Engineering, Universidad Autónoma Metropolitana-Iztapalapa, Apartado Postal 55-535, Iztapalapa, México, D.F. 09340*

ABSTRACT

In this article, a study of the dynamics of the immobilization of mesophilic anaerobic bacteria was conducted. Low-density polyethylene (LDPE) ground to fine particles was used as immobilization support in an inverse fluidized bed. After 135 d of immobilization, the biomass concentration attained was 375 μg protein/g dry plastic support. The specific rate for acetate consumption was 0.008 μmol acetate/μg protein/h. The kinetic parameters of the bioparticles (K_s = 1.067 m*M* and V_{max} = 0.00754 μmol acetate/μg protein/h) were estimated by the initial rate method at different acetate concentrations and were compared to the values obtained by a direct parameter estimation method from the dynamic assays for acetate consumption with an initial substrate concentration of 10 m*M*. The particles were prepared for scanning electron microscopy (SEM) to evaluate the biofilm qualitatively. The SEM studies showed that the surface and cavities of the support were colonized, and the bacterial structures found were bacilli similar to *Methanothrix* as the predominant microorganism and a few groups of *Methanosarcina*-like cocci. The K_s value found is between those reported for the predominant bacterial group immobilized.

Index Entries: Anaerobic bioimmobilization; fluidized bed; plastic support; specific activity; substrate affinity.

INTRODUCTION

The waste-water treatment in fluidized beds has been extensively reported as an efficient operation for high-loading organic rates, with some advantages over the Upflow Ascendent Sludge Blanket (UASB) reactor, such as a great retention of biomass in the reactor, large superficial area for biomass attachment, and the possibility to control biofilm thickness *(1–6)*. However, the cell attachment mechanism to the support materials used in these systems is not fully understood, and considerable scientific work has been done in an attempt to describe the immobilization process to different supports and by different microorganisms in anaerobic conditions *(7–11)*.

*Author to whom all correspondence and reprint requests should be addressed.

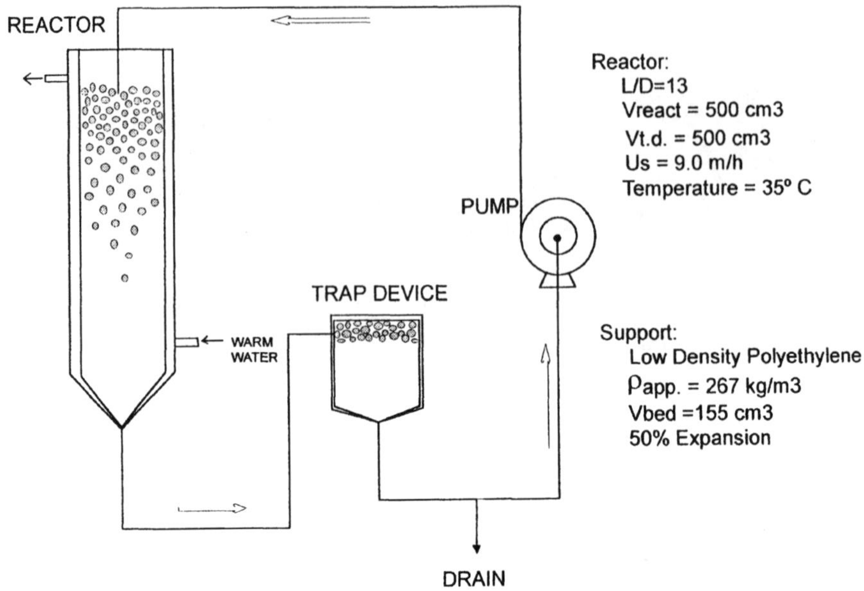

Fig. 1. Schematic diagram of the batch downflow fluidized bed used for bioimmobilization.

The objective of this work was to describe the immobilization dynamics of a mesophilic sludge on low-density polyethylene (LDPE) particles in a downflow fluidized bed similar to that of Go1nzález et al. *(9)*, the specific acetoclastic activity, and the affinity for the substrate of the fully immobilized bioparticles.

MATERIALS AND METHODS

Immobilization System

The immobilization system was a batch downflow fluidized bed of 500 cm^3 (L/D 13) shown as a diagram in Fig. 1. It has a device to trap the plastic particles that flow out of the bed. The support used was LDPE pellets ground to fine particles with a size range distribution from 0.56 to 0.28 mm and with an apparent density of 267 kg/m^3. Approximately 155 cm^3 of support were fluidized at a superficial velocity of 9 m/h, and the bed expansion achieved was 50%. The reactor was operated at 35°C and pH 7.0. Every 2 d the immobilization medium *(12)* containing 1 g acetate/L and 50 mg of volatile suspended solids (VSS)/L of inoculum was replaced by fresh medium.

Inoculum

The inoculum was a mesophilic sludge with 21 g VSS/L with a specific methanogenic activity of 0.42 mmol CH_4/g VSS/h. The sludge was observed in an Olympus BH2-RFCA light and epifluorescence microscope and showed microbial structures similar to *Methanothrix* as the dominant microorganism, some grouped *Methanosarcina*-like cocci, and short bacilli, slightly curved and strongly fluorescent that could be *Methanobacterium*.

Specific Activity Determinations

The biofilm accumulation was assessed through the specific acetoclastic activity of the bioparticles at several intervals of the colonization time (15, 30, 45, 70, 100, and 135 d). All specific activity determinations were made by placing a known amount of wet particles into serologic bottles containing 20 mL of mineral medium *(12)* and 10 m*M* acetate. The acetate was analyzed with standard gas chromatography using an HP 5890 chromatograph with an AT-1000 capillary column at 120°C.

The acetate degradation rate was related to the final biomass content, which was measured by detaching the biofilm by sonication with successive washes in distilled water to quantify directly the total suspended solids (TSS) and the VSS as indicated by the standard methods *(13)*, or in phosphate buffer 0.1*M*, pH 7.0, to determine the protein content by Coomassie brilliant blue G-250 *(14)*. All quantities were related to the amount of dry support used for each determination.

Bioparticles Substrate Affinity

The affinity of the bioparticles for the substrate at concentrations of 1, 2, 3, 4, 5, 10, and 15 m*M* of acetate was estimated at the 135th d and was measured following the same methodology described. All specific activity rates were obtained by means of the initial rate method. The kinetic parameters, K_s and V_{max}, were obtained by the inverse modifications to the Michaelis-Menten model and by an integral method *(15)*. The integral method uses measurements of substrate concentration as a function of time to evaluate the kinetic parameters that fit the concentration measurements according to the linearized equation:

$$K_s \ln (S/S_0) + S_0 - S = V_{max} t \quad (1)$$

where S and S_0 are the substrate concentrations at time t and $t = 0$, respectively.

Sample Preparation for SEM

The bioparticles were prepared following a methodology similar to that of Zellner et al. *(11)*. Prior to fixation, the bioparticles were washed with cacodylate buffer 0.1*M*, pH 7.5 (Sigma, St. Louis, MO) and placed in buffer solution with 3% of glutaraldehyde (25% aqueous solution Grade I, Sigma) for 24 h. After this time, the sample was washed again with buffer solution, dehydrated by increasing ethanol concentrations, and dried with a Labconco Lyph·Lock 4.5 lyophilizer at –50°C and 10 μ of vacuum for 3 to 5 h. The dried sample was sputtered with gold or aluminum with a Sputter-Coater Bal-Tec SCD 050 and observed at 15–25 kV in a Jeol-5200 and a Zeiss DSM 940 A scanning microscopes.

RESULTS

Immobilization Dynamics

The immobilization process was carried out for 135 d. The protein, TSS, and VSS contents attained at this time were: 0.375 mg, 106.0 mg, and 39.75 mg/g of dry support, respectively. Figure 2 shows the acetate degradation pattern at each time during the immobilization period. On the 100th d, a noticeable decrement of the specific activity was detected, owing to an unexpected malfunction of the immobilization system that led to the total leakage of the liquid phase and loss of biomass.

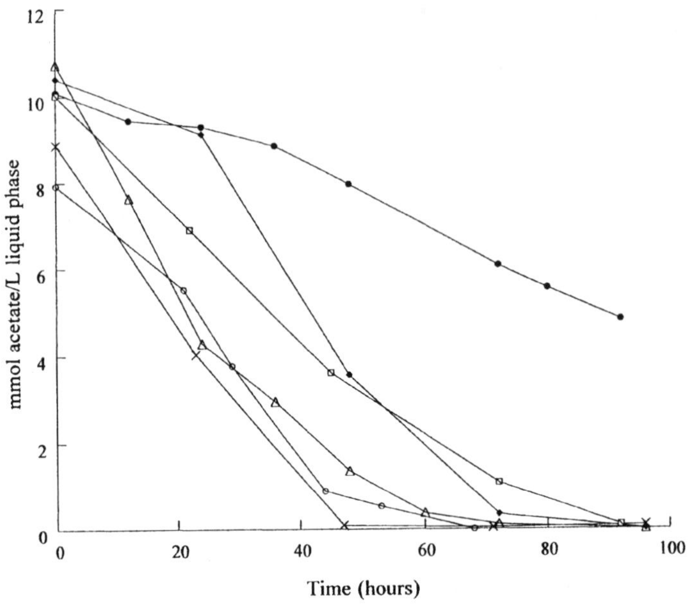

Fig. 2. Acetate degradation pattern at several times of bioimmobilization. ●, 15 d; ○, 30 d; ◆, 45 d; x, 70 d; □, 100 d; △, 135 d.

However, it can be seen that some of the activity was recovered at the 135th d, suggesting that when the biofilm is already attached to the support, the fixation of more microorganisms onto the surface is facilitated, as Meier-Schneiders et al. *(10)* have already found.

The specific acetate degradation rates at each time follow an exponential curve, and the maximum activity was found around 70th d (*see* Fig. 3). It seems that the bioimmobilization process reached a plateau after this time. This is in good agreement with previous reports that deal with batch *(9)* and continuous *(2,3,6,11)* immobilization studies, which found out that about 60 d are necessary to reach full immobilization and a steady-state organic load removal.

In this exponential curve can be distinguished two-phase bioimmobilization dynamics. The first phase corresponds to the biofilm formation, and the second one to a steady-state phase, and thereafter no activity increase is noticed, suggesting a dynamic saturation mechanism. It is not possible to say if the bioimmobilization process is limited by the rate of biomass attachment or by the biomass activity.

Affinity of the Bioparticles for the Substrate

The acetate consumption patterns at different acetate concentrations are shown in Fig. 4. It can be seen that all patterns are similar with a clear tendency to exhaust the substrate. At 15 m*M*, a slight decrement was found in the consumption rate, because there is a delay in the consumption pattern. The Michaelis-Menten-like curve constructed with the specific rates obtained is shown in Fig. 5. The maximum specific rate was found at 10 m*M*, and there seems to be a slight inhibition by the substrate at 15 m*M*.

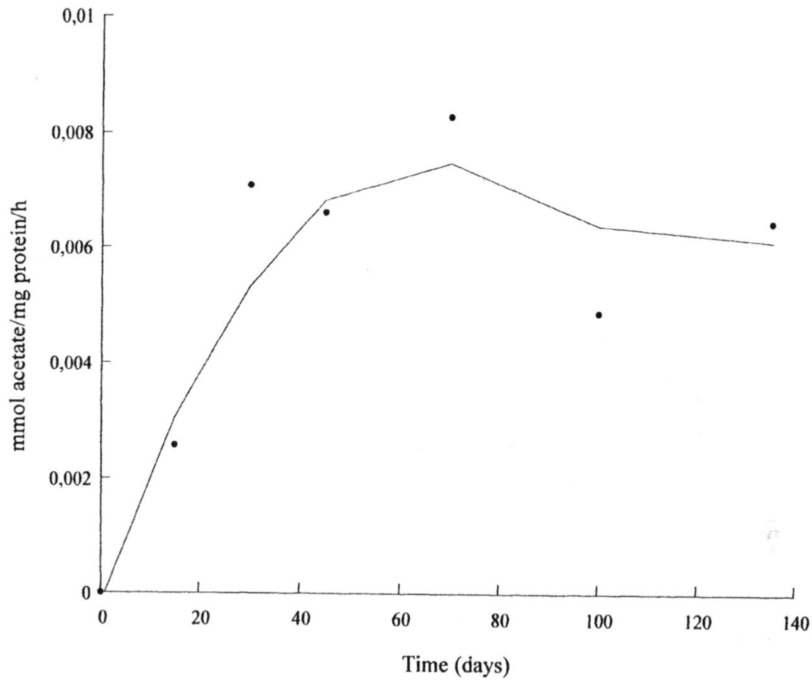

Fig. 3. Specific acetate degradation rates during the immobilization time. Note that the bioimmobilization dynamic follows an exponential behavior reaching a plateau after the 70th day of immobilization.

The kinetic parameters obtained by the Hanes-Woolf model were $K_s = 1.067$ m*M* and $V_{max} = 0.00754$ µmol acetate/µg protein/h, which are in good agreement with the experimental data (*see* Fig. 5). On the other hand, the kinetic parameters obtained by the integral method *(15)* with $S_0 = 10$ m*M* were $K_s = 1.723$ m*M* and $V_{max} = 0.0098$ µmol acetate/µg protein/h.

The V_{max} values obtained are similar, but there is a discrepancy between the K_s values obtained by the two methods that can be explained because the integral method considered only one substrate consumption pattern, corresponding to the highest consumption rate. Nevertheless, the K_s values obtained are in the range for those reported for *Methanothrix*. The V_{max} values are under those reported for this microorganism. It is probably because of the low biomass concentration attached to the support. The comparative values of the kinetic parameters are listed in Table 1.

SEM Studies of the Bioparticles

The SEM studies showed that the support has a furrowed surface with crevices, pleats, and cavities quite suitable for colonization (Fig. 6A,B) and apparently offered a larger surface area. The particles were colonized mainly by *Methanothrix*-like bacteria (Fig. 6C,D) forming long filaments attached to the surface of the media. The biofilm was not dense or thick; it was very superficial. The pleats and crevices were well colonized (not shown), as opposed to the smooth surfaces of the support

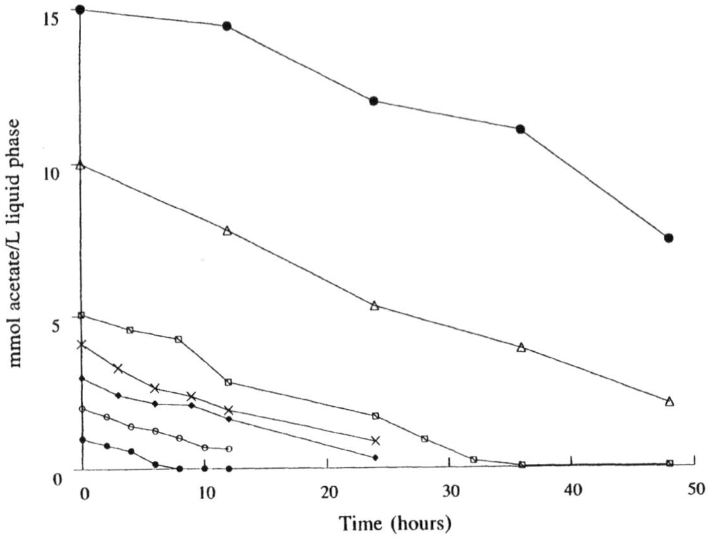

Fig. 4. Acetate consumption patterns of the bioparticles at the 135th d of immobilization. •, acet 1; ○, acet 2; ♦, acet 3; ×, acet 4; □, acet 5; △, acet 10; ●, acet 15.

where the adhesion was poor. The *Methanosarcina*-like bacteria and the small, curved bacilli were scarcely seen on the particles.

DISCUSSION

In spite of the superficial area offered by the support, the amount of biofilm found is low, as reflected in the low acetoclastic activity. Owing to the necessary high superficial velocity used to expand the bed, the shear stress between the particles and the liquid phase produces biofilm detachment *(4)*. Low superficial velocities lead to less shear stresses, so the microorganisms settle on the surface of the support, remaining there for longer times until the irreversible adhesion occurs *(5)*. At these operation conditions, the bed does not expand enough to reach a fluidized state, and, after a time of operation, the bed fractionates and shows "channelization" or channel formation with preferential flow. This is because of the dynamic friction that the particles exert on each other and with the liquid phase because of its irregular shape and rough edge.

These factors finally determine the maximal biofilm thickness and the time to reach a steady-state phase in the bioimmobilization exponential curve. Although González et al. *(9)* found out that the system we studied is quite suitable for immobilization, this work suggests that the cell attachment mechanism and the cell reproduction rate are slow with respect to the detachment by shear stresses at which the particles are subjected in the immobilization system.

The K_s value obtained is typical to that reported earlier for *Methanothrix* bacteria, and this is supported by the SEM studies, which show that the support was colonized by *Methanothrix*-like bacteria as the most abundant acetotrophic methanogen, and the specific activities are assigned to this microorganism. The prevalence of this microorganism is owing to the low acetate concentration used to feed the reactor.

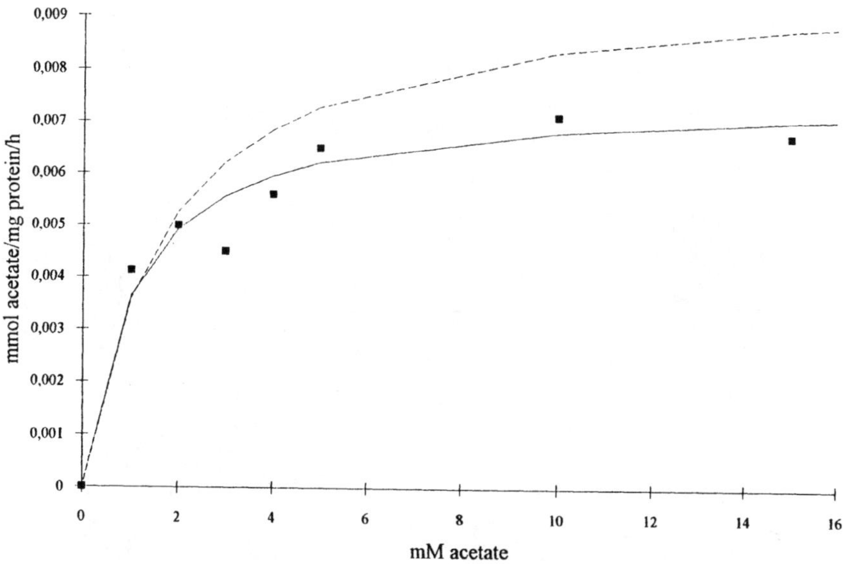

Fig. 5. Michaelis-Menten curve constructed with the kinetic parameter estimated with the Hanes-Woolf model (—) and the integral method (-----). Experimental data shown as filled squares.

Table 1
Kinetic Parameters Obtained by the Inverse Modification to the Michaelis-Menten Model and the Integral Method Compared with Those Reported in the Literature

	V_{max}, mg COD/mg VSS/d	K_s mg COD/L	References
Methanothrix	2.9–6.1	11–30	*(16)*
		45–76	*(17)*
		20	*(18)*
Methanosarcina	5.9–12	320	*(16)*
		320	*(17)*
This work	0.116	68	Hanes-Woolf
	0.15	110.24	Integral method *(15)*

With the evidence of this work and that of González et al. *(9)*, it seems that the main difference between a batch and a continuous immobilization system is that the bioimmobilization dynamics for the batch system is exponential with two stages: biofilm formation and steady-state phases.

Although for the continuous systems, the bioimmobilization pattern is sigmoidal and shows three consecutive stages: the lag, the biofilm production, and the steady-state phases, as Gorris et al. *(2)* already pointed out. The lack of a lag phase in a batch immobilization system is probably the result of the fact that there is no washout of cells from the system because it is closed. The inoculum is accumulated into the bed with subsequent reinoculations, and the microorganisms are present in great numbers since the beginning of the process. Nevertheless, the time to attain a steady-state is similar in both systems.

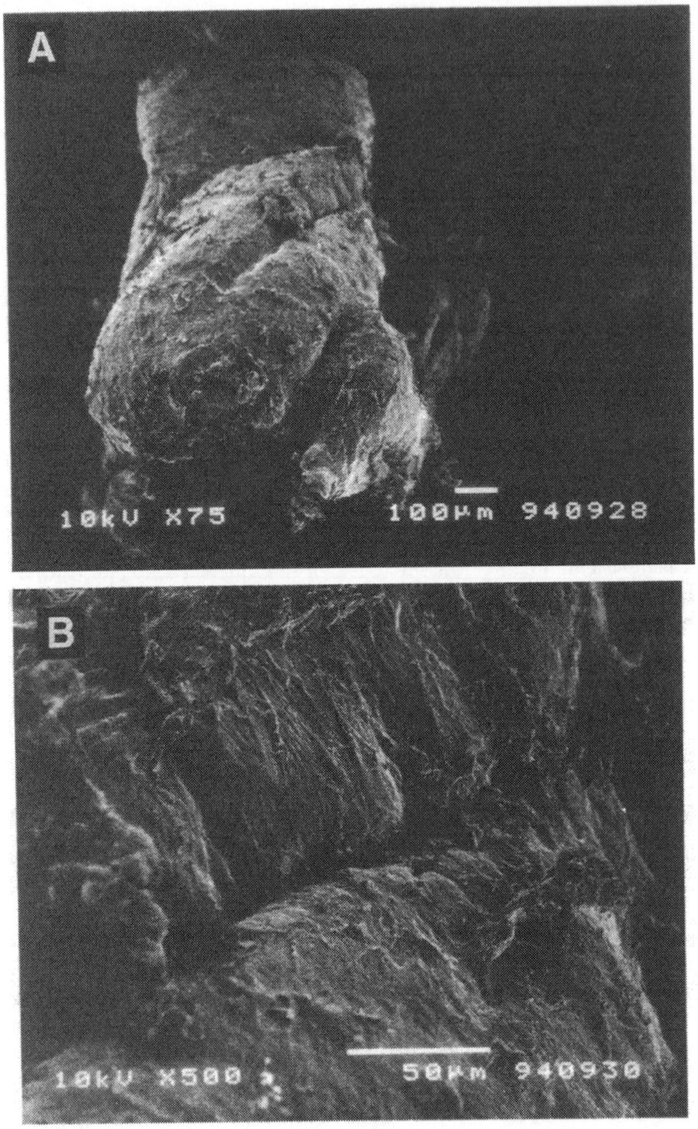

Fig. 6. **(A)** SEM photographs of the LDPE particles used as support, **(B)** a magnification that shows the furrowed surface of the support, **(C)** the *Methanothrix*-like bacteria adhered on a smooth surface of the support, and **(D)** filaments of the same microorganisms attached to a crevice of the particle.

ACKNOWLEDGMENTS

We thank Jacqueline Cañetas from the Institute of Physics, UNAM and Jesús Arenas from the Basic Sciences Division, UAM-I for their technical support in the SEM studies.

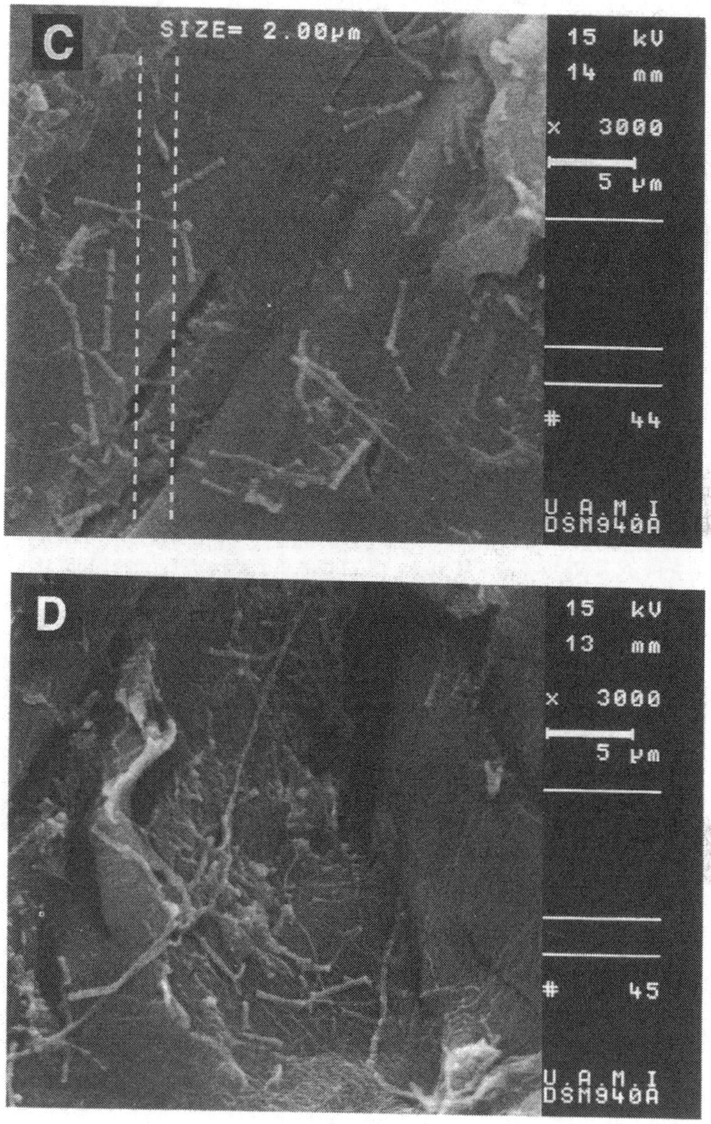

Fig. 6C and D.

REFERENCES

1. Henze, M. and Herremöes, P. (1983), *Water Sci. Technol.* **15,** 1–101.
2. Gorris, L. G. M., Van Deursen, J. M. A., van der Drift, C., and Vogels, G. D. (1989), *Biotechnol. Bioeng.* **33,** 687–693.
3. Binot, R. A., Bol, T., Naveau, H. P., and Nyns, E. J. (1983), *Water Sci. Technol.* **15,** 103–115.
4. Iza, J. (1991), in *Ananerobic Treatment Technology for Municipal and Industrial Wastewater,* Switzenbaum, M. S., ed., International Association on Water Pollution Research and Control, Pergamon, Oxford, UK, pp. 109–132.

5. Guiot, S. (1992), Bioimmobilization in Anaerobic Reactors. *Bioprocesos Anaerobios para el Tratamiento de Efluentes Industriales*. UAM/IMP/ORSTOM. May 6–7, Mexico City.
6. Rubindamayugi, M. S. T., Op Den Camp, H. J. M., Lubberding, H. J., Gijzen, H. J., and Vogels, G. D. (1992), *Water Sci. Technol.* **25(1),** 99–106.
7. Shapiro, M. and Switzenbaum, M. S. (1984), *Biotechnol. Lett.* **6(11),** 729–734.
8. Verrier, D., Mortier, B., Dubourguier, H. C., and Albagnac, W. (1988), in *Proceedings of the 5th International Symposium on Anaerobic Digestion*. Hall, E. R. and Hobson, P. N., eds., Pergamon, Oxford, UK, pp. 61–69.
9. González, G., Ramírez, F., and Monroy, O. (1992), *Biotechnol. Lett.* **14(2),** 149–154.
10. Meier-Schneiders, M., Busch, C., and Diekert, G. (1993), *Appl. Microbiol. Biotechnol.* **38,** 667–673.
11. Zellner, G., Diekman, H., Austermann-Haun, U., and Seyfried, C. F. (1994), *Biotechnol. Lett.* **16(3),** 611–620.
12. Balch, W. E., Fox, G. E., Magrum, L. J., Woese, C. R., and Wolfe, R. S. (1979), *Microbiol. Rev.* **43(2),** 260–296.
13. APHA-AWWA-WPCF. (1988), *Standard Methods for the Examination of Water and Wastewater*, 16th ed. Greenberg, A. E., Trussell, R. R., Clesceri, L. S., and Franson, A. A. H., eds., Port City Press, Baltimore, MD.
14. Sedmark, J. J. and Grossberg, S. E. (1977), *Anal. Biochem.* **79,** 544–552.
15. Counotte, G. H. M. and Prins, R. A. (1979), *Appl. Environ. Microbiol.* **38(4),** 758–760.
16. Pavlostathis, S. G. and Giraldo-Gómez, E. (1991), in *Ananerobic Treatment Technology for Municipal and Industrial Wastewater*, Switzenbaum, M. S., ed., International Association on Water Pollution Research and Control. Pergamon, Oxford, UK, pp. 35–60.
17. Guyot, J. P., Noyola, A., and Monroy, O. (1990), *Biotechnol. Lett.* **12,** 155–160.
18. Kuba, T., Furumai, H., and Kusuda, T. (1990), *Water Res.* **24(11),** 1365–1372.

0273-2289/96/5758—0535$6.75

Continuous Alcoholic Fermentation Process in a Tower Reactor with Recycling of Flocculating Yeast

TERESA C. B. PAIVA,*[,1] SUNAO SATO,[2] ALEXANDRE E. S. VISCONTI,[1] AND LÚCIA A. B. CASTRO[1]

[1]*Departamento de Biotecnologia, Faculdade de Engenharia Química de Lorena, cp. 116, CEP 12.600-000, Lorena, SP, Brazil; and* [2]*Departamento de Tecnologia Bioquímico-Farmacêutica, Faculdade de Ciências Farmacêuticas de Universidade de São Paulo, Brazil*

ABSTRACT

The influence of different substrate concentrations on the performance of a continuous system of alcohol prduction by fermentation using a tower reactor with recycling of flocculating yeasts was investigated. All experiments were carried out using a flocculating yeast strain IR-2, isolated from fermented food, and identified as *Saccharomyces cerevisiae*. Cane sugar juice was used as a substrate with sugar concentrations of 160, 170, 180, 190, and 200 g/L. Constant values of dilution rate, 0.20 h^{-1}, temperature, 30°C, and pH 3.3, were used. The performance of the reactor was observed to be efficient with high substrate concentrations. Maximum productivities of 18 g/L/h, 99% substrate conversion and ethanol concentrations of 90 g/L were obtained using 200 g/L of sugar in the feedstock. For substrate concentrations of 160 g/L, a maximum yield of 0.45 g of ethanol/g of sugar was observed or 90% of the theoretical value.

Index Entries: Tower reactor; flocculating yeast; ethanol.

INTRODUCTION

Special attention has been given to continuous alcoholic fermentation processes because of the advantages they present with respect to the conventional batch fermentation processes.

Recently, several continuous processes have been developed, such as fermentation with the use of immobilized yeast cells *(1)*, flocculating yeast ceils *(2–7)*, or membrane separation *(8)*.

The continuous production of ethanol from different raw materials, using flocculating yeast, in recycling tower reactors, has been widely studied at the laboratory scale *(3,4,9)*, with the view to optimization and future application at the industrial scale. In this work, the performance of a continuous ethanol production process in a tower reactor with recycled flocculating cells was evaluated. The initial

*Author to whom all correspondence and reprint requests should be addressed.

sugar concentrations used ranged from 160 to 200 g/L, the dilution rate was 0.20 h^{-1}, and the raw material was sugar cane juice (widely employed in Brazilian alcohol and sugar mills). The aim was a higher alcohol production with simultaneous reduction of the production residues (stillage), which is a big problem in many alcohol-producing regions.

MATERIALS AND METHODS

Microorganisms

A flocculating yeast strain, *Saccharomyces cerevisiae*, called IR-2, was used, isolated from fermented foodstuffs supplied by the Fermentation Research Institute (Tsukuba-Japan).

Culture

The *S. cerevisiae* cultures were kept in a refrigerator at 4°C, in an agar malt extract (malt extract: 30.0 g/L; peptone: 5.0 g/L; agar-agar: 15.0 g/L). The cultures were subcultivated every month; the test tubes were then incubated at 28°C for 24 h.

Medium

The main source of carbon used was the sugar cane juice, containing: sugar concentrations of 160–200 g/L; 1.0 g/L $(NH_4)_2SO4$; 0.20 g/L KH_2PO_4; 0.1 g/L $MgSO_4 \cdot 7H_2O$; 0.1 g/L $CaCl_2 \cdot 2H_2O$; 0.5 g/L yeast extract. The medium was sterilized in autoclave at 121°C and 1 atm for 45 min, and after cooling, a 40% formaldehyde diluted 1:6000 and mixtured in a penicilin-via-acid solution was added.

Inoculum Preparation

The inoculum was prepared in a two-stage growth process. In the first stage, a 1-mL cellular suspension was used to inoculate 50 mL of culture medium in a 125-mL Erlenmeyer flask with sugar cane juice diluted to 10 g/L and additional nutrient salts. Incubation was done at 30°C with shaking at 130 rpm in a rotational shaker for 24 h. Next, the second growth stage was inoculated: the culture cells were transferred to 500-mL Erlenmeyer flasks containing 200 mL of the culture medium, with sugar cane juice diluted to 50 g/L and additional nutrient salts. The incubation conditions were identical to those of the first stage. The reactor was inoculated with the contents of the Erlenmeyer flasks.

Fermentation Tests

The reactor was inoculated with a concentration of cells of 30 g/L (dry base). The initial phase was of retention and accumulation of yeast cells into the reactor. This was achieved by continuously pumping the sugar cane medium diluted to 50, 100, and 150 g/L sugar concentrations, according to yeast growth. The system remained in operation for 72 h, until a cell concentration of 100 g/L (dry base) was attained. After the accumulation stage, the dilution rate was adjusted to $D = 0.20\ h^{-1}$ and the initial sugar concentration was increased from 160 to 200 g/L. Fermentation was monitored by periodical sampling of the fermenting medium, in order to determine cell concentration and of the effluent for determination of the residual sugar concentration and ethanol. Figure 1 shows a scheme of the reactor used in the experiments.

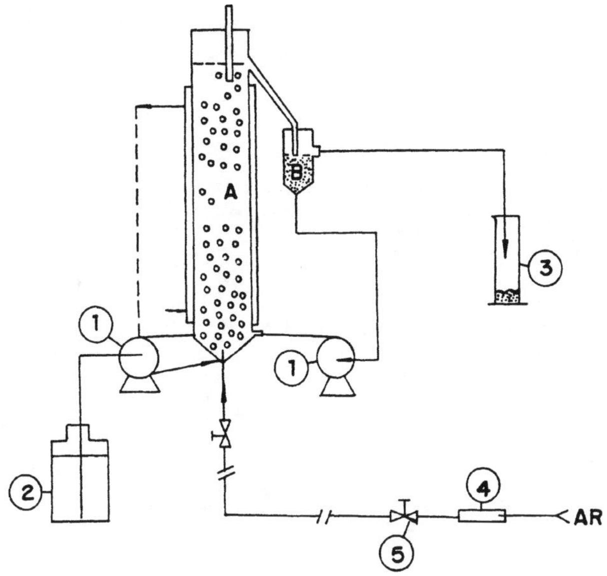

Fig. 1. Schematic diagram of a cell recycling continuous fermentation system using flocculating yeast. A, main reactor 220 mL; B, settler; 1, turbine pump; 2, medium reservoir; 3, product receiver; 4, flow meter; 5, air valve.

RESULTS AND DISCUSSION

Figures 2–6 show the performance data of the system, for each sugar concentration tested. The initial concentration of cells in the reactor was around 100 g/L for all experiments. Duration was approx 30 d, in which no inhibition of cell growth by the substrate concentrations used was observed. When initial sugar concentrations of 200 g/L were used, cell inhibition with partial deflocculation of the cells took place. This is attributed, likely, to an overload of substrate, defined by critical fermentation rate, with a 20% reduction in the cell concentration. According to Bueno Netto et al. *(10)*, the stability of the system is limited when the substrate load attains a critical load near 39–40 g/L/h, resulting in deflocculation and reduction in biomass.

The performance of the bioreactor for this sugar cane juice medium shows some differences when compared to those reported by Kuriyama et al. *(3)*, who used a simulated medium (brown sugar). However, similar cell densities were found around 90–100 g/L. Concerning the flocculating characteristics of the yeast strain *S. cerevisiae* IR-2, no adverse effects were observed.

Regarding the amount of cells lost in the effluent, average values around 6.0 g/L were encountered. It was also observed that this value might vary a lot, depending on the operational conditions used.

Substrate conversion was approx 99% for all tests. These results are shown in Figs. 2–6. Considering that the tolerated sugar losses in the process are in the range of 1–2 g/L, i.e., minimum conversion rate of 98%, it was possible to obtain virtually complete substrate conversion in all experiments.

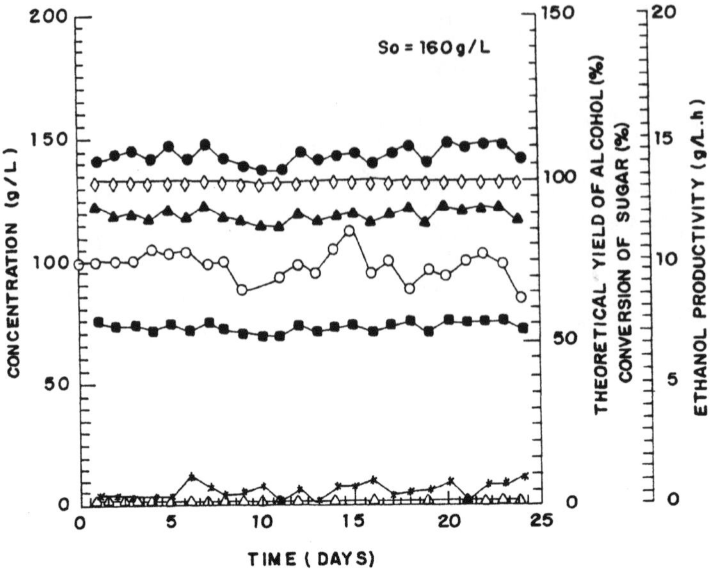

Fig. 2. Performance data in continuous alcohol fermentation with sugar cane juice concentration of 160 g/L. ●, ethanol productivity; ◇, conversion of sugar; ▲, theoretical yield of alcohol; ○, cellular conc. in reactor; ■, ethanol conc.; *, cellular conc. in effluent; △, residual TRS.

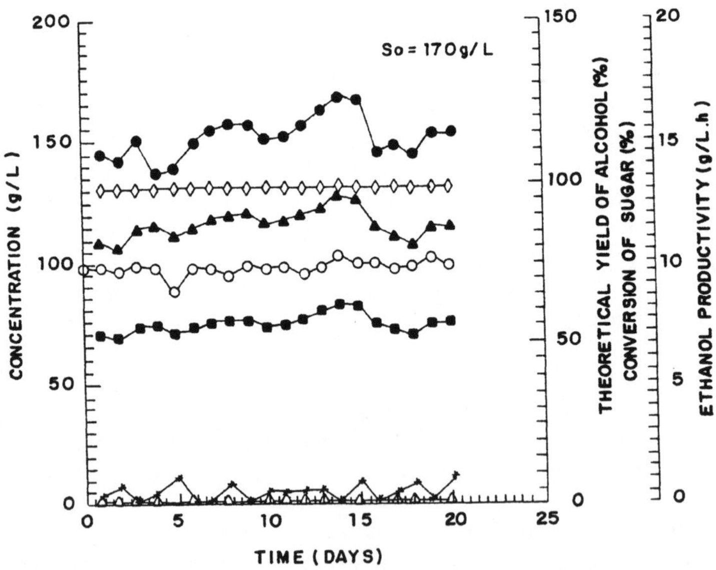

Fig. 3. Performance data in continuous alcohol fermentation with sugar cane juice concentration of 170 g/L. ●, ethanol productivity; ◇, conversion of sugar; ▲, theoretical yield of alcohol; ○, cellular conc. in reactor; ■, ethanol conc.; *, cellular conc. in effluent; △, residual TRS.

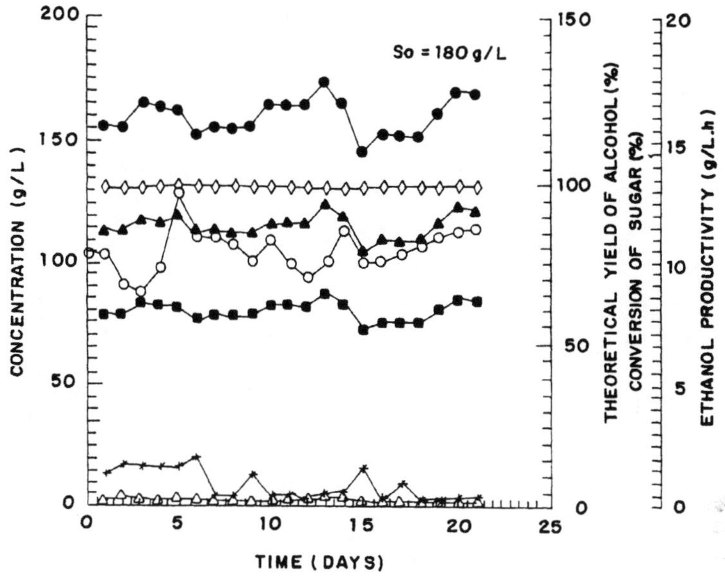

Fig. 4. Performance data in continuous alcohol fermentation with sugar cane juice concentration of 180 g/L. ●, ethanol productivity; ◇, conversion of sugar; ▲, theoretical yield of alcohol; ○, cellular conc. in reactor; ■, ethanol conc.; *, cellular conc. in effluent; △, residual TRS.

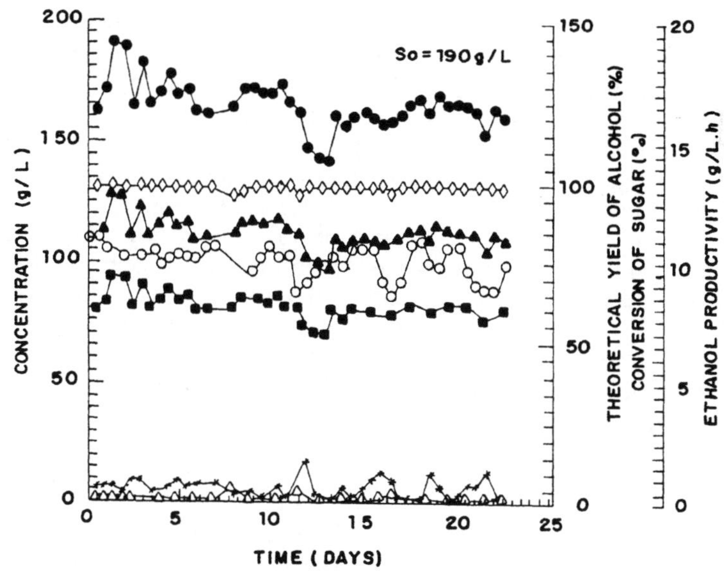

Fig. 5. Performance data in continuous alcohol fermentation with sugar cane juice concentration of 190 g/L. ●, ethanol productivity; ◇, conversion of sugar; ▲, theoretical yield of alcohol; ○, cellular conc. in reactor; ■, ethanol conc.; *, cellular conc. in effluent; △, residual TRS.

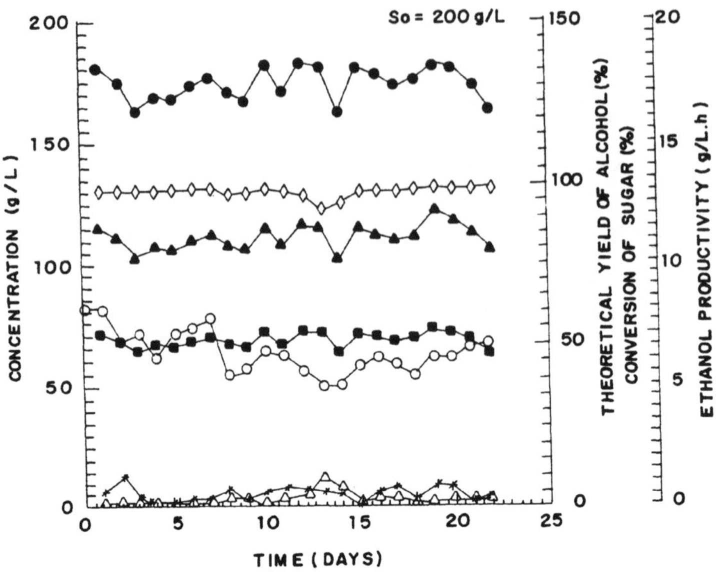

Fig. 6. Performance data in continuous alcohol fermentation with sugar cane juice concentration of 200 g/L. ●, ethanol productivity; ◇, conversion of sugar; ▲, theoretical yield of alcohol; ○, cellular conc. in reactor; ■, ethanol conc.; *, cellular conc. in effluent; △, residual TRS.

Concerning the residual sugar concentration, this tended to increase when an initial concentration of 200 g/L was used. It is interesting to note that even with this increase in residual sugar concentration, no significant variation in the substrate conversion was observed, remaining constant around 99%.

The maximum ethanol concentration attained was 90 g/L with a maximum productivity of 18 g/L/h. The ethanol concentrations and productivity were higher for the fermentations with substrate concentrations of 200 g/L.

Considering the productivity degree of this system and compared to those reported in the literature, it was observed that Kida et al. *(2)* obtained a maximum productivity of 30 g/L/h operating with high dilution rates of 0.5 h^{-1}. The concentration of ethanol produced by these authors was 60 g/L, less than the concentrations reported above of 90 and 81 g/L, respectively. Recently, Kuriyama et al. *(4)* obtained productivities of 28–29 g/L/h with 69–70 g/L of ethanol, using flocculating yeast, a tower reactor, or a conventional mixing reactor, resulting in high ethanol production rates and productivities.

The maximum yield obtained in the process was 0.45 g of ethanol/g sugar or 90% of the theoretical value, when substrate concentrations of 160–170 g/L were used. This yield is considered reasonable when compared to those in the literature *(7,10)*.

Table 1 presents the average values of the concentrations of residual total reducing sugars (TRS) and cells in the reactor and effluent, substrate conversion, yield, and productivity, obtained in a 220-mL tower reactor, at steady state, at the different substrate concentrations used. A linear increase in ethanol productivity and concentration and a linear decrease in yield were observed with the increase in initial sugar concentration.

Table 1
Average Values Obtained in Steady State of the Tests in a 220-mL Tower Reactor

Initial TRS, g/L	Residual TRS, g/L	Cell conc. in the reactor, g/L	Cell conc. in the effluent, g/L	Ethanol conc., g/L	Conversion of sugar, %	Theoretical yield of alcohol, %	Ethanol productivity, g/L/h
160	1.0 ± 0.1	98.3 ± 6.4	6.0 ± 3.3	72.5 ± 1.6	99.3 ± 0.1	88.3 ± 1.9	14.4 ± 0.3
170	1.1 ± 0.2	99.7 ± 2.6	5.7 ± 3.8	77.1 ± 3.5	99.3 ± 0.1	88.3 ± 4.0	15.3 ± 0.8
180	1.8 ± 0.6	104.3 ± 8.9	7.9 ± 6.1	79.6 ± 3.7	99.1 ± 0.4	85.8 ± 3.7	15.9 ± 0.7
190	2.0 ± 0.7	101.6 ± 6.3	6.3 ± 4.1	82.4 ± 5.1	98.9 ± 0.3	84.7 ± 5.2	16.5 ± 1.0
200	4.2 ± 3.3	80.0 ± 11.7	7.1 ± 4.0	86.6 ± 3.5	97.9 ± 1.6	82.6 ± 3.7	17.3 ± 0.7

CONCLUSIONS

A continuous tower reactor, using naturally flocculating yeast, enabled the obtention of high productivity degrees (18 g/L/h) and ethanol concentrations for high sugar concentrations without jeopardizing the process performance. These conditions make a future system scale-up possible, envisaging the industrial use of a simple continuous process (just one reactor), with low installation and operational costs, furthermore allowing reductions up to 10% of the total volume of residues normally produced at typical process conditions.

The fact that high concentrations (average of 100 g/L dry base) of naturally immobilized yeast in the reactor were maintained, which are much superior to the concentrations obtained at present in the alcohol industry, provides evidence of the strong flocculating capacity of the cells used, favoring the continuous process proposed.

ACKNOWLEDGMENTS

The authors wish to thank CNPq-RHAE for the financial aid granted and the Department of Biotechnology of the Faculty of Chemical Engineering at Lorena, where this research was made possible.

REFERENCES

1. Das, D., Gaidhani, N. R., Murari, K., and Gupta, P. S. (1993), *J. Ferment. Bioeng.* **75,** 132–137.
2. Kida, K., Yamadaki, M., Asano, S., Nakata, T., and Sonoda, Y. (1989), *J. Ferment. Bioeng.* **68,** 107–111.
3. Kuriyama, H., Ishibashi, H., Miyagawa, H., Kobayashi, H., and Mikami, E. (1993), *Biotechnol. Lett.* **15,** 4.
4. Kuriyama, H., Ishibashi, H., Umeda, I., Murakami, T., and Kobayashi, H. (1993), *J. Chem. Eng. Japan* **26,** 4.
5. Nam, K. D., Lee, I. K., and Cho, H. H. (1991), *Snop Misaengmul Hakhoenchi, Pusan* **19,** 487–490.
6. Son, S. M., Kim, I. G., and Pyum, Y. R. (1992), *Snop Misaengmul Hakhoechi, Seoul* **20,** 607–613.
7. Tabera, J. and Iznaola, M. A. (1989), *Alimentacion, Equipos Tecnologia, Spain* **8,** 111–117.
8. Luper, S. A., ed. (1986), in *Membrane Separations in Biotechnology,* Marcel Dekker, NY, pp. 161–200.
9. Kuriyama, H., Seiko, Y., Murakami, T., Kobayashi, H., and Sonoda, Y. (1985), *J. Ferment. Bioeng.* **63,** 159–165.
10. Bueno Netto, C. L., Destruhaut, A., and Goma, G. (1985), *Biotechnol. Lett.* **7,** 355–360.

FSC
www.fsc.org
MIX
Papier | Fördert
gute Waldnutzung
FSC® C083411

Kolibri 360 GmbH
Ferdinand-Jühlke-Straße 7
99095 Erfurt, Deutschland
produktsicherheit@kolibri360.de